Mammalian Protein Metabolism

VOLUME II

CONTRIBUTORS TO VOLUME II

JAMES B. ALLISON

GUILLERMO ARROYAVE

MOISÉS BÉHAR

K. L. BLAXTER

D. P. CUTHBERTSON

G. A. J. GOODLAD

A. E. HARPER

C. R. C. HEARD

D. M. HEGSTED

L. E. HOLT, JR.

R. A. McCANCE

H. N. MUNRO

B. S. PLATT

NEVIN S. SCRIMSHAW

S. E. SNYDERMAN

R. J. C. STEWART

FERNANDO VITERI

DONALD M. WATKIN

ELSIE M. WIDDOWSON

MAMMALIAN PROTEIN METABOLISM

Edited by

H. N. MUNRO

DEPARTMENT OF BIOCHEMISTRY
THE UNIVERSITY
GLASGOW, SCOTLAND

and

J. B. ALLISON

BUREAU OF BIOLOGICAL RESEARCH
RUTGERS, THE STATE UNIVERSITY
NEW BRUNSWICK, NEW JERSEY

VOLUME II

1964

ACADEMIC PRESS • New York and London

ACADEMIC PRESS INC.
111 Fifth Avenue, New York, New York 10003

United Kingdom Edition published by
ACADEMIC PRESS INC. (LONDON) LTD.
Berkeley Square House, London W.1

LIBRARY OF CONGRESS CATALOG CARD NUMBER: 63-21397

PRINTED IN THE UNITED STATES OF AMERICA.

List of Contributors

Numbers in parentheses indicate the pages on which the authors' contributions begin.

JAMES B. ALLISON, Bureau of Biological Research, Rutgers, The State University, New Brunswick, New Jersey (41)

GUILLERMO ARROYAVE, Institute of Nutrition of Central America and Panama (INCAP), Guatemala, Central America (523)

MOISÉS BÉHAR, Institute of Nutrition of Central America and Panama (INCAP), Guatemala, Central America (523)

K. L. BLAXTER, The Hannah Dairy Research Institute, Ayr, Scotland (173)

D. P. CUTHBERTSON, Rowett Research Institute, Bucksburn, Aberdeen, Scotland (373)

G. A. J. GOODLAD, Department of Physiology and Biochemistry, St. Salvator's College, University of St. Andrews, Fife, Scotland (415)

A. E. HARPER, Department of Nutrition and Food Science, Massachusetts Institute of Technology, Cambridge, Massachusetts (87)

C. R. C. HEARD, Human Nutrition Research Unit, National Institute for Medical Research, Mill Hill, London, England (445)

D. M. HEGSTED, Department of Nutrition, Harvard School of Public Health, Boston, Massachusetts (135)

L. E. HOLT, JR., Department of Pediatrics, New York University Medical Center, New York, New York (321)

R. A. McCANCE, Department of Experimental Medicine, University of Cambridge, England (225)

H. N. MUNRO, Department of Biochemistry, The University, Glasgow, Scotland (3, 267)

B. S. PLATT, Human Nutrition Research Unit, National Institute for Medical Research, Mill Hill, London, England (445)

NEVIN S. SCRIMSHAW, The Department of Nutrition and Food Science, Massachusetts Institute of Technology, Cambridge, Massachusetts (523, 569)

S. E. SNYDERMAN, Department of Pediatrics, New York University Medical Center, New York, New York (321)

R. J. C. STEWART, Human Nutrition Research Unit, National Institute for Medical Research, Mill Hill, London, England (445)

FERNANDO VITERI, Institute of Nutrition of Central America and Panama (INCAP), Guatemala, Central America (523)

DONALD M. WATKIN, Clinical Research Center, Massachusetts Institute of Technology, Cambridge, Massachusetts (247)

ELSIE M. WIDDOWSON, Department of Experimental Medicine, University of Cambridge, England (225)

Preface

Shortly after World War II, an international meeting was convened to consider nutritional deficiencies affecting populations in the post-war world. The agenda did not include protein deficiency. Since then, a revolution in outlook has taken place, so that in 1960 the Director of the Nutrition Division of the Food and Agricultural Division of the United Nations stated that, in the view of his organization, protein malnutrition is without doubt the most compelling nutritional problem in the underdeveloped countries today (M. Autret, 1961, *in* "Progress in Meeting Protein Needs of Infants and Preschool Children," National Academy of Sciences–National Research Council, Washington). In the more highly developed countries interest in the role of dietary proteins in the promotion of health has also quickened during the past few years.

The past decade has witnessed a revolution in a different aspect of protein studies. In less than ten years, the study of protein biosynthesis has progressed from a state of unproved and contradictory hypotheses to become the seat of intense and fruitful activity. When the dust of battle has settled, current work bids fair to resolving problems as varied as the nature of the inheritance mechanism and the genesis of the malignant cell.

These investigations into the importance of proteins in nutrition and into the mechanism of protein formation have each produced an abundant and specialized literature in which it is difficult for anyone other than the specialist to evaluate recent advances. In consequence, a barrier of non-communication has arisen between experts in each field of inquiry, although in reality they are but opposite sides of the same coin: the doctor faced with a case of protein malnutrition is observing a natural experiment in the inhibition of protein synthesis. Other areas of protein metabolism present a similar rapid growth of knowledge. Changes in protein metabolism are now known to be involved in the action of many hormones. In the field of medicine, the physician observes an expanding series of errors in the metabolism of individual amino acids, and the surgeon continues to explore the significance and the repair of the large dissipation of body protein resulting from injury.

It is thus difficult nowadays for the research worker in any one branch of the subject to comprehend the advances which are taking place on all fronts in protein metabolism. The objective of this treatise is to provide an up-to-date account of current thought in all areas of protein metabolism which will meet the needs of specialists in nutrition, biochemistry, clinical chemistry, medicine, and indeed in all aspects of biology in which mammalian protein metabolism is studied. We are not aware that this has been done in recent years.

The contents of the treatise are presented in three parts. After a short historical account of early studies on protein metabolism, the first part provides a picture of the biochemical mechanisms involved in protein metabolism, the second part deals with the place of proteins in nutrition, the third part gives an account of protein metabolism in diseased states. Although written as a multi-author text, the book is designed to be read in continuity by those who have only a general knowledge of the outlines of protein metabolism, as presented in textbooks of general biochemistry. In order to emphasize this continuity, each main part—biochemical, nutritional, and pathological—is introduced by a short general survey of the area covered. It is hoped that, in this way, the treatise will serve as a source-book for the graduate student who requires an advanced survey of the whole subject, and, at the same time, will interest the specialist who needs an authoritative account of selected areas of protein metabolism.

As editors, we should like to say how fortunate we have been in securing as authors many of the leading investigators in the field of protein metabolism, and we are deeply conscious of the honor they do us in participating in the writing of this book. We should also like to acknowledge with gratitude the help and encouragement of many colleagues and friends who have advised us during the assembly of this book. In particular, it is a pleasure to thank Professor D. McKie of the Department of the History and Philosophy of Science at London University for the advice on the historical introduction; Dr. L. Fowden of University College, London, for information on the amino acids found in plants; Dr. J. G. Black of Unilever for providing us with a survey of literature on protein metabolism in the skin; and Dr. G. Leaf of the University of Glasgow for much general advice on protein metabolism. Dr. Renwick of the Genetics Department at Glasgow University kindly gave advice on some of the genetic points in abnormalities of protein biosynthesis, discussed in the introductory material to Part III, and Mr. Robin Callander, Medical Artist at Glasgow University, provided a number of excellent illustrations. We are much indebted to these and many other friends for the help they have given us.

H. N. Munro
October, 1963 J. B. Allison

Table of Contents

Part II

Nutritional Aspects of Protein Metabolism

AN INTRODUCTION TO NUTRITIONAL ASPECTS OF PROTEIN METABOLISM

H. N. Munro

CHAPTER 12

THE NUTRITIVE VALUE OF DIETARY PROTEINS

J. B. Allison

CHAPTER 13

AMINO ACID TOXICITIES AND IMBALANCES

A. E. Harper

CHAPTER 14

PROTEIN REQUIREMENTS

D. M. Hegsted

CHAPTER 15

PROTEIN METABOLISM AND REQUIREMENTS IN PREGNANCY AND LACTATION

K. L. Blaxter

CHAPTER 16

PROTEIN METABOLISM AND REQUIREMENTS IN THE NEWBORN

R. A. McCance and Elsie M. Widdowson

CHAPTER 17

PROTEIN METABOLISM AND REQUIREMENTS IN THE ELDERLY

Donald M. Watkin

Part III

Pathological Aspects of Protein Metabolism

A GENERAL SURVEY OF PATHOLOGICAL CHANGES IN PROTEIN METABOLISM

H. N. Munro

CHAPTER 18

ANOMALIES OF AMINO ACID METABOLISM

L. E. Holt, Jr., and S. E. Snyderman

CHAPTER 19

PHYSICAL INJURY AND ITS EFFECTS ON PROTEIN METABOLISM

D. P. Cuthbertson

CHAPTER 20

PROTEIN METABOLISM AND TUMOR GROWTH

G. A. J. Goodlad

CHAPTER 21

EXPERIMENTAL PROTEIN-CALORIE DEFICIENCY

B. S. Platt, C. R. C. Heard, and R. J. C. Stewart

CHAPTER 22

CLINICAL ASPECTS OF PROTEIN MALNUTRITION

Fernando Viteri, Moisés Béhar, Guillermo Arroyave, and Nevin S. Scrimshaw

CHAPTER 23

PROTEIN DEFICIENCY AND INFECTIVE DISEASE

Nevin S. Scrimshaw

Mammalian Protein Metabolism

Volume I

PART II

Nutritional Aspects of Protein Metabolism

An Introduction to Nutritional Aspects of Protein Metabolism

H. N. MUNRO

I. Introduction

In this part of our treatise, the chapters provide a survey of what is known about protein requirements under various physiological conditions, special emphasis being placed on the needs of man. Not only is man an important species, but in addition the objectives in feeding farm animals are oriented around production of meat and dairy products, whereas the objectives in the case of man are presumably less materialistic. As an introduction to the subject for the benefit of readers who are not aware of the complexity of the nutritional problems of assessing protein requirements, we shall first survey briefly the background against which these subsequent essays should be examined. The material will be presented in two portions: first, methods of measuring the quantity and quality of protein in the diet; second, general aspects of the assessment of protein requirements of the body.

II. Measurement of Quantity and Quality of Dietary Protein

As discussed in the historical chapter introducing this treatise (Volume I, Chapter 1), the first indication that nitrogenous compounds were essential constituents of the diet was provided at the beginning of the nineteenth century by Magendie. Almost a hundred years later, firm indications of differences in protein quality were established by Willcock and Hopkins, and by Osborne and Mendel, and this led eventually to the identification of essential and nonessential amino acids. With this came the realization that the requirement for protein in the diet is in fact a requirement for amino acids, and that proteins containing certain concentrations of essential amino acids are nutritionally more acceptable than others.

This led logically to the proposition that protein quality can be expressed in terms of the concentrations of essential amino acids in the protein molecule. Thus, to evaluate the status of a diet with regard to its protein content, we require means both of measuring the total amount of protein present in the diet and also of evaluating protein quality.

A. Measurement of the Protein Content of the Diet

The most frequently used method of measuring the amount of protein in a tissue or in a foodstuff is by means of nitrogen assay. The usual procedure is to multiply the N content of the diet by 6.25 to give the quantity of protein present. This at once makes two assumptions which have to be examined in some further detail: First, can we safely assume that all, or nearly all, the nitrogen of the diet is present in the form of protein? If this can be accepted as a reasonable approximation, we then have to consider whether we are justified in using a single factor of 6.25 to convert from nitrogen to protein. Once these points have been discussed, the range of variation in protein content observed in different major foods will be examined. Finally, the place of amino acid analysis as an alternative procedure to nitrogen determination will be considered.

1. The Forms of Nitrogen in the Diet

It is commonly assumed that, for the purposes of nutritional studies, the nitrogen of the diet consists essentially of protein. This assumption is not strictly correct even for animal tissues, and is quite erroneous in the case of many plant foods. From the nutritional point of view, the nonprotein nitrogen constituents of foods can be considered in two groups. The first of these consists of utilizable amino acids present in free form instead of as protein; such amino acids will contribute to the biological value of the food as if they were a protein source and it seems a matter of indifference whether they are free or protein-bound. In addition, however, the nonprotein nitrogen fraction of tissues includes substances which are probably not used by the body as nitrogen sources, and should therefore not be considered as part of the nutritionally available dietary nitrogen; this group embraces such substances as urea, creatine, and nucleic acids in the case of animal foodstuffs, and the nonutilizable amino acids found in plant foods.

Analysis of edible tissues, such as muscle and liver, shows that only a small proportion of the nitrogen, some 1 to 2 %, occurs in the form of free amino acids and small peptides (see Chapter 4), and that a minor proportion of the tissue nitrogen is present as nucleic acid nitrogen and phospholipid nitrogen (e.g., Harrison, 1953): meat muscle also contains appreciable soluble nitrogen in the form of creatine, and the dipeptides carnosine and anserine. In fish muscle the proportion of nonprotein nitrogen is made up of a great variety of compounds which account for 9 to 18 % of the total nitrogen in teleost fishes and a much higher proportion in elasmobranch fishes (Tarr, 1958). Some 95 % of the nitrogen of cow's milk occurs as protein, but only about 80 % in human milk (Morrison, 1952; Hytten, 1954) ; of the 20 % of nonprotein nitrogen, only about one-third is present as free amino acids (Am. Acad. Pediat., 1960).

The free nonprotein nitrogen of plant tissues varies considerably both in quantity and in form. In the first place, the proportion of nonprotein nitrogen can be quite large in some edible parts of plants. Whereas in seeds protein accounts for more than 95 % of the total nitrogen, in leaves the proportion is 80 to 90 %, in stems and roots it is about 60 % and in the potato, the carrot, and other storage organs it is as little as 30 to 40 % of the total nitrogen of the tissue (Kulkarni and Sohonie, 1956; McKee, 1958b; Fowden, 1960). A considerable part of the nonprotein nitrogen of these plant tissues consists of small peptides and free amino acids. In the case of the potato, much of the nonprotein nitrogen is accounted for by glutamine and asparagine (Neuberger and Sanger, 1942) and only low concentrations of most essential amino acids are present, except for comparatively large amounts of arginine and lysine (Hughes, 1958).

In the second place, plant tissues provide a profusion of nonprotein amino acids. About 100 amino acids, only some 20 of which occur in the plant proteins, have been identified to date from plant sources (Steward and Thompson, 1954; McKee, 1958a, 1962; Fowden, 1962). The formulas of some of these nonprotein amino acids are shown in Fig. 1. A few of these are also encountered in animal tissues; for example, γ-aminobutyric acid is found in the brain, and β-alanine occurs in muscle in the form of the dipeptides carnosine and anserine. The distribution of these nonprotein amino acids in plants is quite variable; γ-aminobutyric acid and α-aminoadipic acid are widespread in the plant kingdom, whereas hypoglycin A has so far been encountered only in *Blighia sapida*, a West Indian species.

Although not all of these nonprotein amino acids occur in articles of diet, small amounts of a number of them must inevitably be ingested with plant foods. Some, such as *S*-methylcysteine and its γ-glutamyl-dipeptide, which occurs in quite large amounts in dormant seeds of the green bean (*Phaseolus vulgaris*), are presumably metabolized by conventional pathways in the mammalian body. Others may be excreted unchanged; this has been shown to occur in the case of 5-hydroxypipecolic acid (Fig. 1), which is abundant in dates and whose presence in the urine of a patient who consumed large amounts of dates was on one occasion mistaken for an aminoaciduria of metabolic origin (Fowden, private communication). A form of sickness in Jamaica, with acute loss of liver glycogen, has been attributed to ingestion

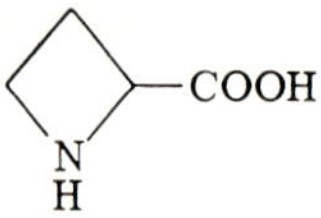

Azetidine-2-carboxylic acid (*Liliaceae*)

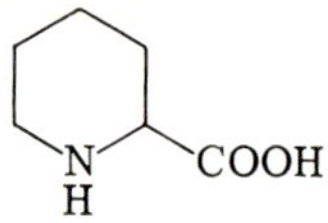

Pipecolic acid (*Legumes*)

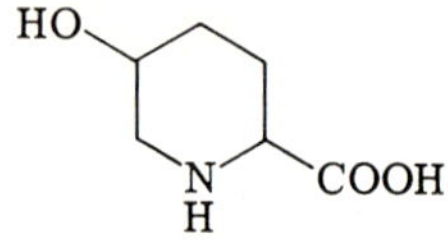

5-Hydroxy-pipecolic acid (*Dates*)

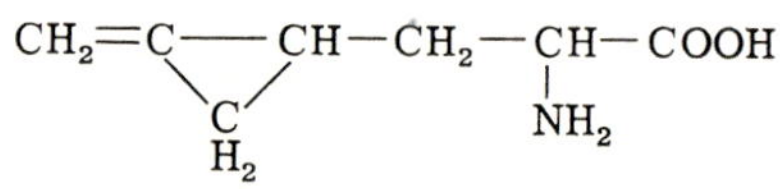

Hypoglycin A (*Blighia sapida*)

FIG. 1. Some nonprotein amino acids occurring in plants.

of hypoglycin A (Fig. 1) present in the fruits of *Blighia*, one of the local plants (Hassal *et al.*, 1954). Another and better known illness of dietary origin is lathyrism, caused by ingestion of the seed of various *Lathyrus* species (especially *L. odoratus*). The toxic factor was originally identified as γ-glutamyl-β-aminopropionitrile; however, this accounts for only the mesenchymal lesions, notably the effects on bone. The neurological symptoms of lathyrism are due to another toxin, possibly β-cyanoalanine (Ressler, 1962); a number of other amino acids and derivatives are present in *Lathyrus* species (Bell and Foster, 1962) and may play a part in producing the toxic effects.

From these considerations, it can be concluded that, in the case of mixed diets, the assumption that dietary N occurs essentially as protein is reasonable though never absolutely correct. On the other hand, when dealing with diets made up mainly of root or tuber crops, such as cassava or potatoes, it is as well to remember that the nitrogen content of the diet is not only low but even so may exaggerate the true protein nitrogen of the food, because of the presence of large proportions of nonprotein nitrogen, some of which may not be available to the body for metabolic purposes.

2. Conversion Factors for Obtaining Protein from Nitrogen

For tables of food composition and in the case of diets, the practice for obtaining the protein content is to multiply the N content by the factor 6.25, on the grounds that most proteins do not deviate seriously from 16 % N content. When such a factor is used, it is best to designate the result as "crude protein," since Jones (1941) showed from a survey of proteins isolated from individual foods that the percentage of N in the total protein mixture for each food could vary from 15.7 in the case of milk to 19 in the case of nuts. For precision, such deviations from the theoretical value of 16 % should be recognized in computing the protein content of a food or a diet. Indeed, in assembling their survey of "The Composition of Foods," McCance and Widdowson (1960) use the factor of 6.25 for all foods except milk or milk products (6.38), cereals (5.7), and gelatin (5.55). Furthermore, in these food tables protein N has been differentiated from nonprotein N in the case of meat, fish, yeast, and coffee extracts, since the nonprotein N of these foods is not necessarily made up of compounds used for metabolic purposes. Other foods contain nonprotein N mainly as amides of amino acids, and are therefore reckoned in with the protein N. In fact, the use of the factor 6.25 under these circumstances may well be misleading since the N content of amides is much higher than 16 %. On the whole, such foods generally contribute little to the total N content of a mixed diet. However, they may assume greater importance when diets predominantly vegetable in nature and of low protein content are under consideration, such as diets consumed in some tropical countries. Thus a sample of cassava flour with a nitrogen content of 0.32%, equivalent to 2% crude protein (N × 6.25), was found on summation of its amino acids to contain only 1.25 % total amino acids; of the 42 % of the total N content not accounted for as amino acid N, two thirds was found as ammonia, only a part of which could have arisen from amides (Close *et al.*, 1953).

3. Range of Variations in Protein Content of Foods

Table I displays the protein concentration per 100 gm of edible portion for some major articles of diet consumed in different parts of the world. The range is wide, from 0.5 % up to 35 % protein. Those at the lower end of the scale are clearly poor sources of protein irrespective of the nutritive value of the protein present.

It is obvious that the amount of protein provided by different foods depends not only on their protein concentration but also on the amount consumed. For this reason, some foods with low concentrations of protein may nevertheless provide useful amounts if consumed in large quantities. There is, however, a limit to this compensatory process, as the following calculations show. In Chapter 14 of this book, Hegsted provides estimates for the minimum intake of protein compatible with maintenance by adult human subjects of N equilibrium, which represents an obvious lower limit of protein intake for survival. From Hegsted's data, it is apparent that, to achieve even N equilibrium, the FAO reference male subject of 65 kg body weight (FAO, 1957a) would need about 30 gm of protein per day, assuming complete utilization of the dietary protein (biological value

TABLE I

PROTEIN CONTENT OF VARIOUS MAJOR FOODSTUFFS[a]

Foodstuff	Protein content	
	As gm per 100 gm edible portion	As percentage of total energy content of food
Sago flour	0.5	0.6
Cassava, fresh	0.7	1.8
Cassava, flour	1.5	1.8
Plantains	1.0	3.1
Sweet potatoes	1.5	5.3
Irish potatoes	2.0	10.7
Yams, fresh	2.0	7.7
Yam flour	3.5	4.4
Milk	3.3	20.6
Rice, polished	7.0	8.0
Maize meal (96% extraction)	9.5	10.5
Sorghum flour	10.0	11.3
Wheat flour (70% extraction)	10.0	11.4
Eggs	13.0	33.0
Groundnuts	15.0	18.1
Fish, lean	17.0	93.0
Beef, lean	19.0	37.6
Beans, kidney	24.0	27.3
Soya bean seed	35.0	36.6

[a] From Platt (1962).

of 100). This amount of protein represents about 4% of the total energy intake of 3200 kcal needed by the FAO reference man. Accordingly, if the diet provides less than 4 % of its energy in the form of protein, the subject can never consume enough of it to assure N equilibrium, since appetite will limit his food intake before he has obtained the necessary amount of protein. Since the biological value of the mixed proteins of the diet will inevitably be less than 100, then the critical concentration of protein calories becomes even higher.

In accordance with this concept of a minimal proportion of dietary energy as protein in order to provide even the minimum level for N equilibrium, the data on the protein content of various foodstuffs have also been computed in Table I to show the percentage of the energy content of the food provided by protein. Clearly, some foods cannot by themselves provide the minimum protein needed for maintenance of the adult, and are even less adequate for the greater demands of growth. For example, two important tropical foods, cassava and plantain, combine a low protein content with a high starch content, and thus have a low protein-calorie ratio. On the other hand, the potato, although almost as low in absolute protein content, contains much less starch and in consequence has a more favorable protein-calorie ratio, so that larger intakes of this food can be tolerated before the limit of appetite is attained. This may be the reason why the Irish of previous centuries, though living mainly on large quantities of potatoes (Woodham-Smith, 1962), nevertheless had a reputation for sound physique.

4. Analysis of Foods for Amino Acid Content

It is apparent from the foregoing discussion that measurement of the protein content of a diet from nitrogen analysis can sometimes be in error, first because of the presence of nonamino acid nitrogen, and second because a single conversion factor from nitrogen to protein is not universally applicable. Such sources of error are not of much moment in assessing mixed Western diets, for which tables of food composition provide a reliable guide to the relatively large quantites of protein provided from animal sources. The errors are likely to be more important in assessing the degree of protein insufficiency of tropical diets, where crops of low nitrogen content, such as cassava, are the main protein source (e.g., Close *et al.*, 1953).

If we accept that the function of protein in the diet is confined to providing amino acids, and notably essential amino acids, these uncertainties can be reduced by determining, in place of crude protein, the amounts of individual amino acids contained in each foodstuff. Automatic amino acid analytical techniques using column chromatographic separation can now provide recoveries of amino acids with errors of less than 3 % and conditions for protein hydrolysis have been worked out which permit satisfactory liberation of amino acids from proteins even in the presence of large amounts of carbohydrate (Dustin *et al.*, 1953). This approach suffers from the defects of expense of equipment and limitation to the number of samples which can be examined at one time. Microbiological assay of amino acids provides an alternative method of estimating the amino acid content of proteins and has the advantage that many samples can be run at one time and that the hydrolysis of the protein need not be so complete as that demanded by column chromatography, with consequent reduction in the risk of destroying amino acids during hydrolysis.

It is clear that information about the amino acid composition of foodstuffs will eventually supersede measurements based on nitrogen content, and already some data of this kind have been assembled (Orr and Watt, 1957). Hughes (1959) has demonstrated that the amino acid content of mixed diets can be accurately predicted from published data of the amino acid composition of foods, but he found that the application of such information to a vegetarian diet was less successful. In addition, as will emerge later in this discussion, a knowledge of the amino acid content of diets is not the whole story, since not all the amino acid content of a food may be available to the body.

B. Measurement of Protein Quality

As discussed in detail in Chapter 12, tests designed to evaluate protein quality are of two kinds. First, the quality of a dietary protein can be tested in relation to some function of the body such as growth or nitrogen balance; such tests do not demand any knowledge of the amino acid composition of the protein. Alternatively, the pattern of amino acids present in the protein can be compared with the pattern required by the recipient; such methods of evaluation of protein quality require precise information about the amino acid composition of the protein and also about the needs of the subject for individual amino acids. These biological and chemical methods of evaluation will now be considered in turn.

1. Biological Methods of Protein Evaluation

The earliest evidence of differences in protein quality were obtained by studies made on growing animals, and this led to the development of the "Protein Efficiency Ratio" (PER) as a means of evaluating protein quality by growth performance (Osborne *et al.*, 1919). This test recognizes the fact that animals receiving diets of low protein quality will show some loss of appetite; therefore, PER relates the gain in body weight to the amount of protein consumed during the period of study. It is usual to incorporate the test protein at a level of 10 % by weight in a synthetic diet providing excess of all other nutrients than protein. For example, a series of cooperative tests carried out with young rats in different laboratories gave average PER values of 2.8 for casein, 2.4 for soy protein, and 0.4 for wheat gluten (Derse, 1960); that is, the young rat under defined conditions will gain 2.8 gm body weight for each gm of casein eaten, 2.4 gm per gm of soy protein and 0.4 gm per gm of wheat gluten.

It is to be noted that the protein efficiency ratio and other tests based on the same principle (net protein ratio, nitrogen growth index) have the disadvantage of being limited to the growing animal and are furthermore unlikely to appeal to those studying the needs of the human infant. These disadvantages do not apply to assessment of protein quality by nitrogen balance methods, which can be used on adult as well as on growing subjects and are applicable to studies on man. The principle involved is illustrated by Fig. 2, which is a diagrammatic representation of the procedure adopted by Karl Thomas in 1909 when he carried out the first determinations of biological value, using himself as subject. When the subject takes a diet

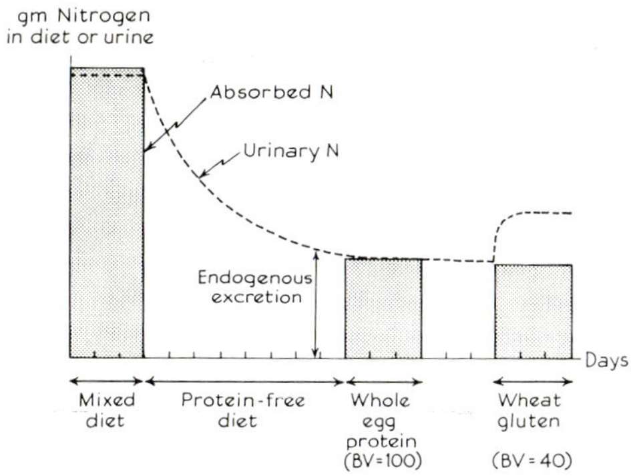

FIG. 2. Schematic representation of the procedure for measuring the Biological Values of dietary proteins by the Thomas-Mitchell method.

containing no protein, his output of nitrogenous compounds in the urine and feces falls but does not cease altogether; this continuing release of nitrogen from the body is sometimes referred to as the "endogenous nitrogen output." Since the subject is receiving no dietary protein, his nitrogen balance (the difference between nitrogen intake in the diet and nitrogen output in the urine and feces) will be negative to the extent of the nitrogen lost in urine and feces on this protein-free diet. Let us now add to the protein-free diet a protein which is (*a*) fully absorbed, (*b*) provides an amount of nitrogen equal to the nitrogen lost in the urine, and (*c*) contains a pattern of amino acids in optimal proportions for utilization by the body. Such a dietary protein will be fully utilized to replace the tissue loss of nitrogen and in consequence none of it will be excreted as extra nitrogen in the urine. This occurs when the protein of whole egg is fed to the rat, since at low intakes virtually 100 % of the absorbed amino acids are utilized for tissue replenishment. If, on the other hand, a protein is added to the protein-free diet which is not of this high quality, such as wheat gluten, some of the absorbed protein N will be excreted as extra urinary nitrogen because the concentrations of the various amino acids in the poor-quality protein differ from the pattern required by the body.

Using nitrogen balance in this way, the "Biological Value" (BV) of a protein (BV = retained N/absorbed N × 100) can be measured in either young animals, adult animals, or in man (Table II). It is apparent that proteins which are really completely retained, such as whole egg protein, will have biological values approximating to 100. Such values are, however, only observed in the adult animal receiving less nitrogen than is needed to bring him into N equilibrium. When larger amounts than this are given, the utilization of the nitrogen falls off progressively. In consequence, determinations of biological value are carried out at suboptimal levels of

TABLE II

BIOLOGICAL VALUES OF SOME PROTEINS DETERMINED BY NITROGEN BALANCE MEASUREMENT ON GROWING RATS, MATURE RATS, AND ADULT MEN[a]

Protein	Biological Values		
	Growing rat	Mature rat	Mature human
Egg albumin	97	94	91
Beef muscle	76	69	67
Casein	69	51	56
Wheat gluten	40	65	42

[a] Cited with permission from Mitchell (1959).

protein intake. At these low levels, some tissues lose a considerable part of their cellular protein, and the biological values obtained under these conditions consequently represent the relative merits of different proteins in the partly protein-depleted animal. It is an assumption to conclude that the same relative scale of protein values will apply in the maintenance of high levels of labile protein stores. Consequently, studies on the growth and N balance of the whole animal must be supplemented by examination of the effect of dietary protein quality on the protein content of individual tissues, especially of those tissues whose protein content is particularly sensitive to the level of protein fed in the diet. In fact, Henry *et al.* (1953, 1961) have reported some data for the relative values of various dietary proteins in causing deposition of labile liver protein, and have demonstrated fair but not complete agreement with the ranking of the same proteins by biological values determined on the whole animal by nitrogen balance methods. More studies are required on the influence of protein quality on the performance of individual tissues of the body.

Finally, it is to be noted that the biological value of a protein is not fixed, but varies with the physiological state of the test animal; thus biological values are higher in protein-depleted animals (Allison, 1955), and can vary from growing to adult animals (Table II).

2. Chemical Methods of Protein Evaluation

Ever since it was first observed sixty years ago that proteins differ in their capacity to support growth, nutritional interest has centered around the 20 or so amino acids that occur in proteins. The process of identifying which of these amino acids have to be provided preformed in the diet was eventually resolved by Rose's studies with rats growing on diets containing mixtures of amino acids in place of protein. When growth was impaired by removing one amino acid from the complete mixture, it was classed as essential (indispensable), whereas those which could be removed without influencing growth rate were nonessential (dispensable). An essential amino acid was therefore defined by Rose (1938) as one which "cannot be synthesized by the animal body from materials ordinarily available at a speed commensurate with the demands for normal growth." In a mixture of 19 amino acids supporting good growth, only 10 were found to be essential, namely, arginine, histidine, isoleucine, leucine, lysine, methionine, phenylalanine, threonine, tryptophan, and valine. Omission of one of these caused immediate and complete cessation of growth, except in the case of arginine, withdrawal of which reduced growth rate to some 70 to 80 % of that obtained with the complete mixture; arginine is nevertheless included in the list of essential amino acids, since its absence is not compatible with optimal growth rate.

Rose's definition also allows for the capacity of α-oxo and α-hydroxy acids corresponding to the essential α-amino acids to act as dietary substitutes for the essential amino acids, since these acids are not "ordinarily available" in the diet. Except in the cases of the oxo and hydroxy acids corresponding to isoleucine, lysine, and threonine, the body can form the corresponding essential amino acid by transamination. This emphasizes that the failure to make essential amino acids lies in the production of the carbon skeleton, and this has been confirmed by feeding C^{14}-sucrose to mice and isolating the amino acids individually from the body proteins: none of the isotope was found in the essential amino acids, with the exception of histidine and the methyl group of methionine, whereas the nonessential amino acids contained an abundance of radioactive carbon (Steele, 1952). The mammal can also derive some of the essential amino acids from the corresponding D forms (see Berg, 1959, for a review).

With regard to the nonessential amino acids, it has to be remembered that two of these (tyrosine and cystine) can be formed in the body from two essential amino acids (phenylalanine and methionine, respectively) by irreversible reactions, and consequently the amounts of tyrosine and cystine in the diet have a bearing on the requirement of the body for phenylalanine and methionine.

The amino acid requirements of man have been determined by N balance procedures which have demonstrated a need for all the amino acids essential to the growing rat except for arginine and histidine. When any of the other eight essential amino acids was omitted from a synthetic mixture, nitrogen balance became negative. This does not necessarily mean that deletion of arginine or histidine is without effect on protein metabolism; Holt *et al.* (1942) have demonstrated a reduced production of spermatozoa by human subjects receiving amino acid mixtures lacking arginine, presumably by affecting production of protamine, a protein with a large content of arginine.

Rose and his associates have also determined the minimum amounts of each essential amino acid needed by the young adult subject to attain N equilibrium (Rose, 1957). Other workers have explored the essential amino acid requirements of subjects and have obtained data differing in one or other respect from those of Rose and also from one another; the reasons for these discrepancies are discussed in Chapter 14. It should be noted that the importance of establishing precise amino acid requirements is that it would provide a pattern of essential amino acid requirements with which to compare the composition of dietary proteins. If the function of dietary protein is to serve as a source of amino acids for the body, the most economical protein will be one which provides a mixture of amino acids, both essential and nonessential, in the proportions needed by the body. If a given dietary protein is, by comparison with the pattern, low in any one of the essential amino acids, this will become the factor limiting the usefulness of that protein: the "limiting amino acid."

The uncertainties of the exact pattern of amino acid requirements make it necessary to consider alternative means of determining the optimum proportions of essential amino acids. Since the protein mixture in whole egg has a Biological Value of 100, the concentrations of essential amino acids present in this protein source are often taken to be present in optimal proportions and are used as a reference pattern. In Table III the concentrations of essential amino acids in an animal protein (beef muscle) and a vegetable protein (wheat gluten) are given as percentages of the concentrations of these amino acids as found in whole egg protein. None of the essential amino acids of beef muscle falls below 69 % of the concentration found in the reference protein; on the other hand, the lysine concentration in wheat gluten is only 33 % of that in whole egg protein; indeed, all the essential amino acids of wheat gluten are less abundant per 100 gm than in the reference protein. Lysine thus becomes the limiting essential amino acid in gluten, with methionine the second most limiting; indeed, these two amino acids are often found in low concentrations in plant proteins. If lysine is the factor limiting the extent to which gluten can be used in the body, it might be predicted that the degree of deficiency of lysine would be a measure of the Biological Value of wheat gluten. The term "Chemical Score" was therefore introduced by Block and Mitchell (1946) to describe the relative concentration of the limiting amino acid in each protein; the Chemical Score of beef protein would thus be 69 and the score for gluten 33. Do these chemical scores agree with the measurement of protein quality by biological tests? Figure 3 shows that proteins of animal origin tend to have a high Biological Value and a high score, whereas those of plant origin have a lower Biological Value and a smaller score. Agreement is not perfect, since proteins such as zein from maize, with a chemical score of

TABLE III

PERCENTAGE OF IDEAL CONCENTRATION OF ESSENTIAL AMINO ACIDS OBSERVED IN A TYPICAL ANIMAL PROTEIN AND A TYPICAL PLANT PROTEIN[a]

Essential amino acid	Percentage of concentration in whole egg protein	
	Beef (%)	Wheat gluten (%)
Histidine	154	94
Isoleucine	84	74
Leucine	87	80
Lysine	141	*33*
Methionine	84	52
Phenylalanine	*69*	89
Threonine	90	57
Tryptophan	89	68
Valine	73	61

[a] Data computed from Allison (1955).

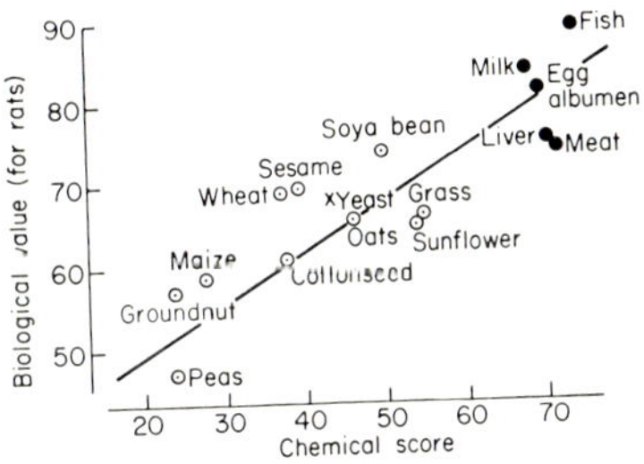

FIG. 3. Correlation diagram of the Biological Value and the Chemical Score for 16 animals protein sources (full circles) and vegetable protein sources (open circles). (From Carpenter, 1951.)

zero because of complete absence of an essential amino acid, nevertheless have some biological value, presumably because they fulfill functions in the body other than supplying amino acids for protein synthesis.

Several attempts have been made to devise improved forms of chemical scoring of protein quality. All eventually depend on the pattern of amino acid requirements of the body which is accepted as the standard for comparison. Not only is this disputed, but all the patterns so far proposed have been measured at minimal levels of amino acid intake to maintain N equilibrium and are thus subject to the criticism made of the biological tests of protein quality, namely, that the optimum pattern for various physiological functions may differ from the optimum pattern for maintenance.

It will be noted that the concept of chemical score assumes that the nutritive value of a dietary protein is determined entirely by its gross content of each amino acid. If peptide sequences within the protein molecule do play a significant part in nutrition, chemical scores will be inadequate for the prediction of protein quality. Some years ago, certain peptide fractions ("strepogenin") in partially hydrolyzed proteins were found to stimulate the growth of bacteria (Sprince and Woolley, 1944). Since the growth rate of rats receiving synthetic amino acid mixtures or on completely hydrolyzed proteins is often inferior to growth on diets containing intact protein (Womack and Rose, 1946), it was surmised that peptides similar to strepogenin might be of importance in animal nutrition. Although some authors (e.g., Woolley, 1946) have considered that partial protein hydrolyzates improved the growth of such animals, many others could find no evidence for a growth-stimulant action of peptides given to animals (Frost and Sandy, 1948; Silber *et al.*, 1949; Maddy and Elvehjem, 1949; Ramasarma *et al.*, 1949; Kodicek and Mistry, 1957). There is evidence that some peptides are absorbed more rapidly into the bacterial cell than their constituent amino acids, and the subsequent hydrolysis of the absorbed peptide then provides the cell with an abundant source of these amino acids (Kihara and Snell, 1955; Rowlands *et al.*, 1957). This mechanism of action of strepogenin on bacterial growth seems more likely than a specific function of peptides in cell metabolism or protein synthesis. We may therefore ignore peptide sequence when assessing protein quality on the basis of chemical composition.

3. Supplementary Effects between Dietary Proteins

Study of the Biological Values of isolated, purified proteins or of single foods is artificial. The amino acids absorbed from the intestine after a meal are a mixture derived from all the proteins present in the meal. The mixed diet of Western countries provides a variety of proteins of both animal and vegetable origin, and the amino acid deficiencies of one type of food tend to be compensated by a surplus in another, so that the biological value of the mixture is better than one would expect from simple averaging of the Biological Values of the individual dietary proteins. For example, Geiger (1948) observed that rats failed to make a satisfactory weight gain on diets containing either 9% wheat gluten or 9% blood protein as the sole protein source. However, growth was resumed when a mixture of 4½% of each protein was present in the diet. As noted above, gluten is particularly deficient in lysine, and the blood protein is mainly deficient in isoleucine; each could therefore compensate to some extent for the other's deficiencies. For this supplementary effect to take place, the two proteins must be consumed in the same meal; thus

Geiger (1948) observed no supplemetary effect when the meals of gluten and blood protein were fed 12 hours apart to the rats. The supplementary relationships between proteins is of immense practical importance in planning for the compensation of dietary deficiencies in areas where protein malnutrition is prevalent. It is clear that this must be done by encouraging the local populations to grow foods containing proteins with different limiting amino acids.

An interesting aspect of amino acid patterns in relation to protein quality is the occurrence of imbalances and toxicities resulting from excess quantities of amino acids present in naturally occurring or synthetic mixtures. These phenomena and their relevance to nutrition are discussed in Chapter 13. The practical effects are probably of most importance with diets low in protein.

4. Availability of Dietary Amino Acids

The amounts of amino acids present in the diet are not necessarily the amounts available to the body. Availability of dietary amino acids can be reduced by (*a*) incomplete digestion and absorption of a general kind, (*b*) the presence of inhibitors of digestive enzymes found in some legumes, (*c*) damage to food proteins and amino acids from heat treatment.

The amounts of amino acids which escape absorption are measured by true digestibility (total fecal N on the diet less fecal N output when a protein-free diet is fed) and there are many data recorded in the literature to show that absorption is usually nearly complete. For example, Henry and Kon (1957) report values for rats fed diets containing 8 % of different proteins; true digestibility was 98 % for egg protein, 97 % for casein, 88 % for soybean, and 86 % for a deteriorated dried skim milk powder. The reasons for incomplete absorption are varied. In mucoproteins, some stretches of peptide chain adjacent to sugar residues have been found to be totally resistant to attack by digestive enzymes (Gottschalk and Fazekas de St. Groth, 1960; Rosevear and Smith, 1961; Johansen *et al.*, 1961). The presence of other nutrients in the meal and particularly the nonutilized cellulose walls of plants are other factors which reduce available nitrogen. It is difficult to assess whether lack of complete absorption affects amino acids selectively or is more general. Several authors (Kuiken and Lyman, 1948; Kuiken, 1952; Gupta and Elvehjem, 1957; Gupta *et al.*, 1958; Borchers, 1962) have attempted to evaluate this by assaying the recoveries of individual dietary amino acids in the feces, but perusal of Chapter 2 will convince the reader that this approach is complicated by mixture of amino acids from the diet with amino acids derived from the endogenous secretion of the gut wall and by the transformations of amino acids by bacteria in the lower parts of the intestine.

It is reasonable to suppose that utilization of dietary protein will be reduced by the presence in the diet of inhibitors of digestive enzymes. The presence of heat-labile trypsin inhibitors in soybeans and other legumes was thought to explain the improvement in utilization of protein of such foods when they were heat treated (Klose *et al.*, 1946). However, recent investigation (Birk and Gertler, 1961) demonstrates that a considerable amount of the trypsin inhibitor can be extracted by chemical means without improving the nutritive value of the soybean protein, and in consequence the beneficial effects of heat treatment on soybean protein may arise for other reasons.

In other instances, the availability of certain amino acids, especially lysine, methionine, and cystine, can be depressed by heat treatment. Depression of the nutritive value of protein by heat was first noted by McCollum and Davis (1915), who autoclaved casein. The causes of damage by heat are varied. The Maillard reaction between sugars and amino acids can produce unavailable products; free ϵ-amino groups of lysine in the intact protein combine under conditions of moist heat to form a linkage that resists enzymic hydrolysis. Loss of available lysine is important, since it is frequently the limiting amino acid in plant proteins. Carpenter (1958, 1960) has taken advantage of the reaction between fluorodinitrobenzene and free ϵ-amino groups to estimate the proportion of lysine molecules in proteins which have been damaged by processing.

The extent of heat damage varies considerably with different foods. Ordinary domestic cooking procedures and commercial canning methods have only trivial effects on the nutritive value of beef protein (Mitchell *et al.*, 1949; Mayfield and Hedrick, 1949a). The baking of bread involves a loss of about 15% of the available lysine (Rosenberg and Rohdenburg, 1951), but more extensive damage has been observed in the nutritive value of a cake mix during baking and especially when slices of the cake were toasted (Block *et al.*, 1946). The greatest hazard arises in the heat processing involved in making milk powder, fish flour, and oilseed meals such as soybean meal (Carpenter, 1958; Mauron, 1961).

These studies on soybean inhibitor and on the effects of heat treatment of proteins demonstrate that nutritive properties can alter without any change in gross amino acid composition. For this reason, amino acid analysis of food proteins can sometimes be an inadequate guide to their nutritive properties and justifies continued use of biological tests of protein quality even when amino acid composition is known. An interesting survey of the various methods by which availability of amino acids can be determined is provided by Mauron (1961). On the whole, availability of amino acids in many foods must be reasonably complete, since the Biological Value and chemical score of proteins generally correlate well (Fig. 3).

Platt *et al.* (1961) have introduced the term "Net Protein Utilization" (NPU) to express in a single index both the digestibility of the protein and the biological value of the amino acid mixture absorbed from the intestine. Whereas biological value is the proportion of the absorbed N retained, NPU represents the proportion of the food N retained (NPU = retained N/food N × 100). In the practical evaluation of diets, this can be taken one step further by combining quality with quantity. For example, if a foodstuff contains 5 % protein with NPU of 40, it will have the same value to the body as a food containing 2 % of protein which has an NPU of 100, i.e., which is completely utilized. Platt and his associates express this concentration of fully utilizable protein as a percentage of the energy content of the food. If, in the above example, the food contained 200 kcal per 100 gm, the concentration of ideal pattern protein present (2 %) would consequently provide 4 % of the total available energy [that is, 2 % × 4 kcal per gm × (100/200)]. This method of expression, termed "Net Dietary-protein Calories per cent," (NDpCal%) recognizes that the requirement of the consumer for energy will limit the amount of any foodstuff which he can eat. Platt *et al.* have computed that, when the NDpCal% of the foodstuff falls below 4.6, it can no longer meet the requirements for protein of the adult; thus, if the average adult human subject is considered to have a requirement for completely utilized protein of 31 gm daily and an energy requirement of 2960 kcal daily, 4.6 % of his energy intake must be in the form of the ideal pattern protein. If the diet contains less than this concentration, he will not consume enough of the diet to meet his protein needs before appetite for energy is satisfied. Such calculations make the assumption that the protein needs of the heavy worker increase in proportion to his extra energy expenditure. This is a debatable point and, if mechanical work does not in fact alter protein requirements, then the critical proportion of calories needed as protein will diminish to 3.4 % at 4000 kcal. It will also be noted that, since the protein requirements of individuals within the population may vary because of inherent metabolic differences (Section III,B), NDpCal% can only be applied to assessing the adequacy of the protein intakes of groups of subjects and not to the intakes of individuals. The studies of Platt and his colleagues provide a practical approach to the problem of expressing by means of a single index the amount of protein available to the body from diets as they are consumed.

III. Assessment of Protein Requirements

For considerably more that a century, it has been known that protein is a necessary constituent of the diet. The history of the study of quantitative needs of the body for protein is set out in Chapter 1 of this series. The earliest estimates were based on observations of the amounts of protein consumed by various groups of the population. Thus, from dietary surveys, Playfair (1865) set down protein needs ranging from 57 gm daily for mere subsistence to 184 gm daily for those undertaking heavy manual work. In 1881, Carl Voit concluded that the average working man had a protein requirement of 118 gm per day, since this was his mean intake. On the basis of studies involving groups of healthy adults, and lasting several months, Chittenden (1905) concluded that protein intake can be safely reduced to 50 gm a day, and that larger intakes are positively dangerous. Although Chittenden's studies were an attempt to determine experimentally the optimum level of protein intake, they were of too short duration to be a reliable guide to the needs of man throughout a lifetime.

The problem of how much protein should be provided at different stages in the life of man is one which is currently exercising many national and international committees. The detailed evidence on which decisions can be reached is considered in the chapters which follow. Here, we shall review briefly the general principles involved. First, what evidence can be adduced from the study of protein metabolism to suggest an optimum level of protein intake? Second, how do age, sex, and other physiological factors influence protein requirements? Lastly, what range of protein intakes have been observed in different groups, and are any of them inadequate?

A. General Principles Involved in Assessing Protein Requirements

Certain aspects of protein metabolism, most of which have been discussed in detail in earlier chapters of this book, have an obvious relevance to the assessment of protein needs.

1. Minimum Protein Intake for Nitrogen Equilibrium

A diet which does not even allow the subject to attain N equilibrium is patently inadequate in protein content. The minimal target which would be acceptable must obviously permit N equilibrium to be achieved. Some investigators also regard this as an adequate level of protein intake, greater intakes being considered an unnecessary luxury. Methods of evaluating the minimal protein intake for N equilibrium are discussed in detail in Chapter 14 by Hegsted, who makes a penetrating analysis of available data. His conclusions need only be summarized here briefly.

It will be recollected that, even when a subject consumes a diet containing no nitrogen, there is a continuous loss of nitrogen from the body, the endogenous N output demonstrated in Fig. 2. Clearly, enough protein must be consumed to replace this unavoidable nitrogen output, and it is therefore important to be able to assess the extent of endogenous losses. Over the range of mammals, endogenous N excretion in the urine approximates to 2 mg per kcal of basal energy (Terroine, 1936) and this can be used to estimate endogenous urinary N for human subjects of various sizes from a knowledge of their basal metabolisms. To this has to be added endogenous fecal N output and also skin losses of nitrogen. The computations made in Chapter 14 show that these various endogenous losses would represent about 30 gm of body protein per day in the case of the FAO reference male subject of 65 kg weight. One may therefore conclude that, given a diet with protein of biological value 100 which is fully absorbed, the subject will come into N equilibrium on 30 gm of protein a day. If the dietary protein is of average biological value 70, it will be proportionately less well utilized and this minimal intake must be increased to 43 gm daily (0.65 gm/kg body weight). It may be noted that, in 1865, Playfair assessed the minimum protein intake for subsistence at 57 gm!

2. *Effect of Higher Protein Intakes on Protein Metabolism*

In practice, the populations of Western countries consume on an average about 80 to 110 gm of protein daily, amounts well above the minimum level for attaining bare N equilibrium. Does the excess have any beneficial effects on the body? Those who claim that it does point to several possible benefits of a high protein intake: (*a*) accumulation of labile protein "stores" in the body, (*b*) greater resistance to infection, and (*c*) perhaps some advantages in response to and recovery from injury.

As seen in Fig. 2, the giving of a protein-free diet does not immediately result in a reduction in N output to the endogenous level; there is an intervening period during which the body loses a considerably greater amount of N. This is considered to represent dissipation of labile body protein, the evidence for the existence of which is considered in Chapter 10. When just enough protein is provided to bring the subject back into N equilibrium (Fig. 2), this amount will not restore the labile protein stores of the body to the levels present in the well-fed subject who, in many cases in Western countries, consumes two or three times the amount of protein needed to bring him into N equilibrium (Table V). Does it matter, then, if the recommended protein intake provides only enough protein to keep the subject in equilibrium and not enough to fill his protein "stores"? Are labile protein stores of any nutritional significance?

Labile protein stores represent one end in a continuous spectrum of tissue protein stabilities toward variations in nutritional conditions. As discussed in Chapters 10 and 12, the protein content of some tissues, such as liver, pancreas, and gut mucosa, changes rapidly in response to variations in the intake of protein; at the other end of the spectrum are tissues like tendon, brain, and heart whose cells resist for long periods the effects of an inadequate protein intake. Thus, a few days of feeding with a protein-free diet can result in the liver cell losing a quarter of its original protein content (Kosterlitz, 1947). This is not to be interpreted as the emptying of an inert store of protein, since the loss of protein is accompanied by changes in the metabolism of ribonucleic acid and of phospholipid in the cell (Munro and Clark, 1960). Instead, it is the overt evidence for a complex readjustment in the mechanism for protein synthesis within these cells, in response to changes in amino acid supply. The effect of this change in tissue protein content will now be considered, first, in relation to the general metabolism of body protein as a whole, and second, in relation to the functional activity of tissues from which such large losses occur on depletion.

Does the presence of labile protein in the body confer benefits by acting as a readily available source of amino acids? Opinion is divided. Holt *et al.* (1962) thought that rats subjected to a protein-free diet were not benefited by a previous high intake of protein, and Samuels *et al.* (1948) reached the same conclusion when studying the effect of the preceding diet on survival of starving rats. On the other hand, Shapiro and Fisher (1962) consider that reserve protein deposited in the chick by feeding a high intake of protein can induce better growth during a subsequent period of low protein intake.

The second consideration is that cells depleted of labile protein and other cytoplasmic constituents may undergo impairment of function. In the case of the liver, it has been shown (Wang *et al.*, 1949) that the capacity to clear bromosulfalein declines in parallel with loss of liver protein. Other functions of the liver, such as the capacity to inactivate estrone (Vasington *et al.*, 1958) and to detoxicate chloroform (L. L. Miller and Whipple, 1940), are impaired in animals receiving a low intake of protein. On a low protein intake, there is a loss of coenzymes and of minerals, and the capacity of the body to retain riboflavin and nicotinic acid becomes reduced (Bro-Rasmussen, 1958). Whether these functional changes are critical to the long-term health of the body as a whole has not been established.

In addition to deposition of labile body protein, there is now considerable evidence that an abundant intake of protein improves resistance to many forms of infectious disease. The evidence for this conclusion is presented in detail in Chapter 23. With regard to the effect of labile protein reserves on the capacity to meet stress situations, Halac (1961) conducted a series of tests on rats which received high protein intakes and were then subjected to severe exercise, to food and water starvation, to X-irradiation, or to convulsions produced by metrazol injection. In none of these tests did the animals on a 64% casein diet show a better performance or survival than rats on a 27% casein diet; however, the conclusion might well be different if rats on a low intake of protein were included in the study. There is another type of stress in which previous level of protein is known to affect the response. Previous dietary intake of protein has a bearing on the catabolic response to injury. After receipt of an injury, there is an outpouring of nitrogen from the body, in which the labile protein stores are dissipated; consequently, the metabolic response to injury is more extensive in subjects previously on a high level of dietary protein (Munro and Chalmers, 1945). It is thus possible that redistribution of amino acids from labile body protein might be of benefit in the healing process, but as yet the biological significance of this catabolic response in promoting the healing of the injured tissue has not been established (see Chapter 19). A type of experimental injury in which labile protein reserves perform a more definite function is the restoration of plasma proteins after their removal by plasmapheresis. If well-nourished dogs are subjected to acute reduction in plasma protein level by plasmapheresis, they can continue to regenerate plasma proteins from labile protein sources within the body; on the other hand, the protein-depleted dog does not have this capacity to provide materials for plasma protein regeneration (Maddon and Whipple, 1940). There is therefore some evidence to suggest that the capacity of the body to deal with certain emergencies is affected by the amount of labile protein in the body.

From these observations, the only firm information to emerge is that the protein content of some organs in the body varies very rapidly with change of diet, but the functional significance of this is difficult to establish. In trying to decide whether maintenance of these very labile tissue proteins should be included in calculations of protein requirements, it is important to recognize that there are two fairly sharply separated phases of nitrogen loss when subjects are given a protein-free diet (see Chapter 10, Section II, C, 2 for evidence). As shown by the nitrogen excretion curve in Fig. 2, there is at first a phase of rapid loss lasting a few days and this corresponds mainly to disappearance of labile protein from the liver and some other viscera. After a few days of protein deprivation, other tissues such as muscle begin to lose protein and become the main source of the irreducible minimum nitrogen loss continuing as long as the protein-free diet is fed. It has been argued in a recent (unpublished) report of W. H. O. and F. A. O.

on protein requirements that this latter phase represents true protein depletion, whereas the initial outpouring of urinary nitrogen is due to a metabolic adaptation of the viscera to lack of amino acids. If this distinction can be sustained, it means that the extra loss of nitrogen at the beginning of protein deficiency (Fig. 2) is not to be regarded as of crucial importance to the organism because the protein involved in the rapid adjustment of these tissues to diet may be considered to be an extension of the free amino acid pool. Once this phase of adjustment of the viscera to the protein-free diet is complete, an intake of protein sufficient to balance the "endogenous" level of loss will prevent true depletion. This is the viewpoint taken in Chapter 14 for the computation of the minimal protein needs of adults. However, only further experimentation can decide whether labile liver and visceral protein really has no function other than as part of an unstable metabolic pool.

3. Energy Intake and Protein Utilization

The interrelationships of energy metabolism and protein metabolism are discussed in detail in Chapter 10 of this treatise. All the experimental evidence conspires to show that the N balance of the adult subject subsisting on an adequate intake of protein is sensitive to changes in energy intake, whether from carbohydrate, fat, or ethanol, irrespective of whether the change is a large one or a small one. Increments in energy intake beyond requirements lead to N retention, whereas removal of energy from the diet results in N loss from the body, the magnitude of this action of energy intake on N balance being about 2 gm N for each 1000 kcal added or subtracted. It may be concluded that an inadequate intake of energy will by itself lead to loss of body protein, and will consequently aggravate protein deficiency in the diet. Presumably populations subjected to chronic insufficiency of energy intake adapt so that the effect on body protein wears off, perhaps through reduced basal energy requirements. For example, Keys *et al.* (1950) reduced the energy intake of 10 human subjects from their normal level of 3500 kcal/day to 1600 kcal/day for 24 weeks. During the entire experiment, nitrogen balance was negative, and this can be attributed to the low level of energy intake, since the moderate reduction in protein intake associated with these submaintenance diets would produce only transient effects on N balance. There was, however, some adaptation to the submaintenance caloric intake, since the average nitrogen balance was −3.8 gm/day during the first 12 weeks but only −1.3 gm/day during the second 12-week period. This adaptation was associated with a diminution in basal metabolism.

Does energy intake affect the biological value of dietary protein? Allison *et al.* (1946) demonstrated that the improvement in N balance caused by adding a standard amount of protein to the diet of dogs was not affected by caloric level until this had been reduced to less than half the animal's energy needs. The converse also occurs, namely that addition of energy to the diet has a beneficial effect on N balance when the diet contains adequate amounts of protein, but at low levels of protein intake this N-retaining action of extra energy is impaired (Munro and Naismith, 1953). From a survey of published data on human subjects, Calloway and Spector (1954) concluded "To the general principle set forth—that on a fixed adequate protein intake, energy level is the deciding factor in nitrogen balance and that with a fixed adequate caloric intake, protein level is the determinant—may be added a corollary. That is, at each fixed inadequate protein intake there is an individual limiting energy level beyond which increasing calories without protein and protein without calories is without benefit." Thus, addition of protein to the diet of an undernourished population may not be fully effective if the energy intake is not adequate, and conversely addition of energy-yielding nutrients may fail to improve N balance if protein intake is limiting.

In addition to the general action of energy-yielding nutrients on the course of protein metabolism, dietary carbohydrate has a special action which appears to result from the temporary deposition of amino acids in muscle for a few hours after each meal containing carbohydrate (Munro *et al.*, 1959). This causes a temporary fall in plasma amino acid level after a meal of carbohydrate which, in the normally nourished subject, is not of importance, but in the subject whose intake of protein is marginal could well aggravate the poor supply of amino acids to key tissues such as the liver. It is to be noted that diets productive of protein malnutrition in infants are often notable for their high carbohydrate content as well as their low protein level; the excessive intake of carbohydrate may be a factor in the liver cell deterioration in such cases.

4. Effect of Changes in Energy Expenditure on Protein Metabolism

The effect of work on protein metabolism has been in dispute ever since the time of Liebig; early opinions on the subject are discussed in Chapter 1, and more recent evidence is considered in Chapter 10 of this treatise. Cathcart (1925) reviewed the literature and concluded that a small but distinct increment in N output occurs after muscular exercise. Nevertheless, there is good evidence that human subjects (Thomas, 1910) and rats (Mitchell and Kruger, 1928) on low-protein intakes can undertake severe exercise without showing a significant increase in endogenous nitrogen excretion. The increase in N output sometimes observed when subjects on a mixed diet are exercised has been attributed by Wilson (1932) to acceleration in the metabolism of food protein, and not to an

TABLE IV

Effect of Absorptive Phase on Response of Protein Metabolism to Exercise[a]

Subject	Level of energy intake	Daily urinary N output (gm) (mean of 4 days' collection)		
		Work before meal	Work after meal	Difference
H.N.M.	Surfeit	9.29	10.09	+0.80
D.P.C.	Surfeit	13.91	14.12	+0.21
D.P.C.	Adequate	14.16	14.66	+0.50
H.N.M.	Adequate	10.71	11.45	+0.74

[a] Data computed from Cuthbertson *et al.* (1937) for 2 subjects carrying out severe exercise on a stationary bicycle either before or shortly after a mixed meal. The diet was constant throughout, except that the energy intake was either adequate or excessive. Urine was collected during the 2 days of heavy work and the first 2 post-work days, the mean output during the 4-day period being averaged in the table; the difference in N excretion due to time of performing the exercise is statistically significant ($P < 0.05$ by the t test).

increase in "endogenous" metabolism. This conclusion is confirmed by studies on sheep (Blaxter, 1963), in which the increased N output of the animals during exercise can be related to the level of protein in the diet. In this connection, also, Cuthbertson *et al.* (1937) obtained evidence that the timing of the work in relation to meals was of importance in obtaining the extra excretion of nitrogen. These authors undertook severe exercise on a bicycle ergometer either shortly before or shortly after a mixed meal. Their data (recalculated in Table IV) show that N output was significantly greater when the work was performed during the period of absorption than when carried out in the postabsorptive period. As discussed above, there appears to be a deposition of amino acids in muscle shortly after each meal containing carbohydrate, and it is possible that muscular work carried out during the absorptive period may impair this action of dietary carbohydrate on protein metabolism.

Whatever may be the metabolic significance of the small changes in protein metabolism observed after experimentally imposed heavy exercise, it is very difficult to judge their relevance to the protein requirements of men undertaking heavy manual labor. Evidence of an increased N output in laboratory work experiments has frequently been obtained in circumstances under which the extra energy cost of the work was not met by adding calories to the basal diet during the work period (e.g., Cathcart and Burnett, 1926), whereas the working man adjusts his caloric intake to meet his energy requirements. Failure to provide energy to cover the work cost in experimental studies may account for part of the extra N excretion observed, through a change in the subject's energy balance. Second, the protein needs of a subject trained to perform manual labor throughout much of his adult life may depend on more complicated factors than those revealed by short bouts of exercise under laboratory conditions. Some authors (Kraut and Müller, 1950; Yoshimura, 1955) point out that seasonal increases in muscular activity and athletic training can increase the need for protein because muscle mass undergoes changes. This still leaves open the question whether the subject who is regularly engaged in heavy labor has an increased need for protein. The problem is an important one, since heavy workers in Western countries automatically select a protein intake which is considerably greater than those of sedentary workers (Table V). One aspect of physical work which indicates more conclusively an increased protein need is the extra sweating which can occur. According to climatic conditions, this may increase nitrogen loss through the skin by 1–2 gm daily and thus increase the protein needs for replacement by as much as 15 gm.

Energy expenditure can also be increased by exposing the subject to cold. In a cold environment, rats on a constant food intake show an increased excretion of nitrogen (Lathe and Peters, 1949; Ingle *et al.*, 1953). However, the change in nitrogen output is not an inevitable consequence of the raised energy metabolism. In short-term studies on fasting human subjects (Voit, 1878) and fasting dogs (Rubner, 1902) and in an investigation of the metabolism of sheep at different planes of nutrition (Graham *et al.*, 1959), the increased heat output evoked by the cold environment came exclusively from the combustion of fat. A recent extensive study on human subjects (Issekutz *et al.*, 1962) shows that, although energy metabolism changes immediately, the increase in N output can be delayed by as much as 2 days after the commencement of exposure to a cold environment, and the increased nitrogen excretion eventually produced persists for several days after returning to the warm environment. The action of cold on protein metabolism probably occurs mainly through an increase in activity of the thyroid gland and adrenal cortex (You *et al.*, 1950; Hardy, 1961), both of which would cause an impairment in nitrogen balance (see Chapter 10). The independence of action of a cold environment on energy requirements and on protein requirements is also suggested by the finding that a diet too low in protein to support survival of rats in a warm environment will nevertheless permit growth of the rats in a cold environment (Andik *et al.*, 1963). Presumably, the increased food intake demanded by energy needs in the cold environment results in a better intake of protein; if protein requirements were increased in proportion to the greater energy needs, growth rate would not be improved in the cold environment despite the larger amount of protein consumed. The fact that the rats grew in the cold shows that protein requirements were not increased in proportion to energy requirements.

5. *Influence of Other Nutrients on Protein Utilization*

Any mineral or vitamin whose presence in the diet is necessary for normal growth and metabolism can be presumed to affect utilization of dietary protein, insofar as deficiency of the essential vitamin or mineral leads to failure of growth and thus of protein deposition. Nevertheless, a convincing picture of the precise role of these nutritional factors in protein utilization is difficult to establish from the scattered observations published in the literature.

The efficiency with which the rat utilizes various food proteins is known to be affected by dietary vitamins. Thus utilization of casein and fibrin improves when the supply of members of the vitamin B complex is increased (Sure and Romans, 1948; Sure, 1950; Mayfield and Hedrick, 1949b). The supply of vitamin B complex may be more critical for protein utilization when the dietary protein is of poor quality; Marfatia and Sreenivasan (1960) observed that growth on diets containing low quality proteins was more seriously depressed by suboptimal intake of the B complex than was growth on diets containing better quality protein sources.

Abnormalities of protein metabolism have been observed in animals made deficient in members of the vitamin B complex. In several instances, enlargement of the liver with deposition of protein has been noted (Guggenheim and Diamant, 1959; Mookerjea and Hawkins, 1960), a picture which is compatible with increased adrenocortical activity. Deficiency of pyridoxine causes an increased excretion of urea (Hawkins *et al.*, 1959). Despite these evidences of altered protein metabolism, Hawkins and his associates found that the capacity of the protein-depleted rat to regenerate liver protein and plasma proteins is unaffected by riboflavin deficiency (Mookerjea and Hawkins, 1960) or by pyridoxine deficiency (Hawkins *et al.*, 1962).

The influence of mineral intake on protein regeneration has also been examined. Cannon *et al.* (1952) found that adequate supplies of potassium were essential for the replacement of body protein in the protein-depleted rat. This was confirmed by Frost and Sandy (1953), who showed that a dietary source of phosphorus was also needed, but deficiency of sodium, calcium, or magnesium in the repletion diet did not affect the capacity of the rat to regenerate protein. Despite this evidence that magnesium is nonessential for protein repletion, Bunce *et al.* (1963) demonstrated that chicks and rats on diets low in magnesium are unable to tolerate high intakes of protein; growth is retarded, there is a reduction in serum protein level, and there are signs specific to magnesium deficiency. It has also been found that the guinea pig requires abundant magnesium in order to utilize amino acid supplements added to a casein diet (Heinicke *et al.*, 1956).

These various observations suggest that protein synthesis is probably not directly affected by deficiencies of vitamins and minerals in the diet. Endocrine function may be more sensitive to nutritional deficiencies of minerals and vitamins, and lead to abnormalities of regulation of protein metabolism, notably, regulation through the adrenal cortex.

B. Physiological Factors Influencing Protein Needs

1. Factors Affecting the Protein Requirements of Adults

The requirements of the adult subject are generally considered to be those for maintenance of tissue protein, and are commonly expressed in gm/kg body weight. For this reason women, who are of smaller mean weight, have a lower total requirement. So far, such adult needs have been given in terms of minimum amounts to achieve N equilibrium; requirements of mixed dietary protein (BV = 70) varying from 0.58 to 0.74 gm/kg body weight for subjects of differing body build have been computed by Hegsted (Chapter 14) to be necessary in order to maintain nitrogen equilibrium. A new W. H. O./F. A. O. report (1964) on protein requirements recommends an additional 10 % to these estimates to allow replacement of nitrogen lost in the course of the minor stresses and infections of ordinary life. This allowance, together with an additional amount for minimal sweating, raises the average protein requirements computed in this report to about 15 % higher than those given in Chapter 14.

It is important to know not only the average amount of protein needed to maintain an adult in N equilibrium, but also the range of variation in this figure encountered within a group of individuals. Sherman (1920) assembled the published data for 109 nitrogen balance experiments, from which he was able to assess the minimum protein intake for attaining N equilibrium on a mixed diet (no account was taken of cutaneous losses). From these data, the average figure for maintenance of equilibrium for a 70-kg man was 44 gm of mixed protein per day (0.64 gm/kg body weight). The range of maintenance requirements extended from 21 gm to 65 gm per 70 kg subject (0.30 to 0.93 gm/kg body weight/day). Sherman himself thought that a large part of this variation was due to differences in the degree of adjustment of the subjects to the low protein intakes used in the various studies. To this can be added differences in the average biological values of the protein in the diets employed to establish protein requirements. This means that inherent differences in the protein needs of the subjects must have been much less than the range of 0.3 to 0.9 gm/kg. Bricker *et al.* (1948) made a careful and prolonged study of the minimal protein requirements for N equilibrium (including cutaneous losses) of 10 college women. The subjects received a single mixed diet on which they showed an average net protein utilization (NPU) of 61 %. In order to achieve N equilibrium, the group required an average intake of 0.55 gm of the protein of this diet per killogram body weight; from their data, the requirement for individual members of the group can be computed to range from 0.41 to 0.69 gm/kg, that is, by 25 % around the mean. This range includes the variability from individual to individual due to differences in digestibility as well as to differences in tissue requirements (Williams, 1959). Data of this kind are important when planning for the protein needs of a whole population, since the requirements of every individual in the group have to be considered, and not just the average needs. In the new W. H. O./F. A. O. report, the range of requirements to meet the needs of 95 % of the population are considered to have a scatter of 20 % around the mean value. Consequently, for mixed dietary protein of which the utilization is 70 % (Biological Value × digestibility) the mean figure recommended is 0.84 gm/kg body weight/day for adults but the range of requirements to cover almost all subjects varies from 0.67 to 1.01 gm/kg body weight. This means that populations not receiving 1 gm/kg may include some subjects whose requirements are not met.

Apart from biological variability in the requirements of individual members of a population, it is also uncertain whether differences in energy expenditure have a significant effect on minimal protein needs. This is an important question, since the energy intake of individuals varies from about 2500 kcal daily for sedentary workers to about 4000 kcal for heavy manual workers, and the evidence to be considered later shows that the heavy workers eat more protein in proportion to their extra energy intake. Do they need this extra protein? This is a matter which, as discussed in an earlier section, needs further study before the question can be conclusively answered.

In contrast to this, some reduction in energy intake and in protein intake tends to occur in old age (Table V) and this raises the question of whether requirements in the elderly differ from those of the younger adult. There is some evidence (Tuttle *et al.*, 1960) that the requirement for essential amino acids rises in the older male, which is contrary to the dietary pattern of a reduced protein intake. It is therefore a matter of some importance to know whether the degenerative diseases of the elderly can be alleviated or prevented by attention to protein intake. The question of protein requirements in the elderly is dealt with in Chapter 17. The limited evidence suggests that, while the average amount of protein needed to bring elderly subjects into N equilibrium may not differ appreciably from the mean figure observed in younger groups of adults, there is probably a greater range of requirements among the elderly.

2. Protein Requirements in Pregnancy and Lactation

The protein requirements of adult women are increased by pregnancy and lactation. During the course of pregnancy, some 800 gm of additional protein accumulate in the body (Thomson and Hytten, 1961). Even at the end of pregnancy, this represents deposition of no more than 5 gm of protein daily, and means some 7 gm of extra dietary protein needed per day. The total output of protein in milk is about 10 gm daily (F. A. O. 1957b), and, in consequence, the additional dietary protein needed for lactation becomes about 15 gm of high quality protein. The validity of such calculations is discussed in detail in Chapter 15 of this book. They imply that pregnancy and lactation do not impose a large additional demand on dietary protein.

3. Protein Requirements in Infancy and Childhood

For the infant, Holt (1959) has suggested that the diet must be sufficient to enable it to withstand the challenges of infancy and of later life. As discussed in Chapter 16, it is commonly accepted that the amount of protein provided by mother's milk, namely, about 2 gm/kg of the infant's body weight, is also the requirement of the young infant, with a requirement of about 1.5 gm protein/kg body weight during the second six months of life. The efficiency of conversion of milk protein to body protein by the young infant is about 50%. In later childhood, when growth rate slows down, the amount of protein needed for maintenance of existing tissue becomes the main factor, and deposition as growth represents a relatively small fraction of the intake. The requirements of the growing child can be computed, as in the case of the adult, by assessing the needs for maintenance of existing tissue, and then adding a quota for deposition of protein in growth; the latter involves an estimate of the efficiency with which dietary protein is used during growth. The hazards of such estimates are discussed in Chapter 14 in connection with the general problem of assessing protein requirements.

4. Factors Affecting the Pattern of Amino Acid Requirements

Human requirements for protein are usually given in terms of gm per kg body weight of some standard protein such as an ideal protein with a biological value of 100 (e.g., the FAO reference protein), or the average protein of a mixed diet with a Biological Value of 70 (Chapter 14).

However, although the amino acid composition of a given protein is fixed, the biological value expresses the efficiency with which the body utilizes the protein as a dietary source of amino acids, and this can alter if the needs of the body for one amino acid, or group of amino acids, change relative to the requirements for other amino acids. Table II shows that the biological values of proteins fed to growing rats can differ in some instances from the values of the same proteins when given to adult rats, the effect in the case of wheat gluten being quite marked. Allison (1955) has demonstrated that many proteins have a higher biological value when tested on protein-depleted dogs than is obtained from dogs in a normal state of nourishment.

These differences in Biological Value reflect differences in amino acid requirements by the body, which may be of two kinds. First, the requirements for a single amino acid may change; if, for example, the amount of lysine needed by the older rat is relatively low, then a lysine-deficient protein such as wheat gluten will show a higher Biological Value in such an animal (Table II). Second, the requirements for essential amino acids as a proportion of the total amino acid intake may change; this is the most likely reason for the improved biological value of almost all proteins tested on the protein-depleted dog.

It is thus apparent that the next step in defining protein requirements will involve subdividing these into three components: (*a*) the requirement for total nitrogen, (*b*) the pattern of essential amino acid requirements, that is, the relative proportions of various essential amino acids needed, and (*c*) the optimum ratio of essential to nonessential amino acids, that is, the proportion of the total N intake which should be provided in the form of essential amino acids.

C. The Observed Range of Protein Intakes

How much protein do different people consume? Does it vary greatly from individual to individual among the adults in a single community? How variable are the protein intakes of different races?

The question of variability within the community is answered by conducting surveys of the actual amounts of food eaten by individuals or by families in the population. Table V presents data obtained from direct weighing of the food intakes of different groups of the adult population of Scotland; most of this information comes from studies made by one team of investigators and has been made available through the kindness of the senior investigator, Dr. Durnin. It is apparent that the average intake of protein varies considerably in different groups of adults, the determining factor being the energy requirements of the subject. Those undertaking heavy work have a high caloric intake and consequently a high protein intake, so that, irrespective of occupational category, the

TABLE V

PROTEIN INTAKES OF GROUPS OF ADULTS IN SCOTLAND, OBTAINED BY WEIGHING ALL FOODSTUFFS CONSUMED DURING A 7-DAY PERIOD

Group[a]	Number in group	Mean age (yr)	Mean energy intake per day (kcal)	Mean protein intake per day		
				Total (gm)	As % of total energy intake	Intake per kg body weight (with range) (gm/kg)
Younger men						
Students (*1*)	9	22	3060	91	11.9	1.28 (1.05–1.46)
Clerks (*2*)	10	28	3040	96	12.6	1.49 —
Coal miners (*2*)	19	36	4030	121	12.0	1.84 —
Older men						
Older clerks (*3*)	8	56	2440	83	13.6	1.24 (1.04–1.52)
Older workers (*4*) (heavy work)	26	60	3430	113	13.2	1.48 (1.24–1.86)
Older workers (*4*) (moderate work)	24	61	2910	90	12.4	1.31 (0.85–1.80)
Retired men (*1*) (living alone)	9	73	2050	72	14.0	1.03 (0.78–1.24)
Younger women						
Technicians (*1*)	18	19	2200	64	11.6	1.18 (0.89–1.65)
Young shop assistants (*5*)	12	21	2220	70	12.3	1.21 (0.99–1.63)
Pregnant women (*6*) (first child)	489	—	2450	75	12.4	1.20 —
Older women						
Older housewives (*7*)	21	60	1940	62	12.7	1.02 (0.69–1.39)
Elderly women (*8*) (living alone)	17	66	1890	62	13.2	1.03 (0.69–1.59)

[a] References: (*1*) Unpublished data (1963) kindly provided by Dr. J. V. G. A. Durnin; (*2*) Garry *et al.* (1955); (*3*) Blake and Durnin (1963); (*4*) Durnin *et al.* (1961b); (*5*) Durnin *et al.* (1957); (*6*) Thomson (1958, 1959); (*7*) Durnin *et al.* (1961c); (*8*) Durnin *et al.* (1961a).

individual consumes about 12 to 14 % of his dietary energy in the form of protein, the proportion tending to increase slightly with advancing age (Table V). This confirms data from many countries showing that, in general, some 10 to 14 % of energy intake comes from protein sources (Cuthbertson, 1940; Dole, 1957); thus, the dominant food drive is for energy, the protein intake being raised or lowered according to the changing demand for calories. Finally, the protein intake per kg of body weight observed in these groups can be compared with the recommended allowances for maintenance of nitrogen equilibrium, assessed at some 0.6 to 0.7 gm mixed dietary protein per kg of body weight for adults

(Chapter 14). None of the groups recorded in Table V had an average intake of less than 1.0 gm protein/kg, and groups undertaking very heavy work (coal miners) selected an intake which was three times the minimum level for equilibrium. However, individual intakes show a considerable range of variation, as low as 0.7 gm/kg among some older subjects. Although there is no evidence of recognizable protein deficiency in Scotland, there may be a proportion of older people who do little more than meet the minimum needs for N equilibrium.

In order to compare the protein intakes of different races, the total amount of protein available in the country can be divided by the number of its citizens to give an estimate of the protein eaten per caput. Such assessments have obvious limitations, especially when a large proportion of the consumers are also producers of food, so that the magnitude of their produce cannot easily be estimated. Nevertheless, the figures for protein intake per head of the population and the proportion of animal protein in the diet are much less in some less fortunate tropical areas than in the Western countries (Table VI). It can be concluded that a considerable number of the population in tropical countries whose average intakes are shown in Table VI must be living at or below the subsistence level for protein intake. In addition, the inhabitants of such countries often suffer from inadequate energy intake which will aggravate the situation by depressing the use made of the meager supply of dietary protein. D. S. Miller and Payne (1961) offer an equation for computing the effects of caloric restriction on the protein values of such diets.

D. Future Approaches to the Study of Protein Requirements

It is apparent that the lower limit of protein needs is fixed by the endogenous N output of the subject on a protein-free diet. It has been

TABLE VI

TOTAL PROTEIN INTAKES PER HEAD OF THE POPULATION AND INTAKES OF PROTEIN FROM ANIMAL SOURCES IN DIFFERENT COUNTRIES[a]

Country	Mean daily protein intake	
	Total protein (gm)	Animal protein (gm)
New Zealand	105	72
Australia	94	61
United States	93	64
Great Britain	87	52
Japan	68	18
India	56	6
Mauritius	46	11

[a] From F. A. O. (1961).

argued that an intake of protein only sufficient to replace this loss will not be optimal for health because the body is prevented from laying down labile protein deposits and because such an intake does not permit maximal resistance to disease. If we allow the thesis that additional protein may be necessary in order to obtain these benefits, we are then faced with the problem of assessing the dietary protein level at which these additional requirements are satisfied. Lack of information about such possible needs stems from the fact that the measurement of nitrogen balance as a criterion for protein requirements has overshadowed the picture. The use of nitrogen balance will indeed indicate the minimum level of protein needed for the functional requirements of major tissues, but it does not cover the possibility that some small organs may require a higher level of protein intake in order to operate with maximal efficiency. This possibility can only be settled by exploration of the function and metabolism of individual tissues and organs at different levels of protein intake, and indeed some work of this kind has already been performed. It has thus been demonstrated that the biological values of dietary proteins assessed from studies on the whole body may not necessarily reflect their relative values for such purposes as regeneration of plasma proteins (Pommerenke *et al.*, 1935) or for the deposition of protein and ribonucleic acid in the cells of some tissues (Allison *et al.*, 1962). This topic receives more attention in Chapter 12.

Eventually, it is to be anticipated that the biochemical features of protein metabolism in different tissues of the body will provide a rational and quantitative explanation for the needs of the body for dietary protein. This information will complement and extend the data obtained on the whole animal which at present provide most of our information about the protein requirements of a species. Indeed, one of the main objectives of this book is to encourage such a marriage between nutritional and metabolic studies.

References

Allison, J. B. (1955). *Physiol. Rev.* **35**, 644.

Allison, J. B., Anderson, J. A., and Seeley, R. D. (1946). *Ann. N. Y. Acad. Sci.* **47**, 245.

Allison, J. B., Wannemacher, R. W., Banks, W. L., Wunner, W. H., and Gomez-Brenes, R. A. (1962). *J. Nutr.* **78**, 333.

Am. Acad. Pediat. (1960). Committee on Nutrition, *Pediatrics* **26**, 1039.

Andik, I., Donhoffer, S., Farkas, M., and Schmidt, P. (1963). *Brit. J. Nutr.* **17**, 257.

Bell, E. A., and Foster, R. Q. (1962). *Nature* **194**, 91.

Berg, C. P. (1959). *In* "Protein and Amino Acid Nutrition" (A. A. Albanese, ed.), p. 57. Academic Press, New York.

Birk, Y., and Gertler, A. (1961). *J. Nutr.* **75**, 379.

Blake, E. C., and Durnin, J. V. G. A. (1963). *Proc. Nutr. Soc.* (*Engl. Scot.*) **22**, i.

Blaxter, K. L. (1963). Personal communication.

Block, R. J., and Mitchell, H. H. (1946). *Nutr. Abstr. Rev.* **16**, 249.
Block, R. J., Cannon, P. R., Wissler, R. W., Steffee, C. H., Straube, R. L., Frazier, L. E., and Woolridge, R. L. (1946). *Arch. Biochem.* **10**, 295.
Borchers, R. (1962). *J. Nutr.* **78**, 330.
Bricker, M. L., Shiveley, R. F., Smith, J. M., Mitchell, H. H., and Hamilton, T. S. (1948). *J. Nutr.* **37**, 169.
Bro-Rasmussen, F. (1958). *Nutr. Abstr. Rev.* **28**, 369.
Bunce, G. E., Reeves, P. G., Oba, T. S., and Sauberlich, H. E. (1963). *J. Nutr.* **79**, 220.
Calloway, D. H., and Spector, H. (1954). *Am. J. Clin. Nutr.* **2**, 405.
Cannon, P. R., Frazier, L. E., and Hughes, R. H. (1952). *Metab., Clin. Exptl.* **1**, 49.
Carpenter, K. J. (1951). *Proc. Nutr. Soc. (Engl. Scot.)* **5**, 243.
Carpenter, K. J. (1958). *Proc. Nutr. Soc. (Engl. Scot.)* **17**, 91.
Carpenter, K. J. (1960). *Biochem. J.* **77**, 604.
Cathcart, E. P. (1925). *Physiol. Rev.* **5**, 225.
Cathcart, E. P., and Burnett, W. A. (1926). *Proc. Roy. Soc.* **B99**, 405.
Chittenden, R. H. (1905). "Physiological Economy in Nutrition." Stokes, New York.
Close, J., Adriaens, E. L., Moore, S., and Bigwood, E. J. (1953). *Bull. Soc. Chim. Biol.* **35**, 985.
Cuthbertson, D. P. (1940). *Nutr. Abstr. Rev.* **10**, 1.
Cuthbertson, D. P., McGirr, J. L., and Munro, H. N. (1937). *Biochem. J.* **31**, 2293.
Derse, P. H. (1960). *J. Assoc. Offic. Agr. Chemists* **43**, 38.
Dole, V. P. (1957). *Am. J. Clin. Nutr.* **5**, 72.
Durnin, J. V. G. A., Blake, E. C., and Brockway, J. M. (1957). *Brit. J. Nutr.* **11**, 85.
Durnin, J. V. G. A., Blake, E. C., Brockway, J. M., and Drury, E. A. (1961a). *Brit. J. Nutr.* **15**, 499.
Durnin, J. V. G. A., Blake, E. C., Allan, M. K., Shaw, E. J., Wilson, E. A., Blair, S., and Yuill, S. A. (1961b). *Brit. J. Nutr.* **15**, 587.
Durnin, J. V. G. A., Blake, E. C., Allan, M. K., Shaw, E. J., and Blair, S. (1961c). *J. Nutr.* **75**, 73.
Dustin, J. P., Czajkowska, M., Moore, S., and Bigwood, E. J. (1953). *Anal. Chim. Acta* **9**, 256.
F. A. O. (1957a). "Caloric Requirements," Nutr. Studies No. 15. Food and Agriculture Organization of the United Nations.
F. A. O. (1957b). "Protein Requirements," Nutr. Studies No. 16. Food and Agriculture Organization of the United Nations.
F. A. O. (1961). "State of Food and Agriculture," Report. Food and Agriculture Organization of the United Nations. Rome.
Fowden, L. (1960). *In* "The Biochemist's Handbook" (C. Long, ed.), p. 990. Spon, London.
Fowden, L. (1962). *Endeavour* **21**, 35.
Frost, D. V., and Sandy, H. R. (1948). *J. Biol. Chem.* **175**, 635.
Frost, D. V., and Sandy, H. R. (1953). *Proc. Soc. Exptl. Biol. Med.* **83**, 102.
Garry, R. C., Passmore, R., Warnock, G. M., and Durnin, J. V. G. A. (1955). *Med. Res. Council, Spec. Rept. Ser.* **289**.
Geiger, E. (1948). *J. Nutr.* **36**, 813.
Gottschalk, A., and Fazekas de St. Groth, S. (1960). *Biochim. Biophys. Acta* **43**, 513.
Graham, N. M., Wainman, F. W., Blaxter, K. L., and Armstrong, D. G. (1959). *J. Agr. Sci.* **52**, 13.
Guggenheim, K., and Diamant, E. J. (1959). *Brit. J. Nutr.* **13**, 67.
Gupta, J. D., and Elvehjem, C. A. (1957). *J. Nutr.* **62**, 313.

Gupta, J. D., Dakroury, A. M., Harper, A. E., and Elvehjem, C. A. (1958). *J. Nutr.* **64,** 259.
Halac, E. (1961). *Am. J. Clin. Nutr.* **9,** 557.
Hardy, J. D. (1961). *Physiol. Rev.* **41,** 521.
Harrison, M. F. (1953). *Biochem. J.* **55,** 204.
Hassal, C. H., Feyle, K., and Feng, P. (1954). *Nature* **173,** 356.
Hawkins, W. W., Leonard, V. G., and Coles, C. M. (1959). *J. Nutr.* **69,** 383.
Hawkins, W. W., Leonard, V. G., and Maxwell, J. E. (1962). *J. Nutr.* **76,** 231.
Heinicke, H. R., Harper, A. E., and Elvehjem, C. A. (1956). *J. Nutr.* **58,** 269.
Henry, K. M., and Kon, S. K. (1957). *Brit. J. Nutr.* **11,** 305.
Henry, K. M., Kosterlitz, H. W., and Quenouille, M. H. (1953). *Brit. J. Nutr.* **7,** 51.
Henry, K. M., Cormack, R. M., and Kosterlitz, H. W. (1961). *Brit. J. Nutr.* **15,** 199.
Holt, L. E. (1959). *J. Pediat.* **54,** 496.
Holt, L. E., Albanese, A. A., Shettles, L. B., Kajdi, C. N., and Wangerin, D. M. (1942). *Federation Proc.* **1,** 116.
Holt, L. E., Halac, E., and Kajdi, C. N. (1962). *J. Am. Med. Assoc.* **181,** 699.
Hughes, B. P. (1958). *Brit. J. Nutr.* **12,** 188.
Hughes, B. P. (1959). *Brit. J. Nutr.* **13,** 330.
Hytten, F. E. (1954). *Brit. Med. J.* **i,** 249.
Ingle, D. J., Meeks, R. C., and Humphrey, L. M. (1953). *Am. J. Physiol.* **173,** 387.
Issekutz, B., Rodahl, K., and Birkhead, N. C. (1962). *J. Nutr.* **78,** 189.
Johansen, P. G., Marshall, R. D., and Neuberger, A. (1961). *Biochem. J.* **78,** 518.
Jones, D. B. (1941). *U. S. Dept. Agr., Circ.* **183** (revised).
Keys, A., Brozek, J., Henschel, A., Mickelsen, O., and Taylor, H. L. (1950). "The Biology of Human Starvation." Vol. I Univ. Minnesota Press, Minneapolis, Minnesota.
Kihara, H., and Snell, E. E. (1955). *J. Biol. Chem.* **197,** 791.
Klose, A. A., Hill, B., and Fevold, H. L. (1946). *Proc. Soc. Exptl. Biol. Med.* **62,** 10.
Kodicek, E., and Mistry, S. P. (1957). *Proc. Nutr. Soc.* (*Engl. Scot.*) **16,** xiii.
Kosterlitz, H. W. (1947). *J. Physiol.* (*London*) **106,** 194.
Kraut, H., and Müller, E. A. (1950). *Biochem. Z.* **320,** 302.
Kuiken, K. A. (1952). *J. Nutr.* **46,** 13.
Kuiken, K. A., and Lyman, C. M. (1948). *J. Nutr.* **36,** 359.
Kulkarni, L., and Sohonie, K. (1956). *Indian J. Med. Res.* **44,** 511.
Lathe, G. H., and Peters, R. A. (1949). *Quart. J. Exptl. Physiol.* **35,** 55.
McCance, R. A., and Widdowson, E. M. (1960). *Med. Res. Council, Spec. Rept. Ser.* **297.**
McCollum, E. V., and Davis, M. (1915). *J. Biol. Chem.* **23,** 247.
McKee, H. S. (1958a). *In* "Handbuch der Pflanzenphysiologie" (W. Ruhland, ed.) Vol. VIII, p. 518. Springer, Berlin.
McKee, H. S. (1958b). *In* "Handbuch der Pflanzenphysiologie" (W. Ruhland, ed.), Vol. VIII p. 564. Springer, Berlin.
McKee, H. S. (1962). "Nitrogen Metabolism in Plants." Oxford Univ. Press, Oxford.
Madden, S. C., and Whipple, G. H. (1940). *Physiol. Rev.* **20,** 194.
Maddy, K. H., and Elvehjem, C. A. (1949). *J. Biol. Chem.* **177,** 577.
Marfatia, U., and Sreenivasan, A. (1960). *J. Nutr.* **70,** 163.
Mauron, J. (1961). *In* "Progress in Meeting Protein Needs of Infants and Children," Publ. No. 843, Natl. Acad. Sci.—Natl. Res. Council, Washington, D. C.
Mayfield, H. L., and Hedrick, M. T. (1949a). *J. Nutr.* **37,** 487.
Mayfield, H. L., and Hedrick, M. T. (1949b). *J. Nutr.* **37,** 475.
Miller, D. S., and Payne, P. R. (1961). *J. Nutr.* **75,** 225.

Miller, L. L., and Whipple, G. H. (1940). *Am. J. Med. Sci.* **199,** 204.
Mitchell, H. H. (1959). *In* "Protein and Amino Acid Nutrition" (A. A. Albanese, ed.), p. 11. Academic Press, New York.
Mitchell, H. H., and Kruger, J. H. (1928). *J. Biol. Chem.* **76,** 55.
Mitchell, H. H., Hamilton, T. S., and Beadles, J. R. (1949). *J. Nutr.* **39,** 413.
Mookerjea, S., and Hawkins, W. W. (1960). *Brit. J. Nutr.* **14,** 231.
Moorjani, M. N., and Subrahmanyan, V. (1950). *Indian J. Med. Res.* **38,** 145.
Morrison, S. D. (1952). "Human Milk." Tech. Commun. No. 18. Commonwealth Bureau of Animal Nutrition, Commonwealth Agric. Bur., Slough, England.
Munro, H. N., and Chalmers, M. I. (1945). *Brit. J. Exptl. Pathol.* **26,** 396.
Munro, H. N., and Clark, C. M. (1960). *Proc. Nutr. Soc.* (*Engl. Scot.*) **19,** 55.
Munro, H. N., and Naismith, D. J. (1953). *Biochem. J.* **54,** 191.
Munro, H. N., Black, J. G., and Thomson, W. S. T. (1959). *Brit. J. Nutr.* **13,** 475.
Neuberger, A., and Sanger, F. (1942). *Biochem. J.* **36,** 662.
Osborne, T. B., Mendel, L. B., and Ferry, E. L. (1919). *J. Biol. Chem.* **37,** 223.
Orr, M. L., and Watt, B. K. (1957). "Amino Acid Content of Foods," Home Economics Research Report No. 4, U. S. Dept. Agric., Washington, D. C.
Platt, B. S. (1962). *Med. Res. Council, Spec. Rept. Ser.* **302.**
Platt, B. S., Miller, D. S., and Payne, P. R. (1961). *In* "Recent Advances in Human Nutrition" (J. F. Brock, ed.), p. 351. Churchill, London.
Playfair, L. (1865). *Med. Times and Gazette* **i,** 459, 485, and 511.
Pommerenke, W. T., Slavin, H. B., Kariher, D. H., and Whipple, G. H. (1935). *J. Exptl. Med.* **61,** 261.
Ramasarma, G. B., Henderson, L. M., and Elvehjem, C. A. (1949). *J. Nutr.* **38,** 177.
Ressler, C. (1962). *J. Biol. Chem.* **237,** 733.
Rose W. C. (1938). *Physiol. Rev.* **18,** 109.
Rose, W. C. (1957). *Nutr. Abstr. Rev.* **27,** 631.
Rosenberg, H. R., and Rohdenburg, E. L. (1951). *J. Nutr.* **45,** 593.
Rosevear, J. W., and Smith, E. L. (1961). *J. Biol. Chem.* **236,** 425.
Rowlands, D. A., Gale, E. F., Folkes, J. P., and Marrian, D. H. (1957). *Biochem. J.* **65,** 519.
Rubner, M. (1902). "Die Gesetze des Energieverbrauchs bei der Ernährungstherapie." Deuticke, Leipzig and Vienna.
Samuels, L. T., Gilmore, R. C., and Reinecke, R. M. (1948). *J. Nutr.* **36,** 639.
Shapiro, R., and Fisher, H. (1962). *Federation Proc.* **21,** 395.
Sherman, H. C. (1920). *J. Biol. Chem.* **41,** 97.
Silber, R. H., Howe, E. E., Porter, C. C., and Mushett, C. W. (1949). *J. Nutr.* **37,** 429.
Sprince, H., and Woolley, D. W. (1944). *J. Exptl. Med.* **80,** 213.
Steele, R. (1952). *J. Biol. Chem.* **198,** 237.
Steward, E. C., and Thompson, J. F. (1954). *In* "The Proteins," 1st ed. (H. Neurath and K. Bailey, eds.), Vol. II, Part A, p. 513. Academic Press, New York.
Sure, B. (1950). *Proc. Soc. Exptl. Biol. Med.* **75,** 300.
Sure, B., and Romans, F. (1948). *J. Nutr.* **36,** 727.
Tarr, H. L. A. (1958). *Ann. Rev. Biochem.* **27,** 223.
Terroine, E. F. (1936). *Quart. Bull. Health Organ. League Nations* **5,** 427.
Thomas, K. (1909). *Arch. Physiol.* p. 219.
Thomas, K. (1910). *Arch. Physiol.* Suppl., p. 249.
Thomson, A. M. (1958). *Brit. J. Nutr.* **12,** 446.
Thomson, A. M. (1959). *Brit. J. Nutr.* **13,** 190.
Thomson, A. M., and Hytten, F. E. (1961). *Proc. Nutr. Soc.* (*Engl. Scot.*) **20,** 76.

Tuttle, S. G., Swendseid, M. E., and Bassett, S. H. (1960). *Federation Proc.* **19,** 11.
Vasington, F. E., Parker, A., Headley, W., and Vanderlinde, R. E. (1958). *Endocrinology* **62,** 557.
Voit, C. (1878). *Z. Biol.* **14,** 80.
Voit, C. (1881). *In* "Handbuch des allgemeinen Stoffwechsels und der Ernährung" (L. Hermann, ed.), Vol. VI, p. 3. Vogel, Leipzig.
Wang, C. F., Hegsted, D. M., Lapi, A., Zamcheck, N., and Black, M. D. (1949). *J. Lab. Clin. Med.* **34,** 953.
W.H.O/F.A.O. (1964). "Protein requirements." (in press).
Williams, R. J. (1959). *In* "Protein and Amino Acid Nutrition" (A. A. Albanese, ed.), p. 45. Academic Press, New York.
Wilson, H. E. C. (1932). *J. Physiol.* (*London*) **75,** 67.
Womack, M., and Rose, W. C. (1946). *J. Biol. Chem.* **162,** 735.
Woodham-Smith, C. (1962). "The Great Hunger: Ireland 1845–9." Hamilton, London.
Woolley, D. W. (1946). *J. Biol. Chem.* **162,** 383.
Yoshimura, H. (1955). *J. Japan. Soc. Food and Nutr.* **7,** 155.
You, S.S., You, R. W., and Sellars, E. A. (1950). *Endocrinology* **47,** 156.

CHAPTER 12

The Nutritive Value of Dietary Proteins

JAMES B. ALLISON

Bureau of Biological Research
and Department of Physiology and Biochemistry,
Rutgers, The State University, New Brunswick, New Jersey

I. Introduction

The plan of this treatise requires that some consideration be given to the determinations of the nutritive value of dietary proteins, but there is little need to repeat in detail the material available in past reviews of this subject (Mitchell, 1950; Allison, 1949, 1955, 1957; Frost, 1959; Albanese, 1959). The following pages, however, will contain a brief summary of some of the methods used to determine nutritive value with emphasis being placed upon more recent advances in procedures and interpretations of results. The significance of nutritive value as it is correlated with the development and maintenance of individual tissues and cells will be considered in these interpretations. An attempt will be made to correlate over-all nutritive value of dietary proteins, commonly determined and reported in the literature, with the impact upon individual tissues, cells, and molecules. The discussion will recognize that the nutritive value and these impacts may vary with the physiological state of the individual. It is hoped that such a presentation of the determination of nutritive value of dietary proteins will focus attention and research upon the need to correlate nutrition more directly with the biochemistry of the individual tissue and cell. Such a focus, which is so essential to modern nutrition, must be considered as a way toward a more complete understanding of the dynamic system which forms the whole animal. Nutrition is an integrating science which must not be allowed either to be too superficial in consideration of the whole animal or too far removed from the whole animal while considering detailed molecular behavior. Since nutritive value is correlated primarily with the essential amino acids, the composition of the various proteins to be considered in this review are recorded in Table I.

Many methods have been proposed for testing the quality of dietary proteins. Most are based on the effects of the dietary protein on the whole animal; Fig. 1 is an attempt to demonstrate the relationships between such tests, which are all basically variants of the observation that the amount and quality of protein in the diet can result in a gain or loss of body substance. Such gain or loss can be identified by a change in body weight or by a body component, often its nitrogen content.

Historically, the first demonstrations of differences in the nutritive quality of dietary proteins were established with growing animals, and they are still the most popular form of test. Consequently, the first section of this chapter will present methods used on growing animals, including indexes based on body weight change, on body nitrogen content, and on changes observed in individual tissues and their components. However, growth is only a part of the function of dietary protein, which must also serve for replacement of body proteins for maintenance. The succeeding section therefore deals with nutritive values of proteins in the adult animal. Both of these sections, on the growing and the adult animal, deal essentially with the study of the quality of isolated proteins. However, the diet as consumed under natural circumstances consists of a mixture of proteins and their utilization is also influenced by intake of energy, vitamins and minerals; consequently, we have next to discuss methods of assessing protein quality under practical conditions. This is followed by a section in which the needs of individual tissues are considered in relation to protein quality; this is most clearly demonstrated by the

TABLE I

AMINO ACID PATTERNS IN VARIOUS DIETARY PROTEIN SOURCES

Protein source	Amino acid[a]												
	Arginine[b]	Histidine[b]	Isoleucine[b]	Leucine[b]	Lysine[b]	Methionine[b]	Cystine	Phenylalanine[b]	Tyrosine	Threonine[b]	Tryptophan[b]	Valine[b]	Nutritive value
Egg albumid	374	151	403	543	450	260	186	371	208	305	72	466	100
Fish protein concentrate	352	—	317	474	549	178	84	231	159	283	62	327	85
Beef	402	203	323	488	537	169	84	243	185	278	63	321	80
Casein	244	190	397	626	506	203	25	338	322	281	60	460	77
Cottonseed flour	702	166	308	236	269	86	102	325	164	221	74	308	60
Peanut flour	705	139	259	442	218	52	90	307	181	174	49	285	56
Wheat gluten	230	124	284	450	126	104	148	315	174	175	48	266	44

[a] Values are expressed as mg/gm N.
[b] Essential amino acids.

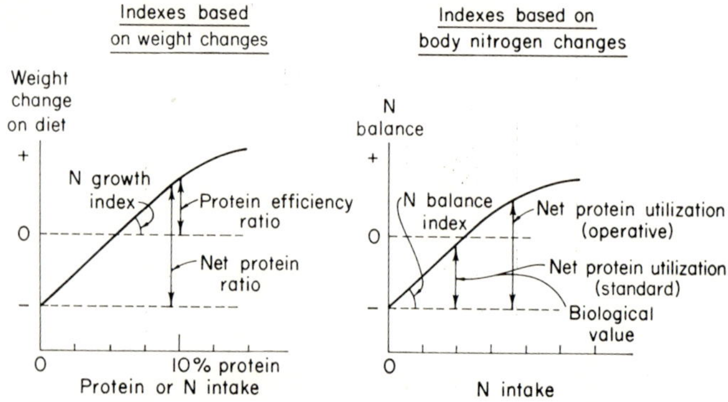

FIG. 1. Procedures for measuring the nutritive values of proteins.

Indexes based on weight changes:

Protein Efficiency Ratio (PER). Gain in body weight divided by amount of protein or nitrogen consumed.

Net Protein Ratio. Difference in final body weight between test group fed the protein in diet and a group receiving a protein-free diet, divided by the amount of protein taken by test group.

Nitrogen Growth Index. The slope of the straight portion of the line relating weight gain to nitrogen intake.

Indexes based on changes in body nitrogen content:

Biological Value (BV). The proportion of absorbed nitrogen retained in the body for maintenance and/or growth.

Net Protein Utilization (NPU). Difference in final carcass nitrogen content between a test group fed the diet with protein and a group fed a protein-free diet, expressed as a percentage of the nitrogen intake of the test group.

(*1*) Net Protein Utilization (standard)—NPU_{st}—is determined under standard conditions at a fixed level of protein intake below maintenance. The NPU_{st} is the same as biological value multiplied by digestibility.

(*2*) Net Protein Utilization (operative)—NPU_{op}—is determined at levels above maintenance requirements and is designed to include uncontrolled factors in the diet.

Nitrogen Balance Index. The slope of the straight portion of the line relating N balance to nitrogen absorbed.

different rates at which tissues lose protein when the diet is deficient in protein, and the rates at which they are repleted when the protein intake is restored. Finally, since differences in dietary protein quality are essentially due to differences in amino acid composition of these proteins, the last section of this chapter will present our knowledge of the ideal amino acid pattern, and will consider whether such a pattern varies with differences in the physiological state of the body.

II. Nutritive Value and Growth

All estimates of the nutritive value of a dietary protein involve a direct or indirect measurement of the retention of nitrogen in the body for maintenance and growth. Maintenance and growth of the whole body is an algebraic sum of gains and losses within the various cells and tissues, and emphasis will be placed upon this algebraic summation in the discussions to follow. As a first approximation, however, this over-all growth of the body, measured as gain in weight, length, or both, is often used as an estimate of gain in body nitrogen. The correlation between gain in weight and gain in nitrogen is particularly good in the rat (Hegsted and Worcester, 1947; Howard *et al.*, 1958; Bender, 1956).

A. Protein Efficiency Ratio

Growth of the rat, as a measure of nutritive value of dietary protein, was put on a quantitative basis by Osborne *et al.* (1919), who introduced the concept of the "Protein Efficiency Ratio" (PER). This ratio is defined as the gain in body weight per gram of protein or of nitrogen consumed. Originally, the ratio was determined at several nitrogen intakes and the maximum PER calculated (Barnes and Bosshardt, 1946), so that the optimum nitrogen intake for maximum gain in weight could be determined. The data recorded in Fig. 2 illustrate the gain in body weight at different nitrogen intakes (*A*) and the Protein Efficiency Ratios calculated from these data (*B*). The rats were fed the diet containing a certain amount of nitrogen from weaning over a period of 28 days (Allison *et al.*, 1959). The dietary proteins used in these experiments were egg albumin (circles with cross bars), casein (open circles),

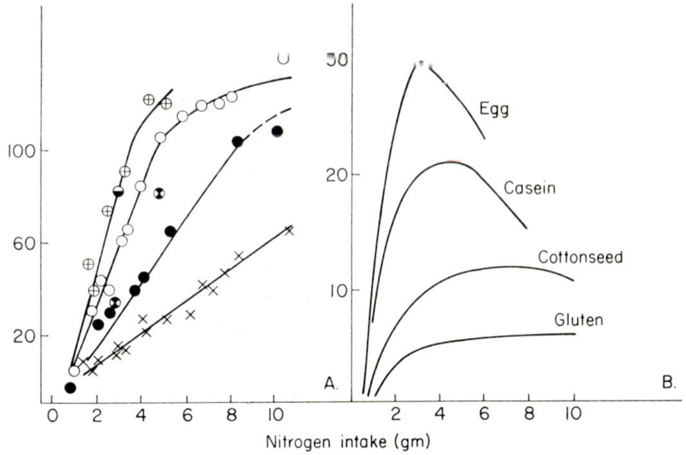

FIG. 2. Growth of rats from weaning over a period of 4 weeks while being fed different quantities of protein in a purified diet. *A*. Gain in body weight in grams at different nitrogen intakes. *B*. Protein efficiency ratios calculated from *A*. ⊕, egg albumin; ○, casein; ●, cottonseed flour; ×, wheat gluten.

TABLE II

GAIN IN BODY WEIGHT OF PUPPIES AND NITROGEN RETAINED FOR GROWTH PER GRAM OF NITROGEN INTAKE

Protein source	Gain in body weight (gm) / N intake (gm)	N retention / N intake
Egg albumin	9.0	0.50
Casein	9.8	0.43
Wheat gluten	8.8	0.18

cottonseed flour (closed circles), and wheat gluten (crosses). These data illustrate the variation in PER with nitrogen intake: the higher the nutritive value of the dietary protein, the greater the effect of intake upon this ratio. Care must be taken, therefore, in the interpretation of PER when only one nitrogen intake is used for purposes of comparing the nutritive value of one dietary protein with another. If a single intake is used the diet should contain approximately 10 % protein to obtain the best comparison between the various protein efficiency ratios.

Although gain in body weight in the rat generally correlates well with gain in body protein, this correlation is not always good and it is especially variable in some animals such as the dog. The data in Table II illustrate the effect of feeding different protein sources in a purified diet on the gain in body weight and on the nitrogen retained by young dogs. The dogs receiving egg or casein proteins in the diet developed into lean, active, muscular animals. Those fed wheat gluten gained in weight similarly to those fed egg or casein but they were obese and inactive, exhibiting some of the apathy associated with protein depletion. The values in the last column of Table II demonstrate that the dogs fed wheat gluten developed only about one-third as much body nitrogen as those fed the egg protein (Allison and Wannemacher, 1957; Allison, 1958a).

B. Nitrogen Growth Index

The relationship between nitrogen intake and body weight gain is curvilinear, possibly semilogarithmic, but the lower part of the curves in Fig. 2A can be considered as essentially linear. The slopes of the lines, unlike PER, are independent of nitrogen intake and, for that reason, may be considered as a more quantitative measure of nutritive value. The slopes of these lines have been called the "Nitrogen Growth Indexes" (Allison, 1959). These indexes, for the data plotted in Fig. 2, are egg albumin, 33; casein, 25; cottonseed flour, 15; and wheat gluten, 7. The data in Table III illustrate the grams of total nitrogen and amino acids available from various dietary protein sources which result in 50 gm

TABLE III

GRAMS OF TOTAL NITROGEN AND AMINO ACID AVAILABLE FROM VARIOUS DIETARY PROTEIN SOURCES WHICH RESULT IN 50 GRAMS GAIN IN BODY WEIGHT PER RAT[a]

	Protein source									Minimum values
	Egg	Casein + methionine	Casein	Fish	Cottonseed + methionine + lysine	Soybean	Wheat gluten + lysine	Cottonseed	Wheat gluten	
Total nitrogen	1.7	1.7	2.5	2.5	2.8	3.0	3.0	4.0	8.0	1.7
Arginine	0.63	0.41	0.61	1.07	1.95	1.35	0.69	2.80	1.8	0.41
Histidine	0.25	0.32	0.47	0.26	0.46	0.74	0.37	0.66	1.0	0.25
Isoleucine	0.68	0.67	0.98	0.79	0.68	0.98	0.85	0.95	2.3	0.67
Leucine	0.93	1.16	1.56	1.18	1.04	1.45	1.35	1.48	3.6	0.93
Lysine	0.76	0.86	1.30	1.37	1.00	1.18	0.80	1.07	1.0	0.76
Phenylalanine	0.63	0.57	0.84	0.57	0.92	0.92	0.95	1.30	2.5	0.57
Tyrosine	0.35	0.55	0.80	0.40	0.48	0.60	0.52	0.68	1.4	0.35
Total aromatic	0.98	1.12	1.64	0.97	1.40	1.52	1.47	1.98	3.9	0.97
Cystine	0.31	0.04	0.06	0.21	0.28	0.33	0.44	0.41	1.1	—
Methionine	0.44	0.62	0.65	0.45	0.64	0.26	0.31	0.34	0.8	—
Total sulfur	0.75	0.66	0.71	0.66	0.90	0.59	0.75	0.75	1.9	0.59
Threonine	0.52	0.48	0.70	0.70	0.62	0.77	0.52	0.88	1.4	0.45
Tryptophan	0.12	0.10	0.15	0.15	0.20	0.26	0.14	0.29	0.4	0.10
Valine	0.80	0.78	1.15	0.81	0.86	0.98	0.80	1.33	2.1	0.80

[a] From Allison (1961).

gain in body weight. The fraction of nitrogen intake required for growth varies with the pattern of amino acids presented to the body by the dietary protein. If it is assumed that the gain in body weight represents a gain in tissue containing 16 % protein, then 50 gm would be equivalent to 8 gm of protein or 1.28 gm of nitrogen. The difference between 1.28 and the minimum 1.7 total nitrogen intake, recorded in the table, could represent the nitrogen needed for maintenance together with some loss through excretion.

C. Biological Value

The "Biological Value" of a dietary protein was defined by Thomas (1909) as the fraction of absorbed nitrogen retained in the body for maintenance and growth. The calculation of Biological Value (BV) therefore, requires an estimation of the amount of nitrogen absorbed into the body and the amount of absorbed nitrogen retained. Absorbed nitrogen (A) can be estimated from Eq. (1):

$$A = I - (F - Fm) \tag{1}$$

where I is nitrogen intake, F is the fecal nitrogen excreted, and Fm is fecal nitrogen of body origin (endogenous). It is customary to feed a protein-free diet to determine Fm. The nitrogen retained is estimated by using the concept of nitrogen balance which is discussed in Chapter 11. Allison and Anderson (1945) demonstrated a curvilinear relationship between nitrogen balance (A) and absorbed nitrogen (B) in the dog, but the curve was essentially linear in the region of negative and low positive balance (Fig. 6A). Similarly, Bricker *et al.* (1945) demonstrated a linear relationship between absorbed nitrogen and nitrogen balance in man and Bricker and Mitchell (1947) reported a linear relationship between these two variables in the rat. If it is assumed that the endogenous excretion of urinary and fecal nitrogen is constant and independent of nitrogen intake, such a linear correlation makes possible the calculation of the Biological Value according to the following equation:

$$\mathrm{BV} = \frac{B - B_0}{A} \tag{2}$$

B in this equation is the nitrogen balance established while feeding protein in the diet, and B_0 the nitrogen balance established during associated periods of feeding a protein-free diet. Biological Values, therefore, should be essentially constant if the determinations are made in the linear portion of the balance curves (Fig. 6A). The significance of the calculation of Biological Value from Eq. (2) will be discussed further in Section III,A (see Table IV).

It will be noted that the relationship between nitrogen intake and gain in body weight is also essentially linear at low nitrogen intakes. An approximate Biological Value for growth, therefore, can be estimated from the Nitrogen Growth Index by assuming a fairly constant composition of the body of the rat. For example, the rats fed egg albumin gained 33 gm in body weight per gram of nitrogen intake. An increase of 33 gm in body weight would be equivalent to an approximate retention of 0.85 gm in nitrogen for growth. Assuming that egg protein nitrogen is 90% absorbed, the BV would be 0.85/0.9 or 0.94, which could be interpreted to mean that 94 % of the absorbed egg nitrogen was retained in the body of the animal for growth. Similar calculations for casein, cottonseed flour, and wheat gluten result in Biological Values of 0.71, 0.51, and 0.20, respectively. These calculations will be in error if the body composition of the rat varies as it grows at different rates with different quantities and qualities of dietary protein intake.

The Biological Value also can differ for maintenance and for growth. The lines drawn through the data obtained for egg and casein can be extrapolated to zero nitrogen intake and will describe the repletion and maintenance of body weight as well as gain. These results can be interpreted to mean that egg and casein yield approximately the same BV for maintenance as for growth. On the other hand, the data for wheat gluten, illustrated in Fig. 2A, experimentally does not continue as a straight line to zero nitrogen intake. Mitchell and Beadles (1950) demonstrated that relatively little lysine is needed for maintenance as compared with the amount needed for growth in the rat so that the Biological Value for wheat gluten is relatively high for maintenance and low for growth. The value of adding lysine to wheat gluten for growth is illustrated in Fig. 2 by the circles with the "hour glass" designs. Rats fed wheat gluten supplemented with lysine grew almost but not quite as rapidly as those fed casein. Possibly, growth of rats fed wheat gluten supplemented with lysine would be somewhat restricted because of the low percentage of essential amino acids in this protein (Harper and Kumta, 1959). The effect of a low percentage of essential amino acids upon nutritive value was illustrated by the experiments reported by Howe *et al.* (1960). These authors demonstrated that a mixture of amino acids, made up to correspond to the FAO reference pattern (see Section VI) and fed to rats resulted in a Nitrogen Growth Index of 12 when the percentage of essential amino acid was 31.1. Retaining the FAO pattern but increasing the percentage of essential amino acids to 56 increased the Nitrogen Growth Index to 19. It is possible also that proteins of high nutritive value supply an excess of the essential amino acids. Bender (1961), for example, found that the essential amino acids of defatted egg could be diluted 15 % by weight with a mixture of dispensable amino acids without altering the nutritive value of the egg protein in the rat. Further dilution, however, with 30 % of the dispensable amino acid mixture reduced the nutritive value.

D. Net Protein Ratio

The relationship between body weight gain and nitrogen intake can be extended into the region of body weight loss when nitrogen intake is decreased to very low values including zero intake (Fig. 1). Bender and Doell (1957) took advantage of this relationship to calculate the "Net Protein Ratio" (NPR). This ratio is the weight loss of a negative control group added to the weight gain of a test group divided by the protein consumed by the latter. The values for NPR avoid another fundamental criticism of PER which is that this ratio makes no allowances for maintenance but assumes that nitrogen intake is only available for growth. NPR and the Nitrogen Growth Index should give the same values if the linear relationship between nitrogen intake and body weight gain can be extended to include body weight loss of animals fed a protein-free diet (see Fig. 1).

E. Net Protein Utilization

Obviously, a more accurate estimate of nitrogen retention in the animal would be obtained by determining carcass nitrogen instead of body weight. Such an estimate of nitrogen retention was proposed by Bender and Miller (1953) (D. S. Miller and Bender, 1955; Bender, 1956). "Net Protein Utilization" (NPU) was defined as the difference in body nitrogen of a group fed dietary protein and a group fed no protein divided by the nitrogen consumed by the test group. This definition can be expressed mathematically as follows (D. S. Miller and Payne, 1961 a,b,c):

$$\mathrm{NPU} = \frac{Cp - C_0}{I} \tag{3}$$

where Cp is carcass nitrogen of animals fed the diet containing protein, C_0 is the carcass nitrogen of animals fed a protein-free diet, and I is the nitrogen intake of animals fed the diet containing protein. The BV as defined by Thomas (1909) could be estimated by dividing NPU by digestibility. Digestibility is calculated by dividing absorbed nitrogen [A in Eq. (1)] by nitrogen intake (I). It was found that the nitrogen to water ratio of the carcass could be estimated from determination of dry matter rather than by an analysis for nitrogen.

F. Tissue Proteins during Growth

1. Impact of Dietary Proteins upon Tissues

The impact of nutrition upon the metabolism of individual cells and tissues of the body needs to be studied in more detail. Published data indicate that this impact may differ from one tissue to another and that it is particularly marked in the liver (Waterlow *et al.*, 1960). The statement was made that growth is the algebraic sum of development of various tissues of the body. The nutritive value, therefore, of a dietary protein for each tissue could vary with the tissue under study and need not correlate with the value determined from gain in body weight.

The ratios between tissue proteins and deoxyribonucleic acid (DNA) are plotted in Fig. 3 (Allison *et al.*, 1962). These data were obtained from the same animals that were used to obtain the gain in body weights plotted in Fig. 2. Thus, the protein in each liver cell may be considered as increasing with an enlargement of dietary nitrogen intake—the greatest increase, however, being observed in the animals fed egg or casein, the minimum in those fed wheat gluten. Similarly, the protein relative to DNA increased in the muscle with a rise in nitrogen intake. The separation between egg or casein and cottonseed flour or wheat gluten was as marked as in the liver—even more so, at low nitrogen intakes. The total

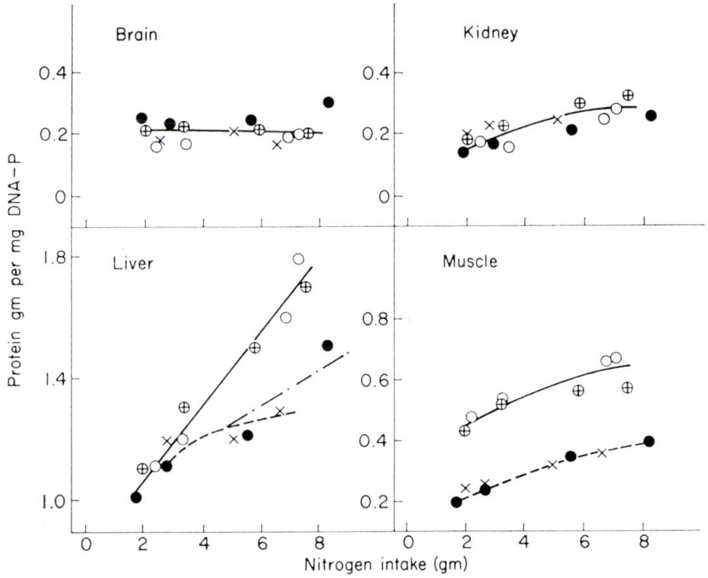

FIG. 3. The effects of nitrogen intake upon tissue proteins relative to DNA in rats fed diets containing the proteins recorded in Fig. 2. (Published with permission of J. Nutrition, Allison *et al.*, 1962.)

protein with respect to DNA also increased with nitrogen intake in the kidney but there was no significant difference between the ratios in animals fed the various protein sources. The ratio of protein to DNA in the brain, on the other hand, was independent of nitrogen intake and was not affected by the over-all nutritional value of the protein. Previous data had demonstrated that the brain of the rat was most resistant to depletion in proteins, followed by the kidney, then the muscle, and finally the liver (Wainio *et al.*, 1959; Allison and Fitzpatrick, 1960). There is need to determine the physiological significance of these variations in sensitivity of the cellular proteins in various tissues, especially as they can be correlated with the concept of "protein reserves." Other biochemical differences associated with the shifts in tissue proteins undoubtedly exist. For example, the lipids migrating electrophoretically with the serum globulins were higher in the rats fed wheat gluten or cottonseed flour than in those fed egg or casein (Allison *et al.*, 1962).

2. RNA, DNA, and Tissue Protein

The liver illustrates a tissue with labile cytoplasmic proteins, most of them components of enzyme systems, which rise or fall in quantity as dietary nitrogen increases or decreases. These labile proteins are often defined as "labile protein reserves" since they contribute amino acids to the metabolic pool for protein anabolism. The impact of protein nutrition, therefore, is especially marked in liver cells (Addis *et al.*, 1936; R. M. Campbell and Kosterlitz, 1948; Munro and Clark, 1960; Muramatsu and Ashida, 1962). Munro and associates have studied extensively the nutritional regulation of protein and ribonucleic acid (RNA) metabolism in the liver cell, their data demonstrating a correlation between the RNA content of the normal liver cell and nitrogen intake and pattern of amino acids. The change in RNA was associated with the magnitude of the endoplasmic reticulum.

The data recorded in Fig. 4 illustrate the changes in DNA, RNA, and in liver protein with the kind and quantity of dietary protein intake expressed as nitrogen intake. These data were obtained with the same animals used for the results recorded in Fig. 2. The RNA phosphorus (RNA-P) relative to a unit weight of liver increased with increasing nitrogen intake and was greatest in animals fed egg and minimum in those fed wheat gluten. Indeed, the RNA-P content of the liver correlated nicely with the nutritive value of the dietary protein, increasing with a rise in

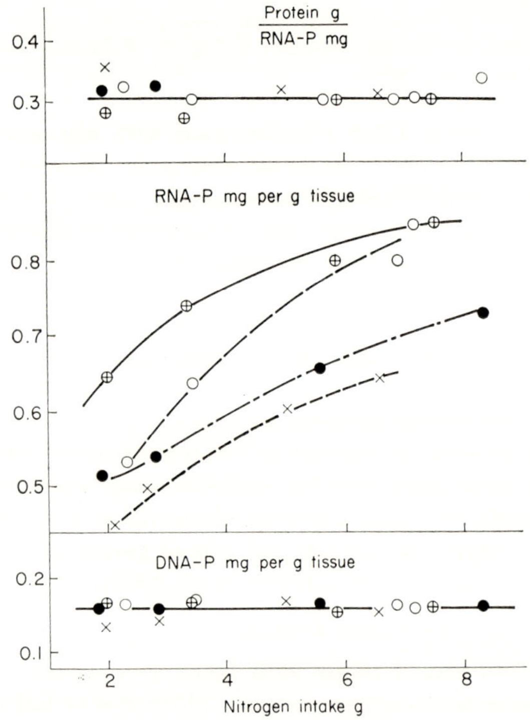

FIG. 4. The effects of nitrogen intake upon liver RNA and DNA. (Published with permission of Federation Proc., Allison *et al.*, 1963.)

the nitrogen growth index. Total liver protein relative to a unit weight of liver also increased with nitrogen intake, being maximum in rats fed egg protein and minimum in those fed wheat gluten. The DNA-P relative to a unit weight of liver, on the other hand, was essentially independent of nitrogen intake and the ratio of protein to RNA is constant, suggesting that there is a close correlation between total RNA in the liver and protein biosynthesis. Indeed, a determination of RNA permits an approximate calculation of liver protein. Data also have been obtained to illustrate a rather close correlation between RNA and the magnitude of the free amino acid metabolic pool in the liver (Allison *et al.*, 1963). Possibly there is a homeostatic mechanism that results in the reduction, for example, in S-RNA (soluble RNA) when the free amino acid pool is reduced due to reduced nitrogen intake or to the presence of an imbalance in amino acid needed for protein biosynthesis.

There are at least two ribonucleases present in the cells. One of these has an optimum activity at pH 5.8, the other at pH 7.8. Ribonuclease (pH 5.8) activity relative to the protein content of the liver decreased with the increased nitrogen intake and was relatively higher in animals fed wheat gluten than in those fed egg proteins. Ribonuclease (pH 7.8) was not altered by nitrogen intake nor nutritive value of the dietary protein. Allison *et al.* (1962) suggested that ribonuclease may play a role in a homeostatic mechanism controlling synthesis of liver protein.

Studies on the growth of the Walker carcinoma 256 demonstrated that rate of growth and protein biosynthesis were correlated with the magnitude of RNA (Allison *et al.*, 1961b). The possible relationship between ribonuclease activity and the growth of the tumor was illustrated by the effect of the phosphoramide drugs. These drugs reduced the growth of the tumor and at the same time increased ribonuclease activity markedly (Zigman and Allison, 1959; Wannemacher *et al.*, 1962). A decrease in RNA was associated with the rise in ribonuclease activity. This association suggested, as indicated before, the possibility that the reduction in RNA may be, in some tissues at least, under the control of ribonuclease activity. Such a hypothesis involving ribonuclease will need much critical study, especially since the sources and kinds of RNA and ribonuclease vary considerably.

III. Nutritive Value and Maintenance

A. Nitrogen Balance

The interrelationship between maintenance and growth can be illustrated, in part, through the use of the concept of nitrogen balance (B) (Allison, 1951) which is defined by the following equation:

$$B = I - (U + F + S) \tag{4}$$

where B is the difference between dietary nitrogen intake (I) and nitrogen excreted in the urine (U), feces (F), and skin (S). The value for S in man may range from 0.1 to 0.4 gm per m^2 per day, but it is neglected in nitrogen balance studies because it is usually relatively small and difficult to measure. The value for S in animals like the dog or rat is believed to be negligible. Nitrogen balance is discussed in detail in Chapter 11.

This concept of nitrogen balance is illustrated in Fig. 5. The cross-hatched area in negative balance illustrates the continual loss of nitrogen from the body of boys at different ages. It is the amount of nitrogen that is needed to maintain equilibrium and it would become constant and maximum when the individual had reached adulthood and stopped growing. The cross-hatched area was calculated by assuming the excretion of endogenous nitrogen to be approximately equal to 2 mg of nitrogen per basal calorie, a value proposed by Terroine (1936), Sorg-Matter (1928), and others. These data are essentially the same as those presented by Hegsted (1957, 1959) and by the Food and Nutrition Board (1959) of the National Research Council. The cross-hatched area, however, will vary according to the magnitude of the endogenous amino acid pool of the body and will be maximum when that pool is filled.

The darkened portion of Fig. 5 estimates the amount of nitrogen needed for growth at different ages. The continual increase in the area for maintenance with time, correlated with the variations in the needs for growth, could be interpreted to mean that the "protein reserves" are low relative to the potential for growth during the rapid development of the infant. The rate of growth of skeletal structures could dominate early develop-

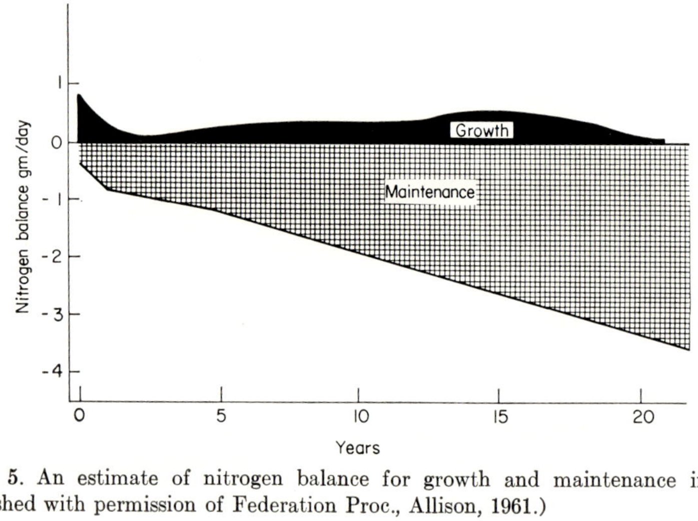

FIG. 5. An estimate of nitrogen balance for growth and maintenance in boys. (Published with permission of Federation Proc., Allison, 1961.)

ment while growth of muscle and other tissues with adequate protein reserves would become more evident following this period. The effects of malnutrition are most marked during rapid growth, possibly with minimum development of reserve proteins. It could be argued, therefore, that with minimum reserves available during periods of rapid growth, a deficiency in nitrogen intake or an inadequate pattern of amino acids could result in growth outstripping the filling of reserves, particularly in the liver and in the muscles. Such an individual could be most susceptible to disease and many of the symptoms of protein-calorie deficiency are made manifest following a period of stress such as a childhood illness (Scrimshaw, 1959). Another period of stress could be associated with growth during adolescence.

B. Biological Value and Nitrogen Balance Index

One of the most useful measurements involving nitrogen balance has been the determination of Biological Value, which is discussed briefly in Section II,C. The marked effect of amino acid imbalance upon nitrogen balance has been illustrated recently by Bressani (1962). His data give added meaning to Biological Value in terms of amino acid balances and emphasize the important relationship between balance and nutritive value. Mitchell and associates have done more than any other group of workers to make the determination of this value quantitative and meaningful (Mitchell, 1929, 1944, 1954, 1959). The relationship between nitrogen intake and nitrogen balance is curvilinear, possibly semi-logarithmic, but the lower part of the curve, in negative and low positive balance, is essentially linear under most experimental conditions (Fig. 6A). The slope of the linear portion of this curve is the Biological Value provided the excretion of body nitrogen (endogenous) is constant and independent of nitrogen intake (Allison and Anderson, 1945; Bricker *et al.* 1945; Bricker and Mitchell, 1947). Thomas (1909), for example, defined Biological Value (BV) as the fraction of absorbed food nitrogen (A) [see Eq. (1)] retained in the body of the animal, a definition that can be described by the following equation:

$$\mathrm{BV} = \frac{A - (U - Um)}{A} \tag{5}$$

where U is the excretion of urinary nitrogen and Um the excretion of urinary nitrogen which could be assigned to endogenous metabolism. $(U - Um)$, therefore, would represent the excretion of urinary nitrogen that came from exogenous sources.

The equivalent of A in Eq. (1) can be substituted into the numerator of Eq. (5), so that

$$(BV)\ (A) = I - (U + F) + (Fm + Um) \tag{6}$$

The expression $I - (U + F)$ is nitrogen balance [see Eq. (4)], hence,

$$(BV)\ (A) = B + (Fm + Um) \tag{7}$$

or

$$B = (BV)\ (A) - (Fm + Um) \tag{8}$$

where $(Fm + Um)$ represents the excretion of endogenous nitrogen by way of the feces and urine, respectively. Thus, if this excretion is constant and independent of nitrogen intake, then the slope of the line relating nitrogen balance to absorbed nitrogen would be the Biological Value (BV) as defined by Thomas (1909). There is some evidence, however, that feeding nitrogen may conserve body nitrogen, thereby reducing excretion of so-called endogenous nitrogen. The addition of methionine to a protein-free diet fed to dogs reduced the excretion of urea nitrogen without affecting the excretion of creatinine (L. L. Miller, 1944; Allison *et al.*, 1947a). Thus, the urea nitrogen arising from the catabolism of amino acids from body nitrogen was reduced by supplementing with methionine. Supplementation of casein with methionine increased the Biological Value calculated according to Eq. (8) to values as high as 1.5, especially in the protein-depleted animal. Values as high as this suggest that feeding nitrogen, under these experimental conditions, changed the excretion of body nitrogen and the shape of the curve so that Eq. (8) is not always valid. Brush *et al.* (1947) also reported unusually high Biological Values determined in partially depleted rats, results which were interpreted to mean that the excretion of body nitrogen was less during the period of nitrogen feeding than when the animals were fed a protein-free diet. Nasset (1956), in a study of nitrogen balance vs. nitrogen intake while feeding mixtures of amino acids to rats, found that the relationship was not always linear in negative nitrogen balance. He emphasized the unstable values of nitrogen balance data obtained while feeding a nitrogen-free diet. Since the relationship between nitrogen balance and nitrogen intake may vary and since it becomes definitely curvilinear in positive balance, Allison *et al.* (1946) suggested that the tangent to the curve at any given nitrogen intake be called the "Nitrogen Balance Index" for that intake. If Eq. (8) is valid for the experimental conditions used, the index would be equal to the Biological Value when calculated from the linear portion of the curve. If not, the index would always be the rate of increase of nitrogen balance with respect to nitrogen intake, a measure of changing anabolism at any given intake.

The index varies with the pattern of amino acids, the caloric intake, the requirements for growth and for maintenance, and the physiological state of the individual. The index is a valuable measure of nutritive value but it does not reveal changes in individual body tissues and it suffers from the experimental difficulties associated with the measurement of nitrogen balances or other estimations of retention of nitrogen in the body.

Some indexes for absorbed nitrogen, determined over the linear portion of the nitrogen balance vs. absorbed nitrogen curves, are recorded in Table IV. These data demonstrate the increased index in animals depleted in body proteins when compared with normal controls. The high value for casein in the adult cat may be the result of a lower requirement for sulfur amino acids than in the dog or the rat. The higher index for wheat gluten in the adult rat compared to the growing rat is believed to be the result of less need for lysine by the adult than by the young rat (Mitchell, 1947). The high value for wheat gluten in the cat may also be associated, at least in part, with a reduced requirement for lysine but prolonged feeding of wheat gluten to the cat does not develop a normal animal. A depleted rat needs just as much lysine for repletion as a young animal needs for growth. These indexes, determined over a short period of time, therefore can be misleading. In general, the pattern of amino acids needed for growth appears to be good for both young and adult since it is required for the development of new tissue in the young and the repletion of depleted body nitrogen in both young and adult.

TABLE IV

NITROGEN BALANCE INDEXES FOR ABSORBED NITROGEN (BV) OVER LINEAR PORTION OF NITROGEN BALANCE VS. ABSORBED NITROGEN CURVES[a]

Subject	Whole egg	Egg white	Beef	Casein	Peanut flour	Wheat gluten
Man	0.94	0.91	0.67	0.68	0.56	0.42[b]
Dog (normal adult)	0.87	1.14	0.77	0.73	0.56	0.44
Dog (depleted adult)	1.06	1.14	—	0.84	—	0.70
Cat (adult)	—	1.00	—	0.93	—	0.74[c]
Rat (adult)	0.82	0.94	0.69	0.59	0.46	0.65[d]
Rat (growing)	0.87	0.97	0.76	0.69	0.54	0.40

[a] Allison (1955).
[b] Hawley *et al.* (1948).
[c] S. A. Miller and Allison (1958).
[d] Mitchell and Beadles (1950).

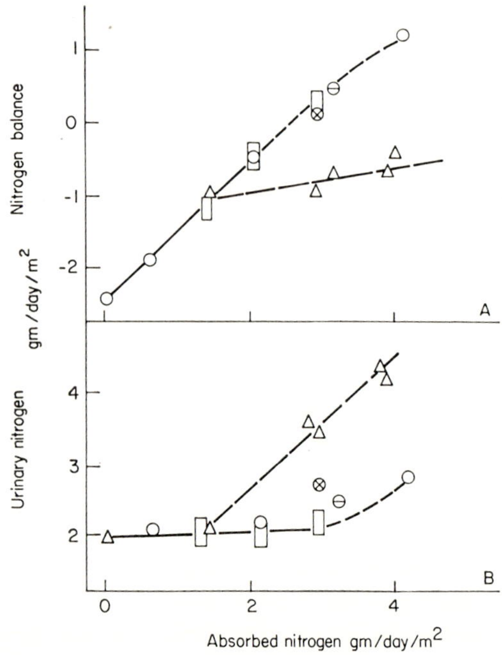

FIG. 6. Absorbed egg albumin nitrogen in gm/day/m^2 plotted against nitrogen balance (*A*) or urinary nitrogen (*B*) in gm/day/m^2 of body surface (see text). (Reproduced by permission of N. Y. Academy of Science, Allison *et al.*, 1946.)

The concept of Nitrogen Balance Index is illustrated by the data plotted in Fig. 6 (Allison *et al.*, 1946). The circles represent average data obtained while feeding an adequate caloric intake of a purified diet containing egg albumin to dogs. The rectangles record similar data obtained while feeding 50 % of this intake of calories; and the triangles, average data obtained while feeding 25 % of an adequate caloric intake. In general, a reduction in caloric intake increased the excretion of urinary nitrogen (Allison, 1956a), the amount of increase depending, in part, upon the body reserves. The data were plotted in Fig. 6, however, starting at a common origin, emphasizing the relationship between nitrogen intake and slope of the line. Under these experimental conditions, the Nitrogen Balance Index was essentially constant in negative and low positive balance when the caloric intake was either adequate or 50 % so. When the calories were reduced to 25 %, the index remained unchanged at low intakes of nitrogen but rapidly approached zero as the intake was increased. Hence, the dog could not be placed in nitrogen equilibium. The increased excretion of urinary nitrogen as the index decreases is illustrated by the data plotted in the lower half of Fig. 6B. This increase is primarily a rise in the excretion of urea, a rise which is correlated with an increased catabolism of amino acids and a reduction in protein anabolism.

The rise in catabolism and reduction in synthesis of certain tissue proteins when the caloric intake is reduced is a complex response to lack of energy intake. Rosenthal and Allison (1951) recognized two responses to a restricted intake of calories. The first response to a caloric restriction was an increase in excretion of urinary nitrogen without altering the Nitrogen Balance Index, a response which was associated with an over-all rise in catabolism of proteins. Continued feeding of a markedly restricted intake of calories depleted the protein stores of the animal. The second response resulted in a marked increase in the excretion of nitrogen which was associated with the reduction in the Nitrogen Balance Index. Some dogs resisted this second response longer than others, a resistance that was correlated with the magnitude of the fat stores and reserve proteins.

Swanson (1959) has written an excellent review of "Food Energy and the Metabolism of Nitrogen." She points out that "whether catabolism or anabolism predominates in the adult animal following caloric restriction depends on the degree to which energy value of the diet has been reduced, the quantity and quality of the dietary protein, the size and nature of the labile body reserves in respect to protein, fat, and glycogen, and in some instances, dietary sources of energy." In the experiments illustrated in Fig. 6 the response to an increase in calories was the same whether carbohydrate or fat was added to the diet. For example, the circle with the bar illustrates the effect of adding sucrose to increase the caloric intake from 25 to 50 % of normal. The circle with the cross bar records similar data obtained by increasing the caloric intake with lard. Under other experimental conditions differences between carbohydrate and fat may be found (Munro, 1951) (see Chapter 10).

C. Summary Chart of Indexes Used for Growth and Maintenance Determinations of Nutritive Values

The relationships between the Nitrogen Growth Index, Protein Efficiency Ratio, and Net Protein Ratio, all dependent upon body weight changes, are recorded in Fig. 1. Similarly, Nitrogen Balance Index (see Section III,B), Biological Value, and Net Protein Utilization, based upon body nitrogen changes, are illustrated in this figure. Net Protein Utilization is labeled (standard), following the suggestion of Platt *et al.* (1961) (see Chapter 21), to differentiate it from measurements made on a curvilinear portion of the curve. As the animal gains in size the maintenance portion of the curve (below 0 nitrogen balance) increases until, in adulthood, maintenance becomes the major feature of nitrogen retention. Positive balance in adult man, for example, is the result of repleting depleted tissues or an increase in muscular or tissue growth which could be called "adult growth." It should be emphasized again that the area for maintenance and the area for growth may not always have as simple a relationship as suggested by these curves. The requirements for maintenance, for example, may differ from those for growth with one dependent upon the other, particularly in the young growing animal.

IV. Mixed Diets

The preceding sections of this chapter have dealt in the main with the study of isolated proteins and the determination of their quality for various functions of the body. However, diets as consumed inevitably consist of mixtures of proteins, which interact not only with one another but also with other nutrients, so that the nutritive value of the protein of a diet has to be considered from the practical point of view in the context of the diet in which it is consumed. This section will consider some aspects of this more complex relationship, and the attempts which have been made to provide indexes of nutritive quality which take cognizance of the other factors in the diet.

A. Supplementation

Given an adequate intake of calories, and also of metabolites other than amino acids, the nutritive value of a dietary protein would be essentially a function of the pattern of indispensable amino acids absorbed into the body, the relative percentage of these amino acids to those considered dispensable, and the nitrogen intake. Supplementation involves an attempt to improve the pattern of absorbed amino acids and the amount of nitrogen intake. Emphasis is placed upon absorption into the body because of the role that rate of digestion may play in influencing the pattern of absorption, particularly a mixture of free and combined amino acids. Howe *et al.* (1960), for example, found that when casein was supplemented with free amino acids to simulate the FAO reference pattern (see Section VI) the Nitrogen Growth Index was the same whether the amino acids were fed free or partially combined in casein. On the other hand, when cottonseed flour supplemented with amino acids to simulate the FAO pattern, was fed to young rats, the Nitrogen Growth Index was lower than when either the cottonseed flour or the mixture of amino acids was fed. Variation in digestion of the cottonseed flour could result in relatively rapid absorption of some of the free amino acids to produce an imbalance within the body.

Improvement in patterns of amino acids provided by the diet is usually sought by mixing protein sources in natural or prepared food stuffs. Semolina, for example, can be made into an alimentary paste with a Nitrogen

Growth Index of 7.2. Obviously, this index is too low to permit proper development of a young animal. Such a paste, however, is usually supplemented with sauces such as tomato sauce plus cheese or meat sauce plus cheese. Adding either one of these sauces to the semolina food increased the Nitrogen Growth Index to 16, an adequate value for growth if the nitrogen intake was sufficiently high. The nitrogen intake, however, was too low for adequate growth when the semolina was supplemented with the sauces used for these experiments (Allison *et al.*, 1959). These results illustrate the importance of not only improving the nutritive value but also the quantity of dietary protein needed to meet the requirements for synthesis of tissue proteins. Testing of the nutritive value of mixed diets should include, if possible, the feeding of sufficient nitrogen to maintain good growth since the Nitrogen Growth Index may be adequate at low nitrogen intakes but may decrease at higher intakes, indicating toxicity. Although such a toxic response is rare, feeding at higher nitrogen intakes to produce a normal animal is always a good check upon the value of a mixed food.

The data so far indicate that if the Biological Value of a protein is less than 60, increasing the intake of that protein may not overcome the amino acid deficiencies but may even cause an imbalance. A protein of lower nutritive value, however, may be an excellent supplementing protein in a mixed diet. Cottonseed flour is an example of a protein that may have a Biological Value of 60 and be sufficient without supplementation to produce a normal rat. Some samples of cottonseed flour, however, may have a lower Biological Value, but in either case, they can be excellent supplementing protein sources. Cottonseed flour, for example, combined with maize, sorghum, and yeast resulted in a mixture with a nutritive value similar to casein (Scrimshaw *et al.*, 1957; Bressani and Scrimshaw, 1961).

This discussion has emphasized the importance of measuring both the quality and quantity of proteins to evaluate a diet. All the methods for estimating the nutritive value of dietary protein for growth can be applied to mixed diets and the over-all utilization of the dietary protein calculated. For example, net protein values, suggested by Mitchell and Carman (1924), are functions of nitrogen retention, digestibility, and protein source (Allison *et al.*, 1946). J. A. Campbell and Chapman (1959) and J. A. Campbell (1960) recognized that the protein value of a food could be considered to be a function of the BV, the percentage of protein in the food and the amount of food consumed in a "reasonable daily intake." A "Protein Rating" for a food was calculated, therefore, by multiplying the PER of the food, determined under standard conditions, by the grams of protein in a "reasonable daily intake."

B. Net Dietary-Protein Value

Platt and Miller (1959) have emphasized some of the variables that can affect the determination of protein values of diets which are eaten by man. Some of these are: (*a*) the complexity of the mixed diets, (*b*) the composition and timing of the meals or snacks together with the total intake, (*c*) the processing and cooking procedure used to prepare the diets, (*d*) the protein-calorie ratio of the diet, (*e*) the adequacy of the caloric intake, and (*f*) the completeness of diet, particularly the level of some vitamins and minerals. Consideration has been given, therefore, to various methods of calculating a "Net Dietary-protein Value" (NDpV) for the food as it is eaten without modification (D. S. Miller and Bender, 1955; Platt and Miller, 1959) (see Chapter 21). This value is a product of quantity and quality of protein as follows:

$$\text{NDpV} = \text{protein concentration} \times \text{Net Protein Utilization (operative)} \qquad (9)$$

Determined under standard conditions in an otherwise complete diet "Net Protein Utilization" has been called NPU (standardized) but if it was determined by feeding a mixed diet as it is consumed by a family or population, the value was designated NPU (operative).

On the basis that people eat to satisfy energy needs, the interdependence of two dietary factors—protein and metabolizable energy—resulted in the expression of NDpV in terms of the calories which could be derived from protein (assuming complete utilization for energy) calculated as a percentage of the total metabolizable energy in the diet. This value, called "Net Dietary-protein Calories per cent" (NDpCal%) is calculated according to the following equation (see Chapter 21):

$$\text{NDpCal\%} = \frac{\text{NDpV} \times 4}{\text{kcal/gm food}} \qquad (10)$$

The relationships between "Protein Score," protein calories per cent, and NDpCal% over a wide range of protein concentrations in the diet have been expressed in the form of a nomograph (Platt *et al.*, 1961; Miller and Payne, 1961a). The NDpCal% can be read off the nomograph at various concentrations of dietary protein. Thus nutritive values of various mixed diets can be expressed relatively as numbers.

D. S. Miller and Payne (1961a) pointed out that nitrogen intake may be diverted into three metabolic pathways: N used for maintenance (I_m), for growth (I_g), and for energy (I_e).

$$I = I_m + I_g + I_e \qquad (11)$$

Allison (1958b) presented data to suggest that in the presence of adequate dietary proteins, the rate of growth of rat tissue is, in general, a function of caloric intake. A competition, however, may develop between the need for growth or maintenance and for energy. Data were presented to demonstrate that the caloric intake could be reduced 50% from normal without altering the rate of retention of nitrogen in adult dogs, but when the caloric intake was reduced to 25% of normal, the dogs could not be placed in nitrogen equilibrium (Fig. 6A). At this low caloric intake, as the nitrogen intake was increased, urea nitrogen excretion increased and the retention of nitrogen was reduced markedly (Allison *et al.*, 1946). However, at low nitrogen intakes on this restricted diet, nitrogen was retained. Similarly, Cox *et al.* (1953) presented data that suggested some protein anabolism under restricted caloric intakes.

It should be emphasized that under all conditions excess dietary protein intake, above the amount that can be used for anabolic purposes, results in an increased excretion of nitrogen and a reduced percentage of dietary nitrogen retained. It is customary to consider the BV or NPU as decreasing in magnitude after the needs for protein anabolism in the body have been met. The terms BV and NPU_{st}, however, could be considered essentially constant for any one protein source, if they are determined at nitrogen intakes below the absolute needs of the body during the feeding of an otherwise complete diet.

C. *Complete Protein*

The concept "Complete Protein" was introduced by Howard *et al.* (1958, 1960) to estimate that fraction of total dietary protein which may be utilized for anabolic purposes. They recognized that the pattern of essential amino acids required for anabolic purposes may depend upon the physiological state (growth, tissue repair, pregnancy, lactation, adult maintenance). For their purpose, however, Complete Protein was represented by the pattern of essential amino acids needed for growth in the rat. The complete protein content of a food or diet may be calculated from the equation:

$$P_c = Q \times \frac{A}{Ar} \qquad (12)$$

where P_c equals percentage of complete protein in food, Q is the percentage of total protein nitrogen × 6.25, A represents the quantity of limiting amino acid in grams per 16 gm nitrogen, and Ar equals the requirements for the limiting of amino acids in grams per 16 gm nitrogen. Calculation of the complete protein requires the establishment of a "reference protein" for growth. Reference patterns of amino acids for growth will be discussed in Section VI.

D. Plasma Amino Acid Ratio

A number of factors influence the concentrations of amino acids in the plasma. To a certain extent the plasma amino acids represent a significant portion of the metabolic pool of free amino acids in the body, a pool which is high when the "protein stores" are full and low when the animal is depleted in those stores. The plasma amino acids also reflect the dietary intake. Denton and Elvehjem (1953, 1954), for example, observed that, in general, the increase in concentration of plasma amino acids after a meal was proportional to the amino acid content of the dietary protein. Longenecker and Hause (1959) demonstrated that changes in concentrations of plasma amino acids after a meal were directly correlated with the amino acid composition of the ingested protein. They calculated a "Plasma Amino Acid Ratio" (PAA) by determining the difference between the concentration in the plasma of each essential amino acid in the fasting animal (18-hr fast) and the amount found after ingestion of a meal (average of five hourly plasma levels). This difference was then divided by the requirement for the essential amino acid under consideration (see Section VI). The ratios for each amino acid were placed in order of magnitude from the lowest to the highest, thereby obtaining a limiting order for the series of essential acids. Using this method, lysine was found to be the first limiting amino acid for wheat gluten when fed to the dog. Similar correlations have been reported for rats and man by Campbell and associates (McLaughlan *et al.*, 1961; Morrison *et al.*, 1961a,b). These authors, in a report to the Committee on Amino Acids of the Food and Nutrition Board of the National Research Council, stated that reproducible responses to ingested protein and good estimates of PAA ratios can be obtained in man. They suggest that there is need for further studies on "(1) the order in which amino acids become limiting in foods, (2) the relation between PAA ratios and protein quality of a variety of proteins, (3) the applicability of the method for evaluating protein in diets and in supplements for children, (4) the clinical significance of differences in response between individual subjects, (5) the applicability of the procedure for measuring availability of amino acids in human foods."

V. Depletion and Repletion

A. Body Proteins as Reserves

Data have been presented in Section III,B to illustrate the apparent increased efficiency of retention of dietary nitrogen often observed in animals depleted in body proteins. A number of explanations for this increased efficiency have been proposed, most of them involving the concept of dynamic protein metabolism and "protein reserves." It is recognized that the body does not store protein or create reserves in the same way as fat is metabolized. There is evidence, however, that some tissue proteins can be reversibly depleted and repleted by a variation in the quantity and quality of dietary proteins. These proteins can contribute to the free amino acid supply of the body, sometimes called the amino acid metabolic pool, and thereby help maintain essential structures and functions involving amino acids when the intake is deficient.

Some tissue proteins are depleted very rapidly when an animal is fed a protein-free or starvation diet. A young rat, for example, will lose a large portion of these so-called "Labile Protein Reserves" after 12 hours on a starvation diet, enough to affect the results of certain pair-feeding experiments. These Labile Reserves represent only a small fraction of the total body protein but the debate concerning the significance and the role of protein reserves is often restricted to these very labile tissue proteins (Allison *et al.*, 1963).

There is, however, no sharp dividing line between these Labile Reserves and the total body protein that can be reversibly depleted and repleted by variation in nitrogen intake. This larger concept of reserves can include 20 % of the body nitrogen and can contribute, it is believed, to the maintenance of essential tissue structures and metabolic functions over a long period of time. An adult dog, for example, fed a good maintenance diet and then placed on a protein-free diet will lose approximately 20–25 % of its body nitrogen over a period of 90 days, nitrogen which can be repleted by feeding an adequate diet (Allison and Wannemacher, 1957). An animal maintained on a less adequate diet, and somewhat depleted in body nitrogen, may be in serious condition within a period of 14 days. Some definitions of protein reserves therefore include all body proteins that can be depleted and repleted. They are thought to contribute amino acids to the metabolic pool available to each tissue for maintenance of essential tissues and functions and to meet stress conditions such as the healing of wounds and the repair of damaged metabolism (Allison and Fitzpatrick, 1960).

These over-all body protein reserves vary in magnitude with nitrogen intake and with nutritive value. The amount of protein per unit of DNA in the liver, muscle, and kidney, for example, increases in the rat with nitrogen intake, thereby approaching a maximum value which represents full reserves (Fig. 3). The quantity of these cellular proteins also varies with the nutritive value of the dietary protein, increasing as the nutritive value increases. Some tissues, like the liver and muscles, are more susceptible to the effect of nutritive value than others like the kidney (Fig. 3). An ordinary determination of nutritive value, using gain in body weight or in body nitrogen, does not reveal imbalances in the distribution of body nitrogen in various tissue cells (Allison *et al.*, 1962).

The needs of the body may vary with the physiological state. The caloric intake for rapid repletion of depleted tissues in the adult, for example, must be increased above normal values. Feeding a protein of low nutritive value may replete the nitrogen lost during depletion but may produce a biochemical imbalance not associated with a good pattern of amino acids. Care must be taken not to overload metabolism with fat, protein, or other dietary constituents, or even drugs. Feeding unnecessarily high protein diets, even to so-called normal individuals, may represent overloading so that damage is done to metabolism. The problem is one of proper balance between nutrients as much as proper quantities.

The loss in weight of young rats, correlated with loss in Labile Protein Reserves can be illustrated by the following data which were taken from a recent paper by Allison *et al.* (1963). The loss in weight of the liver, heart, and kidney in young rats fed the protein-free diet for 3 days is illustrated by the open circles in Fig. 7. The black circles record the weight losses in similar young animals starved for the same period. Starvation caused a greater loss in weight in the liver and heart than feeding the protein-free diet. No such difference was observed during loss in weight of the kidney and there was no effect of feeding either the protein-free diet or of starvation upon the weight of the brain. The following data will correlate in part these weight changes with the magnitude of cellular proteins and the part that these proteins may play as reserves will be discussed.

The effects of starvation in protein on the RNA/DNA and the protein/DNA ratios in the liver are illustrated by the white circles in Fig. 8 while the black circles record similar data obtained from the young animals starved over the period of 3 days. The rapid decrease in RNA and tissue protein with protein or protein-calorie deprivation is illustrated by these data, with the decrease greater in those animals deprived of both protein and calories. Thus, it could be said that the depletion of Labile Protein Reserves was greater in the starved animals than in those fed calories but no protein. It should be emphasized again, however, that there is no sharp dividing line between so-called Labile Protein Reserves and the total body protein available to meet a metabolic need. The starved animals must utilize more of the amino acids for energy than the animals fed calories but in both sets of animals the free amino acid pool is depleted, as illustrated in Figs. 9 and 10. The white bars record the free amino acids in the liver of the control animals fed an 18 % casein diet. The bars with slanted lines illustrate the data obtained from livers in animals fed calories but no protein and the black bars present data involving livers of starved young rats. The greater utilization of the free amino acids for energy

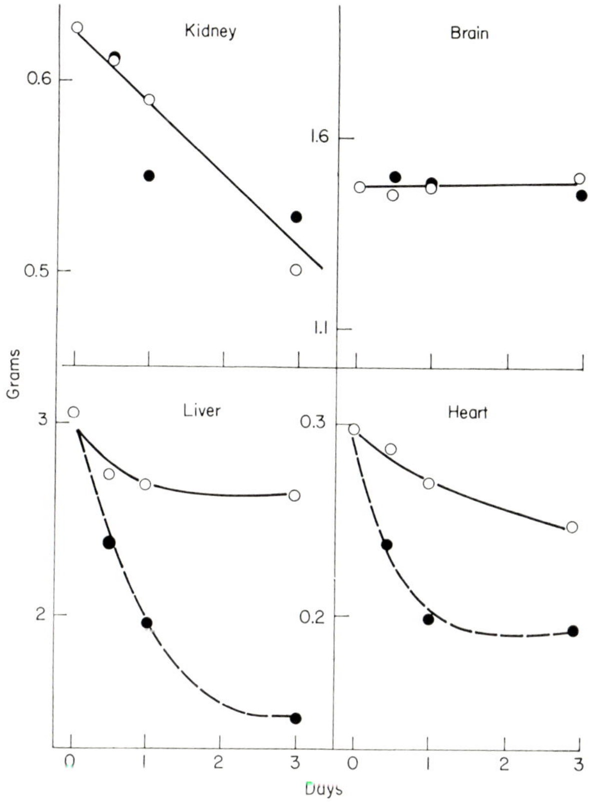

FIG. 7. Loss in weight of various tissues in young rats fed a protein-free diet (open circles) or starved (closed circles) for 3 days.

purposes is suggested by the increased formation of urea and taurine in the starved animals than in those fed calories, recorded in Fig. 10.

The rapid reduction in the free amino acid pool with time is illustrated for methionine in the upper left-hand portion of Fig. 8. It is not surprising, therefore, to find that the uptake of S^{35} by protein in the liver from S^{35}-methionine is greater in the depleted than in the normal animal. The technique used to determine uptake involved giving to the animal a constant dose per body weight of S^{35}-methionine which resulted in a greater concentration of the labeled acid as the free amino acid pool decreased with depletion.

The "labile tissue proteins" are abundant also in the muscle of the body, and the data obtained can be interpreted to mean that the muscles can contribute amino acids to the rest of the body. The initial reaction to depletion in the muscles results in a decrease in the free amino acid pool

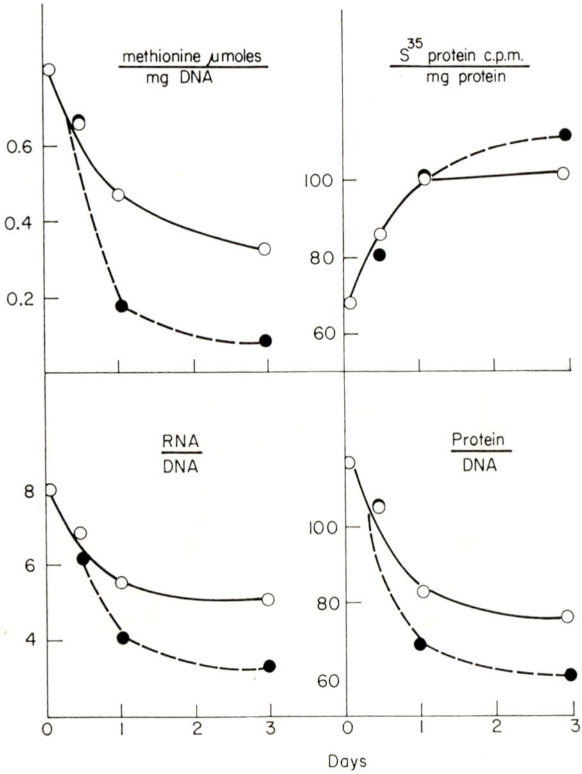

FIG. 8. The effect of feeding a protein-free diet (open circles) or starvation (closed circles) for various time periods upon liver RNA to DNA ratios, protein to DNA ratio, micromoles of free methionine per mg of DNA and S^{35} uptake per mg of liver protein. (With permission of Federation Proc., Allison *et al.*, 1963.)

which is followed, however, with a tendency to increase so that the S^{35} uptake from labeled methionine is at first increased and then decreased as the period of depletion reaches 3 days (Allison *et al.*, 1963). The muscles, therefore, may represent a most important source for free amino acids during periods of deprivation. Arroyave (1962) has reviewed the literature and emphasized the concept that skeletal muscle may be sacrificed during chronic protein deprivation to supply amino acids for synthesis of more "essential" proteins. Thus each tissue must be considered separately and yet as a part of the dynamic whole.

Serum albumin is usually considered as one of the major protein reserves of the body. The reduction of serum albumin in these experiments using young animals is illustrated in Fig. 11. There was no difference in the serum albumin concentration in rats deprived of protein plus calories. The upper

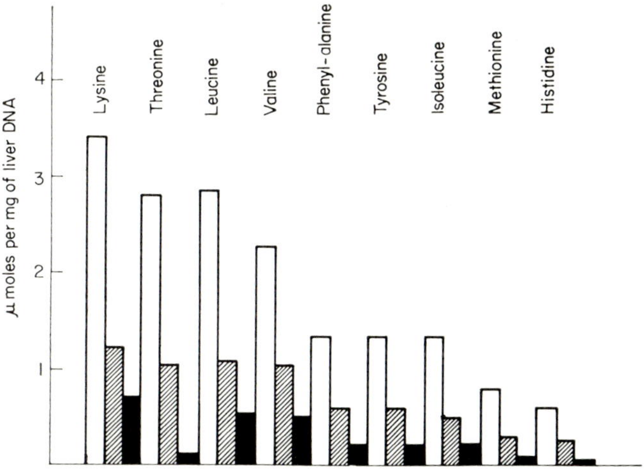

FIG. 9. The micromoles of free essential amino acids per mg of liver DNA in rats fed 18% casein (white bars), protein-free diet (slanted lines) or starvation (black bars) for 3 days. (With permission of Federation Proc., Allison *et al.*, 1963.)

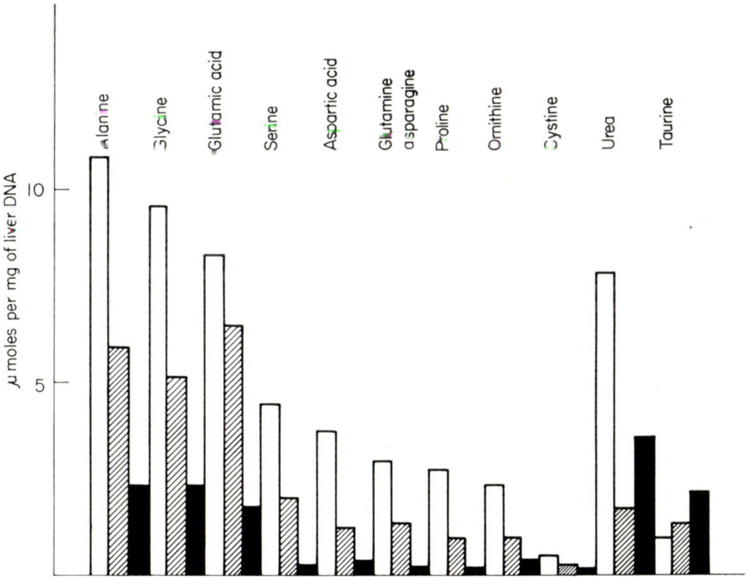

FIG. 10. The micromoles of free nonessential amino acids per mg of liver DNA in rats fed diets recorded in Fig. 9 for 3 days. (With permission of Federation Proc., Allison *et. al.*, 1963.)

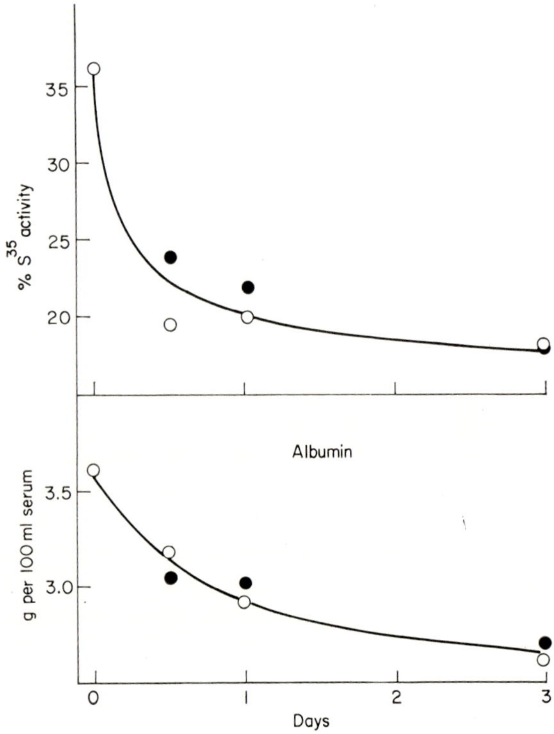

FIG. 11. The effect of feeding a protein-free diet (open circles) or starvation (closed circles) upon serum albumin and upon the percentage of S^{35} activity in serum albumin. (With permission of Federation Proc., Allison *et al.*, 1963.)

part of Fig. 11 records the uptake of S^{35} by albumin from S^{35}-methionine in these animals. These data can be interpreted to mean that the synthesis of albumin is the same in the two groups of animals but is rapidly reduced in both of them with depletion in body proteins.

B. Measurement through Nitrogen Balance

When an individual is fed a diet deficient in proteins, calories, or sometimes other correlated constituents, certain body proteins are depleted. Usually the metabolic pool of amino acids is reduced and the excretion in urea falls so that the body is conserving nitrogen. Arroyave (1962) has used the urea/creatinine ratio in the urine to estimate the adjustments to protein metabolism in children. Less nitrogen is needed to maintain equilibrium and the depleted individual can be placed well into positive balance, thereby demonstrating a capacity to replete (Allison and Wannemacher, 1957). This capacity to replete is illustrated by the data plotted in Fig. 12. These data were obtained by first depleting the dogs by feeding a protein-free diet until they had lost approximately 20 % of their body nitrogen. The black circles record the nitrogen balance produced when the depleted animals were fed 0.2 gm casein nitrogen/day/kg body weight and 80 cal/day/kg for 28 days. This amount of nitrogen produced a small but constant positive balance of approximately 0.05 gm nitrogen/day/kg body weight. The dog was being maintained in the depleted state with a small amount of repletion taking place. The circles with vertical bars illustrate data obtained while feeding the same caloric intake but increasing the casein nitrogen intake to 0.6 gm/day/kg body weight. The slope of the line drawn through these points is a measure of the rate of repletion while the area under the curve records the amount of repletion. The amount of nitrogen retained during this period of repletion was calculated to be 3.2 gm/kg body weight. Increasing the intake to 1 gm/day/kg but with the same caloric intake raised the initial positive balance to 0.6 gm/day/kg, thereby enlarging the area under the curve and the amount of repletion but without altering the slope (Allison and Fitzpatrick, 1960). The open circles record data obtained while feeding 1.12 gm casein nitrogen/day/kg but at a higher caloric intake of 140 cal/day/kg. The higher caloric intake increased markedly the rate of repletion as measured

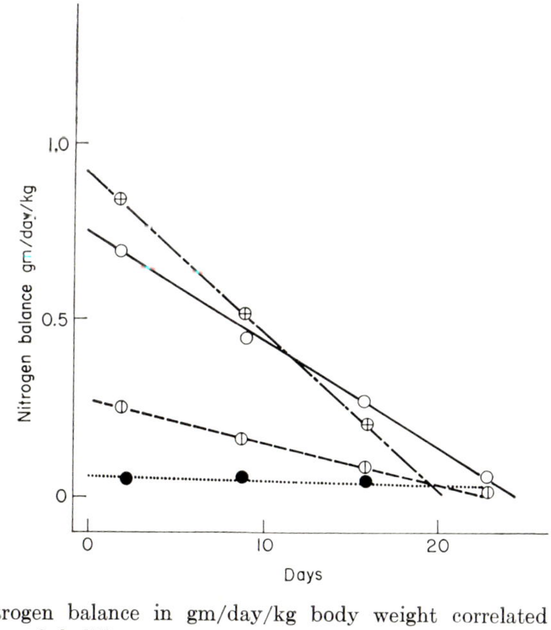

FIG. 12. Nitrogen balance in gm/day/kg body weight correlated with days of repletion in dogs fed different amounts of casein nitrogen and calorie intake (see text) (Allison and Wannemacher, 1957).

by the slope of the line. The increase in rate of repletion by raising the caloric intake may be associated with the higher energy requirements also found during growth in young animals. The nitrogen retained at this higher nitrogen and caloric intake was calculated to be 9.5 gm/kg body weight. The line drawn through the circles with cross bars records the effect of supplementing the 1.12 gm of casein nitrogen with 0.3 % methionine. The rate of repletion was increased by supplementation but the amount of repletion remained at 9.5 gm of nitrogen per kg of body weight.

A similar series of repletion experiments were completed with wheat gluten instead of casein in the purified diet (Allison and Wannemacher, 1957). The rate of repletion, as measured by the slope of the line, was reduced in animals repleted by feeding wheat gluten. Adding lysine to wheat gluten resulted in the same slope and amount of repletion recorded for unsupplemented casein. It is interesting to record here, however, that the Nitrogen Balance Indexes were much higher in the depleted than in normal animals. Casein, for example, which is deficient in sulfur amino acids, had an average Nitrogen Balance Index of 0.74 when fed to the normal dog. This value was increased to 0.84 in the depleted animals. Similarly, the index for wheat gluten was 0.44 when fed to the normal dog but increased to 0.70 when ingested by the depleted animal.

C. Measurement through Body Weight

In an excellent review of methods of measuring the nutritive value of proteins, Frost (1959) discussed the "Rat-Repletion Method" in detail. This assay consists in measuring the gain in body weight which is correlated with the repletion in body nitrogen. The method was developed by Cannon during his classic studies of the effect of protein deficiency on loss of natural resistance to disease (Cannon *et al.*, 1944; L. E. Frazier *et al.*, 1947; Benditt *et al.*, 1947; Wissler *et al.*, 1948). More recently Cannon *et al.* (1954) have reviewed the advantages and applications of the Rat-Repletion Method to the solution of various medical and biological problems. Frost (1959) and Frost and Sandy (1948, 1949, 1951) have studied the method in detail to determine the optimal conditions for depletion and repletion in body proteins. This method, like all others that sum up changes in various tissues into one value, however, is not designed to reveal alterations which take place in tissues and cells. Supplementing the Rat-Repletion Method, for example, with studies involving changes within the tissues of the animal would be most helpful.

D. Tissue Proteins

The concept of a dynamic state, discussed briefly in Section III,A, may be used to explain the loss or gain of nitrogen at different rates from the various tissues or compartments of the body. For example, plasma albumin was easily depleted by feeding a protein deficient diet while some plasma globulins were not. Certain blood proteins took priority over tissue proteins in the protein-depleted animal (Whipple, 1948; Whipple *et al.*, 1947). The cytoplasmic proteins of the intestinal tract and liver were most rapidly depleted (Addis *et al.*, 1936; Kosterlitz and Campbell, 1945; Munro and Clark, 1960). The studies of Wainio *et al.* (1953, 1954, 1959) demonstrated that the total protein and certain enzymes of the liver were most labile in protein depletion while the total protein and the enzyme systems in the brain were very resistant. Similarly, Bargoni *et al.* (1952) reported that unit activities of a number of enzymes in the brain were unchanged by depletion in over-all body proteins. Wainio and colleagues found that the total protein and enzymes of the heart ventricle were more resistant than those of the kidney, skeletal muscle, and spleen. More consideration needs to be given, therefore, to the effects of depletion and repletion upon cellular and molecular moieties. Data have been obtained to indicate that protein depletion in the rat and the dog results in a decrease in certain tissue proteins which can be correlated with a fall in ribonucleic acid and in protein biosynthesis (Wannemacher *et al.*, 1962; Allison *et al.*, 1962; Wannemacher *et al.*, 1963).

The reduction in the activity of some enzyme systems, however, particularly those associated with the "protein reserves" of the liver, were nicely correlated with the nutritive value of dietary proteins as determined by standard methods for nutritive value (Litwack *et al.*, 1952, 1953, 1954; Muramatsu *et al.*, 1954; Muramatsu and Ashida, 1955, 1962). Muramatsu and Ashida (1962) divided enzyme systems of the liver of rats into various types according to their response to the dietary level of protein in the diet. For example, some enzyme systems decreased in activity, others remained unchanged, or increased in activity as the protein intake was reduced. Those that decreased, such as xanthine oxidase, succinic dehydrogenase, and glutamic dehydrogenase, correlated well with the fall in liver protein and the decrease in growth rate of the rats.

E. Plasma Proteins

Numerous papers have presented data correlating depletion in body proteins with changes in serum protein and serum enzyme activities. No attempt will be made to review all of these data, but brief attention will be given to a consideration of the effects of protein depletion and repletion upon serum albumin and globulins (Chow *et al.*, 1945).

The reduction of serum albumin associated with protein deficiency in man has been discussed recently by Scrimshaw and Behar (1961) and Arroyave (1962), in the dog by Allison and Wannemacher (1957), and in the rat by Wannemacher (1961). Preliminary data indicate that the decrease in serum albumin during depletion is the result of a slower synthesis of this protein in the liver (Gitlin *et al.*, 1958; Wannemacher, 1961). Recently, Wannemacher *et al.* (1963) presented data that could be interpreted to mean that there is a decrease in concentration and synthesis of serum albumin and beta globulin in the depleted dog, with evidence that there is a reduction in the rate of catabolism of most serum proteins. Possibly, the concentrations of the various fractions of serum proteins may be dependent primarily upon the rates of synthesis.

These estimates of rates of catabolism and anabolism of serum proteins were based, in part, upon the uptake and loss of amino acids tagged with radioactive atoms such as S^{35}-methionine. These data demonstrated that the specific activity of fraction III (Cohn *et al.*, 1950) classified electrophoretically with the beta globulins, was higher than normal in both tumor-bearing and protein-depleted dogs (Allison *et al.*, 1958). Garrow (1957) suggested that this increased uptake of S^{35} by serum proteins in depleted animals could be explained by the reduction in the metabolic pool of amino acids, thereby increasing relatively the concentrations of the tagged amino acid for synthesis. Data presented by Allison *et al.* (1963) support this explanation.

The effects of depletion of body protein upon serum proteins in the dog are illustrated by the data plotted in Fig. 13. These data were obtained by feeding a protein-free diet to two groups of dogs with 6 animals in each group. One group was fed the diet with 30 % of the calories in the form of lard, the other received the same percentage of calories as fat, but in the form of corn oil. The open circles record the data obtained while feeding the protein-free diet containing lard; the triangles, the data obtained while feeding corn oil. Urine and fecal collections were made throughout the depletion period to estimate the nitrogen lost from the bodies. It was assumed that the over-all content of protein of the animal was approximately 16 % (Allison and Wannemacher, 1961).

The decrease in the albumin-globulin ratio with respect to loss in body nitrogen is illustrated in the lower left portion of Fig. 13. This excellent correlation between loss in nitrogen and the ratio, however, was observed only in laboratory controlled experiments. There was no difference in ratios between the group of dogs fed lard and those fed corn oil. Lard in the protein-free diet did increase the lipid migrating electrophoretically with the alpha and beta globulins, illustrated in the lower right-hand portion of the figure. This increase began after the dogs had lost approximately 10 % of body nitrogen. The rise in lipid was slight in animals fed corn oil. There was an increase also in cholesterol ester and phospholipid phosphorus with loss in body nitrogen in animals fed lard, less so in those fed corn oil

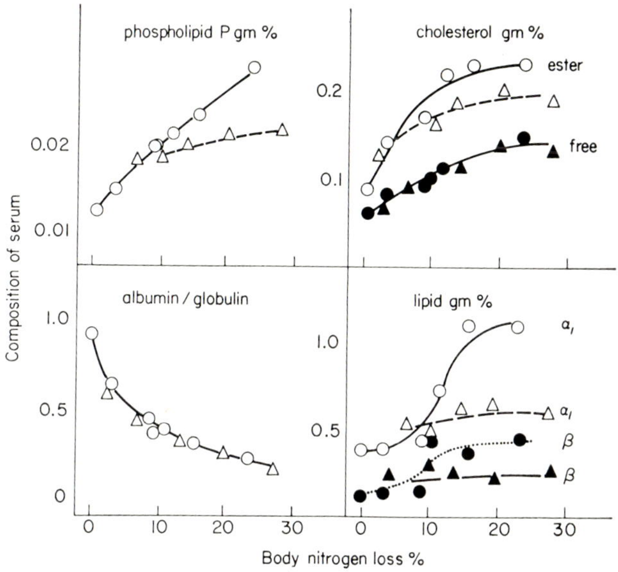

FIG. 13. Correlation between loss in body nitrogen and the serum albumin-globulin ratio and various serum lipid fractions. Circles, protein-free diet with 30 % of the calories as lard; triangles, corn oil (Allison and Wannemacher, 1961).

(see upper half of Fig. 13). There was no rise in these lipid fractions in animals fed a protein-free diet containing 5 % of the calories in the form of either lard or corn oil until the loss in body nitrogen became very high, approximately 18–20 % (Allison and Wannemacher, 1961). These results support the assumption that the protein-depletion lipemia may be the result, in part, of overloading reduced oxidation and transportation mechanisms in the depleted animal.

Children with protein malnutrition seldom exhibit this lipemia of depletion, possibly because of the type of diet they have received. Schwartz and Dean (1957) and Arroyave *et al.* (1959), however, have reported an initial rise in serum lipids during repletion of children with kwashiorkor, followed by a decrease to normal values. Indeed, the initial lipemia of repletion in children followed by a decrease to normal as repletion progresses is sometimes used as a criterion of satisfactory recovery. Recovery from lipemia in children may be the result of repletion of certain oxidative enzyme systems and transport mechanisms, thereby improving the ability to oxidize and utilize lipids.

Consideration must be given to the quality as well as the quantity of dietary protein needed for repletion. Wheat gluten, for example, with a Biological Value of approximately 42 % fed at a high concentration to overcome the lysine deficiency, will not produce normal growth in the young animal nor replete the depleted tissues in the adult without producing some abnormalities. The data recorded in Fig. 14 illustrate some of the problems of repletion related to quality of dietary protein. To obtain these data, adult dogs were fed a purified diet that could be altered in calories and in protein content (Allison and Wannemacher, 1961). The white bars over *C* record results obtained while feeding to normal dogs a purified diet with an intake of 0.6 gm casein nitrogen and 80 cal/day/kg body weight. The bars with horizontal lines over *C′* illustrate data obtained when normal dogs were fed the same diet but with the higher intake of calories and protein used for rapid repletion of the depleted animal (1 gm casein nitrogen and 140 cal/day/kg body weight). Because of excess calories (more than were needed for maintenance of the normal dog) these animals became obese, with a tendency for serum cholesterol ester and phospholipid to increase above normal controls. The black bars over *D* record data obtained after dogs had been depleted approximately 20 % in body nitrogen by feed-

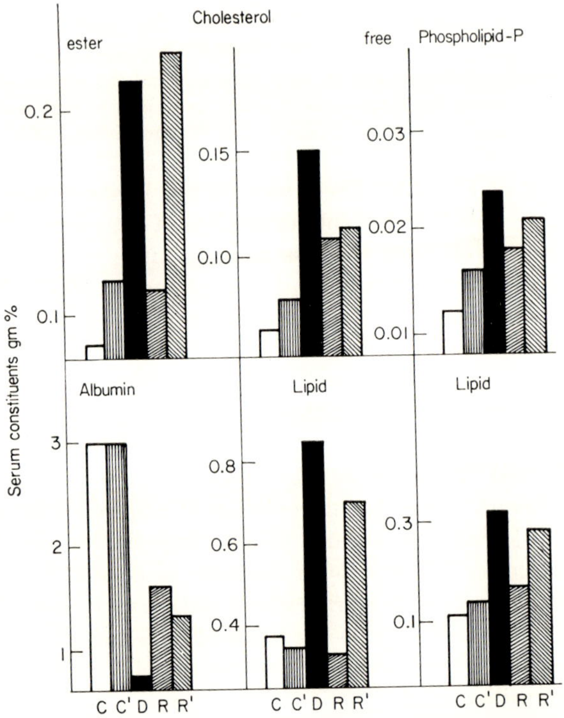

FIG. 14. Effects of repletion with casein or with wheat gluten on serum albumin and serum lipids. *C*, control; *C′*, dog fed excess calories; *D*, depleted dogs; *R*, dogs repleted with casein; *R′*, dogs repleted with wheat gluten (left lipid column refers to alpha globulins, and the right to beta globulins) (Allison and Wannemacher, 1961).

ing a protein-free diet. These data illustrate, as do those in Fig. 13, the marked drop in serum albumin in the depleted animal with the rise in lipid migrating electrophoretically with alpha and beta globulins and the increase in serum cholesterol and phospholipid. The bars with slanted lines over R in Fig. 14 record data obtained after the body protein had been mostly repleted by feeding the purified diet containing 1 gm casein nitrogen and 140 cal/day/kg body weight. Under these conditions of feeding, the lipemia dropped rapidly toward normal values after approximately 40 % of the body nitrogen had been repleted (Allison, 1961). This return to normal did not occur while feeding wheat gluten, even after 90 % of the body nitrogen was apparently repleted. The repletion was estimated through daily measurements of nitrogen retention (see bars with cross-hatched lines over R' in Fig. 14).

The suggestion was made that this lipemia was the result of overloading oxidative and transport mechanisms of the body. There is a tendency for the depleted dog to become somewhat hypothyroid, a condition which will result in marked lipemia in rats fed a high fat diet (Parmer, 1962). A lipemia was also observed in rats made hypertensive by unilateral nephrectomy and then administering deoxycorticosterone acetate with 1 % sodium chloride in the drinking water (Allison *et al.*, 1961a). Data have been obtained to indicate that the rats adapt, in part, to hypertension by developing a degree of hypothyroidism. These data emphasize, as do many others, recorded in various chapters in this treatise, the interrelationship between diet and endocrine balance.

VI. Reference Patterns of Amino Acids

The measurement of nutritive value is thought to estimate how closely the pattern of essential amino acids of a dietary protein approaches the pattern utilized most efficiently for protein synthesis. For example, since the Biological Value of egg proteins approaches 100 %, the pattern of essential amino acids in these proteins is assumed to represent an efficient pattern. Mitchell and Block (1946) and Block and Mitchell (1946–1947) have compared the amino acid content of many foods with whole egg and proposed that the BV of a protein (y) may be estimated approximately from the maximum percentage deficit in the most limiting amino acid (x) by the equation $y = 100 - 0.634x$. A calculation of nutritive value from a consideration of amino acid patterns is often called a "Protein Score." Such scores, however, assume that the pattern of amino acids absorbed into the body will be represented by the chemical analysis of the food. This assumption is not always correct since liberation of amino acids in the intestinal tract through digestion and rates of absorption may alter the pattern from the one calculated from analytical data. Available lysine from the food, for example, may differ markedly from the amount of lysine determined from a chemical analysis. This lack of availability is often the result of food processing (Carpenter, 1958; Mauron, 1961; Bender, 1961).

Oser (1951, 1959) has pointed out that the amino acid most limiting may not determine entirely the nutritive value but that all deficits may contribute. He proposed, therefore, an "Essential Amino Acid Index" (EAA) which correlates the ratios of each essential amino acid in a dietary protein relative to the amount in egg protein. Mitchell (1954) also calculated such a geometric mean of the deficits of all the essential amino acids relative to those in egg, dropping arginine as essential and including tyrosine with phenylalanine and cystine with methionine. Recently, Rosenberg and Eckert (1961) presented an interesting concept of multiple amino acid supplementation of proteins when they suggested an examination of all the limiting amino acids and nonessentials in a single experiment. A multidimensional logarithmic grid of diet composition would be arranged around a completely balanced protein. Their studies on supplementation of rice and bread with lysine and threonine resulted in this suggestion and supported the principle that "multiple amino acid supplementation of a protein should be carried out in such a manner that, according to the needs of the organism, the supplemented amino acids are present in the proper ratio to each other and in balance with the next limiting amino acid or nutrient in the diet" (Rosenberg and Eckert, 1961). They developed a three-dimensional model to determine the amounts of the two limiting amino acids that would give "peak" performance for growth of the rat and to test if the ratio between these amounts is a constant or varies according to the concentrations added. Their data suggest that the ratios between lysine, threonine, and tryptophan should be approximately 1.4:1:0.25 (see Table V). Thus, the problem of supplementation is very complex but whether the solution of the problem involves the calculation of some "Chemical Score" or a more biological approach, reference patterns of amino acids would be most helpful.

The previous discussions of amino acid requirements and nutritive values of dietary proteins have pointed out that the pattern of amino acids needed for growth may differ from the one needed for maintenance and that the optimum patterns may vary with the physiological state of the individual. In general, however, the data suggest that the over-all pattern optimum for growth, would be best also for maintenance and repletion of depleted tissues (Mitchell, 1959). For that reason egg proteins, with the highest nutritive value for growth, have been discussed as an ideal protein to obtain a reference pattern of essential amino acids. For similar reasons, the essential amino acids of human milk have been proposed as a reference for man (Kühnau, 1949; Nehring and Schwerdtfeger, 1961; György, 1959).

TABLE V

ESSENTIAL AMINO ACIDS IN REFERENCE PATTERNS

Amino acid	Reference patterns from				Provisional reference proteins (threonine approximately 4%) based upon			
	Human milk	Whole egg	Rat carcass[a]	FAO	Human milk	Rat carcass[a]	Target[b]	Rat growth[c]
Isoleucine	1.4	1.3	0.9	1.5	5.6	3.5	4.3	4.3
Leucine	1.9	1.8	1.6	1.7	7.6	6.5	7.8	6.8
Lysine	1.4	1.4	1.9	1.5	5.6	7.6	5.2	5.6
Phenylalanine	1.0	1.1	0.9	1.0	4.0	3.7	4.9	4.0
Tyrosine	1.2	0.7	0.8	1.0	4.8	2.9	—	4.0
Total aromatic	2.2	1.8	1.7	2.0	8.8	6.6	—	8.0
Cystine	0.5	0.5	0.4	0.7	2.0	1.5	—	—
Methionine	0.5	0.7	0.4	0.8	2.0	1.7	—	—
Total sulfur	1.0	1.2	0.8	1.5	4.0	3.2	4.7	4.5
Threonine	1.0	1.0	1.0	1.0	4.0	3.9	4.1	4.0
Tryptophan	0.34	0.22	0.2	0.5	1.3	0.8	1.0	1.0
Valine	1.4	1.4	1.4	1.5	5.6	5.5	5.0	5.6

[a] Williams *et al.* (1954).
[b] Bender (1961).
[c] Estimate made using various data.

The Committee on Proteins of the Food and Agriculture Organization of the United Nations, however, decided to seek a provisional reference pattern of amino acids from studies of the requirements in man, which were based primarily upon the classic experiments in adult man by Rose and his associates (1949, 1957). More recent studies by Leverton and colleagues (1956a–d), Swendseid and colleagues (Swendseid and Dunn, 1956; Swenseid *et al.*, 1956), and Clark *et al.* (1957) for maintenance in the adult, and by Holt and Snyderman (1956), and Snyderman *et al.* (1959a,b,c) for the growth of premature infants were used, whenever available, to determine such a pattern of amino acids. The pattern was called "provisional" because the committee recognized the need to study more extensively the requirements of amino acids for growth and maintenance and the effects of various patterns upon the development of different tissues under various physiological states.

The FAO provisional pattern of amino acids is illustrated in Table V, where the ratio of each essential amino acid is calculated with respect to threonine as unity. This ratio was chosen because it revealed a close correlation with similar patterns calculated for the proteins in human milk, and egg, and the rat carcass (Williams *et al.*, 1954). The requirement for tryptophan, however, is higher in the FAO pattern than in the other three. Howe *et al.* (1960) found that the FAO requirement for tryptophan was excessive for growth in the rat. Swendseid *et al.* (1961, 1962) have presented data to indicate that the tryptophan level in the FAO pattern is too high for young men and women. Possibly the reference protein should be calculated using the ratios illustrated in Table V. The problem would be to determine the magnitude of the values in the protein. The FAO provisional reference protein was calculated on the basis of threonine equal to 2.8 gm of this amino acid per 100 gm of protein. This calculated reference protein was shown to be too low in essential amino acids for optimal growth in the rat (Howe *et al.*, 1960). These authors found that 4.1 % of threonine in a mixture of amino acids would result in good growth of this animal. They found also that 1 % of tryptophan was adequate. Data have been obtained in our laboratories to suggest that 350 mg of lysine per gram of nitrogen or 5.6 % lysine gave good growth in the rat. Such figures result in ratios of lysine:tryptophan:threonine of approximately 1.4:0.25:1.0, which are the ratios mentioned previously and in agreement with those in Table V. The data suggest also that the total sulfur amino acid requirement would be approximately 4.5 %. These values are in excellent agreement with the "Target" reference protein proposed by Bender (1961) which is recorded in Table V. This target protein was derived from experiments involving dilution of egg protein with nonessential nitrogen and studies with mixtures of amino acids to simulate egg protein. The reference protein given for rat growth was estimated from over-all data available to us although there is no reason to believe that these estimates are superior to the Target protein of Bender. These two references are quite similar to the analysis for the carcass of the rat (Williams *et al.*, 1954). The higher value for sulfur amino acids in the references than in the carcass analysis could be explained by the varied roles assigned to methionine in intermediary metabolism. The higher value for lysine in the carcass analysis than in the references could be the result of efficient utilization of lysine for protein biosynthesis. Research is needed to determine the relative utilization of the various amino acids in different tissues and pathways in intermediary metabolism as presented in the body from the gastrointestinal tract. Studies of the retention of nitrogen from protein hydrolyzates fed intravenously indicate that the pattern fed through the veins must differ from the one absorbed through the gastrointestinal tract for maximum utilization. Thus, any reference pattern should be labelled provisional and be the subject for further research.

The role of the nonessential amino acids has not been emphasized in this discussion except to point to the need for sufficient nitrogen in the diet to meet the requirements for metabolism. If, for example, the essential amino acids were fed at concentrations which met the needs of protein anabolism and for other specific purposes, additional essential amino acids would increase nitrogen retention by furnishing the material required for synthesis of the nonessentials. The addition of a pattern of nonessential amino acids, however, would result in still greater nitrogen retention (Rechcigl *et al.*, 1957). A single nonessential amino acid may not meet the need as well as a mixture; and may even produce an abnormal response. Excess glycine, for example, produces such a response in the rat. A type of interrelationship between essential and nonessential amino acids may be illustrated by the metabolism of methionine, glycine, and arginine. Glycine and arginine are converted into guanidoacetic acid, which in turn is methylated through reactions converting methionine into creatine. The toxicity of excess dietary methionine can be overcome by adding glycine in the presence of adequate arginine (Allison, 1956b). The toxicity of excess guanidoacetic acid may be prevented by adding sufficient methionine. A proper mixture of methionine and guanidoacetic acid added to casein improved the growth of normal tissues in the tumor-bearing rat (Allison, 1956b) and the repletion of serum albumin in the depleted dog (Allison and Wannemacher, 1957). This type of interrelationship, as well as many others, such as the antagonism which may exist between abnormal mixtures of acids of similar structure, illustrate the great importance of balance (Harper and Kumta, 1959).

It is appropriate to end this discussion of the nutritive value of dietary proteins by emphasizing the importance of balance between all constituents of the diet. Water (Bressani and Braham, 1962), inorganic elements (M. S. Frazier *et al.*, 1956), amino acids (Harper and Kumta, 1959), vitamins (Riggs and Hegsted, 1949; Shils *et al.*, 1949; Allison *et al.*, 1954), and fats and carbohydrates (Munro, 1951; Swanson, 1959) in adequate quantities and proportions are essential to the development and well-being of each tissue through nitrogen metabolism. "The body is well adapted to correct for many dietary errors but not all of them, and it is the important function of the nutritionist to discover and correct those errors which lead to inefficiency or premature death of the living system" (Allison and Wannemacher, 1957).

References

Albanese, A. A. (1959). In "Protein and Amino Acid Nutrition" (A. A. Albanese, ed.), p. 297. Academic Press, New York.

Addis, T., Poo, L. J., and Lew, W. (1936). *J. Biol. Chem.* **115,** 117.

Allison, J. B. (1949). *Advan. Protein Chem.* **5,** 155.

Allison, J. B. (1951). *Federation Proc.* **10,** 676.

Allison, J. B. (1955). *Physiol. Rev.* **35,** 664.

Allison, J. B. (1956a). *Am. J. Clin. Nutr.* **4,** 662.

Allison, J. B. (1956b). *In* "Some Aspects of Amino Acid Supplementation" (W. H. Cole, ed.), p. 1. Rutgers Univ. Press, New Brunswick, New Jersey.

Allison, J. B. (1957). *J. Am. Med. Assoc.* **164,** 283.
Allison, J. B. (1958a). *Voeding* **19,** 119.
Allison, J. B. (1958b). *Ann. N. Y. Acad. Sci.* **69,** 1009.
Allison, J. B. (1959). *In* "Protein and Amino Acid Nutrition" (A. A. Albanese, ed.), p. 97. Academic Press, New York.
Allison, J. B. (1961). *Federation Proc.* **20,** 66.
Allison, J. B. and Anderson, J. A. (1945). *J. Nutr.* **29,** 413.
Allison, J. B., and Fitzpatrick, W. H. (1960). "Dietary Proteins in Health and Di-ease." Thomas, Springfield, Illinois.
Allison, J. B., and Wannemacher, R. W., Jr. (1957). *In* "Amino Acid Malnutrition" (W. H. Cole, ed.), p. 1. Rutgers Univ. Press, New Brunswick, New Jersey.
Allison, J. B., and Wannemacher, R. W., Jr. (1961). *In* "Meeting Protein Needs of Infants and Children," Publ. No. 843, p. 357. Natl. Acad. Sci.—Natl. Research Council, Washington, D.C.
Allison, J. B., Anderson, J. A., Seeley, R. D. (1946). *Ann. N. Y. Acad. Sci.* **47,** 245.
Allison, J. B., Anderson, J. A., and Seeley, R. D. (1947a). *J. Nutr.* **33,** 361.
Allison, J. B., Seeley, R. D., and Ferguson, F. P. (1947b). *J. Biol. Chem.* **171,** 91.
Allison, J. B., Wannemacher, R. W., Jr., and Migliarese, J. F. (1954). *J. Nutr.* **52,** 415.
Allison, J. B., Wannemacher, R. W., Jr., Russell, T., and McCoy, J. R. (1958). *Cancer Res.* **18,** 394.
Allison, J. B., Wannemacher, R. W., Jr., Middleton, E., and Spoerlein, T. (1959). *Food Technol.* **13,** 597.
Allison, J. B., Brande, P. F., Wannemacher, R. W., Jr., Parmer, L. P., and Leathem, J. H. (1961a). *J. Nutr.* **73,** 321.
Allison, J. B., Wannemacher, R. W., Jr., Parmer, L. P., and Gomez-Brenes, R. A. (1961b), *J. Nutr.* **74,** 176.
Allison, J. B., Wannemacher, R. W., Jr., Banks W. L., Wunner, W. H., and Gomez-Brenes, R. A. (1962). *J. Nutr.* **78,** 333.
Allison, J. B., Wannemacher, R. W., Jr., and Banks, W. L., Jr. (1963). *Federation Proc.* **22,** 1126.
Arroyave, G. (1962). *Am. J. Clin. Nutr.* **11,** 447.
Arroyave, G., Viteri, F., and Scrimshaw, N. S. (1959). *Am. J. Clin. Nutr.* **7,** 185.
Bargoni, N., Cafiero, M., DiBella, S., De Mori, E., and Grillo, M.A. (1952). *Experientia* **8,** 206.
Barnes, R. H., and Bosshardt, D. K. (1946). *Ann. N. Y. Acad. Sci.* **47,** 273.
Bender, A. E. (1956). *Brit. J. Nutr.* **10,** 135.
Bender, A. E. (1961). *In* "Meeting Protein Needs of Infants and Children," Publ. No. 843, p. 407. Natl. Acad. Sci.—Natl. Research Council, Washington, D.C.
Bender, A. E., and Doell, B. H. (1957). *Brit. J. Nutr.* **11,** 140.
Bender, A. E., and Miller, D. S. (1953). *Biochem. J.* **53,** vii.
Benditt, E. P., Humphreys, E. M., Straube, R. L., Wissler, R. W., and Steffee, C. H. (1947). *J. Nutr.* **33,** 85.
Block, R. J., and Mitchell, H. H. (1946–1947). *Nutr. Abstr. Rev.* **16,** 249.
Bressani, R. (1962). *J. Nutr.* **78,** 365.
Bressani, R., and Braham, J. E. (1962). *Federation Proc.* **21,** 404.
Bressani, R., and Scrimshaw, N. S. (1961). *In* "Meeting Protein Needs of Infants and Children," Publ. No. 843, p. 35. Natl. Acad. Sci.—Natl. Research Council, Washington, D.C.
Bricker, M. L., and Mitchell, H. H. (1947). *J. Nutr.* **34,** 491.
Bricker, M. L., Mitchell, H. H., and Kinsman, G. M. (1945). *J. Nutr.* **30,** 269.
Brush, M., Willman, W., and Swanson, P. (1947). *J. Nutr.* **33,** 389.

Campbell, J. A. (1960). *J. Agr. Food Chem.* **8,** 323.
Campbell, J. A., and Chapman, D. G. (1959). *J. Can. Dietet. Assoc.* **21,** 51.
Campbell, R. M., and Kosterlitz, H. W. (1948). *J. Physiol. (London)* **107,** 383.
Cannon, P. R., Humphreys, E. M., Wissler, R. W., and Frazier, L. E. (1944). *J. Clin. Invest.* **23,** 601.
Cannon, P. R., Frazier, L. E., and Hughes, R. H. (1954). *J. Lab. Clin. Med.* **44,** 250.
Carpenter, K. J. (1958). *Proc. Nutr. Soc. (Engl. Scot.)* **17,** 91.
Chow, B. F., Allison, J. B., Cole, W. H., and Seeley, R. D. (1945). *Proc. Soc. Exptl. Biol. Med.* **60,** 14.
Clark, H. E., Mertz, E. T., Kwong, E. H., Howe, J. M., and Delong, D. C. (1957). *J. Nutr.* **62,** 71.
Cohn, E. J., Gurd, F. R. N., Sorgenor, D. M., Barnes, B. A., Brown, R. K., Derouaux, G., Gillespie, J. M., Kahnt, F. W., Lever, W. F., Liv, C. H., Helman, D. M., Moulton, R. F., Scamid, K., and Uroma, E. A. (1950). *J. Am. Chem. Soc.* **72,** 465.
Cox, W. M., Jr., Ellingson, R. C., and Mueller, A. J. (1953). *Pediatrics* **11,** 435.
Denton, A. E., and Elvehjem, C. A. (1953). *J. Nutr.* **49,** 221.
Denton, A. E., and Elvehjem, C. A. (1954). *J. Biol Chem.* **206,** 449.
Fisher, H., and Scott, H. M. (1954). *Arch. Biochem.* **51,** 517.
Food and Nutrition Board (1959). "Evaluation of Protein Nutrition," Publ. No. 711, pp. 9–17. Natl. Acad. Sci.—Natl. Research Council, Washington, D.C.
Frazier, L. E., Wissler, R. W., Steffee, C. H., Woodridge, R. L., and Cannon, P. R. (1947). *J. Nutr.* **33,** 65.
Frazier, M. S., Hughes, R. H., Cannon, P. R. (1956). *Am. J. Clin. Nutr.* **4,** 655.
Frost, D. V. (1959) *In* "Protein and Amino Acid Nutrition" (A. A. Albanese, ed., p. 225. Academic Press, New York.
Frost, D. V., and Sandy, H. R. (1948). *Proc. Soc. Exptl. Biol. Med.* **68,** 51.
Frost, D. V., and Sandy, H. R. (1949). *J. Nutr.* **39,** 247.
Frost, D. V., and Sandy, H. R. (1951). *J. Biol. Chem.* **189,** 249.
Garrow, J. S. (1957). *In* "Amino Acid Malnutrition" (W. H. Cole, ed.), p. 14. Rutgers Univ. Press, New Brunswick, New Jersey.
Gitlin, D., Cravioto, J., Frenk, S., Mantano, E. L., Ramos-Galvan, R., Gomez, F., and Janeway, C. A. (1958). *J. Clin. Invest.* **37,** 682.
György, P. (1959). *Federation Proc.* **18,** 9.
Harper, A. E., and Kumta, U. S. (1959). *Federation Proc.* **18,** 12.
Hawley, E. E., Murlin, J. R., Nasset, E. S., Symanski, T. A., Blackwook, M., and Robinson, J. A. (1948). *J. Nutr.* **36,** 153.
Hegsted, D. M. (1957). *J. Am. Dietet. Assoc.* **33,** 225.
Hegsted, D. M. (1959). Symposium on Protein Requirements and Its Assessment in Man, *Federation Proc.* **18,** 1130.
Hegsted, D. M., and Worcester, J. (1947). *J. Nutr.* **33,** 685.
Holt, L. E., Jr., and Snyderman, S. E. (1956). *In* "Some Aspects of Amino Acid Supplementation" (W. H. Cole, ed.), p. 60. Rutgers Univ. Press, New Brunswick, New Jersey.
Howard, H. W., Monson, W. J., Bauer, C. D., and Block, R. J. (1958). *J. Nuir.* 64, 151.
Howard, H. W., Bauer, F. D., and Block, R. J. (1960). *J. Agri. Food Chem.* **8,** 486.
Howe, E. E., Gilfillan, E. W., and Allison, J. B. (1960). *J. Nutr.* **70,** 385.
Kosterlitz, H. W., and Campbell, R. M. (1945). *Nutr. Abstr. Rev.* **15,** 1.
Kühnau, J. (1949). *Angew. Chem.* **61,** 357.
Leverton, R. M., Gram, M. R., Chaloupka, J., Brodovsky, E., and Mitchell, A. (1956a). *J. Nutr.* **58,** 59.

Leverton, R. M., Gram, M. R., Brodovsky, E., Chaloupka, J., Mitchell, A., and Johnson, N. (1956b). *J. Nutr.* **58,** 83.
Leverton, R. M., Johnson, N., Pazur, J., and Ellison, J. (1956c). *J. Nutr.* **58,** 219.
Leverton, R. M., Johnson, N., Ellison, J., Geschwender, D., and Schmidt, F. (1956d). *J. Nutr.* **58,** 341.
Litwack, G., Williams, J. N., Jr., Chen, L., and Elvehjem, C. A. (1952). *J. Nutr.* **47,** 299.
Litwack, G., Williams, J. N., Jr., Fatterpaker, P., Chen, L., and Elvehjem, C. A. (1953). *J. Nutr.* **49,** 579.
Litwack, G., Fatterpaker, P., Williams, J. N., Jr., and Elvehjem, C. A. (1954). *J. Nutr.* **52,** 187.
Longenecker, J. B., and Hause, N. L. (1959). *Arch. Biochem. Biophys.* **84,** 46.
McLaughlan, J. M., Noel, F., Morrison, A. B., and Campbell, J. A. (1961). *Can. J. Biochem. Physiol.* **39,** 1670.
Mauron, J. (1961). *In* "Meeting Protein Needs of Infants and Children," Publ. No. 843, p. 425. Natl. Acad. Sci.—Natl. Research council, Washington, D.C.
Miller, D. S., and Bender, A. E. (1955). *Brit. J. Nutr.* **9,** 383.
Miller, D. S., and Payne, P. R. (1961a). *J. Nutr.* **74,** 413.
Miller, D. S., and Payne, P. R. (1961b). *Brit. J. Nutr.* **15,** 11.
Miller, D. S., and Payne, P. R. (1961c). *J. Nutr.* **75,** 225.
Miller, L. L. (1944). *J. Biol. Chem.* **152,** 603.
Miller, S. A., and Allison, J. B. (1958). *J. Nutr.* **64,** 493.
Mitchell, H. H. (1929). *Bull. Natl. Res. Council (U.S.)* **67,** 84 pp.
Mitchell, H. H. (1944). *Ind. Eng. Chem., Anal. Ed.* **16,** 696.
Mitchell, H. H. (1947). *Arch. Biochem.* **12,** 293–300.
Mitchell, H. H. (1950). *In* "Protein and Amino Acid Requirements of Mammals" (A. A. Albanese, ed.), p. 1. Academic Press, New York.
Mitchell, H. H. (1954). *Abhandl. Deut. Akad. Landwirtsch.* **5,** 279.
Mitchell, H. H. (1959). *In* "Protein and Amino Acid Nutrition" (A. A. Albanese, ed., p. 11. Academic Press, New York.
Mitchell, H. H., and Beadles, J. R. (1950). *J. Nutr.* **40,** 25.
Mitchell, H. H., and Block, R. J. (1946). *J. Biol. Chem.* **163,** 599.
Mitchell, H. H., and Carman, G. G. (1924). *J. Biol. Chem.* **60,** 613.
Morrison, A. B., Middleton, E. J., and McLaughlan, J. M. (1961a). *Can. J. Biochem. Physiol.* **39,** 1676.
Morrison, A. B., McLaughlan, J. M., Noel, F. J., and Campbell, J. A. (1961b). *Can. J. Biochem. Physiol.* **39,** 1681.
Munro, H. N. (1951). *Physiol. Rev.* **31,** 449.
Munro, H. N., and Clark, C. M. (1960). *Proc. Nutr. Soc. (Engl. Scot.)* **19,** 55.
Muramatsu, K., and Ashida, K. (1955). *J. Agr. Chem. Soc. Japan* **29,** 725.
Muramatsu, K., and Ashida, K. (1962). *J. Nutr.* **76,** 143.
Muramatsu, K., Yano, T., and Ashida, K. (1954). *J. Agr. Chem. Soc. Japan* **28,** 984.
Nasset, E. S. (1956). *In* "Some Aspects of Amino Acid Supplementation" (W. H. Cole, ed.), p. 3. Rutgers Univ. Press, New Brunswick, New Jersey.
Nehring, K., and Schwerdtfeger, E. (1951). *Arch. Tierernähr.* **1,** 295.
Osborne, T. B., Mendel, L. B., and Ferry, E. L. (1919). *J. Biol. Chem.* **37,** 223.
Oser, B. L. (1951). *J. Am. Dietet. Assoc.* **27,** 396.
Oser, B. L. (1959). *In* "Protein and Amino Acid Nutrition" (A. A. Albanese, ed.), p. 281. Academic Press, New York.
Parmer, L. P. (1962). "Studies on Hypertension in Rats," Ph.D. thesis, Rutgers University.

Platt, B. S. and Miller, D. S. (1959). *Proc. Nutr. Soc. (Engl. Scot.)* **18,** vii.

Platt, B. S., Miller, D. S., and Payne, P. R. (1961). *In* "Recent Advances in Clinical Nutrition" (J. F. Brock, ed.), p. 351. Little, Brown, Boston, Massachusetts.

Rechcigl, M., Loosli, J. K., and Williams, H. H. (1957). *J. Nutr.* **63,** 177.

Riggs, T. R., and Hegsted, D. M. (1949). *J. Biol. Chem.* **178,** 669.

Rose, W. C. (1949). *Federation Proc.* **8,** 546.

Rose, W. C. (1957). *Nutr. Abstr. Rev.* **27,** 631.

Rosenberg, H. R., and Eckert, R. E. (1961). *In* "Meeting Protein Needs of Infants and Children," Publ. No. 843, p. 451. Natl. Acad. Sci.—Natl. Research Council, Washington, D.C.

Rosenthal, H. L., and Allison, J. B. (1951). *J. Nutr.* **44,** 423.

Schwartz, R., and Dean, R. F. A. (1957). *J. Trop. Pediat.* **3,** 23.

Scrimshaw, N. S. (1959). *Federation Proc.* **18,** 1207.

Scrimshaw, N. S., and Béhar, M. (1961). *Science* **133,** 2039.

Scrimshaw, N. S., Squibb, R. L., Bressani, R., Béhar, M., Viteri, F., and Arroyave, G. (1957). *In* "Amino Acid Malnutrition" (W. H. Cole, ed.), p. 28. Rutgers Univ. Press, New Brunswick, New Jersey.

Shils, M. E., Seligman, H. M., and Goldwater, L. J. (1949). *J. Nutr.* **37,** 227.

Snyderman, S. E., Holt, L. E., Jr., Smellie, F., Boyer, A., and Westall, R. G. (1959a). *A.M.A. J. Diseases Children* **97,** 186.

Snyderman, S. E., Boyer, A., and Holt, L. E., Jr. (1959b). *A.M.A. J. Diseases Children* **97,** 192.

Snyderman, S. E., Norton, P. M., Fowler, D. I., and Holt, L. E., Jr. (1959c). *A.M.A. J. Diseases Children* **97,** 175.

Sorg-Matter, H. (1928). *Arch. Intern. Physiol.* **30,** 126.

Swanson, P. (1959). *In* "Protein and Amino Acid Nutrition" (A. A. Albanese, ed.), p. 195. Academic Press, New York.

Swendseid, M. E., and Dunn, M. S. (1956). *J. Nutr.* **58,** 507.

Swendseid, M. E., Williams, I., and Dunn, M. S. (1956). *J. Nutr.* **58,** 495.

Swendseid, M. E., Watts, J. H., Harris, C. L., and Tuttle, S. G. (1961). *J. Nutr.* **75,** 295.

Swendseid, M. E., Harris, C. L., and Tuttle, S. G. (1962). *J. Nutr.* **77,** 391.

Terroine, E. F. (1936). *Quart. Bull. Health Organ. League Nations* **5,** 427.

Thomas, K. (1909). *Arch. Physiol.* p. 219.

Wainio, W. W., Eichel, B., and Eichel, H. J., Person, F., Estes, F. L., and Allison, J. B. (1953). *J. Nutr.* **49,** 465.

Wainio, W. W., Allison, J. B., Eichel, B., Person, P., and Rowley, G. R. (1954). *J. Nutr.* **52,** 565.

Wainio, W. W., Allison, J. B., Kremzner, L. T., Bernstein, E., and Aronoff, M. (1959). *J. Nutr.* **67,** 197.

Wannemacher, R. W., Jr. (1961). *Proc. Soc. Exptl. Biol. Med.* **107,** 277.

Wannemacher, R. W., Jr., Allison, J. B., Chu, D., and Crossley, M. L. (1962). *Proc. Soc. Exptl. Biol. Med.* **111,** 708.

Wannemacher, R. W., Jr., Russell, T. J., and Allison, J. B. (1963). *J. Nutr.* **80,** 315.

Waterlow, J. C., Cravioto, J., and Stephen, J. M. L. (1960). *Advan. Protein Chem.* **15,** 131.

Whipple, G. H. (1948). "Hemoglobin, Plasma Protein and Cell Protein: Their Production and Interchange," American Lecture Ser. No. 11. Thomas, Springfield, Illinois.

Whipple, G. H., Miller, L. L., and Robscheit-Robbins (1947). *J. Exptl. Med.* **85**, 277.
Williams, H. H., Curtin, L. V., Abraham, J., Loosli, J. K., and Maynard, L. A. (1954). *J. Biol. Chem.* **208**, 277.
Wissler, R. W., Steffee, C. H., Frazier, L. E., Woolridge, R. L., and Benditt, E. P. (1948). *J. Nutr.* **36**, 245.
Zigman, S., and Allison, J. B. (1959). *Cancer Res.* **19**, 1105.

CHAPTER 13

Amino Acid Toxicities and Imbalances

A. E. Harper

Department of Nutrition and Food Science, Massachusetts Institute of Technology, Cambridge, Massachusetts

I. Introduction

A. The Concept of Amino Acid Balance

Amino acid imbalance and toxicity cannot be discussed independent of the basic concept of amino acid balance which underlies this entire subject.

In 1906 Willcock and Hopkins (1906) reported that a supplement of tryptophan prolonged the lives of mice fed a tryptophan-free diet consisting largely of zein. This was the first demonstration of a nutritional requirement for a specific organic compound. Within a short time, Osborne and Mendel (1914) demonstrated specific requirements for lysine and sulfur-containing amino acids as well, and introduced the concept of amino acid balance. From their observations that proteins varied greatly in amino acid composition and in nutritional adequacy, they concluded that the nutritional value of a protein depended upon the proportions of the various indispensable amino acids it contained, that is, upon its amino acid balance (Osborne and Mendel, 1915). Only after all of the amino acids essential for animals had been identified and extensive analyses of proteins for these amino acids had been completed, could the concept of amino acid balance be fully developed.

Efforts to identify the amino acids that are essential dietary constituents culminated in the discovery of threonine by McCoy *et al.* (1935; Rose, 1938). During this same period, information about the relative nutritional value of many proteins had accumulated owing in large measure to the work of Mitchell and his associates (Mitchell, 1950) who used whole-egg proteins as a standard against which to compare the others. From all of this information and from the accumulated information on amino acid composition of proteins, Block and Mitchell (1946–1947) developed the concept of Chemical Score to express the relationship between the nutritional value of a protein and its amino acid composition. In essence, this was a practical application of the concept of amino acid balance. It provided a means of predicting the maximum efficiency of utilization of a protein by comparing the proportions of the various indispensable amino acids in the protein with the proportions in a protein of the highest known nutritional value

As knowledge of the quantitative amino acid requirements of various species of animals accrued the concept was developed still further (Almquist, 1953, 1954; Food and Nutrition Board, 1959). It is now generally accepted that the maximum efficiency with which a protein can be used by an animal as a source of indispensable amino acids depends upon how closely the proportions of the indispensable amino acids it provides correspond, after correcting for digestibility, with the proportional amino acid requirements of the animal. It also depends upon the ratio of dispensable to indispensable amino acids in the protein (Harper and Kumta, 1959). If the amino acid balance of a protein is such that a small amount of of the protein will provide each amino acid in exactly the required amounts, it can be used with great efficiency and will have a high Protein Efficiency Ratio, Biological Value, Chemical Score, or Net Protein Utilization (see Chapter 12). On the other hand, if the amino acid balance of a protein is such that a surplus of several amino acids must be ingested to ensure that the quantitative requirements of one or more of the others will be met, the protein cannot be used with great efficiency and it will have a lower Protein Efficiency Ratio, Biological Value, Chemical Score, or Net Protein Utilization.

The concept of amino acid balance has special significance among nutritional balances for three reasons: first, all of the amino acids needed by higher organisms are obtained together from the proteins of foodstuffs; second, except for minor quantities required for special purposes, all amino acids are used concomitantly for the synthesis of tissue proteins; third, and most important, there is essentially no storage of amino acids in the body. The omission of only one of the indispensable amino acids from the diet of a well-nourished animal or subject results immediately in failure of the body to use the others except as a source of calories; within one day this results, in young growing animals, in loss of appetite and growth failure and, in mature animals, in severely negative nitrogen balance, indicating loss of body proteins (Rose, 1938; Frazier *et al.*, 1947; Lepkovsky, 1948). The effects of omitting one indispensable amino acid from the diet for only a few hours can be detected (Elman, 1939; Geiger, 1947, 1948; Cannon *et al.*, 1947). Omission of most of the other essential nutrients for a single day, and generally for considerably longer, from the diet of an animal that has previously been well-nourished has little immediate effect on growth and food intake and may have none at all for some time, until the body reserves are depleted.

This is not to deny that animals with an inadequate protein intake can use amino acids derived from the breakdown of certain more expendable proteins of the body to maintain those that are less expendable. Many muscle and liver proteins are depleted in animals starved or fed on protein-deficient diets while those of heart and brain are depleted very little (Addis *et al.*, 1936, 1940). Some enzymes fall in activity during starvation (Miller, 1948; Allard and Cantero, 1957) while others may actually increase (Rosen *et al.*, 1959). Therefore, the failure of an animal to maintain weight for even a short time on a diet that is deficient in a single amino acid is not due to an inability to mobilize amino acids readily from tissues; it can be due only to the lack of a reserve supply of amino acids.

Above all, it is this inability of the body to compensate even for a short time for even a slight deficiency of a single amino acid that makes amino acid balance so critical and makes the organism so sensitive to alterations in the amino acid pattern of the diet. If there were no body stores of minerals or vitamins, and if a lack of one mineral or one vitamin prevented efficient utilization of all of the others, mineral balance and vitamin balance would be more critical.

B. Questions Arising from the Concept of Amino Acid Balance

Implicit in the concept of amino acid balance (as it is generally used), is the assumption that, although the efficiency of utilization of an unbalanced protein is low, the surplus of amino acids which cannot be used for protein synthesis by an animal ingesting such a protein exerts no adverse effect. This appears to be true, for the most part, within the limits usually encountered in diets composed of natural foodstuffs. However, observations that amino acid requirements increase as the protein content of a diet is increased (Grau, 1948; Almquist and Merritt, 1950; Grau and Kamei, 1950; Salmon, 1954; Munaver and Harper, 1959) raise a question about the general validity of the assumption. The relationship between dietary level of protein and amino acid requirements can be studied only by increasing the amount of unbalanced protein in the diet. The observations of Grau and others indicate that, if the surplus of amino acids which cannot be used for protein synthesis is great enough or is of a certain composition, not only the efficiency with which the dietary protein is used is low, but the animal's requirement for the amino acid that is in greatest deficit in the diet may be increased.

The extent to which the amino acid pattern of a diet may be unbalanced without affecting amino acid requirements has not been adequately studied. It is difficult, except when the quantity of dietary protein is inadequate to support maximal growth, to assess with confidence whether a treatment has resulted in a small increase in the amino acid needs of an animal (Harper, 1959a; M. A. Morrison *et al.*, 1963). The results of amino acid imbalance studies indicate that even a small, nonutilizable, surplus of certain amino acids can sometimes increase the need for others when the total protein intake is low (Harper, 1959b). A basic question is raised by these observations, that is: when an animal must catabolize a surplus of amino acids, can it catabolize preferentially those provided in excess without also catabolizing a portion of those not provided in excess? The activities of many of the enzymes required for the catabolism of amino acids increase when the diet contains a surplus of protein (Rosen *et al.*, 1959; Schimke, 1962), but whether there is a relationship between these observations and those on the effect of dietary protein content on amino acid requirements has not been resolved.

Another question that arises from the concept of amino acid balance is: how large an excess of an individual amino acid can an organism tolerate? This question aroused interest before the concept of amino acid balance had been well developed (Lewis, 1925). Over the years many observations have indicated that there is a definite limit of tolerance for most amino acids and that this limit depends upon the level of protein in the diet, the levels of certain vitamins such as pyridoxine and niacin, and the age of the animal (Sassen, 1955). Although the quantities of individual amino acids shown to exert adverse effects have usually been well beyond those that could logically be used as dietary supplements, information about the tolerance of animals for amino acids, or for that matter, for any compound in a food or feedstuff, is important from both a practical and a fundamental viewpoint.

The problems of amino acid imbalance and toxicity have thus arisen independently of it, from observations that certain additions of amino acids to diets result in adverse effects. Although historically the concept of amino acid imbalance arose only after some information about toxicity of individual amino acids had accumulated, it is more appropriate to discuss imbalance first because it has been investigated in more detail.

II. Amino Acid Imbalance

A. Development of the Concept

The term "amino acid imbalance" has been used to describe the amino acid pattern of diets containing an excess of an individual amino acid (Hier *et al.*, 1944), but it is applied here to amino acid patterns resembling more closely those of the diets used by Elvehjem, Krehl and their associates (Elvehjem and Krehl, 1955) when they found that the addition of gelatin or threonine to a 9% casein diet lacking niacin resulted in a severe depression in growth rate that was prevented by a supplement of niacin or tryptophan (Table I).

This effect was originally thought to be specific for animals fed on diets deficient in niacin; however, similar adverse effects of amino acid supplements were subsequently observed in studies on the deposition of liver fat in rats fed on low protein diets containing niacin, and a literature survey revealed other such examples from experiments on several species of animals (Harper, 1958). The general type of phenomenon, in which an increase in the quantity of the second most-limiting amino acid in a diet caused a depression in growth rate that was readily prevented by a further supplement of the amino acid that was most limiting is illustrated in

TABLE I

EFFECT OF NIACIN AND TRYPTOPHAN (Try) ON GROWTH OF RATS FED ON DIETS CONTAINING 8% CASEIN AND GELATIN OR THREONINE (Thr)[a]

Additions to 8% casein diet[b]				Weight gain (gm/2 wk)
Gelatin (%)	DL-Thr (%)	Niacin (mg/100 gm)	L-Try (%)	
—	—	—	—	15
6.0	—	—	—	5
6.0	—	2.5	—	27
6.0	—	—	0.1	32
—	0.36	—	—	4
—	0.36	2.5	—	35
—	0.36	—	0.1	32

[a] Adapted from M. A. Morrison *et al.* (1960).
[b] All diets contained 0.3% DL-methionine.

TABLE II

EFFECTS OF AMINO ACID SUPPLEMENTS ON GROWTH OF RATS ON DIETS CONTAINING 8% EGG ALBUMIN[a]

Supplements				Weight gain (gm/2 wk)
DL-Thr (%)	DL-Val (%)	L-Lys·HCl (%)	L-His·HCl (%)	
0.36	0.2	—	—	29
0.36	0.2	0.5	—	19
0.36	0.2	0.5	0.25	28

[a] Adapted from Winje *et al.* (1954).

TABLE III

EFFECT OF AMINO ACID SUPPLEMENTS ON GROWTH OF RATS ON DIETS CONTAINING 87% RICE[a]

Supplements			Weight gain (gm/2 wk)
L-Lys (%)	DL-Thr (%)	Other	
—	—	—	15
0.4	0.5	—	30
0.4	0.5	3% A.A. mix.[b] − Leu	22
0.4	0.5	3% A.A. mix.[b] + Leu	52

[a] Adapted from Deshpande *et al.* (1955).

[b] Indispensable amino acids in proportions in casein.

Table II. Also, Salmon (1954) had observed that the addition of a larger amount of gelatin to a low protein diet caused a depression in growth rate that could be completely prevented only by tryptophan and not by niacin. In further studies (Sauberlich, 1956; Deshpande *et al.*, 1958a), additions to low protein diets of unbalanced proteins or mixtures of amino acids lacking a single indispensable amino acid were found to cause depressions in growth rate, all of which could be prevented by a supplement of the amino acid that was most limiting in the diet (Table III). These growth depressions, which are prevented by small supplements of the limiting amino acid or acids, have been attributed to amino acid imbalances.

The subject of amino acid imbalance has been reviewed previously (Elvehjem and Krehl, 1955; Salmon, 1958; Harper, 1958; Food and Nutrition Board, 1959) and two general approaches have been taken. One has been to attribute all adverse effects from dietary additions of individual amino acids, mixtures of amino acids, or quantities of unbalanced protein, to amino acid imbalances (Salmon, 1958). This can certainly be defended, as was indicated previously (Harper, 1958), because any addition to a diet of amino acids (except a proportional increase in all of them) results in a change in amino acid balance. However, this broad use of the term "amino acid imbalance" implies some type of common basis for all such effects, and while there may be some common elements underlying all adverse effects due to dietary additions of amino acids, there are also differences among them.

The other approach has been to group the various adverse effects in some way that will facilitate an understanding of them (Harper, 1958). To accomplish this, it has been proposed that they be separated into toxic effects, antagonisms, and imbalances. This classification, based on apparent differences in the underlying causes, was originally suggested in the form of a working hypothesis that would probably require modification. Nevertheless, the general hypothesis of distinct groups of adverse effects still appears valid; it is proposed that its use be continued unless evidence of a common basis for all such effects is discovered.

B. Thoughts on the Definition of Amino Acid Imbalance

Originally, adverse effects that resulted from "the addition of a relatively small amount (what might be called a supplemental amount) of an indispensable amino acid, or of a mixture of such amino acids, or of an unbalanced protein to a diet that is low in one or more amino acids . . . and that can be prevented by the concomitant addition of small quantities of the limiting amino acids to the diet" were attributed to amino acid imbalances (Harper, 1958). Subsequently, it became clear that the quantitative restrictions imposed by this definition should be eliminated (Harper, 1959b). The growth rate of rats fed a diet containing enough protein to support maximum growth can be depressed by the addition of a large quantity of a mixture of amino acids or protein lacking completely one indispensable amino acid (Table IV), and this depression can be prevented by a small supplement of the missing amino acid (Harper and

TABLE IV

EFFECT OF GELATIN ON GROWTH OF RATS FED AN ADEQUATE DIET[a]

Diet[b]			Weight gain (gm/2 wk)
Casein (%)	Gelatin (%)	DL-Tryptophan (%)	
12	9	—	80
12	15	—	70
12	15	0.3	83

[a] Adapted from Harper and Kumta (1959).

[b] All diets contained 0.3% DL-methionine.

Kumta, 1959). However, there is no evidence to indicate that the small additions of amino acids that sometimes depress growth rate when the diet is low in protein will do so when the diet is high enough in protein to support maximum growth.

Removing the quantitative restrictions from the definition may, superficially at least, make it appear more difficult to distinguish between amino acid imbalances and toxicities. Fisher *et al.* (1960a) have raised this point in relation to growth depressions caused by additions of dispensable amino acids. However, the critical points that distinguish imbalances from other changes in dietary amino acid pattern which cause adverse effects still remain. These are: (*a*) a diet in which there is an imbalance of amino acids, especially if it is adequate in protein, must contain a marginal amount of at least one indispensable amino acid; and (*b*) adverse effects which are attributed to an amino acid imbalance must be completely prevented by a small supplement of the limiting amino acid or acids.

Adverse effects from large additions of many of the individual amino acids (toxicities) are less severe when either the quantity of protein in the diet or the quantity of the limiting amino acid in the diet is increased (Sassen, 1955; Salmon, 1958; Fisher *et al.*, 1960a; Sauberlich, 1961). They are seldom, however, completely prevented by a supplement of the limiting amino acid in the diet, nor by a high protein intake, particularly if the dietary level of protein and the level of amino acid causing the adverse effect are increased proportionately. In contrast, when the dietary level of protein and the quantity of the amino acid or amino acid mixture causing an imbalance are increased proportionately, a small supplement of the limiting amino acid completely prevents the adverse effect (Table V) and, in fact, as the two are increased proportionately the adverse effect may even be reduced (Table VI).

An amino acid imbalance would then be defined, much as before (Harper, 1959b), as *a change in the proportions of amino acids in a diet which results in a depression in food intake or growth rate that can be completely prevented by a supplement of the indispensable amino acid present in least amount in the diet in relation to the amount required for optimal performance.*

C. Nutritional Effects of Amino Acid Imbalances

1. Specificity of Amino Acids Causing Imbalances

As indicated above, amino acid imbalances are of two types. One type results from addition of a quantity of a protein or an amino acid mixture lacking one indispensable amino acid (Table III) to a diet containing from a low to a moderate amount of protein (an amount that just satisfies

TABLE V

PREVENTION BY TRYPTOPHAN OF GROWTH DEPRESSION CAUSED BY GELATIN IN RATS FED ON DIETS CONTAINING DIFFERENT AMOUNTS OF CASEIN[a]

Diet[b]			Weight gain (gm/2 wk)
Casein (%)	Gelatin (%)	DL-Tryptophan (%)	
6	—	—	22
6	12	—	13
6	12	0.2	51
8	—	—	53
8	15	—	24
8	15	0.2	77

[a] From Elias *et al.* (1959–1961).
[b] All diets contained 0.3% DL-methionine.

TABLE VI

EFFECT OF FIBRIN CONTENT OF DIET ON GROWTH DEPRESSION CAUSED BY METHIONINE (Met) AND PHENYLALANINE (Phe)[a]

Diet			Weight gain (gm/2 wk)
Fibrin (%)	DL-Met (%)	DL-Phe (%)	
6	—	—	35
6	0.2	0.3	22
10	—	—	61
10	0.2	0.3	64
10	0.6	0.9	65

[a] Adapted from Kumta and Harper (1960a).

the requirements for maximum growth (Sauberlich, 1956; Deshpande *et al.*, 1958a; Harper and Kumta, 1959). The creation of imbalances by this procedure is usually quite predictable. The other results from addition of a small amount of amino acid or acids to a diet that is low in protein. Although such an addition does not alter the amino acid balance greatly, it may cause a substantial depression in growth rate (Table I). The amino acid causing the imbalance is usually the one that is second most-limiting for growth (Hankes *et al.*, 1949; Henderson *et al.*, 1953; Rosenberg *et al.*, 1959; Winje *et al.*, 1954) and the imbalance occurs when the protein content of the diet is relatively low. The creation of imbalances by this procedure is not routinely predictable, which immediately raises the question of the specificity of this effect.

Henderson *et al.* (1953) altered the composition of an amino acid diet deficient in tryptophan and niacin to make three different amino acids in turn the second most-limiting. In each case addition of a quantity of the second most-limiting amino acid caused an imbalance that increased the need for niacin or tryptophan (Table VII).

The effects of additions of amino acids other than the one that is second most-limiting have been determined (M. A. Morrison and Harper, 1960) in diets having nearly the same amino acid pattern as one used by Henderson *et al.* (1953). Of the amino acids tested, only threonine, which was second most-limiting in this diet, or combinations of amino acids including threonine, caused severe depressions in growth rate (Table VIII). In a second study (Kumta *et al.*, 1961) in which diets containing fibrin were used, the addition of methionine and phenylalanine, which are slightly less limiting than leucine, isoleucine, valine, and histidine, caused an

TABLE VII

EFFECT OF SECOND LIMITING AMINO ACID ON GROWTH OF RATS FED AN AMINO ACID DIET DEFICIENT IN NIACIN AND TRYPTOPHAN[a]

Diet				Weight gain (gm/4 wk)
Basal	L-Lys (%)	DL-Val (%)	Niacin (mg/100 gm)	
Amino acid mix. − Lys	0.4		—	45
	1.0		—	13
	1.0		2.5	57
Amino acid mix. − Val		0.8	—	23
		2.0	—	9
		2.0	2.5	61

[a] After Henderson *et al.* (1953).

TABLE VIII

SPECIFICITY OF THREONINE IN RETARDATION OF GROWTH OF RATS ON DIETS CONTAINING 8% OR 9% CASEIN AND 0.3% DL-METHIONINE BUT LACKING NIACIN[a]

Diet supplement	Weight gain (gm/2 wk)
None	19–25
0.2% DL-Threonine	7–10
0.4% DL-Phenylalanine	22
0.4% L-Lysine·HCl	20
0.4% DL-Valine	25
0.36% L-Leucine	25
0.36% DL-Valine	22
0.36% DL-Isoleucine	22
0.9% L-Arginine	24

[a] Adapted from M. A. Morrison and Harper (1960).

imbalance comparable to that caused by threonine in casein diets deficient primarily in tryptophan. The effect of methionine and phenylalanine was not specific in this experiment as several other combinations of amino acids also caused depressions in growth rate (Table IX).

One point of difference between these two experiments stands out; in the first example, in which the second most-limiting amino acid was quite specific in causing an imbalance (Table VIII), tryptophan was the most limiting amino acid, threonine was clearly next limiting, and thereafter several others were nearly equally limiting (Harper, 1959c). In the second example, in which there appeared to be relatively little specificity, several amino acids were nearly equally limiting (Kumta and Harper, 1960b). The same conditions held in a study of amino acid supplementation of corn meal in which Rosenberg *et al.* (1960) found that several different amino acids caused imbalances. This raises the possibility that the organism is particularly sensitive to small changes in amino acid balance when the total protein content of the diet is low and that, when such a diet is deficient in two amino acids in sequence, it is particularly sensitive to the balance between the first and second limiting amino acids (Henderson *et al.*, 1953; Harper, 1958; Rosenberg *et al.*, 1959) (Table X). This sensitivity apparently exists over a relatively narrow range of dietary levels of protein (Kumta *et al.*, 1961; Florentino and Pearson, 1962). Kumta *et al.* (1961), for example, caused imbalances with small quantities of methionine and phenylalanine in diets containing 6 % fibrin but when the fibrin content was increased by 2 % or decreased by 2 %, additions of methionine and phenylalanine caused no depression in growth rate or food intake (Table VI).

In contrast to the effects of small additions of one or two amino acids, additions of protein or amino acid mixtures lacking one amino acid cause imbalances regularly and predictably. If the quantity of unbalanced protein or amino acid mixture in a diet is kept constant while the quantity of basal protein is increased, the effect of the unbalanced mixture becomes

TABLE IX

LACK OF SPECIFICITY OF METHIONINE AND PHENYLALANINE IN RETARDATION OF GROWTH OF RATS ON DIETS CONTAINING 6% FIBRIN[a]

Diet supplements	Weight gain (gm/2 wk)
None	33
0.4% DL-Met + 0.6% DL-Phe	17
0.4% L-Arg + 0.6% L-Lys	18
0.4% L-Arg + 0.6 %DL-Thr	20
0.4% DL-Thr + 0.4% DL-Try	19

[a] Adapted from Kumta *et al.* (1961).

TABLE X

EFFECT OF LYSINE AND THREONINE SUPPLEMENTS ON THE GROWTH OF RATS FED A DIET OF 90% RICE[a]

Supplements		Weight gain (gm/5 wk)
L-Lysine·HCl (%)	DL-Threonine (%)	
—	—	57
0.1	—	78
0.1	0.1	112
0.1	0.2	138
0.1	0.3	114
0.2	0.3	151
0.2	0.1	136
0.25	0.1	152
0.3	0.1	131
0.3	0.2	154

[a] After Rosenberg *et al.* (1959).

TABLE XI

EFFECT OF DIETARY PROTEIN ON GROWTH DEPRESSION RESULTING FROM AN IMBALANCE CAUSED BY ADDING 3.5% OF A MIXTURE OF INDISPENSABLE AMINO ACIDS LACKING THREONINE[a]

Diet		Weight gain (gm/2 wk)
Casein (%)	Mixture	
6	−	22
6	+	5
8	−	45
8	+	31
10	−	72
10	+	59
15	−	85
15	+	83

[a] Adapted from Harper (1959b).

proportionately less (Table XI). However, if the quantity of the unbalanced mixture is increased too, the growth rate of animals fed higher levels of the basal protein is also depressed (Table XII). This appears to occur over a range of protein levels until the diet contains some minimal amount of the amino acid in shortest supply with respect to the requirement, an amount that must be somewhat above the accepted minimum requirement.

TABLE XII

EFFECT OF VARIOUS PROPORTIONS OF CASEIN AND GELATIN ON THE GROWTH OF RATS[a]

Diet		"Chemical Score"	Weight gain (gm/2 wk)
Casein (%)	Gelatin (%)		
7[b]	—	100	56
7[b]	6.4	50	61
7[b]	11.5	35	43
9[c]	—	100	77
9[c]	8.2	50	72
9[c]	14.7	35	62
11[d]	—	100	78
11[d]	10.0	50	78
11[d]	18.0	35	66
11[d]	18.0 + 0.2% DL-Try	—	76
11[d]	29.0	25	38
11[d]	29.0 + 0.2% DL-Try	—	70

[a] From Elias *et al.* (1959–1961).
[b] Supplemented with 0.2% DL-Met + 0.1% DL-Thr.
[c] Supplemented with 0.25% DL-Met + 0.1% DL-Thr.
[d] Supplemented with 0.3% DL-Met.

2. Amino Acid Imbalance and Chemical Score

The accepted relationship between the Chemical Score (amino acid balance) of a protein and its nutritional value may be affected by an amino acid imbalance. In experiments on protein mixtures having a wide range of Chemical Scores, the mixtures being prepared from combinations of casein and gelatin (the experiments need to be repeated with other combinations), the rat tolerated a protein mixture with a Chemical Score[1] of 50 without any apparent ill effects. If the Chemical Score fell below 50, food intake and growth were depressed (Table XII). This is not apparent if Biological Value is determined by the classic paired-feeding technique with mature animals (Mitchell *et al.*, 1945), because Block's and Mitchell's (1946–1947) plot of Biological Value versus Chemical Score is a straight line. No adverse effect of the poor amino acid balance would be detected in comparisons involving paired-feeding if the food intake of an animal fed a diet with a poorly balanced pattern of amino acids fell to a point at which the animal consumed only as much protein as it could metabolize effi-

[1] Based on the amount of tryptophan per 16 grams of nitrogen, assuming casein supplemented with methionine and threonine to have a "Chemical Score" of 100. Equal quantities of nitrogen from supplemented casein and gelatin give a "Chemical Score" of 50.

ciently. Measurements of growth and efficiency of nitrogen utilization in experiments in which nitrogen balance and paired-feeding techniques were used to study the effects of an amino acid imbalance support this conclusion (Kumta *et al.*, 1958). However, when food consumption and growth are used as criteria, the performance of an animal fed a diet in which there is an imbalance of amino acids may be much below that predicted from the amino acid pattern of the diet or its Chemical Score.

From the experiments cited above (Table XII) it appears that imbalances should occur primarily when the Chemical Score of a diet is less than 50, yet some diets having Chemical Scores considerably greater than 50 appear to be imbalanced. Examples of imbalances produced by small supplements of amino acids, usually the second most-limiting amino acid, have been discussed (Tables V and IX). In some of these an imbalance occurs with only a small change in the Chemical Score of the protein. In others, such as those caused by a mixture of indispensable amino acids lacking one indispensable amino acid, the change in Chemical Score is greater but is not enough to lower the score to 50 (Table III). It thus appears that, when the protein level is not adequate for maximum growth, food intake can be depressed owing to an imbalance, even though the Chemical Score of the diet exceeds 50.

In some of the examples cited, the proportion of indispensable to dispensable amino acids in the diet has been high, suggesting that imbalances are created more readily under these conditions. When the proportion of indispensable to dispensable amino acids was lower, as in the example in which gelatin was used to create an amino acid imbalance (Table XII), the animal tolerated well a protein mixture with a Chemical Score of 50. The rat also grows well on a diet high in wheat gluten (Munaver and Harper, 1959) which has a Chemical Score of about 30 and a high proportion of dispensable to indispensable amino acids. This suggests that the relationship between Chemical Score and amino acid imbalance is influenced by the ratio of indispensable to dispensable amino acids in the diet and that the lower the ratio, the better a protein of low Chemical Score will be tolerated.

3. Relationships among Dietary Protein Level, Amino Acid Imbalance, and Amino Acid Requirements

The above discussion leads back to the problem of the relationship between dietary level of protein and amino acid requirements. Evidence that amino acid requirements increase as the dietary level of protein is increased was cited in Section I,B. In most of these studies, amino acid requirements were expressed as a percentage of the diet, so it is not clear whether the absolute amount of amino acid required to support a given

TABLE XIII

EFFECT OF WHEAT GLUTEN CONTENT OF DIET ON THE LYSINE REQUIREMENT FOR MAXIMUM GROWTH OF RATS[a]

Diet composition		Weight gain (gm/2 wk)	Lysine intake (gm/2 wk)
Wheat gluten (%)	Total lysine (%)		
30	0.9	73	1.3
30	1.0	83	1.4
47	1.05	71	1.4
47	1.10	79	1.5
53	1.05	70	1.4
53	1.15	76	1.6
53	1.20	79	1.6

[a] Adapted from Munaver and Harper (1959).

rate of growth is increased. In one study (Munaver and Harper, 1959), in which wheat gluten was used, not only was the requirement as a percentage of the diet increased but also the total amount to support maximum growth for two weeks (Table XIII).

The analogy between experiments on the effect of dietary level of protein on amino acid requirements and experiments on amino acid imbalance was drawn earlier (Section I,B). When a mixture of amino acids lacking one amino acid is added to a diet and an imbalance is created, the total amino acid (protein) content of the diet is increased and more of the limiting amino acid, expressed as a percentage of the diet, is required to prevent a depression in growth rate. The absolute amount of amino acid required to overcome the depression may also be increased (Harper, 1959b; Kumta and Harper, 1960a). In most experiments of this type the dietary level of protein has been below that required to support maximum growth, so the values obtained cannot be considered to indicate true requirements. Nevertheless, the general procedure for studying amino acid imbalances is analogous to that used to study the effect of dietary protein level on amino acid requirements. Demonstration of an increase in the absolute amino acid requirements appears to depend on the severity of the imbalance induced, and unless it is fairly severe, the increase in requirement may be so small that it is difficult to demonstrate a significant effect by the growth assay, particularly when the diet is nearly adequate (Harper, 1959a,b). This may account for the lack of effect of protein level on amino acid requirement reported by Bressani and Mertz (1958). If the Chemical Score of wheat gluten of about 30 is low enough for this protein to be considered as one in which there is a naturally occurring amino acid imbalance, the increase in lysine requirement with increasing dietary levels of wheat gluten can be taken as evidence of both an effect of protein level and of amino acid imbalance on amino acid requirement, as was implied by Grau and Kamei (1950) in a study of the lysine requirement of the chick. A similar conclusion regarding casein was drawn by Fisher *et al.* (1960b) when they found the arginine requirement of the chick to be higher with casein than with soybean protein diets. All of this leads to the conclusion that the demonstrations of an increase in amino acid requirements with increasing quantities of protein in the diet are actually examples of the effects of amino acid imbalance.

D. Toward an Explanation of Amino Acid Imbalance

Amino acid imbalances were originally detected in experiments on animals fed *ad libitum,* and this procedure has been used in most of the subsequent studies. The growth rate of animals fed a diet in which there is an imbalance of amino acids is depressed, and as would be expected, the animals eat much less than the more rapidly growing control animals receiving a balanced mixture of amino acids. Depressed food intake has been suggested as an explanation of the depressed growth rate (Waddell, 1958) but since food intake must fall when growth rate is depressed—unless there is a defect in absorption (as in coeliac disease) or metabolism (as in diabetes) which prevents the organism from converting foodstuffs into energy or usable building blocks—this explanation leads only to rephrasing the question "How does an amino acid imbalance cause a depression in growth rate?" to "How does an amino acid imbalance cause a depression in food intake?" Nutritional deficiencies and imbalances that retard growth will depress food intake, but for many reasons. Regardless of the wording of the question, the problem still remains as to how a surplus, and sometimes a very small surplus, of amino acids that cannot be used for protein synthesis can affect adversely an animal that is able to cope effectively with many times as much of a balanced mixture of amino acids.

1. Palatability

One obvious explanation of reduced food intake is that additions of unbalanced proteins or amino acid mixtures that cause imbalances make the diet unpalatable. This explanation is difficult to accept for several reasons.

1. The proteins and amino acids that cause imbalances when added to diets, sometimes in quite small amounts, are readily consumed in much larger amounts when they are added to diets having a satisfactory amino acid balance, e.g., as little as 0.08 % L-threonine depresses the growth and food intake of rats receiving a diet deficient in niacin and tryptophan (Hankes *et al.*, 1949), yet it is readily consumed in quantities up to 1 %

without any adverse effect if tryptophan is included in the diet (Benton *et al.*, 1956b).

2. It is difficult to conceive of 0.05 % L-histidine·HCl improving the palatability of a diet already containing approximately 4 % of an amino acid mixture that includes 0.2 % L-arginine·HCl and 0.6 % L-lysine·HCl (Kumta and Harper, 1960a). Even more difficult to accept is that the palatability of a diet in which an imbalance is created by the addition of 6 % gelatin can be improved equally well by a supplement of 0.1 % DL-tryptophan or 0.0025 % niacin which exerts a sparing effect on tryptophan (M. A. Morrison *et al.*, 1963).

3. Depending on the previous treatment of the animal, an amino acid imbalance may depress food intake within a few hours (Kumta and Harper, 1962) or not for three days (Sanahuja and Harper, 1962).

4. Perhaps most convincing, an amino acid that causes an imbalance and depresses growth and food intake when added to one diet may, when added to another, correct an imbalance and thus stimulate growth and food intake. This is true of threonine, which causes an imbalance corrected by tryptophan when it is added to a niacin-deficient diet containing 8–9 % casein (M. A. Morrison *et al.*, 1960) and which corrects an imbalance caused by adding lysine to a diet consisting largely of rice (Rosenberg *et al.*, 1959). This can be explained only by postulating that threonine can improve the palatability of one diet and reduce the palatability of another.

2. *Rapidity of Response to Amino Acid Imbalance*

If the hypothesis that amino acids causing imbalances depress food intake by reducing the palatability of a diet is not accepted, an alternate hypothesis must be proposed. The most promising alternative would appear to be that depressed food intake and growth rate are the result of some physiological or biochemical response to an imbalanced diet. A logical mechanism, suggested originally by Salmon (1954), is that the surplus of amino acids causing an amino acid imbalance may stimulate amino acid catabolism and, hence, result in some loss of the amino acid that is already limiting or marginal for growth. This explanation would account for depressed growth and food intake and for the observation that the growth depression due to an amino acid imbalance can be prevented only by increasing the level of the limiting amino acid in the diet.

If amino acid catabolism were stimulated, efficiency of protein utilization should be depressed and nitrogen retention should fall. In nitrogen balance experiments, Deshpande *et al.* (1958b) observed a fall in nitrogen retention of rats during the first day after they were fed a diet in which an amino acid imbalance had been created. This was followed immediately by an upward adjustment so that, in longer experiments, although the

TABLE XIV
EFFECT OF AMINO ACID IMBALANCE AND PAIRED-FEEDING ON NITROGEN BALANCE OF PROTEIN-DEPLETED ADULT RATS[a]

Diet	3-Day N intake (mg)	N retained (%)
6% Fibrin[b] (*ad libitum*)	483	60
6% Fibrin + 0.6% DL-Met + 0.9% DL-Phe (*ad libitum*)	228	44
6% Fibrin[b] (pair-fed)	228	33

[a] Adapted from Kumta *et al.* (1958).
[b] Glutamic acid added to make all diets isonitrogenous.

additions of amino acids that caused an imbalance reduced nitrogen retention, the effect was not significantly different from that caused by restriction of the food intake of a control group to the amount eaten by the experimental group (Kumta *et al.*, 1958). Fisher and Shapiro (1961) have reported similar observations in studies on the effect of an imbalance on efficiency of food utilization in chicks. The most striking feature of the nitrogen balance experiments was the rapidity with which food intake was depressed (Table XIV). This occurred within 24 hours and, thus, the food intake of the pair-fed control animals was severely restricted during the remaining days of the balance period—the calorie intake must have been quite inadequate.

Most experiments on amino acid imbalance have been continued for 2–4 weeks, presumably on the assumption that the effects, including any metabolic changes, would gradually become more severe. The alternative assumption, that the fall in food intake might be part of a homeostatic mechanism which limits food intake and, thereby, minimizes the effects of a diet causing an adverse change in the internal environment of the organism, appears to have been overlooked. The rapidity of the fall in food intake when amino acids causing an imbalance are added to a diet suggests: (*a*) that physiological adjustments, which occur within a short time to minimize the effect of an amino acid mixture causing an imbalance, depress food intake so much that nitrogen balance experiments are difficult to interpret; and (*b*) that the most important effects due to ingestion of an amino acid mixture causing an imbalance, may occur within a few hours. The results of short-term nitrogen balance studies (Deshpande *et al.*, 1958b) support this view but the accuracy of such a study in the rat is limited. R. C. Wilson *et al.* (1962) have suggested that the threonine-tryptophan imbalance which involves niacin deficiency may not fit this pattern.

In an effort to determine how rapidly a depression in food intake could be detected, animals were fed a protein-free diet for 4–6 days, were kept without food overnight, then were allowed to eat *ad libitum* a control diet or one to which a mixture of amino acids causing an imbalance had been added. Food intake was measured at short intervals. In some of these experiments, a depression in food intake was evident within 6 hours (Kumta and Harper, 1962). It had previously been observed that the food intake of rats fed *ad libitum* was depressed within 24 hours by a supplement of amino acids causing an imbalance and that animals trained to eat for only 2 hours daily ate normally the first day but much less the second day (Kumta *et al.*, 1958). When rats were protein-depleted and were then fed the different diets *ad libitum* without any intervening period of starvation, they ate normally for between 2 and 3 days, presumably until they were repleted, before showing a depression in food intake (Sanahuja and Harper, 1962).

Since the body does not have a reserve of amino acids which will prevent a depression in growth rate if one amino acid is in short supply for even a few hours (Elman, 1939; Geiger, 1947, 1948; Cannon *et al.*, 1947), a small loss or reduction in the efficiency of utilization of one of the indispensable amino acids immediately after ingestion of a meal could be responsible for the depression in food intake and growth rate due to an amino acid imbalance.

3. Possible Causes of Depressed Food Intake and Growth

A variety of reactions can be postulated which would reduce food intake either directly or indirectly. Apart from palatability, which was discussed earlier (Section II,D,1), delayed stomach emptying, delayed protein digestion, delayed absorption or transport of the limiting amino acid, increased rate of excretion or catabolism of the limiting amino acid, inhibition of protein synthesis, or a direct effect of altered amino acid pattern of the body fluids on the appetite-regulating mechanisms could each account in whole or in part for the effects observed.

a. Delayed Stomach Emptying. The rates of stomach emptying of animals ingesting fibrin diets supplemented with methionine and phenylalanine to produce an imbalance have been measured after feeding them a single meal (Kumta and Harper, 1961). No delay in stomach emptying was observed in this or in other experiments in which other imbalances were studied. Previously such an effect had been reported in animals fed a niacin-deficient low protein diet with threonine added to produce an imbalance (Shahinian *et al.*, 1952), but the animals in that study had been kept on a protein-free diet for two weeks, then were force-fed the diet lacking niacin for 7–10 days, so were severely malnourished. Delayed stomach emptying is apparently not the immediate cause of depression in growth and food intake due to an amino acid imbalance, but when a malnourished animal is force-fed more of a nutritionally unsatisfactory diet than it will eat *ad libitum*, a delay in stomach emptying may result.

b. Depressed Rate of Protein Digestion. A lowered rate of protein digestion, which might result in the limiting amino acid becoming available too slowly to support rapid growth, can also be eliminated as a possible cause of the depressed food intake and growth of animals fed a diet in which there is an imbalance of amino acids; Henderson *et al.* (1953) were able to produce amino acid imbalances using diets in which the protein was replaced by crystalline amino acids.

c. Delayed Amino Acid Absorption and Transport. Competition among amino acids, leading to delayed absorption and transport of the limiting amino acid, has not been investigated as a possible cause of the adverse effects due to amino acid imbalances. However, addition of lysine to a rice diet causes an imbalance corrected by threonine and addition of threonine causes an imbalance corrected by lysine (Rosenberg *et al.*, 1959). Since these two amino acids do not compete with each other for absorption in the intestine (T. H. Wilson, 1962), these imbalances cannot be attributed to delayed intestinal absorption. Also, consideration of the total absorptive capacity of the intestine, which can amount to 600–700 mg of amino acids in the rat in the 2 hours following the feeding of a single meal (Peraino *et al.*, 1959), makes it unlikely that the 400–600 mg of additional amino acids ingested over a 24-hour period by animals fed a diet in which a severe imbalance has been created, would seriously impede absorption. Delhumeau *et al.* (1962) have studied absorption from isolated loops of intestine *in situ* of quantities of mixtures of amino acids comparable to those ingested by animals used in imbalance experiments. They did not detect specific competitions in absorption when an amino acid mixture as unbalanced as zein was used. Christensen (1963) has emphasized the need for detailed study of this subject.

d. Increased Amino Acid Excretion and Catabolism. Increased excretion of amino acids was suggested by Salmon (1954) as a possible cause of the growth depression due to amino acid imbalances. Sauberlich and Salmon (1955) found that the amount of tryptophan excreted by rats fed a tryptophan-deficient diet increased when gelatin was added to produce an imbalance, but they concluded that the increase was too small to account for the growth depression observed. Florentino and Pearson (1962) suggested on the basis of subsequent calculations that Sauberlich and Salmon (1955) might have underestimated the significance of the urinary loss. R. C. Wilson *et al.* (1962) reported that urinary excretion of the limiting amino acid was not changed by an imbalance. In some exploratory experiments (Kumta and Harper, 1960c), histidine excretion was somewhat elevated after feeding a mixture of amino acids that increased the histidine requirement but the effect was not considered significant. The possibility that increased excretion of the limiting amino acid may be a factor contributing to the adverse effects of amino acid imbalances, remains controversial.

The possibility that an amino acid imbalance stimulates amino acid catabolism (Section I,D,2) rather than the excretion of intact amino acids, has been suggested (Harper, 1958; Salmon, 1958). The results of very short nitrogen-balance studies (Deshpande *et al.*, 1958b) support this idea, but longer studies do not (Kumta *et al.*, 1958). However, since the conclusion from the longer studies is based on comparisons with pair-fed controls that were severely restricted in calorie intake, it may not be valid. Results of experiments in which blood urea concentration was measured at intervals over a period of 6.5 hours in rats fed a single meal in which there was an imbalance of amino acids or an isonitrogenous control diet, indicated that an amino acid imbalance caused a sharper and more prolonged elevation of blood urea (Kumta and Harper, 1961). Another mixture of amino acids that caused an imbalance was discovered by this procedure, but results obtained with a series of amino acid mixtures did not reveal a consistent relationship between the effects of the mixtures on blood urea concentration and on food intake and growth. Since blood urea concentration is influenced by many factors, particularly by the relative rates of catabolism of the different amino acids in the mixtures, this technique has obvious limitations. These limitations can probably be circumvented only by studying with isotopes the rates of catabolism of individual amino acids over short intervals with a diet in which an imbalance has been created.

Two reports of the use of isotopic tracers in amino acid imbalance experiments have been published. Although the studies differed in many details, in both C^{14}-labeled tryptophan was administered to rats fed a diet with a deficiency in niacin and tryptophan and an imbalance caused by threonine. Evidence of increased tryptophan catabolism was obtained in one study (Florentino and Pearson, 1962) but not in the other (R. C. Wilson *et al.*, 1962). In both experiments, the rats had been subjected to an amino acid imbalance for 10 days or more, and, therefore, the physiological adjustment resulting in lower food intake had occurred prior to the time of the isotope study. Under such conditions, the results may not represent an initial effect of the imbalance.

e. Depressed Rate of Protein Synthesis. The possibility that the extra amino acids that cause an imbalance stimulate protein synthesis, was suggested by Keoppe and Henderson (1955). Evidence to support this was presented by R. C. Wilson *et al.* (1962) who observed that an imbalance caused by threonine increased the amount of C^{14}-labeled tryptophan incorporated into liver proteins. Interpretation of these observations may be complicated because in starvation and after feeding a protein-free diet incorporation into the liver may increase even though the animal is losing weight (Stenram, 1962). Both Salmon (1954) and Harper (1958) took the opposite view—that increased protein synthesis was unlikely in animals showing a depressed rate of growth. It seemed much more likely that an imbalance would depress the rate of protein synthesis. An effort was made to test this possibility by injecting animals fed either a balanced or an imbalanced diet with enough tryptophan to stimulate synthesis of tryptophan pyrrolase. Although ethionine will inhibit the synthesis of this enzyme, the imbalance failed to do so (Kumta and Harper, 1961). A large excess of leucine, which will severely depress growth rate, also failed to inhibit tryptophan pyrrolase induction; it also did not inhibit liver regeneration in partially hepatectomized rats more than could be accounted for by the reduction in food intake (Spolter and Harper, 1961). The high specificity of the reactions of protein synthesis makes it unlikely, in any event, that direct inhibition of protein synthesis would be the primary cause of reduced food intake and growth depression. There is, of course, no question about total protein synthesis in the body being depressed by an amino acid imbalance because growth rate is depressed, but this is probably secondary to lowered food intake. However, an immediate and excessive tissue uptake of the limiting amino acid when the animal is fed a diet in which there is an imbalance of amino acids, together with increased turnover, could be important steps in the sequence of events leading to depressed food intake.

f. Altered Plasma Amino Acid Pattern. The only clear-cut positive evidence of altered amino acid metabolism as a result of amino acid imbalance comes from plasma amino acid studies (Sauberlich and Salmon, 1955; Kumta and Harper, 1962). Kumta and Harper studied the pattern of blood amino acids in animals fed on diets containing 6 % fibrin and either an amino acid mixture which lacked histidine or a mixture of phenylalanine and methionine which could produce an imbalance. When the former was used, plasma histidine concentration fell very sharply to a very low level (Table XV). This was particularly striking because the control group and the experimental group were pair-fed, so they had exactly the same histidine intake. The mixture of methionine and phenylalanine, which produces a less severe imbalance, one which is corrected by a mixture of leucine, isoleucine, valine, and histidine, did not cause as severe a depression in the plasma concentration of any of the amino acids. It did, however, cause a moderate fall in the concentrations of the

TABLE XV

FALL IN PLASMA HISTIDINE CONCENTRATION DUE TO AMINO ACID IMBALANCE CAUSED BY ADDING TO A 6% FIBRIN DIET A MIXTURE OF INDISPENSABLE AMINO ACIDS LACKING HISTIDINE[a]

Time after feeding (hr)	Plasma histidine concentration (μM/100 ml)	
	6% Fibrin	6% Fibrin + mixture
Fasting	4.9	4.9
1	4.5	1.8
3.5	3.8	1.2
6.5	3.2	0.9

[a] Adapted from Kumta and Harper (1962).

TABLE XVI

EFFECT OF AMINO ACID IMBALANCE DUE TO ADDITION OF DIETARY METHIONINE AND PHENYLALANINE ON PLASMA AMINO ACIDS OF PROTEIN-DEPLETED RATS FED A SINGLE MEAL[a]

Time after feeding (hr)	Diet	Plasma amino acid concentrations (μM/100 ml)			
		Leu	Iso	Val	His
1	Control	6.0	3.7	7.8	8.2
	Imbalance	4.8	4.0	5.6	5.9
3.5	Control	6.2	4.4	8.0	7.0
	Imbalance	4.5	3.8	4.8	5.6
6.5	Control	5.6	4.2	7.6	6.8
	Imbalance	4.2	3.4	4.5	5.5

[a] Adapted from Kumta and Harper (1962).

amino acids required to prevent a growth depression (Table XVI). Similar changes have been observed in threonine concentration in the tissues of rats fed a diet in which an imbalance was created by adding an amino acid mixture lacking in threonine (Leung and Harper, 1962). Sauberlich and Salmon (1955) noted a fall in plasma tryptophan concentration in rats fed *ad libitum* a low-casein diet to which gelatin had been added. The changes in plasma amino acid pattern resembled those seen in animals fed on diets deficient in one amino acid (Wu, 1954; Longenecker and Hause, 1959; Gray *et al.*, 1960; A. B. Morrison *et al.*, 1961; Peraino and Harper, 1962). Some of these diets had a larger surplus of nonutilizable amino acids than those used in studies of amino acid imbalances.

4. Effect of Amino Acid Imbalance on Food Intake and Preference

Since the rapid depression in food intake caused by an amino acid imbalance appeared to be a homeostatic response to a dietary alteration that could affect the "milieu intérieur" of the organism adversely, it was thought that metabolic changes resulting from the imbalance would be exaggerated if food intake could be increased. To accomplish this without force-feeding (which could not readily be done in a way that would simulate feeding *ad libitum* as used in most of the experiments on amino acid imbalances), rats receiving a 6 % fibrin diet, to which methionine and phenylalanine were added to produce an imbalance, were injected with insulin. This stimulated their food intake. It did not, however, result in any obvious physiological defect; the animals grew well and appeared to be equal in all respects to the control animals. This suggested that food intake might be directly affected by the altered dietary pattern of amino acids.

In studies of the nutritional value of protein hydrolyzates, Frost and Sandy (1949) offered groups of animals equal quantities of the various preparations to be assessed and then allowed them to eat a protein-free diet *ad libitum.* They observed that rats given hydrolyzates high in nutritive value ate more of the protein-free diet than those receiving hydrolyzates of low nutritive value. This appeared to provide a simple method for detecting an amino acid imbalance. To test this, protein-depleted animals were provided with two food cups, one containing a measured amount of the balanced diet (6 % fibrin) or the diet supplemented with amino acids to produce an imbalance (6 % fibrin plus amino acid mixture lacking histidine) and the other a surplus of a protein-free diet (Sanahuja and Harper, 1962). The animals were not subjected to starvation before the feeding period, as they had been in earlier studies (Kumta and Harper, 1962). In the initial experiments, the animals receiving the diet with an imbalance of amino acids in it began to show a preference for the protein-free diet after the third day and many of them stopped eating the imbalanced diet althogether (Table XVII). In a series of studies of this type, rats that had been previously protein-depleted, starved, or fed *ad libitum,* all showed a similar preference for a protein-free diet, but the time required for this to become evident varied with the prior treatment of the animals (Sanahuja and Harper, 1963b). Since the animals preferred a protein-free diet that would not support growth to one which, despite an imbalance of amino acids, would support a slow rate of growth, these observations indicated that the regulating mechanism was sensitive to something other than the nutritive value of the diet.

This imbalance was known to cause a rapid fall in the plasma concentration of histidine, the most limiting amino acid in the diet (Kumta and Harper, 1962). It was also known that the plasma amino acid pattern of

TABLE XVII

FOOD INTAKE OF PROTEIN-DEPLETED RATS OFFERED BALANCED OR IMBALANCED DIET SIMULTANEOUSLY WITH PROTEIN-FREE DIET[a]

Day of experiment	Food intake on day indicated (gm)			
	Bal. diet +	Protein-free	Imbal. diet +	Protein-free
1	10.5	1.5	13.5	0.2
2	13.0	0.0	12.0	0.0
3	13.5	0.0	9.5	1.5
5	14.5	0.0	3.8	4.0
8	14.0	0.2	3.8	5.0
12	13.0	0.0	1.5	6.5

[a] Adapted from Sanahuja and Harper (1962).

TABLE XVIII

BLOOD HISTIDINE CONCENTRATION OF RATS FED AN IMBALANCED DIET AND THE SAME DIET SUPPLEMENTED WITH HISTIDINE TO CORRECT THE IMBALANCE[a]

Day of experiment	Plasma histidine concentration (μM/100 ml)		
	Balanced	"Imbalanced"	"Corrected"
Fasting conc.	9.2	9.2	9.2
1	8.5	—	—
2	8.9	2.2	—
3	9.3	1.9	5.9
4	9.1	1.6	5.3

[a] Adapted from Sanahuja and Harper (1963a).

animals fed a protein-free diet for a short time was not abnormal, presumably because the plasma amino acids were derived from the breakdown of tissue proteins. Correction of the imbalance, by supplementing the diet with a small amount of the limiting amino acid, stimulated food intake and growth; resulted in a reversal of the animal's preference for the protein-free diet; and returned the plasma amino acid pattern to normal. Measurement of plasma amino acids indicated that plasma histidine concentration fell during the period when food intake fell (Table XVIII) and that, at the time of the greatest depression in food intake, the ratios of several of the indispensable amino acids to histidine (Table XIX) began to rise (Sanahuja and Harper, 1963a). This association between changes in food intake and in plasma amino acid pattern, which may or may not be indicative of a causal relationship, is of interest in relation to the suggestion by Almquist (1954) that an accumulation of amino acids

TABLE XIX

Ratio of Concentrations of Certain Indispensable Amino Acids in Plasma to That of Histidine during Period When Food Intake Falls after Feeding "Imbalanced" Diet

Day of experiment	Amino acid ratios					
	Balanced		Imbalanced		Corrected	
	Thr/His	Phe/His	Thr/His	Phe/His	Thr/His	Phe/His
Fasting	2.3	1.0	2.3	1.0	2.3	1.0
2	2.5	1.7	21.7	3.4	7.5	1.0
3	4.0	1.0	46.5	7.4	10.5	1.6
4	3.8	0.7	65.6	8.5	14.4	2.0

[a] Adapted from Sanahuja and Harper (1963a).

TABLE XX

Food Intake of Rats Offered Basal (6% Fibrin) Diet and "Corrected" (Basal + Amino Acid Mixture + Histidine) Diets Simultaneously for Two Weeks[a]

Rat number	Food intake (gm/2 wk)			
	Protein-depleted		Non-depleted	
	Basal	Corrected	Basal	Corrected
1	2	178	201	3
2	4	214	196	9
3	3	215	4	210
4	240	4	2	199
5	212	5	1	213
6	2	203	206	3

[a] Sanahuja and Harper (1963b).

that cannot be used for protein synthesis might depress food intake, and the suggestion by Mellinkoff (1957) that plasma amino acid concentrations may influence appetite regulation.

The effect of the dietary amino acid pattern on diet selection is particularly striking. Within 24 hours, rats that have not been protein-depleted will select a protein-free diet in preference to one containing amino acids that cause an imbalance. Supplementing the latter with a small amount of the limiting amino acid to correct the imbalance will result in the animal rejecting the protein-free diet almost immediately. When the balanced diet and the diet in which there is an imbalance of amino acids are offered together in separate containers, the one in which there is an imbalance is rejected; but, if the amino acid balance of the diet is then corrected by a supplement of the limiting amino acid—thus made nutri-

TABLE XXI

FOOD CONSUMPTION OF RATS OFFERED IMBALANCED (6% FIBRIN + AMINO ACID MIXTURE − HISTIDINE) DIET AND CORRECTED (IMBALANCED PLUS 0.05% L-HISTIDINE) DIET SIMULTANEOUSLY[a]

Day of experiment	Percentage of total food from each diet	
	"Imbalanced"	"Corrected"
1	78	22
2	57	43
3	20	80
5	17	83
7	3	97
10	0	100
12	2	98
14	0	100

[a] Sanahuja and Harper (1963b).

tionally superior to the balanced control diet—neither one is preferentially selected over the other (Table XX). Again, if the diet in which there is an imbalance and the "corrected" diet, which differ by only 0.05–0.1 % L-histidine·HCl, are offered together, a preference is not evident as quickly (Table XXI), but within three days the diet in which there is an imbalance of amino acids is largely rejected (Sanahuja and Harper, 1963b).

These results indicate that the rat is amazingly sensitive to the amino acid pattern of a diet and that this sensitivity is independent of the nutritional value of the diet. They also indicate that the rat can in some way distinguish very effectively between two diets differing only slightly in amino acid composition.

5. *Hypothesis Concerning Food Intake Regulation and Plasma Amino Acid Pattern*

Some years ago, Rose (1938) and Frazier *et al.* (1947) emphasized that the initial response to an amino acid deficiency is a depression in food intake. It is also known that animals will consume more of a protein-free diet than of a diet completely deficient in one amino acid (Frazier *et al.*, 1947; Greenstein and Winitz, 1961) and that the plasma amino acid pattern of animals fed an amino-acid-deficient diet is characterized by a very low concentration of the amino acid which limits growth (Longenecker and Hause, 1959; Gray *et al.*, 1960; A. B. Morrison *et al.*, 1961). Here, too, there is an association between altered blood amino acid pattern and appetite depression.

It is not surprising that growth and food intake of animals fed a diet lacking one of the indispensable amino acids are depressed, as such a diet cannot provide the bare essentials for tissue synthesis no matter how much of it is consumed. It is difficult, however, to accept the idea (in fact, there has been considerable resistance to its acceptance) that the growth and food intake of rats fed a diet that will support a moderate rate of growth will be depressed if certain mixtures of amino acids, or under some conditions relatively small amounts of one or two amino acids, are added to such a diet. It is possible that the depression in food intake is triggered by the altered plasma amino acid pattern, which resembles that found after feeding a diet more deficient than the imbalanced diet. Such an alteration in plasma amino acid pattern could result in the organism reacting as it would to a more severe deficiency.

It is possible, with this as a basis, to postulate a sequence of events which can account for the phenomenon of amino acid imbalance. This sequence is presented as an hypothesis based on limited experimental results and some interesting associations. It must be thoroughly tested.

The sequence of events can be envisaged as follows: First, a diet is ingested in which there is an imbalance of amino acids and in which, therefore, there is a surplus of amino acids which cannot be used for protein synthesis. Next, the poorly balanced mixture of amino acids is absorbed from the intestine and the amino acid patterns of blood plasma and tissue fluids are altered to resemble that of the diet. The influx of amino acids then stimulates amino acid metabolism [Koeppe and Henderson (1955) suggested that an amino acid imbalance may stimulate protein synthesis] and the surplus of amino acids which cannot be used for protein synthesis results in particularly high efficiency of utilization of those in shortest supply. High efficiency in utilization of the most limiting amino acid or acids is reflected in the amino acid pattern of the blood, in which the concentration of the amino acid or acids in short supply fall still further (Kumta and Harper, 1962; Sanahuja and Harper, 1962). This fall in the concentration of the most limiting amino acid, alternatively, may result from an increase in amino acid catabolism which could also be stimulated by the surplus of amino acids that cannot be used for protein synthesis. The evidence of increased amino acid excretion (Sauberlich and Salmon, 1955), depressed nitrogen retention (Deshpande *et al.*, 1958b), and elevated plasma urea concentration (Kumta and Harper, 1962) after feeding imbalanced diets, although suggestive of this, is by no means unequivocal. If the altered blood amino acid pattern in some way produces a feeling of satiety, food intake would be depressed and, hence, also protein synthesis and growth. Since food intake is depressed very rapidly in animals fed most diets in which there is an imbalance of amino acids, direct effects of an imbalance on amino acid metabolism may be difficult to separate from the indirect effects of low food consumption. The possibility that the mechanism resulting in depressed food intake may be inde-dependent of the hypothalamus should be explored, because Agnew and Mayer (1956) have found that destruction of the satiety center in the hypothalamus by gold thioglucose treatment did not stimulate the food intake of mice fed a thiamine-deficient diet.

Not only is food intake apparently regulated by some mechanism that is sensitive to the dietary amino acid pattern, but food preference is also. Although there is no direct evidence that blood amino acid pattern is the triggering mechanism, the observations reported above indicate that the rat selects very effectively for a diet that will result in a "normal" plasma amino acid pattern, even if it is nutritionally inferior to another that will lead to a severely altered plasma amino acid pattern, and he is not selective between two diets differing in nutritive value if either can produce a "normal" blood amino acid pattern.

The idea that the depression in food intake and the altered food preference are responses that contribute to homeostasis, is appealing. The food intake-depressing effect of an amino acid-deficient diet has been mentioned and the fact that such a diet is consumed in smaller quantities than a protein-free diet. Ousterhout (1960) has shown that chicks fed a protein-free diet survive longer than those fed on diets deficient in only one of several amino acids. Sidransky and associates have shown that force-feeding a diet deficient in a single amino acid results in severe tissue changes and early death of the rat whereas rats allowed to eat the deficient diet *ad libitum* survive much longer and show no obvious abnormalities (Sidransky and Baba, 1960; Sidransky and Rechcigl, 1962). The depression in food intake is apparently protective under these conditions. R. C. Wilson *et al.* (1962), Deshpande *et al.* (1958a), and Kumta *et al.* (1958) were not able to force-feed rats large amounts of a diet having a severe imbalance of amino acids. Insulin injection stimulated the food intake of rats offered a diet in which there was a moderate imbalance of amino acids without producing any adverse consequences (Kumta and Harper, 1962) but not of rats fed a diet with a more severe imbalance (Sanahuja and Harper, 1961). However, Klain *et al.* (1962) were able to stimulate both growth and food intake of rats fed either kind of diet by keeping the rats at low temperature. Although rats fed *ad libitum* exclusively on diets in which there is an imbalance of amino acids do not suffer the same severe consequences as those fed exclusively on diets completely deficient in an amino acid, the protective response which results in depressed food intake may be triggered by a similar sequence of events.

III. Amino Acids in Excess

An attempt has been made to distinguish between adverse effects from excessive intakes of individual amino acids—or amino acid toxicities, as they are generally called—and amino acid imbalances (Section II,A).

Toxicity is such a general term that any adverse effect resulting from administration of an amino acid might be called a toxic effect, just as it might be attributed to an amino acid imbalance (Section II,A). However, classification and grouping of events and objects having common features is essential for the development of generalizations that facilitate understanding of a subject; therefore, in a previous discussion (Harper, 1958) adverse effects from dietary additions of amino acids which seemed to have common features were grouped as "imbalances" and "antagonisms" whereas those that seemed to be unique and to permit no generalization were grouped together as "toxicity." As more of these effects are found to have common features, further subdivision under this heading, or new groupings independent of it, will be necessary. Fisher *et al.* (1960a,b) have suggested that growth depressions caused by additions of dispensable amino acids may be still another type of imbalance but until these have been studied in more detail they are included here as toxicities.

As early as 1925, H. B. Lewis (1925) reported that a large excess of cystine was toxic. Since then, there have been many reports of toxic effects from feeding amino acids individually in excess. Most of these observations have been cited or reviewed (Almquist, 1951; Sassen, 1955; Harper *et al.*, 1955) so they will not be discussed in detail. Also toxic effects of parenterally administered amino acids will not be discussed because they appear to be attributable mainly to ammonia toxicity (Greenstein and Winitz, 1961) and thus have an entirely different basis from toxic reactions to orally administered amino acids.

A. *Toxicity of Dietary Excesses of Amino Acids*

Studies of excessive dietary intakes of individual amino acids have been sporadic. Excessive intakes of most of the indispensable and several of the dispensable amino acids depress growth under some conditions (Sassen, 1955; Wretlind, 1949; Sauberlich, 1961), but the investigations have been done under such a variety of conditions that meaningful comparisons are difficult. Effects of excesses of only tyrosine, methionine, and leucine have been documented in much detail. Sassen (1955) concluded a review of amino acid toxicity by quoting Almquist (1951) as follows: "Massive doses of any amino acid might be expected to be toxic. The significance and the consistency of this relation remains to be examined." Because work on this subject has remained sporadic during the past 12 years, this statement is as appropriate in 1963 as it was in 1951.

1. *Specific Toxic Effects*

a. Tyrosine Toxicity. The toxicity of a high dietary intake of tyrosine has been well documented (Sullivan *et al.*, 1932; Lillie, 1932; Martin,

1943, 1947; Schweizer, 1947). Levels of tyrosine of from 0.5 to 20 % of the diet have been tested in diets containing from 4 to 18 % protein. Schweizer (1947) found that feeding young rats 1 % or more of tyrosine in a diet containing 12 % protein produced lesions of the eyes characterized by vascularization, erosion, and cloudiness of the corneal epithelium, conjunctivitis, and keratitis. The last can be quite severe, resulting in complete closing of one or both eyes. The toes are also affected, particularly on the front feet. They become swollen, red and encrusted, with histologic examination revealing hyperkeratosis and edema. In severe cases the toes atrophy. The signs may clear up within 2–3 weeks but may reappear later. Mature rats are much more resistant to tyrosine toxicity than young rats.

Martin (1947) has reported that the condition does not develop in pyridoxine-deficient rats and that it is alleviated by a high protein diet and by supplements of cystine and glycine. It is also alleviated by threonine (Becker *et al.*, 1961–1963). The quantities of threonine required to prevent the appearance of the signs exceed the requirement for this amino acid, indicating that the effect of threonine is unrelated to its effect on growth (Table XXII).

The basis for the toxicity is not known. It has been suggested that excessive tyramine formation may be the cause and Bernhart and Zilliken (1959) found a small increase in tyramine excretion after feeding excess tyrosine, but after feeding tyramine in quantities as high as 1 % of the diet no signs resembling tyrosine toxicity were produced. The inclusion of an amine-oxidase inhibitor in a diet containing excess tyrosine did not increase the severity of the signs (Becker, 1961). Phenylalanine in excess does not cause comparable signs.

b. Methionine Toxicity. High mortality among rats fed a diet containing from to 6 to 12 % of DL-methionine in a stock diet was observed by Earle *et*

TABLE XXII

EFFECT OF THREONINE ON GROWTH AND SIGNS OF TOXICITY OF RATS FED EXCESS TYROSINE[a]

Diet			Weight gain (gm/2 wk)	Signs of toxicity (max. of 10)
Casein (%)	L-Tyr (%)	L-Thr (%)		
6	—	—	33	0
6	3	—	7	6
6	3	0.2	11	5
6	3	0.8	19	0

[a] Alam and Harper (1963).

TABLE XXIII
EFFECT OF GLYCINE AND RELATED COMPOUNDS ON GROWTH OF 250-GRAM RATS FED EXCESS METHIONINE[a]

Diet			Change in weight in 20 days
Casein (%)	DL-Met (%)	Glycine (%)	
12	—	—	7
12	4.8	—	−37
12	—	4.8	−1
12	4.8	4.8	−1

[a] After Roth and Allison (1949).

al. (1942). The animals ate poorly, appeared ill, and lost weight. Atrophy of liver and spleen cells was observed. Toxicity of an excessive intake of methionine has since been reported on many occasions (Almquist, 1951; Sassen, 1955). In most studies depressed growth rate, weight loss, and mortality have been used as the criteria of toxicity.

The toxic effect of methionine is greater in pyridoxine-deficient rats (DeBey *et al.*, 1952). It is alleviated in both chicks and rats by adding glycine, combinations of glycine and arginine, or guanidinoacetic acid (Table XXIII) to the high methionine diets (McKittrick, 1947; Roth and Allison, 1949). The effectiveness of these agents is, however, limited (Baron, 1958). The toxicity of methionine is greater in the young than in the mature rat. Histological changes may clear up on prolonged feeding of as much as 4 % of either D- or L-methionine but growth remains retarded (Stekol and Szaran, 1962).

The reason for the toxicity of methionine has not been established. The various agents that alleviate toxicity lead to increased creatinine excretion in the urine which suggests that a methyl acceptor is one of the limiting factors in the metabolism of an excess of methionine. The limited effectiveness of methyl acceptors in preventing the toxicity indicates that there may also be an enzymatic defect in the young rat, as suggested by Stekol and Szaran (1962).

c. Other Amino Acid Toxicities. Growth-depressing effects of high dietary levels of many individual amino acids have been reported (Sassen, 1955; Wretlind, 1949; Sauberlich, 1961). Relatively few studies in histopathology of the effects of excessive intakes of individual amino acids have been reported, and seldom have specific signs of toxicity been observed. Examples of the growth-depressing effects are shown in Table XXIV.

The quantity required to cause toxicity varies greatly among the in-

TABLE XXIV
EFFECTS OF HIGH DIETARY LEVELS OF VARIOUS AMINO ACIDS ON GROWTH OF RATS[a]

Addition	Weight gain (gm/wk)	
	6% Casein	40% Casein
No additions	20	44
5% DL-Threonine	15	
5% L-Lysine	13	
2% DL-Methionine	12	20
5% L-Histidine	5	
5% DL-Tryptophan	4	28
5% DL-Phenylalanine	3	43

[a] Adapted from Sauberlich (1961).

dividual amino acids (Wretlind, 1949; Sauberlich, 1961) indicating, as Russell *et al.* (1952) suggested, that the basis for the effect is probably dependent on the structure and metabolism of the particular amino acid. Methionine, tyrosine, tryptophan, and histidine, which enter into many metabolic pathways, appear to be the most toxic. Isoleucine and valine, which are readily oxidized completely and are not metabolized by a wide variety of pathways, are less toxic (Salmon, 1958). The others fall variously in between, roughly in the sequence: leucine, phenylalanine, arginine, lysine, and threonine—although there is not complete agreement about this (Wretlind, 1949; Sauberlich, 1961).

2. *Relationships between Amino Acid Toxicity, Amino Acid Imbalance, and Adequacy of Diet*

In general, the toxic effect of a given quantity of an individual amino acid is less when the diet is adequate than when it is inadequate, probably because the rapidly growing animal is better able to metabolize an excess of many of the individual amino acids than a slower growing animal on a deficient diet. There appears, however, to be a definite limit to the amount of several amino acids that can be metabolized efficiently even by an animal fed an adequate diet.

The growth depression due to an excess of an individual amino acid is greater when the diet is low in protein, indicating that the organism can tolerate an excess of one better when it has an adequate supply of all of the others (Grau and Kamei, 1950; Harper *et al.*, 1955; Fisher *et al.*, 1960a,b; Sauberlich, 1961; Becker, 1961). However, there have been few studies of amino acid toxicity in which the dietary level of protein and the toxic level of the amino acid have been increased proportionally. In one

TABLE XXV

EFFECT OF DIFFERENT LEVELS OF PROTEIN AHD HISTIDINE ON GROWTH OF RATS[a]

Diet		Weight gain (gm/2 wk)
Casein (%)	Histidine·HCl (%)	
6	—	25
6	2.0	10
9	—	41
9	3.0	25
12	—	70
12	4.0	44
12	2.0	58

[a] Adapted from Becker (1961).

such study (Becker and Harper, 1961), when the proportion of histidine to casein was kept constant, toxicity occurred with each level of protein (Table XXV). Also, severe growth depressions due to excesses of methionine and tryptophan have been observed in animals fed an abundance of protein (Sauberlich, 1961).

The relationship between the quality of the protein in the diet and the effect of an excess of histidine has been studied using a diet which contained 6 % casein and was most deficient in methionine and next in threonine (Becker, 1961). The addition of 2 % L-histidine·HCl to this basal diet depressed growth. Methionine supplementation stimulated the growth of the group receiving excess histidine so that it equalled that of the control group receiving no additional histidine. However, methionine also stimulated the growth of the control group, so, if the group receiving additional methionine is considered to be the appropriate control, the growth depression due to excess histidine was about the same as it was when neither group received methionine. When threonine was then added as well as methionine, again the growth of both the control and the group receiving excess histidine was stimulated but again, although threonine supplementation stimulated the growth of the group receiving excess histidine so that it equalled that of the control group receiving methionine, the growth rate was not equal to that of the control group receiving both threonine and methionine (Table XXVI). Also, in a study by Benton *et al.* (1956b) in which a supplement of the most limiting amino acid, threonine, alleviated a growth depression due to an excess of 3 % phenylalanine, threonine was much less effective when the phenylalanine level was raised.

These results resemble some of those described in the discussion of amino acid imbalances (Section II,A), but the differences between them should be emphasized. The most comparable imbalance experiment is that of the

TABLE XXVI
EFFECT OF ADDITIONS OF LIMITING AMINO ACIDS ON GROWTH-DEPRESSING EFFECT OF EXCESS HISTIDINE[a]

Diet				Weight gain (gm/3 wk)
Casein (%)	L-His·HCl (%)	DL-Met (%)	DL-Thr (%)	
6				27
6	2			12
6	2	0.2		24
6		0.2		56
6	2	0.2	0.54	47
6		0.2	0.54	69

[a] Adapted from Becker (1961).

type exemplified in Table I, in which threonine alone, added to an 8 % casein diet lacking niacin, does not depress growth; in contrast, excess histidine added to a 6 % casein diet does depress growth (Table XXVI). A supplement of methionine, the most limiting amino acid in both basal diets, stimulates the growth of rats receiving excess histidine (Table XXVI) but depresses the growth of those fed the 8 % casein diet with threonine added (Table I). Addition of threonine, the second limiting amino acid in the diet containing excess histidine, stimulates growth (Table XXVI); similarly, addition of tryptophan, the second limiting amino acid in the diet imbalanced by the addition of threonine, stimulates growth (Table I). However, the growth of the group receiving both methionine and threonine with excess histidine is still less than that of the control group receiving methionine and threonine (Table XXVI), whereas the growth of the group in the imbalance study receiving methionine and tryptophan with extra threonine is greater than that of the group receiving only methionine and tryptophan (Table I).

These observations indicate that, although amino acid toxicities may be reduced in severity by improving the quality of low protein diets with supplements of the limiting amino acid, a growth depression is not prevented, whereas, in conditions classed as imbalances, improving the quality of the low protein diet with a supplement of the most limiting amino acid completely prevents the growth depression and usually stimulates growth.

3. *Toward an Explanation of Amino Acid Toxicity*

It seems unlikely, just from the differences in the quantities of individual amino acids which are required to produce toxic effects, that there is a common basis underlying them. Nor is a common basis evident from the few procedures found to alleviate certain individual toxicities. Extra glycine, glycine and arginine, or guanidinoacetic acid alleviate methionine toxicity even though glycine is synthesized in the body; extra threonine alleviates tyrosine toxicity, but the quantity required exceeds that required for maximum growth by a factor of at least two. How toxic an excess of an individual amino acid is, probably depends upon what the metabolic products of the amino acid are and how readily these can be eliminated. Those giving rise to a large number of metabolic products appear to be most toxic. The greater toxicity of most amino acids in low protein diets and in diets deficient in vitamins, such as pyridoxine, riboflavin, and cobalamine, suggests that some of the enzymes required for catabolism of amino acids may become limiting in organisms fed on deficient diets. Evidence that the activities of many of the enzymes of amino acid metabolism in the liver increase in animals fed on high protein diets, has been mentioned (Section I,A).

As with amino acid imbalances, food intake is depressed by amino acid toxicity. Also, as with imbalances, it seems unlikely that an excess of an amino acid does this by reducing the palatability of a diet. If lowered palatability is assumed to be important, it is necessary to postulate that excess glycine, which alleviates methionine toxicity (Roth and Allison, 1949), improves the palatability of excess methionine, and that excess threonine, which alleviates tyrosine toxicity, improves the palatability of excess tyrosine. This seems particularly improbable when it is considered that glycine (Sassen, 1955) and threonine (Sauberlich, 1961) themselves depress growth when added in comparable quantities to other diets.

Plasma amino acid studies have not provided as suggestive evidence for mechanisms of amino acid toxicity as they have for amino acid imbalances. The main effect observed is a very high plasma level of the amino acid in excess (Hier, 1947; Richardson *et al.*, 1953; Sauberlich, 1961; Alam and Harper, 1963; Holt, 1962). Plasma concentrations of some other amino acids are depressed when one amino acid is provided in excess, but there is little evidence of a unique pattern of the type seen after feeding imbalanced diets. With toxicities, as well as imbalances, it seems quite possible that appetite depression is a homeostatic response which prevents an undue change in the internal environment. If, as was postulated for amino acid imbalances, a large increase in the ratio of one indispensable amino acid to another in body fluids may induce a feeling of satiety, it is possible that the high ratio in the plasma of the amino acid in excess in the diet to that of the others (provided in normal amounts) may also induce such an effect. However, the severe signs due to excess tyrosine indicate that, although depressed food intake may be a protective response to an excess of an amino acid, which results in depressed growth rate, there must be some other basis for the lesions which develop.

B. Amino Acid Antagonism

Two examples of adverse effects of dietary excesses of individual amino acids appear to be unique. These are effects of an excess of leucine (Harper *et al.*, 1955) and an excess of lysine (Jones, 1962a). These are classified as amino acid antagonisms because excess leucine depresses the utilization of the structurally similar isoleucine and valine when the latter two are not limiting for growth (Spolter and Harper, 1961), and excess lysine depresses the utilization of arginine. An amino acid antagonism differs from an imbalance in that the growth depression caused by an imbalance is prevented by a supplement of the most limiting amino acid, whereas the growth depression due to an antagonism is not prevented by a supplement of the most limiting amino acid. The distinction between amino acid antagonism and toxicity is that the former is alleviated by structurally similar amino acids whereas the latter is not. That some of the effects now attributed to toxicity may eventually prove to be a result of antagonism is not precluded nor is the possibility that some amino acids may exert both antagonistic and toxic effects, depending upon the quantities fed. The antagonism between lysine and arginine has been demonstrated since this topic was last reviewed (Harper, 1958; Salmon, 1958).

1. Types of Antagonisms

a. Leucine-Isoleucine and Valine Antagonism. The antagonism between leucine and isoleucine and valine has been studied in some detail in rats fed a 9 % casein diet. The addition of 3 % or more of leucine to this diet causes a growth depression that is not prevented by the most limiting amino acid, threonine, but is largely prevented by additional isoleucine and valine (Table XXVII). When the basal diet is made adequate by supplementing with other amino acids, an excess of isoleucine and valine will cause a growth depression that is prevented by additional leucine

TABLE XXVII

GROWTH-DEPRESSING EFFECT OF AN EXCESS OF LEUCINE AND ITS PREVENTION BY ISOLEUCINE AND VALINE[a]

Diet				Weight gain (gm/2 wk)
Casein (%)	L-Leu (%)	DL-Iso (%)	DL-Val (%)	
9	—	—	—	36
9	3	—	—	4
9	3	1.2	—	16
9	3	1.2	1.2	31

[a] Adapted from Benton *et al.* (1956a).

TABLE XXVIII

EFFECT OF LEUCINE IN PREVENTING GROWTH DEPRESSION DUE TO EXCESS ISOLEUCINE AND VALINE[a]

Diet[b]				Weight gain (gm/2 wk)
Casein (%)	DL-Iso (%)	DL-Val (%)	L-Leu (%)	
9	0.5	0.5	—	54
9	0.5	0.5	0.5	52
9	2.0	0.5	—	30
9	2.0	0.5	0.5	50
9	0.5	2.0	—	36
9	0.5	2.0	0.5	58

[a] Adapted from Benton *et al.* (1956a).

[b] All diets contained DL-thr, 0.6%; L-lys, 0.4%; L-his, 0.2%; DL-phe, 0.3%; L-arg, 0.17%.

even though leucine is not the limiting amino acid (Table XXVIII). An excess of leucine has also been shown to depress the weight gain of pregnant rats and to reduce the number of surviving litters (Cohlan and Stone, 1961). Of interest in relation to the occurrence of a leucine-isoleucine and valine antagonism is the observation of Ousterhout (1960) that chicks survive for a shorter time when they are fed a diet lacking only isoleucine or valine (18–19 days) than when they are fed a diet lacking leucine, isoleucine, and valine (34 days). Also, Rose (1957) found that human subjects do not tolerate a diet lacking isoleucine as well as they do diets lacking other amino acids. A similar antagonism has also been demonstrated in cell cultures (Paul, 1959).

b. Lysine-Arginine Antagonism. A comparable relationship between lysine and arginine has been demonstrated by Jones (1962a). Excess lysine apparently increases the arginine requirement of the chick and a similar effect can be demonstrated in the rat (Jones, 1962b).

c. Antagonisms among D-*Amino Acids.* A. Wretlind (1956) has observed that D-leucine inhibits the conversion of D-valine to the L-form in the rat. Rechcigl *et al.* (1958) have observed that the utilization of D-leucine is inhibited by DL-norleucine.

2. Nutritional Effects of Amino Acid Antagonism

Addition of 2 % L-leucine to an isoleucine-deficient diet increased the requirement of the rat for isoleucine (Harper *et al.*, 1955) and addition of 2 % lysine to a diet marginal in arginine increased the requirement of the chick for arginine (Jones, 1962a). Further, the growth of rats fed on diets high in leucine and low in isoleucine was stimulated when the leucine content of the diets was reduced (Bender, 1960; DeMuelenaere and Harper, 1961).

The corn protein, zein, is very high in leucine; thus, if the antagonism between leucine and isoleucine is of practical importance, it should be so in diets consisting largely of corn or corn products. Sauberlich *et al.* (1953) found that, of a variety of amino acid supplements, only those containing isoleucine stimulated appreciably the growth of rats fed a corn diet. Experiments by Benton *et al.* (1955) confirmed this observation but did not resolve the question of whether part of the response to isoleucine was due to the high leucine content of corn. Subsequently, Rosenberg *et al.* (1960) reinvestigated this problem in detail and came to the conclusion that isoleucine was one of four amino acids that are nearly equally third limiting in corn diets and that the response to isoleucine could be explained without postulating a leucine-isoleucine antagonism. The amino acid balance of corn is such that several amino acids are limiting for the rat fed a diet of corn and imbalances among these may be produced readily by amino acid supplementation. It is therefore difficult to distinguish among simple deficiencies, imbalances, and antagonisms in corn diets supplemented with amino acids. As Rosenberg *et al.* (1960) suggested, this problem deserves further study.

In growth experiments in which the availability of isoleucine in corn proteins was found to be low (Deshpande *et al.*, 1957), leucine-isoleucine antagonism was probably a complicating factor. In subsequent studies in which the total leucine content of the diet was decreased, much higher values were obtained for the availability of isoleucine from corn (DeMuelenaere *et al.*, 1961). Also, in studies of amino acid imbalances in fibrin diets, Kumta *et al.* (1961) found that as little as 0.6% L-leucine could cause an antagonism. So, although it is questionable whether the leucine content of corn is high enough to cause an antagonism, addition of leucine to a diet can increase the need of an animal for isoleucine and removal of leucine from such a diet can improve growth.

Since an amino acid antagonism, like an imbalance, depresses food intake, the growth of an animal fed a diet in which there is enough of an excess of an amino acid to cause an antagonism, may be less than would be predicted from the Chemical Score of the diet (Bender, 1960), as was indicated above (Section II,C,2) for animals fed diets in which amino acid imbalances occur.

3. *Toward an Explanation of Amino Acid Antagonisms*

The same mechanisms can be proposed to explain amino acid antagonisms as were suggested to explain amino acid imbalances (Section II,D).

a. Palatability. If palatability is reduced by an excess of leucine, the poor palatability caused by addition of as much as 5 % leucine must be prevented by small supplements of isoleucine and valine. This seems unlikely since in other experiments leucine prevented growth depressions caused by isoleucine or valine (Benton *et al.*, 1956a).

b. Stomach Emptying. The dry matter from a control diet and that from a diet containing excess leucine was found to empty from the stomach at nearly the same rate. Nitrogen emptied a little more quickly from the stomachs of rats receiving excess leucine but no more so than from the stomachs of rats fed a high leucine diet supplemented with isoleucine and valine, which largely prevented the adverse effect of excess leucine (Rogers *et al.*, 1962).

c. Digestion. Excess leucine does not depress growth by inhibiting protein digestion because the antagonism between leucine and isoleucine and valine can be demonstrated in diets containing free amino acids instead of protein (Spolter and Harper, 1961).

d. Absorption. Kamin and Handler (1952) found that isoleucine absorption from the intestine was depressed when leucine was administered with isoleucine. Similar observations were made in both an *in vivo* system and an *in vitro* system by Hagihira *et al.* (1960). This depressed absorption could contribute to the lowered plasma concentrations of isoleucine and valine observed by Rogers *et al.* (1962) after administering an excess of leucine to rats. The very great absorptive capacity of the small intestine argues against this being a major effect *in vivo*. Even though the absorption of isoleucine and valine may be delayed somewhat by an excess of leucine (particularly in the upper part of the small intestine) as more leucine is absorbed, the ratio of leucine to isoleucine and valine will fall and competition for absorptive sites should diminish as the food residue passes farther down the intestine. As indicated above (Section II,D,3,*c*) Delhumeau *et al.* (1962) found little evidence of specific competitions involving leucine and isoleucine when they studied intestinal absorption of quantities of amino acids resembling those ingested by animals fed a low protein diet high in leucine.

e. Nitrogen Balance. If an excessive intake of leucine stimulated amino acid catabolism, this should be evident from nitrogen balance studies. However, excess leucine caused no more reduction in the percentage of ingested nitrogen retained than would be expected from the depression it caused in food intake. Restricting the food intake of an isonitrogenous control group to that of a group receiving excess leucine caused a similar reduction in nitrogen retention (Spolter and Harper, 1961).

f. Urinary Excretion. Beyer *et al.* (1947) and Kamin and Handler (1951) have found that excess leucine depresses the tubular reabsorption of isoleucine, which could contribute to a decrease in the efficiency of utilization of isoleucine. However, preliminary experiments in which excess leucine was included in a diet containing C^{14}-isoleucine as a tracer, indicated that the extra leucine had little effect on urinary excretion of C^{14} in the intact animal (Tannous *et al.*, 1963).

g. Protein Synthesis. The possibility was considered that leucine might act as antagonist of isoleucine and valine in protein synthesis (much in the way that synthetic amino acid analogs do) and lead to decreased incorporation of isoleucine and valine into tissue proteins or to the formation of inactive enzymes. It seems improbable that such is the case because excess leucine does not inhibit the synthesis of tryptophan pyrrolase following injection of tryptophan, nor does it inhibit the regeneration of liver protein in partially hepatectomized rats (Table XXIX) more than would be expected from the depression it causes in food intake. Restricting the food intake of a control group to the amount consumed by a group receiving excess leucine caused a comparable inhibition of the liver regeneration (Spolter and Harper, 1963).

h. Blood Amino Acid Pattern. The leucine concentration of the blood plasma is greatly elevated in rats fed an excess of leucine in a diet containing 9% casein. The threonine concentration is low; since threonine is the most limiting amino acid in the diet, it would be expected to show the greatest fall from the fasting value, as it does when the basal diet is fed. However, when an excess of leucine is added, isoleucine and valine each show a greater percentage fall than does threonine (Table XXX), indicating (as do the growth experiments) that an excess of leucine in some way affects isoleucine and valine utilization (Rogers *et al.*, 1962). A similar change occurs in the pattern of free amino acids in muscle and some other organs; the pattern in the liver is not so greatly affected (Tannous *et al.*, 1963).

i. Food Intake. As with amino acid imbalances, the antagonism caused by an excess of leucine depresses food intake within 24 hours (Spolter

TABLE XXIX

EFFECT OF EXCESS LEUCINE ON 4-DAY PROTEIN REGENERATION IN PARTIALLY HEPATECTOMIZED RATS[a]

Access to food	Diet			Weight change (gm)	Liver regenerated (gm)	Protein regenerated (mg)
	L-Leu (%)	DL-Iso (%)	DL-Val (%)			
Ad libitum	5	0.3	0.3	+2	1.14	434
Ad libitum	5	—	—	−13	0.82	340
Pair-fed	5	0.3	0.3	−15	0.55	325

[a] Adapted from Spolter and Harper (1963).

TABLE XXX

EFFECT OF EXCESS LEUCINE ON PLASMA CONCENTRATIONS OF INDIVIDUAL AMINO ACIDS OF RATS FED A SINGLE MEAL[a]

Amino acid	Per cent of fasting concentration (average of 3-, 5- and 8-hr values)	
	9% Casein	9% Casein + 5% L-leu
Leucine	77	240
Threonine	73	83
Lysine	100	83
Isoleucine	80	37
Valine	93	37

[a] Adapted from Rogers *et al.* (1962).

TABLE XXXI

EFFECT OF INSULIN INJECTIONS ON FOOD INTAKE OF RATS FED EXCESS LEUCINE[a]

Leucine content (%)	Weight gain of surviving rats (gm/2 wk)		Mortality due to insulin (%)
	Saline	Insulin (1 unit/day)	
3	27	39	40
5	11	29	80
7	2	25	85

[a] Adapted from Spolter and Harper (1961).

and Harper, 1961). As in the studies on amino acid imbalances, an effort was made to stimulate food intake in order to increase the severity of the antagonism. When the level of leucine was low (3 %), insulin injections stimulated the food intake of most of the animals, and their growth increased considerably. As the level of leucine was increased from 3 % to 5 % and 7 %, the mortality rate from insulin injection increased; however, those that could eat enough food to prevent hypoglycemic shock grew better than animals receiving leucine but no insulin but not as rapidly as control animals receiving no leucine (Table XXXI). Benton (1962) has also observed that the growth depression is prevented if animals are force-fed a diet containing 3 % leucine.

These observations, like those from amino acid imbalance studies, indicate that a diet having an abnormal pattern of amino acids can depress food intake even though the animal can tolerate such a diet and grow reasonably well on it. This suggests that food intake may be depressed due to the altered dietary pattern of amino acids, as a result of some homeostatic mechanism, even when the change in the pattern is not great enough to cause a direct adverse effect. Most of the animals fed the diet containing the highest quantity of leucine were unable to eat enough to prevent hypoglycemic shock when they were injected with insulin, which suggests that this is a very effective mechanism.

The association between the change in plasma amino acid pattern and the depression in food intake following ingestion of a diet high in leucine, raises the question here as in the imbalance studies as to whether the change in the proportion of leucine to isoleucine and valine in the plasma (an increase in the former and a fall in the latter) may in some way lead to a feeling of satiety which depresses food intake (Rogers *et al.*, 1962). The possibility that the hypothesis discussed in relation to amino acid imbalance (Section II,D,5) might, with minor modifications, also apply to amino acid antagonisms, would seem to be worth exploring.

IV. Summary and Conclusions

A. Classification of Adverse Effects of Amino Acids

As has been emphasized, all adverse effects resulting from additions of amino acids can be called toxic effects if growth depression is accepted as evidence of toxicity. They can also all be attributed to imbalances if any change in amino acid pattern that causes an adverse effect is defined as an imbalance. However, there appears to be enough evidence for refinement of the classification of these effects to propose that:

a. The term "imbalance" be applied to those changes in the amino acid pattern of a diet that cause a growth depression which is completely prevented by a small supplement of the limiting amino acid or acids.

b. The term antagonism be applied to those changes in the amino acid pattern of a diet that cause a growth depression which is largely or completely prevented by a small supplement of a structurally similar amino acid or acids not originally limiting in the diet.

c. The effects of excessive intakes of individual amino acids that do not fit the above categories be grouped under the general term toxicity until some meaningful basis for grouping them in another way is developed.

B. Effects of Imbalances and Antagonisms on Amino Acid Requirements

The effects of amino acid imbalances and antagonisms are prevented by increasing the amounts of either the limiting or the structurally related amino acids in the diet, respectively. Therefore, it can be said that imbalances and antagonisms increase amino acid requirements. The extent to which the requirement for maximum growth is increased may be small and difficult to measure accurately, mainly owing to the nature of the dose-response curve. But, the increase in the amount required to support a given growth level when the protein content of the diet is suboptimal, is usually quite readily measured.

Low toxic levels of amino acids can also be considered to increase the need of an animal or subject for the limiting amino acid because the toxic effect of a given level of an amino acid is usually less severe in a more rapidly growing, better nourished animal. However, when the toxic level is raised high enough, an increase in the amount of the limiting amino acid will not restore normal growth.

C. Practical Significance of Adverse Effects of Amino Acids

It is important to emphasize before concluding this discussion of adverse effects that the accumulated evidence does not in any way contradict the original conclusions of Willcock and Hopkins (1906) and Osborne and Mendel (1914, 1915), which have been so amply confirmed and reinforced by almost 60 years of investigation, that diets deficient in one or more amino acids are improved by supplementing them with appropriate quantities of the amino acids present in the diets in inadequate amounts. It is also important to emphasize, however, as is indicated by studies on the effect of protein level on the amino acid requirements of animals, that the "law of limiting factors" or the "law of minimum" applies to amino acid supplementation. This was recently re-emphasized very effectively by the observations of Rosenberg *et al.* (1959, 1960). Both the studies on adverse effects and those by Rosenberg *et al.* emphasize further that it is possible to depress food intake and growth by certain types of alterations in the amino acid balance of a diet so that not only is a point reached at which no further response is obtained upon adding another increment of an amino acid but a point of overbalance (Rosenberg *et al.*, 1959) or imbalance may be reached and the response becomes negative. From a practical view point these are perhaps the most important conclusions, and by no means novel ones, to be drawn from all of the studies on adverse effects of amino acids.

D. Basis for Adverse Effects

The bases for the adverse effects attributed to amino acid imbalances, antagonisms, and toxicities have not been established. The most consistent and general finding is that the amino acid pattern of the blood and body fluids is frequently more out of balance than the pattern of the diet producing the adverse effect. Food intake is rapidly depressed by all of these conditions, so that it is difficult to distinguish clearly between direct metabolic effects and the indirect effects of low food intake.

E. Amino Acid Balance, Food Intake, and Homeostasis

The rapid depression in food intake caused by amino acid imbalances, antagonisms, and toxicities is postulated to be a homeostatic response, possibly initiated by the severely altered balance of amino acids in the blood and other body fluids, which inhibits the animal from ingesting an excessive amount of a severely unbalanced diet. Force-feeding a severely unbalanced diet can result in degeneration of certain organs and even early death of an animal (Sidransky and Baba, 1960; Sidransky and Rechcigl, 1962). Reduction of food intake caused by moderate imbalances may be a response to a protective mechanism which is so sensitive to changes in the composition of body fluids that it sometimes responds to changes produced by the ingestion of diets that can actually be tolerated.

References

Addis, T., Poo, L. J., and Lew, W. (1936). *J. Biol. Chem.* **115,** 111 and 117.

Addis, T., Lee, D. D., Lew, W., and Poo, L. J. (1940). *J. Nutr.* **19,** 199.

Agnew, L. R. C., and Mayer, J. (1956). *Nature* **177,** 1235.

Alam, S. Q., and Harper, A. E. (1963). Unpublished observations.

Allard, C., and Cantero, A. (1957). *Exptl. Cell. Res.* **13,** 69.

Almquist, H. J. (1951). *Ann. Rev. Biochem.* **20,** 314.

Almquist, H. J. (1953). *Arch. Biochem. Biophys.* **46,** 250; **48,** 482.

Almquist, H. J. (1954). *Arch. Biochem. Biophys.* **52,** 197.

Almquist, H. J., and Merritt, J. B. (1950). *Proc. Soc. Exptl. Biol. Med.* **73,** 136.

Baron, H. (1958). *J. Nutr.* **64,** 229.

Becker, R. V. (1961). M.S. Thesis, University of Wisconsin.

Becker, R. V., and Harper, A. E. (1961). Unpublished work.

Becker, R. V., Stucki, W. P., Alam, S. Q., and Harper, A. E. (1961–1963). Unpublished observations.

Bender, A. E. (1960). *Clin. Chim. Acta* **5,** 1.

Benton, D. A. (1962). Personal communication.

Benton, D. A., Harper, A. E., and Elvehjem, C. A. (1955). *Arch. Biochem. Biophys.* **57,** 13.

Benton, D. A., Harper, A. E., Spivey, H. E., and Elvehjem, C. A. (1956a). *Arch. Biochem. Biophys.* **60,** 147.

Benton, D. A., Harper, A. E., Spivey, H. E., and Elvehjem. C. A. (1956b). *Arch. Biochem. Biophys.* **60,** 156.

Bernhart, F. W., and A. Zilliken. (1959). *Arch. Biochem. Biophys.* **82,** 462.

Beyer, K. H., Wright, L. D., Skeggs, H. R., Russo, H. F., and Shaner, C. A. (1947). *Am. J. Physiol.* **151,** 202.

Block, R. J., and Mitchell, H. H. (1946–1947). *Nutr. Abstr. Rev.* **16,** 249.

Bressani, R., and Mertz, E. T. (1958). *J. Nutr.* **65,** 481.

Cannon, P. R., Steffee, C. H., Frazier, L. E., Rawley, D. A., and Stepto, R. C. (1947). *Federation Proc.* **6,** 390.

Christensen, H. N. (1963). *Federation Proc.* **22,** 1110.

Cohlan, S. Q., and Stone, S. M. (1961). *J. Nutr.* **74,** 93.

DeBey, H. J., Snell, E. E., and Baumann, C. A. (1952). *J. Nutr.* **46,** 203.

Delhumeau, G., Pratt, G. V., and Gitler, C. (1962). *J. Nutr.* **77,** 52.
DeMuelenaere, H. J. H., and Harper, A. E. (1961). Unpublished experiments.
DeMuelenaere, H. J. H., Chen, M-L., and Harper, A. E. (1961). Unpublished observations.
Deshpande, P. D., Harper, A. E., Quiros-Perez, F., and Elvehjem, C. A. (1955). *J. Nutr.* **57,** 415.
Deshpande, P. D., Harper, A. E., Collins, M., and Elvehjem, C. A. (1957). *Arch. Biochem. Biophys.* **67,** 341.
Deshpande, P. D., Harper, A. E., and Elvehjem, C. A. (1958a). *J. Biol. Chem.* **230,** 327.
Deshpande, P. D., Harper, A. E., and Elvehjem, C. A. (1958b). *J. Biol. Chem.* **230,** 335.
Earle, D. P., Smull, K. and Victor, J. (1942). *J. Exptl. Med.* **75,** 179 and 191.
Elias, L. G., Kumta, U. S., and Harper, A. E. (1959–1961). Unpublished work.
Elman, R. (1939). *Proc. Soc. Exptl. Biol. Med.* **40,** 484.
Elvehjem, C. A., and Krehl, W. A. (1955). *Borden's Rev. Nutr. Res.* **16,** 69.
Fisher, H., and Shapiro, R. (1961). *J. Nutr.* **75,** 395.
Fisher, H., Griminger, P., Leveille, G. A., and Shapiro, R. (1960a). *J. Nutr.* **71,** 213.
Fisher, H., Shapiro, R., and Griminger, P. (1960b). *J. Nutr.* **72,** 16.
Florentino, R. F., and Pearson, W. N. (1962). *J. Nutr.* **78,** 101.
Food and Nutrition Board (1959). "Evaluation of Protein Nutrition," Publ. No. 711. Natl. Acad. Sci.—Natl. Res. Council, Washington, D. C.
Frazier, L. E., Wissler, R. W., Steffee, C. H., Woolridge, R. L., and Cannon, P. R. (1947). *J. Nutr.* **33,** 65.
Frost, D. V., and Sandy, H. R. (1949). *J. Nutr.* **39,** 427.
Geiger, E. (1947). *J. Nutr.* **34,** 97.
Geiger, E. (1948). *J. Nutr.* **36,** 813.
Grau, C. R. (1948). *J. Nutr.* **36,** 99.
Grau, C. R., and Kamei, M. (1950). *J. Nutr.* **41,** 89.
Gray, J. A., Olsen, E. M., Hill, D. C., and Branion, H. D. (1960). *Can. J. Biochem. Physiol.* **38,** 435.
Greenstein, J. P., and Winitz, M. (1961). "Chemistry of the Amino Acids," Vol. I, p. 286. Wiley, New York.
Hagihira, H., Ogata, M., Takedatsu, N., and Suda, M. (1960). *J. Biochem.* (*Tokyo*) **47,** 1.
Hankes, L. V., Henderson, L. M., and Elvehjem, C. A. (1949). *J. Biol. Chem.* **180,** 1027.
Harper, A. E. (1958). *Ann. N. Y. Acad. Sci.* **69,** 1025.
Harper, A. E. (1959a). *Federation Proc.* **18,** 104.
Harper, A. E. (1959b). *J. Nutr.* **68,** 405.
Harper, A. E. (1959c). *J. Nutr.* **67,** 109.
Harper, A. E., and Kumta, U. S. (1959). *Federation Proc.* **18,** 1136.
Harper, A. E., Benton, D. A., and Elvehjem, C. A. (1955). *Arch. Biochem. Biophys.* **57,** 1.
Henderson, L. M., Koeppe, O. J., and Zimmerman, H. H. (1953). *J. Biol. Chem.* **201,** 697.
Hier, S. W. (1947). *J. Biol. Chem.* **171,** 813.
Hier, S. W., Graham, C. E., and Klein, D. (1944). *Proc. Soc. Exptl. Biol. Med.* **56,** 187.
Holt, L. E., Jr (1962). Personal communication.

Jones, J. D. (1962a). *Federation Proc.* **21,** 1.
Jones, J. D. (1962b). Personal communication.
Kamin, H., and Handler, P. (1951). *Am. J. Physiol.* **164,** 654.
Kamin, H., and Handler, P. (1952). *Am. J. Physiol.* **169,** 305.
Klain, G. J., Vaughan, D. H., and Vaughan, L. N. (1962). *J. Nutr.* **78,** 359.
Koeppe, O. J., and Henderson, L. M. (1955). *J. Nutr.* **55,** 23.
Kumta, U. S., and Harper, A. E. (1960a). *J. Nutr.* **70,** 141.
Kumta, U. S., and Harper, A. E. (1960b). *J. Nutr.* **71,** 310.
Kumta, U. S., and Harper, A. E. (1960c). Unpublished experiments.
Kumta, U. S., and Harper, A. E. (1961). *J. Nutr.* **74,** 139.
Kumta, U. S., and Harper, A. E. (1962). *Proc. Soc. Exptl. Biol. Med.* **110,** 512.
Kumta, U. S., Harper, A. E., and Elvehjem, C. A. (1958). *J. Biol. Chem.* **233,** 1505
Kumta, U. S., Elias, L. G., and Harper, A. E. (1961). *J. Nutr.* **73,** 229.
Lepkovsky, S. (1948). *Advan. Food Res.* **1,** 105.
Leung, P. M-B., and Harper, A. E. (1962). Unpublished observations.
Lewis, H. B. (1925). *J. Biol. Chem.* **65,** 187.
Lillie, R. D. (1932). *Public Health Rept.* (*U.S.*) **47,** 83.
Longenecker, J. B., and Hause, N. L. (1959). *Arch. Biochem. Biophys.* **84,** 46.
McCoy, R. H., Meyer, C. E., and Rose, W. C. (1935). *J. Biol. Chem.* **112,** 283.
McKittrick, D. S. (1947). *Arch. Biochem.* **15,** 133.
Martin, G. J. (1943). *Arch. Biochem.* **1,** 397.
Martin, G. J. (1947). *J. Am. Pharmacol. Assoc.* **36,** 187.
Mellinkoff, S. (1957). *Ann. Rev. Physiol.* **19,** 193.
Miller, L. L. (1948). *J. Biol. Chem.* **172,** 113.
Mitchell, H. H. (1950). *In* "Protein and Amino Acid Requirements of Mammals" (A. A. Albanese, ed.), p. 11. Academic Press, New York.
Mitchell, H. H., Hamilton, T. S., Beadles, J. R., and Simpson, F. (1945). *J. Nutr.* **29,** 13.
Morrison, A. B., Middleton, E. J., and McLaughlan, J. M. (1961). *Can. J. Biochem. Physiol.* **39,** 1675.
Morrison, M. A., and Harper, A. E. (1960). *J. Nutr.* **71,** 296.
Morrison, M. A., Reynolds, M. S., and Harper, A. E. (1960). *J. Nutr.* **72,** 302.
Morrison, M. A., Reynolds, M. S., and Harper, A. E. (1963). *J. Nutr.* **80,** 441.
Munaver, S. M., and Harper, A. E. (1959). *J. Nutr.* **69,** 58.
Osborne, T. B., and Mendel, L. B. (1914). *J. Biol. Chem.* **17,** 325.
Osborne, T. B., and Mendel, L. B. (1915). *J. Biol. Chem.* **20,** 351.
Ousterhout, L. E. (1960). *J. Nutr.* **70,** 226.
Paul, J. (1959). *J. Exptl. Zool.* **142,** 475.
Peraino, C., and Harper, A. E. (1962). *Arch. Biochem. Biophys.* **97,** 442.
Peraino, C., Rogers, Q. R., Yoshida, M., Chen, M.-L., and Harper, A. E. (1959). *Can. J. Biochem. Physiol.* **37,** 1475.
Rechcigl, M., Jr., Loosli, J. K., and Williams, H. H. (1958). *Science* **127,** 1051.
Richardson, L. R., Blaylock, L. G., and Lyman, C. M. (1953). *J. Nutr.* **51,** 515.
Rogers, Q. R., Spolter, P. D., and Harper, A. E. (1962). *Arch. Biochem. Biophys.* **97,** 497.
Rose, W. C. (1938). *Physiol. Rev.* **18,** 109.
Rose, W. C. (1957). *Nutr. Abstr. Rev.* **27,** 631.
Rosen, F., Roberts, N. R., and Nichol, C. A. (1959). *J. Biol. Chem.* **234,** 476.
Rosenberg, H. R., Culik, R., and Eckert, R. E. (1959). *J. Nutr.* **69,** 217.
Rosenberg, H. R., Rohdenburg, E. L., and Eckert, R. E. (1960). *J. Nutr.* **72,** 415.

Roth, J. S., Allison, J. B. (1949). *Proc. Soc. Exptl. Biol. Med.* **70**, 327.
Russell, W. C., Taylor, M. W., and Hogan, J. M. (1952). *Arch. Biochem. Biophys.* **39**, 249.
Salmon, W. D. (1954). *Arch. Biochem. Biophys.* **51**, 30.
Salmon, W. D. (1958). *Am. J. Clin. Nutr.* **6**, 487.
Sanahuja, J. C., and Harper, A. E. (1961). Unpublished work.
Sanahuja, J. C., and Harper, A. E. (1962). *Am. J. Physiol.* **202**, 165.
Sanahuja, J. C., and Harper, A. E. (1963a). *Am. J. Physiol.* **204**, 686.
Sanahuja, J. C., and Harper, A. E. (1963b). *J. Nutr.* **81**, 363.
Sassen, A. (1955). *Acta Gastro-Enterol. Belg.* **12**, 944.
Sauberlich, H. E. (1956). *J. Nutr.* **59**, 353.
Sauberlich, H. E. (1961). *J. Nutr.* **75**, 61.
Sauberlich, H. E., and Salmon, W. D. (1955). *J. Biol. Chem.* **214**, 463.
Sauberlich, H. E., Chang, W.-Y., and Salmon, W. O. (1953). *J. Nutr.* **51**, 623.
Schimke, R. T. (1962). *J. Biol. Chem.* **237**, 459.
Schweizer, W. (1947). *J. Physiol.* (*London*) **106**, 167.
Shahinian, S. S., Ebisuzaki, K., Kring, J. P., Williams, J. N., Jr., and Elvehjem, C. A. (1952). *Proc. Soc. Exptl. Biol. Med.* **80**, 146.
Sidransky, H., and Baba, T. (1960). *J. Nutr.* **70**, 463.
Sidransky, H., and Rechcigl, M., Jr. (1962). *J. Nutr.* **78**, 269.
Spolter, P. D., and Harper, A. E. (1961). *Am. J. Physiol.* **200**, 513.
Spolter, P. D., and Harper, A. E. (1963). *Arch. Biochem. Biophys.* **100**, 369.
Stekol, J. A., and Szaran, J. (1962). *J. Nutr.* **77**, 81.
Stenram, U. (1962). *Z. Zellforsch.* **58**, 107.
Sullivan, M. X., Hess, W. C., and Sebrell, W. H. (1932). *Public Health Rept.* **47**, 75.
Tannous, R. I., Rogers, Q. R., and Harper, A. E. (1963). Unpublished experiments.
Waddell, J. (1958). *In* "Processed Plant Protein Foodstuffs" (A. M. Altschul, ed.), p. 307. Academic Press, New York.
Willcock, E. G., and Hopkins, F. G. (1906). *J. Physiol.* (*London*) **35**, 88.
Wilson, R. C., Wortham, J. S., Benton, D. A., and Henderson, L. M. (1962). *J. Nutr.* **77**, 142.
Wilson, T. H. (1962). "Intestinal Absorption." Saunders, Philadelphia, Pennsylvania.
Winje, M. E., Harper, A. E. Benton, D. A., Boldt, R. E., and Elvehiem C. A. (1954). *J. Nutr.* **54**, 155.
Wretlind, K. A. J. (1949). *Acta Physiol. Scand.* **17**, Suppl. 59.
Wretlind, A. (1956). *Acta Physiol. Scand.* **36**, 119.
Wu, A. (1954). *J. Biol. Chem.* **207**, 775.

CHAPTER 14

Protein Requirements

D. M. Hegsted

Department of Nutrition, Harvard School of Public Health. Boston, Massachusetts

I. Introduction

The estimation of nutritional requirements of man, whether of protein or other nutrients, poses a number of problems, the most general and basic one being the definition of the over-all objectives to be achieved by good nutrition in man. The common statements that we are trying to achieve "optimum health" or "buoyant health," while true, are not particularly useful unless some criteria for their measure are available. In such statements there is usually the implication that high levels of intake are required to develop this optimum health status. It is reasonably clear that when one considers the nutritional requirements of farm animals or of laboratory animals, the objectives are not those of the animal, but those imposed by man. In general, for farm animals, the objective is simply the most economic production of milk, meat, eggs, or other products under the conditions in which the animals are to be raised. For meat-producing animals, the most rapid growth possible resulting in a high quality carcass is usually the most economic procedure since the maintenance requirements are thereby minimized. This may, of course, be modified by the prices and availability of various kinds of feeds and it may be more economic to produce animals on pasture than by feeding concentrates. It seems reasonably clear that a low plane of nutrition is advantageous during the growth period for animals to be kept to adult stages for breeding or productive purposes (Lagervall, 1958). The point to be made, without attempting to examine the information available, is that although the data may be difficult to obtain, the objectives are clear and given sufficient time, energy, and money, the relative value of different nutritional regimes can be assessed.

In man the objectives are presumably a long life with minimal disease and maximal intellectual capacity. We assume that it should be a "happy life" and that the life must be worth living, although it is apparent that conditions must be very bad indeed before life is not worth living. It must be noted that size and rate of growth per se are of little or no direct consequence except as they may be causally related to more serious objectives. If one assumes that the various objectives, such as being happy and living a long time, are not necessarily related to the same kind of diet, then we are presumably forced to place a value judgment on the relative merits of them. The well-known experiments of McCay and his co-workers (McCay *et al.*, 1935, 1939) demonstrated that very serious restriction of food, serious enough to interfere markedly with growth, not only allowed animals to live much longer, but the animals "were physiologically younger at the same chronological age of the controls, and the incidence of practically all the common causes of death of rats was greatly diminished." Ross (1959, 1961) has produced confirmatory data and also suggests that some protein restriction may contribute to a longer life in rats. Experiments with more primitive organisms (Lansing, 1952) also suggests that when growth ceases, senescence begins. Observations of these kinds become of increasing importance in a modern society where suspicion continues to grow that many of the major causes of death, such as atherosclerotic heart disease, may be in part related to the diet and to overnutrition, rather than undernutrition, and may begin early in life.

It must be noted that the experiments of McCay and others do not duplicate a free society. Given a choice, the animals clearly prefer to eat more and then die sooner. Whether they would modify this choice if they had the intellectual equipment to realize the end result is doubtful. It is difficult to visualize severe dietary restriction in an affluent society, even though it might contribute to longevity. Nevertheless, the nutritional dilemma of our time may be whether we can have our cake and eat it too.

An optimistic point of view is that as we understand those factors which contribute to longevity, mental achievement, happiness, etc., we will be able to obtain them all. The point must be clearly made, however, that many of the experimental criteria designed to test nutritional adequacy may have little relevance to the problem we are trying to solve. These include such criteria as maximum rates of growth, or maximum number of young produced. Even susceptibility to infection may be of little consequence when better methods are available to deal with infections. This is not to say that much cannot still be done with the techniques available, but certainly much more pertinent approaches are required.

II. Technical Problems

A major technical problem in the determination of nutritional requirements is that the animal studied must be in a state where a response can be obtained to a change in the intake of the nutrient of interest. In nitrogen balance studies with adult animals, for example, it is well recognized that the adequately fed animal is in nitrogen equilibrium (losses through the skin ignored) and that this condition must prevail over a wide range of intakes above the minimal need. When the nitrogen intake is lowered, there is a short lag period during which the animal is in negative nitrogen balance, and the equilibrium is then achieved at the new level of intake if it is not too low. Thus, nitrogen balance studies will probably not yield information relevant to nitrogen requirements unless the nitrogen intake is the critical factor. On the other hand, it can be argued with some justification that if nitrogen or any other nutrient is limiting in the diet, the animal is, *ipso facto*, not a normal animal. Improved and more sensitive methods will improve the situation, but it is probable that some extrapolation or interpolation will be necessary to derive the requirements for optimal nutrition from such data collected on undernourished animals.

Figure 1 represents the general kind of response curves obtained. The problem is to determine, with as much accuracy as possible, the lowest

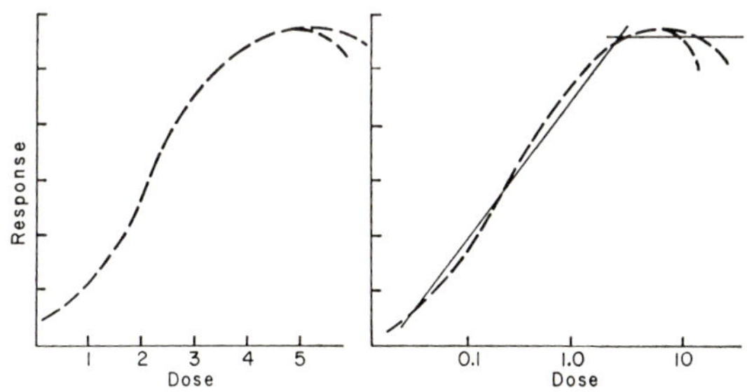

FIG. 1. Hypothetical dose-response curves.

level of intake which produces the maximal response, whether this be nitrogen retention, gain in weight, serum levels, long life, or any other appropriate criterion. If the mean response of groups of animals or subjects fed selected levels are compared by ordinary statistical tests, it will be apparent that as one approaches the optimum response the differences between groups will become smaller and smaller. Thus, by this method, "most of the data [in the response curve] are discarded in making the critical test, at exactly the point at which large groups are required to show the small differences to be significant, and the most likely error is that one will conclude that the groups are not different, when in fact, they may be. This type of error will invariably lead to a low estimate of the requirement. A low estimate is considered a more serious error than the less likely overestimation" (Hegsted, 1948).

In the past 10 to 15 years many authors have used statistical methods to define the curve which best represents the response to different doses, rather than simply comparing the response of different groups. For most nutrients the log dose-response curve appears to be essentially linear. The suggestion (Hegsted, 1948) that the intersection of the line representing the linear response with the point of maximal response might be taken as the best estimate of the minimal requirement has been widely used. Unfortunately, it is often not easy to determine whether the response is, indeed, proportional to the dose upon an arithmetic or logarithmic scale. Somewhat typical data are shown in Fig. 2 taken from the original paper of Osborne *et al.* (1919) in which they proposed the use of "Protein Efficiency" in the assessment of the nutritive value of proteins. Lactalbumin

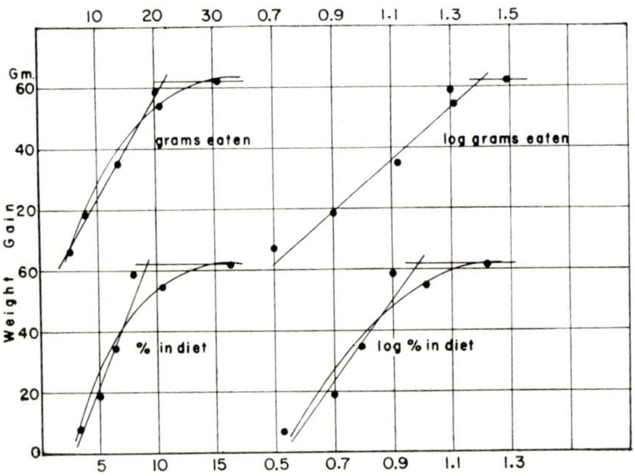

FIG. 2. Gain in weight of rats fed different levels of lactalbumin from the paper of Osborne *et al.* (1919). Same data plotted with lactalbumin content expressed in different ways. Curves drawn freehand.

was provided as the protein source for groups of rats at 3.3, 4.9, 6.2, 7.9, 10.2, and 16.2 % of the diet. Food intake and growth over a limited period of time were measured. The gain in weight has been plotted in Fig. 2 against the per cent of protein in the diet, and the grams of protein eaten as well as the logarithms of these two parameters. Somewhat typically, the animals were perverse enough to allow those at the 7.9 % level to grow slightly more (59 gm) than those at the 10.2 % level (54 gm). Those at the 16.2 % level gained 61 gm. Clearly there is no statistical difference between these three groups. Nevertheless, the estimate of the requirement might vary a great deal, depending upon the whimsy of the investigator who draws the curves.

Most of the investigators who have reported responses to various levels of protein or amino acid intake in experimental animals have assumed that the response was linear to the dose on an arithmetic scale. Others (Almquist, 1959) assume it may be semilogarithmic. Generally, the data are insufficient in order to determine which scale is more appropriate. Replotting the data from many papers will, however, indicate that in many instances the fit is as good or possibly better when plotted as logarithms. This is a matter of some substance since a semilogarithmic curve indicates a constantly decreasing efficiency of the protein fed as the level is raised, that is, a constantly decreasing Biological Value (see Section V,A,3). Also the relatively large error in the estimate of the mimimal requirement is emphasized in the semilogarithmic plot. The dose-response curve is expected to show a decline at high levels of intake where adverse effects are caused by high doses. Most nutritional investigators assume that there is a rather broad region of intake which is safe before adverse effects become manifest. It is somewhat disturbing to find that some investigators studying amino acid and protein requirements have concluded that quadratic equations of the type $Y = a + bx - cx^2$ represent the best fit of the data. Schwartz *et al.* (1958), for example, estimated the lysine requirement of chicks from the first derivative of such an equation fitted to the growth data. It may be suspected that such an equation represents the composite of two effects, one positive and the second negative. This author finds such an approach less satisfying, since the more complicated the equation the more difficult it is to attach intelligible interpretations to the various components of the curve. It does, however, serve as a warning as to the probable inadequacies of simpler approaches. The papers of Rosenberg *et al.* (1960a,b) also indicate the experimental sophistication which may be necessary to determine the ideal pattern of amino acids in a diet.

With experimental animals a considerable number of tests have been used with greater or lesser success as indicators of protein deficiency or as assay methods of protein quality. Unfortunately, only two methods appear to be relevant to the problem of protein requirements in man; these are growth rate and nitrogen balance. A consideration of some of the problems these methods present is given in the appropriate sections of this chapter, but it is clear that improved and sensitive methods reflecting the adequacy of protein intake, particularly functional tests, would be most helpful.

III. Protein Requirements of Adults

A. Endogenous Urinary Nitrogen

The most basic premise in the calculation of the minimal protein requirements is that the minimal urinary nitrogen excretion (endogenous nitrogen) is proportional to the basal metabolic rate. This was originally demonstrated by Terroine and Sorg-Matter (1927) and further developed by Sorg-Matter (1928), Smuts (1935), Brody (1945), and others. The compilation of the data upon the basal metabolic rate from many species (Brody, 1945) indicates that the rate of caloric expenditure in the basal state is proportional to $wt^{0.72}$. As shown in Fig. 3 the minimal nitrogen excretion over a considerable range of weight and in different species appears to be proportional to weight to the same power. It is assumed, therefore, that both basal metabolic rate and the endogenous nitrogen excretion

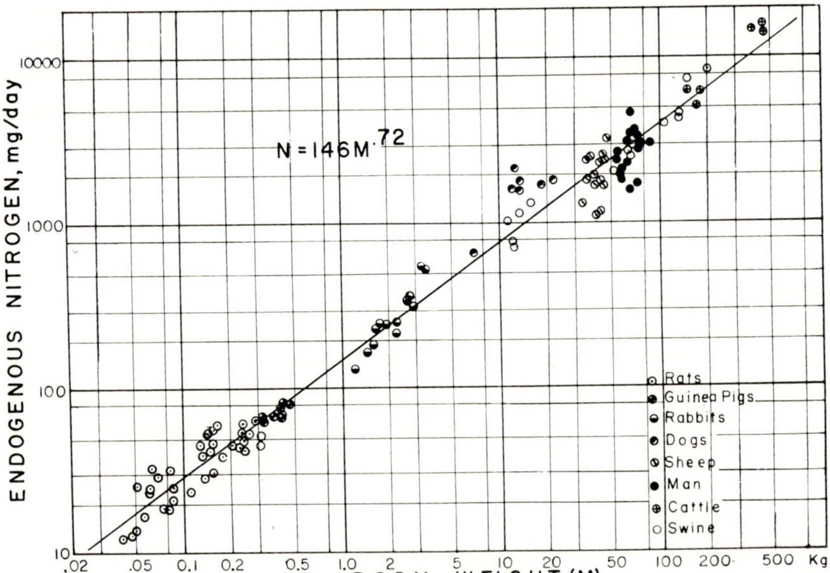

FIG. 3. Endogenous urinary nitrogen as a function of body weight in mature animals. Drawn from data tabulated by Brody (1945).

are related to the same factor, presumably the amount of metabolically active tissue. While the correlation in Fig. 3 appears excellent, it should be noted that the spread of the data around the line is considerable and the difference between a 50-kg man and a 70-kg man would not be distinguishable. Thus, there is still some question as to whether differences in weight within a species necessarily follow the same relationship. Some of the data presented by Brody (see Brody, 1945, Fig. 14.2) indicate the possibility that they may not. Also there is the question in man as to whether $wt^{0.72}$ or basal metabolic rate (BMR) calculated according to the DuBois formula, which considers both height and weight, are the most appropriate measures for estimating the endogenous nitrogen. We assume that there is a direct relationship between BMR and endogenous excretion regardless of the relationship of either of these to weight and body conformation. Thus, measured metabolic rates should be preferable when they are available, although $wt^{0.72}$ is expected to give essentially similar figures.

When an animal or human subject is fed a nitrogen-free diet, the urinary nitrogen level falls rapidly and after a week or two reaches a substantial plateau. Although Fig. 3 indicates that most of the data center around a value of 2 mg of nitrogen per basal calorie, the value is subject to the usual biological variation and is influenced to some degree by the extent of protein depletion. For example, in the human subjects studied by Bricker *et al.* (1945) the minimum excretion was found to be 1.4 mg N per basal calorie. The value of 2 mg is, therefore, not an absolute minimum and is reached without severe protein depletion.

The importance of this parameter is clear when one considers that, by definition, the absorbed nitrogen from a protein of Biological Value (BV) of 100 is 100 % efficient in replacing these obligatory urinary losses. Thus, on the average, 2 mg of absorbed nitrogen per basal calorie supplied by such a protein, plus sufficient to meet other obligatory losses, will result in N equilibrium. An individual with a basal metabolic rate of 60 calories per hour or 1440 calories per day would have a minimum requirement of the highest quality protein of 18 gm per day to replace the urinary losses (2 mg N $\times$ 6.25 $=$ 12.5 mg protein/kcal; 12.5 $\times$ 1440 $=$ 18 gm protein). Since Biological Value is a measure of the efficiency with which a protein is able to replace these losses, the amounts required when proteins of lower Biological Value are fed may be calculated. In the example above, if the dietary protein had a BV of only 70 %, then the amount needed would be proportionately increased, 18/0.7 $=$ 25.7 gm.

B. Fecal Nitrogen

Consideration of Biological Value does not include fecal losses, either those of endogenous origin or undigested dietary protein. The origin of the fecal nitrogen is not known with certainty. When animals are fed a nitrogen-free diet a small amount of fecal nitrogen is produced. It is commonly assumed that the larger amounts occurring in the feces when diets containing proteins are fed represent undigested protein. The extent to which this assumption is true remains to be determined. From the work of McCance *et al.* (1947, 1948) it appears that differences in fecal nitrogen when different kinds of bread were fed were probably mostly of endogenous origin. Fecal nitrogen in animals is influenced by the bulk of food consumed, among other othings (Mitchell, 1942), and it is reasonable to assume that variations in the composition of the diet may affect both the composition and the amount of the secretions into the gut and the disposition of the protein and nitrogenous constituents they contain. Studies with isotopically labeled proteins will apparently be required to determine the relative contribution of dietary and endogenous sources under various conditions.

Regardless of the source of the fecal nitrogen, it represents a loss which must be covered by the dietary protein. It has been commonly assumed that fecal losses are of the order of 10 % of the intake and on this assumption one would add 10 % to the estimates derived above. The requirement of 18 gm when the protein is of Biological Value 100 would thus be raised to 20 gm and the requirement would be 28 gm when the protein had a BV of 70 % (Food and Nutrition Board, 1959). It is probable that this estimate is too low when diets low in protein are fed. Many proteins appear to have a digestibility of approximately 90 %, but fecal losses are greater than 10 % of the protein fed. Hegsted *et al.* (1946), for example, found that the total fecal losses were from 0.8 to 1.1 gm per day on diets which supplied about 4.5 gm of nitrogen. Evidence on children presented later also suggests that 10 % is too low and this might be more logically raised to 20 % in making minimal estimates.

C. Losses in Skin

The other loss of body nitrogen which occurs in the adult is through the skin and by growth of hair and nails. Skin losses include both the nitrogen in sweat and sloughed-off cells. These losses are difficult to measure and have rarely been considered in nitrogen balance studies. Mitchell and Edman (1962) have recently reviewed the limited evidence in this field. In their own studies, Mitchell *et al.* (1949) found the nitrogen in sweat to be about 360 mg during an 8-hour day under minimal sweating conditions. Losses ten times this high, amounting to 22 % of the total nitrogen excretion, were recorded in a hot, humid environment that produced profuse sweating. These men were eating a high protein diet and the relationship between intake and losses through the skin does not appear to have been investigated.

Mitchell (1949) found in long-term balance studies with 3 young men consuming 84 gm of protein per day that there was a consistent positive nitrogen balance amounting to 1.379 gm of nitrogen per day. Such data lead him (Mitchell and Edman, 1962) to conclude that "dermal excretion of nitrogen . . . may account for 0.38 grams of nitrogen . . . the remainder of approximately 1 gm of nitrogen . . . equivalent to 0.56 gm per m^2 may be considered as being retained for the renewal of the integumental losses in desquamated cells and the extrusion of hair and nails. . . ." Others (Wallace *et al.*, 1958; Wallace, 1959) have argued that the continuous positive balances commonly observed in studies of this kind are largely the result of technical errors. The intake is likely to be slightly overestimated since the diet may be incompletely consumed and the excretion slightly underestimated since complete collection is difficult. Thus, if the intake and excretion were each 10 gm of nitrogen per day, a 2 % overestimate of the first and a similar underestimate of the second would give data indicating a retention of approximately 0.4 gm per day. It is probable that extremely careful balance studies must be done in order for the intake to be done with this degree of exactness. However, it is not the fact that there are errors in the measurements which is distressing, but the fact that the errors tend to be cumulative and become large when the diet is high in nitrogen. It is probably impossible to demonstrate the magnitude of these errors. Incidentally, it is of much interest that in infants fed milk, in which the accurate assessment of intake should be easy as compared to mixed diets, the nitrogen retention figures are often consistent with theoretical calculations (see Section V,A,2). Also it should be noted that nitrogen balance becomes inherently less exact as the intake is raised, since one is interested in a small difference obtained by subtracting two large values. Certainly balance data on various nutrients, calcium, for example (Duncan, 1958), have often yielded data which appear unrealistic in terms of total body content. Macy's (1942) data upon children indicate retentions of protein which are as large as the total weight gain. The problem, therefore, is not whether or not nitrogen is lost through the skin, but the magnitude of this loss and the relation of it to protein intake. The men studied by Mitchell with large losses in sweat were consuming a high protein diet. It would be unusual if the adaptive mechanisms of the body did not lower these losses when lower levels of protein are fed. It must be agreed, however, that there are requirements for the losses through the skin and for integumental growth.

If one assumes that 18 to 20 gm of egg protein (Biological Value 100) is a reasonable estimate of the minimal protein requirement and if one calculates the amounts of essential amino acids supplied by this quantity of the protein, these amino acids will be seen to be supplied in amounts considerably in excess of Rose's estimated requirements. Since Rose's estimates of requirements are thought to be high (see Section VI,A) and since protein with a Biological Value of 100 should provide a protein of optimal quality, there is a discrepancy here which requires examination. Mitchell and Edman (1962) suggest that this discrepancy might be accounted for by a consideration of the amino acid content of the epidermis. They assumed that 0.56 gm of nitrogen were lost per square meter of body surface and that this had the composition of human epidermis as compiled by Block and Weiss (1956). They believe that the sum of the amino acids in 20 gm of egg protein (to meet endogenous losses) and those calculated as being required to meet epidermal losses are a more reasonable estimate of the total need. The figures thus derived are considerably above previous estimates of amino acid requirements. It will be apparent, however, that this procedure can be questioned since the Biological Value of a protein as usually measured would be influenced by the amino acid requirements for integumental growth even though the losses through the skin were not actually measured. Mitchell suggests that minimal estimates of nitrogen requirements would be raised by approximately 27 % by proper consideration of these losses. What allowances should be made for differences in Biological Value is unclear, but it does not seem that the amino acid needs of each tissue can be considered independently.

D. Estimates of Protein Needs

Ignoring the question of the nitrogen requirements for integumental growth for the moment, there is much experimental evidence on adult subjects which is consistent with a minimal requirement of protein equivalent to 2 mg of nitrogen per basal calorie with corrections for Biological Value. The total literature need not be reviewed here. Sherman *et al.* (1920) reviewed the literature in 1920. The protein requirements to maintain nitrogen balance were found to vary from 21 to 65 gm per day with an average of 44 gm per day for a 70-kg man. In Sherman's own studies, largely of cereal grains, he derived a mean value of 0.5 gm/kg/day as the requirement for balance. This value with an allowance for safety was given as the basis for the original Recommended Daily Allowance for the Food and Nutrition Board (1945) of 1.0 gm/kg/day. Since the Recommended Allowance also assumed that the diet would supply a considerable proportion of the protein from animal sources, thus assuring a reasonably high Biological Value, the Recommended Allowance would appear to be some two to three times the estimated minimal need. Bricker *et al.* (1945) found the minimal needs to be 22.4 gm/day when the protein was supplied from milk, 38.7 from white flour, 23.4 from soy flour, 27.5 from a mixture of soy and white flour, and 25.4 from mixed foods. The average basal metabolism of the 9 subjects used was 1300 calories per day. Conversion of these values to those of a 70-kg man, assumed to have a surface area of 1.8 m^2 and a basal metabolic rate of 1650 calories per day raises them by about 25 %. Calculated on a similar basis, Hegsted *et al.* (1946) obtained a value of 32.4 gm/70 kg man/day on a diet of mixed vegetables and 27 gm when approximately one-third of the protein was supplied as meat.

The general approach outlined here was used by a committee of the F.A.O. (1957) and the Committee on Amino Acids of the Food and Nutrition Board (1959) in deriving estimates of the minimal protein needs. The former estimated the minimum adult need as 0.35 gm/kg/day and the latter between 0.31 and 0.34, depending upon body size; in Table I are presented re-estimates. The fecal losses have been raised to 20 % of the intake, which would then yield the estimates shown in column 8. If one then adds the additional estimate for losses through the skin according to Mitchell and Edman (1962), the estimate in column 11 is obtained. Further

TABLE I

ESTIMATES OF ADULT PROTEIN REQUIREMENTS

Height (cm)	Weight (kg)	Surface area (m^2)	Basal metabolism (kcal/day)	Endogenous N (gm/day)	Protein equivalent of endogenous N (gm/day)	Requirement corrected for fecal losses		Addition for skin losses, etc., as protein (gm/day)	Total requirement, protein of BV = 100		Total requirement, protein of BV = 70	
						Gm/day	Gm/kg/day		Gm/day	Gm/kg/day	Gm/day	Gm/kg/day
1	2	3	4	5	6	7	8	9	10	11	12	13
Men												
155	50	1.46	1384	2.77	17.3	20.8	0.42	5.1	25.9	0.52	37.0	0.74
155	60	1.58	1498	3.00	18.8	22.6	0.38	5.5	28.1	0.47	40.1	0.67
165	60	1.66	1574	3.15	19.7	23.6	0.39	5.8	29.4	0.49	42.0	0.70
165	70	1.77	1678	3.36	21.0	25.2	0.36	6.2	31.4	0.45	44.9	0.64
175	70	1.84	1744	3.49	21.8	26.2	0.37	6.4	32.6	0.47	46.6	0.67
175	80	1.96	1858	3.72	23.3	28.0	0.35	6.9	34.9	0.44	49.9	0.62
Women												
150	45	1.36	1191	2.38	14.9	17.9	0.40	3.4	21.3	0.47	30.4	0.68
150	55	1.43	1253	2.51	15.7	18.8	0.34	3.6	22.4	0.41	32.0	0.58
160	50	1.50	1314	2.63	16.5	19.8	0.40	3.8	23.6	0.47	33.7	0.67
160	60	1.62	1419	2.84	17.8	21.4	0.36	4.1	25.5	0.43	36.4	0.61
170	55	1.63	1428	2.86	17.9	21.5	0.39	4.1	25.6	0.47	36.7	0.67
170	65	1.75	1533	3.07	19.2	23.0	0.35	4.4	27.4	0.42	39.1	0.60

Column 3—Based upon DuBois' formula.

Column 4—Assuming 39.5 kcal/m^2 for men and 36.5 kcal/m^2 for women.

Column 5—2 mg N per basal calorie.

Columns 7 and 8—20% added to endogenous N to account for fecal losses.

Column 9—Assuming 0.56 gm N (3.5 gm protein) per m^2 for men and 0.4 gm N (2.5 gm protein) per m^2 for women.

correction, based upon an assumed Biological Value of only 70 %, would yield the minimum needs shown in column 13. These would appear to be the highest estimates of the minimal needs that can be reasonably justified with the evidence available today. It is probable that they may be higher than actual needs but, as has been stressed, care should be exercised that dietary recommendations are not based upon serious underestimates of the actual need.

The minimal estimates derived by the F.A.O. Committee were arbitrarily increased by 50 % to provide a "safe practical allowance." The values of the Committee on Amino Acids would have to be raised considerably more than this to account for the Recommended Daily Allowance of the Food and Nutrition Board. Such large arbitrary corrections clearly indicate that the nutrition fraternity is unwilling to take the estimates of the minimal protein needs very seriously. It is somewhat ludicrous to go through the effort of calculating the requirements in detail and then double or triple the value obtained, since the "guesstimate" added is as large as the estimate. The factual data become a kind of scientific window dressing which lends authority to the final figure. On the other hand, there are cogent arguments at this time for reasonably large allowances for safety. Practically nothing is known of the variability in the requirements of individuals, or the cause of them. Data available upon amino acid requirements suggest that there may be a 100 % difference between the needs of individuals, but these differences include the technical error of the estimate which must be large. The average diet, like the average man, is a statistical concept and corrections for Biological Value based upon the national diet penalize the regions which have diets of lower value as well as the poorer economic groups. Recommendations as to the appropriate intakes must be higher than the average minimal estimate; but one of the greatest needs in putting a scientific base upon such recommendations is information upon the variation in individual requirements.

On the other hand, there is little or no clinical evidence of protein deficiency in adults consuming diets considerably lower in proteins than the amounts recommended. An intake of 2400 calories as white flour would supply approximately 60 gm of protein. This compares with the estimated requirement of Bricker *et al.* (1945) for women consuming white flour of 58.6 gm per day. This estimate included an allowance for "adult growth" of 6 gm of protein. Since the white flour was found to have a Biological Value of 41 %, it is doubtful that diets of poorer quality than this will be found in practice. Thus, the conclusion that "It is most unlikely that protein deficiency will develop in healthy adults on a diet in which cereals and vegetables supply adequate calories" (Hegsted *et al.*, 1946) still appears to be valid. The concern over proteins generated by the studies upon protein-calorie malnutrition and kwashiorkor has probably overemphasized the practical importance of protein.

IV. Protein Requirements in Pregnancy and Lactation

Minimal protein requirements are obviously increased during pregnancy and lactation. A newborn infant weighing 3.5 kg and containing approximately 2% nitrogen in the body represents approximately 450 gm of protein which must be synthesized principally during the last trimester of pregnancy. Over a 90-day period approximately 5 gm of additional protein would thus be stored per day. Growth of accessory tissues raise this value by some undetermined amount. A similar calculation for breast milk based upon the production of a liter a day containing 1.2% protein gives an increased requirement of 12 gm per day. These calculations assume, of course, a maximal efficiency of the conversion of the dietary protein into tissue or milk protein; and little, if anything is known of the efficiency of this conversion. Data are available which indicate that high protein intakes during pregnancy increase the likelihood of a favorable outcome (Tompkins, 1955; Hillman, 1960). On the other hand, the apparently normal birth weight of many infants in underdeveloped areas (Brock and Autret, 1952) and apparently excellent milk production (Walker, 1958) argue against the need for unusually high protein intakes. The F.A.O. recommendation is that 10 gm additional protein (plus correction for Biological Value) be allowed during pregnancy and 30 gm during lactation. The Recommended Daily Allowance provides 20 additional gm during pregnancy and 40 gm during lactation. The general attitude is one of relative generosity during an apparently critical period, but there is little scientific basis for the actual amounts recommended (see Chapter 15).

V. Protein Requirements of Infants and Children

A. *Infants*

1. *Theoretical Estimates*

Although few, if any, studies have been done to determine experimentally the minimal protein needs of children and infants, the minimum requirement can be calculated using the approach outlined for adults. The Biological Value of proteins is similar although possibly not identical (Block and Mitchell, 1946–1947; Mitchell, 1950) for either adult or young animals, thus indicating that the proportions of amino acids required are similar. Terroine (1936) also found that the minimal nitrogen excretion per basal calorie in young animals was similar to that of adults. Thus, the maintenance need, that is, the requirement to maintain the tissue already formed, may be calculated from information upon body size and basal metabolism. To this value the estimated requirement for growth must be added, and this can be estimated from the rates of growth. Finally, additions may be made for fecal losses, losses through the skin, etc.

In the calculation made by Hegsted (1957) the maintenance need was estimated as equivalent to 2 mg N per basal calorie as in adults. Values for the basal metabolic rate for infants and children were taken from a variety of sources (Hegsted, 1957). The growth requirement was estimated from the growth charts of Stuart (no date) and Stuart and Meredith (1946) assuming that the tissue gained contained 18 % protein. Finally, as in adults, an additional 10 % was allowed for fecal losses. The results of such calculations are shown in Fig. 4. The Committee on Amino Acids (Food and Nutrition Board, 1959) arrived at similar estimates. The estimates of the F.A.O. (1957) are shown in Fig. 5. It is assumed that a somewhat similar approach was used in deriving these curves, but the basis of the calculations was not disclosed. The curves are quite different from

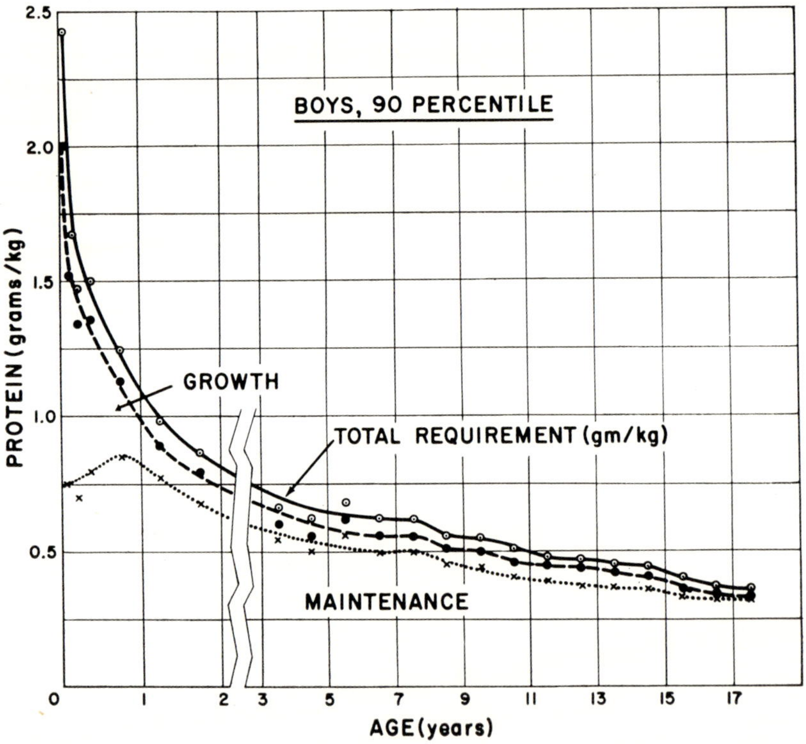

FIG. 4. Estimates of minimal protein requirements according to Hegsted (1957).

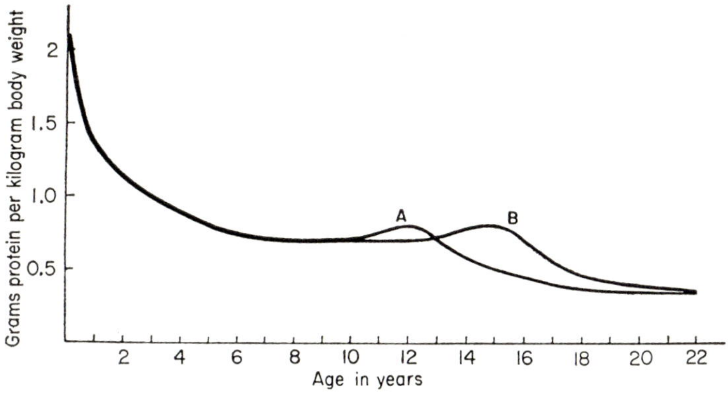

FIG. 5. Estimates of minimal protein requirements according to F.A.O. (1957). Curve A: girls; curve B: boys.

those of Hegsted, particularly for children between 3 and 10 years of age and for adolescents.

It is known that the infant is born with a low total body nitrogen content, approximately 2 % compared to 3 % in the adult. The rate at which "chemical maturity" develops in most infants is unknown and presumably depends upon the diet, but it is generally believed to be complete within the first year of life (Holt *et al.*, 1960). The desirable rate at which chemical maturity should be reached is also unknown. In view of these uncertainties, the additional nitrogen requirements they impose were not included in the calculation of Hegsted. However, it is apparent that to increase the body content of a 3.5-kg infant 1 % would require 35 gm of nitrogen or approximately 220 gm of protein. Spread over a one-year period this would be 0.6 gm of protein per day, or assuming an average weight of 6 kg during this period, an additional 0.1 gm/kg/day. Thus chemical maturity is not an important factor in computing protein requirements.

2. Experimental Evidence

These calculations agree rather well with the data of Fomon and May (1958) in infants who were fed pasteurized breast milk during the first 6 months of life. The average intake during the first 1.5 months of life yielded 147 kcal/kg/day and 2.4 gm protein/kg/day. Between 4.5 and 6 months of age the intake was 94 kcal/kg/day and 1.5 gm protein/kg/day. The nitrogen retained during the first 1.5 months was equivalent to 1.125 gm protein/kg/day, whereas our estimate during the first 2 months would be 1.2 gm/kg/day. Between 4.5 and 6 months of age the average retention would be 0.26 gm/kg/day. Our estimate was approximately 0.5 at 4.5 months, falling to 0.27 at 9 months. Considering the fact that during the first 6 months the intake per kilogram of body weight is rapidly falling and that the nitrogen retention is undoubtedly closely related to the actual gain in weight, the estimates obtained by Hegsted and the actual data of Fomon and May must be considered as essentially the same.

Some differences should be noted, however; the children studied by Fomon and May had a mean gain of 42 gm per gram of nitrogen retained, thus indicating an actual protein content of the gain of 15 % rather than 18 %. It is probable that the retentions per gram gain are smaller during the first months of life than later. Also, the average fecal nitrogen in Fomon's children amounts to approximately 15 % of the dietary intake and in one was as high as 19 %. It seems likely that the value of 10 % assumed by Hegsted is a little too low.

One may obtain an estimate of the disposition of the protein consumed by Fomon's infants if the maintenance need is estimated from Fig. 3. Utilizing this value plus the nitrogen retained as the total amount utilized, and comparing this with the nitrogen absorbed (total intake minus fecal nitrogen)[1], one obtains an efficiency of utilization or Biological Value during the first 1.5 months from the protein of breast milk of approximately 85 %. That is, 85 % of the nitrogen absorbed can be accounted for either as that required for maintenance or for growth at an intake of 1.125 gm/kg/day. In this connection it is of interest to compare the values obtained by Snyderman *et al.* (1960), who fed amino acid mixture prepared to duplicate the composition of breast milk or the FAO reference pattern. These patterns are quite similar except that the breast milk provides 150 % as much leucine, and only 74 % as much tryptophan and 46 % as much methionine as the FAO pattern. The over-all result with the two mixtures when fed to premature infants was essentially the same. These authors' figures show that the urinary nitrogen excretion while the infants were receiving the equivalent of 2 gm of protein/kg/day from these sources was, in most instances, practically equivalent to the estimated endogenous nitrogen, and in a few instances was somewhat lower. Adding the estimated maintenance requirement and the nitrogen retained together accounts for essentially all of the nitrogen fed (93 to 117 %). The Biological Value of these mixtures thus appears to be, for all practical purposes, 100. The relative intake of these infants is somewhat lower than those studied by Fomon.

[1] It should be noted that this is "apparent absorption" rather than "true absorption," the latter usually being measured by subtracting the so-called metabolic nitrogen or fecal nitrogen obtained when a nitrogen-free diet is fed. The metabolic nitrogen was not taken into account in Hegsted's calculations and this may in part account for the larger values obtained here.

Whether this accounts for the better utilization, or whether breast milk is somewhat less well utilized than an equivalent mixture of amino acids is unknown. In any event, the excellent efficiency is of interest and supports the validity of estimates assuming complete utilization.

On the other hand, between 4.5 and 6 months of age, when the infants in Fomon's group received about 1.5 gm/kg/day, the efficiency of utilization similarly calculated falls to only 60 %. Since there is no reason to assume that infants of this age are inherently less efficient, it is likely that this decrease in Biological Value is due to an excess intake. Our estimate of the minimal need at this age is between 0.5 and 0.24 gm/kg/day.

It would be of interest to know the Biological Value of breast milk determined by conventional rat assay. Little data appear to be available. Holt (1959) quotes an unpublished report of Platt to the effect that the net protein utilization of breast milk is 95.4 compared to 81.5 for cow's milk. Biological Values would be somewhat higher since they are based upon absorbed protein and maintenance requirements are considered.

Fomon (1961) also studied 12 infants who received a prepared formula which supplied 3.3 gm protein/kg/day during the first 1.5 months and 2.5 gm/kg/day between 4.5 and 6 months. Nitrogen retentions, as expected, were slightly higher than on breast milk and were equivalent to 1.32 gm protein/kg/day and 0.64 gm/kg/day during the two periods. The efficiency of utilization was much less than for breast milk and it may be assumed that this was in large part due to the higher protein intake. Comparative data at equal intakes and particularly at lower intakes would be of major interest. Comparing the gain in weight with the nitrogen retained, the tissue formed appeared to contain 19 % protein as compared to 15 % in the breast-fed infants. Fomon points out that no known advantage can be assigned to these higher nitrogen retentions. Finally, it must be noted that these data obtained on infants (contrary to much or all of the data on older children and adults) do not indicate great errors in the balance technique. That is, the nitrogen retention figures appear quite reasonable when compared to the gain in weight. It has been noted that this was true also of the data supplied by Albanese (1947) and Holmes (1946). The greater accuracy which can be obtained in measuring formula diets as opposed to mixed diets is quite obvious and may provide an adequate explanation.

3. Efficiency at Optimal Intakes

As has been noted in Section II, if the response to dietary protein is proportional to the logarithm of the intake, there would be a continuing fall in the efficiency of utilization as higher levels are fed and maximal rates of growth are approached. There is reason to believe that this might be the case, since the log-dose response curve is so ubiquitous in nature. A teleological argument was advanced (Hegsted, 1959) that this might be true, particularly with proteins of lower Biological Value. As higher levels of dietary protein are provided, and if these are not of optimal amino acid composition, the organism must develop mechanisms to dispose of the excess amino acids. If one assumes that these disposal mechanisms are not completely efficient in sorting out those amino acids in short supply from those in excess, some of the amino acids in short supply and presumably limiting growth would also be destroyed. The biological efficiency of the dietary protein would thus decline. Miller and Payne (1961a,b), in attempting to derive methods for protein assay which account for both the Biological Value and the level of protein in the diet, have assumed, in fact, that there is a continuous fall in efficiency of utilization as the dietary level is increased above the maintenance requirement. They further assume that the fall in efficiency is such that all proteins reach zero utilization when 54 % of the dietary calories are supplied as protein. A critique of their overall proposal is not appropriate here and has been made by Njaa (1962). Miller and Payne base their proposal upon their own data on three proteins fed at varying levels to rats and data supplied by Forbes *et al.* (1958). It may be noted that neither of these sources is entirely adequate for the purposes under discussion here, which are changes in efficiency as optimal intakes for maximum growth are approached. The data of Miller and Payne are largely from animals which received relatively high intakes, as compared to requirements, and the data of Forbes *et al.* (1958) were obtained with restricted food intakes. If growth is not allowed the efficiency of utilization must fall as the intake is raised. More appropriate data are those of Barnes *et al.* (1946). Figure 6 was constructed

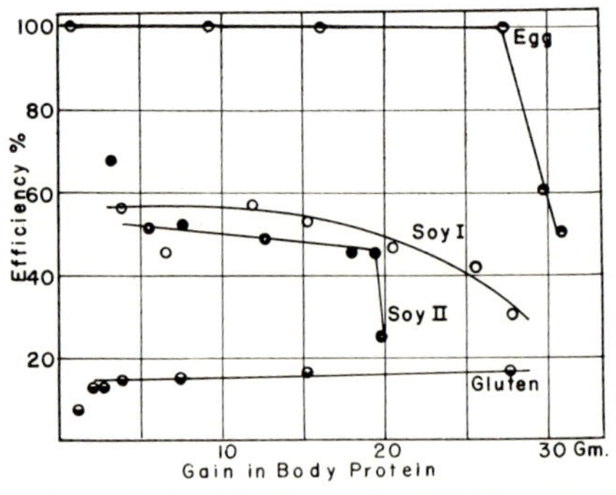

FIG. 6. Calculated efficiency of utilization of available protein for the formation of body protein in rats fed levels of intake which allowed different rates of growth. Available protein calculated as intake less fecal losses and the estimated maintenance requirement (see text). Original data from Barnes *et al.* (1946).

from these data. The efficiency of utilization was calculated from the average size of the animals (assuming the maintenance requirement shown for animals of different size in Fig. 3) and from the gain in total body protein. For the animals fed egg protein, essentially all of the protein absorbed can be accounted for by these two requirements. It is apparent that there was little tendency for the efficiency of utilization of egg protein and wheat gluten to fall as the protein and weight gain increased. Of course, once the maximum gain was reached, the provision of additional egg protein resulted in a precipitous fall in the efficiency of utilization. The choice of soy flour as the other examples may have been unfortunate, particularly since maximal growth was not obtained with the second sample. One may suspect that growth inhibitors of various kinds (see Liener, 1958) may not have been completely eliminated from these preparations. Clearly insufficient data are available and more are needed since this is an important problem in estimating optimal needs of protein. The fact that high efficiencies were obtained by Fomon and May (1958) and by Snyderman *et al.* (1962) in infants growing at normal rates suggests, at least, that a large allowance for inefficient utilization is not justified at this time, contrary to the conclusion of Miller and Payne.

B. Children

Few, if any, data can be found which bear directly upon the minimal protein requirements of children. Macy (1942) conducted 593 balance studies upon 29 children between 4 and 12 years of age. The mean nitrogen intake at 4 years of age was 9.58 gm per day (60 gm protein) and increased to approximately 13 gm (81 gm protein) at 10 to 12 years of age. The mean nitrogen retentions for each age group ranged from 5 to 8 % of the intake, equivalent to 2.9 to 6.6 gm protein retained per day. Assuming the tissue formed contained 18 % protein these retentions would be the equivalent of 16 to 37 gm gain per day. These values are greatly in excess of the actual gain of children. Even if one accepts the proposition discussed earlier that there is a substantial apparent nitrogen retention due to losses through the skin and the growth of skin, hair, and nails, it is difficult to see the relationship between these values and minimal protein requirements. They emphasize again the probability that slight overestimates of consumption and underestimates of excretion may result in apparently substantial retentions.

Estimates of the protein requirement based upon rates of growth and maintenance requirements (Hegsted, 1957) emphasize the relatively small amount of protein deposited daily due to growth in the human species. Even during the so-called "growth spurt" at adolescence the increments due to growth appear small compared to the total intake or requirement.

It has been suggested that very high intakes of protein, 15 % of the total calories, may represent the requirement during adolescence. According to Johnston (1953), working with tuberculous children and adolescents, optimal performance requires intakes of this order of magnitude. It is apparently true that even minor infections or stresses of various kinds (see Chapter 23) may induce slight negative nitrogen balances. Whether such effects contribute substantially to the over-all protein requirements remains to be demonstrated.

The standards proposed by F.A.O. (1957) for the estimation of minimal protein needs at various ages are shown in Fig. 5. While it is assumed that this curve was derived from considerations similar to those discussed in the previous section, the report does not specifically describe the derivation of the curve. The requirement for very young infants is slightly lower than the estimate of Hegsted, and is presumably based upon estimates of the protein supplied by breast milk. The estimate for older children and particularly for adolescents is much higher than that derived by Hegsted. These are presumably compromise figures arrived at by attempts to satisfy the points of view of various committee members. These values, like that for adults, are to be corrected for Biological Value of the diet consumed and a 50 % allowance for safety provided in converting them into recommended intakes.

VI. Amino Acid Requirements

Information upon the amino acid requirements of man is also largely concentrated upon those of adults and very young infants. The classic work of Rose *et al.* (1954a,b; 1955a–e; Rose and Wixom, 1955a,b) deals with young men, the work at Nebraska (Leverton *et al.*, 1956a–e), Wisconsin (Jones *et al.*, 1956) and California (Swendseid and Dunn, 1956; Swendseid *et al.*, 1956) with young women, and Holt and his co-workers (Pratt *et al.*, 1955; Snyderman *et al.*, 1955, 1959a–d; Holt, 1960) have studied young infants. More recently Nakagawa *et al.* (1960, 1961a,b, 1962) deal with the amino acid requirements of boys 10 to 12 years of age. The conclusions of these various groups are summarized in Table II. Since much of the work has been previously summarized (Food and Nutrition Board, 1959) it is more useful to consider problems raised by the data available and the methods used rather than present the details of these studies.

A. *Accuracy of the Estimates*

The primary questions that arise in considering the information available are the accuracy of the estimates which have been made and the applicability of the results to the general problem of protein nutrition. All of the studies or nearly all, have been done with diets very low in natural

TABLE II

ESTIMATES OF MINIMAL REQUIREMENTS OF INDISPENSABLE AMINO ACIDS

Amino acid	Infants[a] (mg/kg)	Adult female[b] (mg/day)	Adult male[b] (mg/day)	Children[c] (mg/day)
Histidine	34	—	—	—
Isoleucine	126	450	700	1000
Leucine	150	620	1100	1500
Lysine	103	500	800	1600
Methionine	45[d]	350	1100[e], 200[d]	800[e]
Total S AA	—	550	1100, 1010	—
Phenylalanine	90[f]	220	1100, 300[f]	800
Total aromatic AA	—	1120	1100, 1400	—
Threonine	87	305	500	1000
Tryptophan	22	157	250	250
Valine	105	650	800	900

[a] Taken from Holt *et al.* (1960).
[b] Taken from Food and Nutrition Board (1959).
[c] Nakagawa *et al.* (1960, 1961a,b, 1962).
[d] In presence of adequate cystine.
[e] No cystine supplied.
[f] In presence of adequate tyrosine.

proteins in which all, or practically all, of the nitrogen has been supplied as amino acids and other purified nitrogen sources. This has been necessary since proteins cannot be found which are sufficiently low in certain of the essential amino acids. Because of the expense and difficulty of the procedure, relatively few subjects have been studied and the periods of observation have been rather short. Similarly, only a limited number of levels of amino acid addition have been tested. In all of the studies nitrogen balance has been the principal criterion used, although with young infants weight gain has been useful.

The data upon amino acid requirements in man are somewhat unique in comparison to much other nutritional data with man, and most of the data upon nutritional requirements of animals, in that attempts have been made to titrate the requirements of each individual studied. Thus, the final value selected by Rose (1957) as the minimal daily requirement was "the smallest amount capable of inducing *distinctly positive* nitrogen balance in the *individual requiring the greatest amount.*" Leverton *et al.* (1956a) defined the requirement for nitrogen balance as "the amount of an amino acid which will keep a subject in a zone of nitrogen equilibrium, where the difference between intake and excretion is within 5 percent." Thus, Rose required a distinctly positive balance, whereas Leverton assumed an excretion within 5 % of the total nitrogen intake was not different from equilibrium. However, the value selected as the minimal requirement by Leverton was also the amount required by the individual with the highest apparent requirement by their criteria. Holt's criteria were "satisfactory nitrogen retention and weight gain" (Holt *et al.*, 1960). With one exception, that of an infant who appeared to have an unusually high leucine requirement, these workers have also selected the amount required by those infants with the highest requirement. Nakagawa *et al.* (1960, 1961a,b, 1962) do not specifically state their criteria, but it appears to be the estimate of the amount required for nitrogen equilibrium and relatively few subjects or levels of intake were studied.

The reasons for selecting these end points are fairly obvious. As was mentioned in discussing protein requirements, an underestimate of the need can be considered a more serious error than an overestimate. When few subjects are studied and rather large differences are found, the average value may have little significance. On the other hand, it should be clearly recognized that much of the material upon which nutritional requirements are based is presumably an average value. This is almost uniformly true of estimates of the nutritional requirements of animals where groups of animals are used and the estimates are based upon the average response of the group. The selection of values based upon those subjects with the highest apparent requirements places great emphasis upon the data from one or a very few individuals. The variation encountered is undoubtedly due to both biological variation and technical error, the latter including not only the familiar errors in determining nitrogen balance, but any factor, such as the influence of the previous diet, which may influence the result obtained. For example, the data upon the tryptophan requirement of adult men (Rose *et al.*, 1954b) is based upon three individuals whose requirements were estimated to be 0.15, 0.15, and 0.25 gm per day. An inspection of the data presented in this study leaves little doubt that the third subject responded differently from the other two, but that the reasons for this difference are, of course, in doubt. The selection of the highest value appears logical but prevents statistical evaluation of the accuracy of the estimated requirement or comparison with other data.

It can be argued that an appropriate means of evaluating the data upon amino acid requirements would be simply to plot the dose against the response and estimate in the usual way the amount required to maintain N equilibrium. In Fig. 7 the data tabulated by Jones *et al.* (1956) on the nitrogen balance in young women who received various levels of lysine intake have been plotted. These data are selected simply as a convenient example since the crude values are easily obtained from the publication. The regression line, determined by the method of least squares, is $Y = 1.56X - 0.81$ where Y is nitrogen balance and X is the lysine intake in

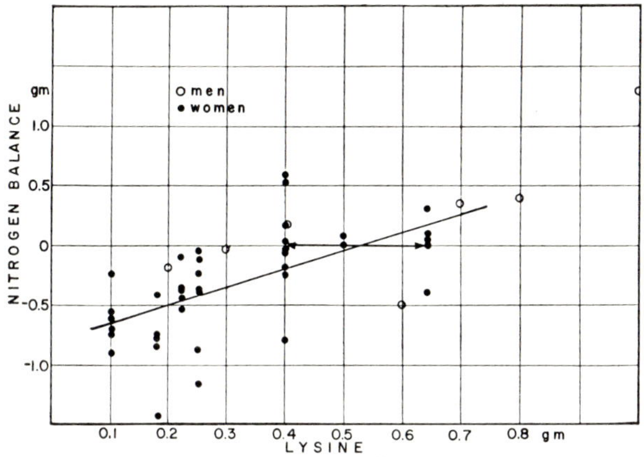

FIG. 7. Relationship between lysine intake and nitrogen balance. Data on women from Jones *et al.* (1956). Data on men from Rose *et al.* (1955b). Regression line calculated only from data of Jones *et al.*

grams. Y is zero when X is 0.52 and this level of lysine intake would be the best estimate of the mean lysine requirement to maintain nitrogen equilibrium. Surprisingly enough, this is not much different from the estimate of Jones *et al.* which was "between 0.4 and 0.5 grams per day." The approximate variance of X when X is 0.52 was calculated according to the method of Scheffé (1960). The standard error of the estimate is ± 0.058 gm. Two standard errors on either side of the mean estimate are shown in Fig. 7 by the solid line drawn through the point $Y = 0$, $X = 0.52$. The limits are 0.4 to 0.64. Thus the mean requirement can be estimated to fall within these limits with a probability of 0.05. It is important to note that this corresponds to a standard error of the mean and does not indicate the range in requirements of individuals utilized in the study.

We have also plotted in Fig. 7 the data supplied by Rose *et al.* (1955b) upon the nitrogen balance in two presumably typical subjects fed several levels of lysine. Rose concluded that the minimum daily requirement of lysine was 0.8 gm per day, and that in 6 men studied the requirement ranged from 0.4 to 0.8 gm per day. It is not apparent from the values plotted in Fig. 7 that the requirements of lysine for men are particularly different from those of the women studied by Jones *et al.* In this connection it is of interest that whereas in the work referred to above there appeared to be no relationship between body size and amino acid requirements, Clark *et al.* (1960) found a highly significant correlation between the lysine requirements of men and their body weight. According to their paper, the lysine requirement of the 10 men studied varied from 400 to 1200 mg per day with a mean requirement of 750. In 10 women the lysine requirement varied from 300 to 700 mg per day with a mean value of 550. The correlation between body size and lysine requirement was not significant in the women studied because of two aberrant values, but even with these two eliminated the correlation was much less than in men. It should be noted that the correlation made was with a selected lysine requirement and that, as is typical of much of the literature in this field, the crude protocols demonstrating how the requirement was estimated are not shown. More disturbing from the point of view of the general problem of estimating amino acid requirements is the demonstration by Clark *et al.* (1960) of the great difference in nitrogen balance at equal levels of lysine intake when the order of feeding was changed. Undoubtedly the values in the literature are in part dependent upon the length of the feeding trial and the previous degree of nitrogen depletion, but we do not know sufficient about the effect to make corrections which might in part eliminate such sources of error.

Contrary to the impression gained from Fig. 7 that the differences in lysine requirements of men and women so far reported may be in large part due to interpretive decisions of the investigator, the data upon isoleucine requirements give exactly the opposite impression. In Fig. 8 the nitrogen balance data published by Swendseid and Dunn (1956) at different

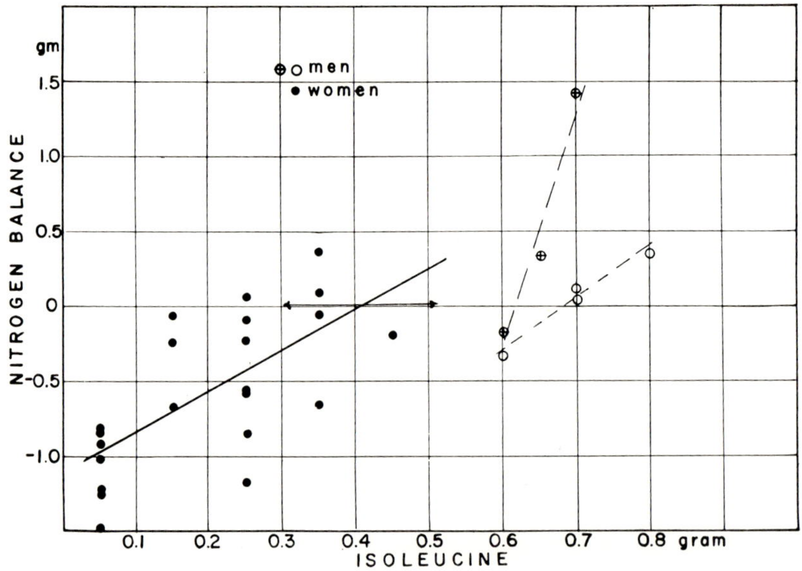

FIG. 8. Relationship between isoleucine intake and nitrogen balance. Regression line calculated only on data for women from Swendseid and Dunn (1956). Data on two men from Rose *et al.* (1955d).

levels of isoleucine intake have been plotted, together with the data upon two men from the paper of Rose *et al.* (1955d). The regression line, calculated only from the data of Swendseid and Dunn, indicates a mean requirement of 404 mg/day. This compares to their estimate of 450 mg/day which was based upon the conclusion that three subjects required 250 mg/day, that three required 350 mg/day and that one subject had a requirement of 450 mg/day. As was the case with the lysine requirements the calculation of the mean requirement by this method does not appear to seriously underestimate the requirement. The standard error of the estimate calculated as described above for the lysine data indicates the probable limits ($P = 0.05$) as between 0.31 and 0.50 gm per day. Again, this is an estimate of the accuracy of the mean figure and not an estimate of the range of the requirements of individuals. It is apparent, however, that the data obtained with men by Rose *et al.* (1955d) are very poorly represented by the regression line. Rose gave the requirement of the four men studied as 0.7, 0.7, and 0.65 gm per day. Whether the great difference in the apparent response of men and women is a sex difference or due to differences in technique is problematical, but the data clearly indicate the need for further study.[2]

These examples suffice to show that amino acid requirements are not known with particular accuracy, nor are the causes of the variability found well defined. It would appear that considerable information might be obtained by fuller exploitation of the crude data upon nitrogen balance in the various experiments, much of which has not been made generally available. At the least, such a study might demonstrate those areas in which the data seem most insecure. Although Clark *et al.* (1960) did not find that expressing requirements on the basis of surface area, metabolic rate, or weight to the 0.72 power was an improvement over the correlation between requirement and weight, there is a need to define the most suitable parameter since there is considerable reason to believe that protein requirements are most closely related to the basal metabolic rate.

B. Caloric Requirements

Rose *et al.* (1954a) concluded that the caloric requirements to maintain nitrogen equilibrium were considerably higher when their subjects consumed the amino acid-containing diet than when whole protein was provided. This appears to have been a more or less general finding (Jones *et al.*, 1956; Pratt *et al.*, 1955) although different authors appear to have been variously impressed with whether this was, or was not, a striking finding. The paper of Linkswiler *et al.* (1958) is therefore of considerable

[2] Since this was written a more extensive analysis of the data on amino acid requirements of adults has been presented [Hegsted (1963)].

interest. In studies designed to determine the availability of amino acids to man by comparing equal levels of intake provided either by proteins or as free amino acids, they fed two different purified diets. These diets were similar except that in one the mixture of essential amino acids amounts to 14.9 gm, whereas in the other only 8.4 gm was included. In the latter diet 19.2 gm of glutamic acid was given. Both diets contained 18.8 gm of glycine and 28.3 gm of diammonium citrate as sources of nonessential nitrogen. On the first diet the subjects required 45 to 47 kcal/kg/day to maintain weight. With the second diet only 30 to 40 kcal/kg/day were required. The latter subjects were started at the higher level of intake but the amount of food had to be decreased to prevent weight gains. The authors speculate that the difference in caloric need might have been related to the intake of glutamic acid. Metta *et al.* (1960) failed to demonstrate a greater caloric need in rats which received comparable amino acid intakes as free amino acids or as protein. This is another area which deserves more direct study since the unusual caloric requirements often observed raise questions as to the applicability of the studies to general amino acid requirements.

C. *Nonessential Nitrogen*

It is known that glycine alone is not an effective source of nonessential nitrogen (Rose *et al.*, 1949; Lardy and Feldott, 1950; Rechcigl *et al.*, 1957). The data so far available suggest that relatively simple mixtures are as adequate as mixtures of most of the nonessential amino acids. However, the level and kind of nonessential nitrogen will undoubtedly receive more study in the future. Nearly all of the data available on adult man have been collected with total nitrogen intakes of 8 to 13 gm per day, equivalent to protein intakes of 50 to 80 gm. The inclusion of minimal amounts of essential amino acids in such diets gives a very small proportion of essential amino acids compared to the amounts which occur in natural proteins. A considerable amount of data, particularly with chicks, suggests that the individual amino acid requirement is not independent of the total protein supplied by the diet. Thus, as an example, Grau (1948) concluded that the lysine requirement of chicks increased as the total protein content of the diet was raised. Forbes *et al.* (1955) concluded that the isoleucine requirement was 2.6% of the dietary protein, irrespective of the total nitrogen content of the diet. Grau and Carroll (1958) concluded that "there is abundant evidence that the requirement for a particular amino acid is significantly influenced by the levels of other amino acids present in the diet." Bressani and Mertz (1958) found that the total lysine requirement was not changed once sufficient protein was supplied to allow maximum growth in rats. The studies of Swendseid *et al.* (1959, 1960) on man led to the conclusion that the minimal amount of egg protein required to maintain nitrogen balance was probably due to the total nitrogen content rather than the essential amino acids it contains. However, they could not show that the inclusion of large amounts of nonessential amino acids raised the requirement for the essential amino acids supplied as egg protein. It is uncertain, therefore, whether the relatively large amounts of nonessential nitrogen included in the diets have any effect upon the published estimates of the requirements of the essential amino acids in man.

On the other hand, the calculation of the amounts of essential amino acids supplied by minimal quantities of egg protein, for example, show that the essential amino acids are supplied in excess of the estimates made by Rose and others. The calculations shown in the report of the Committee on Amino Acids (Food and Nutrition Board, 1959) indicate that nearly twice the supposed minimum requirement of all the essential amino acids is provided by the amount of egg protein required to maintain N equilibrium. Snyderman *et al.* (1962) have apparently shown that this is true of milk in studies done with young infants. The level of milk in the diet was decreased to the point that satisfactory nitrogen retention and growth no longer occurred. The addition of glycine at this point resulted in prompt resumption of growth and retention of normal amounts of nitrogen. Thus they conclude that the limiting factor in milk protein is the total nitrogen rather than a particular essential amino acid. At first glance this may appear to be no more than an interesting fact concerning the relative amounts of essential and nonessential amino acids. However, it is apparent that if one or more of the essential amino acids are not the limiting factors then the reasons for differences in the Biological Value of proteins become completely obscure. That is to say, since the nonessential amino acids are expected to be essentially interchangeable, differences in Biological Value and the rather close correlation between essential amino acid content and Biological Value (Block and Mitchell, 1946–1947) are not understandable if total nitrogen is the limiting factor. Data on rats (Bender, 1961) are contrary to the conclusion of Snyderman *et al.* This publication should stimulate considerable work since it questions the basic approach to the estimates of protein and amino acid requirements.

D. Availability of Amino Acids

One may logically assume that the amino acids supplied in free form are completely available to the organism. In view of the fact that proteins may not be completely digested, the availability of the amino acids in proteins might be a factor affecting amino acid requirements and may make the results obtained with purified diets containing pure amino acids not applicable to natural diets. From animal feeding trials, it is clear that unavailability of lysine is an important practical problem (Carpenter *et al.*, 1962) and arginine, tryptophan, and cystine may be destroyed or rendered unavailable (Liener, 1958). Insufficient direct data are available to determine the importance of these phenomena in man although there is no reason to believe that man has capacities for utilization of amino acids that other animals do not. Heat treatment of human diets may be generally less severe than is often true of materials which are utilized in animal feeding (Heller *et al.*, 1961). The studies of Linkswiler *et al.* (1958, 1960a,b) on the availability of the valine, threonine, and isoleucine from corn do not demonstrate that the requirements of these amino acids are higher when they are supplied as cereal. These results must be considered in the light of the general inaccuracy of the technique discussed above. Since lysine may be made unavailable with only moderate processing (Calhoun *et al.*, 1962) and since lysine is apparently a limiting amino acid in most cereal diets, more information upon this amino acid is greatly needed.

It should be noted that increases in fecal nitrogen over the nitrogen excretion upon a nitrogen-free diet are generally interpreted as evidence of undigested protein. However, considerable amounts of nitrogen including amino acids are secreted into the gut (Nasset, 1962). The amount and the disposition of this nitrogen are unknown, but are undoubtedly influenced by the diet fed, the bacterial flora, etc. It is more than likely that some of the amino acids appearing in the feces are synthesized in the gut either from the endogenous nitrogen or from the dietary nitrogen. Thus the measure of the fecal amino acids as a measure of availability is open to serious question (Kuiken and Lyman, 1948; Watts *et al.*, 1959a,b). Adequate studies of digestion will probably require the use of proteins in which the nitrogen is isotopically labeled.

E. Amino Acid Patterns

Nearly all of the data upon the amino acid requirements of experimental animals have been obtained with young and growing animals. It seems apparent that in growing animals the absolute requirements are largely dictated by the rate at which new tissue or new body protein is being formed. Comparisons based upon the percentage required in the diet or the requirement per unit of body weight do not bear directly upon estimates of the requirement of man. The Biological Values of various proteins determined with a number of species, including man, are at least of similar magnitude and, as first shown by Block and Mitchell (1946–1947) and more extensively by Mitchell (1954), are rather closely correlated with their content of essential amino acids. The implication is that there is probably a general optimal pattern of amino acids required which is similar for different species and at different stages of life. Much has been made, therefore, of the pattern of amino acids in proteins. Definition of this pattern together with knowledge of the over-all protein requirements would permit the assessment of the amino acid adequacy of diets or prediction of amino acid requirements. The F.A.O. report suggested an ideal pattern based largely upon the data available upon the essential amino acid requirements of adult human beings.

The expression of a pattern is most easily visualized when the amino acids are expressed as ratios of one to the other. This is most logically and usually done by comparison to tryptophan, the amino acid required in the smallest amounts. The proportions of amino acids required when compared to the tryptophan requirement for different ages and sexes of human beings and some other species are shown in Table III. The list is not inclusive of all the data in the literature, but is intended to be demonstrative of problems which still remain. The ratios contained in egg protein, assumed to approach the ideal natural protein with a Biological Value of 100, are included for comparison.

Inspection of the table apparently leads to two opposite conclusions,

TABLE III

PATTERNS OF AMINO ACID REQUIREMENTS

Amino acid	Adult male[a]	Adult female[a]	Boys[b]	Infants[c]	Young rat (Rose)[d]	Young rat (Rama Rao)[e]	Young chick[d]	Egg protein[f]
Histidine	—	—	—	1.5	2.0	1.7	1.5	
Isoleucine	2.8	2.9	4	5.7	4.0	3.7	3.0	4.0
Leucine	4.4	3.8	6	6.8	2.5	4.7	7.0	5.3
Lysine	3.2	3.2	6.4	4.7	5.0	6.0	5.0	3.7
Methionine	4.4, 0.8	2.2	3.2	2.0	2.0	1.0	2.3	1.8
Cystine	3.2	1.3	—	—	—	2.3	—	—
Total S AA	4.4, 4.0	3.5	—	—	—	3.3	—	3.2
Phenylalanine	4.4, 1.2	1.4	3.2	4.1	3.5	2.8	4.0	3.5
Tyrosine	4.4	5.7	—	—	—	2.0	—	2.6
Total aromatic AA	4.4, 5.6	7.1	—	—	—	4.8	—	—
Threonine	2.0	1.9	4.0	4.0	2.5	3.3	3.0	2.9
Tryptophan	1.0	1.0	1.0	1.0	1.0	1.0	1.0	1.0
Valine	3.2	4.1	3.6	4.8	3.5	3.7	4.0	4.3

[a] From Food and Nutrition Board (1959).
[b] Compiled from reports of Nakagawa (1960, 1961a, b, 1962).
[c] From Holt *et al.* (1960).
[d] From Almquist (1959).
[e] Rama Rao *et al.* (1959, 1960, 1961).
[f] Calculated from F.A.O. (1957).

depending upon the inspector. The conclusions of the F.A.O. Committee and the N.R.C. Committee on Amino Acids was that there was agreement, so that an "ideal pattern" could logically be proposed. On the other hand, the proportion of leucine required varies from approximately 3 to 7, of valine from 3 to 5, of lysine from 3 to 6, etc., in different studies. A variation of 100 % could be viewed as frank disagreement, and, as has already been emphasized, there is no logical means of deciding which figures are more meaningful.

Some of the difficulties are fairly obvious. The calculations of these ratios place much emphasis upon the value of the tryptophan requirement. There is no reason to believe that this value is more stable or better known than others. It is likely that it is less definite than certain other values in some of the studies. In this connection it is of interest that Bender (1961) attempted a direct determination of the relative adequacy of each of the amino acids at the level they occur in egg protein. An amino acid mixture was made based upon the composition of egg protein. Each essential amino acid was in turn decreased to half this level and the diet fed with the expectation that if the original level was just optimal, half the level would produce half as much growth as the complete mixture. By this test, he concluded that lysine, the sulfur amino acids, leucine, and isoleucine were indeed supplied at just the amounts required, but that all other essential amino acids were supplied in excess of the needs. It is of particular interest that Bender found that tryptophan is present in egg protein at 160 % of the amount required. If this is a more valid estimate of tryptophan requirement, the ratio of the requirements of other essential amino acids to tryptophan would be markedly changed from that in egg protein. Along these same lines, it is of interest that Bender (1961) found that the predicted utilization based upon Chemical Score (most limiting amino acid, according to the FAO pattern) failed when most of the essential amino acids are individually limiting to a serious degree. When most of the essential amino acids were supplied at less than 50 % of the presumed requirement, the utilization was better than predicted. This would indicate that there may be mechanisms which conserve specific amino acids in short supply. Only when valine or the sulfur amino acids were limiting was the utilization proportional to the amount supplied. From the practical point of view these findings may not be important since prediction of utilization from the Chemical Score appeared relatively satisfactory when the Chemical Score was above 50 %. Few diets will be found which have a lower score than this. It should be noted that when diets of very low Biological Values are encountered the amino acid requirements are not likely to be met by increased protein intake but an improvement in the Biological Value by protein supplementation is indicated (Amino Acid Committee, see Food and Nutrition Board, 1959).

Scrimshaw, Bressani, and co-workers (Scrimshaw *et al.*, 1958; Bressani *et al.*, 1958, 1960) fed cereal diets to young children in which glutamic acid and glycine were added to provide a total intake of nitrogen equivalent to either 3 gm protein/kg/day in one study, or 2 gm in another. The corn diet was estimated to have a Biological Value of only 13 %, as judged by the F.A.O. recommendations. Stepwise addition of the limiting amino acids were made and the nitrogen balance determined. In the corn diet tryptophan was judged the most limiting and, as expected, addition of tryptophan improved nitrogen balance. Further improvement was also obtained by the addition of lysine, the next most limiting amino acid. Valine and the sulfur amino acids were apparently the next most limiting followed by isoleucine and threonine. The addition of methionine markedly interfered with nitrogen retention. The further addition of valine had little effect, but addition of isoleucine improved the nitrogen balance. When the methionine addition was decreased to half the amount, nitrogen balance was better than when the total amount recommended was provided. Generally similar data were obtained upon the wheat diet, the most limiting amino acids improving balance but the addition of methionine had an adverse effect.

Clearly these studies raise questions both as to the appropriateness of the mixture of amino acids recommended by the F.A.O. Committee and the prediction of Biological Value from the amino acid composition. One might conclude that the amount of methionine recommended is much too high, but this would imply that egg protein and other such high quality proteins contain excess methionine. No other evidence is available to support this conclusion. It is more likely that excess methionine has adverse effects when certain other amino acids are limiting. Since the preceding chapter is devoted to amino acid imbalance, this cannot be discussed here. However, these studies demonstrate that evaluation of the nutritive value on the basis of the single most limiting amino acid can be a seriously inadequate method, and that amino acid imbalance may be of practical consequence.

Comparisons of the patterns of amino acids shown in Table II for adults and children would suggest that so-called "ideal" patterns might be different for different ages. Some support for this can be found in the rat studies. Williams *et al.* (1954), Price *et al.* (1953), and Grau (1947) demonstrated a reasonable constancy in the amino acid composition of the carcasses of animals of different ages. Following the suggestion of Mitchell (1950) one might therefore assume that some estimate of the requirement for the different amino acids could be made from estimates of the rate of deposition of the amino acids in the carcass. Hartsook and Mitchell (1956) derived estimates of the methionine requirements of rats and concluded that the methionine and cystine requirements expressed as per cent of the diet decreased exponentially from 0.98 to 0.26 % with age. The requirement of the two together, expressed as per cent of the dietary protein, increased from 7 to 8 %. Forbes and Rao (1959) in similar studies concluded that the lysine and tryptophan requirements decreased exponentially from 6.0 to 0.77 and from 0.6 to 0.3, respectively, when expressed as a percentage of the dietary protein requirement. The lysine to tryptophan ratios would thus be 10 for the young rats and 2.3 in old rats. Mitchell's (1950) basic suggestion was that if the requirements of one essential amino acid were known at various ages, the rest could be calculated, since the ratios in the carcass are essentially constant. This appears reasonable, but we do not actually know that the efficiency of the utilization of each amino acid is the same at different ages.

Several attempts have been made to compare the efficiencies of the pattern of amino acids proposed by FAO and those occurring in egg or other proteins. Howe *et al.* (1960) found that the FAO mixture was nutritionally superior to one patterned after casein, as might be expected, but also found that decreasing the tryptophan content of the FAO pattern by 50 % did not impair its nutritive value. These studies were done with weanling rats. Swendseid *et al.* (1961, 1962) found an egg pattern to be slightly superior to the FAO pattern in nitrogen balance studies in young men. Egg protein itself may have been slightly superior to either mixture of amino acids. Generally similar results were reported by Kirk *et al.* (1962) in comparisons of a pattern based upon milk with the FAO pattern. Swendseid *et al.* (1961) also reported that lowering the tryptophan content of the FAO mixture by 20 % did not decrease the efficiency of the mixture measured by nitrogen balance in young men. In studies on 7 premature infants by Snyderman *et al.* (1960) a mixture patterned after that of human milk was found to give similar results to that obtained with the FAO pattern.

The interpretation of these studies is most difficult. It must be recognized that the studies done with man are obviously the most relevant to man, but usually have a very poor and undetermined degree of accuracy. Thus the method may be insufficient to measure the differences expected. Studies with animals may or may not be more accurate, although they can be made so, but suffer from the fact that the results may not be applicable to human nutrition.

VII. Summary

The basic position that the minimum protein needs are closely correlated with basal metabolic rate appears to have been generally accepted. This proposition is, however, based largely upon interspecies comparisons which may obscure intraspecies comparisons. The extent of individual differences in requirements and whether these are related to metabolic rate has had little study. Not only are there little data available on this point, but it appears unlikely that those currently available on man are sufficiently accurate to permit a decision.

A second major question is the efficiency of utilization of dietary proteins at levels of intake which are adequate. No entirely valid approach to determine this in adult man appears to be available. The evidence from animals is controversial. In young children the limited data indicate that maximum or near maximum efficiency of utilization can be achieved in rapidly growing infants. No large allowance can be logically justified on this basis but much more information is required. The estimates derived by Hegsted (1957) would thus appear to be as well justified as can be made at this time with perhaps two modifications. The fecal nitrogen excretion was estimated to be approximately 10 % of the intake. This appears to be too low and did not consider the so-called "metabolic nitrogen." A more reasonable estimate, particularly when the protein intake is low, would be of the order of 15 to 20 %. Also no allowance was made for losses through the skin and the growth of the integument. Some allowance is undoubtedly justified, but it is probable that the estimate made by Mitchell is on the high side.

It seems quite clear that the data so far available upon the amino acid requirements are characterized by a considerable, but unknown, degree of inaccuracy. A number of the possible reasons and unanswered questions have been discussed. It appears to this author that too much reliance has been placed upon the values available in arriving at practical dietary standards.

References

Albanese, A. A. (1947). *Advan. Protein Chem.* **3**, 227.

Almquist, H. J. (1959). *In* "Protein and Amino Acid Nutrition" (A. A. Albanese, ed.), p. 349. Academic Press, New York.

Barnes, R. H., Bates, M. J., and Maack, J. E. (1946). *J. Nutr.* **32**, 535.

Bender, A. E. (1961). *In* "Meeting Protein Needs of Infants and Children," Publ. No. 843, p. 407. Natl. Acad. Sci.—Natl. Res. Council, Washington, D. C.

Block, R. J., and Mitchell, H. H. (1946–1947). *Nutr. Abstr. Rev.* **16**, 249.

Block, R. J., with the cooperation of Weiss, K. W. (1956). "Amino Acid Handbook: Methods and Results of Protein Analysis." Thomas, Springfield, Illinois.

Bressani, R., and Mertz, E. T. (1958). *J. Nutr.* **65**, 481.

Bressani, R., Scrimshaw, N., Behar, M., and Viteri, F. (1958). *J. Nutr.* **66**, 501.

Bressani, R., Wilson, D. L., Behar, M., and Scrimshaw, N. (1960). *J. Nutr.* **70**, 176.

Bricker, M., Mitchell, H. H., and Kinsman, G. M. (1945). *J. Nutr.* **30**, 269.

Brock, J. F., and Autret, M. (1952). World Health Organization Monograph Series No. 8, Geneva.

Brody, S. (1945). "Bioenergetics and Growth." Reinhold, New York.
Calhoun, W. K., Hepburn, F. N., and Bradley, W. B. (1962). *J. Nutr.* **70,** 337.
Carpenter, K. J., Morgan, C. B., Lea, C. H., and Parr, L. J. (1962). *Brit. J. Nutr.* **16,** 451.
Clark, H. E., Yang, S. P., Walton, W., and Mertz, E. T. (1960). *J. Nutr.* **71,** 229.
Duncan, D. L. (1958). *Nutr. Abstr. Rev.* **28,** 695.
F.A.O. (1957). "Protein Requirements," Nutr. Studies No. 16, Food and Agricultural Organization of the United Nations.
Fomon, S. J. (1961). *Pediatrics* **28,** 347.
Fomon, S. J., and May, C. D. (1958). *Pediatrics* **22,** 101.
Food and Nutrition Board (1945). "Recommended Dietary Allowances." Reprint and Circular Series No. 122. Natl. Acad. Sci.—Natl. Res. Council, Washington, D. C.
Food and Nutrition Board (1959). "Evaluation of Protein Nutrition," Bull. No. 711. Natl. Acad. Sci.—Natl. Res. Council, Washington, D. C.
Forbes, R. M., and Rao, T. (1959). *Arch. Biochem. Biophys.* **82,** 348.
Forbes, R. M., Vaughan, L., and Norton, H. W. (1955). *J. Nutr.* **57,** 593.
Forbes, R. M., Vaughan, L., and Yohe, M. (1958). *J. Nutr.* **64,** 291.
Grau, C. R. (1947). *J. Biol. Chem.* **170,** 661.
Grau, C. R. (1948). *J. Nutr.* **36,** 99.
Grau, C. R., and Carroll, R. W. (1958). *In* "Processed Plant Protein Foodstuffs" (A. M. Altschul, ed.), p. 153. Academic Press, New York.
Hartsook, E. W., and Mitchell, H. H. (1956). *J. Nutr.* **60,** 173.
Hegsted, D. M. (1948). *J. Nutr.* **35,** 399.
Hegsted, D. M. (1957). *J. Am. Dietet. Assoc.* **33,** 225.
Hegsted, D. M. (1959). *Federation Proc.* **18,** 1130.
Hegsted, D. M. (1963). *Federation Proc.* **22,** 1424.
Hegsted, D. M., Tsongas, A. G., Abbott, D. B., and Stare, F. J. (1946). *J. Lab. Clin. Med.* **31,** 261.
Heller, B. S., Chutkow, M. R., Lushbough, C. H., Siedler, A. J., and Schweigert, B. S. (1961). *J. Nutr.* **73,** 113.
Hillman, R. W. (1960). *In* "Modern Nutrition in Health and Disease," 2nd ed. (M. G. Wohl and R. S. Goodhart, eds.), p. 957. Lea & Febiger, Philadelphia, Pennsylvania.
Holmes, J. O. (1946). *In* "Nutrition and Chemical Growth in Childhood" (I. G. Macy, ed.), Vol. II, p. 1304. Thomas, Springfield, Illinois.
Holt, L. E., Jr. (1959). *J. Pediat.* **54,** 496.
Holt, L. E., Jr., György, P., Pratt, E. L., Snyderman, S. E. and Wallace, W. M. (1960). "Protein and Amino Acid Requirements in Early Life." N. Y. U. Press, New York.
Howe, E. E., Gilfillan, E. W., and Allison, J. B. (1960). *J. Nutr.* **70,** 385.
Johnston, J. A. (1953). "Nutritional Studies in Adolescent Girls and Their Relation to Tuberculosis." Thomas, Springfield, Illinois.
Jones, E. M., Baumann, C. A., and Reynolds, M. S. (1956). *J. Nutr.* **60,** 549.
Kirk, M. C., Metheny, N., and Reynolds, M. S. (1962). *J. Nutr.* **77,** 448.
Kuiken, K. A., and Lyman, C. M. (1948). *J. Nutr.* **36,** 359.
Lansing, A. I., ed. (1952). "Cowdry's Problems of Ageing, Biological and Medical Aspects." Williams & Wilkins, Baltimore, Maryland.
Lagervall, M. (1958). *Ann. Roy. Agr. Coll. Sweden* **24,** 1.
Lardy, H. A., and Feldott, G. (1950). *J. Biol. Chem.* **186,** 85.
Leverton, R. M., Gram, M. R., Chloupka, M., Brodovsky, E., and Mitchell, A. (1956a). *J. Nutr.* **58,** 59.

Leverton, R. M., Johnson, N., Pazur, J., and Ellison, J. (1956b). *J. Nutr.* **58**, 219.
Leverton, R. M., Gram, M. R., Brodovsky, E., Chaloupka, M., Mitchell, A., and Johnson, N. (1956c). *J. Nutr.* **58**, 83.
Leverton, R. M., Johnson, N., Ellison, J., Geschwender, D., and Schmidt, F. (1956d). *J. Nutr.* **58**, 341.
Leverton, R. M., Ellison, J., Johnson, M., Pazur, J., Schmidt, F., and Geschwender, D. (1956e). *J. Nutr.* **58**, 355.
Liener, I. E. (1958). *In* "Processed Plant Protein Foodstuffs" (A. M. Altschul, ed.), p. 79. Academic Press, New York.
Linkswiler, H., Geschwender, D., Ellison, J., and Fox, H. M. (1958). *J. Nutr.* **65**, 441.
Linkswiler, H., Fox, H. M., and Fry, P. C. (1960a). *J. Nutr.* **72**, 397.
Linkswiler, H., Fox, H. M., and Fry, P. C. (1960b). *J. Nutr.* **72**, 389.
McCance, R. A., and Walsham, C. M. (1948). *Brit. J. Nutr.* **2**, 26.
McCance, R. A., and Widdowson, E. M. (1947). *J. Hyg.* **45**, 49.
McCay, C. M., Crowell, M. F., and Maynard, L. A. (1935). *J. Nutr.* **10**, 63.
McCay, C. M., Maynard, L. A., Sperling, G., and Barnes, L. L. (1939). *J. Nutr.* **18**, 1.
Macy, I. G., ed. (1942). "Nutrition and Chemical Growth in Childhood," Vol. I. Thomas, Springfield, Illinois.
Metta, V. C., Firth, J. A., and Johnson, B. C. (1960). *J. Nutr.* **71**, 332.
Miller, D. S., and Payne, P. R. (1961a). *J. Nutr.* **74**, 413.
Miller, D. S., and Payne, P. R. (1961b). *J. Nutr.* **75**, 225.
Mitchell, H. H. (1942). *J. Animal Sci.* **1**, 159.
Mitchell, H. H. (1949). *Arch. Biochem.* **21**, 335.
Mitchell, H. H. (1950). *In* "Protein and Amino Acid Requirements of Mammals" (A. A. Albanese, ed.), p. 17. Academic Press, New York.
Mitchell, H. H. (1954). *Wiss. Abhandl. Deut. Akad. Landwirtschaftswiss.* **V2**, 279.
Mitchell, H. H., and Edman, M. (1962). *Am. J. Clin. Nutr.* **10**, 163.
Mitchell, H. H. Hamilton, T. S., and Haines, W. T. (1949). *J. Biol. Chem.* **178**, 345.
Nakagawa, I., Takahashi, T., and Suzuki, T. (1960). *J. Nutr.* **71**, 176.
Nakagawa, I., Takahashi, T., and Suzuki, T. (1961a). *J. Nutr.* **73**, 186.
Nakagawa, I., Takahashi, T., and Suzuki, T., (1961b). *J. Nutr.* **74**, 401.
Nakagawa, I., Takahashi, T., Suzuki, T., and Kobayashi, K. (1962). *J. Nutr.* **77**, 61.
Nasset, E. S. (1962). *J. Nutr.* **76**, 131.
Njaa, L. R. (1962). *Brit. J. Nutr.* **16**, 185.
Osborne, T. B., Mendel, L. B., and Ferry, E. L. (1919). *J. Biol. Chem.* **37**, 223.
Pratt, E. L., Snyderman, S. E., Cheung, M. W., Norton, P., and Holt, L. E., Jr. (1955). *J. Nutr.* **56**, 231.
Price, W. A., Jr., Taylor, M. W., and Russel, W. C. (1953). *J. Nutr.* **51**, 413.
Rama Rao, P. B., Metta, V. C., and Johnson, B. C. (1959). *J. Nutr.* **69**, 387.
Rama Rao, P. B., Metta, V. C., Norton, H. W., and Johnson, B. C. (1960). *J. Nutr.* **71**, 361.
Rama Rao, P. B., Norton, H. W., and Johnson, B. C. (1961). *J. Nutr.* **73**, 38.
Rechcigl, M., Jr., Loosli, J. K., and Williams, H. H. (1957). *J. Nutr.* **63**, 177.
Rose, W. C. (1957). *Nutr. Abstr. Rev.* **27**, 631.
Rose, W. C., and Wixom, R. L. (1955a). *J. Biol. Chem.* **216**, 763.
Rose, W. C., and Wixom, R. L. (1955b). *J. Biol. Chem.* **217**, 95.
Rose, W. C., and Wixom, R. L. (1955c). *J. Biol. Chem.* **217**, 997.
Rose, W. C., Smith, L. C., Womack, M., and Shane, M. (1949). *J. Biol. Chem.* **181**, 307.
Rose, W. C., Coon, M. J., and Lambert, G. F. (1954a). *J. Biol. Chem.* **210**, 331.

Rose, W. C., Lambert, G. F., and Coon, M. J. (1954b). *J. Biol. Chem.* **211,** 815.
Rose, W. C., Leach, B. E., Coon, M. J., and Lambert, G. F. (1955a). *J. Biol. Chem.* **213,** 913.
Rose, W. C., Borman, A., Coon, M. J., and Lambert, G. F. (1955b). *J. Biol. Chem.* **214,** 579.
Rose, W. C., Coon, M. J., Lockhart, H. B., and Lambert, G. F. (1955c). *J. Biol. Chem.* **215,** 101.
Rose, W. C., Eades, C. H., Jr., and Coon, M. J. (1955d). *J. Biol. Chem.* **216,** 225.
Rose, W. C., Wixom, R. L., Lockhart, H. B., and Lambert, G. F. (1955e). *J. Biol. Chem.* **217,** 987.
Rosenberg, H. R., Rohdenburg, E. L., and Eckert, R. E. (1960a). *J. Nutr.* **72,** 415.
Rosenberg, H. R., Rohdenburg, E. L., and Eckert, R. E. (1960b). *J. Nutr.* **72,** 423.
Ross, M. H. (1959). *Federation Proc.* **19,** 1190.
Ross, M. H. (1961). *J. Nutr.* **75,** 197.
Scheffé, H. A. (1960). "The Analysis of Variance." Wiley, New York.
Schwartz, H. G., Taylor, M. W., and Fisher, H. (1958). *J. Nutr.* **65,** 25.
Scrimshaw, N. S., Bressani, R., Behar, M., and Viteri, F. (1958). *J. Nutr.* **66,** 485.
Sherman, H. C., Gillett, L. H., and Osterberg, E. (1920). *J. Biol. Chem.* **41,** 97.
Smuts, D. B. (1935). *J. Nutr.* **9,** 403.
Snyderman, S. E., Pratt, E. L., Cheung, M. W., Norton, P., and Holt, L. E., Jr. (1955). *J. Nutr.* **56,** 253.
Snyderman, S. E., Holt, L. E., Jr., Smellie, F., Boyer, A., and Westall, R. G. (1959a). *A.M.A. J. Diseases Children* **97,** 186.
Snyderman, S. E., Boyer, A. and Holt, L. E., Jr. (1959b). *A.M.A. J. Diseases Children* **97,** 192.
Snyderman, S. E., Norton, P., Fowler, D. I., and Holt, L. E., Jr. (1959c). *A.M.A. J. Diseases Children* **97,** 175.
Snyderman, S. E., Prose, P. H., and Holt, L. E., Jr. (1959d). *A.M.A. J. Diseases Children* **98,** 459.
Snyderman, S. E., Holt, L. E., Jr., and Boyer, A. (1960). *J. Nutr.* **72,** 404.
Snyderman, S. E., Holt, L. E., Jr., Dancis, J., Roitman, E., Boyer, A., and Balis, M. E. (1962). *J. Nutr.* **78,** 57.
Sorg-Matter, H. (1928). *Arch. Intern. Physiol.* **20,** 126.
Stuart, H. C. (no date). Percentile Chart for Measurement Boys-Girls, Anthropometric Chart. Children's Medical Center, Boston, Massachusetts.
Stuart, H. C., and Meredith, H. V. (1946). *Am. J. Public Health* **36,** 1365.
Swendseid, M. E., and Dunn, M. S. (1956). *J. Nutr.* **58,** 507.
Swendseid, M. E., Williams, I., and Dunn, M. S. (1956). *J. Nutr.* **58,** 495.
Swendseid, M. E., Feeley, R. J., Harris, C. L., and Tuttle, S. G. (1959). *J. Nutr.* **68,** 203.
Swendseid, M. E., Harris, C. L., and Tuttle, S. G. (1960). *J. Nutr.* **71,** 105.
Swendseid, M. E., Watts, J. H., Harris, C. L., and Tuttle, S. G. (1961). *J. Nutr.* **75,** 295.
Swendseid, M. E., Harris, C. L., and Tuttle, S. G. (1962). *J. Nutr.* **77,** 391.
Terroine, E. F. (1936). *Quart. Bull. Health Organ. League Nations* **5,** 427.
Terroine, E. F., and Sorg-Matter, H. (1927). *Arch. Intern. Physiol.* **29,** 121.
Tompkins, W. T. (1955). *In* "Modern Nutrition in Health and Disease" (M. G. Wohl and R. S. Goodhart, eds.), p. 885. Lea & Febiger, Philadelphia, Pennsylvania.
Walker, A. R. F. (1958). *Ann. N. Y. Acad. Sci.* **69,** 989.
Wallace, W. M., Weil, W. B., and Taylor, A. (1958). *Ciba Found. Colloq. Aging* **4,** 116.
Wallace, W. M. (1959). *Federation Proc.* **18,** 1125.

Watts, J. H., Booker, L. K., McAfee, J. W., Williams, E. G., Wright, W. G. and Jones, F. (1959a) *J. Nutr.* **67,** 483.
Watts, J. H., Booker, L. K., McAfee, J. W., Graham, D. C. W., and Jones, F. (1959b). *J. Nutr.* **67,** 497.
Williams, H. H., Curtin, L. V., Abraham, J., Loosli, J. K., and Maynard, L. A. (1954). *J. Biol. Chem.* **208,** 277.

CHAPTER 15

Protein Metabolism and Requirements in Pregnancy and Lactation

K. L. BLAXTER

The Hannah Dairy Research Institute, Ayr, Scotland

I. Introduction

The great number and diversity of mammalian species mean that any short account of mammalian protein metabolism during pregnancy and lactation must be selective. An attempt has been made in this chapter to draw general conclusions about protein needs during pregnancy and lactation, and in doing this much detailed information has, of necessity, been neglected. Furthermore, since a definitive account of lactation has recently been published in the book by Kon and Cowie (1961), slightly more emphasis has been given to protein metabolism during pregnancy than to its metabolism in lactation. The approach adopted throughout is largely comparative; at the outset the limitations of this approach or rather the limitations involved in arguments from a particular species to the multitude of mammalian kind are examined.

II. Rates of Protein Synthesis in Pregnancy and Lactation

During pregnancy and more especially during lactation, the net rate of protein synthesis by the adult mammal is greatly increased. In several species it is possible to make a quantitative assessment of the net rates of protein synthesis in pregnancy and lactation and compare them with the rates of synthesis which occur in the early stages of postnatal growth. This has been done in Table I where daily rates of net synthesis have been expressed in terms of the amounts of N retained in growth, or in pregnancy, or secreted in milk during lactation. For comparative purposes these rates have been expressed per kilogram of weight, although other reference bases, such as surface area or metabolic body size (body weight raised to the power 0.73), could have been used. The table shows, first, that in each species considered (as might be expected) the rate of protein synthesis per kilogram of weight is smaller in the pregnant animal than it is in her rapidly growing young. Considering the animal as a whole, pregnancy results in a rate of synthesis of new protein which is considerably less than that attained in the early stages of postnatal growth. Second, with the apparent exception of man, Table I shows that lactation implies a rate of synthesis of new protein which is three-to tenfold that noted in pregnancy. The rate of synthesis per kilogram of body weight in all species is, however, less in lactation than in early life. The apparent exception of man is largely due to the inclusion of a large term for augmentation of the N content of the tissues of the mother during pregnancy. If the synthesis of pregnancy in women is taken to be the protein deposited in the uterus plus that contributing to the growth of the breast, then the net rate of synthesis is reduced to a third, and the apparent exception disappears.

The comparison of rates of protein secretion in milk with other rates of protein metabolism as shown in Table I ignores the fact that the synthesis of the proteins of milk takes place in the mammary gland. In rats, the weight of the mammary gland is about 7 % of body weight (see review by Linzell, 1959) and in the cow about 5 % of body weight. The rate of synthesis of protein N/kg of mammary tissue in all the species listed in Table I greatly exceeds the rate of synthesis achieved per kilogram of soft tissue in juvenile growth. It seems probable that the rate of protein

TABLE I

THE RATES OF PROTEIN SYNTHESIS (DEPOSITION) DURING EARLY POSTNATAL GROWTH, THE LATE STAGES OF PREGNANCY, AND DURING FULL LACTATION

Period	Rat[a]	Sheep[b]	Man[c]	Pig[d]	Ox[e]
Early postnatal growth					
Gm N retained/day	0.040	11	1.02	5	25
Mean body weight (kg)	0.013	7	4.0	2.2	40
Mg N retained/day/kg	3077	1571	255	2272	625
Pregnancy					
Gm N retained/day	0.086	—	3.3	8.6	32
Mean body weight (kg)	0.20	60	60	200	400
Mg N retained/day/kg	430	—	55	34	80
Lactation					
Gm N secreted/day	0.57	21	2.27	84	100
Mean weight (kg)	0.20	60	60	200	400
Mg N secreted/day/kg	2850	350	38	420	250

[a] Rat. *Postnatal growth:* The increase in the N content of a single newborn rat from birth to 15 days was calculated from the data of Spray (1950) and Widdowson and Spray (1951) to be 595 mg or 40 mg/day. The mean weight of the baby rat during the 15-day period was taken to be 13 gm. *Pregnancy:* The gain of N during pregnancy as determined by Boyne *et al.* (1953) in their slaughter experiments (1.8 gm) was used. This was divided by 21, the number of days of experiment. *Lactation:* In the absence of direct measurement, the daily gain by an individual baby rat as determined by Spray (1950) has been multiplied by 100/70, assuming the retention of N to be 70% and again by 10 which is the modal number of young born to the rat (Blaxter, 1961).

[b] Sheep. *Postnatal growth:* From the equation weight gain (lb in 28 days) = 0.18 milk intake (lb in 28 days) + 2.5 (Wallace, 1948), a mean intake of 2.25 kg milk/day and a protein content of the gain in weight of 16.2%. The mean weight has been taken to be 7 kg. *Lactation:* The milk yields of 2.25 kg/day noted by Wallace (1948) have been taken. The N content of milk has been taken to be 0.91 gm/100 ml, corresponding to 6% protein (N × 6.38), since this is a more representative value than the 5.4% protein given by Ling *et al.* (1961) for sheep milk.

[c] Man. *Postnatal growth:* Mean % retention of N of breast fed infants taken to be 45% (Clements, 1949). For a mean intake of 1000 ml mature milk containing 0.227 gm N/100 ml (Macy and Kelly, 1961) daily retention is 1.02 gm. *Pregnancy:* The mean retention of N in the tenth lunar month has been taken from the compilation of metabolic results made by Hunscher *et al.* (1933). This includes gain of N by extra-uterine tissues; the actual reproductive gain is only 1 gm/day. *Lactation:* See under postnatal growth.

[d] Pig. *Postnatal growth:* The N retentions noted by Ludvigsen and Thorbek (1959, 1960) in artificially reared pigs have been used. *Pregnancy:* The amounts of N which Mitchell *et al.* (1931) found to be retained by pigs in the last week of pregnancy have been used. *Lactation:* The highest value for milk found by Lodge (1959) for the daily milk yield of sows given high levels of food (10 kg) has been used together with an N content of 0.84 gm/100 ml.

[e] Ox. *Postnatal growth:* The maximal daily retention noted by Blaxter and Wood (1952) has been used. Higher values were, however, noted by Soxhlet (1878) but the calves were much larger. *Pregnancy:* The N retentions noted in the terminal stages of pregnancy by Jakobsen (1957) have been used. *Lactation:* A yield of 4 gal (18.2 kg) has been adopted; it is not a maximal figure. The N content is taken to be 0.55 gm/100 ml (Macy and Kelly, 1961).

synthesis in the alveolar mammary cell exceeds that of any other cell type in the body with the exception of the regenerating liver cell. Richardson and Folley (1954) in this regard have computed from histological work that the average alveolar cell of the goat secretes 1.5×10^{-9} μg casein/second. The number of molecules of casein synthesized per cell per second computed on the basis of an average molecular weight for casein is of the order of 10^{-19} moles/second. This is a remarkable rate of synthesis.

III. The Comparative Biology of Pregnancy and Lactation

Table I also shows that net rates of protein synthesis vary appreciably from species to species and, in particular, that on average the rat synthesizes about ten times as much protein per unit time per unit body weight as does man, irrespective of whether it is growing, pregnant, or lactating. Comparative aspects of protein metabolism as they concern endogenous N excretion, synthesis of protein in the liver, and N retention in the body as a whole have been commented on by Munro in Chapter 10. When comparative aspects of protein metabolism in pregnancy and lactation are considered, it is evident at once that the number of animal species available for such a study is sadly limited. Quantitative data are available for few animals in their feral stage and with lactation, studies are mostly limited to those species which can be milked artificially. Indeed, with the addition of the goat, the mouse, and the rabbit, the species given in Table I are the only ones in which pregnancy and lactational metabolism have been studied in any quantitative detail.

This state of affairs necessarily raises problems regarding the general applicability of information obtained with a single species. How far can a concordance of quantitative and qualitative experimental work obtained with a few species be expressed in more general terms as applicable to "pregnancy" or "lactation"? A study of the protein metabolism of the pregnant rat must provide some information applicable to species other than rats, but such a general interpretation must take into account any peculiarities of the rat with respect to its reproductive behavior. It is here that certain broad generalizations about pregnancy and lactation have value, for they enable the reproductive performance of a single species to be set in the perspective of all species. An analogous approach was employed by Benedict (1938) when he used the so-called surface area law linking the metabolic rates of different species to their body size as the point of departure in describing the metabolism of individual species.

A. Comparative Aspects of Pregnancy

A generalization about pregnancy, probably first given quantitative expression by D'Arcy Thompson (1952), is that the larger a species the

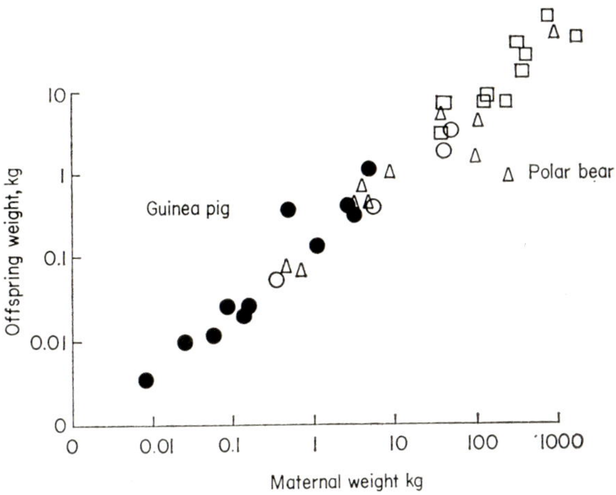

FIG. 1. The relation between birth weight of offspring and weight of mother in those mammals for which records are available of both weight and gestation period. The guinea pig is clearly an anomalous creature and so is the polar bear. Key: ○, primates; ●, rodents; △, carnivores; □, ungulates.

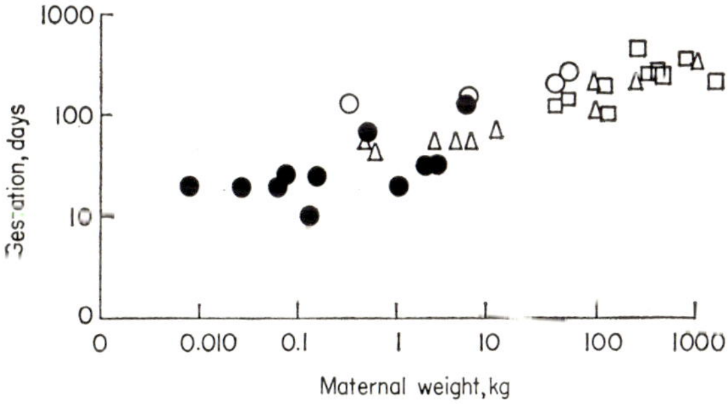

FIG. 2. The relation between gestation period and body weight of the mother in the same species given in Fig. 1. The functions of body weight with which pregnancy varies are for primates $wt^{0.128}$, for rodents $wt^{0.199}$, for carnivores $wt^{0.171}$, and for ungulates $wt^{0.203}$. The general equation in the text for all species is $wt^{0.253}$. Key: ○, primates; ●, rodents; △, carnivores; □, ungulates.

smaller, relatively speaking, are the offspring at birth. Another generalization is that the duration of pregnancy does not increase proportionally with the size of a species. The first generalization was tested by Leitch *et al.* (1959) by an appeal to recorded weights of mammals and their newborn young, and these data, together with Kenneth's (1943) compilation of records of the gestation periods of mammalian species, enable a test to be made of the second generalization. The results are given in Figs. 1 and 2.

Broadly speaking, the weight of the newborn young increases not directly with the weight of a species but with a fractional power of weight according to the equation

$$F = 0.54\ W^{0.832} \tag{1}$$

where F and W are weights of offspring (F) and mother (W) in grams. Similarly, the duration of pregnancy in days (D) increases with increasing species size according to the equation.

$$D = 12\ W^{0.252} \tag{2}$$

The two equations imply than on average an animal weighing 100 gm can be expected to produce an offspring weighing 25 % of her own weight in 38 days and an animal weighing 100 kg to produce offspring weighing 7.8 % of her own weight in 218 days. The crude rate of synthesis of offspring during pregnancy R can be defined as the weight of offspring at birth divided by the time taken to produce it, and from the above equation is

$$F/D = R = 0.046\ W^{0.58} \tag{3}$$

This equation may be compared with the interspecies generalization that the basal rate of heat production in kcal/24 hr (Q) is

$$Q = 70.5\ W^{0.73} \tag{4}$$

and indicates that with increasing body size the crude rate of the synthetic processes involved in pregnancy decline faster than does maternal metabolism. In general, the metabolic strain imposed by pregnancy as measured by the ratio of rate of production of the offspring relative to the basal metabolic rate of a species is less in the larger species.

With this general information about pregnancy in the mammalia as a class, it is possible to examine pregnancy in particular species. This has been done in Table II where the deviations of each of the common species from the interspecies generalization have been given. The table shows, and this was indeed evident in Fig. 2, that man, a primate, has a relatively long gestation for its size, the ungulates average ones, and rodents usually short ones. Ungulates have heavy offspring relative to their size, rodents average ones, and man a very small one. The crude rate of synthesis of offspring expressed relative to the interspecies values can be obtained by dividing the third column by the sixth in Table II and shows than man is unique in having a crude rate less than half that of the interspecies value, while the rabbit has one which is threefold. In Fig. 1, two species have been indicated by name, the laboratory guinea pig and the polar bear. These represent extremes of reproductive performance as judged by the birth weight of their offspring relative to the interspecies mean. There is some

TABLE II

SOME QUANTITATIVE ASPECTS OF PREGNANCY IN THOSE MAMMALIAN SPECIES FOR WHICH MOST EXPERIMENTAL OBSERVATIONS ARE AVAILABLE

Species	Weight of offspring at term			Length of pregnancy		
	Predicted from an inter-species generalization[a] (kg)	Observed (kg)	Observed/predicted	Predicted from an inter-species generalization[b] (days)	Observed (days)	Observed/predicted
Man	5.1	3.2	0.6	192	273	1.4
Ox	24.7	34.0	1.4	309	279	0.9
Pig	13.9	9.5	0.7	260	115	0.4
Rat	0.044	0.044	1.0	46	21	0.4
Sheep	5.1	7.3	1.4	192	150	0.8
Mouse	0.008	0.006	0.8	27	20	0.7
Rabbit	0.44	0.40	0.9	92	31	0.3

[a] See Eq. (1).
[b] See Eq. (2).

doubt, however, whether these extremes are fully comparable, for in the polar bear, as in certain other carnivores, implantation of the fertilized ovum is delayed, and the duration of pregnancy is probably not measured best by the interval from effective coitus to parturition (Courrier and Baçless, 1955). Even so, it appears that generalizations about such simple measures as the length and outcome of pregnancy based on information obtained with a few species are often difficult to support because of the considerable interspecies variation that occurs. It follows that considerable care must be taken in drawing general conclusions about protein metabolism in pregnancy from comparable limited information.

B. Comparative Aspects of Lactation

Probably the best known of the interspecies generalizations made about lactation is that of Bunge (1898). Bunge stated that the composition of the milk of different species was directly related to the rate of growth of their young, because he found that those species with the most rapidly growing young also produced milk containing the most protein and mineral matter. This hypothesis has been discussed elsewhere (Blaxter, 1961), where it was shown that the comparisons Bunge made were complicated by the fact that the species he considered to grow most rapidly also tended to be small. It was argued that, since the capacity of the stomach of animals of different species probably varies more closely with body weight than does their metabolic demand, which varies with their surface area, it

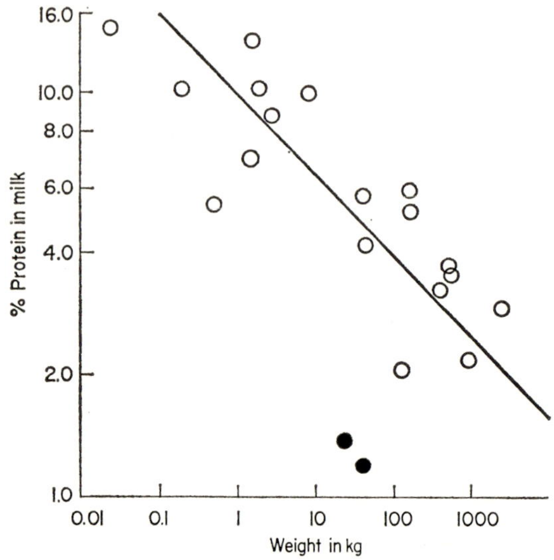

FIG. 3. The relation between the protein content of milk and body weight in 20 mammalian species for which data have been found. Values for primates (man and orangutan) are shown separately. The data were obtained from collections by W. L. Davies (1936), J. G. Davis and McDonald (1953), Brody, (1945), Blaxter (1961), and the following additional sources: Neuhaus (1959): horse; Simon (1959): elephant; Bollinger and Gross (1960): opossum; Jørgensen (1960): mink. Data on body weight of the rhinoceros were not traced; the protein content of its milk is 3.1% (Grzimek, 1957).

would be expected, purely on dimensional grounds (Rashevsky, 1960), that small species would produce more concentrated milks than large ones. Figure 3, in which the protein content of milk is plotted against body size for different nonaquatic mammalian species, illustrates this dimensional argument. If metabolism varies with body weight raised to the power 0.73 and stomach capacity varies directly with weight, it would be expected that the *concentration* of nutrients in milk would vary with $W^{0.73}/W = W^{-0.27}$. The observed exponent of -0.2, which is the slope of the best fitting line in Fig. 3, though subject to error, provides some confirmation of this argument. It further follows from the dimensional argument that the total *amount* of energy or of protein secreted in the milk of different species will vary with the same fractional power of body weight as does energy metabolism. The experiments of Brody and Nisbet (1938), which included quantitative studies with rats, goats, and cows, support this corollary to the dimensional argument and Rüsse's (1961) quantitative studies with dogs when computed to be in comparable terms to those of Brody and Nisbet are in agreement, too.

TABLE III

THE PROTEIN CONTENT OF THE NEWBORN, THE DURATION OF LACTATION RELATIVE TO PREGNANCY, AND THE LENGTH OF PREGNANCY AND WEIGHT OF OFFSPRING RELATIVE TO EXPECTATIONS BASED ON BODY SIZE

Species	Protein in newborn (%)		Relative duration of lactation[a]	Values relative to body size	
	Actual	Fat-free		Length of pregnancy	Weight of offspring
Ox	15.9	16.6	0.2	0.9	1.4
Sheep	15.8	16.2	0.4	0.8	1.4
Man	11.9	14.1	0.6	1.4	0.6
Mouse	12.5	12.8	1.0	0.7	0.8
Pig	11.3	11.4	0.5	0.4	0.7
Rabbit	11.1	11.3	1.0	0.3	0.9
Rat	10.8	10.9	1.3	0.4	1.0

[a] The data are those sources listed by Blaxter (1961) together with some derived from sources listed in the footnotes to Table I. The relative duration of lactation is the ratio days of lactation/days of pregnancy; the relative lengths of pregnancy and and weight of offspring are those derived in Table II.

A further series of generalizations about lactation which have been adopted by many without question is that those species which produce the most mature young either produce less milk or have shorter lactations than closely related species producing immature young. As first suggested by Moulton (1923), the best estimate of the maturity of an animal at birth is probably the protein content of its fat-free body. Table III summarizes for the same species listed in Table II the protein content of the fat-free body at birth and the length of lactation relative to the length of pregnancy. In addition, it gives the length of pregnancy and the weight of offspring relative to expectations based on body size. Table III shows that species which are immature at birth tend to suckle longer, but that immaturity at birth cannot be ascribed simply to a shortness of the gestation period. Even here, exceptions abound, as is evident in the considerable variation both in Table III and Fig. 3. Furthermore, simplifying assumptions do no reflect underlying regulatory mechanisms which are simple. Lactation length and total lactation yield, for example, are markedly affected by the duration of the stimulus of suckling. Continual suckling prolongs lactation (Selye *et al.*, 1934) and it is possible, by continually replacing the litter of a foster rat by new young, to prolong her lactation for a year or more (Bruce, 1958).

As with pregnancy, so with lactation, the great diversity of its expression in different species and the absence of any clear-cut generalization about its quantitative aspects makes it difficult to argue from the results obtained with a single species to the mammalia as a class. This difficulty, however, should not obscure the outstanding features of lactation in the mammalia. All mammals secrete a fluid containing lactose in concentrations which do not vary from species to species by more than ±2 % about a mean of 5 % in fat-free milk. Lactation in all mammals involves secretion of the proteins casein and β-lactoglobulin and a characteristic fat distinct from depot fat. No other cells in the body produce these characteristic products.

IV. Pregnancy

A. Fetal Growth

It is usually assumed by those dealing with the over-all pattern of growth that fetal growth, by analogy with bacterial growth, is some sort of multiplicative or autocatalytic process best represented by an exponential equation. This is probably true of the very early stages when the fertilized zygote forms the morula by successive division. During morphogenesis the same multiplicative model may be true. It seems probable, however, that after this time growth is of a different nature. Huggett has often drawn attention to the lack of deviation from linearity of the relation between crown-rump length and age of fetuses of different species, including whales (Huggett and Frazer, 1959), a finding which is not consistent wih multiplicative or exponential growth. Stephenson (1962) has recently examined the alternative view that fetal growth represents linear additions to existing material along growing surfaces, that is, the growth pattern follows a cubic law. This model would result in the linear relation noted by Huggett; Stephenson in fact found a greater conformity for the cubic equations than for an exponential equation in relating body weight to age in fetal sheep. It is of some interest, too, that analysis of data on gestation in pigs (Mitchell *et al.*, 1931) showed that total ash deposition varied with time raised to the power 3.2, and N deposition with time raised to the power of 2.5. The theoretical value for the cubic relationship is 3.0 and the low exponent for N suggests that a simple surface accretion model for protein growth is very much of an approximation.

Many mathematical analyses of data on the relation between the N content of the fetus and adnexa and fetal age have been made. Some equations used to describe such data have been exponential functions, some power functions, other quadratic and cubic equations. For the reasons given above, no attempt has been made to re-express data on N deposition in pregnancy in a form common to all species. Such a step might well anihilate interesting species differences or alternatively generate spurious ones. Instead the data have been presented in their original form.

Figure 4 shows the N deposition in the uterus of the cow and of the sow

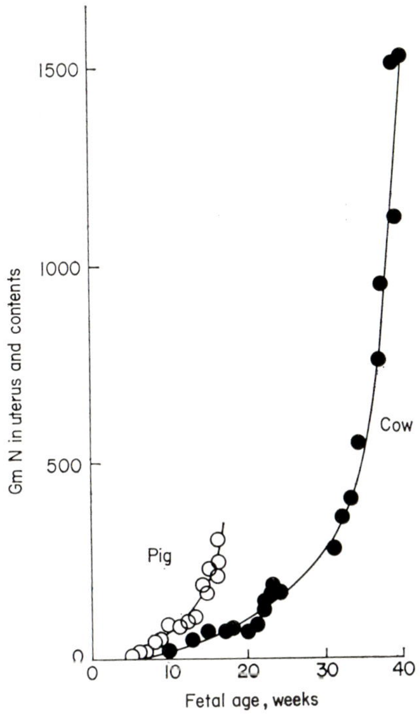

FIG. 4. The total N content of the uterus and its contents in pregnant sows and pregnant cows, plotted against time. The data are recomputed from the observations of Mitchell *et al.* (1931) and of Jakobsen (1957).

plotted on the same scale. This figure shows that, although the cow produces about 1500 gm N in conception products in 273 days and the sow about 270 gm N in 115 days; that is, the crude rate of synthesis by the cow is 5 gm N/day and in the sow 2.3 gm N/day, the uterus of the cow in the first 115 days of her pregnancy contains less N than that of the sow. The deposition of N takes place primarily in the second half of pregnancy in both species. This is brought out in Fig. 5, where the same results given in Fig. 4 have been plotted with the abscissa scaled to length of pregnancy. Comparable data relating to laboratory animals and man could be cited in addition to these for pig and cow. They also reveal that protein deposition in pregnancy occurs mainly during the second half of pregnancy and then at an ever-increasing rate.

B. Nitrogen Deposition in the Placenta and Fetus

When the products of conception are considered separately, it is evident that growth of weight of the placenta occurs rapidly at first and that in many species the weight of the placenta does not change very much in the

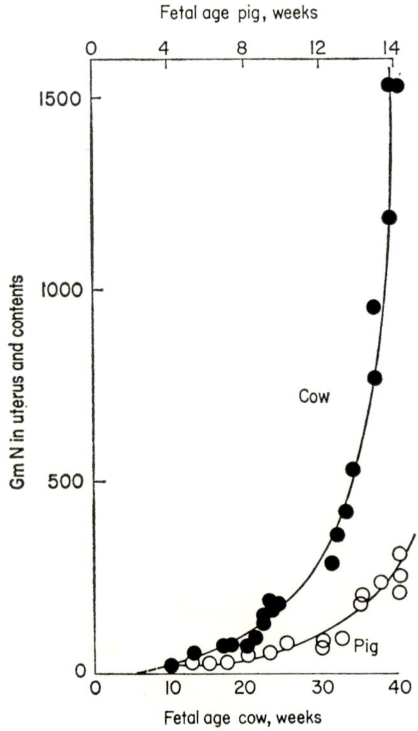

FIG. 5. The same data as in Fig. 4 replotted on a time scale with unit value conception to pregnancy.

last third of pregnancy. In some species, notably the guinea pig (Flexner and Pohl, 1941a,b), the placenta continues to increase in weight throughout pregnancy. In the ewe, goat, and rabbit, the weight of the placenta changes little after the 70th day (ewe, goat) or 21st day (rabbit) (Barcroft, 1948). The N concentration in the placenta tends, however, to increase slightly during pregnancy so that a small increase in N content of the placenta is noted during pregnancy in several species. Figure 6 shows data for the cow and for woman, which indicate a small rise in N content throughout. At term the placenta accounts for as much as 20 % of the total N in the uterus, as in woman (Widdowson and Spray, 1951), or as little as 6 %, as in the cow (Jakobsen, 1957). With most species, the placenta accounts for 10–15 % of the total products of conception at term.

The growth of the fetus and the deposition of protein in the fetus largely account for the massive increase in the N content of uterine contents during the last half of pregnancy. The weight of the fetus increases at a less rapid rate than does its protein content, for as the fetus grows the concen-

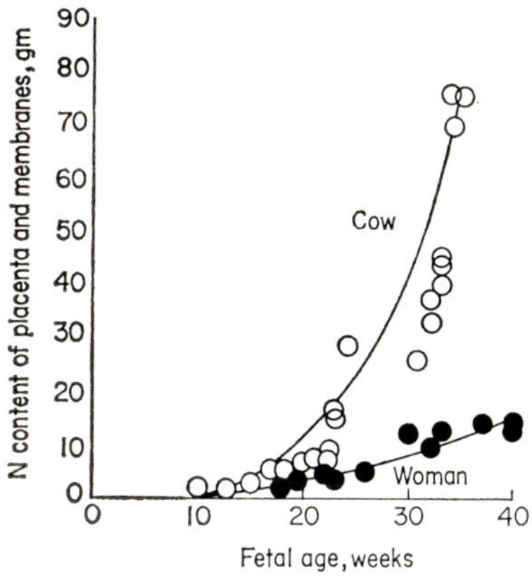

FIG. 6. The N content of the placenta in cow and woman. Note the marked increase in the cow and the slight increase in woman. The latter occurs despite the so-called placental regression and erosion of late pregnancy. The data were recomputed from those of Jakobsen (1957) and Widdowson and Spray (1951).

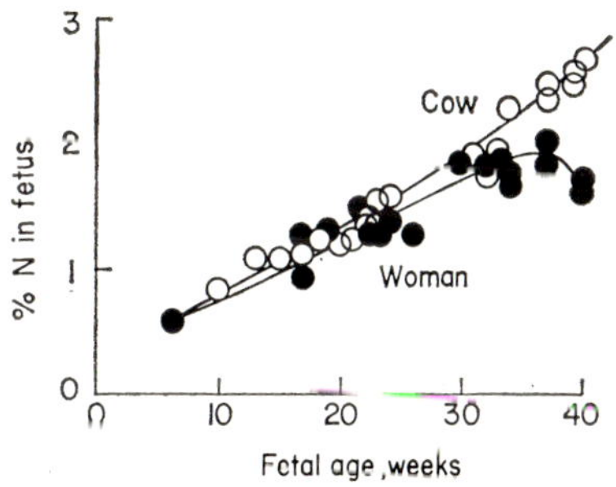

FIG. 7. The percentage of fetal N in the cow and in woman. The fall in percentage N in woman after about 30 weeks is due to concomitant deposition of fat. The data are those of Jakobsen (1957) and Widdowson and Spray (1951).

tration of N in it increases. Figure 7, for example, shows the increase in N concentration in fetal cow and fetal man, two species with the same gestation time. In both species the N concentration is initially very low, less than 1 %. By the 200th day it has increased twofold. With the cow the increase continues subsequent to the 200th day, but with man a fall in concentration occurs, largely due to the concomitant deposition in the fetus of about 16 % of fat. In cows, fat deposition is only 4 % of fetal weight. Again, the results given in Fig. 7 appear general to many species.

C. Differential Growth of the Fetus

That growth of the fetus necessarily involves differential growth of parts must have been evident in the eighteenth century when the homuncu-

TABLE IV

WEIGHT OF TISSUES AND ORGANS IN SHEEP FETUSES AT DIFFERENT STAGES OF PREGNANCY[a]

Tissue	Weight of tissue (gm)				
	Age of fetus (days)				
	72	97	112	120	140
Skeleton	39	230	288	476	780
Muscle	92	229	330	482	765
Liver	9	45	63	92	79
Gut	3	11	27	40	89
Brain	4	19	36	57	47
Heart	2	8	15	21	34
Kidney	2	10	14	16	42
Lung	6	36	68	91	99
Blood	20	65	170	170	500
Skin	9	74	220	356	590
Weight of fetus	180	740	1440	2100	3400

[a] The data are those of Carlyle (1945), recomputed by Barcroft (1946) and given here in a slightly abridged form.

lus theory of development was laid to rest. An immense amount of morphological study and of biochemical and physiological investigation has been made in which the course of development of different organs and tissues has been described. Although this differential growth involves changes in N metabolism, it is impossible here to do more than touch upon these aspects. To illustrate the extent of differential growth of parts of the fetus, the results obtained by Carlyle (1945) with sheep are given in Table IV.

This table shows that during growth some organs and tissues of the sheep fetus develop more rapidly than others. Thus, at 72 days the muscle is twice as heavy as the skeleton but at term it is about the same weight. The liver at 72 days of fetal age is three times the weight of the gut, but at term the gut is the heavier organ. Liver and skin weigh the same at 72 days but by the 140th day the skin weighs nearly eight times as much as the liver. The same general pattern of development is seen in other species, and immediately predicates the possibility of sequential changes in the needs for amino acids, dependent on the differential rates of formation of the different proteins of tissues.

D. Composition of Fetal Tissues

1. Muscle

The tissue of the body that contains the most N in adult life is the muscle. Dickerson and Widdowson (1960) and Widdowson and Dickerson

TABLE V

THE DISTRIBUTION OF PROTEIN FRACTIONS IN FETAL AND ADULT MUSCLE OF MAN AND PIG

Nitrogen source	Per cent at following stages:			
	Man			
	0.35[a]	0.54[a]	1.0[b]	≫1.0[c]
Nonprotein N	1.2	1.7	2.4	3.0
Sarcoplasmic protein N	3.6	3.7	3.9	6.7
Fibillar protein N	5.7	8.7	10.9	19.9
Extracellular protein N	0.6	1.8	3.8	1.4
Collagen N	—	1.3	2.8	1.0
Total protein N	9.7	13.5	18.5	27.5
	Pig			
	0.40[a]	0.79[a]	1.0[b]	≫1.0[c]
Nonprotein N	1.1	2.0	2.6	4.3
Sarcoplasmic protein N	4.0	2.4	2.2	8.0
Fibrillar protein N	2.5	6.1	6.5	19.8
Extracellular N	0.4	1.0	2.3	1.1
Collagen	0.4	1.2	1.3	1.2
Total protein N	7.2	9.4	11.0	28.4

[a] The stage of pregnancy is the age of fetus divided by the gestation period in days.

[b] At birth.

[c] In adult life.

(1960) have summarized analytical studies of muscle and liver in developing pig and human embryos. Fetal muscle differs in histological appearance from newborn muscle and adult muscle. In the fetus, muscle nuclei are large, the fibers are small and widely separated by extracellular material. At birth the fibers are still small but more closely packed, while the adult muscle fibers are much larger and the nuclei few and small. These changes seen on histological section are reflected in changes in the proportions of sarcoplasmic, fibrillar, and extracellular protein N. Some results are summarized in Table V. It is clear that the fibrillar proteins of muscle increase at a greater rate than the sarcoplasmic proteins during pregnancy in both species. The fibrillar protein fraction isolated by the Cambridge workers using the method of D. S. Robinson (1952) includes deoxyribonucleoprotein, and is not homogeneous, so the increase in the fibrillar protein itself must be greater than Table V suggests. Analysis of fetal tissue for nucleic acids (Davidson and Waymouth, 1944a; Davidson, 1950) shows that fetal tissue not only has a higher water content than adult tissue but also a higher total nucleic acid content. The high nucleo-

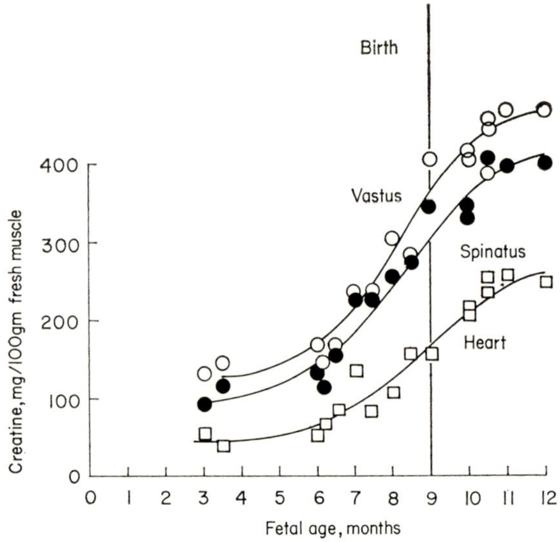

FIG. 8. The creatine content of the muscles of calves during pregnancy and the first 3 months of lactation. The data are those of Blaxter *et al.* (1953).

protein content of embryonic cattle muscle was indeed commented on by Kossel as early as 1882.

The results in Table V also show that the extracellular proteins—elastin, collagen, and reticulin—increase markedly during pregnancy; indeed at birth they account for 18 % (man) or 17 % (pig) of the total N of muscle. These percentage values are four to five times those noted at maturity. Nonprotein N tends to make up a slightly higher proportion of the total N in fetal tissue than in adult tissue, no doubt because of the higher water content of the former. The creatine content of muscle increases markedly during late pregnancy in the cow, as shown in Fig. 8, where it is evident that the small differences in concentration in somatic muscle groups is established very early in pregnancy and that the heart, though functional at an early age, follows a similar pattern (Blaxter *et al.*, 1953). These conclusions are true both on a fresh weight and on a dry weight basis. Creatine concentration probably reflects very closely the growth of the contractile proteins of the muscle cell.

2. *Liver*

Similar observations to those for muscle tissue can be made for other tissues of the body. With the fetal liver, an organ which has been subject to much study, its metabolism is higher than that of any other fetal tissue as judged by its O_2 consumption/gm dry weight. In sheep, the metabolism of the fetal liver declines by nearly 50 % from the 70th day of fetal life to birth (Carlyle, 1945). The nucleic acids of fetal rat liver show marked changes during development. The concentration of DNA is relatively constant on a dry basis from the 16th to the 19th prenatal day and then falls to a third by the 21st day. This fall is part of the maturation process since it continues, albeit at a slower rate, after birth. RNA-P falls far less rapidly so that the ratio RNA-P/DNA-P changes from a value of 0.9 at the 16th day of fetal life to 2.9 at the 40th day of extrauterine life (Geschwind and Li, 1949). Wirth and Schreier (1956) confirmed the fall in the DNA content of fetal liver, but found a steady rise in the RNA content. Studies on the incorporation of 2-C^{14}-glycine into RNA showed the RNA incorporation to exceed that of the DNA incorporation (Schreier *et al.*, 1956).

Birth is necessarily associated with entirely different demands on liver function and it is not surprising that marked differences are found in enzyme concentrations in fetal as compared with adult liver tissue. Dawkins (1959) has studied some of these changes and has shown that the chief differences between the liver cell of the fetus and of the adult lies in the mitochondrial fraction; from analytical data and measurements of the activity of enzymes of the respiratory chain he concluded that the individual liver cell contains fewer mitochondria. Part of this may stem from the presence of hematopoietic tissue in the fetal liver and from glycogen stores. Histochemical studies of the fetal liver cells, however, show only a few identifiable mitochondria (Payne, 1957). Nemeth and Dickermann (1960) have shown that the $NADPH_2$ (reduced nicotinamide adenine dinucleotide phosphate) concentration is much lower in fetal liver than in adult liver, which suggests a marked reduction in those enzymic reactions, mainly synthetic, associated with a high $NADPH_2$ concentration. Adult enzymic function does not necessarily take long to develop. Nemeth (1959), for example, showed that tryptophan pyrrolase (tryptophan peroxidase) activity was increased within hours of birth. The enzymes of the respiratory chain, however, take longer to develop. In the rat, mature values for cytochrome oxidase are not reached until 3 days of life.

3. Blood

The plasma proteins of the fetus and the newborn are distinct from those of the mother. The fetal plasma protein fetuin (α-glycoprotein), first discovered by Pederson (1945), is present in many species and γ-globulin can be completely absent. Fetuin is a mucoprotein with a molecular weight of 48,400 containing 74 % by weight of polypeptide, the carbohydrate consisting of galactose, mannose, glucosamine, galactosamine, and sialic acid (Spiro, 1960). It cannot be detected in maternal blood. In blood of the fetal calf in the 6th week of fetal life, the plasma albumin is less than

10 % of the total plasma protein and it rises to 30 % of the total by the 20th week and to 45 % by the 40th week. The β-globulins tend to be higher early in pregnancy than later and the concentration of the α-globulin fraction, which includes fetuin together with those α-globulins homologous with those of adult blood, falls during pregnancy from 65 to 35 % of the total protein. In the fetal goat (Barboriak *et al.*, 1958), albumin increases from 6 % on the 46th day to 45 % on the 135th day and fetuin follows virtually the opposite course. A similar inverse relationship between fetuin and serum albumin occurs in the sheep.

4. Amino Acid Composition of the Fetus

The above brief survey of the changing proportions and concentrations of the N-containing compounds of three tissues during pregnancy implies that the proportions in which amino acids are deposited in the fetus may well vary from time to time. This is quite apart from changes in rates of amino acid turnover which could equally well be predicated from the changes in enzymic activities. Analysis of mouse embryos suggests that this conclusion is correct (Singer *et al.*, 1956); it was found that the ratios of lysine:leucine:tryptophan changed with age. With changes in the collagen content of the embryo as great as those discussed above, changes in the hydroxyproline content of the fetus with age can be envisaged. No doubt growth of hair in those species born with a pelage necessarily implies a greater demand for cystine in the later stages than the earlier stages of pregnancy. Table VI summarizes the results of amino acid analyses made by Jakobsen (1957) with cattle fetuses. The table indicates that the amino acid composition of the fetus changes with its age and that the free amino acids of maternal blood contain amounts of the basic amino acids in excess of those deposited, but less of certain of the non-essentials and of tryptophan. From the data on which Table VI was based, Jakobsen calculated that in the last 5 weeks of pregnancy the cow deposits 257 gm lysine in the fetus, or about 7 gm lysine each day. The daily deposition of this essential amino acid during pregnancy serves to place pregnancy demands for amino acids in their true perspective. The lactating cow producing 25 kg milk each day secretes 55 gm lysine in milk. Pregnancy demands for this amino acid are thus only a fraction of lactation demands, which emphasizes the point made earlier that lactation involves greater rates of protein synthesis than occur even in the late stages of pregnancy.

E. Transfer of Nitrogen to the Fetus during Pregnancy

Measurement of the concentration of N-containing compounds in the arterial supply and venous drainage from the uterus provides a measure of the compounds utilized to synthesis of the proteins of the fetus, of the

TABLE VI

AMINO ACID COMPOSITION OF CATTLE FETUSES AND OF THE FREE AMINO ACIDS IN MATERNAL BLOOD (GM/16 GM N)[a]

Amino acid	Fetus		Free amino acids of maternal blood
	20 Weeks	40 Weeks	
Histidine	1.63	1.47	4.7
Isoleucine	2.80	2.94	5.0
Leucine	7.77	6.53	7.4
Lysine	5.81	7.00	8.8
Methionine	1.43	1.72	1.9
Phenylalanine	3.92	3.24	2.6
Threonine	3.99	3.70	3.1
Tryptophan	1.0	1.1	0.5
Valine	4.04	4.20	9.0
Arginine	5.24	6.73	8.5
Tyrosine	2.89	2.78	2.1
Proline	6.78	7.90	7.8
Serine	3.95	4.36	5.7
Glutamic acid	12.20	13.20	8.1
Aspartic acid	8.23	8.19	—
Glycine	6.33	10.73	7.1
Alanine	6.21	6.14	7.6

[a] From Jakobsen (1957).

placenta, and to provide for the hypertrophy of the uterine wall. Setnikar (1957) Setnikar and Barron (1958), and Jakobsen (1957) have carried out such experiments with pregnant goats. Their experiments agree in that they show that the venous blood draining the pregnant uterus contains less amino N than the arterial blood, but that the plasma protein N undergoes no diminution. Jakobsen's results suggest some uptake of small peptides. His analyses of individual amino acids by the Moore and Stein technique show that the different amino acids are taken up in amounts which vary with the amino acid concerned. Setnikar and Barron found no significant correlation of the arteriovenous differences in amino acid concentration and stage of pregnancy. Since the N demands increase as pregnancy advances, this implies that uterine blood flow increases in direct proportion to the rate of intrauterine protein synthesis. This is in agreement with direct flow measurements made with sheep (Barcroft, 1946). Setnikar calculated from his own measurements of uterine flow of blood and arteriovenous differences that the amino acids of maternal blood were a sufficient source of N to meet the demands for synthesis of new protein.

Simultaneous measurement of concentrations of N-containing compounds in fetal and uterine vessels gives information on the mechanism of passage across the placenta. Many such measurements have been made. They show, in confirmation of the early observations of Van Slyke and Meyer (1913), that the concentration of free amino acid N in fetal vessels is higher than in maternal vessels (see review by Hagermann and Villee, 1960). This implies that the passage across the placenta is an active process. Experiments with guinea pigs (Dancis *et al.*, 1958), in which during late pregnancy the placenta is of the most simple type in terms of the number of cellular barriers interposed between fetal and maternal blood, show that infusion of L-histidine into the maternal circulation results in an increase in its concentration in maternal plasma. The concentration in fetal plasma also rose but at an ever-decreasing rate. Furthermore, there is evidence of discrimination by the human placenta against the D-isomer of histidine, for on infusion into the maternal blood it passes much more slowly to the fetal blood than does L-histidine (Page *et al.*, 1957). Placental transfer of thyroxine and triiodothyronine has been studied by Myant (1958), among others, using the amino acids labeled with I^{131}. He showed that, in the rabbit given labeled thyroxine, the ratio of concentration in fetal to maternal serum varied from 0.05 before the 20th day to 0.3 at term. Removal of the thyroid of the fetus by decapitating it *in utero* had no effect on the transfer. Transfer of triiodothyronine was more rapid than of thyroxine.

The above results show that amino acids pass the placenta and some pass against concentration gradients. Their use as a substrate for synthesis of protein in the fetus has been studied by Dancis *et al.* (1957). Human fetal tissue was incubated with glycine-2-C^{14}. In fetal liver, evidence was obtained of incorporation of labeled glycine into electrophoretically identifiable fractions of the plasma proteins, with the sole exception of γ-globulin.

As far as the major end product of amino acid catabolism, urea, is concerned, it seems certain that the concentration of urea is the same in fetal as in maternal blood. This is true of the goat (de Bella, 1958), the pig (Zweymuller *et al.*, 1959), and man (McGaughey *et al.*, 1959), and no doubt of other species. The urea concentration of the allantoic fluid is above that of the blood and its composition in the sheep at different stages of pregnancy shows that an increasing addition of fetal urine occurs throughout pregnancy (McDougall, 1949). Urea concentrations in amniotic fluid were also found to be higher than in blood and transport of urea across the chorioamniotic membrane probably occurs (McGaughey *et al.*, 1959).

F. Transfer of Intact Protein to the Fetus in Utero

In some species the offspring acquires immune proteins from its mother while still *in utero*. In other species this transfer is mainly through ingestion of immune proteins in the colostrum. Colostral transfer of immunity is usually associated with the more complex type of placentation, notably, the syndesmochorial placenta, as found in sheep and cattle. Here, no passage of immune proteins *in utero* has been detected (Lovell and Rees, 1961). In the rabbit and guinea pig it has been shown (Brambell *et al.*, 1949, 1959), however, that the transfer of immune proteins takes place through the yolk sac placenta and recent work shows this transfer to be a highly selective and specific process (Brambell *et al.*, 1960). With primates there is some disagreement about the method of transfer. Two hypotheses have been put forward. In the first, the transfer of intact protein is thought to be from maternal blood to amniotic fluid and from amniotic fluid either to lung or gut of the embryo and thence to its blood. In the second, a specific and highly selective transfer of protein across the placenta is predicated.

The passage of serum proteins labeled with I^{131} from maternal blood of woman and of rhesus monkeys to amniotic fluid has been demonstrated (Dancis *et al.*, 1961; Bangham *et al.*, 1958). Passage of the serum globulins is more rapid than that of serum albumin. Furthermore, the composition of amniotic fluid suggests that it is a dialyzate across the fetal membrane (Abbas and Tovey, 1960). Labeling amniotic fluid with albumin tagged with I^{131} showed that the monkey fetus ingests amniotic fluid. Little labeled protein, however, appears in fetal blood, which implies that the first method of transfer referred to above is not important. It has been suggested (Aršavskij, 1959) that amniotic fluid is an important source of N for the fetus, and that a considerable part is played by protein ingested by this route in fetal nutrition. Quantitative information is, however, lacking.

G. Nitrogen Deposition in Maternal Tissues, including the Wall of the Uterus and the Mammary Gland, during Pregnancy

It is common belief that the retention of materials in the body during pregnancy normally exceeds the amounts attributable to the growth of the products of conception and enlargement of the organs of reproduction. The general argument that pregnancy stimulates the mother to store more N than is present in the fetus and adnexa (here called extrareproductive storage of N) was first made by Bar and Daunay (1905) in experiments with dogs. A more qualified and basically teleological argument has been put forward that those species which lactate abundantly lay in stores during pregnancy, while those species that do not milk well pass on more nutrients to their fetuses before they are born (Spray, 1950). This argument was based on observations that pregnant rats and mice retained more N and mineral elements in pregnancy than they transmitted to the fetus and adnexa, whereas no comparable storage was noted in the guinea pig. A third, and again a teleological argument, is that the pregnant mother stores

N in her own tissues early in pregnancy to meet the demands made later in pregnancy by the fetus (Campbell *et al.*, 1953). A basis for the belief that pregnancy leads to a maternal storage of protein resides in the demonstration of an endocrine function of the placenta and the transfer of hormones from the fetus to the maternal organism (Amoroso, 1952). This aspect will be touched on later. A further basis is that maternal weight gains exceed the weight gained by reproductive structures. This maternal gain is, however, in part due to changes in maternal hydration and blood volume.

Since an allowance for maternal N storage considerably affects estimates of protein needs in pregnancy, the evidence for maternal storage of N in a number of species is presented in some detail. Most information is available for the rat, more general information is available for farm animals and for man, while with most other species information is fragmentary.

1. Rat

Herring (1920) was probably the first to make an extensive study of the extrauterine tissues of the rat during pregnancy. Since then, numerous investigations have been made and most, but not all, show an increase in "extrauterine" body weight in pregnant compared with nonpregnant animals. Part of this is due to changes in hydration of the body (Bond, 1958); not all is due to a change in protein or fat content. Extrauterine storage of N in rats has been studied, first, by deducting from the N retentions measured in pregnancy the N content of the newly born litter, and, second, by a comparison of analyses of the carcasses of animals at parturition with those of nonpregnant animals. Using the former technique, Morse and Schmidt (1944) found the extrauterine change in N retention over 21 days of pregnancy in the rat to vary from a maximum of +2.08 gm to a minimum of −1.60 gm. The extrauterine loss of N was associated with a low consumption of food. In these experiments there was certainly no obligatory storage of N in extrauterine tissues. By analysis of carcasses, Poo *et al.* (1940) noted, however, a considerable maternal storage with some tendency for the storage to increase with increasing protein content of the diet up to a protein content of 27 %. Spray (1950) similarly observed the considerable maternal storage of 1.4 gm N, using the carcass analysis technique. When, however, Boyne *et al.* (1953) repeated Spray's work and combined it with N balance studies, their results failed to demonstrate any maternal storage whatever. These latter results are summarized in Table VII.

The above results (others could be cited), show that N retention in the extrauterine tissues of the rat during pregnancy is not invariably positive. It can amount to a storage of nearly 5 gm N in 21 days as in experiments

TABLE VII

EXPERIMENTS ON N RETENTION IN THE RAT IN PREGNANCY[a]

	Gm/N in 21 days	
	Pregnant rats	Control rats
N metabolism experiment		
Intake	11.974	10.473
Retention	4.067	2.401
Difference = total retention (*a*)	1.666	
Carcass N determinations		
Increase in N content of maternal body by analyses of carcass	−0.163	
Increase in N content of products of conception by analysis of carcass (*b*)	1.820	
Estimate of maternal storage from N balance (*a* − *b*)	−0.160	

[a] Data of Boyne *et al.* (1953).

TABLE VIII

MEAN DAILY N METABOLISM BY PREGNANT RATS GIVEN DIFFERENT DIETS[a]

Diet characteristics	Second week of pregnancy		Third week of pregnancy	
	N intake (kg/kg)	N retained (kg/kg)	N intake (kg/kg)	N retained (kg/kg)
Basal (16% protein and 3.7 kcal/gm)	1612	199	1581	423
Protein increased to 25%	3204	431	2945	610
Energy increased to 4.6 kcal/gm by substitution of fat for carbohydrate	1747	377	1632	456

[a] Data of Pike *et al.* (1954).

of Poo *et al.* (1940), or can constitute a loss of over 2 gm. The cause of this variation is mainly nutritional. Thus Bourdel and Jacquot (1960), by giving three different amounts of protein to pregnant rats, showed that although the N content of the uterine contents was unaffected, the total N retention in gestation varied from 59 to 354 % of the N content of uterine contents. A more informative, though less complete, experiment was made by Pike *et al.* (1954) as summarized in Table VIII. It is evident from this experiment that in the 2nd week of pregnancy, when structural organization of the embryo is taking place, additional energy resulted in an increase in total N retention. This N could hardly be in fetal structures and must have represented an increase in extrauterine deposit. Late in pregnancy, when a considerable increase in fetal mass occurs, no such effect was present. This suggests that protein supplies were then limiting.

In rats and mice, removal of the fetuses from the uterus at mid-term does not terminate pregnancy provided the placentas remain viable. Gestation proceeds, as judged by the absence of estrus, maintenance of the corpora lutea of pregnancy, development of the mammary gland, relaxation of the pubic symphysis at the 21st day, and delivery of the uterine contents (Klein, 1935; Newton, 1935). Such preparations are useful because they permit study of the maternal aspects of pregnancy uncomplicated by growth of the fetuses. Brooksby and Newton (1938) showed that the gain in weight of pregnant mice evident early in pregnancy persisted after fetal ablation and this was confirmed by Campbell *et al.* (1953) with rats. Bourdel and Jacquot (1960) have summarized their experiments and those of others and have shown that after fetal ablation maternal N retention, if already established, continues and that this anabolic function of the placenta can be ascribed to prolactin.

During pregnancy in rats, voluntary intake of food of adequate nutritive value increases over that observed in controls (Cole and Hart, 1938). The increase has a fairly rapid onset and amounts to about 30 % This no doubt accounts for a large part of the increase in extrauterine N retention observed during early pregnancy. In the last 5 days of pregnancy, however, food intake can fall rapidly to about 60 % of that observed in pregnancy as a whole (Morrison, 1956). The continuation of an extrauterine N retention in late pregnancy after fetal ablation and the demonstration that it is of the same magnitude as that observed in early pregnancy could be due to nothing more than a continuation of the high food intakes established in early pregnancy.

While the extrauterine tissues of the rat taken together need not necessarily retain N in pregnancy, there is considerable evidence that even when the body of the rat as a whole is losing N, some organs can gain N. The liver increases markedly in both size and total N content (Souders and Morgan, 1957) in normal pregnancy. Smaller increases can occur in the weights of the kidney and the gut (Boyne *et al.*, 1953); some endocrine glands such as the adrenal may regress, others increase in weight. In addition, the uterus itself increases in size and growth of the mammary gland occurs.

The increase in the size of the liver of the pregnant rat is associated with a reduction in glycogen content, an increase in water content, a slight rise in DNA content, smaller than that predicted on the basis of the increased body size of the pregnant animals, and a large rise in the RNA content, which is disproportionate to the rise in phospholipid and protein content (Davidson and Waymouth, 1944b; Campbell and Kosterlitz, 1949). In nonpregnant animals the RNA content of the liver cell varies directly with the protein N of the cell. In pregnancy, however, the increase in RNA is greatly in excess of any increase associated with its increased protein content. Campbell and Kosterlitz's data on the increase in the DNA content of the liver indicate that an increase in cell number takes place. Only about 60 % of the nuclei in the liver are those of hepatocytes, but measurement of collagen N in pregnant livers indicates that no connective tissue proliferation occurs (Harkness and Harkness, 1956a) which suggests that these changes in the composition of the liver represent a true hypertrophy of the liver cells with but small increase in their number.

The cause of the "excess RNA" content of the liver during pregnancy was investigated by Campbell *et al.* (1953). It is dependent on the presence of a viable placenta. One fraction of the excess RNA varied with the protein content of the liver cell as in nonpregnant animals. The other varied linearly with the energy intake of the rat and also with the weight of the placentas. Doubling food intake from 4 to 8 gm/100 gm body weight increased the ratio RNA-P/DNA-P from 5.3 to 5.5. Doubling the weight of the placentas from 3 to 6 gm increased the ratio from 5.3 to 5.7. Together, these two findings suggest that both a placental factor and the increase in food intake that accompanies pregnancy are responsible for this increase in what presumably is the synthetic activity of the liver with respect to protein.

An increase in the N content of the mammary gland and of the uterine wall is necessarily included in the definition of extrauterine N deposition. Griffith and Turner (1961), using the DNA concentration in fat-free dry mammary gland tissues of the rat as an index of its growth, found no significant growth over virgin controls on the 5th day of pregnancy and a threefold increase by the 1st day of lactation. Moon (1962) noted a doubling of the DNA content of the abdominal inguinal glands by the 20th day of pregnancy. Some 60–70 % of the mammary gland development recorded on the 14th day of lactation had occurred during pregnancy in both these studies. Similar studies of nucleic acids in the mammary gland have been made by Greenbaum and Slater (1957) and by McLean (1958). In its transition from the quiescent to the lactating state, increases in enzyme concentrations in mammary tissue occur, reflecting its increased metabolic status, and all involve an increase in protein deposition. The effects of estrogens and progesterone in stimulating duct growth and in promoting alveolar cell proliferation have been reviewed by Jacobson (1961).

A third organ that hypertrophies in the maternal organism is the uterine wall. Analyses of endometrium and myometrium together show that this is associated with an increase in both cell size and cell number (Needham and Cawkwell, 1957). Using the DNA content of the tissue as an index of cell number, the amount of RNA per cell increases sevenfold and the total N per cell increases fourfold during pregnancy (Wakid and Needham, 1960). In addition a marked increase in collagen takes place. The uterus increases in weight from 0.47 gm in controls to 2.76 gm on the 19th–22nd day, and the total collagen from 13 mg to 64 mg (Harkness and Harkness, 1954). Collagen growth depends on distention of the uterine horn (Harkness and Harkness, 1956b).

Growth of the uterus is rapidly initiated by estrogen injection in the ovariectomized animal, and considerable work has been done in describing the sequence of enzymic changes which occur when this tissue starts growing rapidly. Scott and Lisi (1960) have summarized in tabular form the measured changes in the rat uterus after estradiol administration. The sequence is plain. Increased capilliary permeability is noted at 1 hour and the first enzymic changes observed are those concerned with amino acid incorporation into proteins; they are noted at the 5th hour. Increased incorporation of uridine into RNA occurs at the 12th hour and an increase in RNA and DNA content occurs at about the 21st hour. An increase in collagen content in response to estrogen administration occurs still later (Harkness *et al.*, 1957).

The uterus of the rat involutes very rapidly at parturition. The rate of dissimilation of the collagen of rat uterus is represented by a half-life of 24 hours (R. D. Harkness and Moralee, 1956). Woessner (1962) has suggested that the process involves an enzymic (probably histiocytic) solubilization of the collagen, followed by an action of lysosomal acid cathepsins to form peptides which are subsequently degraded by peptidases.

2. Pig

Reproduction in the pig has been reviewed by Duncan and Lodge (1960), who pay some attention to extrauterine storage of N. Table IX summarizes

TABLE IX

MATERNAL N RETENTION DURING PREGNANCY AND THE N CONTENT OF THE PRODUCTS OF CONCEPTION AT TERM

Authority	N retention of sow in pregnancy	N in conception at term	N retention (%) in extrauterine tissue
R. E. Evans (1929)	1439	233	16
Mitchell *et al.* (1931)	813	235	29
Lenkeit *et al.* (1955a, b)	1879	341	18
	1266	290	23

the mean results of three sets of experiments in which N retention over the whole of pregnancy was measured. When compared with the amount of N in the uterine contents at term, there is no doubt that an extrauterine storage occurred. In Mitchell's experiments the pregnancy studied was the first and the increased N retention must therefore include a component represented by normal body growth (Mitchell *et al.*, 1931). In Evan's experiments the products of conception were estimated (R. E. Evans, 1929), while in Lenkeit's they were determined (see Lenkeit and Gütte, 1955; Lenkeit *et al.*, 1955a,b). In addition to these experiments Penzes (1961) analyzed the organs of pregnant and nonpregnant sows, and found evidence of increases in weight during pregnancy for many tissues. This was mainly due to an increased hydration, for the concentration of N fell.

The above evidence that extrauterine retention in sows can occur is comparable to that obtained in rats, in that it provides no information about whether or not it is obligatory. As far as the reproductive tissues of the mother are concerned, the uterine wall certainly increases in N content. In Mitchell's experiments, the weight of the uterine wall increased from 86 gm in the nonpregnant animal to 2950 gm at term, representing a total increase of 54 gm N. In addition, N is deposited in the mammary glands. No data on the magnitude of this have been found, but Moustgaard (1956) has estimated this from unpublished work by de Villiers to be 70 gm N in the last 14 days of pregnancy. Mammary growth commences much earlier than this, so the estimate may well be low.

Many experiments have shown that weight gains of pregnant sows can vary enormously. For example, Smith (1960) noted a range from 14 to 51 kg in experiments in which different amounts of food were given. The gain in weight of the uterus and its contents is usually about 18–20 kg. Thus, even if both mammary growth and tissue hydration are ignored, a loss of the protein and fat of maternal tissues must have occurred in the sows given little food. Similar conclusions can be drawn from the experiments of Dean and Tribble (1961). As with the rat, so with the sow, the major determinant of whether or not a gain of protein occurs in extrauterine and extrareproductive tissues is the amount of food given. A slight increase in N excretion of the pregnant sow has been noted (Hennig, 1959).

3. Sheep

Sheep are kept under a wide range of environmental and nutritional conditions. In some they gain appreciable weight, in others they lose weight, so much so that the ewe at the end of her pregnancy weighs less than at the beginning although she may well be carrying two lambs. This is well recognized in practice and the problems of attaining adequate weight gains in pregnancy by provision of food have understandably attracted more attention than whether or not pregnancy increases weight gains and N deposition (W. Thomson and Aitken, 1959). In the experiments summarized by Thomson and Aitken, the range of change in weight during pregnancy was from a loss of 14 lb to a gain of 40 lb.

Rombauts (1959a,b) has, however, reported experiments in which pregnant and nonpregnant ewes were given the same food and N retention measured. In one experiment the pregnant ewes consumed 18 % more food and the N retentions by the pregnant ewes exceeded those of the nonpregnant by 203 gm. A further experiment with 8 ewes resulted in a retention of 248 gm. Both values exceed the amounts of N deposited in the uterus and mammary glands, which amounts to about 110 gm N. Rombauts (1959a) showed by carcass analyses that the increase in N retention was due to increased amounts in the skeletal muscles with appreciable amounts in the liver, kidneys, spleen, pancreas, and digestive tract. There was no hypertrophy of the heart.

Wallace's (1948) experiments in which ewes were given widely different amounts of food during pregnancy show that the reduction of maternal organ and tissue weight in undernutrition and the increase in weight following overnutrition is not the same for all organs and tissues. When body weight at term is reduced to 77 % of that in mid-pregnancy, the skeleton, the nervous system, and the blood are reduced to about 90 % of mid-term values. Muscle, the main repository of protein, is reduced in weight, however, to 57 % and the liver to 71 % of mid-term values. In overnutrition, when body weight at term increases to 142 % of that in mid-pregnancy, the weights of the bones and nervous tissue do not increase by more than 8 %. Blood increases by 38 %, muscle mass by 29 %, and liver weight by 55 %. These results show that any increase in liver weight that may occur as a result of pregnancy is not maintained in the face of nutritional deprivation of the mother. Furthermore, liver weight in the pregnant ewe and indeed the total protein of the liver, for Wallace gives analytical data, is more sensitive to nutritional manipulation than is the large protein mass of the muscles. P. Robinson's (1948) data shown in Fig. 9 are of interest in this regard. They show very clearly not only the magnitude but the rapidity of the effects of a change in nutritional level on liver weight in the nonpregnant sheep. For these reasons it seems probable that a considerable part of the changes in maternal N retention noted by Rombauts and discussed above could be due to the increased appetite of the ewe during pregnancy. This increase in appetite is not, however, a constant one. In late pregnancy voluntary intake of food declines (Graham and Williams, 1962), the declines being greatest for those ewes with the heaviest uterine contents.

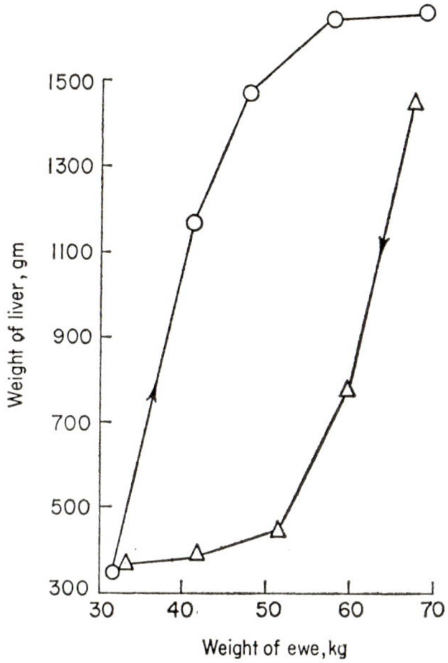

FIG. 9. The effect on liver weight of a reduction of body weight by underfeeding well-fed ewes (△——△) and of an increase of body weight by overfeeding underfed ewes (○——○). Replotted from the data of P. Robinson (1948) to show that the weight of the liver is particularly sensitive to a change in nutritional level. These results confirm in part work with rats which shows that some of the increase in liver N in pregnancy can be attributed to the greater food intake in pregnancy.

4. *Cow*

Jakobsen's (1957) study of pregnancy in the cow included the determination of N retention in pairs of identical twins, one of which was pregnant. His results are slightly difficult to interpret since in two experiments only were the twins given the same amount of protein. These two experiments made on the 157th and 251st day of pregnancy (gestation in the cow is normally 273 days) showed that the daily retentions of N by the pregnant twin were 1.1 gm and 0.6 gm greater than by the nonpregnant twin. It can be computed that the uterine deposition was at least 20 gm N daily in these animals. These results show that, when given the same amount of food as a pregnant cow, a nonpregnant cow will retain very nearly the same amount of N as one which is depositing large amounts of N in the uterus.

Van Es (1961), in a series of experiments with pregnant cows, determined body N retention and the retention of carbon. From Jakobsen's (1957) results on the composition of the uterine contents in pregnancy, he calculated the amounts of N associated with the growth of the fetus and associated structures. In late pregnancy all 8 of his cows gained N, and in

5 of them the gain of N was greater than could be accounted for by the products of conception. The total gain of C recorded by Van Es enables an estimate to be made of fat retention. In the 5 cows which made "non-reproductive gains", 166 gm protein and 1033 gm fat were retained per day. Retention of protein in reproductive structures was 114 gm and, since the fat content of the fetus and adnexa is small, it follows that about 50 gm protein (8 gm N) and nearly 1 kg fat were retained in maternal extra-reproductive tissues. The large amounts of fat accompanying the extra-reproductive deposition of protein does not suggest that pregnancy in the cow specifically augments N retention. The ratio of protein/fat retained is in fact smaller than is normally expected in a mature cow gaining weight.

Jakobsen in his work determined the increase in the N content of the udder of the cow during her first pregnancy and found that deposition obeyed this equation

$$\text{N (gm)} = 25.2\, e^{0.0236(t - 175)} \qquad (5)$$

where t is age in days and the term $(t - 175)$ indicates that growth does not begin until the 6th month of pregnancy. Equation (5) states that at term the N in the udder is about 250 gm. This may be compared with the N in the uterus at term which is about 1200 gm, and the uterine wall at term, which also contains about 250 gm N.

No information regarding specific changes in the N metabolism of the liver of the cow during pregnancy has been traced.

5. *Man*

Garry and Stiven in 1936 reviewed the information available up to that time relating to the N content of the products of conception in woman and concluded that at term the fetus accounted for 70 gm N, the placenta 18 gm, and the fetal fluids 1 gm, that is, a total of 89 gm. They further estimated the N deposited in the breast and uterine wall as 17 gm and 39 gm, respectively, concluding that pregnancy in woman was associated with the deposition of 135–145 gm N in reproductive structures. More recent analytical work, notably, that of Widdowson and Spray (1951) and summaries of diverse analytical data given by Huggett (1941) and by Clements (1949), are in quantitative agreement with Garry and Stiven's analysis over 25 years ago. Garry and Stiven also summarized the results of numerous N balance trials made with pregnant women and A. M. Thomson and Hytten (1960) have brought this summary up-to-date. Nitrogen retention by women (mostly American) during pregnancy amounts to 200–400 gm, implying that the uterine contents account for at the most 70 % and at the least 30 % of the total maternal retention. Conversely, a considerable extrauterine storage occurs which is not accounted for by growth of the breast or uterine wall. It is largely on the basis of this evidence that the teleological argument that woman stores protein in her tissues during pregnancy to provide a "reservoir for lactation" has been widely accepted. It was thought to receive support from the fact that body weight gains in pregnancy commonly exceed those accounted for by the fetus, adnexa, and estimates of breast weight.

Subdivisions of the gain in weight by woman during pregnancy have been made by several workers. Leitch (1957), for example, calculated that for an average woman (54 kg, first pregnancy) weight gain was 12.5 kg, of which 5.0 kg consisted of the product of conception and 7.5 kg of maternal tissue. A gain of 1.5 kg protein (240 gm N) in maternal tissues was stated to occur. This included breast tissue protein and protein of the uterus wall which from Garry and Stiven's summary probably amounts to 56 gm N, leaving a true maternal gain of 184 gm N. Leitch points out that an increase of 60 gm circulating hemoglobin occurs during pregnancy, leaving about 175 gm N to be accounted for. This amount is about the same as that noted from the N balance studies. No information about its distribution is given.

A similar subdivision was made by Thomson and Hytten (1960). The 7.5 kg maternal store was divided into 3.4 kg plasma and extravascular water, a component which is lost rapidly after parturition, and 4.1 kg representing growth of breast and uterus. It will also include about 0.25 kg red blood cells. Taking the N content of this tissue as 2.5 %, the maternal storage would amount to 102 gm N, of which 56 gm must consist of breast and uterus tissue, leaving a true maternal storage of 46 gm, that is, about a quarter that given by Leitch and noted in N retention trials.

Later work by Thomson and Hytten (1962) suggest that even this estimate of a true maternal tissue retention of only 46 gm N may be too high. Direct determination of body water by dilution techniques indicates that, if protein is stored it must be stored dry, a conclusion which is untenable. Furthermore, recent studies of N retention in late pregnancy in normal women indicate gains of 2–5 gm N in the last week of pregnancy. The value of 2 gm is not much different from expectation based on the rate of fetal N deposition and the range is certain evidence of considerable variation (Lister, 1961).

These results on body composition in pregnancy cast considerable doubt on the validity of the long-term N retention trials which resulted in the estimates of 100–300 gm N retained by women over and above the gain due to conception. Nehring (1957), and other workers in East Germany concerned with animal experiments, have shown that large systematic errors can arise in N metabolism studies amounting to overestimation of retention by 13 to 23 % and that these can be obviated only by very careful work in which losses of N as NH_4 and as dust from scurf and hair are obviated.

The percentage errors need not be large to account for a maternal storage of 100 gm N by pregnant women. The total N taken in during the pregnancy of a woman consuming 80 gm protein each day is about 3.5 kg. A retention of 100 gm N could be accounted for by a systematic error of 3 % in the collection of the end products of N metabolism. Cutaneous losses of N alone could account for a spurious storage of more than 70 gm N if Darke's (1960) results are correct, and losses of NH_4 N in our experience, admittedly with a less cooperative animal, the sheep, can amount to 2–3 % of the total N loss unless careful precautions are taken.

During pregnancy in woman, plasma volume increases by about 40 %, a maximum occurring in the 32nd week (Lund, 1951), whereas the red cell mass increases by 17 % (Hytten and Duncan, 1956). The concentration of hemoglobin in the blood thus decreases and this physiological fact appears to be the genesis of the belief that anemia is a common deficiency disease of pregnancy. Iron therapy will increase hemoglobin concentration in most pregnant women, necessitating a further demand for amino acids (Hytten and Duncan, 1956). In any event, the physiological increase in total circulating hemoglobin necessarily implies a demand for N on the part of the maternal organism, as distinct from reproductive structures. Its magnitude, however, is small.

6. *Conclusions on Maternal Extrareproductive Storage of N in the Mammalia*

In the rat it is certain that an increase in the deposition of protein and of nucleic acids in extrauterine tissues occurs during pregnancy. The uterine wall and mammary gland account for only part of this deposition, for an important deposition occurs in the liver. The total N content of the maternal body, however, does not necessarily increase, implying an ability of the rat to translocate amino acids from sites other than those at which N deposition occurs. There is no evidence of a specific storage of protein which the rat makes to meet the demands of an ensuing lactation. Rather, the over-all pattern suggests that any changes in the N content of extrareproductive tissues represents an increase in the synthetic capacity of the organ concerned which is determined by the enhanced demand for amino acids on the part of the fetus and, more important perhaps, by the enhanced supply of substrates in the form of an augmented food intake.

With the pig, although N retention data, which for reasons given when discussing N retention in man may be suspect, suggest it can occur, there is nothing to suggest that it is obligatory. The same appears to be true of the sheep in which, as with rats, maternal storage of protein is associated with an increased food intake. With cows, there is no evidence of any specific augmentation of N retention, and in man, evidence of such augmentation from N retention studies is not in agreement with other more direct estimates of the composition of body gains. Limited data for other species tend to support the conclusion that there is no obligatory change in maternal N metabolism accompanying pregnancy. It follows that minimal N requirements during pregnancy can be measured by the amounts of N retained in the uterus and its wall and in the developing mammary gland, with at the most a very minor allowance for deposition in the tissues of the mother.

This conclusion implies that protein needs in pregnancy are met when the maternal protein mass is maintained constant at approximately the same level as that pertaining before pregnancy. It does not follow, however, that this is necessarily an ideal state of affairs either for the mother or for the offspring, and the subsequent section deals with the effect of increasing or decreasing maternal stores on the well-being of the fetus and on subsequent lactational performance of the mother.

H. Effects of Maternal Nutrition on Fetal Growth and on Subsequent Lactation

In dealing with storage of N by maternal tissues other than those directly concerned with reproduction, it was emphasized that pregnancy could proceed and lactation be initiated whether the maternal tissues were gaining or losing N. It was concluded, on the tacit assumption that it is undesirable to deplete body tissues of their protein or to lay down unwanted reserves, that net protein needs in pregnancy are simply the minimal amounts which are retained. When converted into dietary terms, maternal needs would thus be given by the expression:

$$\text{Daily protein requirement during pregnancy as an addition to normal adult needs} = \frac{\text{Daily storage of protein in the uterus, the uterine wall, and the mammary gland}}{\text{Biological value of dietary protein expressed as a decimal} \times \text{True digestibility of protein expressed as a decimal}}$$

The assumption, which is really implicit in this argument, is that maintenance of maternal body protein close to prereproductive levels and the making of no allowance for a maternal protein gain have no effects on the subsequent well-being of the offspring. If pregnancy proceeds to an uneventful termination, lactation is successful and plentiful and the offspring are not placed at physiological risk by under- or overfeeding the mother during pregnancy and thereby altering her total body protein, then the assumption is tenable. Whether the mother gains or loses her own protein is immaterial on this hypothesis: the fetus is safeguarded from such dietary hazards and more or less lives a life of its own. If, however, fetal development or subsequent lactation is enhanced by ensuring that the mother's own tissues increase in mass, then the assumption is untenable. A quantitative examination of the relationship between the change in maternal protein mass and the success of reproduction as a whole is therefore desirable.

One extreme point of view about the effect of prenatal nutrition on the offspring was expressed by Wilkerson and Gortner (1932), who studied N deposition in pig embryos. They concluded: "In spite of the constant indirect communication with the maternal organism, mammalian embryos follow certain definite and fixed chemical courses during development which appear to be governed by the inherent nature of the embryo itself and the specific selective absorption of the placenta rather than any variations in maternal nutrition. . . ." With man, Thomson and Hytten (1960) clearly reached the same conclusion when they concluded that "Most authorities now agree that within a wide range of dietetic conditions, the human foetus will grow normally, if necessary at the expense of the maternal tissues." At the other extreme is the view based on experimental work with sheep that "Competition for available nutrients exists between the foetus and its mother when these are in short supply, . . . within the foetus, as within the mother's own body, some tissues and organs are penalized more than others" (Wallace, 1948).

These two points of view may well reflect differences in the behavior of the different species considered, but it is probable that this explanation is not the entire one. Questions of degree of undernutrition are involved: sheep commonly receive insufficient food, man and pigs do not.

It is difficult in many respects to separate effects of protein deficiency from those due to an over-all shortage of food. Both states of undernutrition lead to a reduction in maternal reserves of protein, and indeed of fat, for voluntary intake of food declines in protein deficiency despite an adequate calorific value of the diet per gram. Lack of energy in the diet probably results in a greater depletion of protein reserves than does mild deficiency of protein in the diet. In this regard, starvation of rats for 40 hours on the 9th day of pregnancy results in 28 % of the young having congenital malformations (Runner and Miller, 1956). Fasting later in pregnancy reduces birth weight in the rat by 40 % (Warkany, 1945; Kalter and Warkany, 1959). Fasting in the rabbit in late pregnancy not only results in reduction in fetal weight but prematurity of birth (Jonen, 1931). When protein-free diets are given during late pregnancy, the results depend on the length of time for which they are given. Seegers (1937), for example, found that even 2 days of specific undernutrition before term reduced the average weight of young rats but that living young could be born after 13 days subsistence on the diet, although they weighed about 20 %

less than normal and contained only 12.7 mg N/gm body weight instead of the usual 15.2 mg/gm. Feeding a diet deficient in an essential amino acid (a gelatin diet) had comparable effects. If a protein-free diet is given to the rat for more than 13 days before term, the chances of survival of the young are very slight.

The most extensive study of the effect of protein nutrition on pregnancy and lactation in the rat were carried out over six succeeding generations of rats by Slonaker (1938). Five diets ranging in protein content from 10.3 to 26.3 % protein were used and the results show that the average birth weight of the young was greatest for the diet containing 18 % protein. The dams of this group produced, however, the smallest litters, which may well have been fortuitous; Slonaker's results and conclusions can be criticized in this regard (Russel, 1948). Even so, his results do show that mortality of the young from birth to weaning was greatest in those rats given the diet containing least protein and, equally important, that very high dietary levels did not result in optimal performance.

Experiments with pigs, as reviewed by Duncan and Lodge (1960), show that low protein diets reduce the weight of the offspring and may reduce the number born. If minimal protein needs during the last part of pregnancy are assessed simply as the sum of endogenous losses and the growth of the reproductive structures, then it may be calculated that the protein content of a diet in which the protein has a Biological Value of 70 and a true digestibility of 85 % is about 14 % of the dry matter of the ration. There is no doubt that viable litters are born under these circumstances, but the evidence from practical trials suggest that heavier piglets are obtained when 15 to 17.5 % protein is given (Stevenson and Ellis, 1957).

With ewes the evidence is overwhelming that the birth weight and viability of the lamb at birth is profoundly modified by the nutritional status of the mother during pregnancy. Much of this work has been summarized by W. Thomson and Aitken (1959) and it shows a clear gradation of offspring weight with total food intake, or equally of offspring weight with the loss or gain of weight by the ewe during her pregnancy.

There is considerable evidence that the food supply during pregnancy affects subsequent lactational performance and that loss of maternal tissue weight in pregnancy results in diminution of lactation. This has been noted in rats (Perisse and Salmon-Legagneur, 1960), sheep (Thomson and Aitken, 1959: review), and cattle (Burt, 1956: review). The physiological mechanisms involved all appear to be related to the growth of the mammary gland, most of which is completed at the time of parturition. There is further evidence that undernutrition during pregnancy may result in a depression of the protein content of the milk (Rook, 1961: review), but the greatest effect is a quantitative one on the amount of secretion.

I. *Optimal Protein Nutrition during Pregnancy*

The precise definition of an allowance of protein optimal for pregnancy is clearly difficult. Offspring size and vigor, recovery of the mother from the strain of parturition, and her subsequent lactational performance all appear to be functions of the food intake during pregnancy and hence of the change in the protein content of the extrareproductive structures of the maternal body. Even so, much of the evidence which shows that the outcome of pregnancy is affected by changes in maternal nutrition has been obtained by observing declines in reproductive performance on reduction of the intake of food or of the protein in the diet. There is little evidence to show that overt overnutrition with respect to calories or protein increases performance, and what evidence there is suggests that any difference between minimal needs of protein and optimal allowances is quite small. It can, in fact, be argued that for a population of animals in which minimal needs are likely to vary from individual to individual, optimal allowances are probably best estimated as those which ensure that only a few individuals receive less than sufficient to meet their requirements, inferring that the remainder receives more. From estimates of variation in the protein needs of individuals, a safety margin defined in statistical terms as an amount which ensures that only one animal in 50 receives insufficient protein would entail giving about 10% more protein than the minimum requirement calculated (see p. 205). This approach, which combines the analytical approach with estimates of variability, does not define an optimal allowance of protein in terms of a desirable average N retention in extrareproductive structures. Rather, it defines the optimum in terms of a low probability of occurrence of subminimal levels of protein nutrition in the population as a whole.

V. Lactation

A. *Quantitative Aspects of Lactation*

The amount of protein secreted in milk each day is largely determined by the volume of milk secreted, for the protein concentration in milk is subject to relatively little variation. The values given in Table I for the amount of N secreted in milk by different species were average ones, and they cover an enormous individual variation. In cows, total lactation milk yields (lactation lasts 300 days or less) can vary from less than 100 kg to more than 800 kg. In sows, studies summarized by Duncan and Lodge (1960) show a variation in total lactation yield in 8 weeks from 84 kg to 498 kg. In women, Morrison's (1952) summary suggests a variation in yield in mid-lactation from less than 500 gm/day to about 3500 gm/day. These ranges exclude cases of agalactia, which are common in both sows and women, but not in cows. Nor do they include the maximal daily variation which occurs. During the course of lactation, milk secretion increases to a maximum and then declines; the maximum daily yield is usually about 30–40 % greater than the average daily yield. It is remarkable that such interindividual variation, amounting to a 6- to 10-fold range in the provision of food for the developing young, should occur, especially when it is considered that in the intrauterine stages of development variation in nutrient supply as judged by birth weight is so small.

There are many reasons for this variation, or rather much of the variation can be explained in terms of concomitant variation in the age, parity, body weight, nutritive status, and genetic constitution of the mother, by variation in the suckling stimulus provided by the young, and by external factors affecting the duration of lactation. These are not discussed here: several chapters in Kon and Cowie's (1961) book deal with these aspects. What is important to note is that the amount of protein secreted by an individual animal in its milk can depart very markedly from any mean estimate applicable to that species. This fact implies that protein requirements for the support of lactation in a species, if expressed in terms of the average quantity required each day, are not very meaningful. An example is the FAO recommendation that 30 gm of additional protein is required by the average woman to support lactation in addition to that in her normal diet (F.A.O., 1957). Average milk yield was assumed to be 850 ml/day. Some women secrete four times as much, some only a quarter of this amount. To provide more precise estimates of the protein needs of individuals, either cognizance must be taken of the amount of milk produced, or an alternative approach adopted in which it is assumed that the animal is able to regulate its food intake or energy consumption according to its needs. This latter approach, which is dealt with on page 215, is the only one applicable to animals in which milk protein yield is not readily measured.

B. *The Proteins of Milk*

Understandably, most investigational work concerned with the isolation and identification of the proteins of milk has ben done with cow milk. The classical fractionation methods of Rowland (1938) divided the proteins of cow milk into five fractions: casein, lactalbumin, lactoglobulin, serum protein, and a proteose-peptone fraction. A review of electrophoretic studies made up to 1956 by a committee dealing with nomenclature of milk proteins showed that the casein fraction could be subdivided into three distinct fractions, α-, β- and γ-casein; that from the remaining proteins, an α-lactalbumin, β-lactoglobulin, and a serum albumin could be crystallized, and immune euglobulins and pseudoglobulins could be separated. Electrophoresis indicated that the proteose-peptone fraction contained three fractions (Jeness *et al.*, 1956). By 1960, when a similar review was made

(Brunner *et al.*, 1960), it was realized that α-casein was heterogeneous, consisting of two main fractions, and that the β-lactoglobulin consisted of two distinct proteins, designated lactoglobulin A and lactogobulin B. Although these two fractions are under genetic control (Aschaffenburg and Drewry, 1957), they are remarkable since they do not appear to be immunologically distinct (Gough and Jeness, 1962). There is evidence, however, that each one is heterogeneous (Ogston and Tombs, 1957). A number of other minor proteins have been isolated and it now appears that the proteins of the milk of the cow consist of a complex and highly integrated system of a number of caseins and their aggregates, caseins being defined as acid-precipitable phosphoproteins, together with a system in the whey remaining after precipitation of the casein which contains at least eight major proteins and many minor ones, including lactoglobulins, immune proteins, and those proteins derived from the blood or from alveolar cells.

Some work, not in such comparable detail, has been done with milks of man and goat, and such studies are of interest from a comparative point of view. Investigations by Deutsch (1947) and by Sloan *et al.* (1961) have together embraced a very wide range of mammalian species. The latter electrophoretic studies suggest that in rodents, lagomorphs, and marsupials only one casein, corresponding to the α-casein of the cow, is present. In primates and some rodents two caseins are present, one corresponding to the α-casein of the cow and the other to a slower moving component not easily described as homologous to the cow β-casein. In most species studied, both an α- and β-casein are found, and it is only in the cow that the small γ-component is seen. The whey proteins from the milks of species from more primitive orders contain fewer distinct proteins than do those from the more advanced ones. It seems that in evolution the unique proteins of milk, the caseins, are present in all therian mammals, and that these caseins are usually but not invariably present in complex mixtures. In this regard, casein has been found in the milk of Echidna, a monotreme (Marston, 1926). This means that the origin of casein is phylogenetically remote, since therian animals and the monotremes have no common mammalian ancestor, their most recent common ancestry being in an early therapsid reptile. The evolutionary significance of the caseins may well be closely linked to the nurture of animals with endoskeletons of pseudoapatite, since casein is a phosphoprotein which readily forms complexes with calcium and is thus an excellent vehicle for transport of bone-forming elements from mother to young.

C. The Synthesis of the Proteins of Milk

Blackwood (1932), working at this institute with cows, was the first to demonstrate that the free amino acids of blood plasma were taken up from arterial blood by the mammary gland in amounts which she concluded could account for a substantial proportion of the protein synthesized. Her experiments, however, involved but small differences in the concentration of amino N in arterial and venous blood supplying the gland; in view of this the alternative proposal was made (W. R. Graham, Jr., *et al.*, 1938) that plasma proteins were the main precursor of the proteins of milk. These conclusions were based on equally small A/V differences in plasma protein concentrations. The early studies in this field have been reviewed by Barry (1961), and it is evident that until the application of isotopic labeling of amino acids to the problems of protein synthesis, few advances were made.

There is now strong evidence (see reviews by Barry, 1958, 1961, and by Folley, 1961) that the major portion of at least four of the essential amino acids to be found in the casein, lactalbumin, and β-lactoglobulin of milk are derived from a pool of amino acids in the mammary gland which is in equilibrium with the free amino acids of blood. The same is true of seven nonessential amino acids. The amino acids studied to date are lysine, methionine, tyrosine, valine, glutamine, glutamic acid, glycine, proline, serine, asparagine, and aspartic acid. Casein is a phosphoprotein with at least 40 %, and probably more, of the phosphorus present in monoester groups, presumably combined with serine. Phosphoserine is not, however, incorporated into casein as such, but the serine is phosphorylated after its incorporation (Kumar *et al.*, 1960). There is every reason to believe that the intracellular stages of protein synthesis in mammary gland from amino acids conforms to the general pattern of activation, transfer to S-RNA (soluble RNA) assembly on the RNA templates of endoplastic reticulum as seen in other protein-synthesizing tissues (see Folley, 1961, for review).

The immune globulins of milk and the small amount of serum albumin in milk, in contradistinction to the caseins, lactalbumin, and β-lactoglobulins, arise from the corresponding proteins in the blood (Larson, 1958). In early lactation in the cow, the globulin fraction can account for about 50 % of the total protein. From histological studies Dixon *et al.* (1961) concluded that the acinar cells of the mammary epithelium transport serum protein from the blood during the colostral phase of milk secretion. These same cells produce milk protein during lactation.

D. The Amino Acid Composition of Milk Proteins

The complexity of the protein distributions of milks of different mammals necessarily implies that the amino acid composition of the total protein secreted in milk may well vary from species to species. Analyses of defined casein fractions from human milk and cow milk show that homologous proteins do not have the same amino acid composition. Human κ-casein, for example, contains more cystine and fewer basic amino acids than does cow κ-casein (Alais and Jollès, 1962), and the whole caseins of cow, sheep,

TABLE X

AMINO ACID COMPOSITION OF THE MIXED PROTEINS OF ANIMAL MUSCLE AND OF THE MILKS OF FOUR SPECIES

Amino acid	Animal muscle[a]	Cow milk[b]	Human milk[c]	Rat milk[d]	Pig milk[e]
Arginine	7.2 ± 0.9	4.3	4.2	3.4	5.7
Histidine	2.1 ± 0.2	2.6	2.3	2.5	2.2
Lysine	7.6 ± 1.0	7.5	6.7	5.3	7.4
Tyrosine	3.1 ± 0.3	5.3	—	3.9	2.1
Tryptophan	1.2 ± 0.2	1.6	1.8	4.6	—
Phenylalanine	4–5	5.7	3.9	3.9	3.5
Cystine	1.1 ± 0.3	1.0	—	4.0	—
Methionine	3.2 ± 0.3	3.4	1.1	1.9	1.4
Serine	5.7	—	—	—	—
Threonine	5.3 ± 0.4	4.5	5.0	4.5	3.5
Leucine	8.0	11.3	9.3	5.5	8.0
Isoleucine	6.3	8.5	5.9	4.7	4.2
Valine	5.8	8.4	7.0	5.0	5.0
Glutamic acid	15	—	—	20.2	—
Aspartic acid	—	—	—	5.8	—
Glycine	4	2.3	—	1.7	—
Proline	6	—	—	7.8	—

[a] Block and Bolling (1947, 1951).
[b] Block and Mitchell (1946–1947).
[c] Morrison (1952).
[d] Glass (1957).
[e] Beacon and Bowland (1951).

and goat also show significant differences, cow milk containing half the number of alanine residues that are in sheep and goat milk (Jollès *et al.*, 1961). In cow milk, α- and β-caseins vary with respect to their content of proline, tryptophan, and tyrosine (Gordon *et al.*, 1949, 1950). Similarly, wide differences occur between the different proteins of milk; β-lactoglobulin, for example, contains much more lysine than casein. The amino acid composition of the major proteins of cow milk has been summarized by Henry (1957).

From a nutritional point of view it is probably more informative to deal with the amino acids present in the mixed proteins of the milk of a species, bearing in mind that some variation due to variation in the proportions of different proteins in milk is necessarily encompassed. Table X summarizes data on the amino acid composition of the mixed proteins of the milk of different species, and, for comparative purposes, the composition of the mixed proteins of animal muscle, that is, of the major protein mass of the maternal organism.

Table X shows that the four milks listed contain less lysine and arginine and about the same amount of histidine as muscle proteins. This suggests that diets adequate in basic amino acids for growth are also qualitatively adequate for lactation. For the aromatic amino acids, tyrosine, tryptophan, and phenylalanine, with the exception of the high value for tryptophan in rat milk, their content in the proteins of milk probably differs little from that in muscle. Milks do not contain very different amounts of sulfur-containing amino acids than do muscle proteins; they contain about the same amount of threonine, but tend to contain more of the leucines than muscle proteins.

E. Biological Values of Proteins in Lactation

From these analyses, all of which are subject to within-species variation as well as to analytical error, it would appear that the protein secreted in milk probably contains slightly less basic amino acid and slightly more of the leucines than does mixed muscle protein. If it is acceded that one of the determinants of the Biological Value of a protein is the proportion of "essential" amino acids retained in the body or secreted in milk relative to those supplied in the diet, then Biological Values of protein in lactation should be the same or very similar to those for growth. This general statement can be, and indeed has been, criticized (Ericson, 1961). Although the amino acid composition of the body of all species is very comparable, the requirements of amino acids for weight maintenance vary from species to species and so Biological Values could well vary according to the proportional demands of net synthesis and maintenance. This does not, however, detract from the hypothesis stated above in which two functions, both involving net synthesis, are compared. The Biological Value (BV) of a protein for the joint function of maintenance and lactation can be defined as an expression which entails measurement of the N secreted in the milk, a

$$BV = \frac{100 \times \left\{ \begin{array}{l} \text{body retention of N + milk N + endogenous N} \\ \quad + \text{metabolic N in feces} \end{array} \right\}}{\text{Truly digestible N of diet}} \qquad (6)$$

feat which is only accomplished with ease in animals kept for milk production. With milking cows, an examination of all published N retention studies (120 experiments in all) showed that the Biological Value of protein in lactation was 74.8 with a standard deviation of ±4.6 (Blaxter, 1958). The relative constancy of the Biological Value, as indicated by the coefficient of variation, is to be expected in view of the massive conversion of dietary to microbial protein in the rumen of the cow, so that differences in Biological Value, as between proteins, are small in ruminants. With growing cattle, recalculation of data obtained by Swanson and Herman (1943) also gave a Biological Value of 75, while a more recent analysis of 38 further experiments (McDonald and Preston, 1962) gave a mean

Biological Value in growing calves of 73.7 ± 4.8. There thus appears to be no difference of any magnitude in the Biological Value of protein for lactation in the cow and for growth in the cow, a finding in agreement with the hypothesis outlined above. With lactating sows, Werner and Hennig (1960) obtained a Biological Value of a skimmed cow milk protein of 80–98 (range of 7 experiments). With growing rats the Biological Value of unheated skim milk protein varies from 81–92 (Henry, 1957). This again suggests that Biological Values of proteins for lactation probably differ little from those for growth.

No data obtained with laboratory species have been found which refute this argument. Russell (1948) summarized data which showed that in rats, supplements of S-containing amino acids improved lactational performance when casein diets were given. Similar results were later obtained by Nelson and Evans (1958). All these experiments show the S-containing amino acids are the limiting amino acids when casein is the source of N for the lactating rat. S-containing amino acids are also the first to be limiting for growth of the rat when casein is the N source (Block and Mitchell, 1946–1947). The only quantitative data available show, however, that the lactating rat may require proportionately slightly more cystine and methionine when given a casein diet than does a growing one (Nelson and Evans, 1961). The comparison, however, is not entirely valid for reasons given by Nelson and Evans, although it accords with the high cystine content of the rat milk given in Table X. It thus appears as a generalization, to which future experimentation may provide small exceptions, as in the instance of the rat, that the Biological Value of dietary protein for lactation probably differs little from that for juvenile growth.

F. The Effect of Energy and Protein Deficiencies on Lactation

It is well recognized in rats (see review by Nelson and Evans, 1961), sheep (see review by W. Thomson and Aitken, 1959), cattle (see reviews by Reid, 1956; Burt, 1957; and Rook, 1961), man (see review by Hytten and Thomson, 1961), and in pigs (see review by Duncan and Lodge, 1960) that the amount of milk secreted depends on the energy supplied in the diet. In the cow a reduction in the energy intake leads to a reduction in the protein of milk, largely due to a reduction in its casein content. In a recent experiment, Rook and Line (1961) found that total N in milk was reduced from 0.52 to 0.48 % by underfeeding. Similar results have been obtained with pigs (Lodge, 1959) in which the protein content of the milk was reduced from 6.1 to 5.5 % when the amount of diet was reduced by 26 %. Underfeeding with respect to energy thus produces effects on total protein secretion which are greater than those deduced from the reduction in milk volume alone.

When the dietary supply of protein for cows is reduced, small depressions occur in the N content of the milk (Rowland, 1946) while large increases in N supply lead to small increases in N content, mostly due to nonprotein N (Rook, 1953). Milk yield is not affected very much by small changes in the cow's protein nutrition (Bartlett and Blaxter, 1947). In ewes, a reduction in milk yield may occur in protein deficiency, but the growth of wool, which accounts for a considerable part of the N retention of sheep, is depressed by protein undernutrition both in pregnancy and lactation (W. Thomson and Aitken, 1959). Lenkeit and Gütte (1957) have found that protein deficiency in sows leads to a reduction of yield. With women, data on effects on yield of milk are not very reliable (Hytten and A. M. Thomson, 1961), although a small decrease in yield probably occurs in mild deficiency. The total N content of milk tends to increase with increasing N content of human diets (Karmarkar *et al.*, 1959). No effects on either yield or total N content of human milk were, however, noted by Deb and Cama (1962).

These results show that, in protein deficiency, a variety of species have the ability to make good a lack of protein in their diet from their body tissues, and that for short periods of time, and for mild deficiency, no great reduction in the amount of protein secreted occurs.

G. Generalized Protein Requirements in Lactation

It is not intended to give a detailed account of experiments which relate to the protein requirements of different animal species during lactation. Rather, an attempt has been made from the sparse information available to draw generalizations about the protein needs of lactation in the mammalia as a class. From any such interspecies generalizations there are bound to be exceptions, and it is hoped that the general statements enable a placing of species differences in their true perspective. One interesting example, that of man, is dealt with in detail.

To generalize protein needs in lactation necessarily means that protein needs for the maintenance of the body must be considered too. From the Terroine and Sorg-Matter law (1927), it appears that the sum of the endogenous losses of N and the metabolic fecal N of adult mammals is about 2.3 mg N/basal kilocalorie of heat produced. If dietary protein is assumed to have a Biological Value of 70 and the average specific dynamic action of food is assumed to be of 15 %, then the ratio of digested protein energy to total metabolizable calories in diet required to maintain any adult mammal is 9.8 kcal digested protein/100 kcal metabolizable energy. A statement of protein requirements in terms of a ratio of digested protein energy to metabolizable energy is preferred to a statement in terms of percentage concentration of digested protein in the dry matter of a diet because of the wide variations between classes of mammals in the energy value of their habitual dietaries. Herbivora, for example, can eat food with a metabolizable energy per gram of dry food as low as 2 kcal/gm, while omnivora can consume food with more than twice as much metabolizable energy per unit weight. The expression of protein intake as a proportion of total calories is discussed in the introductory chapter to this part of the treatise. In that chapter, the convention used by Platt and his colleagues of expressing food protein calories as a percentage of *ingested* calories is given. For the reasons given above, the loss of energy due to specific dynamic action will raise the ratio when expressed in relation to *metabolizable* calories.

The validity of this relationship can be checked by particular examples. In man, 9.8 % digested protein calories/100 kcal metabolizable energy corresponds to an allowance of 53 gm protein daily for adult man with an energy requirement of 3000 kcal/day.* This value is lower than the conventional standard of 1 gm protein/kg body weight (ca., 70 gm) but higher than noted in many critical experiments in which protein needs have been accurately determined (Bricker *et al.*, 1945). With cattle weighing 454 kg (1000 lb) with a metabolizable energy requirement of 12,000 kcal, the relationship corresponds to an allowance of 214 gm digested protein/day. Current British recommendations are 272 gm, American ones are 300 gm, Scandinavian ones 220–320 gm, and in German-speaking countries requirements are usually taken to be 0.5 gm/kg body weight, corresponding to 227 gm. With rats weighing 200 gm and a basal metabolism of 25 kcal/day, the generalization corresponds to an allowance of 397 mg protein/day. Burroughs *et al.* (1940) found 80 mg N (500 mg protein) to result in small retentions of N in adult rats. Results with rats, men, and cows thus support the generalization.

During lactation a considerable augmentation of the demand for both energy and protein takes place. The ratio of protein calories to total calories in the milk of 20 mammalian species, excluding aquatic ones and man, is 22.2 with a standard deviation of ±3.8 (Blaxter, 1961), indicating that two thirds of all nonaquatic mammalian species produce milks in which the ratio of protein calories:total calories lies within the small range of 18.4 to 26.0. If it is now assumed that the Biological Value of dietary protein in lactation is on an average 70 % and the specific dynamic action of food associated with secretion of milk is 30 %, then the specific protein

* This corresponds to an intake of 7% of the ingested calories as protein calories and if the dietary protein is assumed to have a biological value of 70, becomes approximately 5% of ideal reference protein per 100 calories consumed. This is the order of magnitude of protein requirements computed in the introductory chapter to this section and also in Chapter 14.

requirement for lactation, expressed as a dietary ratio, will be 22.2 % of the metabolizable energy.

With protein needs for the maintenance of all species assessed as 9.8 % and with protein needs for the specific function of lactation assessed as 22.2 % of the energy supply, then it is obvious that protein needs expressed as dietary concentrations exceed 9.8 % but cannot exceed 22.2 %, and that the main determinant of the protein content of the diet for the support of lactation in a species depends primarily on the rate of milk secretion relative to the size (and hence maintenance cost) of the species concerned.

Since animals tend to increase their voluntary food intake during lactation in proportion to the increased energy demands, an approximation to their minimal protein needs can be obtained from measurements of food intake. In general, if the food intake necessary to maintain a nonlactating nonpregnant animal is x gm and the food intake of the same animal when lactating is Lx gm, the protein requirement expressed as the percentage of digested protein energy in the total metabolizable energy will be

$$\frac{9.8 + (L - 1) \times 22.2}{L} \qquad (7)$$

The nonpregnant rat eats 12–15 gm food daily and at the peak of lactation consumes up to 60 gm daily with an average of about 40 gm (Murray, 1941; Kennedy, 1953), that is, L in the above equation is maximally 4 and the digestible protein requirement is thus maximally 19.1 kcal/100 kcal. In this regard, Slonaker (1938) concluded that 18 % of dietary calories as protein was optimal for lactation in the rat and Russell (1948) concluded from an extensive review of the literature that 17–19 % protein in the diet would meet requirements for lactation. Goettsch (1960) suggested 16.3 % protein was the absolute minimum. With cows, maintenance intakes of metabolizable energy are about 12,000 kcal and in lactation food intake in terms of metabolizable energy can increase by up to three-fold. The optimal concentrations of protein in this instance is 14.8 kcal digested protein/100 kcal metabolizable energy. This value agrees with practical recommendations when calculated to the basis of digestible protein in dry food.

It appears from these two examples that protein needs for lactation in the mammalia can be generalized and that the interspecies range of variation in requirements expressed as dietary concentrations of protein is almost entirely related to the intensity of milk secretion in the species concerned. This broad generalization makes the protein demands of lactation in woman a remarkable exception. Woman is an exception to the generalization because the ratio of protein energy to total energy in human milk is about 9 kcal/100 kcal. This ratio is to date lower than has been recorded in any other species. Even in the monkey (Macaca mulatta) the ratio is 16 kcal/100 kcal, which is within the range of values found in other non-aquatic mammals (Van Wagenen *et al.*, 1941; D. E. Evans, 1959). Making the same generalization as that employed in dealing with mammals other than man, it is evident that protein needs in lactating women will vary between 9.8 kcal/100 kcal metabolizable energy, that is the maintenance need,* and 9.0 kcal/100 kcal, the lactation need. This means that, in contradistinction to other mammalian species, protein needs of woman expressed as dietary concentrations do not increase during lactation above those capable of maintaining her when neither pregnant nor lactating. It would seem that increases in the amount of the same type of diet should be adequate for lactation in women. This conclusion is contrary to the recommendations of the British Medical Association that the need for protein increases from 11 % of the total calories, which is a liberal allowance, to 14 % during lactation (Brit. Med. Assoc., 1950). The recommendations on protein needs made by the F.A.O. (1957) include the statement "women who normally consume liberal quantities of protein need not consume additional protein". This is in agreement with the conclusions reached above, and tends to confirm the supposition that woman is unique among the mammalia in requiring a diet no richer in protein than normal to support lactation. In this regard, modern dietary studies show that women do not increase their intake of protein rich foods more than they do of fatty or starchy foods during lactation (A. M. Thomson, 1959), nor is there evidence from the food habits and taboos of primitive people that there is any increased emphasis given to protein-rich foods during lactation.

References

Abbas, T. M., and Tovey, J. E. (1960). *Brit. Med. J.* **I**, 476.

Alais, C., and Jollès, P. (1962). *Nature* **196**, 1098.

Amoroso, E. C. (1952). *In* "Physiology of Reproduction," 3rd ed. (F. H. A. Marshall and A. S. Parkes, eds.), Vol. 2, p. 28. Little, Brown, Boston, Massachusetts.

Aršavskij, I. A. (1959). *Zh. Obshch. Biol.* **20**, 104; see *Nutr. Abstr. Revs.*

Aschaffenburg, R., and Drewry, J. (1957). *Nature* **180**, 376.

Bangham, D. R., Hobbs, K. R., and Terry, R. J. (1958). *Lancet* **II**, 351.

Bar, P., and Daunay, R. (1905). *J. Physiol. Pathol. Gén.* **7**, 832.

Barboriak, J. J., de Bella, G., Setnikar, I., and Krehl, W. A. (1958). *Am. J. Physiol.* **193**, 89.

Barcroft, J. (1948). "Researches on Prenatal Life." Thomas, Springfield, Illinois.

Barry, J. M. (1958). *Proc. Roy. Soc.* **B149**, 380.

Barry, J. M. (1961). *In* "Milk: The Mammary Gland and Its Secretion" (S. K. Kon and A. T. Cowie, eds.), Vol. 1, p. 389. Academic Press, New York.

Bartlett, S., and Blaxter, K. L. (1947). *J. Agr. Sci.* **37**, 2.

Beacon, S. E., and Bowland, J. P. (1951). *J. Nutr.* **45**, 419.

Benedict, F. G. (1938). *Carnegie Inst. Wash. Publ.* **503**.

* Equivalent to 5% of caloric intake as ideal reference protein as discussed earlier.

Blackwood, J. H. (1932). *Biochem. J.* **26,** 722.

Blaxter, K. L. (1958). *Proc. Conf. on Sci. Principles of Feeding Farm Livestock, Brighton, England,* p. 21. (Farmer & Stock Breeder Pub. Ltd; London.)

Blaxter, K. L. (1961). *In* "Milk: The Mammary Gland and Its Secretion" (S. K. Kon and A. T. Cowie, eds.), Vol. II, p. 305. Academic Press, New York.

Blaxter, K. L., and Wood, W. A. (1952). *Brit. J. Nutr.* **6,** 1.

Blaxter, K. L., Brown, F., and MacDonald, A. M. (1953). *Brit. J. Nutr.* **7,** 105.

Block, R. J., and Bolling, D. (1947). "Amino Acid Composition of Proteins and Foods," 1st ed. Thomas, Springfield, Illinois.

Block, R. J., and Bolling, D. (1951). "Amino Acid Composition of Proteins and Foods," 2nd ed. Thomas, Springfield, Illinois.

Block, R. J., and Mitchell, H. H. (1946–1947). *Nutr. Abstr. Rev.* **16,** 249.

Bollinger, A., and Gross, R. (1960). *Australian J. Sci.* **22,** 292.

Bond, C. F. (1958). *Endocrinology* **63,** 285.

Bourdel, G. and Jacquot, R. (1960). *In* "Amino Acids, Peptides and Proteins," Vol. 4. A.E.C., Soc. Chim. Org. et Biol., Paris.

Boyne, A. W., Chalmers, M. I., and Cuthbertson, D. P. (1953). *Z. Physiol. Chem.* **295,** 424.

Brambell, F. W. R., Hemmings, W. A., Henderson, M., Parry, H. J., and Rowlands, W. T. (1949). *Proc. Roy. Soc.* **B136,** 131.

Brambell, F. W. R., Hemmings, W. A., and Oakley, C. L. (1959). *Proc. Roy. Soc.* **B150,** 312.

Brambell, F. W. R., Hemmings, W. A., Oakley, C. L., and Porter, R. R. (1960). *Proc. Roy. Soc.* **B151,** 478.

Bricker, M., Mitchell, H. H., and Kinsmann, G. M. (1945). *J. Nutr.* **30,** 269.

Brit. Med. Assoc. (1950). Report of Committee on Nutrition.

Brody, S. (1945). "Bioenergetics and Growth." Reinhold, New York.

Brody, S., and Nisbet, R. (1938). *Univ. Missouri Agr. Expt. Sta., Res. Bull.* **285.**

Brooksby, J. B., and Newton, W. H. (1938). *J. Physiol. (London)* **92,** 136.

Bruce, H. M. (1958). *Proc. Roy. Soc.* **B149,** 421.

Brunner, J. R., Ernstrom, C. A., Hollis, R. A., Larson, B. L., Whitney, R. McL., and Zittle, C. A. (1960). *J. Dairy Sci.* **43,** 901.

Bunge, H. (1898). "Lehrbuch der physiologischen Chemie," 4th ed. Leipzig.

Burroughs, E. W., Burroughs, H. S., and Mitchell, H. H. (1940). *J. Nutr.* **19,** 363.

Burt, A. W. A. (1956). *Dairy Sci. Abstr.* **18,** 884.

Burt, A. W. A. (1957). *Dairy Sci. Abstr.* **19,** 435.

Campbell, R. M., and Kosterlitz, H. W. (1949). *J. Endocrinol.* **6,** 171.

Campbell, R. M., Innes, I. R., and Kosterlitz, H. W. (1953). *J. Endocrinol.* **9,** 68.

Carlyle, A. (1945). *J. Physiol. (London)* **104,** 34P.

Clements, F. W. (1949). "Infant Nutrition, Its Physiological Basis." Williams & Wilkins, Baltimore, Maryland.

Cole, H. H., and Hart, G. H. (1938). *Am. J. Physiol.* **123,** 589.

Courrier, R., and Bacless, M. (1955). *Rappts. 3ie Rênunion Endocrinologistes Langue Française* (Masson, Paris).

Dancis, J., Bravermann, N., and Lind, J. (1957). *J. Clin. Invest.* **36,** 398.

Dancis, J., Olsen, G., and Folkaart, G. (1958). *Am. J. Physiol.* **194,** 44.

Dancis, J., Lind, J., Oratz, M., Smolens, J., and Vara, F. (1961). *Am. J. Obstet. Gynecol.* **82,** 167.

D'Arcy Thompson, W. (1952). "On Growth and Form," 2 vols. Cambridge Univ. Press, London and New York.

Darke, S. J. (1960). *Brit. J. Nutr.* **14,** 115.

Davidson, J. N. (1950). "Biochemistry of Nucleic Acids." Methuen, London (2nd ed., Wiley, 1953).
Davidson, J. N., and Waymouth, C. (1944a). *Biochem. J.* **38,** 39.
Davidson, J. N., and Waymouth, C. (1944b). *Biochem. J.* **38,** 379.
Davies, W. L. (1936). "The Chemistry of Milk." Chapman & Hall, London.
Davis, J. G., and McDonald, F. J. (1953). "Richmond's Dairy Chemistry," 5th ed. Chas. Griffin, London.
Dawkins, M. J. R. (1959). *Proc. Roy. Soc.* **B154,** 284.
Dean, B. T., and Tribble, L. F. (1961). *Univ. Missouri Agr. Expt. Sta., Res. Bull.* **774.**
Deb, A. K., and Cama, H. R. (1962). *Brit. J. Nutr.* **16,** 65.
de Bella, G. (1958). *Yale J. Biol. Med.* **30,** 368.
Deutsch, H. F. (1947). *J. Biol. Chem.* **169,** 437.
Dickerson, J. W. T., and Widdowson, E. M. (1960). *Biochem. J.* **74,** 247.
Dixon, F. J., Weigle, W. O., and Vasquez, J. S. (1961). *Lab. Invest.* **10,** 216.
Duncan, D. L., and Lodge, G. A. (1960). *Commonwealth Bur. Animal Nutr. Tech. Commun.* **21.**
Ericson, L. E. (1961). *Acta Physiol. Scand.* **52,** 90.
Evans, D. E. (1959). *Dairy Sci. Abstr.* **21,** 277 (Review).
Evans, R. E. (1929). *J. Agr. Sci.* **19,** 752.
F. A. O. (1957). "Protein Requirements." F. A. O. Nutr. Studies No. 16. Food and Agriculture Organization of the United Nations.
Flexner, L. B., and Pohl, H. A. (1941a). *J. Cellular Comp. Physiol.* **18,** 49.
Flexner, L. B., and Pohl, H. A. (1941b). *Am. J. Physiol.* **132,** 594.
Folley, S. J. (1961). *Dairy Sci. Abstr.* **23,** 511.
Garry, R. C., and Stiven, D. (1936). *Nutr. Abstr. Rev.* **5,** 855.
Geschwind, I., and Li, C. H. (1949). *J. Biol. Chem.* **180,** 467.
Glass, R. L. (1957). See D. E. Evans (1959).
Goettsch, M. (1960). *J. Nutr.* **70,** 307.
Gordon, W. G., Semmett, W. F., Cable, R. S., and Morris, M. (1949). *J. Am. Chem. Soc.* **71,** 3293.
Gordon, W. G., Semmett, W. F., and Bender, M. (1950). *J. Am. Chem. Soc.* **72,** 4282.
Gough, P., and Jeness, R. (1962). *J. Immunol.* **89,** 511.
Graham, N. M. and Williams, A. J. (1962). *Australian J. Agr. Sci.* **13,** 894.
Graham, W. R., Jr., Peterson, V. E., Houchin, O. B., and Turner, C. W. (1938). *J. Biol. Chem.* **122,** 275.
Greenbaum, A. L., and Slater, T. F. (1957). *Biochem. J.* **66,** 155.
Griffith, D. R., and Turner, C. W. (1961). *Proc. Soc. Exptl. Biol. Med.* **106,** 448.
Grzimek, B. (1957). *Milchwissenschaft* **12,** 231.
Hagermann, D. D., and Villee, C. A. (1960). *Physiol. Rev.* **40,** 313.
Harkness, M. L. R., and Harkness, R. D. (1954). *J. Physiol.* (*London*) **123,** 492.
Harkness, M. L. R., and Harkness, R. D. (1956a). *J. Physiol.* (*London*) **134,** 135.
Harkness, N. L. R., and Harkness, R. D. (1956b). *J. Physiol.* (*London*) **132,** 492.
Harkness, M. L. R., Harkness, R. D., and Moralee, B. E. (1957). *J. Physiol.* (*London*) **135,** 270.
Harkness, R. D., and Moralee, B. E. (1956). *J. Physiol.* (*London*), **132,** 502.
Hennig, A. (1959). *Arch. Tierernähr.* **9,** 29.
Henry, K. M. (1957). *Dairy Sci. Abstr.* **19,** 604.
Herring, P. T. (1920). *Brit. Med. J.* **II,** 886.
Huggett, A. St. G. (1941). *Physiol. Rev.* **21,** 438.
Huggett, A. St.G., and Frazer, J. F. D. (1959). *J. Physiol.* (*London*) **146,** 21P

Hunscher, H. A., Donelson, E., Nims, B., Kenyon, F., and Macy, I. G. (1933). *J. Biol. Chem.* **99,** 507.

Hytten, F. E., and Duncan, D. L. (1956). *Nutr. Abstr. Rev.* **26,** 855.

Hytten, F. E., and Thomson, A. M. (1961). *In* "Milk: The Mammary Gland and Its Secretion" (S. K. Kon and A. T. Cowie, eds.), vol. II, p. 3. Academic Press, New York.

Jacobsohn, D. (1961). *In* "Milk: The Mammary Gland and Its Secretion" (S. K. Kon and A. T. Cowie, eds.), Vol. I, p. 127. Academic Press, New York.

Jakobsen, P. E. (1957). *299de Beretng. Forsøgslab. Kobenhavn.*

Jeness, R., Larson, B. L., McMeekin, T. L., Swanson, A. M., Shitnah, C. H., and Whitney, R. McL. (1956). *J. Dairy Sci.* **39,** 536.

Jørgensen, G. (1960). *Dansk Pelsdgravel* **23,** 37.

Jollès, P., Alais, C., and Jollès, J. (1961). *Biochim. Biophys. Acta* **51,** 309.

Jonen, P. (1931). *Z. Geburtshilfe Gynäkol.* **101,** 50.

Kalter, H., and Warkany, J. (1959). *Physiol. Rev.* **39,** 69.

Karmarkar, M. G., Chokshi, H. R., Shah, V. K., and Ramakrishnan, C. V. (1959). *J. Pediat.* **55,** 481.

Kennedy, G. C. (1953). *Proc. Roy. Soc.* **B140,** 578.

Kenneth, J. H. (1943). *Bur. Animal Breeding, Edinburgh, Tech. Commun.* **5.**

Kelin, M. (1935). *Arch. Anat. Microscop.* **31,** 397.

Kon, S. K., and Cowie, A. T., eds. (1961). "Milk: The Mammary Gland and Its Secretion." Academic Press, New York.

Kossel, A. (1882). *Z. Physiol. Chem.* **7,** 7.

Kumar, K. S., Sundararajan, T. A., and Sarma, P. (1960). *J. Biol. Chem.* **235,** 679.

Larson, B. L. (1958). *J. Dairy Sci.* **41,** 1033.

Leitch, I. (1957). *Brit. J. Nutr.* **16,** 38.

Leitch, I., Hytten, F. E., and Billewicz, W. Z. (1959). *Proc. Zool. Soc. London* **133,** 11.

Lenkeit, W., and Gütte, J. O. (1955). *Z. Tierernähr. Futtermittelk.* **10,** 94.

Lenkeit, W., and Gütte, J. O. (1957). *Z. Tierernähr. Futtermittelk.* **12,** 58.

Lenkeit, W., Gütte, J. O., and Streuter-Petermoller, A. (1955a). *Z. Tierernähr. Futtermittelk.* **10,** 228.

Lenkeit, W., Gütte, J. O., Warneke, W., and Kirchoff, W. (1955b). *Z. Tierernähr. Futtermittelk.* **10,** 351.

Ling, E. R., Kon, S. K., and Porter, J. WG. (1961). *In* "Milk: The Mammary Gland an Its Secretion" Vol. II, p. 195. Academic Press, New York.

Linzell, J. L. (1959). *Physiol. Rev.* **39,** 534.

Lister, U. M. (1961). *J. Obstet. Gynaecol. Brit. Commonwealth* **68,** 405.

Lodge, G. A. (1959). *J. Agr. Sci.* **53,** 177.

Lovell, R., and Rees, T. A. (1961). *In* "Milk: The Mammary Gland and Its Secretion" S. K. Kon and A. T. Cowie, eds.), Vol. II, p. 363. Academic Press, New York.

Ludvigsen, J., and Thorbek, G. (1959). *311 de Beretn. Forsøgslab. København.*

Ludvigsen, J., and Thorbek, G. (1960). *320 de Beretn. Forsøgslab. København.*

Lund, C. J. (1951). *Am. J. Obstet. Gynecol.* **62,** 947.

McDonald, I., and Preston, T. R. (1962). Private communication.

McDougall, E. I. (1949). *Biochem. J.* **45,** 397.

McGaughey, H. S., Corey, E. L., Scoggin, W. A., Bibbitt, O. B., and Thornton, W. N. (1959). *Am. J. Obstet. Gynecol.* **78,** 844.

McLean, P. (1958). *Biochim. Biophys. Acta* **30,** 303.

Macy, I. G., and Kelly, H. J. (1961). *In* "Milk: The Mammary Gland and Its Secretion" (S. K. Kon and A. T. Cowie, eds.), Vol. II, p. 265. Academic Press, New York and London.
Marston, H. R. (1926). *Australian J. Exptl. Biol. Med. Sci.* **3,** 217.
Mitchell, H. H., Carroll, W. E., Hamilton, T. S., and Hunt, G. E. (1931). *Univ. Illinois Agr. Expt. Sta. Bull.* **375.**
Moon, R. C. (1962). *Am. J. Physiol.* **203,** 939.
Morrison, S. D. (1952). *Commonwealth Bur. Animal Nutr., Tech. Commun.* **18.**
Morrison, S. D. (1956). *J. Physiol.* (*London*) **133,** 167.
Morse, L. M., and Schmidt, C. L. A. (1944). *Proc. Soc. Exptl. Biol. Med.* **46,** 472.
Moulton, C. R. (1923). *J. Biol. Chem.* **57,** 79.
Moustgaard, J. (1956). *Proc. 3rd Intern. Congr. Animal Reproduction, Cambridge,* 1956, p. 12.
Murray, G. M. (1941). *Onderstepoort J. Vet. Sci. Animal Ind.* **16,** 331.
Myant, N. B. (1958). *J. Physiol.* (*London*) **142,** 329.
Needham, D. M., and Cawkwell, J. M. (1957). *Biochem. J.* **65,** 540.
Nehring, K. (1957). *Biochem. Z.* **328,** 549.
Nelson, M. M., and Evans, H. M. (1958). *Proc. Soc. Exptl. Biol. Med.* **99,** 723.
Nelson, M. M., and Evans, H. M. (1961). *In* "Milk: The Mammary Gland and Its Secretion" (S. K. Kon and A. T. Cowie, eds.), Vol. II, p. 137. Academic Press, New York.
Nemeth, A. M. (1959). *J. Biol. Chem.* **234,** 2921.
Nemeth, A. M., and Dickermann, H. (1960). *J. Biol. Chem.* **235,** 1761.
Neuhaus, U. (1959). *Z. Tierzücht. Züchtungsbiol.* **73,** 370.
Newton, W. H. (1935). *J. Physiol.* (*London*) **84,** 196.
Ogston, A. G., and Tombs, M. P. (1957). *Biochem. J.* **66,** 399.
Page, E. W., Glendening, M. B., Margolis, A., and Harper, H. A. (1957). *Am. J. Obstet. Gynecol.* **73,** 589.
Paton, D. N. (1903). *Lancet* **II,** 21.
Payne, J. M. (1957). *J. Pathol. Bacteriol.* **73,** 141.
Pederson, K. O. (1945). "Ultracentrifugal Studies on Serum and Serum Fractions." Almqvist & Wiksell, Upsala.
Penzes, L. (1961). *Allattenyesztes* (*Budapest*) **10,** 175; see *Nutr. Abstr. Rev.* **32,** 161.
Perisse, J., and Salmon-Legagneur, E. (1960). *Arch. Sci. Physiol.* **14,** 105.
Pike, R. L., Suder, H. B., and Rose, N. L. (1954). *J. Nutr.* **52,** 297.
Poo, L. J., Lew, W., Lee, D. D., and Addis, T. (1940). *Proc. Soc. Exptl. Biol. Med.* **19,** 505.
Rashevsky, N. (1960). "Mathematical Biophysics: Physicomathematical Foundations of Biology," 3rd ed., Vol. 2. Dover Publ., New York.
Reid, J. T. (1956). *Cornell Univ. Agr. Expt. Sta. Mem.* **344.**
Richardson, K. C., and Folley, S. J. (1954). *Nature* **174,** 828.
Robinson, D. S. (1952). *Biochem. J.* **52,** 621.
Robinson, P. (1948). *J. Agr. Sci.* **38,** 345.
Rombauts, P. (1959a). *Compt. Rend. Soc. Biol.* **153,** 556.
Rombauts, P. (1959b). *Compt. Rend.* **248,** 1859.
Rook, J. A. F. (1953). *Proc. Nutr. Soc.* **12,** viii.
Rook, J. A. F. (1961). *Dairy Sci. Abstr.* **23,** 251.
Rook, J. A. F., and Line, C. (1961). *Brit. J. Nutr.* **15,** 109.
Rowland, S. J. (1938). *J. Dairy Res.* **9,** 47.
Rowland, S. J. (1946). *Dairy Ind.* **11,** 656.
Rüsse, I. (1961). *Zentr. Vet. Med.* **8,** 252.

Runner, M. N., and Miller, J. R. (1956). *Anat. Record* **124**, 437.
Russell, F. C. (1948). *Commonwealth Bur. Animal Nutr., Tech. Commun.* **16**.
Schreier, K., Wirth, K., and Hack, W. (1956). *Z. Physiol. Chem.* **304**, 186.
Scott, D. B. M., and Lisi, A. G. (1960). *Biochem. J.* **77**, 52.
Seegers, W. H. (1937). *Am. J. Physiol.* **119**, 474.
Selye, H., Collip, J. B., and Thomson, D. L. (1934). *Endocrinology* **18**, 237.
Setnikar, I. (1957). *Boll. Soc. Ital. Biol. Sper.* **33**, 1636.
Setnikar, I., and Barron, D. H. (1958). *Quart. J. Exp. Physiol.* **43**, 284.
Simon, K. J. (1959). *Indian Vet. J.* **36**, 500.
Singer, E. J., Hochstrasser, H., and Cerecedo, L. R. (1956). *Growth* **20**, 229.
Sloan, R. E., Jeness, R., Kenyon, A. L., and Regehr, E. A. (1961). *Comp. Biochem. Physiol.* **4**, 47.
Slonaker, J. R. (1938). *Am. J. Physiol.* **123**, 526.
Smith, D. M. (1960). *New Zealand Agr. Res.* **3**, 745.
Souders, H. J., and Morgan, A. F. (1957). *Am. J. Physiol.* **191**, 1.
Soxhlet, F. (1878). *Ber. Ver. Stat. Wien no. 1.*
Spiro, R. G. (1960). *J. Biol. Chem.* **235**, 2860.
Spray, C. M. (1950). *Brit. J. Nutr.* **4**, 354.
Stephenson, S. K. (1962). *Nature* **196**, 1070.
Stevenson, J. W., and Ellis, N. R. (1957). *J. Animal Sci.* **16**, 877.
Swanson, E. W., and Herman, H. A. (1943). *Univ. Missouri Agr. Expt. Sta., Res. Bull.* **372**.
Terroine, E. F., and Sorg-Matter, H. (1927). *Arch. Intern. Physiol.* **29**, 121.
Thomson, A. M. (1959) *Brit. J. Nutr.* **13**, 190.
Thomson, A. M., and Hytten, F. E. (1960). *Brit. J. Nutr.* **19**, 5.
Thomson, A. M., and Hytten, F. E. (1962). Private communication.
Thomson, W., and Aitken, F. C. (1959). *Commonwealth Bur. Animal Nutr., Tech. Commun.* **20**.
Van Es, A. J. H. (1961). Proefschrift, Wageningen.
Van Slyke, D. D., and Meyer, G. M. (1913). *J. Biol. Chem.* **16**, 197.
van Wagenen, G., Himwich, H. E., and Catchpole, H. R. (1941). *Proc. Soc. Exptl Biol. Med.* **48**, 133.
Wakid, N. W., and Needham, D. M. (1960). *Biochem. J.* **76**, 95.
Wallace, L. R. (1948). *J. Agr. Sci.* **38**, 93.
Warkany, J. (1945). *Vitamins Hormones* **3**, 73.
Werner, A., and Hennig, A. (1960). *Z. Tierphysiol. Tierernähr. Futtermittelk.* **15**, 143.
Widdowson, E. M., and Dickerson, J. W. T. (1960). *Biochem. J.* **77**, 30.
Widdowson, E. M., and Spray, C. M. (1951). *Arch. Disease Childhood* **26**, 205.
Wilkerson, V. A., and Gortner, R. A. (1932). *Am. J. Physiol.* **102**, 153.
Wirth, K., and Schreier, K. (1956). *Z. Physiol. Chem.* **304**, 182.
Woessner, J. F., Jr. (1962). *Biochem. J.* **83**, 304.
Zweymüller, E., Widdowson, E. M., and McCance, R. A. (1959). *J. Embryol. Exptl. Morphol.* **7**, 202.

CHAPTER 16

Protein Metabolism and Requirements in the Newborn

R. A. McCance and Elsie M. Widdowson

Department of Experimental Medicine, University of Cambridge, England

If we concede at the outset that nature intended the fertilized ovum to grow to maturity at a rate which was limited only by its genetic inheritance and the interests of the species we are provided at once with a frame of reference within which to consider the protein metabolism and requirements of the newly born. The highest growth rates of rats, however, and the greatest stature of individuals can only be achieved when the newborn are suckled in groups of three or less, and these litters may not be large enough to ensure the successful survival of the species under natural conditions (Widdowson and McCance, 1960; Widdowson and Kennedy, 1962). Furthermore, the work of McCay *et al.* (1939) and others (Silberberg and Silberberg, 1955; Ross, 1959; Berg and Simms, 1960; Comfort, 1960) showed that a slow rate of growth due to controlled but extensive undernutrition may lead to considerable prolongation of a rat's life in the laboratory, but in some experiments few survived to demonstrate this, and probably none would have reached maturity in the natural state. Experimentally, therefore, the interests of the individual can be made to conflict with those of the species, but man-made conditions of this kind will not be considered in this chapter.

In all mammals the newborn and suckling periods of life are characterized by an anabolic tendency which will never be equaled later. This can be demonstrated even when there is not enough food available to allow the anabolism, i.e., growth, to take place. McCance and Strangeways (1954), for example, showed that the starving newborn infant derived only some 4 to 5 % of its total calorific requirements from protein breakdown, whereas the healthy adult male utilized protein for about four times as much of his metabolic requirements. Normally, however, when food is provided, growth dominates the picture and creates the great demand for proteins, minerals, and calories which characterize the newborn period.

It should go on to record at this point that there is little, if any, evidence that growth hormone plays any part in this. The somatic development of decapitated rat and rabbit fetuses (Jost, 1954; Bearn, 1961) proceeds almost normally, and the same is true of anencephalic monsters. The growth of hypophysectomized rats and of infants with functional disorders of the pituitary only begins to falter after the period of life which is being considered in this chapter, and which may be defined as limited to the age of suckling.

I. Growth and Mother's Milk

Bunge's (1898) well-known table, as presented in revised form by Blaxter (1961), related the rate of growth of the young to the percentage of protein in the mother's milk. This generalization, far-sighted enough when it was made, requires some modification today because it failed to take two things into account. Bunge did not consider the volume of the milk secreted by the mother and taken by the young, nor did he consider the kcal/protein ratio in the milk. Both of these are important. The kilocalories, for instance, derived from protein vary from 9 % of the total in human milk to 20 % in cow's milk and, further, up to 25 % in milks of the rat and pig. Some milks, moreover, for example those of the rat and reindeer, contain so little carbohydrate that some of the protein in the milk may have to be utilized for gluconeogenesis (Hahn *et al.*, 1961) and thus not be available for growth.

It may be possible to improve upon mother's milk as a food for newborn animals, but this has yet to be proved, and until it has been done the proteins and calories provided by mother's milk may be taken to represent satisfactory amounts for her young. Table I shows an approximation to the number of kilocalories necessary for the newborn of four mammalian species during the time it takes them to double their birth weights at good rates of growth. The figures have been arrived at from a knowledge of the calorie value of the milks, and of the amounts of milk taken. The sources of the information have been Spray (1950) and Kennedy (1956) for the rat; Spray and Widdowson (1951) and Blair *et al.* (1963) for the pig; Blaxter (1952, 1961, 1962), Blaxter and Wood (1952), and Gonzales-Jimenez and Blaxter (1962) for the calf; Fomon and May (1958a) and Macy and Kelly (1961) for man. To some extent the figures presented

TABLE I

CALORIE INTAKES OF NEWBORN ANIMALS OF 4 SPECIES WHILE THEY DOUBLE THEIR BIRTH WEIGHTS ON MOTHER'S MILK

Species	Birth weight (gm)	Time to double (days)	Mean weight (gm)	Total intake (kcal)	Kcal intake/ gm gain	Kcal intake/ gm mean wt	Kcal intake/gm mean wt/day	Kcal intake/gm mean $wt^{0.71}$/day
Rat	5	5	7.5	15	3.0	2.0	0.40	0.72
Pig	1,400	7	2,100	6,500	4.6	3.1	0.44	4.08
Calf	35,000	60	52,500	370,000	10.6	7.1	0.12	2.75
Man	3,500	180	5,250	114,000	32.5	22.0	0.12	1.45

have been matters of judgment and they should on no account be regarded as final. They stand to be revised and corrected as knowledge expands. The four species have very different birth weights and growth rates, as can be seen from Table I, columns 1 and 2. The rat, for example, weighs 5 gm at birth and takes 5 days to double this, whereas the calf weighs 35,000 gm and takes 60 days. The human infant is in quite a peculiar position, for its weight is only one-tenth that of a calf at birth, yet it takes three times as long to double it. During the 5 days in which the rat doubles its birth weight it requires 15 kcal, while the piglet, which weighs much more at birth but takes very little longer, needs 6500 kcal. The calf, which is much larger and takes much longer, requires 370,000 kcal and the human infant—smaller, but a very slow grower—114,000. The species differ much less among themselves when their calorie requirements are compared per gram of gain in weight or gram of mean weight, as has been done in columns 5 and 6. This shows at once the saving in calories brought about by rapid growth. The human "engine" has to be provided with fuel for 180 days—not for 5 only as in the case of the rat—and requires about 10 times as much energy either per gram of weight gained or per gram of mean weight. When time is brought into the calculation the order of the species is changed and the calf and the human baby have a smaller calorie requirement than the pig or rat and the order is changed again when the comparison is made per day per gram of mean $weight^{0.71}$.

Table II shows how these figures for the total calorie intake of the human infant at the third month of its life are related to its "basal" requirements and to the metabolic requirements of the newborn infant and the adult which are much better known. The sources of information have been Karlberg (1952), McCance and Strangeways (1954), and Smith (1959). The requirements of the infants are always higher than those of the adults per unit of body weight and there is a considerable rise from birth to the age of 3 months. The basal rate at birth is well known to be lower than

TABLE II

CALORIE INTAKES OF MEN AND OF INFANTS AT BIRTH AND AT 3 MONTHS OF AGE

Calorie intake	Adults[a]		Infants			
			At birth[b]		3 months old[c]	
	Basal	*Total*	*Basal*	*Total*	*Basal*	*Total*
Kcal/day	1680	3200	168	322	340	630
Kcal/gm/day	0.024	0.046	0.048	0.092	0.065	0.120
Kcal/gm $wt^{0.71}$/day	0.61	1.16	0.51	0.98	0.78	1.45

[a] Weight: 70,000 gm.
[b] Weight: 3500 gm.
[c] Weight: 5250 gm.

that of adults per unit of "surface area." By 3 months it is higher, and the total metabolic requirements at 3 months now estimated for the first time seem to bear the same relationship to the basal requirements as others have found at or soon after birth.

Munro (see Chapter 10) has computed that the total calorie requirements of the adult rat in confinement and of man are of the order of 0.190 and 0.046 kcal/gm body weight/day, respectively. These are nearer equal per unit of surface area if this is taken to be $wt^{0.71}$, and amount to 0.88 kcal/day for the rat and 1.16 kcal/day for man. The total calorie requirements of the newborn rat amount to 0.40 kcal/gm/day (Table I). This is higher than the requirements of the adult rat or of the newborn infant on the same basis but when the total requirements of the newborn rat are expressed per unit of $weight^{0.71}$ the figure is only 0.72, lower than the one for the adult rat and only half that for the human infant and less still than those for the pig and calf. No complete reason can be given for these rates as yet, or for the total calorie requirements of the human infant as compared for example with those of the calf on the basis of surface area. The latter seem high from what is known of the surface area and metabolic rate of the two adults. The rat is almost poikilothermic at birth and it doubles its birth weight long before it has acquired much thermal control of its internal environment. Its warmth comes from its mother and therefore it expends practically no energy on keeping warm and its requirements must be correspondingly less. This may not be the whole explanation, but the converse can be offered for the high metabolic rate of the calf. This newborn animal has almost perfect control of its body temperature at birth, and at environments around 0°C must expend a great deal of energy in maintaining it. The whole matter is very obscure at present and is not made easier by the way in which the energy expenditures of newborn animals may change in the days or weeks following birth (McCance,

TABLE III

PROTEIN INTAKES OF NEWBORN ANIMALS OF 4 SPECIES WHILE THEY DOUBLE THEIR BIRTH WEIGHTS ON MOTHER'S MILK

Species	Birth weight (gm)	Time to double (days)	Mean weight (gm)	Total intake (gm)	Intake/ gm gain (gm)	Intake/ gm mean wt (gm)	Intake/ gm mean wt/ day (gm)
Rat	5	5	7.5	0.75	0.15	0.10	0.020
Pig	1,400	7	2,100	400	0.28	0.19	0.027
Calf	35,000	60	52,500	16,800	0.48	0.32	0.005
Man	3,500	180	5,250	1,900	0.54	0.36	0.002

1959). The oxygen consumption of the newborn lamb, for example, rises three to four times between its last days *in utero*, when it has an oxygen consumption in keeping with its mother's surface area, and the second day of its life when it has one more in line with its own (Dawes and Mott, 1958, 1959). Pigs have not been investigated *in utero*, but their oxygen consumption also rises rapidly in the first days after birth (Mount, 1958), and, as already stated, the basal metabolic rate of the human infant is below that of the adult per unit of surface area at birth and rises rapidly in the first 24 hours (Cross *et al.*, 1963) and then more slowly, reaching a level higher than that of the adult in a few months.

The protein intakes which will enable these four species to double their birth weights at "target" rates of growth are given in Table III. The figures have been arrived at from the composition of the milks and the sources of information already quoted. Since the proteins were derived from the same food as the calories, their relationship to them is fixed by the composition of the four milks. Rats require 0.75 gm of protein for a gain of 5 gm in weight in as many days, while man demands 1,900 and the calf 16,800, but the intake per gram of gain or per gram of mean weight is not so very different and varies inversely as the rate of growth. Thus in man the protein intake per gram of gain is about four times that of the rat. Nevertheless, when the requirements are expressed per gram of mean weight per day, the human baby appears to be very economical. Were he not so, his requirements would be enormous. Protein requirements are generally thought of in terms of so many gm/kg/day. The human infant on this basis requires 2, the rat 20, and the pig 27.

It is the expenditure of energy that creates the demand for calories, and if the incoming ratio of protein to calories is fixed, as it is in the systems under consideration, part of the protein intake may be dissipated to meet the requirements for energy. In older animals on a satisfactory diet every additional kcalorie introduced into the food in the form of carbohydrate or fat leads to the retention of about 2 mg of extra nitrogen (see Chapter

10) and it would be interesting to know to what extent this applies to the newborn period. Extra carbohydrate can make little difference to the nitrogen retention of the pig or puppy in the first day or two of life, for they already retain about 90 % of the nitrogen from their mother's milk for purposes of growth (McCance and Widdowson, 1956a, 1958a), but it might make a difference before they have doubled their birth weights. In the human baby, however, which retains about 50 % of the nitrogen from breast milk at 6–8 days of age (Slater, 1961) and an average of about 22 % over the first 6 months, there is more room for an increase. The usual "experiment" in man has not been to increase the intake of carbohydrate or calories and to maintain the same intake of protein, but to increase the protein by substituting a cow's milk formula for human milk and to keep the calorie intake constant (see, for example, Kagan *et al.*, 1961). The effects of this have been thoroughly investigated and discussed by May (1959), Holt (1959), and Slater (1961). Babies on the cow's milk formulas retain more N but they do not grow, i. e., gain weight, any faster than babies fed on human milk, and the most reasonable conclusion is that the additional N retained leads, while it lasts, to more rapid maturation of the soft tissues and of the skeleton. The value of protein preparations other than milk and the effect of various heat treatments of milk on N retention and on the protein requirements of infants have been investigated. They are a little outside the scope of this article but the papers of F.A.O. (1957), Fomon and May (1958a,b), Fomon *et al.* (1958), Fomon (1959, 1961), Macy and Kelly (1961) and Fomon and Owen (1962) should be consulted by those who are interested. Substitutes for mother's milk are always being sought after by the animal husbandmen and those who are studying infant nutrition in the tropics and other undeveloped parts of the world. These "ersatz" preparations may contain milk proteins, fish meal, or vegetable proteins of which soy meal or cottonseed meal are the most popular. Work on man has been published by Dean (1953), Trowell *et al.* (1954), Fomon and May (1958b), Fomon (1959), and Snyderman *et al.* (1961); on pigs by Bellis (1957), Peo *et al.* (1957, 1958), Hays *et al.* (1958), Lodge and Lucas (1959), and Subcommittee (1959); and on cattle by Loosli (1950), Lofgreen *et al.* (1951), and Brisson *et al.* (1957). This list is by no means exhaustive.

II. Amino Acid Requirements

After Rose (1957) had demonstrated that some 10 of the amino acids could not be synthesized by the growing rat, and that 8 were essential for man, it was clear that a shortage of any one of these in the food might limit the rate of growth. This led to a flood of work on the amino acid "make-up" of individual proteins and of foods, and a reassessment of many of the diets

TABLE IV

AMINO ACID REQUIREMENTS OF HUMAN INFANTS

Amino acid	Requirement for optimum growth as determined by experiment (mg/kg/day)	Reference	Quantity provided by human milk at a protein intake of 2 gm/kg/day (mg/kg/day)
Histidine	30	Snyderman *et al.* (1959c)	42
Isoleucine	90	Hansen (1961)	122
Leucine	150	Hansen (1961)	196
Lysine	100	Snyderman *et al.* (1959b)	140
Phenylalanine	90	Snyderman *et al.* (1955)	74
Threonine	60	Pratt *et al.* (1955)	98
Tryptophan	22	Hansen (1961)	32
Valine	95	Snyderman *et al.* (1959a)	130

used for animals which had only been possible before in terms of total protein. Most of this work, however, has been done on animals outside the newborn period (Chick, 1940, 1942; Widdowson and McCance, 1954; Hutchinson *et al.*, 1956a,b, 1958, 1959, 1960; Rowett Report, 1962). Pigs require the same 10 amino acids as rats (Hanson, 1958) and, if they are fed at the correct balance and level, a 30-lb weanling will grow well on a ration containing 11.4 % of protein equivalent, only 7.5 % of which need be the essential amino acids (Beeson, 1958).

A group at Bellevue Hospital, New York, has made a special study of suckling infants (Snyderman *et al.*, 1959b). Their technique has been to take infants of low birth weight when they were about 6 weeks old and growing well and to provide them with a synthetic diet containing 18 L-amino acids as their sole of protein. A full rate of growth, as determined from the "grid" (Dancis *et al.*, 1948) was obtained when this mixture formed 12 % of the total calorie intake. The amino acid under test was next removed from the mixture and the N made up with glycine. The essential amino acid was then restored stepwise till full growth was resumed as judged by gain of weight, the percentages of proteins in the serum, and the retention of N. Table IV shows the results which these workers obtained, the references to the original papers, and the amounts of these amino acids which would be provided by human milk at a protein intake of 2 gm/kg/day. In every case but one, human milk provides more [but not a great deal more] than the amounts determined by experiment and the comparisons provide some food for thought (see later). The newborn infant does not require to be supplied with arginine for optimum growth. This is an essential amino acid for turkeys and chickens (Klose *et al.*,

1938; Ewing, 1947; Snyderman *et al.*, 1959d; N.A.S. Report, 1960), but mammals vary in their ability to get on without it. Young cats and weanling pigs grow—but only slowly—if it is not provided. Mice do not seem to need it (Snyderman *et al.*, 1959d). Albanese (1953) has suggested that if babies are fed on diets containing insufficient lysine some substitution of arginine for this amino acid may take place in the serum proteins. With the knowledge of today this would seem unlikely and it has not been confirmed—not in pigs at any rate (Bender and Doell, 1962).

Valuable information and generalizations can be obtained by applying to the individual amino acids calculations similar to those which have already been made about calories and total proteins (Tables I and II). This is made possible by reason of the fact that (*a*) the percentage of the individual amino acids in the total body proteins of the rat and pig are known both at birth and weaning (Williams *et al.*, 1954; Cunha, 1957) and a few figures are available for the human baby (Widdowson *et al.*, 1951), (*b*) the amino acids in human milk have been well documented (Macy and Kelly, 1961; Bigwood, 1963), and (*c*) the essential amino acids in sow's colostrum and milk are known (Braude, 1954; Lucas and Lodge, 1961).

Table V sets out: (*a*) The percentage of all essential and some non-essential amino acids in the proteins of the mammalian body. Only one figure has been given for each of them because the composition of the body proteins of one mammal seems—as one would expect—to be very like that of another (Beach *et al.*, 1943), and there is no evidence that the small differences between them would alter any of the conclusions that are to be drawn

TABLE V

AMINO ACID INTAKES AND SOMATIC INCREMENTS WHILE THE NEWBORN BABY AND THE PIGLET DOUBLE THEIR BIRTH WEIGHTS

	Arg	Hist	Ileu	Leu	Lys	Met	Phe	Thr	Try	Val	Tyr	Cy_2S_2	Glu
Amino acids in body as % total protein	6.3	2.5	3.6	6.8	7.5	1.8	3.8	3.8	0.8	5.8	2.6	1.2	11.6
Amino acids in human milk as % total protein	4.2	2.1	6.2	9.8	7.0	1.3	3.7	4.9	1.6	6.5	5.1	2.8	19.0
Increment of amino acids (gm) in baby[a]	25.8	10.2	14.8	28.0	30.7	7.4	15.6	15.6	3.3	23.8	10.7	4.9	47.5
Total intake of amino acids (gm) taken as milk[b]	80	40	118	186	133	25	70	93	30	124	97	53	362
Amino acids in sow's milk as % of total protein	5.9	2.3	4.3	8.2	7.6	1.4	3.6	3.6	—	5.1	—	1.4	—
Increment of amino acids (gm) in piglet[a]	10.1	4.0	5.8	10.9	12.0	2.9	6.1	6.1	1.3	9.3	4.2	1.9	18.6
Total intake of amino acids (gm) in sow's milk[b]	23.5	9.2	17.2	32.8	30.4	5.6	14.4	14.4	—	20.4	—	5.6	—

[a] As birth weight is doubled and 410 gm protein (baby) or 160 gm protein (piglet) is gained.

[b] Needed to double birth weight in 6 months (baby) or 7 days (piglet).

from these deliberations. (*b*) The percentage of the same amino acids in the proteins of human milk, and of sow's milk so far as they are available. The similarities again are more apparent than the differences and there is nothing in the amino acid make-up of these milk proteins to suggest that differences between them have anything much to do with the rate of growth of the organisms they were designed to feed. It is evidently the growth capacity of the animal, the protein/calorie ratio of the milk and the amount of protein ingested per kilogram of body weight which allows the various species to grow at their characteristic rates on mother's milk. (*c*) The increments of the various amino acids in the baby and in the pig while they double their birth weights. These increments are dependent upon the estimates for those of protein, which must only be considered as approximations at present. They are based upon analytical figures, where known, and on the knowledge that as the healthy newborn doubles its weight the percentage of protein in the lean body mass goes up but, with the increments of fat, the percentage of the lean body mass in the body as a whole goes down. (*d*) The total intake of amino acids which, taken as mother's milk, will enable the baby and the piglet to double their birth weights in 6 months and 7 days, respectively. The amounts given for the pig may be too low, for pig's colostrum contains over 17 % of protein and the percentage must remain high for some days.

If the figures for the human baby are considered first it is evident that the proteins of human milk provide a handsome margin of all the essential amino acids over the quantities which are likely to be found in the body proteins at 6 months of age. The surplus varies for different amino acids, which may be important, and the comparisons already made in Table IV show that these surpluses may be more apparent than real and that individual surpluses may reflect individual biochemical necessities. The figures for pigs are not quite so complete as those for the human infant but they are complete enough to establish an important point. Comparison of the last two lines shows at once that sow's milk provides a much smaller surplus of amino acids. This result might have been anticipated from the figures given in Table III for the protein intake per gram of weight gained by the pig and baby, for the amount of protein in the body bears a reasonably constant relationship to the body weight. It may be forecast that rat's milk will provide an even smaller surplus for the newborn rat, and cow's milk one for the calf which will be somewhere between those for the baby and the piglet.

In considering all these estimates it must always be remembered that they are at best averages. There is no doubt from all the work which has been carried out on man that individuals even from birth will vary widely from the mean both in their intakes and requirements (Widdowson, 1947;

TABLE VI

THE ENERGY INTAKE PER GRAM OF PROTEIN USED FOR GROWTH BY THE NEWBORN OF 4 SPECIES WHILE THEY DOUBLE THEIR BIRTH WEIGHTS ON MOTHER'S MILK

	Rat	Pig	Calf	Man
Total intake of energy (kcal)	15	6,500	370,000	114,000
Total intake of protein (gm)	0.75	400	16,800	1,900
Total protein used for growth (gm)	0.56	160	4,200	410
Per cent of protein intake used for growth	75	40	25	22
Energy value (kcal)/gm of protein used for growth	27	40	90	280

Beal, 1953). Animals and birds have their individual idiosyncracies, no less than man (Kinnersley *et al.*, 1928), and Cunha (1957) has indeed shown that individual piglets vary widely in their requirements for lysine and other amino acids.

It was pointed out earlier in this chapter that, since the protein/calorie ratio of milk is a fixed one, some of the protein may have to be dissipated to meet the calorie requirements if these are high. The same thing would eventuate, however, if the net rate of protein synthesis were very low and this seems to be quite as important. Table VI sets out some of the essential relationships in the four species. The calorie intakes and the protein intakes while doubling the birth weights have been given before. Less is known about the increments of body protein in the calf than in the other three species; the figure is very much an approximation but will suffice. The low protein/calorie ratio of human milk is well known. In spite of this the percentage of the protein used for growth is also low and the expenditure of calories for maintenance and growth per gram of protein so used is far higher than it is for the calf, piglet, or rat. There is evidently some connection between the small percentage of the protein used for growth, the low protein/calorie ratio and the high calorie intake per gram of protein incorporated into the baby's body. Yet the figures given in Table IV suggest that, although there is this large dissipation of protein for purposes of energy in man, a smaller intake of the essential amino acids might not meet the baby's requirements.

III. Molecular Protein Metabolism

One of the most urgent requirements of the newly born is protection against the bacterial invasions which it must immediately face. Owing to the inability of the fetus and newborn animal to produce antibodies with the necessary dispatch, it has been ordained by nature that temporary

TABLE VII

TIME OF TRANSMISSION OF PASSIVE IMMUNITY IN MAMMALS[a]

Species	Transmission of passive immunity	
	Prenatal	Postnatal
Ox, goat, sheep	0	+++ (36 hr)
Pig	0	+++ (36 hr)
Horse	0	+++ (36 hr)
Dog	+	++ (10 days)
Mouse	+	++ (16 days)
Rat	+	++ (20 days)
Guinea pig	+++	0
Rabbit	+++	0
Man	+++	0

[a] From Brambell (1958).

protection shall be afforded to the newborn animal by the transmission of antibodies from the mother. These antibodies are always proteins and usually γ-globulins and therefore they must be reckoned as one of the prior protein requirements of the young animal. They are transmitted in various ways and at various times; Table VII, which is taken from an article by Brambell (1958), shows when these times are. Since this chapter is concerned with the newborn animal the species in which transmission takes place before birth will not be considered in any detail, but they must be mentioned because the human baby is one of them.

There are three routes by which these antibodies are transmitted. In the rabbit, guinea pig, and to some extent in the rat, the antibodies reach the fetus by passing through the walls of the yolk sac. In primates this route is closed because the yolk sac is vestigial and functionless, and transmission is now known to be through the placenta (Bangham *et al.*, 1958). The evidence goes to show that transmission through this organ is an active process, or at any rate a highly selective one, for the concentration of antibodies and γ-globulins may be higher in cord blood than in maternal blood (Barr *et al.*, 1949; Kekwick, 1959). There may even be some internal regulation of γ-globulin levels in the circulating fetal blood. Transmission is by the gut in all species which acquire their passive immunity only after birth, but the gut is also the organ which transmits most of the rat's immune bodies from mother to young. Table VII shows that the process goes on for a very variable length of time. In the pig and calf it is confined to the first day or two of life and the colostrum is the prime source of the antibodies and globulins. In rats and mice the transmission may go on with diminishing intensity throughout suckling. Transmission can only be successful after birth so long as the digestive glands in the stomach remain functionless; the colostrum of the cow and pig contains a trypsin inhibitor which helps to reduce the rate of digestion in the duodenum of these species (Nordbring and Olsson, 1958a,b). After 24 hours or so in the pig the cells of the intestine which absorb the intact protein molecules become blocked and no further absorption takes place (Payne, 1962). There is evidence in rats that the administration of cortisone and deoxycorticosterone can bring the transmission to an end and that this is associated with the appearance of an alkaline phosphatase in the duodenum (Brambell, 1961). The functional connection here is not yet clear, and there may in fact be no connection (Payne, 1962). It would be helpful to know whether this enzyme has already appeared in the duodenum by the time of birth in the young of animals such as man in which the transmission after birth is negligible.

The absorption of these proteins may be highly but variably selective (Hemmings, 1960; Brambell, 1961). Species differ in their capacity to absorb the antibodies of other species and also molecules of differing size. The rat, for example, will not absorb the cholinesterase molecule, which is not a true γ-globulin and migrates between the β- and γ-globulins in the electrophoretic field (Halliday and Mihailović, 1961). It is also highly selective in other ways (Halliday and Kekwick, 1960–1961). In nature a second site of selection may complicate the picture, as it does in the rat's mammary gland (Biswas, 1961). Antibodies to *Brucella abortus*, moreover, generated in the mother hedgehog, locate themselves with the β-globulins of her serum while those to *Salmonella pullorum* run with the γ-globulins. The former are not transmitted so freely either into the milk or from the gut of the young and consequently the immunity achieved to *Brucella abortus* is small (Morris and Baldwin, 1962). In the horse, on the contrary, most of the immune bodies seem to be β-globulins and it is these, not the γ-globulins, which are preferentially absorbed by the newborn animal (Kekwick, 1959). Unlike the rat and the other animals so far mentioned, the calf is not highly selective and will absorb many proteins freely including, for example, gelatin. The cow's mammary gland, however, concentrates γ-globulin from serum to colostrum by a factor of 13 which ensures the presentation of plenty of γ-globulins to the absorbing surface (Pierce, 1961a). These differences make it very difficult to formulate any general theory as to the mechanism of absorption and the amounts absorbed. It has been suggested that the γ-globulins enter the cells of the endoderm by a process of pinocytosis (Pierce, 1961a), but all do not agree with this (Payne, 1962). There must, moreover, be some further structurally specific mechanism by which the protein molecules are passed out of the cells. What this may be is not yet known, and is a problem waiting for some molecular biologist to solve.

The calf has provided us with other valuable information about the absorption of these macromolecules. These pass from the cells of the gut into the lymph and reach the general circulation through the thoracic duct (Comline *et al.*, 1951). Their absorption, moreover, is greatly facilitated by an organic substance of a protein nature, and by the glucose-6-phosphate and inorganic phosphate normally present in colostral whey and milk whey. Sodium chloride solutions tend to inhibit absorption (Balfour and Comline, 1962).

In the past, the absorption of these proteins has mostly been followed by studying the concentrations of antibody titer, but other methods of marking the proteins are now being used and it is becoming clear that the quantities absorbed may be considerable. Bangham and Terry (1957) found that 7.5 % of the γ-globulins they had administered were in the serum of the young rat 3 hours later. Nordbring and Olsson (1958a,b) gave 7 gm of immune globulins to a newborn pig and found 17 % of it in the serum. When serum was given 10 % of the γ-globulins were absorbed and also 4 to 5 % of the β-globulins and albumin. McCance and Widdowson (1959a) found that during the first 24 hours of suckling the amount of circulating protein in the newborn pig rose from 1.2 to 4.0 gm. Most of this protein was globulin. More may have reached the circulation but this alone must have represented a considerable proportion of the total protein absorbed because the weight of the body had only gone up by 67 gm, of which not more than 14 gm is likely to have been protein. In the calf the γ-globulins may rise from less than 4 to 30 % of the total circulating protein and the β-globulins to twice their original level within a day or two of birth (Kekwick, 1959).

Once they are absorbed the fate of all these macromolecules is not yet fully known, but Pierce (1960, 1961a,b) has shown that the transient proteinuria of the newborn calf is due to the β-lactoglobulin it has absorbed. This protein is synthesized in the udder and carries no passive immunity, so its loss is of little consequence, and other proteins of low molecular weight may accompany it into the urine—gelatin among them if it has been administered experimentally. The horse would make an interesting study in this connection, since in this species the immune bodies seem to be associated with the β-globulins of the maternal serum (Kekwick, 1959). The concentration of diphtheria antitoxin may fall to half the figure in cord blood by the tenth day of a baby's life, but thereafter the half-life seems to be of the order of 30 days (Barr *et al.*, 1949; Wiener, 1951). Dixon *et al.* (1952) and Dixon *et al.* (1953) have shown that the half-life of homologous γ-globulins in the adults of a number of species varies inversely with the expected metabolic rate from 21 days in cattle to 1.9 days in mice. They found, however, that the half-life was longer in children (20 days) than in adults (13 days), and Schreier and his colleagues (Plückthun and Schreier, 1960; Daffner and Schreier, 1961; Schreier *et al.*, 1961; Schreier,

1962) have added another complication. The incorporation of labelled amino acids into proteins goes on more rapidly in the fetus and newborn than it does in the adult and the half-life of the circulating proteins is also shorter. It is difficult to see at present how these isolated and contradictory findings will be reconciled with the length of time the passive immunity of the newborn must last. If the half-life of the immune globulins in mice is only 2 days, there is good reason why the absorption of immune bodies should go on so long after birth in these animals. When the immune globulins do break up it may be assumed that the constituent amino acids will enter the amino acid pool and participate in the general metabolism of the body.

IV. Immaturity of Organs

Many animals are born with a body which is still structurally immature, and organs which are functionally undeveloped. The volume of the extracellular fluids, for example, is relatively large, and the well-being of the newborn baby depends upon the maintenance of this. Two organs with a more direct bearing on protein metabolism are the gut and the kidney. As already pointed out, it is the specific immaturity of the former which is so vital to the newborn of many species because only the immature organ will allow the absorption of the γ-globulins to go forward and thus provide the young animal with the passive immunity it so badly needs. There are, however, no signs of immaturity about the digestion of the casein and lactalbumin in mother's milk, and the absorption of them as amino acids. This may seem curious since the digestion of protein in the stomach must be relatively slight at this age and some of the carbohydrate-splitting enzymes are certainly not functional at birth. The human infant digests the proteins in the milks of other species if they have been suitably prepared (Fomon, 1959, 1961; Fomon and Owen, 1962). Smith (1959) gives a wide range of figures for the absorption of protein by the newborn infant. Some of the variations may be technical in origin, but Slater's (1961) averages for babies 1 week old were 83.5 ± 5.3 % of the intake for breast-fed infants and 88 ± 2.3 % for bottle-fed ones. These figures are within the range for adults living on foods containing a reasonable quantity of roughage; they give some idea also of the individual variations.

V. Renal Function, Food, and Growth

Kerpel-Fronius and his Hungarian colleagues were probably the first to draw attention to the functional weaknesses of the kidneys in newborn animals (Kramár, 1926, 1927; Kerpel-Fronius and Leövey, 1931; Kerpel-Fronius, 1932, 1935). Their primary interests were in salt metabolism and the hydration of the body, but they showed that young puppies could not concentrate their urine as well as adult dogs when they were given evaporated milk or meat, and that in the osmotic diuresis which followed they became dehydrated and hypertonic, with high concentrations of urea and chlorides in their body fluids. In 1941 McCance and Young showed that the urea clearances and also the sodium and chloride clearances were lower in the first fortnight of an infant's life than they were in adults when the comparison was made on the basis of surface area. They showed also that at small, comparable, urine volumes there was always a lower concentration of osmolar material in the urine of the newborn than in that of the adult. This paper drew attention to the subject and it was followed by others. The most recent recruit seems to have been James (1963), who has been studying the matter over the first hours of extrauterine life. The findings in infants were compared with those in adults although the investigators did not always make a study of the latter themselves. Most of the investigations were made after the administration of water and, therefore, at good rates of urine flow. Much of this work has been reviewed by McCance (1948, 1950, 1962), Heller (1951), McCance and Widdowson (1952, 1954a), Barnett and Vesterdal (1953), and Smith (1959). It has been extended, moreover, in other directions and it is now known that many amino acids escape reabsorption in the newborn infant and are lost in the urine and that this is grossly exaggerated by a vitamin D deficiency (Barlow and McCance, 1948; Jonxis, 1957; Smith, 1959). The position may be summarized by saying that the kidneys of the newborn animal are much less able to maintain a stable "internal environment" than those of an adult when they are presented with a similar, if unphysiological, comparable task (McCance and Widdowson, 1957, 1958b; McCance, 1959). There is, however, evidence that in some limited respects the kidney of the newborn animal may be a better organ of excretion than the placenta had been toward the end of gestation, for the concentration of creatinine commonly falls after birth in the serum of the newborn pig (Zweymüller *et al.*, 1959) and of the guinea pig (Boylan *et al.*, 1958). Figure 1 nevertheless shows that urea, which crosses the placenta freely, is not maintained at the concentration found at birth in at least three species even when the kidney has been subjected to no stress (McCance and Otley, 1951), and following a prolonged and difficult labor when the breakdown of body protein may be much increased for 24 hours or more, the concentration of urea in the body fluids may rise to a considerable height (McCance and Widdowson, 1954a,b, 1956b; McCance, 1959). The same thing may happen to starving newborn piglets if they are made to breathe a gas mixture containing 8 % oxygen and 10 % carbon dioxide (Glauser *et al.*, 1962) and to a much greater extent if they are exposed to cold (McCance and Widdowson, 1959b).

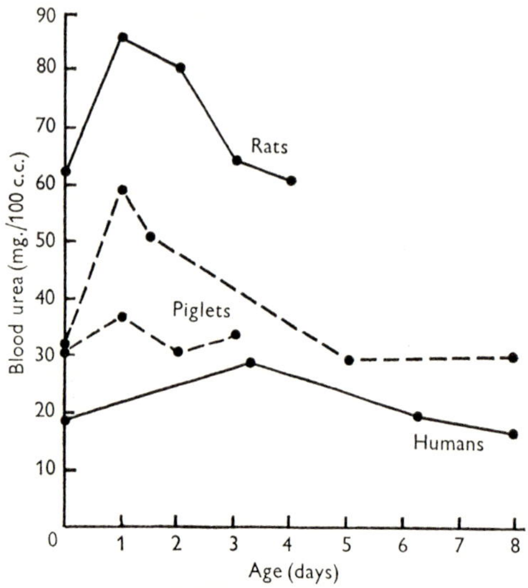

FIG. 1. Changes in the concentration of urea in the blood during the early days of life. (From McCance and Otley, 1951.)

There is, therefore, something of a paradox here between the high protein requirements of the newborn and the functional immaturity of its kidney. However, much of the protein nitrogen in the food of a freely growing animal is incorporated directly or indirectly into tissue protein and never appears in a form to be excreted by the kidney. In some newborn animals the γ-globulins in the colostrum and later milk are absorbed as such and are only broken down to smaller molecules and metabolized some time later. Thus the kidneys of newborn animals are relieved of much of the work which they would be called upon to do in adult life were the intake of protein correspondingly high. Minerals in the food are absorbed also and retained in a growing organism in the tissues and bones. Edwards (1824) fully recognized the overwhelming importance of growth in relationship to every function of the body: "Dans la jeunesse tout parait tendre au développement et à l'accroissement du corps." It was emphasized by Stransky and Bálint (1920) as an aspect of renal function, but the possible implications of these wise remarks were not appreciated at the time and have only recently been demonstrated experimentally (McCance and Widdowson, 1956a). Nine-tenths of the protein nitrogen in the food of the newborn pig and puppy was found to be incorporated into the tissues, and it never presented itself to the kidney for excretion as urea (McCance and Widdowson, 1956a, 1958a). The same is true to a lesser, but still to a surprising, extent for the more slowly growing human infant (Slater, 1961). Immature, therefore, as the kidney of the newborn may appear to be when some such test as the urea clearance or the excretion of para-aminohippuric acid is applied to it, it is quite capable of maintaining the blood urea and creatinine within normal limits, provided growth is taking place as it should and there is no forced degradation of tissue protein.

The stabilizing effect of growth on the internal environment has been demonstrated in other ways. Cell growth demands potassium as well as amino acids, and animals which were being paralyzed by the rise in the concentration of potassium in their sera due to the 2-hourly administration of potassium chloride were protected if the potassium salt was given with milk, which allowed the animals to grow (McCance and Widdowson, 1958b). Water is toxic to puppies in amounts which are correct if they are given with milk solids so that the animals can incorporate the water into their expanding cells (McCance and Widdowson, 1958a). Growth will always act in this way if the animal is capable of growth. If the animal is given no food (Widdowson and McCance, 1959), or if the food contains gelatin instead of milk proteins or too much salt (McCance and Widdowson, 1957), the stability of the internal milieu of the newborn animal is threatened at once. The whole principle of feeding in the newborn period, and indeed later, should be to saturate the growth requirements of the animal in question if rapid growth is desired, but at the same time to provide the excretory organs with the minimum amount of work by ensuring that the food shall contain the smallest possible surplus of amino acids, potassium, sodium, chloride, or indeed anything else which the kidney will have to excrete. It is only by this integration of food intake, growth, and renal function that the newborn animal survives.

References

Albanese, A. A. (1953). *J. Biol. Chem.* **200,** 787.

Balfour, W. E., and Comline, R. S. (1962). *J. Physiol. (London)* **160,** 234.

Bangham, D. R., and Terry, R. J. (1957). *Biochem. J.* **66,** 579.

Bangham, D. R., Hobbs, K. R., and Terry, R. J. (1958). *Lancet* **II,** 351.

Barlow, A., and McCance, R. A. (1948). *Arch. Disease Childhood* **23,** 225.

Barnett, H. L., and Vesterdal, J. (1953). *J. Pediat.* **42,** 99.

Barr, M., Glenny, A. T., and Randall, K. J. (1949). *Lancet* **II,** 324.

Beach, E. F., Munks, B., and Robinson, A. (1943). *J. Biol. Chem.* **148,** 431.

Beal, V. A. (1953). *J. Nutr.* **50,** 223.

Bearn, J. G. (1961). *Nature* **192,** 875.

Beeson, W. M. (1958). *J. Animal Sci.* **17,** 1119.

Bellis, D. B. (1957). *Proc. Nutr. Soc. (Engl. Scot.)* **16,** 98.

Bender, A. E., and Doell, B. H. (1962). *Brit. J. Nutr.* **16,** 191.

Berg, B. N., and Simms, H. S. (1960). *J. Nutr.* **71,** 255.

Bigwood, E. J. (1963). *Compt. rend. recherches. No.* **30.**

Biswas, E. R. I. (1961). *Nature* **192,** 883.

Blair, R., Diack, J. R. B., and MacPherson, R. M. (1963). *Brit. J. Nutr.* **17,** 19.

Blaxter, K. L. (1952). *Brit. J. Nutr.* **6,** 12.

Blaxter, K. L. (1961). *In* "Milk, the Mammary Gland and Its Secretion" (S. K. Kon and A. T. Cowie, eds.), Vol. II, p. 305. Academic Press, New York.
Blaxter, K. L. (1962). Personal communications.
Blaxter, K. L., and Wood, W. A. (1952). *Brit. J. Nutr.* **6,** 1.
Boylan, J. W., Colbourn, E. P., and McCance, R. A. (1958). *J. Physiol. (London)* **141,** 323.
Brambell, F. W. R. (1958). *Biol. Rev. Cambridge Phil. Soc.* **33,** 488.
Brambell, F. W. R. (1961). *Proc. Roy. Soc. Med.* **54,** 992.
Braude, R. (1954). *In* "Progress in the Physiology of Farm Animals" (J. Hammond, ed.), p. 40. Butterworths, London.
Brisson, G. J., Cunningham, H. M., and Haskell, S. R. (1957). *Can. J. Animal Sci.* **37,** 157.
Bunge, G. (1898). "Lehrbuch der physiologischen Chemie," 4th ed. Leipzig.
Chick, H. (1940). *Lancet* **II,** 511.
Chick, H. (1942). *Lancet* **I,** 405.
Comfort, A. (1960). *Proc. Nutr. Soc. (Engl. Scot.)* **19,** 125.
Comline, R. S., Roberts, H. E., and Titchen, D. A. (1951). *Nature* **167,** 561.
Cross, K. W., Hill, J. R., and Rahimtulla, K. A. (1963). *J. Physiol. (London)* **166,** 34P
Cunha, T. J. (1957). "Swine Feeding and Nutrition." Wiley (Interscience), New York.
Daffner, H. W., and Schreier, K. (1961). *Clin. Chim. Acta* **6,** 195.
Dancis, J., O'Connell, J. R., and Holt, L. E., Jr. (1948). *J. Pediat.* **33,** 570.
Dawes, G. S., and Mott, J. C. (1958). *J. Physiol. (London)* **143,** 7P.
Dawes, G. S., and Mott, J. C. (1959). *J. Physiol. (London)* **146,** 295.
Dean, R. F. A. (1953). *Med. Res. Council, Spec. Rept. Ser.* **279.**
Dixon, F. J., Talmage, D. W., Maurer, P. H., and Deichmiller, M. P. (1952). *J. Exptl. Med.* **96,** 313.
Dixon, F. J., Maurer, P. H., and Deichmiller, M. P. (1953). *Proc. Soc. Exptl. Biol. Med.* **83,** 287.
Edwards, W. F. (1824). "De l'influence des agents physiques sur la vie." Crochard, Paris.
Ewing, W. R. (1947). "Poultry Nutrition." Little & Ives, New York.
F. A. O. (1957). "Protein Requirements," Nutr. Studies No. 16. Food and Agricultural Organization of the United Nations.
Fomon, S. J. (1959). *Pediatrics* **24,** 577.
Fomon, S. J. (1961). *In* "Progress in Meeting Protein Needs of Infants and Preschool Children," Publ. No. 843, p. 343. Natl. Acad. Sci.—Natl. Res. Council, Washington, D.C.
Fomon, S. J., and May, C. D. (1958a). *Pediatrics* **22,** 101.
Fomon, S. J., and May, C. D. (1958b). *Pediatrics* **22,** 1134.
Fomon, S. J., and Owen, G. M. (1962). *Pediatrics* **29,** 1005.
Fomon, S. J., Thomas, L. N., and May, C. D. (1958). *Pediatrics* **22,** 935.
Glauser, E. M., McCance, R. A., and Widdowson, E. M. (1962). *J. Physiol. (London)* **161,** 313.
Gonzalez-Jimenez, E., and Blaxter, K. L. (1962). *Brit. J. Nutr.* **16,** 199.
Hahn, P., Koldovsky, O., Melechar, V., and Novák, M. (1961). *Nature* **192,** 1296.
Halliday, R., and Kekwick, R. A. (1960–1961). *Proc. Roy. Soc.* **B153,** 279.
Halliday, R., and Mihailović, R. (1961). *Proc. Roy. Soc.* **B153,** 541.
Hansen, J. D. L. (1961). *In* "Recent Advances in Human Nutrition" (J. F. Brock, ed.), p. 267. Churchill, London.
Hanson, L. E. (1958). *J. Animal Sci.* **17,** 1029.

Hays, V. W., Speer, V. C., and Catron, D. V. (1958). *J. Nutr.* **65,** 63.
Heller, H. (1951). *Arch. Disease Childhood* **26,** 195.
Hemmings, W. A. (1960). *Nature* **186,** 399.
Holt, L. E. (1959). *J. Pediat.* **54,** 496.
Hutchinson, J. B., Moran, T., and Pace, J. (1956a). *Nature* **178,** 46.
Hutchinson, J. B., Moran, T., and Pace, J. (1956b). *Proc. Roy. Soc.* **B145,** 270.
Hutchinson, J. B., Moran, T., and Pace, J. (1958). *Nature* **181,** 1733.
Hutchinson, J. B., Moran, T., and Pace, J. (1959). *Brit. J. Nutr.* **13,** 151.
Hutchinson, J. B., Moran, T., and Pace, J. (1960). *J. Sci. Food Agr.* **11,** 576.
James, L. S. (1963). Personal communication.
Jonxis, J. H. P. (1957). *Ergeb. Inn. Med. Kinderheilk.* **8,** 171.
Jost, A. (1954). *Cold Spring Harbor Symp. Quant. Biol.* **19,** 167.
Kagan, B. M., Felix, N., Molander, C., and Hill, J. (1961). *Am. J. Diseases Children* **102,** 746.
Karlberg, P. (1952). *Acta Paediat.* **41,** Suppl. 89.
Kekwick, R. A. (1959). *Advan. Protein Chem.* **14,** 231.
Kennedy, G. C. (1956). *J. Endocrinol.* **16,** 9.
Kerpel-Fronius, E. (1932). *Monatsschr. Kinderheilk.* **51,** 400.
Kerpel-Fronius, E. (1935). *Z. Kinderheilk.* **57,** 489.
Kerpel-Fronius, E., and Leövey, F. (1931). *Arch Kinderheilk.* **94,** 9.
Kinnersley, H. W., Peters, R. A., and Reader, V. (1928). *Biochem. J.* **22,** 276.
Klose, A. A., Stokstad, E. L. R., and Almquist, H. J. (1938). *J. Biol. Chem.* **123,** 691.
Kramár, E. (1926). *Jahresber. Kinderheilk.* **114,** 356.
Kramár, E. (1927). *Jahresber. Kinderheilk.* **115,** 289.
Lodge, G. A., and Lucas, I. A. M. (1959). *In* "Scientific Principles of Feeding Farm Live Stock," p. 125. Farmer & Stock-Breeder Publs., London.
Lofgreen, G. P., Loosli, J. K., and Maynard, L. A. (1951). *J. Animal Sci.* **10,** 171.
Loosli, J. K. (1950). "Recommended Nutrient Allowances for Domestic Animals, Part 3; Dairy Cattle," Agr. Board Report. Natl. Acad. Sci.—Natl. Res. Council, Washington, D.C.
Lucas, I. A. M., and Lodge, G. A. (1961). Commonwealth Bur. Animal Nutr. Tech. Bull. No. 22. Commonwealth Agr. Bureaux, Farnham Royal, Slough, England.
McCance, R. A. (1948). *Physiol. Rev.* **28,** 331.
McCance, R. A. (1950). *Am. J. Med.* **9,** 229.
McCance, R. A. (1959). *Arch. Disease Childhood* **34,** 361, 459.
McCance, R. A. (1962). *In* "Renal Disease" (D. A. K. Black, ed.), p. 157. Blackwell, Oxford.
McCance, R. A., and Otley, M. (1951). *J. Physiol. (London)* **113,** 18.
McCance, R. A., and Strangeways, W. M. B. (1954). *Brit. J. Nutr.* **8,** 21.
McCance, R. A., and Widdowson, E. M. (1952). *Lancet* **II,** 860.
McCance, R. A., and Widdowson, E. M. (1954a). *Arch. Disease Childhood* **29,** 488.
McCance, R. A., and Widdowson, E. M. (1954b). *Arch. Disease Childhood* **29,** 495.
McCance, R. A., and Widdowson, E. M. (1956a). *J. Physiol. (London)* **133,** 373.
McCance, R. A., and Widdowson, E. M. (1956b). *In* "Modern Views on the Secretion of Urine" (F. R. Winton, ed.), p. 217. The Cushny Memorial Lectures. Churchill, London.
McCance, R. A., and Widdowson, E. M. (1957). *Acta Paediat.* **46,** 337.
McCance, R. A., and Widdowson, E. M. (1958a). *J. Physiol. (London)* **141,** 81.
McCance, R. A., and Widdowson, E. M. (1958b). *J. Physiol. (London)* **141,** 88.
McCance, R. A., and Widdowson, E. M. (1959a). *J. Physiol. (London)* **145,** 547.
McCance, R. A., and Widdowson, E. M. (1959b). *J. Physiol. (London)* **147,** 124.

McCance, R. A., and Young, W. F. (1941). *J. Physiol. (London)* **99,** 265.
McCay, C. M., Maynard, L. A., Sperling, G., and Barnes, L. L. (1939). *J. Nutr.* **18,** 1.
Macy, I. G., and Kelly, H. J. (1961). *In* "Milk, the Mammary Gland and Its Secretion" (S. K. Kon and A. T. Cowie, eds.), Vol. II, p. 265. Academic Press, New York.
May, C. D. (1959). *Pediatrics* **23,** 384.
Mount, L. E. (1958). *J. Physiol. (London)* **142,** 57P.
Morris, B., and Baldwin, R. W. (1962). *Proc. Roy. Soc.* **B155,** 551.
N. A. S. Report (1960). "Nutrient Requirements of Poultry," Publ. No. 827. Natl. Acad. Sci.—Natl. Res. Council, Washington, D.C.
Nordbring, F., and Olsson, B. (1958a). *Acta Soc. Med. Upsaliensis* **63,** 25.
Nordbring, F., and Olsson, B. (1958b). *Acta Soc. Med. Upsaliensis* **63,** 41.
Payne, L. C. (1962). Unpublished communication at a Gordon Conference on Food and Nutrition, New London, New Hampshire.
Peo, E. R., Hays, V. W., Ashton, G. C., Speer, V. C., Liu, C. H., and Catron, D. V. (1957). *J. Nutr.* **62,** 465.
Peo, E. R., Hays, V. W., Ashton, G. C., Speer, V. C., Liu, C. H., and Catron, D. V. (1958). *J. Nutr.* **65,** 475.
Pierce, A. E. (1960). *Nature* **188,** 940.
Pierce, A. E. (1961a). *Proc. Roy. Soc. Med.* **54,** 996.
Pierce, A. E. (1961b). *J. Physiol. (London)* **156,** 136.
Plückthun, H., and Schreier, K. (1960). *Clin. Chim. Acta* **5,** 59.
Pratt, E. L., Snyderman, S. E., Cheung, M. W., Norton, P., and Holt, L. E. (1955). *J. Nutr.* **56,** 231.
Rose, W. C. (1957). *Nutr. Abstr. Rev.* **27,** 631.
Ross, M. H. (1959). *Federation Proc.* **18,** 1190.
Rowett Report (1962). Annual Report No. 18 on Animal Nutrition and Allied Sciences, Rowett Research Inst., Bucksburn, Aberdeen, Scotland.
Schreier, K. (1962). *Monatsschr. Kinderheilk.* **110,** 290.
Schreier, K., Porath, U., Schneider, M., Weber, M., Mehra, K., and Plückthun, H. (1961). *Clin. Chim. Acta* **6,** 205.
Silberberg, M., and Silberberg, R. (1955). *Physiol. Rev.* **35,** 347.
Slater, J. E. (1961). *Brit. J. Nutr.* **15,** 83.
Smith, C. A. (1959). "The Physiology of the Newborn Infant," 3rd ed. Thomas, Springfield, Illinois.
Snyderman, S. E., Pratt, E. L., Cheung, M. W., Norton, P., and Holt, L. E. (1955). *J. Nutr.* **56,** 253.
Snyderman, S. E., Holt, L. E., Smellie, F., Boyer, A., and Westall, R. G. (1959a). *A.M.A. J. Diseases Children* **97,** 186.
Snyderman, S. E., Norton, P. M., Fowler, D. I., and Holt, L. E. (1959b). *A.M.A. J. Diseases Children* **97,** 175.
Snyderman, S. E., Prose, P. H., and Holt, L. E. (1959c). *A.M.A. J. Diseases Children* **98,** 459.
Snyderman, S. E., Boyer, A., and Holt, L. E. (1959d). *A.M.A. J. Diseases Children* **97,** 192.
Snyderman, S. E., Boyer, A., and Holt, L. E. (1961). *In* "Progress in Meeting Protein Needs of Infants and Preschool Children," Publ. No. 843, p. 331. Natl. Acad. Sci.—Natl. Res. Council, Washington, D.C.
Spray, C. M. (1950). *Brit. J. Nutr.* **4,** 354.
Spray, C. M., and Widdowson, E. M. (1951). *Brit. J. Nutr.* **4,** 332.
Stransky, E., and Bálint, A. (1920). *Jahresber. Kinderheilk.* **93,** 350.

Subcommittee (1959). "Nutrient Requirements of Domestic Animals, Part 2: Swine," Agr. Board Report. Natl. Acad. Sci.—Natl. Res. Council, Washington, D.C.

Trowell, H. C., Davies, J. N. P., and Dean, R. F. A. (1954). "Kwashiorkor." Edward Arnold, London.

Widdowson, E. M. (1947). *Med. Res. Council, Spec. Rept. Ser.* **257.**

Widdowson, E. M., and Kennedy, G. C. (1962). *Proc. Roy. Soc.* **B156,** 96.

Widdowson, E. M., and McCance, R. A. (1954). *Med. Res. Council, Spec. Rept. Ser.* **287.**

Widdowson, E. M., and McCance, R. A. (1959). *Acta Paediat.* **48,** 383.

Widdowson, E. M., and McCance, R. A. (1960). *Proc. Roy. Soc.* **B152,** 188.

Widdowson, E. M., McCance, R. A., and Spray, C. M. (1951). *Clin. Sci.* **10,** 113.

Wiener, A. S. (1951). *J. Exptl. Med.* **94,** 213.

Williams, H. H., Curtin, L. V., Abraham, J., Loosli, J. K., and Maynard, L. A. (1954). *J. Biol. Chem.* **208,** 277.

Zweymüller, E., Widdowson, E. M., and McCance, R. A. (1959). *J. Embryol. Exptl. Morphol.* **7,** 202.

CHAPTER 17

Protein Metabolism and Requirements in the Elderly

Donald M. Watkin

Clinical Research Center, Massachusetts Institute of Technology, Cambridge, Massachusetts

I. Introduction

Strictly speaking, a study in man of any parameter as a function of age per se is impossible. Aging in man at least is not merely the passing of years, but also the accumulation of insults. These insults result from injury, infection, emotional and physical stress, malnutrition, genetic inheritance, randomly distributed biochemical errors, overindulgence, and so on. No two individuals have ever traversed 65 years of living to acquire them in exactly the same quantity and quality. It is not surprising, therefore, to find considerable disagreement in the literature over what is really known about protein metabolism in the aging adult. The appraisal is made even more difficult by the small magnitude of changes per unit time in the aging process and by the relative grossness of the techniques which have so far been applied to measuring such changes in man (Watkin, 1958).

II. Plasma Proteins

A relatively simple, economical method of evaluating nutritional status with respect to protein is the measurement of total serum proteins and of their principal fractions. While this approach assesses gross deficiencies due to primary or secondary protein malnutrition with ease (Holt *et al.*, 1962), it has less utility in studying the aging process. From a purely descriptive viewpoint, several studies (Morgan *et al.*, 1955; Herbeuval *et al.*, 1955; Karel *et al.*, 1956; Eastman, 1962; Acheson and Jessop, 1962) are in agreement that albumin concentrations fall while globulin displays a compensatory rise. The pathogenesis of these changes is a matter of speculation with reduced efficiency of protein synthesis by the liver being favored. More direct measures of rates of albumin and globulin synthesis have been applied to studies in infants with kwashiorkor (Cohen and Hansen, 1962) and have suggested that globulin synthesis may have a higher priority than albumin synthesis and may actually lead to further decrease in availability of amino acids for albumin synthesis. Similar studies in the aged have yet to be satisfactorily performed. The success in raising low albumin concentrations in adults of all ages by improvements in quantity and quality of dietary protein is well established (Patek and Post, 1941; Stieglitz, 1950). Since this improvement may be observed in either primary or secondary protein malnutrition, both of which may be present in minor degrees in survey groups of the aged, such evidence has been of little value in deciphering the riddle of the aging process in man.

III. Nitrogen Balance

A series of studies by workers in St. Louis is often quoted (Kountz *et al.*, 1947, 1948, 1951, 1953) to support the contention that aged individuals may require as much as two grams of protein per kilogram of body weight per day to maintain nitrogen equilibrium, although the average minimum requirement of the group as a whole was 0.7 gm/kg body weight/day. This implies a somewhat higher average requirement than has been found in young subjects and a very considerable variation in the minimum requirement of individual elderly subjects. By way of contrast the less-publicized studies of Schulze (1954, 1955) led that author to believe that the minimum requirements for the aged were actually less than those for the young. Studies by Watkin *et al.* (1950) in elderly hypertensive subjects fed the rice-fruit diet indicated that nitrogen equilibrium could be achieved on approximately 0.35 gm rice protein/kg body weight/day. Studies by Bogdonoff *et al.* (1953) led them to conclude that elderly subjects had minimum protein requirements equivalent to those of young subjects and that the aged subjects were able to respond to the anabolic stimulus of increased dietary protein about as effectively as the young.

Horwitt (1953) was unable to observe over a 3-month period any difference between the young and the aged in their ability to achieve positive nitrogen balances on diets containing 6.5 gm nitrogen daily. These observations led him to express serious doubts as to the value of the nitrogen balance technique in determining theoretical requirements as functions of age.

IV. Anabolic Agents

Watkin *et al.* (1955), in an effort to find out whether administration of an androgen could produce a further anabolic stimulus when superimposed on the stimulus of increased protein, studied the nitrogen, potassium, calcium, and phosphorus balances in 8 elderly men on high and low protein diets before, during, and after androgen administration. The studies revealed that these 8 men maintained an average nitrogen balance of 0.64 gm/day while receiving an average protein intake of 0.79 gm/kg/day. On an average protein intake of 1.63 gm/kg/day they maintained an average nitrogen balance of 2.28 gm/day. Administration of the androgen induced an additional retention of 1.24 gm/day on the low protein diet, and 1.63 gm/day on the high protein diet. An analysis of variance performed to detect differences between the 4 men fed a high calorie and the 4 fed a low calorie diet during these studies revealed no effect of the added calories or of individual variation but emphasized the marked influence of protein. The retention on the low protein regimen induced by the androgen was smaller than that induced by high protein feeding alone. These studies demonstrated the ability of aged men to maintain nitrogen equilibrium on modest protein intakes and to respond to the anabolic stimulus of a higher protein diet. This observation, together with the numerous unpleasant side effects accompanying hormone administration, suggested that androgen therapy, while perhaps justifiable in certain cases, is not a substitute for an adequate intake of dietary protein by the aged.

Other investigators have studied the effectiveness of various hormones on nitrogen retention in man (Reifenstein and Albright, 1947; Kountz, 1951; Ackermann *et al.*, 1954; Bogdonoff *et al.*, 1954; and Kountz *et al.*, 1959). These studies have revealed a greater effectiveness of male than of female hormones. The problems of side effects (Watkin, 1962a) and the inconveniences involved in taking medication mitigate against hormone administration as a general practice in dealing with the aged.

V. Requirement and Allowance

Studies such as these have led to the conclusions (Watkin, 1964) that groups of the aged when fed diets containing proteins of high Biological Value have average requirements similar to those of groups of young subjects and that there is little need to recommend for groups of the healthy aged more than the National Research Council recommended allowance of 1 gm/kg/day. These conclusions spring mostly from frustrations imposed by the complexity of the problem and the magnitude of interindividual variability created by the singularity of each man's accumulation of insults. Advocates of higher protein intakes have applied the "allowance" principle (N. A. S. Report, 1958) to the problem. This principle assumes that the protein intake must cover the requirement not only of the average individual but also of all those individuals who fall within at least two standard deviations of the mean. In view of the rise in interindividual variability with advancing age, the allowance would increase with age in proportion to the increment in standard deviation. Advocates of a common allowance for young and old assume that special situations will receive special treatment and, therefore, should not be used in weighting the recommendations for the aged any more than for the young. The social, economic, and political implications of either approach are controversial topics.

VI. Protein Withdrawal

Certain investigations in man have attempted to resolve uncertainties by utilizing protocols designed to magnify differences in protein metabolism between the young and the aged. Among these have been those conceived to answer the problem of whether the aged can respond as adequately as can younger subjects to the stress of protein withdrawal.

Before examining the data on protein withdrawal in middle-aged and elderly men, however, it is important to review some available information from studies conducted in young adults. Such data have been provided by Yoshimura (1961), who has reported studies on two healthy 23-year-old male subjects whose protein intake was abruptly changed from 1.25 to 0.57 gm protein/kg/day and 1.27 to 0.56 gm protein/kg/day, respectively. The low protein regimen in the first subject included animal protein with a protein score of 96. The low protein regimen in the second subject included vegetable protein with a score of 65. The first subject reached a point of equilibrium with respect to nitrogen balance 6 days after reduction in protein intake. In the 6 days, he lost about 4 gm nitrogen. The second subject required 14 days to attain equilibrium during which he lost about 12 gm nitrogen. While ingesting the low vegetable protein diet the second subject demonstrated a marked reduction of both basal metabolic rate and protein-bound iodine. The diet of the first subject, after 87 days of low animal protein intake, was abruptly changed to the control regimen. Within 9 days he had reached nitrogen equilibrium after storing approximately 15 gm nitrogen. The diet of the second subject, after only 22 days of low vegetable protein intake, was abruptly changed to a "recovery diet" containing 1.51 gm protein/kg body weight. On the seventh day (the last day reported), following the abrupt increase in dietary protein, the second subject still had a positive nitrogen balance of over 1 gm although he had already stored approximately 20 gm nitrogen.

Yoshimura's work demonstrates the marked influence of protein quality on the response of young subjects to the stress of dietary protein reduction. It points out the necessity of specifying exactly the quality of protein used in studying any nondietary parameter (such as age or a given disease) of protein metabolism in man. This point may be emphasized further by noting the marked differences in protein requirement for nitrogen balance plotted for three proteins of varying quality by an Expert Committee of the Food and Agriculture Organization (F.A.O., 1957). A similar plot by Yoshimura (1961) emphasizes the importance of the duration of adaptation time on a given protein in determining the Biological Value of that protein. Thus the regression line for week 3 on a given regimen shows far less protein required for nitrogen equilibrium than the line for week 2.

In a more detailed series of investigations, Silverstone *et al.* (1964) studied changes induced by altering the quantity of dietary protein of high Biological Value from medium (7.0 gm nitrogen/day) to low (4.3 gm nitrogen/day) to high (13.3 gm nitrogen/day). Three subjects, with ages averaging 47 years, represented late maturity, and four subjects, with ages averaging 73, represented old age.

The average time required to attain minimal nitrogen loss after the change from the medium to the low protein diet was 10 days for the elderly subjects and 14 days for the middle-aged men. The cumulative nitrogen balances of the elderly and middle-aged groups were computed both for the entire time on a given diet and for the first 5 days on a given diet. Prior to receiving the medium protein regimen, all subjects were given a standard institutional diet which, as presented, contained an average of 82 gm protein of which an average of 60 gm protein were consumed (Watkin *et al.*, 1954). In changing from this regimen to the diet containing 44 gm protein, the elderly men lost 10.6 gm nitrogen in 13 days as opposed to 3.7 gm for the middle-aged group; in the first 5 days, the elderly group lost 6.2 gm compared with 4.4 gm for the middle-aged group. In changing from the 44 gm to the 27 gm daily protein intake, losses of 12.5 and 13.3 gm were observed in 25 days in the elderly and middle-aged group, respectively; losses of 3.9 and 5.1 gm nitrogen were observed in the first 5 days in the elderly and middle-aged, respectively. When the dietary protein was raised from 27 gm to 83 gm daily, nitrogen retentions of 67.8 gm and 60.7 gm were observed in 30 days in the elderly and middle-aged, respectively; retentions of 19.8 and 16.6 gm in the first 5 days were observed in the elderly and middle-aged, respectively. Evaluations of the differences in cumulative nitrogen balances between elderly and middle-aged groups on any given diet revealed no statistically significant differences.

A comparison of these studies with those of Yoshimura (1961) suggests that greater differences exist between men in their twenties and elderly men than between middle-aged and elderly men. Yoshimura's subject receiving a diet containing animal protein reached equilibrium on shifting to a lower intake in 6 days with a loss of about 4 gm nitrogen. The older individuals in the Silverstone *et al.* (1963) study required from 10 to 14 days to reduce their nitrogen losses to a minimum and even then were unable to store nitrogen.

The longer time required to reach equilibrium and the far greater loss of nitrogen observed in Yoshimura's young man ingesting a vegetable protein diet resemble closely the times and losses observed in the older men receiving a better quality protein in small quantities. The question may well be asked, therefore, if the Biological Value of a given protein is in effect lower when tested in an elderly individual than when similarly tested in a young adult. A positive answer would suggest agewise differences in the over-all metabolism of amino acids.

Couch *et al.* (1960) (during one phase of a study of a water-soluble, chemically defined diet) have observed the response of a 70-year-old man to complete nitrogen withdrawal while he consumed a 2800-calorie diet complete in all essential nutrients except nitrogen. This subject was observed on this regimen for a period of 12 days and well within that time had decreased his urinary nitrogen excretion to approximately 2.5 gm/day, a figure comparable to that observed in Yoshimura's young man on a low animal protein diet and to that reported in normal young adults used as controls for studies of patients with hyperthyroidism (Lauter, 1922). It must be noted that this 70-year-old subject was receiving 2800 calories, almost exclusively from carbohydrate, while attaining these low levels of urinary nitrogen excretion. This study provides evidence under even more rigorous conditions that the aged individual can respond adequately to the stress of protein withdrawal. However, it also points out the importance of dietary calories (and possibly the source of these calories) in measuring any parameter of protein metabolism in man. Also to be mentioned is the importance of previous dietary intake and whether this has been adequate in the quality as well as quantity of protein. The diet habits of this upper middle-class income bracket subject were superior throughout his lifetime to those of institutionalized subjects. He therefore represents a patient with quite adequate stores and an enzyme system well-equipped for adaptation. Also he may show the beneficial effect of no protein as opposed to some protein which is either *de facto* or "because of subject age" low in Biological Value.

VII. Amino Acids

To this point protein has been discussed with only broad and nonspecific reference to the quality of protein or, in other words, the amino acid composition of the protein. Relatively few studies of the amino acid requirements of man have been performed in the elderly age group. Two excellent studies, however, are worthy of careful examination. Tuttle *et al.* (1957) found that men over 50 receiving quantities of amino acids, both in synthetic L forms and as contained in egg protein, and substantially in excess of the amounts needed to maintain nitrogen equilibrium in most young adults, were in negative nitrogen balance. This suggested to them that the quantitative requirements for one or more of the essential amino acids are increased in older men or that the supplementary source of nitrogen (in this case glycine) may not be so readily available to them for synthesis of nonessential amino acids as in the case of young adults. In later studies with mixtures of amino acids (Tuttle *et al.*, 1960; Swendseid and Tuttle, 1961), it was found that the minimum intakes of methionine and of lysine needed to maintain elderly subjects in nitrogen equilibrium were more than twice the level needed by young adults. Tuttle *et al.* (1959) have also studied the effect of total nitrogen intake on the requirement for essential amino acids in a similar group of elderly men. In this study they demonstrated an increase in the essential amino acid requirement when the total nitrogen was increased from 7 to 15 gm daily. The interpretation of these findings is not clear in the light of contrasting findings in other studies in the aged with whole proteins (Bogdonoff *et al.*, 1953, 1954; Watkin *et al.*, 1955) and similar studies in younger individuals where egg protein was employed as the source of essential amino acids (Swendseid *et al.*, 1959). Obviously, amino acid mixtures differ markedly from whole proteins. More calories are required to achieve the same level of nitrogen balance reached with isonitrogenous whole protein diets (Rose *et al.*, 1954). The source of nonessential nitrogen may have a marked effect on the final balance (Tuttle *et al.*, 1959). These studies emphasize the importance of studying the relation of nonessential amino acids and of total nitrogen to the requirement for essential amino acids. These ratios can be studied only by using synthetic diets, since in any natural protein the ratio remains constant regardless of intake level. Ratios of essential to nonessential amino acids in plasma on various dietary intakes are being evaluated (Swendseid *et al.*, 1963). Comparisons between these ratios in the young and in the aged may aid in the interpretation of the data mentioned above.

The work of Tuttle *et al.* (1957, 1959, 1960) represents one effort to date to determine if the amino acid requirement of the aged adult differed from that of a young adult. Several other groups have observed that older individuals consumed diets which were inadequate in balanced protein. For example, Mertz *et al.* (1952) showed by a determination of the amino acid composition of self-selected diets that older women who ingested acceptable amounts of protein nitrogen were in negative nitrogen balance because of a limited intake of methionine. Albanese *et al.* (1957) studied the self-selected diets of elderly women residing in a home catering to women from environments with better than adequate dietaries. They found that these women frequently selected diets low in animal protein and that the diets had low lysine/tryptophan ratios, averaging 4.5, whereas human tissue proteins have lysine/tryptophan ratios close to 6.0. When supplements of 600 to 750 mg lysine daily were added to the diets of 20 women, 8 of the 20 showed positive changes in nitrogen balance. In the same study, the authors also observed convalescent patients on self-selected diets and found again a low intake of meat proteins. Supplementation of the diets with 600 to 900 mg lysine daily brought about positive changes in nitrogen balance in 60 % of the 39 patients studied.

Investigations such as these leave little doubt that supplementation of a diet containing a poorly balanced protein will improve the performance of the individual consuming that diet. However, they do not imply that the aged person has a requirement due to age per se which differs dramatically from that of younger adults. The studies of Tuttle *et al.* (1957, 1959, 1960) suggest that some true differences may exist, but studies such as those of Mertz *et al.* and Albanese *et al.* merely indicate that the self-selection pattern of food by older individuals differs from ideal and by implication that on the average it differs from ideal by a greater margin than does the pattern selected by the young.

In summary, several related conclusions about the protein needs of elderly subjects can be drawn from the available evidence. (*a*) Balance studies have not demonstrated an unequivocal need by the average elderly subject for a raised protein intake in order to achieve nitrogen equilibrium, although it is probable that the scatter of individual requirements for equilibrium is much greater than among young subjects. (*b*) In spite of the general similarity of requirements for dietary protein in the young and the old, the needs for essential amino acids may well increase with age. (*c*) This conclusion is compatible with evidence obtained on depleted subjects which suggests a reduced Biological Value of proteins fed to elderly subjects. This would occur if the need for essential amino acids rose without a change in the need for total dietary nitrogen, thus demanding dietary protein containing a higher proportion of essential amino acids.

VIII. Digestion and Absorption

Systematic changes in aging man with respect to protein digestion and absorption have been sought in numerous studies of gastric, pancreatic, and intestinal function. Achlorhydria has been reported (in the clinical absence of Addisonian anemia or gastric cancer) to range from 9.8 to 35.4 % in the over-60 age group (Rafsky and Weingarten, 1947; D. T. Davies and James, 1930; Kopelowitz, 1941; Polland, 1933; and Bloomfield and Polland, 1933). Trypsin and pepsin activity are said to be diminished (Ivy *et al.*, 1950; Lascalea, 1959). Decreased secretion of pancreatic enzymes has been demonstrated (Necheles *et al.*, 1942; Meyer *et al.*, 1940). Preliminary observations mentioning a "high percentage" of deficient absorption of proteins labeled with radioactive isotopes among old men have been reported (Salvini and Grandonico, 1962). While these studies are interesting, their pragmatic effect on nutritional status is doubtful if clinical malabsorption syndromes are excluded. Chinn *et al.* (1956), using I^{131}-albumin, were unable to show any significant differences in protein digestion and absorption between young and old. The excess capacity of gastrointestinal enzyme systems and absorption mechanisms for proteins and amino acids is sufficient to provide a wide safety margin despite small decrements observed in presumably normal groups.

IX. Loss of Cell Function vs. Cell Death

Reduction in performance of an organ or organ system is commonly found in normal aging. For example, in the kidney there are linear decrements in filtration rate, effective renal plasma flow and transfer maxima for a variety of compounds between youth and old age (D. F. Davies and Shock, 1950; Miller *et al.*, 1952; Watkin and Shock, 1955). Much of this reduction is attributable to actual tissue loss due to the cumulative effect of individual cell death. Further evidence is found in the agewise reductions in total body water without agewise reductions in oxygen consumption per unit of body water (Shock *et al.*, 1963). The possibility remains that aging may also be associated with decrements in cellular metabolism in remaining tissue. On the theory that any diminution in metabolic activity would be reflected by decreases of specific enzyme concentrations in proportion to DNA (deoxyribonucleic acid) content of tissue, Barrows (1956, 1960), Barrows *et al.* (1958, 1962), and Falzone *et al.* (1959) have measured D-amino acid oxidase, succinoxidase, pseudocholinesterase, cathepsin, pyrophosphatase, alkaline phosphatase, and acid phosphatase in liver, kidney, and heart in young and old rats. Except for cathepsin, which did increase in senescence, the enzyme activities per unit of DNA measured in the tissues studied showed no marked changes.

Barrows and Raeder (1961), by measurement of the same enzyme concentrations in the liver, kidneys, and hearts of young and old rats subjected to protein deprivation and protein refeeding, could find little evidence for impairment of protein synthesis with senescence. On the contrary, the increased cathepsin concentrations in liver suggested an increased rate of protein turnover with age. To examine this possibility further, these authors measured rates of disappearance of S^{35}-labeled methionine from the liver, kidney, heart, and skeletal muscle of young and old rats. Although the differences were not statistically significant, the rate of protein breakdown tended to be slightly higher in the tissues of senescent animals. The report of an agewise increase in tryptophan peroxidase activity (Rivlin and Knox, 1959) in rat liver and the demonstration of increased RNA (ribonucleic acid) content in cells of aged mice (Wulff *et al.*, 1962) provide ample cause for further investigation.

Studies of isolated mitochondria (Barrows *et al.*, 1960; Fletcher and Sanadi, 1961) have shown no age-related decrements in succinoxidase activity or turnover rates. However, significant decrements have been observed in homogenates prepared from the whole tissues, thus suggesting a loss of mitochondria from cells rather than changes in mitochondrial function in senescence.

Data now at hand suggest that even the maximum agewise decrement in enzyme function (less than 15 % when based on DNA content of tissue) cannot account for whole organ functional decrements of up to 60 % (Shock, 1960). Although more precise data on changes in enzyme activity must await better methods of counting the number of cells per unit volume or weight of tissue, evidence now available is sufficient to place particular emphasis on the role cell death plays in functional impairment associated with aging.

X. Deoxyribonucleic Acid and Ribonucleic Acid

The mechanisms underlying cell death have been examined more extensively with advances in knowledge of the roles of DNA and RNA in protein synthesis. Wulff *et al.* (1962), after comparing data on H^3-cytidine incorporation into RNA in young and old mice with data on RNA content of liver, dorsal root ganglia, and Purkinje cells, have evolved a hypothesis to explain cell death. Starting from the observations that precursor uptake increased in liver and dorsal root ganglia cell nuclei but decreased in Purkinje cell nuclei while RNA content followed the same pattern, they observed that tracer uptake increased more, or decreased less, with age than RNA content, thereby producing a ratio of tracer uptake to RNA content which increased with age.

In explanation of these findings, they theorize: (*1*) With aging, sites for RNA synthesis sustain damage related to continual use. (*2*) These sites produce defective messenger RNA which in turn produces enzymes incapable of normal function. (*3*) The consequent accumulation of substrate molecules results in de-repression of RNA synthesis where this is possible (e.g., in liver cells but not in Purkinje cells). In other words, the maintenance of a functional level of enzymes in the presence of impaired RNA production necessitates increased production of RNA. (*4*) Defective RNA is less stable than normal RNA, resulting in an increase in the ratio of tracer uptake to RNA content.

The accumulation of somatic mutations with age is suggested as the basic cause for the production of the defective RNA. Changes resembling aging, produced by ionizing radiation (Dougherty, 1961), probably have a similar basis. As long as the de-repressed RNA synthesis supplies adequate amounts of functional enzymes, cells survive. When "correct" enzymes are no longer produced in adequate amount, cell function deteriorates and cells die. The hypothesis is of further interest, since it is also compatible with those observations (Barrows and Raeder, 1961) suggesting increased protein turnover with age.

Hypotheses such as this are hopeful indicators that new approaches to research in gerontology may gradually dispel man's ignorance of the basic aging process. An understanding of how aging may occur at the molecular level is essential to conduct, participate in, or even observe research now under way.

Medvedev (1962) has reviewed the roles DNA and RNA may play in the process of aging. He lists the following primary factors as responsible for errors in the synthesis of nucleic acids and proteins: (*1*) random errors in replication of RNA (from DNA); (*2*) radiation; (*3*) changes in the ratio of substrates (during the synthesis of RNA and proteins); (*4*) incorporation of analogs and products of side reactions; (*5*) microdenaturation, (*6*) the influence of free radicals; (*7*) microemission of heat and radiant energy; and (*8*) blocking of active groups. He points out that, in nature, two different mechanisms for the reproduction of specific polymers exist: one, capable of producing the specific polymer (DNA) with almost perfect accuracy; the other (producing RNA) lacking this capacity and hence being incapable of supporting an infinite life for a complex biochemical organism, since exact reproduction of all compounds would be required forever. The primary factors already mentioned act on DNA and RNA in different ways. DNA synthesis is relatively sheltered. In addition, negative characteristics resulting from errors in DNA synthesis are eliminated by natural selection; positive characteristics are spared for subsequent reproduction. Unlike this phylogenetic process, the ontogenetic process retains and accumulates molecules (altered RNA and protein) damaged during synthesis. Since RNA is more readily altered than DNA, the changes in its structure create the conditions for the aging process.

Medvedev believes that proteins are the most changeable component of a living system. They select all errors in DNA-RNA synthesis and contribute additional variants of their own. Proteins require precise functional specificity for the functions they perform. Since they cannot replicate themselves exactly, they are a limiting factor in ontogenesis. These changes in protein form the molecular basis for structural changes in living tissue. They therefore account for breakdowns in structure and function and thus lead eventually to death of the individual.

Medvedev suggests that, since only a fraction of available DNA is involved in the synthesis of cellular RNA, considerable RNA is synthesized autonomously without regulation by DNA. The small amount of DNA intimately involved in cellular RNA and protein synthesis can only furnish the cell with models, not with templates and adaptors. Medvedev proposes that age changes are associated with a gradual weakening of this hereditary control of synthesis. Regulation of this aging mechanism may be achieved through elimination of causes of change and by creating conditions unfavorable to the production of negative effects. One potential method of regulating the accumulation of molecular changes would involve blocking, isolating, and selectively eliminating altered templates, leaving the unchanged templates as the only determinants of cellular synthesis. Another possibility would introduce into old cells soluble RNA from young cells, bringing about reactions with unchanged templates and increasing the production of unaltered proteins. The binding of amino acid analogs to RNA molecules might also prove an effective means of inhibiting altered syntheses or pathological syntheses as in tumor growth. The rebuilding of age-altered proteins and nucleic acids might also be aided by the pool of activated peptide and polynucleotide blocks which could be utilized to replace defective parts in the intermediate steps of synthesis. A major hope continues to be that a method can be found which will maintain the governing role of DNA on protein synthesis for a time period longer than that currently in practice.

XI. Concluding Remarks and Recommendations

This review clearly leaves the impression that the surface of a very complex problem—that of determining the isolated effect of aging on the metabolism of protein—is only now being scratched. In one sense this is due to the lack of good data on human protein metabolism in modern terms in any age category. In another it is due to the lack of adequate experimental design and adequate subject selection in the studies which have been undertaken. Mention has already been made of the importance of the Biological Value of the protein under study, of the caloric content of the diet and of the lifelong diet history of the individual. Stress must also be placed on the necessity of acquiring experimental subject groups who are homogeneous in terms of general health, body size, and lifetime nutritional history as well as age. The presence and the activity of a disease have a marked influence on protein metabolism (Watkin, 1961, 1962b). The importance of acquiring truly young, i.e., 20–30-year-old, subjects for study under conditions identical to those used in investigating the old was emphasized in comparing the Japanese studies in young men with those done in the United States. Application to clinical investigations of the methodology of the animal experimenter and the molecular biologist will reveal wide vistas for future research. Several types of this methodology have already been applied. One, for example, is the study of rates of incorporation of amino acid nitrogen into protein. Such a study (Tschudy *et al.*, 1959) has been performed using N^{15}-labeled aspartic acid in a patient with Hodgkin's disease. In this patient, while her disease was inactive, the turnover of the metabolic pool and the rate of incorporation of amino acid nitrogen into protein were directly proportional to the product of protein and calories in the diet. These changes appeared long before evidence of nitrogen retention or weight gain appeared. Subsequently, while her disease was active, this patient, consuming a diet minimal in protein and calories, showed turnover rates of the nitrogenous metabolic pool and of the nitrogen incorporation into protein equal to those observed while she was fed a high-calorie, high-protein diet. Nitrogen balance data suggest that such increases result only in increased energy and protein metabolism without net nitrogen retention. Another such study (Sharp *et al.*, 1957) in which N^{15}-tagged yeast was fed to two young (one male, one female, both age 24) and four older subjects (all female, average age 66) revealed retentions of absorbed N^{15} during the 5 days subsequent to the tagged feeding of 57.6 % in the young and 49.1 % in the older subjects. Basing their calculations on N^{15} absorbed, the authors found the amount of nitrogen (including both fed and recycled) used for protein synthesis to be 0.284 gm/kg/day in the young and 0.204 gm/kg/day in the older subjects. They also found a mean N^{15} half-life of 86 days in the young subjects and 61 days in the older. They deduced that the turnover of N^{15} (protein, nitrogen) in the young was about one and one-half times that in the older subjects. They concluded that physiologic aging is accompanied by a slowing down in the rates of protein synthesis.

Similar studies performed in healthy young and old subjects on comparable diets with respect to calories and protein quantity and quality would provide information on the pool turnover and nitrogen incorporation rates as a function of age. They could answer for man the question posed by conflicting reports of increased and decreased protein turnover in aged rats and mice.

Data from rate studies such as these must be interpreted with the knowledge that they apply principally to the labile proteins. Thompson and Ballou (1956), using tritium as a tracer in rats, found long half-lives in certain tissues. Collagen, with a half-life of 1000 days, when once laid down, is not turned over at all. This work suggests an additional problem of molecular aging in those tissues which are in an essentially nondynamic state.

Another avenue of investigation would be the determination of the effective Biological Values of various proteins in young and aged men. Should these studies reveal that subject age per se does alter the effective Biological Value of a given protein, studies using chemically defined, water-soluble diets offer an effective, quantitative method of ascertaining which amino acids are particularly required by the aged. Such information could be combined with additional research at the molecular level to identify the biochemical changes which we now define as the aging process.

An additional means of examining the variable of age with respect to protein nutrition lies in the response of the aging organism to the anabolic stimulus of daily strenuous exercise. Studies performed in younger individuals have revealed striking changes in body composition during prolonged periods of strenuous physical work (Watkin and Mueller, 1959; Watkin *et al.*, 1963). Stratified studies in decade groups aged 20 through 90 indicate marked composition changes with advancing age which are correlated temporally with gradual decrements in activity (Shock *et al.*, 1963). Similar studies performed in both young and old men, especially if performed in conjunction with the tracer studies and the chemically defined diet studies mentioned above, could provide a rapid and yet physiological means of identifying significant differences between young and old.

Further research in man into the role of environmental factors subject to human control (nutrition being the outstanding example) in preventing or permitting adaptation to the accumulation of defects resulting from faulty RNA production is indicated. Systematic integration of useful Russian-language publications (Medvedev, 1962) with reports from Western-bloc countries should aid speeding work in this field.

These examples are mentioned more to cite the attitude or approach needed rather than to suggest specific projects. What is required is care, patience, painstaking work, modern technology, and personnel devoted to the task for substantial periods of time. Such an approach will in the long run prove economical both of time and money, since it will permit the orderly accumulation of knowledge and the avoidance of speculation which so readily plays into the hands of food faddists, unethical advertisers, and other parasites who feed on the gullibility of poorly informed aged populations.

The requirement of dietary protein for the elderly is truly protean.

Until the aging process is better defined and until the diseases of man are literally eliminated, broadcast recommendations with respect to dietary management of the aged have little meaning. The prevention of disease through dietary programing in youth and middle age affords the most practical means for the moment of seeking the fountain of youth. For today's aged, individualized medical care, part of which includes the recommendation and promotion of dietary regimens appropriate to the circumstances, has no substitute. Nonetheless, research is bringing understanding which in turn will bring an end to this unhappy, nihilistic era. Aging need not be a necessary consequence of life; the investigation of protein nutrition and metabolism is unquestionably the most promising tool by which man may escape the now not-so-inevitable.

References

Acheson, R. M., and Jessop, W. J. E. (1962). *Gerontologia* **6,** 193.

Ackermann, P. G., Toro, G., Kountz, W. B., and Kheim, T. (1954). *J. Gerontol.* **9,** 450.

Albanese, A. A., Higgons, R. A., Orto, L. A., and Zavattaro, D. N. (1957). *Geriatrics* **12,** 465.

Barrows, C. H., Jr. (1956). *Federation Proc.* **15,** 954.

Barrows, C. H., Jr. (1960). *In* "The Biology of Aging" (B. L. Strehler, ed.). Symposium No. 6, p. 116. Am. Inst. Biol. Sci., Washington, D.C.

Barrows, C. H., Jr., and Raeder, L. M. (1961). *J. Gerontol.* **16,** 321.

Barrows, C. H., Jr., Yiengst, M. J., and Shock, N. W. (1958). *J. Gerontol.* **13,** 351.

Barrows, C. H., Jr., Falzone, J. A., Jr., and Shock, N. W. (1960). *J. Gerontol.* **15,** 130.

Barrows, C. H., Jr., Raeder, L. M., and Falzone, J. A., Jr., (1962). *J. Gerontol.* **17,** 144.

Bloomfield, A. L., and Polland, W. S. (1933). "Gastric Anacidity: Its Relation to Disease." Macmillan, New York.

Bogdonoff, M. D., Shock, N. W., and Nichols, M. P. (1953). *J. Gerontol.* **8,** 272.

Bogdonoff, M. D., Shock, N. W., and Parsons, J. (1954). *J. Gerontol.* **9,** 262.

Chinn, A. B., Lavik, P. S., and Cameron, D. B. (1956). *J. Gerontol.* **11,** 151.

Cohen, S., and Hansen, J. D. L. (1962). *Clin. Sci.* **23,** 351.

Couch, R. B., Watkin, D. M., Rosenberg, L. E., Smith, R. R., Winitz, M., Otey, M. C., Birnbaum, S. M., and Greenstein, J. P. (1960). *Federation Proc.* **19,** 13 (Abstract).

Davies, D. F., and Shock, N. W. (1950). *J. Clin. Invest.* **29,** 496.

Davies, D. T., and James, T. G. I. (1930). *Lancet* **II,** 899.

Dougherty, T. F. (1961). *Federation Proc.* **20,** Suppl. 8, 3.

Eastman, G. (1962). *J. Am. Geriat. Soc.* **10,** 633 and 638.

Falzone, J. A., Jr., Barrows, C. H., Jr., and Shock, N. W. (1959). *J. Gerontol.* **14,** 2.

F.A.O. (1957). "Protein Requirements," Nutr. Studies No. 16, p. 7. Food and Agricultural Organization of the United Nations.

Fletcher, M. J., and Sanadi, D. R. (1961). *J. Gerontol.* **16,** 255.

Herbeuval, R., Cuny, G., Manciaux, M., and Hansen, J. (1955). *In* "Old Age in the Modern World," Gerontol. Congr., p. 574. Williams & Wilkins, Baltimore, Maryland.

Holt, L. E., Jr., Halac, E., Jr., and Kajdi, C. N. (1962). *J. Am. Med. Assoc.* **181,** 699.

Horwitt, M. K. (1953). *J. Am. Dietet. Assoc.* **29,** 443.

Ivy, A. C., Grossman, M. I., and Bachrach, W. A. (1950). "Peptic Ulcer" McGraw-Hill (Blakiston), New York.
Karel, J. L., Wilder V. M., and Beber, M. (1956). *J. Am. Geriat. Soc.* **4**, 667.
Kopelowitz, J. C. (1941). *J. Missouri State Med. Assoc.* **38**, 55.
Kountz, W. B. (1951). *Ann. Internal Med.* **35**, 1055.
Kountz, W. B., Hofstatter, L., and Ackermann, P. G. (1947). *Geriatrics* **2**, 173.
Kountz, W. B., Hofstatter, L., and Ackermann, P. G. (1948). *Geriatrics* **3**, 171.
Kountz, W. B., Hofstatter, L., and Ackermann, P. G. (1951). *J. Gerontol.* **6**, 20.
Kountz, W. B., Ackermann, P. G., Kheim, T., and Toro, G. (1953). *Geriatrics* **8**, 63.
Kountz, W. B., Kheim, T., Toro, J., Ackermann, P. G., and Toro, G. (1959). *J. Am. Geriat. Soc.* **7**, 757.
Lascalea, M. C. (1959). *Excerpta Med., Sect. XX* **2**, 419.
Lauter, S. (1922). *Deut. Arch. Klin. Med.* **139**, 46.
Medvedev, Zh. A. (1962). *In* "Biological Aspects of Aging" (N. W. Shock, ed.), p. 255. Columbia Univ. Press, New York.
Mertz, E. T., Baxter, E. J., Jackson, L. E., Roderuck, C. E., and Weis, A. (1952). *J. Nutr.* **46**, 313.
Meyer, J., Spier, E., and Neuwelt, F. (1940). *A. M. A. Arch. Internal Med.* **65**, 171.
Miller, J. H., McDonald, R. K., and Shock, N. W. (1952). *J. Gerontol.* **7**, 196.
Morgan, A. F., Murai, M., and Gillum, H. L. (1955). *J. Nutr.* **55**, 671.
N. A. S. Report (1958). "Recommended Daily Dietary Allowances, Revised 1958." Publ. No. 589. Natl. Acad. Sci.—Natl. Res. Council, Washington, D.C.
Necheles, H., Plotke, F., and Meyer, J. (1942). *Am. J. Digest. Diseases* **9**, 157.
Patek, A. J., Jr., and Post, J. (1941). *J. Clin. Invest.* **20**, 481.
Polland, W. S. (1933). *A. M. A. Arch. Internal Med.* **51**, 903.
Rafsky, H. A., and Weingarten, M. (1947). *Gastroenterology* **8**, 348.
Reifenstein, E. C., Jr., and Albright, F. (1947). *J. Clin. Invest.* **26**, 24.
Rivlin, R. S., and Knox, W. E. (1959). *Am. J. Physiol.* **197**, 65.
Rose, W. C., Coon, N. G., and Lambert, G. F. (1954). *J. Biol. Chem.* **210**, 331.
Salvini, L., and Grandonico, F. (1962). *In* "Medical and Clinical Aspects of Aging" (H. T. Blumenthal, ed.), p. 54. Columbia Univ. Press, New York (Abstract).
Schulze, W. (1954). *Z. Altersforsch.* **8**, 64.
Schulze, W. (1955). *In* "Old Age in the Modern World," Gerontol. Congr., p. 127. Williams & Wilkins, Baltimore, Maryland.
Sharp, E. S., Lassen, S., Shankman, S., Hazlet, J. W. and Kendis, M. S. (1957). *J. Nutr.* **63**, 155.
Shock, N. W. (1960). *In* "Aging . . . Some Social and Biological Aspects" (N. W. Shock, ed.). Publ. No. viii, p. 241. Am. Assoc. Advance. Sci., Washington, D.C.
Shock, N. W., Watkin, D. M., Yiengst, M. J., Norris, A. H., Gaffney, E. W., Gregerman, R. L., and Falzone, J. A. (1963). *J. Gerontol.* **18**, 1.
Silverstone, F. A., Watkin, D. M., and Shock, N. W. (1964). In preparation.
Stieglitz, E. J. (1950). *J. Am. Med. Assoc.* **142**, 1070.
Swendseid, M. E., and Tuttle, S. G. (1961). *In* "Progress in Meeting Protein Needs of Infants and Preschool Children," Publ. No. 843, p. 323. Natl. Acad. Sci.—Natl. Res. Council, Washington, D.C.
Swendseid, M. E., Feeley, R. J., Harris, C. L., and Tuttle, S. G. (1959). *J. Nutr.* **68**, 203.
Swendseid, M. E., Griffith, W. H., and Tuttle, S. G. (1963). *Metab., Clin. Exptl.* **12**, 96.
Thompson, R. C., and Ballou, J. E. (1956). *J. Biol. Chem.* **223**, 795.
Tschudy, D. P., Bacchus, H., Weissman, S. S., Watkin, D. M., Eubanks, M., and White, J. (1959). *J. Clin. Invest.* **38**, 892.

Tuttle, S. G., Swendseid, M. E., Mulcare, D., Griffith, W. H., and Bassett, S. H. (1957). *Metab., Clin. Exptl.* **6,** 564.

Tuttle, S. G., Swendseid, M. E., Mulcare, D., Griffith, W. H., and Bassett, S. H. (1959). *Metab., Clin. Exptl.* **8,** 61.

Tuttle, S. G., Swendseid, M. E., and Bassett, S. H. (1960). *Federation Proc.* **19,** 11 (Abstract).

Watkin, D. M. (1958). *Ann. N.Y. Acad. Sci.* **69,** 902.

Watkin, D. M. (1961). *Am. J. Clin. Nutr.* **9,** 446.

Watkin, D. M. (1962a). *In* "The Physician and the Total Care of the Cancer Patient," p. 135. Am. Cancer Soc. New York.

Watkin, D. M. (1962b). *Am. J. Clin. Nutr.* **11,** 406.

Watkin, D. M. (1964). *In* "Nutrition: A Comprehensive Treatise" (E. W. McHenry and G. H. Beaton, eds.), Vol. III. Academic Press, New York, in preparation.

Watkin, D. M., and Mueller, P. S. (1959). *Federation Proc.* **18,** 167 (Abstract).

Watkin, D. M., and Shock, N. W. (1955). *J. Clin. Invest.* **34,** 969 (Abstract).

Watkin, D. M., Froeb, H. F., Hatch, F. T., and Gutman, A. B. (1950). *Am. J. Med.* **9,** 441.

Watkin, D. M., Lawry, E. Y., Mann, G. V., and Halperin, M. (1954). *J. Clin. Invest.* **33,** 874.

Watkin, D. M., Parsons, J. M., Yiengst, M. J., and Shock, N. W. (1955). *J. Gerontol.* **10,** 268.

Watkin, D. M., Das, J. B., and McCarthy, M. C. (1963). *Federation Proc.* **22,** 260 (Abstract).

Wulff, U. J., Quastler, H., and Sherman, F. G. (1962). *Proc. Natl. Acad. Sci. U.S.* **48,** 1373.

Yoshimura, H. (1961). *Federation Proc.* **20,** 103.

PART III

Pathological Aspects of Protein Metabolism

A General Survey of Pathological Changes in Protein Metabolism

H. N. MUNRO

I. Introduction

Although many diseases cause changes in protein metabolism a detailed catalogue of all these abnormalities is not appropriate to a book of this type. The purpose is rather to consider how far our present understanding of protein metabolism in the normal mammal can be exploited to illuminate changes observed in disease states. Accordingly, this introductory survey will attempt to classify and consider the known pathological changes in protein metabolism in a way which emphasizes their relationship to normal metabolic processes. As shown in Table I, the available information about protein metabolism in disease has been grouped under six headings: (*1*) abnormalities of protein metabolism in the alimentary tract; (*2*) changes in intermediary metabolism arising from disease; (*3*) conditions directly affecting the mechanism of protein synthesis; (*4*) diseases causing changes in the protein metabolism of individual organs; (*5*) the relationship of protein intake to the occurrence and course of disease; (*6*) the effects of treatment on protein metabolism.

In addition to making a general survey of these topics, certain special-

TABLE I

A Classification of Pathological Changes in Protein Metabolism[a]

I. Abnormalities of protein metabolism in the alimentary tract

A. Diseases causing impaired utilization of dietary protein
    1. Malabsorption due to gastric disease
        Gastrectomy
    2. Malabsorption due to pancreatic disease
        Pancreatectomy
        Congenital fibrosis
        Chronic pancreatitis
        Pancreatic carcinoma
    3. Malabsorption due to intestinal disease
        Anatomical lesions
        Idiopathic steatorrhea, sprue, and celiac disease
        Conditions affecting individual amino acids (Hartnup disease, cystinuria)
        Intestinal infections
B. Diseases causing increased secretion of protein into the tract
    1. General protein loss
        Tuberculosis of intestine
        Ulcerative colitis
        Carcinoma
    2. Plasma protein loss
        Gastric carcinoma
        Hypertrophic gastritis
        Regional enteritis
        Ulcerative colitis
        Idiopathic steatorrhea
        Cardiac failure
        Intestinal lymphangiectasia

II. Changes in intermediary metabolism associated with disease

A. Anomalies of amino acid metabolism
    1. Familial disorders of amino acid metabolism
        Hyperglycinemia
        Primary hyperoxaluria
        Hyperprolinemia
        Hydroxyprolinemia
        Marfan's syndrome
        Anomalies of aromatic amino acid metabolism (albinism, alcaptonuria, familial dysautonomia, familial goitrous cretinism, phenylketonuria, tyrosinosis, tyrosyluria)
        Essential histidinemia
        Anomalies of tryptophan metabolism (Hartnup disease, congenital tryptophanuria)
        Anomalies of urea formation (hyperammonemia, citrullinemia, argininosuccinemia)
        Oculo-cerebro-renal disease
        Anomalies of sulfur amino acid metabolism (cystinosis, cystathionemia, homocystinemia, cystinuria)
        Anomalies of branched chain amino acid metabolism (maple syrup urine disease, leucine-induced hypoglycemia)
        β-Aminoisobutyric aciduria
    2. Nonfamilial disorders of amino acid metabolism

B. General effects of disease on intermediary protein metabolism
   1. Disturbances of endocrine action
      Catabolic hormone action (adrenocorticoids, thyroid hormones, estrogens, progesterone)
      Anabolic hormone action (growth hormone, androgens, insulin)
   2. Conditions leading to stress
      Infection
      Injury
      Undernutrition
      Climatic extremes
      Severe exertion
      Nervous overstimulation
   3. Caloric undernutrition and overnutrition
      Starvation
      Obesity
   4. Malignant tumor cachexia

III. Conditions directly affecting the mechanism of protein synthesis

A. Changes in DNA nucleotide sequence or loss of DNA, due to spontaneous alteration, misreplication, irradiation, chemical mutagens, or infection with DNA-viruses
   1. Affecting structural genes, with consequent change in amino acid sequences in proteins
   2. Affecting rate-controlling and suppressor genes
B. Changes at the postgene level
   1. Alterations in feedback control of rate of synthesis
   2. RNA-virus infection leading to synthesis of a foreign protein

IV. Changes in the protein metabolism of individual organs and tissues

A. Diseases affecting the plasma proteins
   1. Familial defects in synthesis
      Hypogammaglobulinemia
      Analbuminemia
   2. Appearance of previously undetected proteins
      C-reactive protein
      Myeloma proteins
   3. Changes in amount of fractions caused by nutritional deficiencies or by other diseases
B. Diseases of the liver
   1. Hepatic failure
   2. Hepatic coma
C. Diseases of the kidney
   1. Uremia
      Renal azotemia
      Prerenal azotemia
   2. The aminoacidurias
      Renal aminoacidurias
      Prerenal aminoacidurias
   3. Proteinuria
      Glomerular proteinuria (acute nephritis, nephrotic syndrome, toxemia of pregnancy)
      Tubular proteinuria
      Bence-Jones proteinuria

TABLE I—*Continued*

D. Disorders of other systems
  1. Diseases affecting the lymphoreticular system
    Infective diseases
    Autoimmune diseases
    Macroglobulinemia
    Myeloma
  2. Effects of disease on collagen metabolism
    Hydroxyprolinuria
  3. Protein metabolism in muscle disorders
    Muscular atrophy
    Muscular dystrophy

V. Protein intake and disease

A. Protein deficiency as a cause of disease
  1. Without caloric deficiency (kwashiorkor)
  2. With caloric deficiency (marasmus)
  3. Secondary to malabsorption or excessive protein loss
B. Excessive protein intake as a factor in the occurrence of disease
  1. In experimental hypercholesterolemia
  2. In experimental hypertension
C. Influence of protein intake on the response to disease
  1. Infective diseases
  2. Injury
  3. Hypertension
  4. Renal disease

VI. Effect of treatment on protein metabolism

A. Dietary treatment
B. Drugs

[a] Diseases named are a representative but not exhaustive list.

ized topics are of such intrinsic interest that they have been selected for more detailed development in later chapters. These individual chapters deal with abnormalities of amino acid metabolism (Chapter 18), the response of protein metabolism to injury (Chapter 19), protein metabolism in tumor-bearing animals (Chapter 20), the experimental study of protein malnutrition (Chapter 21), the clinical picture of protein malnutrition (Chapter 22), and the influence of protein intake on resistance to infective disease (Chapter 23).

II. Abnormalities of Protein Metabolism in the Alimentary Tract

Two types of defect can be associated with abnormalities of protein metabolism in the gastrointestinal tract. First, protein digestion and absorption may be faulty, so that utilization of dietary proteins becomes impaired. Second, the secretion of proteins into the digestive tract in the form of enzymes, mucus, and plasma proteins may increase under pathological circumstances. Each group will be considered separately, although some diseases (e.g., idiopathic steatorrhea) can be associated with both diminished utilization of dietary protein and increased secretion of endogenous protein into the tract.

A. Diseases Causing Impaired Utilization of Dietary Protein

As discussed in detail in Chapter 2, digestion and absorption of dietary protein by the nonruminant mammal depends on several factors:

1. After a meal, the stomach acts as a holding organ, regulating the transfer of food into the small intestine at a rate which ensures efficient digestion and absorption of nutrients. This balance can be upset by gastric disease, and notably by surgical removal of the stomach, or a significant part of it.

2. Normal digestion proceeds to the stage of complete resolution of dietary protein into free amino acids by enzymes secreted in the digestive juices or present in the intestinal mucosal cells. Absence of one or more of these digestive enzymes consequently can be a cause of incomplete utilization of food protein. This type of impairment occurs in severe pancreatic disease, for example.

3. The absorption of free amino acids by the mucosal cells of the small intestine is an active process involving enzymically controlled concentration mechanisms in the cells of the mucosa. Defective absorption of a general type can occur from degenerative changes in the mucosal cells, as in idiopathic steatorrhea, or it may be specific to absorption of single amino acids or groups of amino acids, as in subjects with cystinuria and with Hartnup disease, who show intestinal as well as renal defects.

The methods of testing for impairment of digestion and absorption of food protein are varied. Estimation of fecal nitrogen output is an obvious first line of attack in examining incomplete protein utilization. In subjects not suffering from gastrointestinal disease and receiving a normal diet, the daily fecal N output usually ranges from 0.5 to 2 gm (Table II), the intake of dietary protein having little effect on fecal N output (Reifenstein *et al.*, 1945). Table II displays some findings in representative gastrointestinal diseases; in general, conditions causing an increased output of N also cause a greater excretion of fat. Comparison of the losses of N and of fat suggests that the latter is somewhat more sensitive to gastrointestinal disturbance. For example, Doig and Girdwood (1960) report that, in a series of 20 patients studied after gastric operation, 4 showed raised fecal N output and 9 had a raised fat output; in 16 subjects with anatomical lesions of the small intestine, the figures were 6 and 12, respectively, and for those with sprue-type steatorrhea, fecal N was elevated in 16 out of 30

TABLE II

REPORTED FECAL NITROGEN AND FAT OUTPUTS OF SUBJECTS WITH ABNORMALITIES OF THE GASTROINTESTINAL TRACT

Group	Number in group	Mean Fecal Output (gm/day) (with range)		Reference
		Nitrogen	Fat	
Control subjects				
Normal subjects	11	1.7 (0.8–2.5)	4.1 (1.8–6.7)	Wollaeger *et al.* (1947)
Normal subjects	6[a]	1.6 (0.8–2.5)	2.9 (1.2–5.0)	Fourman *et al.* (1948)
Normal subjects	20	— (0.6–1.9)	— (2.0–7.5)	Henke *et al.* (1961)
Nongastrointestinal cases	5[b]	1.5 (0.5–3.1)	3.3 (0.5–6.8)	W. T. Cooke *et al.* (1953)
Nongastrointestinal cases	25	1.5 (0.5–2.3)	4.6 (2.0–7.2)	Doig and Girdwood (1960)
Stomach				
Pernicious anemia	25	1.4 (0.7–2.2)	4.1 (2.0–6.9)	Doig and Girdwood (1960)
Pancreas				
Total pancreatectomy	2	6.0 (4–8)	42.0 (36–48)	Waugh *et al.* (1946)
	1[c]	3.7 (2.0–5.4)	25.9 (13–42)	W. T. Cooke *et al.* (1953)
Pancreatic carcinoma	7	3.4 (2.1–5.7)	31.8 (6.6–79.4)	Henke *et al.* (1961)
Chronic pancreatitis	14	1.7 (0.7–2.7)	10.0 (2.4–46.3)	Henke *et al.* (1961)
Biliary tract				
Obstructive jaundice	15	0.9 (0.5–2.1)	15.0 (5–24)	Atkinson *et al.* (1956)
Hepatitis (acute)	8	1.5 (1.1–2.0)	15.1 (7–29)	Gross *et al.* (1950)
Small intestine				
Partial resection				
Minimal	6	3.0 (1.0–5.8)	5.7 (2.4–14.2)	Kalser *et al.* (1960)
Extensive	5	2.2 (1.3–3.0)	11.6 (8.0–15.0)	Kalser *et al.* (1960)
With removal of ileocecal valve	14	6.8 (2.6–10.0)	42.8 (8–97)	Kalser *et al.* (1960)
Idiopathic steatorrhea	3[d]	2.3 (1.3–4.4)	21.1 (2–56)	Bassett *et al.* (1939)
	7	3.2 (2.0–5.1)	18.3 (12–24)	W. T. Cooke *et al.* (1953)
	4	3.4 (2.3–4.3)	21.3 (13–34)	Crane and Neuberger (1960a)
Hookworm	10	3.1 (2.2–5.1)	—	Darke (1959)
Large intestine				
Ulcerative colitis	3[e]	6.0 (3.0–9.3)	7.4 (5.0–11.2)	Welch *et al.* (1937)
Endogenous output[f]				
Normal subjects	2	0.4 (0.4–0.5)	—	G. B. Wallace and Salomon (1909)
Cancer of colon	2	1.8 (1.7–2.0)	—	
Tuberculous ulceration	2	3.6 (3.1–4.2)	—	

[a] The range is for 18 observations on the 6 subjects.
[b] The range is for 60 observations on the 5 subjects.
[c] The range is for 7 observations on the subject.
[d] The range is for 19 observations on the 3 subjects.
[e] The range is for 6 observations on the 3 subjects
[f] All subjects on a diet of carbohydrate.

cases, but fecal fat was raised in 25 of these patients. There are two notable exceptions to the general relationship of raised fat output accompanying raised N output (Table II). In obstructive jaundice and in hepatitis, fat excretion is raised without a concomitant increase in N output, whereas in ulcerative colitis (and presumably in some other diseases of the large intestine), the output of N in the feces is markedly raised due to the ulceration, but fat excretion is minimally augmented.

A combined test of protein digestion and amino acid absorption is obtained when a standard protein is fed and the changes in the amino nitrogen level in the peripheral blood are measured (Van Slyke and Meyer, 1912; West *et al.*, 1946; Anfanger and Heavenrich, 1949; Althausen and Uyeyama, 1954; Clemetson and Churchman, 1955) or the levels of individual amino acids are measured in the plasma (Richmond and Girdwood, 1962). Digestion and absorption can also be assessed by feeding labeled proteins. Crane and Neuberger (1960a) used N^{15}-labeled yeast protein to study the effect of iodiopathic steatorrhea on rate of protein digestion and absorption; the effect of this condition was assessed by measuring the N^{15} content of blood amino acids and urea, and the N^{15} content of urinary urea and ammonia. They also assessed the relative changes in digestion and in absorption in this disease by feeding hydrolyzed yeast protein. Efficiency of protein utilization has been more commonly measured isotopically by feeding I^{131}-labeled proteins, and estimating the loss of radioactivity in the feces or the blood levels of I^{131} (Chinn *et al.*, 1952, 1953; Lavick *et al.*, 1952; Frazer *et al.*, 1955; Shingleton *et al.*, 1955; Freeark *et al.*, 1957; Polachek *et al.*, 1959). Fecal loss of I^{131} is not regarded as satisfactory, but blood levels of radioactivity can be successfully used in the study of malabsorption, although they appear to be less sensitive than similar tests in which I^{131}-labeled fat is fed (Shingleton *et al.*, 1955; Polachek *et al.*, 1959). It would appear from these studies, and from general experience with measurement of fecal N and fecal fat in digestive diseases, that utilization of dietary fat is more sensitive than utilization of dietary protein to abnormalities of the gastrointestinal tract. Finally, tests of absorption have been devised in which single amino acids are fed and the blood level is examined (Heath and Fullerton, 1935; Erf and Rhoads, 1940; Harper and Uyeyama, 1948; Steele and Le Bovit, 1951; Butterworth *et al.*, 1958). However, loading with large doses of single amino acids is hardly a test performed under physiological circumstances.

To supplement these direct tests of protein digestion and absorption, it is possible to sample the digestive juices and the contents of the alimentary tract, and to assess the activities of different enzymes responsible for the digestion of proteins. For example, Lundh (1962) uses a specific synthetic substrate, benzoylarginine ethyl ester, to test for trypsin activity in samples of duodenal contents. This estimation is not invalidated by the presence of peptides from food protein, so that the response to a standard test meal stimulus can be assessed.

When we come to consider the findings obtained by the various tests of efficiency of protein utilization in different pathological conditions of the alimentary tract, it is apparent that clean-cut distinctions as to mechanism cannot be sustained. The so-called "malabsorption syndrome" includes the majority of diseases of interest in this field, and is characterized by loss of weight, usually by excessive output of fecal fat (steatorrhea) and by signs of vitamin and mineral deficiencies, due to poor absorption of these dietary components. The malabsorption syndrome is most frequently due to pathological conditions of the small intestine, but also results from defects in function of the pancreas and hepatobiliary system (Doig and Girdwood, 1960). A detailed list, with classification, of diseases causing malabsorption of nutrients is provided by Spencer (1960).

1. *Malabsorption Due to Gastric Disease*

Simple atrophy of the gastric secretory apparatus does not appear to prejudice protein digestion, since cases of pernicious anemia do not exhibit increased N loss in the feces (Doig and Girdwood, 1960). On the other hand, fecal N output is increased in a fifth of subjects following gastrectomy (Doig and Girdwood, 1960) and nitrogen balance studies (Everson, 1952) suggest that some of these subjects have impaired utilization of protein. The changes in protein digestion after gastrectomy have been explored by Lundh (1958), who aspirated the contents from various parts of the small intestine after a meal containing I^{131} serum albumin. In normal subjects, the stomach controls release of food into the intestine, so that digestion of protein and absorption are completed in the upper part of the small intestine. In cases of total gastrectomy, the radioactive meal was rapidly distributed along the whole length of the small intestine and its rate of removal from the gut was appreciably slower than in normal subjects, an effect attributed to lack of HCl stimulus to pancreatic juice secretion. However, Richmond and Girdwood (1962) found that, following a standard meal of casein, cases of partial gastrectomy showed a higher initial rise in blood amino acid levels than do normal subjects; this suggests that gastrectomy does not result in any reduction in the rate of release of amino acids from food proteins. Crane and Neuberger (1960a) mention also that a case of steatorrhea resulting from partial gastrectomy showed no delay in absorption of N^{15}-labeled yeast protein. Direct delivery of food by tube into the jejunum causes increased fecal N losses (Riegal *et al.*, 1947).

2. Malabsorption Due to Pancreatic Disease

Since pancreatic juice contains a number of important enzymes involved in the hydrolysis of food proteins, pancreatic diseases might be expected to lead to impaired digestion of protein. The degree of impairment observed with the various tests for protein digestion show that this is dependent on the extent of pancreatic damage. After complete pancreatectomy, fecal N outputs of 4 to 8 gm daily have been recorded (Waugh *et al.*, 1946). Such patients can attain N equilibrium at moderately high levels of protein intake, although administration of pancreatic extracts improve N balance under these conditions (Jordan, 1960).

Fecal N output is also raised in adenocarcinoma of the pancreas but is above the normal range in less than half the patients with chronic pancreatitis (Henke *et al.*, 1961). The degree of impairment of absorption of I^{131}-protein in patients with pancreatic disease is also correlated with the extent of the lesion (Polachek *et al.*, 1959). After a standard meal of casein, the blood levels of several amino acids undergo a rise which tends to be less in patients with pancreatic disorders than in normal subjects; however, the mean values for the group of pancreatic cases were not significantly less (Richmond and Girdwood, 1962).

In children with cystic fibrosis of the pancreas, there is not the usual rise in blood amino nitrogen level after feeding protein, whereas there is an increase in blood amino nitrogen content of the usual proportion after giving a protein hydrolyzate (West *et al.*, 1946; Anfanger and Heavenrich, 1949). This finding is compatible with a defect in digestion of protein but a normal absorptive mechanism for amino acids.

3. Malabsorption Due to Intestinal Disease

a. Anatomical Lesions. In a proportion of patients who have undergone removal of segments of the small intestine, there is an increased output of N in the feces (Doig and Girdwood, 1960), especially when the ileocecal valve is also excised (Kalser *et al.*, 1960). The impairment in protein nutrition in such cases is also demonstrated by a reduction in plasma albumin level (Kalser *et al.*, 1960).

b. Idiopathic Steatorrhea. Malabsorption is a prominent feature of the group of diseases to which sprue, celiac disease and idiopathic stearrhoea belong, the clinical features of which have been recently summarized by Frazer (1962). Here, we shall refer to them generally as idiopathic steatorrhea without distinction between adult forms (idiopathic steatorrhea and tropical sprue) and types found in childhood (celiac disease). The general effect of this disease on protein nutrition is demonstrated by the frequency of low plasma albumin levels (Bassett *et al.*, 1939; Forshaw, 1957; Crane and Neuberger, 1960a).

As the name steatorrhea suggests, increased fat output in the feces is usually a prominent feature of such cases. In a proportion of cases, fecal N output is also raised. Thus Bassett *et al.* (1939) found increased values in 2 out of 3 cases, and the outputs were raised in all 4 cases studied by Crane and Neuberger (1960a). In a series of 30 cases of idiopathic steatorrhea, Doig and Girdwood (1960) report excessive fecal N in 16, whereas fecal fat output was abnormal in 25 of the subjects. Long-term studies on individual cases show that the N output varies with different phases of the disease (Bassett *et al.*, 1939). In some cases, output of fat and of nitrogen in the feces are closely correlated, in others no relationship can be discerned (Bassett *et al.*, 1939). In a large series of observations on idiopathic steatorrhea, W. T. Cooke *et al.* (1953) demonstrated a significant but rather low correlation between nitrogen output and fat output. The effect of protein intake on fecal N also varies in different patients; Bassett *et al.* (1939) report normal fecal output when the intake of one subject was increased from 62 gm protein to 115 gm protein daily, but Crane and Neuberger (1960b) found in one severe case that increasing protein intake from 78 gm to 175 gm daily resulted in an increment in fecal N output from 5.6 gm to 12.6 gm daily. In studies with N^{15}-labeled yeast protein in the diet, Crane and Neuberger (1960b) found that excretion in the feces of N^{15} was augmented in this disease, but less than might be anticipated if all the extra fecal N was due to unhydrolyzed food protein; it was therefore suggested that these cases have an increased endogenous loss of N into the gut, a view confirmed by studies with I^{131}-polyvinylpyrrolidone (see later).

After a standard meal of casein, the blood levels of four amino acids showed normal increments in cases of sprue (Richmond and Girdwood, 1962). The absorption of glycine was not abnormal in early cases of idiopathic steatorrhea (Drew *et al.*, 1947), but was impaired in more advanced cases (Erf and Rhoads, 1940; Butterworth *et al.*, 1958). The absorption of urea is delayed (Frazer *et al.*, 1952), as is that of glucose (Frazer *et al.*, 1952) and of water (Higgins *et al.*, 1957). Crane and Neuberger (1960a, b) have analyzed the effect of idiopathic steatorrhea on absorption of N^{15}-labeled whole yeast protein and yeast protein hydrolyzate by measuring fecal N output, blood amino acid and urea levels, and urinary urea and ammonia levels of N^{15}. Absorption after feeding the whole protein was considerably retarded, but even the hydrolyzate was more slowly absorbed than observed in normal control subjects. It was consequently concluded that the defect in idiopathic steatorrhea is a combination of impaired digestion of protein and impaired absorption.

The nature of the lesion in idiopathic steatorrhea is in keeping with this conclusion. Biopsy of the intestinal mucosa shows that there is absence of, or clubbing of, the villi and disorganization of the columnar epithelium covering them, the main site of the changes being the upper jejunum (Shiner and Doniach, 1959; Rubin *et al.*, 1962). During remission, the intestinal picture returns to a more normal appearance and N^{15} absorption from yeast protein rises to normal values (Crane and Neuberger, 1960a). Since the intestinal mucosa is the site of manufacture of some of the digestive enzymes and also is the site of absorption, damage to the mucosa will affect both phases of protein utilization by the digestive tract.

The occurrence of idiopathic steatorrhea is related to intake of wheat gluten. Wartime observations in Holland showed that patients with celiac disease improve on diets free from wheat bread, which led to identification of the intolerance of these subjects to wheat gluten (Dicke, 1950). Removal of gluten from the diet causes remission in most cases of celiac disease and in many of idiopathic steatorrhea (Frazer, 1962). The condition recurs on reintroducing gluten, or peptic and tryptic digests of gluten, but not with complete hydrolyzates of gluten; a peptide fraction responsible for this effect has been isolated from tryptic digests of gluten (van Roon and Haex, 1960; van Roon *et al.*, 1960; Haex *et al.*, 1962). The occurrence of the disease principally in the mucosa of the upper small intestine is apparently due to high concentrations of gluten peptides present in that part of the gut, but digestion prevents their occurrence lower down; instillation of gluten by tube into the lower ileum leads to typical lesions there (Rubin *et al.*, 1962). This demonstrates that the intestinal lesion is secondary to the dietary factor, and not vice versa. Nevertheless, normal subjects are not intolerant to gluten, indicating some peculiarity of the mucosa in cases of idopathic steatorrhea. Frazer (1962) discusses other possible dietary causes of the disease, such as oxidized fats, and comments on the nature of the damage caused by the gluten.

c. Intestinal Diseases Affecting Individual Amino Acids. A specific effect on availability of tryptophan to the body occurs in patients with carcinoid tumors of the intestine. These arise from the argentaffin cells of the mucosa, so called from their capacity to bind silver stains. It has been demonstrated (Erspamer, 1954) that this cell type contains 5-hydroxytryptamine, a tryptophan derivative. Carcinoids occur particularly in the region of the appendix and, by releasing large amounts of 5-hydroxytryptamine into the blood, produce widespread symptoms (MacDonald, 1956). The amine is excreted in large quantities as 5-hydroxyindoleacetic acid in the urine which is thus of diagnostic value (Sjoerdsma *et al.*, 1956). The requirement of these tumors for tryptophan can account for as much as 60% of the daily intake of this amino acid (Sjoerdsma *et al.*, 1956) and the frequency of pellagra in patients with carcinoids has been attributed to shortage of tryptophan for conversion to nicotinic acid, deficiency of which leads to pellagra.

Abnormalities of absorption of individual amino acids from the intestine have also been described. In Hartnup disease, there is a defect in renal tubule reabsorption of several amino acids, including tryptophan, and a similar defect in absorption of tryptophan from the gut has been observed in these patients (Milne *et al.*, 1960). Large doses of tryptophan cause excretion of this amino acid in the feces of such cases, part being transformed to indole derivatives which may then be absorbed and account for the appearance of indole compounds in the urine. As in the case of carcinoid tumors, the defect in tryptophan metabolism is associated with pellagra.

A similar defect in renal transport and intestinal transport occurs in cases of cystinuria (Milne *et al.*, 1961). In this condition, the cells of the renal tubules and the intestinal mucosa show an incapacity to transport cystine, lysine, arginine, and ornithine, concentration of which occurs by a common mechanism (Chapter 4). Thus feeding of large doses of lysine to cystinuric subjects causes fecal excretion of arginine by competition at the site of absorption. Unabsorbed amino acids can be metabolized by the intestinal bacteria to the corresponding amines which may be absorbed and excreted in the urine: agmatine from arginine, cadaverine from lysine, and putrescine from ornithine.

d. Intestinal Infections. Diarrhea causes loss of nutrients, including nitrogenous compounds. Thus Holt *et al.* (1915) observed that severe diarrhea in infants could raise fecal N output from 8 to 25 % of dietary N intake. Probably both decreased absorption of food protein and increased loss of endogenous protein contribute to the augmented fecal N output in infective diarrhea. Infestation with parasitic worms does not usually cause a significant increase in fecal N output (Bray, 1953; Holmes and Darke, 1959) though very heavy hookworm infection has been observed to do so (Darke, 1959).

B. Diseases Causing Increased Secretion of Protein into the Tract

The protein digested and absorbed from the alimentary tract includes not only the protein of the diet but also proteins secreted into the intestinal tract. These consist of the enzymes of the digestive juices, mucin secreted into the gut, and leakage of plasma proteins. The magnitude of this endogenous secretion of protein amounts, in the case of rats on a protein-free diet, to the equivalent of a diet of 10% protein content, and with diets containing protein, the contribution rises (Twombly and Meyer, 1961). Mosenthal (1911), who studied dogs with Thiry-Vella fistulas, computed that this loss of nitrogen into the intestine equalled one third of the daily dietary intake. The contribution of endogenous protein in the case of man is difficult to assess, but Spencer (1960) assesses it at 35 gm daily, rather more than one third of the normal protein intake. Like the dietary protein, most of this must be reabsorbed into the body, presumably as free amino acids. A small part of the endogenous secretion is lost, for the feces of subjects on protein-free diets contain about 0.5 gm N per day (Mendel and Fine, 1912; G. B. Wallace and Salomon, 1909). This endogenous content increases in certain diseases of the intestine (Table II), such as tuberculous ulceration and malignant tumors, the latter being confined to the large intestine and thus below the absorbing area. Similarly in ulcerative colitis, a chronic inflammatory condition of the large intestine, there is loss of protein into the large intestine (Welch *et al.*, 1937) through oozing of blood from the ulcerated areas, sloughing of tissue, and also leakage of plasma proteins (Steinfeld *et al.*, 1960).

Particular interest has focused recently on the amount of plasma protein lost into the gut, both in normal subjects and in patients with gastrointestinal disease. It has been possible to demonstrate that hypoalbuminemia in certain diseases is due to excessive gastrointestinal loss of protein, namely, regional enteritis and ulcerative colitis (Steinfeld *et al.*, 1960), giant hypertrophy of the gastric mucosa (Citrin *et al.*, 1957), gastric neoplasm (M. Schwartz and Jarnum, 1959), and the malabsorption syndrome (Parkins, 1960). In addition, however, there remain a series of subjects with low plasma albumin levels resulting from gastrointestinal losses unassociated with any of these conditions. For these the term "idiopathic hypercatabolic hypoproteinemia" has been suggested. Waldmann *et al.* (1961) report the investigation of 20 cases referred to them with this diagnosis; 2 were identified as cases of analbuminemia, 2 were secondary to cardiac disease, and 12 were found at operation to have disturbances of lymphatic vessels in the small intestine, a condition for which the term "intestinal lymphangiectasia" has been proposed. The albumin was considered to leak from the distended lymphatics of these patients into the intestinal lumen, where it was rapidly digested and the reabsorbed amino acids were returned to the liver to be used in the manufacture of more plasma albumin. This process appears to tax the liver beyond its capacity for albumin synthesis, so that the plasma level of this protein falls below normal.

The recognition of idiopathic hypercatabolic hypoproteinemia and the study of the intestinal defect has provided some technical problems. Such cases, when injected with I^{131}-albumin, show a much reduced half-life for plasma albumin (e.g., 4 days instead of 17 days in the series reported by Waldmann *et al.*, 1961), with an augmented excretion of I^{131} in the feces (in the above series, 0.78 % per day instead of a normal level of 0.01 %). However, this fecal excretion is presumably only a small fraction of the total loss into the intestine, since much of the I^{131} leaking into the intestine will be reabsorbed again when the albumin is digested. Consequently, Gordon (1959) has introduced the use of a polymer with a similar molecular weight to that of albumin, namely, polyvinyl pyrrolidone, which can be labeled with I^{131} and injected intravenously; on secretion into the intestine, it is not affected by the digestive processes and the I^{131} recovered in the feces is taken to represent the extent of leakage into the gut. Thus, in the series of patients seen by Waldmann *et al.* (1961) for idiopathic hypoalbuminemia, the excretion of I^{131}-polymer in the feces during a 4-day period was 8 % of the dose given, whereas normal subjects excrete less than 1 % during this period. However, Jeejeebhoy and Coghill (1961) point out several disadvantages of using a synthetic polymer in place of albumin. They use I^{131}-albumin and, in order to overcome reabsorption of iodine into the blood, administer an ion-exchange resin by mouth which traps the free iodide in the gut and ensures its excretion in the feces. However, as McFarlane points out in Chapter 8, juices entering the alimentary tract will carry radioactive free iodide from the pool of body iodide which is made radioactive by albumin breakdown, and thus will contribute to the radioactivity recovered in the feces after feeding the resin. Consequently, this method will overestimate the degree of leakage into the gut.

III. Changes in Intermediary Protein Metabolism Associated with Disease

As emphasized in the previous section, the evidence as a whole suggests that dietary protein is absorbed from the alimentary tract in the form of free amino acids. The subsequent fate of these amino acids is described in Volume 1. From the point of view of classifying pathological changes in protein metabolism, we can divide abnormalities of intermediary metabolism into those affecting the fate of individual amino acids (such as the inborn errors of amino acid metabolism), and those having a general effect on the course of protein metabolism.

A. Anomalies in the Metabolism of Individual Amino Acids

Diseases affecting the fate of individual amino acids are discussed in detail in Chapter 18. The utilization of a single amino acid can be specifically altered if there is a familial abnormality in the production of an enzyme involved in its metabolism, such as occurs in alcaptonuria. Abnormalities of amino acid metabolism can also occur from pathological processes of nonfamilial type, such as the aminoaciduria of lead poisoning. Many of the recognized familial and nonfamilial anomalies of amino acid metabolism are associated with excretion of the amino acid or a metabolite in excessive quantities in the urine.

B. General Effects of Disease on Intermediary Protein Metabolism

In health, proteins and other nitrogenous body constituents are maintained at a constant level by an integrated series of reactions in which the amino acids derived from the protein of the diet are distributed to the tissues for anabolic purposes and the unwanted amino acids undergo catabolic reactions leading to excretion of their nitrogen. As discussed in detail in Chapter 10, the balance between anabolism and catabolism is influenced by dietary factors other than protein, notably, level of energy intake and, within the body, by the amounts of various hormones secreted from the endocrine glands (Fig. 1). Pathological processes may therefore affect the general course of protein metabolism at several points. They may act directly on tissue protein metabolism (e.g., destruction of tissue by X-irradiation), they may alter endocrine activity and thus bring about changes in hormonal control over protein metabolism (e.g., stress diseases, through overactivity of the adrenal cortex), they may reduce available intracellular energy level, resulting in a variety of changes affecting the fate of amino acids (e.g., in diabetes mellitus), or they may influence protein metabolism by altering food intake, with consequent changes in the amounts both of protein and of energy ingested per day.

Thus, in a variety of ways, disease states can alter the balance of competing pathways to favor either anabolic processes leading to an increment in the total nitrogen content of the body, or catabolic processes resulting in a reduction in body nitrogen. This change from the normal steady state is recognized by alterations in nitrogen balance, i.e., the difference between N intake and N output, which becomes positive when nitrogen is being retained, and negative when loss of body protein is induced. It should be noted that, as an indicator of change in protein metabolism, nitrogen balance has certain limitations. First, it is the algebraic sum of gains and losses in different tissues of the body. For example, administration of cortisone causes loss of nitrogen from the body, coming mainly from the protein of the carcass, but this obscures a gain in the protein content of the liver and other viscera during the period that the body as a whole is losing protein (Trémolières *et al.*, 1955). Nitrogen balance determinations tell us nothing about these shifts in body protein distribution, which are particularly liable to influence the protein content of the viscera on account of the presence of labile protein there (see Chapter 10). Second, the body can in some instances achieve a new equilibrium after a period of nitrogen gain or nitrogen loss, even although the original pathological process is still operative. Consequently, balance studies made at this stage may reveal no deviation in nitrogen balance from normal, even although the protein content of the body is abnormal as the result of

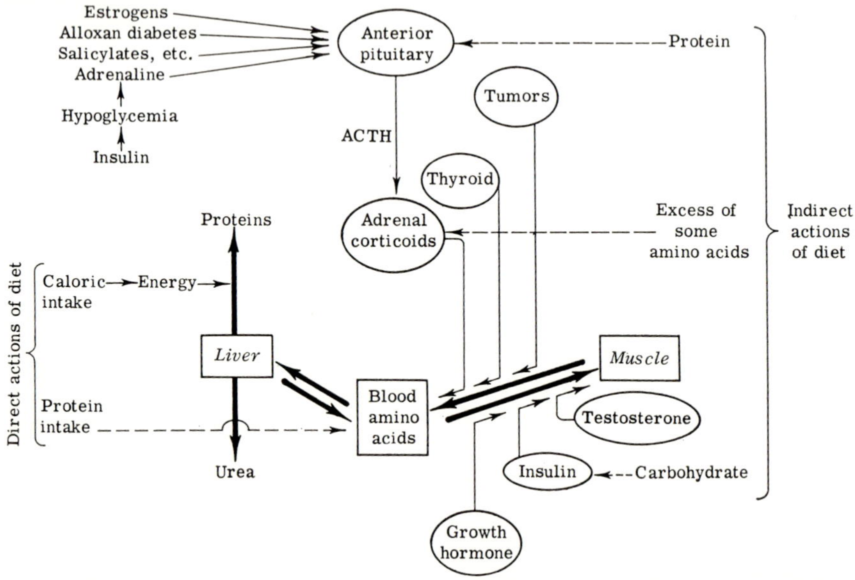

FIG. 1. Hormonal and dietary factors influencing the course of protein metabolism. Hormones affecting protein metabolism are classified as anabolic or catabolic according to whether nitrogen balance is improved or depressed. Not all tissues participate equally in the change. (*1*) The improvement in nitrogen balance caused by anabolic hormones is associated with preferential deposition of amino acids in muscle. (*2*) The negative nitrogen balance caused by catabolic hormones is accompanied by loss of protein from the carcass, but there is often a gain in the protein content of some viscera, such as the liver. These differences in the response of the carcass and viscera are mainly evident in short-term experiments.

The dietary supply of protein and of energy-yielding nutrients has a direct effect on the course of protein metabolism. In addition, diet can influence hormonal control of protein metabolism. (*1*) Protein intake influences the functional activity of the pituitary-adrenal system and affects its sensitivity to stress-producing agents. (*2*) Some amino acids have a (direct?) action on the activity of the adrenal cortex. (*3*) Dietary carbohydrate causes deposition of amino acids in muscle through an insulin-dependent mechanism.

The evidence for these actions of diet and hormones on protein metabolism are discussed in Chapter 10, from which this diagram is taken (Fig. 25).

previous gains or losses. This aspect of the interpretation of nitrogen balance data is seen especially in protein-depleted subjects who need very little food protein in order to achieve N equilibrium and who store nitrogen when receiving a normal intake of protein (F.A.O., 1957).

The technical aspects of nitrogen balance determinations have been discussed in some detail by Reifenstein *et al.* (1945), and the limitations have been assessed by W. M. Wallace (1959), by Waterlow *et al.* (1960), and in Chapter 11 of this treatise. Three comments are worth bearing in mind when evaluating the significance of nitrogen balance data. (*1*) Nitrogen balance as commonly performed means the difference between dietary N intake and the excretion of N in the urine and feces. However, other channels of N loss from the body are not entirely negligible. The daily loss from the skin as sweat and desquamated cells amounts to at least 1.0 gm N per day in adult man (Mitchell and Edman, 1962), and in skin diseases with extensive desquamation can attain 3 gm N or more daily (Salomon and von Noorden, 1907). It can be computed from published data (Gillett *et al.*, 1918) that during the menstrual period the loss of protein in the menstrual flow represents 0.3–0.6 gm N daily. Under pathological conditions, oozing from wounds and burns can be responsible for considerable losses of N and in severe chronic bronchitis the sputum can contain up to 1 gm of N daily (Eaton, 1963). (*2*) Errors involved in measuring N intake and N output all favor a positive N balance (Wallace, 1959). Thus N intake tends to be overestimated because of unconsumed portions of diet, and N output tends to be underestimated due to unrecognized N losses arising from neglect of the various minor channels of N output listed above, and also because of failure to make complete collections of excreta. (*3*) The circumstances under which N balance is measured have to be adequately standardized. Among conditions at the time of the determination which will affect N balance, it is important to pay attention to the adequacy of caloric intake and of water intake, and to avoid psychological disturbances which can cause overproduction of adrenocortical secretion. In addition, the observer must recognize that several days are required before the body will adjust to a new level of protein intake (see Chapter 10). Methods of presenting nitrogen balance data are fully discussed by Reifenstein *et al.* (1945).

Changes in protein metabolism can be conveniently examined in one readily accessible tissue, the blood. Blood amino nitrogen level is responsive to a variety of factors but, even in the experimental study of hormonal action on animals under standardized conditions, the interpretation of changes in amino N level has to be made with caution (Russell, 1955). Measurement of individual free amino acid concentrations in the plasma, for which some representative values for different species are given in Table III, may sometimes prove more informative. For example, Swendseid *et al.* (1963) have demonstrated that subjects fed on a low intake of protein for a month undergo a reduction in essential amino acid N content of plasma but show an increase in nonessential amino acid N, so that the ratio of essential to nonessential amino acids diminishes. This appears to reflect a similar change in the relative amounts of free amino acids in the tissues, as judged by results reported by Wiss (1949) for liver concentrations of free amino acids in rats receiving low and high intakes

TABLE III

CONCENTRATIONS[a] OF FREE ESSENTIAL AMINO ACIDS IN THE PLASMA OF DIFFERENT SPECIES, MOSTLY OBTAINED IN THE POSTABSORPTIVE STATE

Amino acid	Young rat[b]	Adult rat[c]	Cat[d]	Adult dog[e]	Adult dog[f]	Child[g]	Adult human[h]	Adult human[i]	Adult human[j]
Arginine	17	39	28	25	23	11	16	15	15
Histidine	6	7	14	8	13	10	14	12	12
Isoleucine	7	11	8	9	8	11	13	9	7
Leucine	11	19	16	12	15	18	19	17	13
Lysine	69	49	28	18	29	21	22	27	23
Methionine	6	9	4	5	8	3	5	4	3
Phenylalanine	6	9	9	8	8	12	10	8	9
Threonine	8	34	14	18	25	16	17	14	14
Tryptophan	—	12	—	12	14	9	13	—	—
Valine	12	21	24	16	20	29	27	29	23
Total essential amino N	14.7	20.5	15.4	12.7	15.2	14.2	15.6	14.8	13.0

[a] In μg/ml.

[b] By ion-exchange method on young rats receiving 8% casein with 0.3% methionine and not apparently fasting (Swendseid *et al.*, 1962).

[c] By microbiological assay on 8 fasting rats of 200–250 gm (Schurr *et al.*, 1950).

[d] By ion-exchange method on one cat, presumably unfasted (Tallan *et al.*, 1954).

[e] Mean of reported values by microbiological assay on radial vein blood of 4 dogs before feeding experiment (Denton and Elvehjem, 1954).

[f] Mean of reported values by ion-exchange method on 3 dogs fasting after control diet (Longenecker, 1961).

[g] By ion-exchange method on 20 infants, presumably not fasting (Holt, 1963).

[h] By microbiological assay on 24 fasting subjects (Johnson and Bergeim, 1951).

[i] By ion-exchange method on 5 postprandial subjects (Stein and Moore, 1954).

[j] By ion-exchange method on 6 fasting subjects (Frame, 1958).

of protein. Changes in the pattern of free amino acids in the plasma may therefore provide some insight into availability of amino acids within the cells of the body. Other nonprotein nitrogen constituents of blood, notably urea, also provide information of value in examining general changes in protein metabolism due to disease.

The effect of disease on turnover of body proteins can be assessed by giving amino acids labeled with N^{15} and examining the urinary excretion of the isotope, from which some idea of the amounts synthesized into body protein can be computed; the applications and limitations of such techniques are discussed in Chapter 7. The turnover of plasma proteins can be fairly readily investigated in man and in animals by procedures summarized in Chapter 8. When coupled with estimates of the total amounts of plasma proteins, both intravascular and extravascular, a well-documented picture of the metabolic state of this group of proteins can be obtained. Plasma protein metabolism is affected nonspecifically by a variety of diseases, frequently with reduction in albumin concentration and no change or even a rise in globulin level; consequently, changes in plasma protein concentration and in albumin-globulin ratio provide information about deviations in protein metabolism which is relatively simple to obtain.

There is an extensive literature concerning the effects of disease on the course of protein metabolism. Reports on the nitrogen balance of patients have been published for more than a century. In 1897, Atwater and Langworthy made a complete analysis of the literature on this subject, and von Noorden (1907) and Lusk (1928) provided later compilations. Observations on blood nonprotein levels in disease extend over more than fifty years, and much of this field was reviewed in detail in 1946 by Peters and Van Slyke (1946). It is not the intention here to bring this vast field up to date, but rather to select some of the more interesting metabolic changes which have been encountered, and relate them to the normal course of protein metabolism in the healthy subject.

1. Protein Metabolism in Disturbances of Endocrine Action

As pointed out by Ingle (1951) and also in Chapter 9 of this book, the effects of hormones on protein metabolism depend on the state of the recipient. Insufficiency of thyroid secretion causes the stunted growth of the cretin, normal growth is associated with the euthyroid state, and overactivity of the thyroid leads to tissue destruction and loss of body protein. Clearly, optimal amounts of thyroid hormones favor nitrogen retention.

Acute administration of hormones to normal subjects and to animals (Fig. 1) can lead to nitrogen loss from the body (adrenocorticoids, thyroid hormones, estrogens, progesterone) or nitrogen retention by the body (testosterone, growth hormone, insulin). Details of the metabolic picture after giving each of these hormones are discussed in Chapters 9 and 10, and from the evidence given there the following points emerge.

1. The catabolic hormones, with the exception of progesterone, have all been shown to cause *deposition* of protein and RNA in the liver, concurrently with *loss* of protein from the carcass (Fig. 1). This may simply occur because the flood of amino acids coming from breakdown of peripheral tissues provides an abundant supply for liver protein synthesis, but there is also evidence suggesting that, in the case of thyroxine and cortisone, a direct action on liver protein metabolism may be involved. Thyroid hormones and cortisone each increase the turnover of plasma albumin, an effect which is not surprising if protein synthesis in the liver is stimulated by these hormones.

2. The anabolic hormones, testosterone, growth hormone, and insulin, each favor deposition of protein in muscle, partly in labile form. Acute administration of the anabolic hormones thus causes a redistribution of body protein which favors the carcass and is thus opposite in direction to the action of the catabolic hormones (Fig. 1).

Clinical observations support the experimental studies. In the case of the thyroid gland, data reported for N balance on a normal diet show in general a moderate positive N balance in subjects with myxedema and losses of N from subjects with thyrotoxicosis (Janney and Isaacson, 1918; Boothby *et al.*, 1925; Lauter and Jenke, 1925; Krauss, 1926; Crispell *et al.*, 1957; Krane *et al.*, 1956; Dow and Stanbury, 1960). P. B. Cook *et al.* (1959) report slight positive N balances in 7 of 9 thyrotoxic cases, but considerable previous weight loss suggests that most of these subjects must have passed through a phase of negative balance.

The catabolic effect of corticosteroid overproduction is evidenced by the nitrogen loss caused by stress, as discussed below. Prolonged administration of corticoids to human subjects, or continued administration of ACTH, eventually cease to evoke a negative nitrogen balance (Kountz *et al.*, 1953). This may account for the observation that patients with general overactivity of the adrenal cortex (Cushing's syndrome) quite frequently do not show the expected negative nitrogen balance due to adrenocorticoid oversecretion (Albright, 1942). Female patients with a virilizing tumor of the adrenal cortex, and thus exposed to the anabolic effects of endogenously formed androgens, show a positive nitrogen balance (Skoog and Adams, 1957); this agrees with the anabolic action of androgens in animals.

Crispell and his colleagues have studied protein metabolism in endocrine diseases with the use of N^{15}-labeled compounds. The difficulties of interpretation of N^{15} excretion data have been emphasized in Chapter 7, and the conclusions drawn should therefore be viewed with caution. In a pituitary dwarf, Crispell *et al.* (1954a) found that the free amino acid pool and the rate of protein synthesis were reduced, as judged by studies with administered N^{15}-glycine and N^{15}-urea. The giving of growth hormone to this subject led to a small improvement in N balance, but a considerable increase in the free amino acid pool and in rate of N^{15} incorporation into body protein, findings which have recently been confirmed (Haak *et al.*, 1962). By the same technique, patients with adrenocortical overactivity (Cushing's syndrome) were also found to have a reduced rate of protein synthesis, which was considered by the authors to indicate an antianabolic basis for the negative nitrogen balance associated with adrenocortical oversecretion (Parson *et al.*, 1952; Crispell *et al.*, 1954b). However, in patients with myxedema in positive nitrogen balance, N^{15} studies again showed a reduced rate of protein synthesis (Crispell *et al.*, 1956); return to the euthyroid state under treatment with thyroid hormone eventually restored protein synthesis to normal, but one normal subject who received dosage with thyroid hormone and thus became hyperthyroid, underwent a *reduction* in N^{15} incorporation into body protein. Thus excessive amounts of both adrenocortical hormones and of thyroid hormones apparently reduced uptake of N^{15} by body protein as a whole.

Studies of I^{131}-albumin metabolism show that turnover is accelerated by hyperthyroidism induced by administration of thyroxine or triiodothyronine (Rothschild *et al.*, 1957; Iber *et al.*, 1958) and by adrenocortical hormone treatment (Rothschild *et al.*, 1958, 1961; Grossman *et al.*, 1960). Patients with adrenocortical overactivity due to Cushing's syndrome also show an accelerated turnover of plasma albumin (Sterling, 1960). These findings are contrary to the observations on whole body protein synthesis made with N^{15} by Crispell and his colleagues, as reported above, in which abnormally high thyroid or adrenocortical activity *reduced* the uptake of N^{15} into the body proteins. This underlines that the metabolic behavior of individual proteins may differ from that of body protein as a whole, a point already emphasized in connection with shifts of body protein between carcass and viscera under the influence of hormones. In addition, neither the N^{15} nor the I^{131}-albumin method is established beyond criticism as a procedure for measuring turnover (see Chapters 7 and 8).

Diabetes mellitus represents a metabolic disorder in which the implications of a defect in carbohydrate utilization can be worked out from general principles. The controlled diabetic can be maintained in N equilibrium, and on a protein-free diet does not excrete more endogenous N than normal subjects, unless ketosis is also present (Lauter and Jenke, 1925). In severe ketosis, large increments in N excretion occur (Bulger *et al.*, 1925), and excretion of ammonia in the urine parallels the degree of ketosis (Van Slyke and Palmer, 1920). A comprehensive account of the derangement of protein metabolism in diabetes has recently appeared (E. E. Gordon, 1962).

2. Protein Metabolism in Diseases Associated with Stress

Stress has been defined as "a deviation from the normal resting condition in extensive regions of the body" (Selye, 1950). Situations causing stress include (*a*) infection, (*b*) injury, (*c*) undernutrition, (*d*) climatic extremes, (*e*) severe exertion, and (*f*) nervous overstimulation. The principal general responses of the body to these stressors are believed by Selye to occur partly through the autonomic nervous system, but especially through increased activity in the pituitary-adrenal axis (Fig. 1). As discussed above, increased secretion of adrenocortical hormones results in an adverse effect on N balance, but some organs, such as the liver, accumulate protein rather than lose it (Fig. 1).

A standard dose of cortisone causes the same N loss from the body and protein deposition in liver at all levels of protein intake (Silber and Porter, 1953; Goodlad and Munro, 1959). However, it has been demonstrated under several experimental conditions that the sensitivity of the pituitary-adrenal axis to stressors varies with the protein content of the diet (see Chapter 10 for a review of the evidence). Thus agents which cause little or only moderate adrenal stimulation at low intakes of protein may have a much augmented effect when the animal is receiving a diet rich in protein. This does not appear to be due to a difference in sensitivity of the adrenal cortex to ACTH at different levels of protein intake; the weight of evidence suggests that protein intake may influence the amount of ACTH discharged from the pituitary gland in response to the stressor (Munro *et al.*, 1962). Consequently, the disturbance in protein metabolism caused by a stress situation may vary in proportion to the level of protein in the diet.

We shall now examine the changes in protein metabolism induced by the various stress situations listed above.

a. Infection. Febrile illness causes a loss of N from the body which is associated with a rise in energy output, an increase in adrenocortical activity and a diversion of amino acids to the synthesis of antibodies, notably in chronic infections. Lusk (1928) provides an extensive analytical review of the literature on nitrogen balance in fever. In studies on experimental animals receiving a constant food intake [e.g., those of Staehelin (1904), on dogs infected with trypanosomes], loss of body protein was greater than would be predicted from the caloric deficit induced by the fever. This toxic destruction of body protein is also seen in febrile patients receiving low-protein diets (Kocher, 1914). With N^{15}-glycine, Reiss (1959) has demonstrated that fasting rats given a streptococcal pneumonia undergo an accelerated rate of body protein degradation but show an accumulation of protein in the viscera, as compared with fasting control rats, which is reminiscent of the action of cortisone (Fig. 1). Studies on rats infected with malaria show that utilization of dietary protein is impaired by infection (Platt *et al.*, 1961).

b. Injury. The effects of physical injury on protein metabolism are reviewed in Chapter 19. After an injury, such as a fracture, there is an increase in N output of considerable magnitude and duration. This response still occurs in adrenalectomized rats maintained on minimal doses of cortisone (Campbell *et al.*, 1954), but is absent in animals not receiving the hormone, thus pointing to a "permissive" role of the adrenal hormones in this response to a stress situation (Ingle, 1951). The magnitude of the N loss after injury is related to the level of protein in the diet, being very slight in protein-depleted animals (Munro and Chalmers, 1945; Calloway *et al.*, 1955). This suggests that the extensive catabolic response provoked by injuring a well-nourished animal may represent dissipation of labile body protein stores. However, the metabolic status of the animal after injury is not typical of depletion of labile ("stored") protein. After a fracture or other severe injury to the rat, there is no diminution in liver protein content (Calloway *et al.*, 1955; Fleck and Munro, 1963) and the plasma protein level and the rate of degradation of plasma albumin are unaffected (Fleck and Munro, 1963). It would seem probable that the carcass (muscle?) donates the protein involved in the catabolic response to injury. Apparently this carcass material is depleted by feeding a protein-deficient diet for a short period, or else the effector mechanism is sensitive to protein level in the diet, as for example, the pituitary-adrenal axis is known to be. The exact role of the adrenal cortex in the catabolic reaction to injury is a difficult one to assess and the evidence is discussed in detail in Chapter 19.

c. Undernutrition. Severe undernutrition on a qualitatively normal diet is recognized as a cause of adrenocortical overactivity (Selye, 1950). However, N balance is sensitive to even small changes in energy intake (Chapter 10). It is therefore difficult to assess the point at which changes in adrenocortical activity are involved in the response of protein metabolism to undernutrition.

d. Climatic Stress. Exposure to cold has been much used as a means of inducing stress, with consequent adrenocortical enlargement. Exposure of the rat to a cold environment is well known to cause an increased urinary N output (Lathe and Peters, 1949; You *et al.*, 1950; Ingle *et al.*, 1953; Beaton, 1963). A cold environment also induces an increase in heat output. The two responses can, however, be dissociated. Inadequately clad human subjects exposed to the cold show an immediate increment in energy output due to loss of body fat (Voit, 1878), but the loss of body protein may be delayed by as much as 2 days (Issekutz *et al.*, 1962). Furthermore, on returning the subject to a warm environment, the increased N excretion may persist for several days (Issekutz *et al.*, 1962). These findings point to involvement of a hormonal mechanism, and indeed the response of protein metabolism to the cold is greatly diminished by thyroidectomy or adrenalectomy (You *et al.*, 1950). Extremes of heat can also cause a rise in N output (Graham *et al.*, 1959).

e. Severe Exertion. Severe muscular exertion is recognized as a form of stress (Selye, 1950). Nevertheless, a sudden bout of heavy exercise causes no more than transient and slight effects on N output (Cathcart, 1925; Lusk, 1928).

f. Psychological Overstimulation. The pituitary-adrenal axis is sensitive to the stress of excitement. A rise in blood corticosteroid level lasting a few hours results from anesthetizing rats with ether (Kitay, 1961) or with pentobarbital (Yates *et al.*, 1961), and a somewhat more prolonged elevation occurs in rats handled every day (Munro and Steele, 1963). Variations in corticosteroid output of rats in response to handling has led to the conclusion that a proportion of a colony is hypersensitive to stress (Yates *et al.*, 1961). There are also differences in the temperament of mice, reflected in their ability to compete in the society of other mice; those of lower social status undergo adrenal enlargement when taken from isolation into communal cages (Davis and Christian, 1957). It is, however, not known whether these various examples of adrenal instability to transient psychological stress are sufficient to induce recognizable changes in protein metabolism. Thus, Fleck and Munro (1963) used anesthesia as a control procedure in their studies on the metabolic response to fracture and observed no adverse effect of N balance, despite the evidence that ether can induce transient release of considerable amounts of corticosterone into the blood (Kitay, 1961). However, it has recently been found that prolonged anxiety states cause negative nitrogen balance in man (Scrimshaw, 1963).

From this evidence, and also from the discussion of the immediate action of a cold environment and of severe exercise on protein metabolism, it would appear that the stress must be considerable and of a significant duration in order to cause a recognizable change in body protein metabolism.

3. Protein Metabolism in Caloric Undernutrition and Overnutrition

Experimental study has established the sensitivity of protein metabolism to caloric intake (see Chapter 10 for a review of the evidence). The effect is not uniform in all tissues; in acute experiments on animals, changes in energy intake affect the protein content of some organs more than others, the liver being particularly sensitive (Munro and Naismith, 1953; Rosenthal and Allison, 1956). In view of the many observations made on experimental animals, it would therefore be anticipated that deviations in energy intake below or above an adequate level would have corresponding effects on the protein content of the body, and this is borne out by observation.

The effects of undernutrition on protein metabolism are illustrated by the extensive experiments of Keys *et al.* (1950), who also review the literature. Adult human volunteers were given diets providing about half their normal energy and protein intakes for a period of 24 weeks. The protein level, though low, was such that adaptation of N balance to the reduced nitrogen intake should have occurred rapidly; in fact, N balance remained negative throughout the period of undernutrition, and this can in consequence be attributed largely to the low caloric intake and not the protein level fed. During the latter part of the experiment, the dissipation of body protein and body fat diminished to about one third of the initial rate of loss, due to adaptation to caloric undernutrition.

Forced feeding of high caloric intakes to human subjects for short periods leads to retention of nitrogen (B. Krug, 1894; Cuthbertson *et al.*, 1937; Cuthbertson and Munro, 1937; Passmore *et al.*, 1955; Plough *et al.*, 1956). Some protein is laid down in the adipose tissue (Thomas, 1962), but it is likely that other organs mainly share in the nitrogen retention (Chapter 10). If this nitrogen retention continued indefinitely, obesity would be associated with considerable increase in body protein content. In experimental obesity this is not necessarily so; the obesity of hyperglycemic mice results in decreased body protein, whereas gold thioglucose-induced obesity produces a slight increase in total body protein (Mayer, 1960).

When the obese subject is given a low caloric intake in order to effect weight reduction, his body protein appears to be protected by his excess of body fat. Several authors (Dapper, 1894; Evans and Strang, 1931; Keeton and Dickson, 1933; Keeton and Bone, 1935; Sainsbury and Smith, 1937; Kekwick and Pawan, 1957; Young *et al.*, 1958) have demonstrated that obese subjects can maintain N equilibrium on caloric intakes which would cause severe negative N balances in subjects with a normal body fat content.

4. Protein Metabolism in Malignant Tumor Cachexia

The presence of a malignant tumor in the body of a human subject or animal results in distortion of normal protein metabolism. This occurs because "the tumour both concentrates metabolites to the detriment of the host and releases substances into the circulation which produce structural and functional changes in the tissues of the host" (Begg, 1958). A detailed review of these effects is presented in Chapter 20.

Changes observed in the protein metabolism of the tumor-bearing animal or human subject illustrate strikingly the limitations of nitrogen balance studies. Watkin (1961) has demonstrated that patients with non-neoplastic disease can usually be brought into positive balance in metabolic unit studies, whereas under the same conditions patients with active neoplastic disease can display nitrogen equilibrium, positive balance, or negative balance. Patients with rapidly advancing tumors can often achieve positive N balance in spite of a negative caloric balance, owing to the capacity of the tumor to trap amino acids (Waterhouse *et al.*, 1951; Terepka and Waterhouse, 1956). Experimental work on animals shows that these observations on the over-all balance of nitrogen obscure redistributions of body protein (Fenninger and Mider, 1954). Not only is there a translocation of nitrogen from the carcass to the growing tumor, but there is also a phase during which the liver of the host gains protein while the carcass (muscle) is losing nitrogen (Sherman *et al.*, 1950; Stewart and Begg, 1953; Clark and Goodlad, 1960). Since the tumor-bearing animal exhibits adrenocortical enlargement (Begg, 1958), it might be thought that overproduction of adrenocorticoids is responsible for this redistribution of body protein (Fig. 1), but other evidence discussed in Chapter 20 indicates that the picture is more complicated than this interpretation would warrant.

The metabolism of plasma proteins is disturbed in patients and in animals bearing malignant tumors. Plasma albumin level is reduced, even when the tumor is still localized (Mider *et al.*, 1950), and an increase in albumin turnover has been observed very soon after implantation of experimental tumors into rats (Hradec, 1958). A study of N^{15}-aspartic acid metabolism by a patient with an inactive lymphosarcoma has been reported by Tschudy *et al.* (1959). Incorporation of the label into body protein was stimulated by raising both dietary protein and energy levels simultaneously, a response which also occurred when the patient was again studied during an active phase of her malignant condition; however, variations in incorporation of N^{15} did not correlate well with N balance changes (Watkin, 1961).

IV. Conditions Directly Affecting the Mechanism of Protein Synthesis

A defect in the normal mechanism of protein formation can express itself in several forms. First, since all enzymes are proteins, they are vulnerable to defects in formation. In practice, some of these deletions must be incompatible with survival and are not encountered, but an ever-expanding series of inherited metabolic diseases due to defects in enzyme formation are coming to light (see J. B. Stanbury *et al.*, 1960, for a comprehensive survey). Second, the amount and structure of nonenzymic proteins may be changed, as in the case of the partial failure of protein production in analbuminemia (Humphrey, 1960) and in hypogammaglobulinemia (Pearce and Perinpanayagam, 1957), and the structural variants of hemoglobin (see Baglioni, 1963, for a recent comprehensive review). Third, the fact that viruses, which are nucleoprotein particles containing DNA or RNA, can multiply in mammalian cells indicates that the virus can divert the normal mechanism of protein formation into channels leading to the production of a new, foreign protein.

Recent studies on the mechanism of protein synthesis makes it a practical possibility to explore and partly identify the nature of deviations from normal protein formation. The current picture of the mechanism of protein synthesis in animal cells is summarized in Fig. 2, and described in detail in Chapter 6. Since each protein molecule has a definite sequence

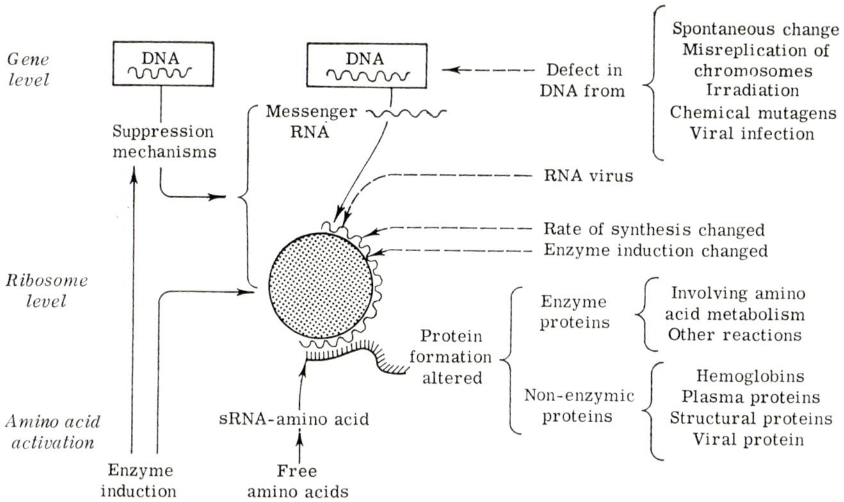

FIG. 2. Diagram to illustrate abnormalities in the mechanism of protein biosynthesis.

of amino acids, protein synthesis requires a precise mechanism for placing each amino acid in the peptide chain. This mechanism is provided by the sequence of nucleotides in the RNA template molecules on which the amino acids are assembled during protein formation. This nucleotide sequence is, in turn, a reflection of a complementary sequence of nucleotides in the DNA of the gene present in the chromosome. Since the main site of protein formation is the system of ribosomal particles present in the cell cytoplasm, the information provided by the DNA within the nucleus must in some manner be passed out to the cytoplasmic ribosomes and this is now considered to be transmitted through a series of messenger RNA molecules, one type for each protein species formed in the cell.

The capacity of a cell to produce a protein does not necessarily mean that it will do so. Thus, fetal hemoglobin differs in structure from adult hemoglobin; the evidence suggests that cells making fetal hemoglobin can later make adult-type hemoglobin, the type of hemoglobin formed being determined at the gene level (Ingram, 1961). The cell thus has the capacity to suppress the passage of the message from the gene to the ribosomes through a mechanism which is itself genetically determined.

Control of protein synthesis can also be exercised at the postgene level. In the case of the liver enzyme tryptophan pyrrolase, which is a porphyrin enzyme oxidizing the amino acid tryptophan, the amount of enzyme present adapts to the amount of tryptophan available in two stages, first by conversion of the heme-free apoenzyme to the active holoenzyme, and then by increase in the *total* amount of enzyme protein through increased protein synthesis (Greengard and Feigelson, 1961). Induction of tryptophan pyrrolase by its substrate can therefore be prevented by inhibiting protein synthesis at the ribosome level with puromycin (Nemeth and de la Haba, 1962). The amount of tryptophan pyrrolase in the liver is also increased by administration of cortisone; unlike induction of the enzyme by tryptophan administration, this action of cortisone is prevented by prior treatment with actinomycin D, which interferes with synthesis of RNA and thus of messenger (Greengard and Acs, 1962). There would consequently appear to be two mechanisms by which synthesis of tryptophan pyrrolase is controlled, one involving the messenger level and the other the ribosome level. This model system suggests means of exploring defects in the control of protein synthesis occurring in disease.

These recent observations on the mechanism of protein formation show that defects in synthesis can arise from several causes (Fig. 2). First, one can identify changes occurring at the gene level. This can take the form of an alteration in the DNA nucleotide code of a structural gene which results in errors in the sequence of amino acids in the protein formed. This alteration in the DNA nucleotide sequence may arise spontaneously, from irradiation damage, or from the action of chemical mutagens; in addition, viruses containing DNA can impose their own nucleotide code and thus control formation of virus protein. When the nucleotide sequence in a structural gene has been changed, the alteration in the amino acid pattern of the corresponding protein may or may not be of significance for its function. For example, in the case of sickle-cell anemia, a single amino acid residue out of the 280 amino acids in hemoglobin is changed, namely, a glutamic acid residue present in normal adult hemoglobin is replaced by a valine residue. This is sufficient to affect the solubility of the hemoglobin molecule and to cause the symptoms of the disease; other changes in the amino acid sequence of hemoglobin can apparently be tolerated without recognizable pathology resulting (Ingram, 1961).

Genetic control over protein synthesis need not involve the structure of the protein molecule, but can occur through changes in suppressor mechanisms (Monod, 1962). For example, a double dose of the "Bombay" gene results in a suppression of enzymes for the synthesis of the ABH blood-group mucopolysaccharides (Race and Sanger, 1962). If, as suggested by Ingram (1961), the same red cell series can at different times make fetal and adult hemoglobin, there must be mechanisms for controlling the disappearance of one and the appearance of the other. Errors in suppressor mechanisms may be responsible for the appearance of inherited diseases late in life (e.g., Huntingdon's chorea).

At the postgene level, a change in protein formation can occur from substitution of a spurious message from an RNA-containing virus. The effect of RNA-virus infection on protein synthesis can be explored experimentally by suppressing the natural messenger RNA synthesis with actinomycin and then presenting the cell with viral RNA (e.g., Baltimore and Franklin, 1962). It is also apparent that rate of protein formation can be controlled at the ribosomal level, as has been shown to be the case with induction of liver tryptophan pyrrolase by tryptophan administration. Pitot and Morris (1961) found that the Morris 5123 hepatoma had little tryptophan pyrolase activity, and that the enzyme in the tumor cell was no longer inducible with tryptophan *or* cortisone, suggesting induction defects of a multiple nature in this tumor. Studies of this kind show the complexity and interest of this approach to protein synthesis in the malignant cell.

V. Changes in the Protein Metabolism of Individual Organs and Tissues

A. Diseases Affecting the Plasma Proteins

The plasma contains a large number of separable protein components, all of which are subject to variation in amount and metabolism through the incidence of disease. Comprehensive surveys of plasma protein metabolism in health and in disease are available (Putman, 1960; Waldenström, 1962).

Changes in plasma proteins may occur through familial defects in synthesis, and are thus specific examples of errors in the mechanism of protein synthesis. It would appear that complete deletion of a normal plasma protein component has not been unequivocally demonstrated. Thus, even in cases of so-called analbuminemia, some plasma albumin can still be detected (Humphrey, 1960). On occasion, new proteins can be detected in the plasma, such as C-reactive protein and myeloma proteins (Petermann, 1960).

Several reviews on the effects of acquired disease on plasma protein metabolism are available (e.g., Petermann, 1960; Wewalka, 1962). The most striking changes occur in the albumin fraction, which is depressed by many diseases and by nutritional deficiencies. On the other hand, the individual globulin fractions tend to increase in pathological conditions.

B. Diseases of the Liver

Since the liver is responsible for transforming amino nitrogen into urea, total hepatectomy results in accumulation of free amino acids in the blood. This increase has also been noted clinically in diseases causing extensive liver cell destruction, such as acute liver atrophy (Peters and Van Slyke, 1946). Although the blood amino acid levels of patients with chronic liver damage (cirrhosis) tend to remain elevated for an abnormally long period after a test meal of protein (Richmond and Girdwood, 1962), the differences are not constant enough to be of diagnostic value. Even in the terminal stages of hepatic failure, with onset of cerebral intoxication (hepatic coma), accumulation of amino acids in the plasma is not a constant feature (Maclagan, 1957; Iber *et al.*, 1957).

A considerable body of evidence shows that blood ammonia levels are elevated in cases of advanced liver damage (Maclagan, 1957), but it seems improbable that ammonia by itself is the direct or only cause of hepatic coma (Sherlock, 1961). The level of ammonia in the blood is the resultant of several factors (see review by Bessman, 1959, for evidence): (*a*) Ammonia is released by the action of intestinal bacteria on food proteins (Webster *et al.*, 1958) and on urea present in the secretions entering the gut (Walser and Bodenlos, 1959). (*b*) Ammonia is produced by the kidney from glutamine and amino acids. (*c*) There is uptake and release of ammonia by the peripheral tissues. (*d*) Ammonia in the blood is transformed by the liver into urea through the arginine-ornithine cycle. Therapy of hepatic coma has included attempts to reduce the high blood ammonia concentration due to failure of the liver to remove it. First, reduction of protein intake and inhibition of the gut flora with antibiotics will limit to a considerable extent the formation of ammonia, and indeed these procedures are part of the treatment of hepatic coma. In 1953, Walshe suggested that administration of glutamic acid might lead to formation of glutamine with the excess ammonia, and thus detoxicate it; he reported promising results in the treatment of hepatic coma. However, extensive controlled trials have not shown any impressive advantage of this treatment (Alexander *et al.*, 1955) though it may, however, have some value in arresting impending coma (Bessman, 1959). It was then thought that stimulation of urea synthesis by arginine administration would be beneficial, since Greenstein and his colleagues had demonstrated that rats could be protected against toxic doses of amino acids or of ammonium citrate by administered arginine, an effect associated with a fall in blood ammonia level and a rise in blood urea concentration (Gullino *et al.*, 1956; Greenstein *et al.*, 1956; Du Ruisseau *et al.*, 1956). Although Najarian and Harper (1956) claimed improvement with this form of therapy in hepatic coma, Reynolds *et al.* (1958) were unable to confirm this. The complexity of equating animal studies with clinical experience on man is illustrated by the experiments of Barak *et al.* (1962) on blood ammonia levels in dogs whose livers were made functionally inefficient by an Eck-fistula bypass. Glutamic acid was able to detoxicate ammonium salts administrated to these dogs, but arginine was ineffective. However, if blood ammonia level was raised by giving glycine parenterally, this form of ammonemia was susceptible to treatment with arginine.

In addition to formation of urea, the liver participates in protein metabolism by formation of most of the plasma proteins. Consequently, changes in plasma protein metabolism are a feature of liver damage. An extensive review of the relationship of liver disease to plasma protein pattern has been provided by Wewalka (1961).

C. Diseases of the Kidney

The function of the kidney in protein metabolism (Chapter 11) is to provide for the excretion of the nitrogenous end products from the plasma while at the same time retaining free amino acids and proteins present in the plasma. Under pathological conditions the kidney may fail in each of these respects, so that the excretory nitrogenous products may be retained (uremia), abnormal amounts of amino acids may be lost in the urine (aminoaciduria), or protein may be excreted (proteinuria).

1. Uremia

In chronic renal failure, there is insufficiency of glomerular filtration and urea and other nitrogenous waste-products accumulate in the blood (renal azotemia). The general toxic effects of uremia are not directly due to accumulation of urea, and are presumably due to some associated aspect of excretory failure (De Wardener, 1961). Before the onset of uremia, changes in nitrogen excretion in chronic renal failure can be investigated by the urea clearance test; capacity to clear urea may diminish to 20 % of normal before a distinct elevation of blood urea level occurs (Peters and Van Slyke, 1946).

In addition to renal disease, increments in blood nonprotein nitrogen level can occur from other causes, grouped together as prerenal azotemia (Fishberg, 1937). Elevations in blood nonprotein nitrogen are recognized as common features of intestinal obstruction (McQuarrie and Whipple, 1919), dehydration due to diarrhea (H. Schwarz and Kohn, 1921), diabetic coma (Bulger *et al.*, 1925), shock (Aub and Wu, 1920), and severe infection, such as pneumonia (Peters and Van Slyke, 1946). Accumulation of nitrogenous end products under these circumstances is presumably the result of diminished urine filtration, aggravated by the increased breakdown of tissue protein which is a frequent feature of these conditions.

2. The Aminoacidurias

The urine normally contains small quantities of a wide range of amino acids (see Chapter 5) the amounts excreted being little influenced by protein intake (Chapter 11). Abnormally large quantities may occur from defects in reabsorption by the kidney tubules (*renal* aminoacidurias), sometimes in association with defects in reabsorption of other metabolites.

Aminoaciduria of *prerenal* origin occurs when there is an excessive level of the amino acid in the plasma (see Chapter 18).

3. Proteinuria

There are about 7 gm protein/100 ml blood plasma. This is not completely retained by the glomerular membrane, so that the filtrate in Bowman's capsule contains 10–20 mg protein/100 ml fluid (De Wardener, 1961). Since the volume filtered per day through the glomerulus is nearly 200 liters, this would lead to considerable loss of protein in the urine if it were not recovered. In fact, the normal 24-hour excretion of protein by man does not exceed 50 mg (Rigas and Heller, 1951) so that the protein of the glomerular filtrate must be almost wholly absorbed. Several investigators (Rather, 1952; Sellers *et al.*, 1954; Oliver *et al.*, 1954) have studied the mechanism of tubular reabsorption and have demonstrated uptake of protein by the cells of the proximal tubule. In the rat, reabsorption of protein by the tubules can be depressed by potassium deficiency (Morrison and Gardner, 1961). Shuster *et al.* (1963) consider that the tubules can also excrete protein.

Proteinuria is usually due to abnormal permeability of the glomerular membrane to plasma proteins; however, it has recently been recognized that the normal small amount of protein in the glomerular filtrate may fail to be reabsorbed and give rise to a tubular type of proteinuria. Thus, cases have been described with proteinuria associated with aminoaciduria of the type due to a tubular reabsorption defect (S. W. Stanbury and Macaulay, 1957; Creeth *et al.*, 1963; Kazantzis *et al.*, 1963). This suggests that the proteinuria in these subjects is also due to a defect in reabsorption during passage of the filtrate down the tubule, a view which is strengthened by the finding (Butler *et al.*, 1962; Creeth *et al.*, 1963) that the pattern of proteins excreted by these patients differs considerably from the spectrum observed in proteinuria due to a glomerular defect. Instead of the usual predominance of plasma albumin in the urinary protein seen in glomerular proteinuria, patients with tubular proteinuria excrete relatively small amounts of albumin and large amounts of α_2- and β-globulins; furthermore, the mean molecular weight is less than that of albumin, being about 45,000. By starch-gel electrophoresis, a series of protein species peculiar to the urine of such cases can be identified (Butler *et al.*, 1962).

However, most patients with proteinuria suffer from a defect in the glomerular membrane that permits passage of plasma proteins into the filtrate in abnormal amounts, especially those of relatively small molecular weight such as albumin and α_1-globulin (De Wardener, 1961). In patients with the nephrotic syndrome (a general name for a clinical condition of chronic proteinuria, hypoproteinemia, and edema arising as the result of a wide variety of causes), the predominant proteins in the urine are usually albumin and α_1-globulin which are thought to filter through the leaky glomerulus; the demonstration histologically of changes in structure of the tubules suggests that some form of tubular defect may also be associated with this syndrome (De Wardener, 1961). In acute nephritis, the pattern of proteinuria is very similar to the spectrum of proteins present in the plasma; this is not surprising, since even whole red cells are excreted in the urine in this disease. In patients with the malignant condition multiple myelomatosis, there is passage of a peculiar group of low-molecular-weight proteins (mean 43,000) into the urine (Putnam and Stelos, 1953). The condition has long been recognized by the precipitation of protein which occurs on heating the urine to 45°–55°C, and then redissolves as the temperature is increased (Bence-Jones protein). This form of proteinuria is discussed in Chapter 20.

The effect of chronic proteinuria on general protein metabolism has been discussed by several authors (Peters and Van Slyke, 1946; Blaine, 1954; Squire, 1956; Milne, 1957; Drabkin *et al.* 1962). The loss of protein in the urine of the nephrotic patient is commonly 10–15 gm/day, and represents a significant call on dietary protein requirements. Blaine (1954) has pointed out that patients with nephrosis in consequence often have a body protein deficit. Since the loss of protein in the urine is usually mainly confined to albumin and α_1-globulin, these components of the plasma protein spectrum are the chief ones to show a reduction in level in nephrotic cases; this leads in turn to a fall in colloid osmotic pressure and edema develops. Studies on nephrotic patients using labeled plasma protein precursors (Kelly *et al.*, 1950; Spector, 1954) are compatible with an increased rate of plasma protein formation in this syndrome, and this conclusion is supported by direct investigation of plasma albumin synthesis in slices of liver prepared from nephrotic rats (Marsh and Drabkin, 1958b). Drabkin and his colleagues (Drabkin and Marsh, 1955; Marsh and Drabkin, 1958a,b; Drabkin *et al.*, 1962) have made extensive use of the rat with experimental nephrosis to demonstrate the rechanneling of liver protein synthesis which leads to increased formation of *all* plasma proteins made in the liver at the expense of manufacture of other liver proteins. As discussed in Chapter 8, the proteinuria does not evoke a compensatory reduction in fractional rate of albumin breakdown, and the rate may even be accelerated in severe nephrosis in children (Gitlin *et al.*, 1956).

Proteinuria is also a feature of the moderate and severe forms of toxemia of pregnancy; mild forms of pre-eclamptic toxemia are recognized by hypertension only, severe forms by hypertension with proteinuria, and eclampsia is marked by convulsions (Nelson, 1955). The proteinuria of eclampsia contains an unusually large proportion of γ-globulin and of α_2-

globulin, while in pre-eclampsia only the former is disproportionately represented by comparison with the spectrum of proteins in the plasma (Lorincz *et al.*, 1961). In toxemia, the normal positive nitrogen balance of pregnancy may become slightly negative even when an adequate diet is fed (Mukherjee and Govan, 1951), but Lister (1961) failed to find a negative balance in patients who were not undernourished. Because of the increased production of glomerular filtrate during pregnancy, reabsorption of histidine by the tubules is incomplete, and histidinuria is a constant feature of pregnancy; in toxemia of pregnancy, glomerular filtration rate diminishes and the histidinuria disappears (Harper and Doolan, 1963).

D. Disorders of Other Systems

1. Diseases Affecting the Lymphoreticular System

The lymphoreticular (reticuloendothelial) system is responsible for synthesis of antibodies which circulate mainly in the γ-globulin fraction of the plasma. Changes in the γ-globulin fraction arise from three groups of conditions: (*a*) The appearance of an antibody in response to an infective disease; in the case of chronic infections, the amount of γ-globulin fraction may be considerably increased. (*b*) An increase in γ-globulin content of the plasma in diseases considered to be due to autoimmunization of the patient to breakdown fragments of his own tissues; these conditions were formerly grouped under the term "collagen diseases" (rheumatoid arthritis, lupus erythematosis, and scleroderma) but more recent investigation shows that a much wider spectrum of diseases may arise from such autoimmune sensitization (I. R. Mackay and Burnet, 1963). (*c*) An increase in the γ-globulin fraction from pathological overproduction of plasma cells (myeloma) or of lymphocytes (macroglobulinemia); these neoplastic cells contribute new globulin peaks to the electrophoretic pattern of the plasma, and other specific properties may appear, such as precipitation of the globulin on cooling (cryoglobulin) which can lead to obstruction of the small cutaneous blood vessels when the skin is chilled. Comprehensive reviews of variations in the γ-globulin fraction in disease are provided by Putman (1960) and by Waldenström (1962).

Two questions of general interest are raised by the appearance of new globulins in the plasma under these circumstances. First, are these globulins really new proteins? In the case of antibodies formed by the host in response to bacterial infection, it is tempting to conclude that the lymphoreticular system has had to construct a protein to suit the situation without previous templates of that type being available. Such a concept would imply that the DNA code can be altered by the bacterial antigen. Burnet (1959) has proposed an ingenious alternative, namely, that there are lymphocytes of various strains, each with the capacity to turn out different types of protein; when an antigen is introduced into the body, the lymphocyte species capable of producing a protein which can act as an antibody to that particular antigen proliferates so that a clone of cells forms and makes large amounts of the appropriate γ-globulin. On this basis, myeloma and macroglobulinemia represent neoplastic clones of plasma cells and of lymphocytes, respectively.

The second aspect of overproduction of globulins is the effect this may have on the economy of protein metabolism. Chronic infective disease, with increased globulin formation, is common in countries where protein malnutrition is rampant, and the production of additional globulin may be a significant factor in the demand for dietary protein. In the case of myeloma globulins, the daily turnover can be as much as 36 gm per day (Berson and Yalow, 1957), implying diversion of a considerable part of the daily amino acid turnover of the body. Other studies on the rate of production and turnover of globulins are reviewed by Putnam (1960).

2. Effects of Disease on Collagen Metabolism

The peculiar metabolic stability of much of the body collagen is discussed in Chapter 7. Slack (1954) has investigated the incorporation of C^{14}-glycine into muscle protein and collagen of rat limbs subjected to denervation combined with dislocation, a procedure which produces gross atrophy of the limb structures. Despite the obvious loss of tissue from atrophy, the incorporation of glycine into both tissues was not reduced and in consequence the disuse atrophy was considered not to be due to impaired protein synthesis, but to increased protein removal.

Since hydroxyproline is an amino acid peculiar to collagen and elastin, the excretion of this substance in the urine has been taken with some justification as a measure of connective tissue metabolism (see Chapter 7 for discussion). Most of the hydroxyproline is excreted in the form of peptides, some of which have sequences of amino acids similar to those found in collagen (Meilman *et al.*, 1963). The excretion of hydroxyproline is low in pituitary dwarfism and is elevated in acromegaly and in some bone diseases (Dull *et al.*, 1962); it is also high in Marfan's syndrome (Sjoerdsma *et al.*, 1958).

It is convenient to mention here that the metabolism of aromatic amino acids may be abnormal in the collagen diseases (rheumatoid arthritis, scleroderma, etc.), which are forms of autoimmune disorder. Urinary excretion of hydroxyphenyllactic acid and other partial degradation products of phenylalanine and tyrosine have been reported in rheumatoid arthritics (Nishimura *et al.*, 1956; McCormick *et al.*, 1962). Similarly, excretion of intermediates of tryptophan breakdown, such as kynurenine and hydroxyanthranilic acid, has been recorded in cases of rheumatoid arthritis and scleroderma (Price *et al.,* 1957; McMillan, 1960; Bett, 1962; Speira, 1962). In patients with rheumatic fever, the urine contains appreciable quantities of 5-methoxytryptamine, a metabolite of tryptophan (Haddox and Saslaw, 1963).

3. Protein Metabolism in Muscle Disorders

Padieu (1959) observed a decrease in labeling of muscle protein with C^{14}-glycine when injected into animals after section of the nerve supply to muscle; however, there was also a reduction in labeling of the free glycine pool of the muscle cells, and it seems improbable that these observations mean more than a decreased rate of entry of glycine into the denervated muscle, perhaps from a change in blood supply. Slack (1954) found no difference in glycine uptake in atrophic limbs.

Wasting of muscle is the cardinal feature of the muscular dystrophy group of diseases. In the inherited muscular dystrophy of mice, study with labeled amino acids suggests a more rapid turnover of muscle protein (Simon *et al.,* 1958; J. Kruh *et al.,* 1960; Dreyfus *et al.,* 1962).

VI. Protein Intake and Disease

Intake of protein and disease processes are interrelated in several ways. First, an inadequate protein intake can by itself be the cause of pathological changes. Second, it can also increase the susceptibility of the subject to some non-nutritional diseases (e.g., to bacterial infection). Third, the level of protein in the diet may determine the way in which the subject reacts to a pathological condition (e.g., to injury). Finally, the disease may alter the patient's needs for dietary protein, usually in the form of an increased requirement to offset a tendency to go into negative N balance. Each of these aspects of the interrelationship between protein intake and disease will now be examined in turn, but it should be pointed out that they are not necessarily separable. For example, protein deficiency can increase susceptibility to infection, it can affect the metabolic response of the body once the infection has become established, and the infection may increase the demands for dietary protein during its course and in convalescence.

A. Influence of Protein Intake on the Incidence of Disease

Too low an intake of protein can undoubtedly result in gross pathological changes especially in children. It also increases susceptibility to other diseases, notably to infections, which are often poorly resisted. It is less certain that a high intake of protein predisposes to any diseases. Precise information about the incidence of disease in relation to protein intake is of considerable importance in determining optimum protein needs for health.

1. Protein Deficiency as a Cause of Disease

Since protein deficiency is a widespread form of malnutrition, it has been intensively studied experimentally on animals. A comprehensive account of the long-term effects of protein deficiency in species other than man is given in Chapter 21 of this book. In the human subject, protein deficiency can occur in three contexts: (*1*) a lack of dietary protein without concurrent insufficiency of energy intake (kwashiorkor); (*2*) protein deficiency together with inadequacy of energy intake (marasmus); (*3*) secondary protein deficiency, from malabsorption or from excessive body protein losses, as in nephrosis. The clinical picture of protein malnutrition in man is presented in Chapter 22.

2. Excessive Protein Intake as a Factor in the Occurrence of Disease

While protein deficiency is the commonest abnormality of protein intake, the high levels of protein consumed in Western countries could conceivably be a factor in the frequency of degenerative diseases such as atherosclerosis in these areas. While the quantity and type of fat consumed by different communities have been suspected to be major factors in the incidence of atherosclerosis, Yudkin (1957) has demonstrated that the frequency of coronary heart disease in different countries and in different classes of the same community cannot be closely correlated with intake of any one dietary nutrient. A multiplicity of factors may thus be involved, of which a high intake of protein could be one. The frequency of high blood cholesterol levels in patients with coronary heart disease has led to intensive study of dietary factors which influence this blood constituent. In experimental animals, however, a large intake of protein generally leads to a *lowering* of blood cholestrol concentration (Moyer *et al.*, 1956, Nath *et al.*, 1958; Barnes *et al.*, 1959; Stamler *et al.*, 1958; De Groot, 1959; Leveille *et al.*, 1960). In man, plasma cholesterol level has been found to increase in proportion to protein intake by some workers (Olsen, 1958; Albanese *et al.*, 1959) but not by others (Keys and Anderson, 1957; Leveille *et al.*, 1962). In the case of the cholesterol-fed rabbit, we have found that deposition of cholesterol in various organs and the intensity of the atherosclerotic changes in the aorta are greatly intensified by a diet containing excess of protein (Munro *et al.*, 1964). Clarkson *et al.* (1962) report a similar worsening of aortic atherosclerosis in cholesterol-fed pigeons when fed a high level of protein.

As pointed out earlier in this chapter (Section III,B,2), a high intake of protein can accentuate the response to a stressor which causes release of ACTH from the anterior pituitary lobe. This interrelationship may determine whether or not pathological change occurs in stress situations. Thus, hypertension produced in rats by injections of crude anterior pituitary extracts is prevented by protein depletion (Berman *et al.*, 1945). Although hypertension produced in the rat by unilateral nephrectomy and administration of deoxycorticosterone with abundant sodium chloride was found by Prado *et al.* (1947) *not* to be affected by a reduction in protein intake from 30 to 15 % of the diet, Allison *et al.* (1961) observed that animals subjected to this procedure and kept on a completely protein-free diet did not develop a high blood pressure. The hypertension induced in rats by subtotal nephrectomy without deoxycorticosterone treatment was also found by Handler and Bernheim (1950a) to be dependent on level of protein intake; subsequent studies (Handler and Bernheim, 1950b) suggested that the nephrectomized animals receiving a low protein diet failed to develop hypertension because of impaired adrenocorticotropic hormone secretion on this diet. This would appear to be a clear instance in which the effect of a stressor on pituitary response was determined by the level of protein in the diet. In dogs with hypertension due to bilateral nephrectomy, Kolff and Page (1955) found that the hypertension was only slightly aggravated by a high intake of protein, but there was an appreciable increase in the severity of degeneration in the arterial wall.

Recently, Mellinkoff *et al.* (1962) have shown that experimental polyarthritis in rats was ameliorated by a very high intake of protein, though several factors, such as the effect of the diet on growth, make interpretation of this result difficult.

All of these examples of the effect of protein intake on vascular degenerative diseases were obtained in experimental animals. There is no evidence to show whether they have any relevance to the corresponding diseases in man. Nevertheless, the experimental findings indicate that a relationship between the incidence of some human diseases and the level of dietary protein should be anticipated, notably in conditions involving the pituitary-adrenal axis or in conditions affecting those tissues which are considerably influenced by the activities of this axis, such as lymphoid tissue. It might, for example, turn out that autoimmune diseases, whose recognition is comparatively recent, occur more frequently in populations with a particular level of protein intake.

B. *Influence of Protein Intake on the Response to Disease*

In deciding to recommend a given level of protein in the diet, it is not sufficient to compute the minimum quantity which will maintain the average subject in nitrogen equilibrium. It is also desirable to provide him with the maximum resistance to attack by disease. It may be that a high level of protein intake will improve his chances with one type of disease (e.g., bacterial infections), but diminish his survival in another condition (e.g., renal failure). Decisions about dietary policy may therefore have to make a choice between alternative evils.

1. Protein Intake and the Course of Infective Diseases

This is an important topic, since protein malnutrition and infective disease coexist in many parts of the world. The evidence in favor of a relationship is detailed in Chapter 23. It would appear that most bacterial diseases are more effectively overcome by the host whose protein intake has been adequate, than by the protein-depleted subject. There is thus a synergism between bacterial infections and protein deficiency which accounts for much of the high infant mortality in areas where there is an inadequate intake of protein.

2. Protein Intake and the Response to Injury

The suppression of the catabolic response to injury in protein-depleted subjects has already been alluded to in an earlier part of this chapter (Section III,B,2). It is still uncertain whether the catabolic response is a desirable feature of the reaction to injury, and thus it is not known whether its disappearance in the depleted subject is to be regretted. There is certainly evidence that the tensile strength of scar tissue is less if the wound is made in a protein-depleted animal and satisfactory healing can consequently be delayed (Kobak *et al.*, 1947). Further discussion of the significance of the catabolic response to injury can be found in Chapter 19.

3. Protein Intake and Treatment of Hypertension

Evidence already considered shows that the protein content of the diet can influence the incidence of hypertension in experimental animals. Other factors, such as caloric intake and salt intake, also enter into the etiology of experimentally induced hypertension (Handler and Bernheim, 1950a,b). The amelioration of hypertension in man has been achieved by reduction in salt intake and by a combination of caloric and protein restriction obtained by feeding a diet of rice (Kempner, 1944). This diet consists of enough rice to provide daily some 20 gm protein, 0.3 gm NaCl, and about 1000 kcal, to which another 1000 kcal can be added as sugar. Kempner considered that his patients could attain nitrogen equilibrium on this diet, although his diet seems too low in protein for such to be possible. His observation that hypertension is greatly reduced on the rice diet has been confirmed, and has been shown to be due essentially to the low salt intake (Dunlop, 1950). One unfavorable aspect of the regimen which may be attributable to the low protein content of the diet is a deterioration in renal function, which is often already poor in such cases; return to a normal diet restores renal function (Dunlop, 1950). Nowadays, hypotensive drugs are used to control hypertension; it might be interesting to know whether they are more effective if a relatively low protein intake is taken.

4. Protein Intake and Treatment of Renal Disease

Renal failure can lead to inability to excrete the end products of protein metabolism and thus to uremia. Consequently, means of reducing protein catabolism form part of the treatment of uremia, though this has become of less importance as the use of dialysis with the artificial kidney has grown in popularity as a means of controlling blood nonprotein nitrogen levels.

The principles on which dietary treatment of uremia are based can be deduced from our present knowledge of factors regulating protein metabolism (Chapter 10). In addition to reducing the intake of protein, the patient is given a diet of high caloric content from carbohydrate sources (Borst, 1948; Bull *et al.*, 1949; Merrill, 1955; Bluemle *et al.*, 1956), which in recent studies has been supplemented with anabolic steroids (Blagg and Parsons, 1960; Wolthuis, 1961). The high intake of carbohydrate and the anabolic agents each reduce the destruction of body protein by mechanisms already discussed (Fig. 1), and thus limit the amount of nitrogen to be excreted each day.

The use of a completely protein-free regimen eventually leads to extensive depletion of body protein and weakens the patient's resistance, and in consequence some physicians give a normal diet and dialyze the patient more frequently with the artificial kidney (Elliott *et al.*, 1960). By feeding a limited amount of protein of high Biological Value, combined with anabolic steroid treatment, nitrogen balance can be improved without significant increase in urea production (Blagg *et al.*, 1962; Lawson *et al.*, 1962).

C. Protein Requirements in Disease

Even in the case of healthy adults, there is disagreement about how to assess protein requirements. Many investigators believe that requirements are fully met by an amount of protein just sufficient to bring the subject into nitrogen equilibrium (see Chapter 14), whereas others consider that a larger intake is desirable in order to ensure deposition and maintenance of labile protein "reserves" in the body (see Chapter 12). The minimum intake of protein needed to achieve nitrogen equilibrium can be estimated directly by varying the level of dietary protein until it is just adequate to achieve equilibrium. It can also be estimated indirectly from the amounts of endogenous metabolic nitrogen lost through the urine (U_m), feces (F_m), and skin (S) when the subject is receiving a protein-free diet (Chapter 14).

$$\text{Endogenous N loss} = (U_m + F_m + S) \tag{1}$$

It is argued that, if the subject is to achieve nitrogen equilibrium, his protein intake must be sufficient to replace these obligatory losses. If the dietary protein is completely absorbed and fully utilized (Biological Value and Net Protein Utilization = 100), the subject's requirements for nitrogen equilibrium would thus be met by an amount of protein equivalent to his endogenous output of nitrogen, that is, endogenous N × 6.25. In practice, dietary protein is not utilized with this efficiency and the amount of protein needed will increase inversely as the Net Protein Utilization (NPU) falls.

$$\text{Minimal requirement} = (U_m + F_m + S) \times 6.25 \times \frac{100}{\text{NPU}} \tag{2}$$

For example, it has been computed in Chapter 14 that, for an adult man of 175 cm weighing 70 kg, U_m is 3.5 gm nitrogen, F_m is 0.7 gm, and S is 1.0 gm. In consequence, the amount of protein of NPU 70 needed to replace these losses will be 47 gm, or 0.7 gm protein/kg body weight.

These calculations do not take labile protein "reserves" into account. The normal adult receiving a Western diet retains some body protein over and above the minimum needed to replace endogenous losses. This retention is due to protein deposition which is extensive in a few organs such as the liver. When protein is completely withdrawn from the diet, this labile protein is rapidly excreted and the subject comes down to an "endogenous" level of N output. The amount of labile body protein lost in this way does not exceed 3–5 % of total body protein; if the protein-free diet is continued, however, the body loses further protein, and Allison (Chapter 12) considers that about 20 % of body protein can be thus dissipated in maintaining essential tissues during depletion, and in consequence he considers it as a "reserve" of protein. If a requirement (R) for maintenance of tissue reserves is included, the equation for protein needs thus becomes:

$$\text{Optimal requirement} = (U_m + F_m + S + R) \times 6.25 \times \frac{100}{\text{NPU}} \tag{3}$$

Unfortunately, the level of protein intake needed for deposition of satisfactory "reserves" is indefinite, whereas the minimum protein intake for nitrogen equilibrium can be estimated as described above with some precision. Consequently, the latter approach provides a more practical means for examining quantitatively whether a given disease affects protein requirements.

Pathological processes can alter protein requirements in three ways:

(1) They can change the amount of endogenous nitrogen excreted through one or more of the existing channels, i.e., they can alter U_m or F_m or S. For example, hyperthyroidism increases endogenous nitrogen excretion in the urine (Mukherjee and Mitchell, 1949). Some gastrointestinal diseases increase nitrogen output in the feces; finally, the increased desquamation which occurs in some skin diseases will augment the cutaneous loss of nitrogen.

(2) Disease can create new channels (N) of nitrogen loss. These include protein losses in the exudate from burned areas of skin, in the urine of patients with kidney disease, and in the sputum coughed up by cases of chronic bronchitis. Diversion of circulating amino acids into tumor tissue can also represent a new means of loss of body nitrogen.

(3) The effect of the disease process on protein requirements may sometimes vary with the level of protein intake. For example, the nitrogen lost after an injury is small in the case of animals receiving a protein-free diet, but increases as the protein intake (I) is raised (Munro and Chalmers, 1945). It would be necessary to include a term ($U \times I$) to express this effect of protein intake on urinary N output. Again, the fecal nitrogen output of subjects with malabsorption will become increasingly abnormal as protein intake (I) is raised. In the case of such diseases, one must therefore add to the estimate of minimal requirements, a term ($F \times I$) which expresses the increase in fecal output in relation to protein intake.

A complete factorial analysis of protein requirements in disease can thus be constructed:

Protein requirements =

$$(U_m + F_m + S + R + N + U \times I + F \times I) \times 6.25 \times \frac{100}{\text{NPU}} \quad (4)$$

The necessary information to make this full calculation is not usually available, but some of the factors which change the protein requirements of patients are known. In the light of these, we can make rough estimates of the magnitude of the changes in protein requirements caused by the main diseases which make such demands, namely, gastrointestinal conditions, endocrine disturbances, infections, injuries, and renal diseases. In so doing, it should be appreciated that the severity of a disease can be very variable. Moreover, requirements can change from the acute to the chronic phase of the disease in the same patient. This is probably due to a fall in endogenous needs following protein depletion by the disease which gives a false impression that requirements are being more easily met. This erroneous picture would be corrected if a satisfactory method could be devised for computing the requirement (R) which at one end of its spectrum represents maintenance of labile "reserves" and at the other provides for repletion of body protein losses. In the depleted subject, the fall in endogenous N output would be balanced by a larger figure, R for repletion of "reserves." A detailed review of the problems associated with measuring the protein needs of patients is given by Levenson and Watkin (1959).

Diseases of the gastrointestinal tract can cause an increased output of nitrogen in the feces (Table II). Whether this arises from an increment in endogenous secretion of protein into the gut or from malabsorption of dietary protein, the increased nitrogen loss in the feces represents an increased demand on dietary protein. For example, a series of cases of ulcerative colitis examined by Welch *et al.* (1937) excreted on average 6 gm N/day in the feces, i.e., 4.5 gm N more than the output of normal subjects. In order to replace this loss, the subject would need to consume an additional 40 gm of dietary protein (NPU = 70); the minimal protein requirement of a 70 kg subject with ulcerative colitis thus becomes 1.2 gm/kg.

Diseases affecting intermediary metabolism will influence protein requirements in proportion to the effect they have on nitrogen balance. Catabolic hormones such as corticosteroids (Goodlad and Munro, 1959) and thyroxine (Deuel *et al.*, 1928) increase nitrogen output even of subjects receiving no dietary protein. In the case of thyroid hormone excess, the increase in endogenous urinary N output is proportional to the increase in basal metabolism (Mukherjee and Mitchell, 1949). If this applies to other channels of endogenous loss, we can conclude that a patient with an increase of 50 % in basal metabolism representing a moderately severe case of thyrotoxicosis will also have a minimal protein requirement 50 % greater than normal, i.e., he will need about 1.0 gm protein/kg body weight. The diabetic patient is also likely to require more dietary protein; Stowers (1963) quotes an estimate of 1.0 gm protein/kg body weight for adults.

Diseases of stress are sometimes associated with a negative nitrogen balance, infection and injury being the most important examples. The nutritional needs of the injured are considered in detail in Chapter 19. The amount of protein needed to control or suppress the catabolic response to injury depends on the type of injury, the phase of the reaction, and the level of protein in the preceding diet. Estimates of the protein intake needed to prevent the nitrogen lost in the acute phase after injury (Riegal *et al.*, 1947) or burning accidents (Blocker *et al.*, 1955; Soroff *et al.*, 1961) are variously put at 2 to 4 gm protein/kg body weight. A type of injury which makes prolonged demands on protein requirements is the production of a discharging loop of bowel (ileostomy, colostomy) from which as much as 20 gm of protein may be lost daily for long periods (Nuguid *et al.*, 1961).

Such patients may thus have a continuing additional protein requirement 50 % greater than the normal minimal needs for meeting endogenous metabolism.

In renal disease, proteinuria often represents a continuing loss of 10 to 15 gm of protein daily (Blaine, 1954). In addition, endogenous nitrogen metabolism appears to be raised in cases of chronic renal failure. In a series of 5 such cases, Herndon *et al.* (1958) found that endogenous urinary non-protein nitrogen output was increased about 50 %. This increased loss of nitrogen is probably due to the toxic action of renal failure on tissue metabolism; this seems likely since Callen and Limazi (1950) observed that red cell turnover is increased in such patients.

These figures serve to show the very considerable increase in minimal protein requirements for equilibrium which can occur in several major diseases. Failure to recognize and provide for this requirement will lead to protein depletion. Further study of protein requirements in disease is obviously desirable.

VII. Effects of Treatment on the Course of Protein Metabolism

A. Dietary Treatment

In a wide variety of diseases, the patient's diet during treatment differs from the food intake of a healthy subject of similar build. A comprehensive survey of diets imposed on patients for therapeutic purposes is given by Davidson and Passmore (1963). We have also noted in earlier sections the use of low calorie diets in obesity and low protein diets in uremic subjects. Since an insufficient intake of energy or of protein can each cause loss of body protein, it is important to recognize these actions of the diets used as part of the treatment. In the case of obesity, it has been pointed out (Section III,B,3) that loss of body protein is minimal if the protein intake of the low calorie diet is kept at a high level.

In addition, the patient's food intake may be affected by impairment of appetite associated with his disease, as for example, in cases of dyspepsia. Since calories from protein usually form a constant proportion of most self-selected diets (Dole, 1957), a reduction in total food intake will usually mean a reduced consumption of protein. In consequence, the patient with impaired appetite can undergo some degree of protein depletion from insufficient protein intake aggravated by inadequacy of caloric intake. If this situation is prolonged, the patient may be sufficiently depleted of body protein to prejudice his chances of recovery. This situation has been particularly recognized by the surgeons, who have proposed the use of parenteral feeding in selected cases (Elman, 1954). Procedures for the administration of adequate calories and amino acids by the intravenous route have been reviewed by Geyer (1960). Evaluation of diets consumed by patients under treatment probably deserves more extensive exploration. Waterlow (1963) has recently discussed the recognition of marginal protein malnutrition; the use of plasma essential to nonessential amino acid ratios might also be helpful in recognizing this state (Swendseid *et al.*, 1963).

Methods of repleting the protein-depleted patient have been the subject of much investigation in connection with infantile malnutrition. From experimental studies on animals, and also on human subjects, it is apparent that the picture in repletion is dominated by both protein intake and caloric intake, each of which can be a limiting factor in restoration of body protein (Benditt *et al.*, 1948a,b; Calloway and Spector, 1954; Forsyth *et al.*, 1955; Waterlow, 1961; see also Chapters 10 and 12).

B. Drugs

Some drugs are known to affect protein metabolism. Changes in protein metabolism are to be expected in patients treated with anabolic or catabolic hormones; the nature of the changes evoked by these has already been discussed. In addition, some drugs influence nitrogen balance, commonly by stimulating the pituitary-adrenal axis to release adrenocorticoid hormones. Thus salicylates are known to have this action (Way *et al.*, 1962). The increase in N output after inhalation anesthetics (Inami, 1931), presumably again occurs through stimulation of the pituitary-adrenal axis; brief exposure to ether is insufficient to change N balance significantly (Fleck and Munro, 1963). Administration of iodides have also been found to increase nitrogen output (Grabfield *et al.*, 1923). On the other hand, there is a substantial nitrogen retention on giving quinine or quinidine (von Boeck, 1871; Loewi, 1907; Howard *et al.*, 1963); since there is a simultaneous decrease in carbohydrate utilization, these drugs thus have an anabolic action similar to that of growth hormone. The action of many other drugs on protein metabolism has been examined; the earlier literature is summarized by Atwater and Langworthy (1897) and by Loewi (1907), and more recent studies by Peters and Van Slyke (1946). Undoubtedly much information is scattered in the literature, a survey of which could provide interesting points of departure for the study of metabolic interference with the normal course of protein metabolism.

References

Albanese, A. A., Higgons, R. A., Lorenze, E. J., and Orto, L. A. (1959). *Geriatrics* **14,** 237.

Albright, F. (1942). *Harvey Lectures* **38,** 123.

Alexander R. N., Berman, E., and Balfour, D. C. (1955). *Gastroenterology* **29,** 711.

Allison, J. B., Brande P. F., Wannemacher, R. W., Parmer, L. P., and Leathem, J. H. (1961). *J. Nutr.* **73,** 321.

Althausen, T. L., and Uyeyama, K. (1954). *Ann. Internal Med.* **41,** 563.

Anfanger, H., and Heavenrich, R. M. (1949). *Am. J. Diseases Children* **77,** 425.

Atkinson, M., Nordin, B. E. C., and Sherlock, S. (1956). *Quart. J. Med.* **25,** 299.

Atwater, W. O., and Langworthy, C. F. (1897). "A Digest of Protein Metabolism Experiments." U.S. Dept. Agr., Washington, D.C.

Aub, J. C., and Wu, H. (1920). *Am. J. Physiol.* **54,** 416.

Baglioni, C. (1963). *In* "Molecular Genetics" (J. H. Taylor, ed.), Part I, p. 405. Academic Press, New York.

Baltimore, D., and Franklin, R. M. (1962). *Biochem. Biophys. Res. Commun.* **9,** 388.

Barak, A. J., Humoller, F. L., Mahler, D. J., and Holthaus, J. M. (1962). *Gastroenterology* **43,** 35.

Barnes, R. H., Kwong, E., Pond, W., Lowry, R., and Loosli, J. K. (1959). *J. Nutr.* **69,** 269.

Bassett, S. H., Keutmann, E. H., Hyde, H. van Z., Van Alstine, H. E., and Russ, E. (1939). *J. Clin. Invest.* **18,** 101.

Beaton, J. R. (1963). *Can. J. Biochem. Physiol.* **41,** 139, 149, 161, and 171.

Begg, R. W. (1958). *Advan. Cancer Res.* **5,** 1.

Benditt, E. P., Humphreys, E. M., Wissler, R. W., Steffee, C. H., Frazier, L. E., and Cannon, P. R. (1948a). *J. Lab. Clin. Med.* **33,** 257.

Benditt, E. P., Woolridge, R. L., and Stepto, R. (1948b). *J. Lab. Clin. Med.* **33,** 269.

Berman, D., Hay, E. C., and Selye, H. (1945). *Can. Med. Assoc. J.* **54,** 69.

Berson, S. A., and Yallow, R. S. (1957). *J. Lab. Clin. Med.* **49,** 386.

Bessman, S. P. (1959). *Advan. Clin. Chem.* **2,** 135.

Bett, I. M. (1962). *Ann. Rheumatic Diseases* **21,** 63.

Blagg, C. R., and Parsons, F. M. (1960). *Lancet* **II,** 577.

Blagg, C. R., Parsons, F. M., and Young, G. A. (1962). *Lancet* **I,** 608.

Blaine, J. D. (1954). *Clin. Sci.* **13,** 567.

Blocker, T. G., Levin, W. C., Nowinski, W. W., Lewis, S. R., and Blocker, V. (1955). *Ann. Surg.* **141,** 589.

Bluemle, L. W., Potter, H. P., and Elkinton, J. R. (1956). *J. Clin. Invest.* **35,** 1094.

Boothby, W. M., Sandiford, I., Sandiford, K., and Slosse, J. (1925). *Ergeb. Physiol.* **24,** 728.

Borst, J. G. G. (1948). *Lancet* **I,** 824.

Bray, B. (1953). *Brit. J. Nutr.* **7,** 3.

Bulger, H. A., Peters, J. P., Lee, C., and Murphy, C. F. (1925). *A.M.A. Arch. Internal Med.* **36,** 857.

Bull, G. M., Joekes, A. M., and Lowe, K. G. (1949). *Lancet* **II,** 229.

Burnet, F. M. (1959). "The Clonal Selection Theory of Acquired Immunity," Abraham Flexner Lecture, 1958. Vanderbilt Univ. Press, Nashville, Tennessee.

Butler, E. A., Flynn, F. V., Harris, H., and Robson, E. B. (1962). *Clin. Chim. Acta* **7,** 34.

Butterworth, C. E., Santini, R., and Perez-Santiago, E. (1958). *J. Clin. Invest.* **37,** 20.

Callen, I. R., and Limazi, L. R. (1950). *Am. J. Clin. Path.* **20,** 3.

Calloway, D. H., and Spector, H. (1954). *Am. J. Clin. Nutr.* **2,** 405.

Calloway, D. H., Grossman, M. I., Bowman, J., and Calhoun, W. K. (1955). *Surgery* **37,** 935.

Campbell, R. M., Sharp, G., Boyne, A. W., and Cuthbertson, D. P. (1954). *Brit. J Exptl. Pathol.* **35,** 566.

Cathcart, E. P. (1925). *Physiol. Rev.* **5,** 225.

Chinn, A. B., Lavik, P. S., Stitt, R. M., and Buckaloo, G. W. (1952). *New. Engl. J. Med.* **247,** 877.

Chinn, A. B., Lavik, P. S., Babb, L. I., Buckaloo, G. W., Stitt, R. M., and Abbott W. E. (1953). *J. Lab. Clin. Med.* **42,** 377.
Citrin, Y., Sterling, K., and Halsted, J. A. (1957). *New. Engl. J. Med.* **257,** 906.
Clark, C..M., and Goodlad, G. A. J. (1960). *Brit. J. Cancer* **14,** 327.
Clarkson, T. B., Prichard, R. W., Lofland, H. B., and Goodman, H. O. (1962). *Circulation Res.* **11,** 400.
Clemetson, C. A. B., and Churchman, J. (1955). *J. Obstet. Gynaecol. Brit. Empire* [N.S.] **62,** 390.
Cook, P. B., Nassim, J. R., and Collins, J. (1959). *Quart. J. Med.* **28,** 505.
Cooke, W. T., Thomas, G., Mangall, D., and Cross, H. (1953). *Clin. Sci.* **12,** 223.
Crane, C. W., and Neuberger, A. (1960a). *Brit. Med. J.* **II,** 815.
Crane, C. W., and Neuberger, A. (1960b). *Brit. Med. J.* **II,** 888.
Creeth, J. M., Kekwick, R. A., Flynn, F. V., Harris, H., and Robson, E. B. (1963). *Clin. Chim. Acta* **8,** 406.
Crispell, K. R., Parson, W., and Hollifield, G. W. (1954a). *J. Clin. Invest.* **33,** 924.
Crispell, K. R., Parson, W., and Harden, G. (1954b). *J. Clin. Invest.* **33,** 342.
Crispell, K. R., Parson, W., and Hollifield, G. W. (1956). *J. Clin. Invest.* **35,** 164.
Crispell, K. R., Williams, G. A., Parson, W., and Hollifield, G. (1957). *J. Clin. Endocrinol. Metab.* **17,** 221.
Cuthbertson, D. P., and Munro, H. N. (1937). *Biochem. J.* **31,** 694.
Cuthbertson, D. P., McCutcheon, A., and Munro, H. N. (1937). *Biochem. J.* **31,** 681 and 689
Dapper, C. (1894). *Beitr. Lehre Stoffwechsel gesund. u. krank. Mensch.* **2,** 65.
Darke, S. J. (1959). *Brit. J. Nutr.* **13,** 278.
Davidson, S., and Passmore, R. (1963). "Human Nutrition and Dietetics," 2nd Ed. Livingstone, Edinburgh and London.
Davis, D. E., and Christian, J. J. (1957). *Proc. Soc. Exptl. Biol. Med.* **94,** 728.
De Groot, A. P. (1959). *Nature* **184,** 903.
Denton, A. E., and Elvehjem, C. A. (1954). *J. Biol. Chem.* **206,** 449.
Deuel, H. J., Sandiford, I., Sandiford, K., and Boothby, W. M. (1928). *J. Biol. Chem.* **76,** 407.
De Wardener, H. E. (1961). "The Kidney," 2nd ed. Churchill, London.
Dicke, W. K. (1950). M.D. Thesis, University of Utrecht.
Doig, A., and Girdwood, R. H. (1960). *Quart. J. Med.* **29,** 333.
Dole, V. P. (1957). *Am. J. Clin. Nutr.* **5,** 72.
Dow, E. C., and Stanbury, J. B. (1960). *J. Clin. Invest.* **39,** 885.
Drabkin, D. L., and Marsh, J. B. (1955). *J. Biol. Chem.* **212,** 623.
Drabkin, D. L., Marsh, J. B., and Braun, G. A. (1962). *Metab., Clin. Exptl.* **11,** 967.
Drew, R., Dixon, K., and Samuel, E. (1947). *Lancet* **I,** 129.
Dreyfus, J. C., Kruh, J., and Schapira, G. (1962). *In* "Protein Metabolism," Symposium, Leyden (F. Gross, ed.), p. 326. Springer, Berlin.
Dull, T. A., Causing, L., and Henneman, P. H. (1962). *J. Clin. Invest.* **41,** 1355.
Dunlop, D. M. (1950). *Brit. J. Nutr.* **4,** 225.
Du Ruisseau, J. P., Greenstein, J. P., Winitz, M., and Birnbaum, S. M., (1956). *Arch. Biochem. Biophys.* **64,** 355.
Eaton, J. C. (1963). Personal communication.
Elliott, W., Hill, M., Kerr, D. N. S., and Ashcroft, R. (1960). *Postgrad. Med. J.* **36,** 230.
Elman, R. (1954). *Surgery* **36,** 1175.
Erf, L. A., and Rhoads, C. P. (1940). *J. Clin. Invest.* **19,** 409.
Erspamer, V. (1954). *Pharmacol. Rev.* **6,** 425.

Evans, F. A., and Strang, J. M. (1931). *J. Am. Med. Assoc.* **97,** 1063.
Everson, T. C. (1952). *Intern. Abstr. Surg.* **95,** 209.
F.A.O. (1957). "Protein Requirements," Nutr. Studies No. 16. Food and Agriculture Organization of the United Nations.
Fenninger, L. D., and Mider, G. B. (1954). *Advan. Cancer Res.* **2,** 229.
Fishberg, A. M. (1937). *Bull. N. Y. Acad. Med.* **13,** 710.
Fleck, A., and Munro, H. N. (1963). *Metab., Clin. Exptl.* **12,** 783.
Forshaw, J. W. B. (1957). *Lancet* **II,** 720.
Forsyth, B. T., Shipman, M. E., and Plough, I. C. (1955). *J. Clin. Invest.* **34,** 1653.
Fourman, L. P. R., Higgins, G., Quelch, P., O'Brien, J. R. P., and Witts, L. J. (1948). *Clin. Sci.* **7,** 121.
Frame, E. G. (1958). *J. Clin. Invest.* **37,** 1710.
Frazer, A. C. (1962). *Proc. Nutr. Soc.* (*Engl. Scot.*) **21,** 42.
Frazer, A. C., French, J., Thomas, G., and Thompson, M. D. (1952). *Clin. Sci.* **11,** 141.
Frazer, A. C., McWeeny, D. J., Pover, W. F. R., Ross, C. A. C., and Sammons, H. G. (1955). *Rec. Trav. Chim.* **74,** 362.
Freeark, R. J., Kozoll, D. D., and Meyer, K. A. (1957). *Surgery* **41,** 268.
Geyer, R. P. (1960). *Physiol. Rev.* **40,** 150.
Gillett, L. H., Wheeler, L., and Yates, A. B. (1918). *Am. J. Physiol.* **47,** 25.
Gitlin, D., Janeway, C. A., and Farr, L. E. (1956). *J. Clin. Invest.* **35,** 44.
Goodlad, G. A. J., and Munro, H. N. (1959). *Biochem. J.* **73,** 343.
Gordon, E. E. (1962). *In* "Clinical Diabetes Mellitus" (M. Ellenberg and H. Rifkin, eds.). McGraw-Hill, New York.
Gordon, R. S. (1959). *Lancet* **I,** 325.
Grabfield, G. P., Alpers, B. J., and Prentiss, A. M. (1923). *J. Pharmacol.* **22,** 393.
Graham, N. M., Wainman, F. W., Blaxter, K. L., and Armstrong, D. G. (1959). *J. Agr. Sci.* **52,** 13.
Greengard, O., and Feigelson, P. (1961). *J. Biol. Chem.* **236,** 158.
Greengard, O., and Acs, G. (1962). *Biochem. Biophys. Acta* **61,** 652.
Greenstein, J. P., Winitz, M., Gullino, P., Birnbaum, S. M., and Otey, M. C. (1956). *Arch. Biochem. Biophys.* **64,** 342.
Gross, J. B., Comfort, M. W., Wollaeger, E. E., and Power, M. H. (1950). *Gastroenterology* **16,** 140.
Grossman, J., Yalow, A. A., and Weston, R. E. (1960). *Metab., Clin. Exptl.* **9,** 528.
Gullino, P., Winitz, M., Birnbaum, S. N., Cornfield, J., Otey, M. C., and Greenstein, J. P. (1956). *Arch. Biochem. Biophys.* **64,** 319.
Haak, A., Kassenaar, A. A. H., and Querido, A. (1962). *In* "Protein Metabolism," Symposium, Leyden (F. Gross, ed.), p. 150. Springer, Berlin.
Haddox, C. H., and Saslaw, M. S. (1963). *J. Clin. Invest.* **42,** 435.
Haex, A. J. C., Seeder, W. A., de Jong, J., and Guis, B. (1962). *Gastroenterologia* **97,** 149.
Handler, P., and Berheim, F. (1950a). *Am. J. Physiol.* **160,** 31.
Handler, P., and Bernheim, F. (1950b). *Am. J. Physiol.* **162,** 368.
Harper, H. A., and Doolan, P. D. (1963). *Clin. Chem.* **9,** 19.
Harper, H. A., and Uyeyama, K. (1948). *Proc. Soc. Exptl. Biol. Med.* **68,** 296.
Heath, C. W., and Fullerton, H. W. (1935). *J. Clin. Invest.* **14,** 475.
Henke, W. J., Vacca, J. B., Van Goidsenhoven, G. E., and Knight, W. A. (1961). *Gastroenterology* **41,** 233.
Herndon, R. F., Freeman, S., and Cleveland, A. S. (1958). *J. Lab. Clin. Med.* **52,** 235.
Higgins, J. A., Lee, P. R., Scholer, J. F., Reitemeier, R. J., Code, C. F., and Wollaeger, E. E. (1957). *J. Clin. Invest.* **36,** 265.

Holmes, E. G., and Darke, S. J. (1959). *Brit. J. Nutr.* **13,** 266.
Holt, L. E. Jr. (1963). Unpublished data.
Holt, L. E., Courtney, A. M., and Fales, H. L. (1915). *Am. J. Diseases Children* **9,** 213.
Howard, R. P., Ells, H. A., and Furman, R. H. (1963). *Metab., Clin. Exptl.* **12,** 359.
Hradec, J. (1958). *Brit. J. Cancer* **12,** 290.
Humphrey, J. H. (1960). *Nature* **187,** 304.
Iber, F. L., Rosen, H., Levenson, S. M., and Chalmers, T. C. (1957). *J. Lab. Clin. Med.* **50,** 417.
Iber, F. L., Nassau, K., Plough, I. C., Berger, F. M., Meroney, W. H., and Freemont-Smith, K. (1958). *J. Clin. Invest.* **37,** 1442.
Inami, R. (1931). *Tohoku J. Exptl. Med.* **17,** 39.
Ingle, D. J. (1951). *Recent Progr. Hormone Res.* **6,** 159.
Ingle, D. J., Meeks, R. C., and Humphrey, L. M. (1953). *Am. J. Physiol.* **173,** 387.
Ingram, V. M. (1961). "Hemoglobin and Its Abnormalities." Thomas, Springfield, Illinois.
Issekutz, B., Rodahl, K., and Birkhead, N. C. (1962). *J. Nutr.* **78,** 189.
Janney, N. W., and Isaacson, V. I. (1918). *A.M.A. Arch. Internal. Med.* **22,** 174.
Jeejeebhoy, K. N., and Coghill, N. F. (1961). *Gut* **2,** 123.
Johnson, C. A., and Bergeim, O. (1951). *J. Biol. Chem.* **188,** 833.
Jordan, G. L. (1960). *In* "Surgical Diseases of the Pancreas" (J. M. Howard and G. L. Jordan, eds.), p. 533. Lippincott, Philadelphia, Pennsylvania.
Kalser, M. H., Roth, J. L. A., Tumen, H., and Johnson, T. A. (1960). *Gastroenterology* **38,** 605.
Kazantzis, G., Flynn, F. V., Spowage, J. S., and Trott, D. G. (1963). *Quart. J. Med.* **32,** 165.
Keeton, R. W., and Bone, D. D. (1935). *A.M.A. Internal Med.* **55,** 262.
Keeton, R. W., and Dickson, D. (1933). *A.M.A. Arch. Internal Med.* **51,** 890.
Kekwick, A., and Pawan, G. L. S. (1957). *Metab., Clin. Exptl.* **6,** 447.
Kelley, V. C., Ziegler, M. R., Doeden, D., and McQuarrie, I. (1950). *Proc. Soc. Exptl. Biol. Med.* **75,** 153.
Kempner, W. (1944). *N. Carolina Med. J.* **5,** 125.
Keys, A., and Anderson, J. T. (1957). *Am. J. Clin. Nutr.* **5,** 29.
Keys, A., Brožek, J., Henschel, A., Mickelsen, O., and Taylor, H. L. (1950). "The Biology of Human Starvation," Vol. I. Univ. of Minnesota Press, Minneapolis, Minnesota.
Kitay, J. I. (1961). *Endocrinology* **68,** 818.
Kobak, M. W., Benditt, E. P., Wissler, R. W. and Steffee, C. H. (1947). *Surg., Gynecol. Obstet.* **85,** 751.
Kocher, R. A. (1914). *Deut. Arch. Klin. Med.* **115,** 106.
Kolff, W. J., and Page, I. H. (1955). *Am. J. Physiol.* **181,** 580.
Kountz, W. B., Ackermann, P. G., and Khiem, T. (1953). *J. Clin. Endocrinol.* **13,** 534.
Krane, S. M., Brownell, G. L., Stanbury, J. B., and Corrigan, H. (1956). *J. Clin. Invest.* **35,** 874.
Krauss, E. (1926). *Deut. Arch. Klin. Med.* **150,** 13.
Krug, B. (1894). *Beitr. Lehre Stoffwechsel gesund. u. Krank. Mensch.* **2,** 83.
Kruh, J., Dreyfus, J. C., Schapira, G., and Gey, G. (1960). *J. Clin. Invest.* **39,** 1180.
Lathe, G. H., and Peters, R. A. (1949). *Quart. J. Exptl. Physiol.* **35,** 55.
Lauter, S., and Jenke, M. (1925). *Deut. Arch. Klin. Med.* **146,** 323.
Lavick, P. S., Mathews, L. W., Buckaloo, G. W., Lemm, F. J., Spector, S., and Friedell, H. L. (1952). *Pediatrics* **10,** 667.

Lawson, L. J., Blainey, J. D., Dawson-Edwards, P., and Tonge, S. M. (1962). *Brit. Med. J.* **II,** 293.
Leveille, G. A., Feigenbaum, A. S., and Fisher, H. (1960). *Arch. Biochem. Biophys.* **86,** 67.
Leveille, G. A., Sauberlich, H. E., Powell, R. C., and Nunes, W. T. (1962). *J. Clin. Invest.* **41,** 1007.
Levenson, S. M., and Watkin, D. M. (1959). *Federation Proc.* **18,** 1155.
Lister, U. M. (1961). *J. Obstet. Gynaecol. Brit. Commonwealth* **68,** 19.
Loewi, O. (1907). *In* "Metabolism and Practical Medicine" (C. von Noorden, ed.), Vol. III, p. 1061. Transl. I. W. Hall, Heinemann, London.
Longenecker, J. B. (1961). *In* "Progress in Meeting Protein Needs of Infants and Preschool Children," Publ. No. 843, p. 469. Natl. Acad. Sci.—Natl. Research Council, Washington, D.C.
Lorincz, A. B., McCartney, C. P., Pottinger, R. E., and Li, K. H. (1961). *Am. J. Obstet. Gynecol.* **82,** 252.
Lundh, G. (1958). *Acta Chir. Scand.* Suppl. 231, 1.
Lundh, G. (1962). *Gastroenterology* **42,** 275.
Lusk, G. (1928). "The Science of Nutrition," 4th ed. Saunders, Philadelphia, Pennsylvania.
McCormick, J. N., Robinson, R., Smith, P., and Day, J. (1962). *Ann. Rheumatic Diseases* **21,** 79.
MacDonald, R. A. (1956). *Am. J. Med.* **21,** 867.
Mackay, I. R., and Burnet, F. M. (1963). "Autoimmune Diseases." Thomas, Springfield, Illinois.
Maclagan, N. F. (1957). *In* "Biochemical Disorders in Human Disease" (R. H. S. Thompson and E. J. King, eds.), p. 47. Academic Press, New York.
McMillan, M. (1960). *J. Clin. Pathol.* **13,** 140.
McQuarrie, I., and Whipple, G. H. (1919). *J. Exptl. Med.* **29,** 397.
Marsh, J. B., and Drabkin, D. L. (1958a). *J. Biol. Chem.* **230,** 1063.
Marsh, J. B., and Drabkin, D. L. (1958b). *J. Biol. Chem.* **230,** 1073.
Mayer, J. (1960). *Am. J. Clin. Nutr.* **8,** 712.
Meilman, E., Urivetzky, M. M., and Rapoport, C. M. (1963). *J. Clin. Invest.* **42,** 263.
Mellinkoff, S. M., Pearson, C. M., Wood, F. D., Eshelman, O., and Dixon, W. (1962). *Am. J. Clin. Nutr.* **10,** 398.
Mendel, L. B., and Fine, M. S. (1912). *J. Biol. Chem.* **11,** 5.
Merrill, J. P. (1955). "The Treatment of Renal Failure." Grune & Stratton, New York.
Mider, G. B., Alling, E. L., and Morton, J. J. (1950). *Cancer* **3,** 56.
Milne, M. D. (1957). *In* "Biochemical Disorders in Human Disease" (R. H. S. Thompson and E. J. King, eds.), p. 177. Academic Press, New York.
Milne, M. D., Crawford, M. A., Girao, C. B., and Loughridge, L. W. (1960). *Quart. J. Med.* **29,** 407.
Milne, M. D., Asatoor, A. M., Edwards, K. D. G., and Loughridge, L. W. (1961). *Gut* **2,** 323.
Mitchell, H. H., and Edman, M. (1962). *Am. J. Clin. Nutr.* **10,** 163.
Monod, J. (1962). *In* "Basic Problems in Neoplastic Disease" (A. Gellhorn and E. Hirschberg, eds.), p. 218. Columbia Univ. Press, New York.
Morrison, A. B., and Gardner, K. D. (1961). *Nature* **190,** 96.
Mosenthal, H. O. (1911). *J. Exptl. Med.* **13,** 319.
Moyer, A. W., Kritchevsky, D., Logan, J. B., and Cox, H. R. (1956). *Proc. Soc. Exptl. Biol. Med.* **92,** 736.

Mukherjee, R., and Mitchell, H. H. (1949). *J. Nutr.* **37,** 303.

Mukherjee, C. L., and Govan, A. D. T. (1951). *J. Obstet. Gynaecol. Brit. Empire* [N.S.] **58,** 701.

Munro, H. N., and Chalmers, M. I. (1945). *Brit. J. Exptl. Pathol.* **26,** 396.

Munro, H. N., and Naismith, D. J. (1953). *Biochem. J.* **54,** 191.

Munro, H. N., and Steele, M. H. (1963). Unpublished data.

Munro, H. N., Hutchison, W. C., Ramaiah, T. R., and Neilson, F. J. (1962). *Brit. J. Nutr.* **16,** 387.

Munro, H. N., Forbes, W., and Steele, M. H. (1964). *Proc. Nutr. Soc. (Engl. Scot.)*, in press.

Najarian, J. S., and Harper, H. A. (1956). *Am. J. Med.* **21,** 832.

Nath, N., Harper, A. E., and Elvehjem, C. A. (1958). *Arch. Biochem. Biophys.* **77,** 234.

Nelson, T. R. (1955). *J. Obstet. Gynaecol. Brit. Empire* [N.S.] **62,** 48.

Nemeth, A. M., and de la Haba, G. (1962). *J. Biol. Chem.* **237,** 1190.

Nishimura, N., Shibata, Y., Yasui, M., and Okamoto, H. (1956). *Proc. Japan. Acad.* **32,** 697.

Nuguid, T. P., Bacon, H. E., and Boutwell, J. (1961). *Surg. Gyn. Obstet.* **113,** 733.

Oliver, J., MacDowell, M., and Lee, Y. C. (1954). *J. Exptl. Med.* **99,** 589.

Olsen, R. E. (1958). *Diabetes* **7,** 202.

Padieu, P. (1959). *Bull. Soc. Chim. Biol.* **41,** 57.

Parkins, R. (1960). *Lancet* **II,** 1366.

Parson, W., Crispell, K. R., and Ebbert, A. (1952). *J. Clin. Invest.* **31,** 548.

Passmore, R., Meiklejohn, A. P., Dewar, A. D., and Thow, R. K. (1955). *Brit. J. Nutr.* **9,** 20 and 27.

Pearce, K. M., and Perinpanayagam, M. S. (1957). *Arch. Disease Childhood* **32,** 422.

Petermann, M. L. (1960). *In* "The Plasma Proteins" (F. W. Putnam, ed.), Vol. II, p. 310. Academic Press, New York.

Peters, J. P., and Van Slyke, D. D. (1946). "Quantitative Clinical Chemistry," 2nd ed., Vol. I: Interpretations. Williams & Wilkins, Baltimore, Maryland.

Pitot, H. C., and Morris, H. P. (1961). *Cancer Res.* **21,** 1009.

Platt, B. S., Miller, D. S., and Payne, P. R. (1961). *In* "Recent Advances in Human Nutrition" (J. F. Brock, ed.), p. 351. Little, Brown, Boston, Massachusetts.

Plough, I. C., Iber, F. L., Shipman, M. E., and Chalmers, T. C. (1956). *Am. J. Clin. Nutr.* **4,** 224.

Polachek, A. A., Cope, C. B., Williard, R. F., and Barnes, F. W. (1959). *Gastroenterolgy* **37,** 38.

Prado, J. P., Dontigny, P., and Selye, H. (1947). *Proc. Soc. Exptl. Biol. Med.* **66,** 446.

Price, J. M., Rukavina, J. G., Mendelson, C., and Johnson, S. A. M. (1957). *J. Invest. Dermatol.* **29,** 289.

Putnam, F. W., ed. (1960). "The Plasma Proteins." Vols. I and II. Academic Press, New York.

Putnam, F. W., and Stelos, P. (1953). *J. Biol. Chem.* **203,** 347.

Race, R. R., and Sanger, R. (1962). "Blood Groups in Man," 4th ed. F. A. Davis, Philadelphia, Pennsylvania.

Rather, L. J. (1952). *Medicine* **31,** 357.

Reifenstein, E. C., Albright, F., and Wells, S. L. (1945). *J. Clin. Endocrinol.* **5,** 367.

Reiss, E. (1959). *Metab. Clin. Exptl.* **8,** 151.

Reynolds, T. B., Redeker, A. G., and Davis, P. A. (1958). *Am. J. Med.* **25,** 359.

Richmond, J., and Girdwood, R. H. (1962). *Clin. Sci.* **22,** 301.

Riegal, C., Koop, C. E., Drew, J., Stevens, L. W., and Rhoads, J. E. (1947). *J. Clin. Invest.* **26,** 18.

Rigas, D. A., and Heller, C. G. (1951). *J. Clin. Invest.* **30,** 853.
Rosenthal, H. L., and Allison, J. B. (1956). *J. Agr. Food Chem.* **4,** 792.
Rothschild, M. A., Bauman, A., Yalow, R. S., and Berson, S. A. (1957). *J. Clin. Invest.* **36,** 422.
Rothschild, M. A., Schreiber, S. S., Oratz, M., and McGee, H. L. (1958). *J. Clin. Invest.* **37,** 1229.
Rothschild, M. A., Oratz, M., Wimer, E., and Schreiber, S. S. (1961). *J. Clin. Invest.* **40,** 545.
Rubin, C. E., Brandborg, L. L., Flick, A. L., Phelps, P., Parmentier, C., and Van Neil, S. (1962). *Gastroenterology* **43,** 621.
Russell, J. A. (1955). *Federation Proc.* **14,** 696.
Sainsbury, R., and Smith, M. C. (1937). *J. Home Econ.* **29,** 468.
Salomon, H., and von Noorden, C. (1907). *In* "Metabolism and Practical Medicine" (C. von Noorden, ed.), Vol. III, p. 746. Transl. I. W. Hall, Heinemann, London.
Schurr, P. E., Thompson, H. T., Henderson, L. M., Williams, J. N., and Elvehjem, C. A. (1950). *J. Biol. Chem.* **182,** 39.
Schwartz, M., and Jarnum, S. (1959). *Lancet* **I,** 327.
Schwarz, H., and Kohn, J. L. (1921). *Am. J. Diseases Children* **21,** 465.
Scrimshaw, N. S. (1963). Private communication.
Sellers, A. L., Griggs, N., Marmorston, J., and Goodman, H. C. (1954). *J. Exptl. Med.* **100,** 1.
Selye, H. (1950). "Stress." Acta Inc., Montreal.
Sherlock, S. (1961). *Gastroenterology* **41,** 1.
Sherman, C. D., Morton, J. J., and Mider, G. B. (1950). *Cancer Res.* **10,** 374.
Shiner, M., and Doniach, I. (1959). *Proc. World Congr. Gastroenterol., 1958* **1,** 586.
Shingleton, W. W., Wells, M. H., Baylin, G. J., Ruffin, J. M., and Saunders, A. (1955). *Surgery* **38,** 134.
Shuster, S., Jones, J. H., and Flynn, G. (1963). *Brit. J. Exptl. Pathl.* **44,** 145.
Silber, R. H., and Porter, C. C. (1953). *Endocrinology* **52,** 518.
Simon, E. J., Lessell, I., Gross, C. S., and Milhorat, A. T. (1958). *Federation Proc.* **17,** 311.
Sjoerdsma, A., Weissbach, H., and Udenfriend, S. (1956). *Am. J. Med.* **20,** 520.
Sjoerdsma, A., Davidson, J. D., Udenfriend, S., and Mitoma, C. (1958). *Lancet* **II,** 994.
Skoog, W. A., and Adams, W. S. (1957). *J. Clin. Endocrinol. Metab.* **17,** 364.
Slack, H. G. B. (1954). *Clin. Sci.* **13,** 155.
Soroff, H. S., Pearson, E., and Artz, C. P. (1961). *Surg. Gyn. Obstet.* **112,** 159.
Spector, W. G. (1954). *Clin. Sci.* **13,** 1.
Spencer, R. P. (1960). "The Intestinal Tract." Thomas, Springfield, Illinois.
Speira, H. (1962). *Arthritis Rheumat.* **5,** 658.
Squire, J. R. (1956). *Am. J. Clin. Nutr.* **4,** 509.
Staehelin, R. (1904). *Arch. Hyg.* **50,** 77.
Stamler, J., Rich, R., and Katz, L. N. (1958). *Circulation Res.* **6,** 442.
Stanbury, J. B., Wyngaarden, J. B., and Fredrickson, D. S., eds. (1960). "The Metabolic Basis of Inherited Disease." McGraw-Hill, New York.
Stanbury, S. W., and Macaulay, D. (1957). *Quart. J. Med.* **26,** 7.
Steele, B. F., and Le Bovit, C. B. (1951). *J. Nutr.* **45,** 235.
Stein, W. H., and Moore, S. (1954). *J. Biol. Chem.* **211,** 915.
Steinfeld, J. L., Davidson, J. D., Gordon, R. S., and Greene, F. E. (1960). *Am. J. Med.* **29,** 405.
Sterling, K. (1960). *J. Clin. Invest.* **39,** 1901.
Stewart, A. G., and Begg, R. W. (1953). *Cancer Res.* **13,** 556 and 560.
Stowers, J. M. (1963). *Nutr. Abstr. Rev.* **33,** 1.

Swendseid, M. E., Hickson, J. B., and Friedrich, B. W. (1962). *J. Nutr.* **78,** 115.
Swendseid, M. E., Griffith, W. H., and Tuttle, S. G. (1963). *Metab., Clin. Exptl.* **12,** 96.
Tallan, H. L., Moore, S., and Stein, W. H. (1954). *J. Biol. Chem.* **211,** 927.
Terepka, A. R., and Waterhouse, C. (1956). *Am. J. Med.* **20,** 225.
Thomas, L. W. (1962). *Quart. J. Exptl. Physiol.* **47,** 179.
Trémolières, J., Derache, R., and Lowy, R. (1955). *Ann. Nutr. Aliment.* **9,** 179.
Tschudy, D. P., Bacchus, H., Weissman, S., Watkin, D. M., Eubanks, M., and White, J. (1959). *J. Clin. Invest.* **38,** 892.
Twombly, J., and Meyer, J. H. (1961). *J. Nutr.* **74,** 453.
van Roon, J. H., and Haex, A. J. C. (1960). *Gastroenterologia* **94,** 227.
van Roon, J. H., Haex, A. J. C., Seeder, W. A., and de Jong, J. (1960). *Experientia* **16,** 209.
Van Slyke, D. D., and Meyer, G. M. (1912). *J. Biol. Chem.* **12,** 399.
Van Slyke, D. D., and Palmer, W. W. (1920). *J. Biol. Chem.* **41,** 567.
Voit, C. (1878). *Z. Biol.* **14,** 57.
von Boeck, H. (1871). *Z. Biol.* **7,** 420.
von Noorden, C., ed. (1907). "Metabolism and Practical Medicine." Transl. I. W. Hall, Heinemann, London.
Waldenström, J. (1962). *In* "Thannhauser's Textbook of Metabolism and Metabolic Disorders," 2nd ed. (N. Zöllner and S. Estren, eds.), Vol. I. p. 413. Transl. S. Estren, Grune & Stratton, New York.
Waldmann, T. A., Steinfeld, J. L., Dutcher, T. F., Davidson, J. D., and Gordon, R. S. (1961). *Gastroenterology* **41,** 197.
Wallace, G. B., and Salomon, H. (1909). *Med. Klin.* **5,** 579.
Wallace, W. M. (1959). *Federation Proc.* **18,** 1.
Walser, M., and Bodenlos, L. J. (1959). *J. Clin. Invest.* **38,** 1617.
Walshe, J. M. (1953). *Lancet* **I,** 1075.
Waterhouse, C., Fenninger, L. D., and Keutmann, E. H. (1951). *Cancer* **4,** 500.
Waterlow, J. C. (1961). *J. Trop. Pediat.* **7,** 16.
Waterlow, J. C. (1963). *Proc. Nutr. Soc. (Engl. Scot.)* **22,** 601.
Waterlow, J. C., Cravioto, J., and Stephen, J. M. L. (1960). *Advan. Protein Chem.* **15,** 131.
Watkin, D. M. (1961). *Am. J. Clin. Nutr.* **9,** 446.
Waugh, J. M., Dixon, C. F., Clagett, O. T., Bollman, J. L., Sprague, R. G., and Comfort, M. W. (1946). *Proc. Staff Meetings Mayo Clin.* **21,** 25.
Way, E. L., Taylor, J., Old, L. J., and George, R. (1962). *J. Endocrinol.* **24,** 7.
Webster, L. T., Davidson, C. S., and Gabuzda, G. J. (1958). *J. Lab. Clin. Med.* **52,** 501.
Welch, C. S., Adams, M., and Wakefield, E. G. (1937). *J. Clin. Invest.* **16,** 161.
West, C. D., Wilson, J. L., and Eyles, R. (1946). *Am. J. Diseases Children* **72,** 251.
Wewalka, F. (1961). *Wien. Z. Inn. Med. Grenzg.* **42,** 85.
Wewalka, F. (1962). *Wien. Z. Inn. Med. Grenzg.* **43,** 309.
Wiss, O. (1949). *Helv. Chim. Acta* **32,** 1344.
Wollaeger, E. E., Comfort, M. W., and Osterberg, A. E. (1947). *Gastroenterology* **9,** 272.
Wolthuis, F. (1961). "Balance Studies on Protein Metabolism in Normal and Uraemic Men." M.D. Thesis, University of Amsterdam.
Yates, F. E., Leeman, S. E., Glenister, D. W., and Fallman, M. F. (1961). *Endocrinology.* **69,** 67.
You, S. S., You, R. W., and Sellars, E. A. (1950). *Endocrinology* **47,** 156.
Young, C. M., Brown, A. M., Empey, E. L., and Turk, D. (1958). *J. Nutr.* **64,** 203.
Yudkin, J. (1957). *Lancet* **II,** 155.

CHAPTER 18

Anomalies of Amino Acid Metabolism

L. E. Holt, Jr., and S. E. Snyderman

Department of Pediatrics, New York University Medical Center, New York, New York

I. Introduction: Classification of Anomalies of Amino Acid Metabolism

In a broad sense all metabolic anomalies may be regarded as anomalies of amino acid metabolism, involving as they do the production of enzymes, which are proteins and must differ in their amino acid patterns. Only a limited segment of this vast field can be covered in a brief review. We shall confine ourselves to a group of disorders characterized by altered concentrations of amino acids in the body fluids, the tissues, and the excreta. The development of modern analytical techniques has led to a phenomenal increase of knowledge of these disorders. More than 40 conditions are now known to be associated with an increased excretion of amino acids in the urine. These include: (*1*) the *renal aminoacidurias* caused by abnormalities of tubular transport in which plasma levels are normal, in some of which the reabsorption defect affects other substances as well, and (*2*) the *prerenal* (*or overflow*) *aminoacidurias* in which a generalized metabolic disturbance gives rise to abnormal concentrations of amino acids in the plasma. This classification is not entirely satisfactory. In several conditions there is both a renal and a prerenal disturbance, as is the case in cystinosis and Hartnup disease. The further study of some diseases in the purely renal group is bringing to light evidence of anomalous transport of amino acids elsewhere. Thus, in cystinuria, apparently a pure renal aminoaciduria affecting the four diamino acids—cystine, lysine, arginine, and ornithine—Milne and his co-workers (1961) have obtained evidence that absorption of these amino acids from the intestine is affected in much the same way as reabsorption from the renal tubule. Some of the renal aminoacidurias are associated with specific clinical syndromes as yet unexplained by any generalized metabolic anomaly, which may well be due to a more widespread anomaly of amino acid metabolism. There are also anomalies of amino acid metabolism characterized by altered concentrations of metabolites of amino acids rather than of the amino acids themselves.

In the present review we shall confine ourselves to tabulating the renal aminoacidurias and referring the reader to appropriate literature. Our discussion will be devoted to the more generalized anomalies in which there appears to be an enzyme block affecting a metabolic pathway of one or another amino acid. Some of these conditions are associated with a characteristic symptomatology, mental defect being found in a number of them. Others are asymptomatic being of interest chiefly to the student of metabolism or of heredity.

A current classification of the anomalies of amino acid metabolism is given in Table I.

II. The Congenital Anomalies

A. Hyperglycinemia

The first patient with this interesting anomaly was studied by Childs, Nyhan, and their co-workers at Johns Hopkins (Childs *et al.*, 1961; Nyhan *et al.*, 1961, Nyhan and Childs, 1962). The patient was a child of 2 years who since birth had exhibited episodes of vomiting, lethargy, and ketosis, and was mentally retarded. He was found to have hypogammaglobulinemia, a tendency to neutropenia and, at times, to thrombocytopenia. A striking chemical finding was marked elevation of glycine in the urine, plasma, and spinal fluid. The hyperglycinemia was attributed to a block in the pathway of glycine to serine which may have been complete.

The symptoms could be related to the level of protein ingested. A generous protein intake was poorly tolerated, the intolerance being traced to the amino acid leucine, loading with which produced exceptionally high levels of leucine in the plasma and attacks of ketosis. Clinical benefit was obtained by reducing the protein intake to the minimum; this was accompanied by disappearance of the neutropenia and a reduction in the hyperglycinemia. Other instances of this disorder have been reported by Nyhan *et al.* (1963) and Cochrane and his co-workers (1963) in the United States and by Tada *et al.* (1963) in Japan.

B. Primary Hyperoxaluria and Oxalosis

These terms refer to a metabolic anomaly characterized by the overproduction of endogenous oxalate and its excessive excretion in urine. This results in progressive deposition of calcium oxalate in the urinary tract with symptoms of renal colic, hematuria, calculus formation, and eventually death from nephrocalcinosis and renal failure. The term "oxalosis" is applied to a small proportion of these cases in which there is deposition of oxalate throughout the body; these cases are likely to give symptoms earlier, sometimes in late infancy, and to terminate before adolescence; diagnosis can be made by demonstration of oxalate crystals in the bone marrow (Burke *et al.*, 1955; Dunn, 1955).

This metabolic anomaly is to be distinguished from "secondary oxaluria" caused by diet which may lead to calculus formation but not to nephrocalcinosis. Excessive ingestion of high oxalate foods such as rhubarb may bring this about; in other instances no such history can be obtained. Oxalate stone formers are found in all countries but are particularly common in the "stone belt" of eastern Europe—an area including the Ukraine, eastern Czechoslovakia, and Hungary (Grossman, 1938). Such cases may yet be endogenous, caused by some dietary ingredient which affects enzyme systems concerned in oxalate metabolism. It has been suggested that these individuals are heterozygotes of primary hyperoxaluria, but there is little evidence to support this contention.

Endogenous oxalate may be formed either from ascorbic acid or from glycine or serine through the glyoxylic pathway (Fig. 1) (Crawhall *et al.*, 1959). The former route appears to be of lesser importance, for loading with ascorbic acid causes relatively little increase in oxaluria unless it is given

TABLE I

CONDITIONS IN WHICH RENAL AMINOACIDURIA HAS BEEN REPORTED

Generalized	Associated tubular defect
Fanconi syndrome (*1*)	Glucose, P
Cystinosis (*1*[a])	Glucose, P
Oculo-cerebro-renal dystrophy (Lowe)	Glucose, P
Hepatolenticular degeneration (*2*)	Glucose, P, uric acid
Galactosemia (*3*)	P?
Poisons: lead (*4*), cadmium (*5*), uranium (*6*), lysol (*7*), oxalic acid, maleic acid (*8*), succinic acid (*9*), butyrobetaine (*10*), sodium fluoride, dinitrobenzene (*11*)	Glucose?, P?
Rickets (vitamin D deficiency) (*12*)	P
Rickets (resistant—some cases) (*13*)	P
Scurvy (*14*)	—
Congenital tubular renal acidosis (Lightwood syndrome) (*15*)	—
Hartnup syndrome	—
Premature and newborn infants (*16*)	Galactose and lactose
Lactosuria and aminoaciduria of infancy (*17*)	Lactose
Mental defect with quadriplegia and ichthyosis (*18*)	—
Familial hypolipidemia (*19*)	Indoluria
Severe potassium depletion (*20*)	—
Epilepsy (*21*)	—
Salicylate intoxication (*22*)	—
Congenital fructose intolerance (*23*)	—
Thalassemia (*24*)	—
Sickle cell anemia (*24*)	—
Osteogenesis imperfecta (*25*)	—
Kwashiorkor—treated (*26*)	—
ACTH and glucocorticoid intoxication (*27*)	—
Pernicious anemia (*28*)	—
Progeria (*29*)	—
Renal disease (*30*)	—
Specific	**Amino acid chiefly affected**
Cystinuria (*31*)	Cystine, ornithine, lysine, arginine
Cadmium poisoning (*5*)	Threonine, serine
Glycinuria (*32*)	Glycine
Pregnancy (*33*)	Histidine
Glucoglycinuria (*34*)	Glycine
Joseph's syndrome (*35*)	Proline
Cerebromacular degeneration (*36*)	Histidine, L-methylhistidine, anserine, carnosine
Hereditary pancreatitis (*37*)	Lysine and leucine
Pancreatitis (*37*)	Leucine

[a] References: (*1*) Fanconi, 1936; Dent, 1947; (*1*[a]) Worthen and Good, 1958. (*2*) Uzman and Hood, 1952. (*3*) Cusworth *et al.*, 1955. (*4*) Wilson *et al.*, 1953. (*5*) Clarkson and

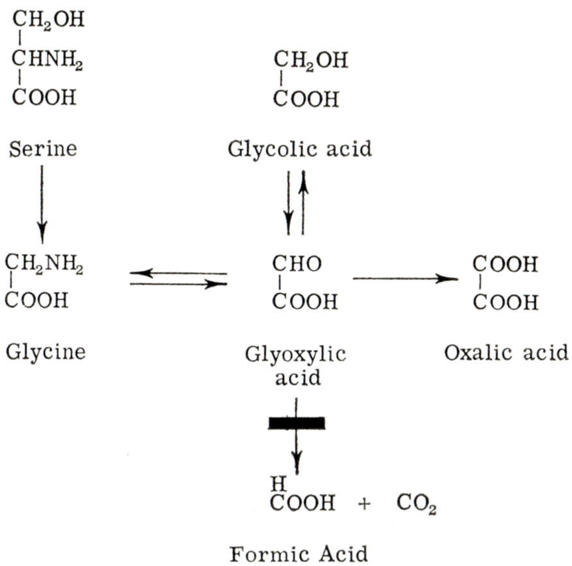

FIG. 1. The block in primary hyperoxaluria.

intravenously. Glyoxylic acid seems to be the immediate precursor of most of the endogenous oxalate (Elder and Wyngaarden, 1959). Frederich and her co-workers (1962) have recently pinpointed a block in the glyoxylic pathway in two siblings with primary hyperoxaluria. The intravenous administration of carboxyl-labeled glyoxalate C^{14} resulted in a respiratory excretion of $C^{14}O_2$ which was only 20% of normal. Both patients also excreted abnormal quantities of glycolic acid. The disease appears to be inherited as a recessive. The parents of the subjects mentioned above exhibited the metabolic defect in a minor degree, suggesting a heterozygote state.

Therapeutic attempts to control the disease have met with little success. The avoidance of high oxalate foods is certainly indicated. Archer *et al.*

Kench, 1956. (*6*) Rothstein and Berk, 1949. (*7*) Spencer and Franglen, 1952. (*8*) Harrison and Harrison, 1954. (*9*) Emslie-Smith *et al.*, 1956. (*10*) Hosein *et al.*, 1960. (*11*) Parkes and Neill, 1953. (*12*) Jonxis *et al.*, 1952. (*13*) Visakorpi, 1960. (*14*) Jonxis and Huisman, 1954. (*15*) Harrison *et al.*, 1958. (*16*) Sereni *et al.*, 1955; Dustin *et al.*, 1955. (*17*) Darling *et al.*, 1960. (*18*) Williams and Tang, 1960. (*19*) Hooft *et al.*, 1962. (*20*) Davidson *et al.*, 1960. (*21*) Millichap and Ulrich, 1962. (*22*) Andrews *et al.*, 1961. (*23*) Froesch *et al.*, 1959. (*24*) Choremis *et al.*, 1959. (*25*) Chowers *et al.*, 1962. (*26*) Schendel *et al.*, 1959. (*27*) Borden *et al.*, 1952; Martin *et al.*, 1954; Ronzoni *et al.*, 1953. (*28*) Weaver and Neill, 1954. (*29*) Steinberg *et al.*, 1957. (*30*) Wallace and Jones, 1960; Woolf and Giles, 1956. (*31*) Dent *et al.*, 1954; Frimpter *et al.*, 1961; Milne *et al.*, 1961. (*32*) de Vries *et al.*, 1957. (*33*) Dignam and Harper, 1954. (*34*) Käser *et al.*, 1961. (*35*) Joseph *et al.*, 1958. (*36*) Bessman and Baldwin, 1962. (*37*) Gross *et al.*, 1957.

(1957, 1958) obtained some reduction in the oxaluria by protein restriction and by the administration of benzoate to limit the available glycine, but the effect was only temporary. Sodium hydroxymethane sulfonate, a potent inhibitor of oxalate formation from glyoxylate *in vitro* is presently being evaluated as a therapeutic agent.

C. Hyperprolinemia

This condition was described by Schaefer *et al.*, (1962) in 4 siblings, some of whom presented evidence of renal disease, deafness, and cerebral dysfunction. A subsequent examination of the family pedigree, however, led to the conclusion that the association of the metabolic anomaly and these symptoms was coincidental. The metabolic anomaly is, however, an interesting one since it has brought to light new information about renal tubular reabsorption. The individuals affected with this anomaly had elevated plasma proline levels, varying from 6 to 20 mg/100 ml, in contrast to the normal which rarely exceeds 3 mg. Values for other amino acids were within normal limits. However, three amino acids were found in excess in the urine—hydroxyproline and glycine as well as proline itself. It was shown that the normal subject when loaded with proline secretes an excess of these three amino acids in the urine, indicating that they compete for a single reabsorption mechanism.

The block responsible for the hyperprolinemia was attributed to a defect of the enzyme proline dehydrogenase which converts proline to pyrrolidine carboxylic acid; the latter substance could not be demonstrated in the plasma or urine of affected subjects.

D. Hydroxyprolinemia

Except in the newborn infant (Fowler *et al.*, 1957) free hydroxyproline is present in the blood and urine in such small amounts as to be detectable only in trace amounts or not at all, even by the most sensitive column chromatographic procedures (Oye, 1962). A single instance of a metabolic anomaly characterized by marked hydroxyprolinemia and hydroxyprolinuria was described by Efron *et al.* (1962) in a mentally defective 11-year-old girl. The condition was not appreciably affected by the intake of proline or hydroxyproline. Since there was mental deficiency without hydroxyprolinemia in the family the association between the mental defect and the chemical anomaly may well have been fortuitous. The condition is sharply to be distinguished from the "bound" hydroxyprolinemia seen in patients with Marfan's syndrome (Prockop and Sjoerdsma, 1961).

E. Marfan's Syndrome (Arachnodactyly)

An increased excretion of bound hydroxyproline has been found in Marfan's syndrome. This is not reduced by lowering the protein intake. An increased degradation of collagen in this disease is regarded as the cause of this specific aminoaciduria. Hydroxyproline excretion may be increased in normal people after a large intake of gelatin (see also Chapter 7).

F. Anomalies of Aromatic Amino Acid Metabolism

1. Albinism

Albinism is a relatively common inborn metabolic error, its incidence varying from 1:5,000 to 1:25,000. It is seen in all races and in many species of animals. It is a hereditary disorder of melanin formation caused by a deficiency of the enzyme tyrosinase in the melanocytes of the skin, hair, and eyes (Knox, 1958).

Albinism exists in three clinical varieties: it may be *generalized*, affecting the entire melanocyte system; it may be *partial*, affecting parts of the skin and hair; or it may be *ocular*, the skin and hair being unaffected. The pattern of inheritance differs in these three types. The generalized type is transmitted in virtually all pedigrees as an autosomal recessive gene, the partial cutaneous type as a dominant, and the ocular type as a sex-linked recessive, affecting males only but transmitted by female carriers.

The generalized albino has an unusually pale skin and hair which is exceptionally blond. The skin is unusually susceptible to ultraviolet radiation; it will burn, but not tan or freckle, and adults exposed to excessive radiation tend to develop precancerous keratoses and carcinomatous lesions. Occasional pigmented nevi may, however, be encountered in these subjects and there are instances in which malignant melanoma has developed in them. The iris is blue gray and semitransparent, the pupil occasionally pink but more often dark; the fundus is unpigmented or reveals only a minimum of pigment. Photophobia is always marked, central vision is poor, and there is nearly always a horizontal nystagmus. A variety of other developmental defects are seen with somewhat greater frequency than normal in the albino. Protection of the skin and eyes from light is a serious problem in these subjects.

Partial albinism affecting the skin or hair but not the eyes is not a therapeutic problem. There may be unpigmented patches of skin in various parts of the body. Frequently there is a localized patch of white hair, most commonly a forelock (Cooke, 1952).

Ocular albinism may affect either the retinal pigment or the iris and is associated with photophobia, nystagmus, and poor vision. The female carrier of this trait can be identified by abortive changes seen on fundascopic examination.

That albinism is a metabolic enzyme defect rather than a structural anomaly was suspected by Garrod (1908) but has only recently been established. Observations with the electron microscope have shown that melanocytes in normal numbers can be identified in their usual locations and that they contain granules which appear to differ from the normal only in their lack of melanin (Barnicot and Birbeck, 1958). The enzyme tyrosinase can be demonstrated in normal skin melanocytes by incubating skin slices with C^{14}-labeled tyrosine which induces radioactive melanin; this fails to occur in the albino (Fitzpatrick, 1960). The location of the metabolic block in albinism is shown in Fig. 2.

Tyrosinase is necessary for the formation of dopa (dihydroxyphenylalanine) from tyrosine and it greatly accelerates the formation of dopa quinone from dopa (Cromartie and Harley-Mason, 1957). The effect of the blocked pathway on other pathways of tyrosine metabolism has not yet been studied.

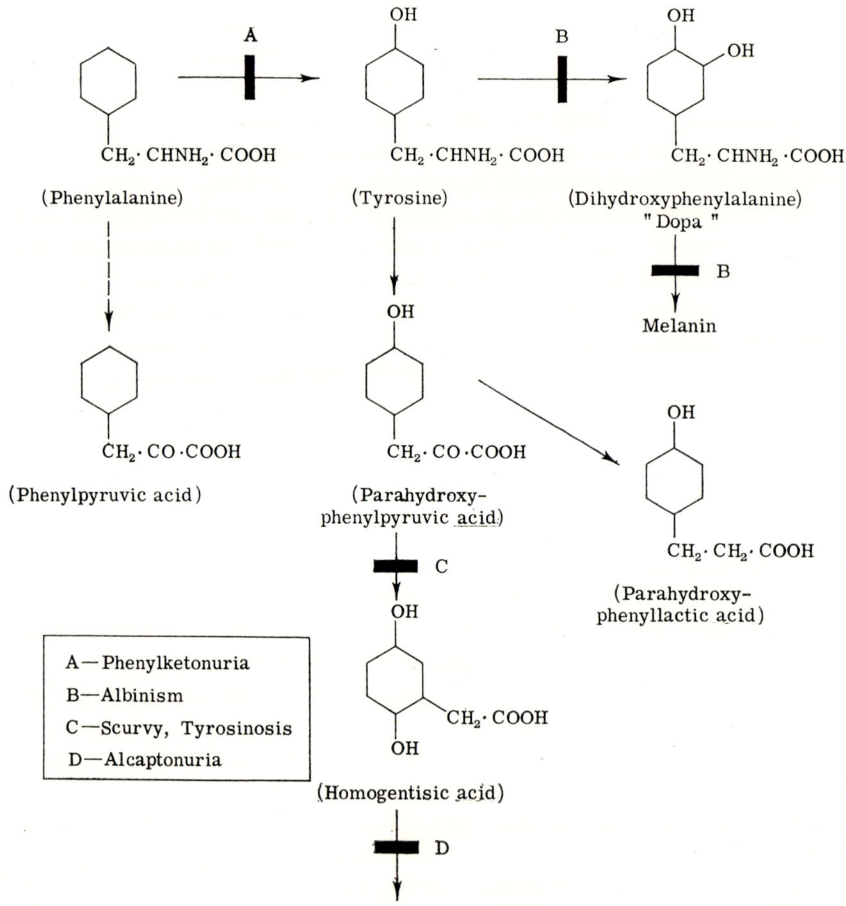

FIG. 2. Some metabolic blocks in the metabolism of the aromatic amino acids.

2. Alcaptonuria

This rare but well-studied anomaly is caused by absence of homogentisic acid oxidase, an enzyme normally present in the liver and kidney which converts homogentisic acid to maleyl acetoacetate, a step in the further degradation of tyrosine (see Chapter 5). The metabolic block (Fig. 2) results in the excretion of homogentisic acid in the urine, and although this substance is readily cleared by the kidney, some of it appears to be retained undergoing oxidation and polymerization to a specific type of melanin pigment, *homogentisic acid melanin,* which in the course of years accumulates in connective tissues.

Alcaptonuria has been detected in the first few days of life from an examination of the urine or, in some instances, from discoloration of the diaper. Symptoms, however, make their appearance only in adult life in the form of discoloration of connective tissues (ochronosis) and by a characteristic type of arthritis (ochronotic arthritis).

The presence of homogentisic acid in the urine may be discovered in a specimen of urine that is allowed to stand for some hours; it gradually darkens, beginning at the surface. This change, resulting in melanin formation, is favored by an alkaline reaction; it may not occur in an acid urine or if the urine contains sufficient ascorbic acid. In some instances discoloration of the diaper attracts attention to the abnormality. The disorder has been mistaken for diabetes because of the copper reduction; alcaptonuric urine will also reduce silver in a photographic film emulsion. Several quantitative tests are available. Its excretion varies with the intake of aromatic amino acids, and hence with the protein intake.

Ochronosis is seldom detected before the age of 20 but becomes increasingly conspicuous with advancing age. It may be observed first in the sclera as a localized bluish-black patch. In other instances it is first noticed in the cartilages of the ears and nose. The pigmentation may be noticed in the sweat and may cause a brownish discoloration of the skin, particularly in the axillae and genital areas. At operation or autopsy the bluish-black pigmentation is more conspicuous; it is most notable in cartilaginous structures which may be almost black in elderly persons.

Arthritis may attract attention to the disease before any anomaly of pigmentation is noted. Its onset is usually in middle age. Most commonly affected are the hip, knee, shoulder, and spine. The severity of the condition varies greatly in different subjects. The roentgenological findings in the spine are very characteristic—these being narrowing and calcification of the intervertebral discs. An ochronotic nephrosis has also been described.

Alcaptonuria is a familial disorder nearly always transmitted by an autosomal recessive gene (Knox, 1958). No test for the detection of heterozygote carriers has yet been devised. A low-protein diet is logical since it reduces the production of homogentisic acid but its effectiveness has not as yet been demonstrated in long-term studies. Ascorbic acid will correct the experimental alcaptonuria produced by tyrosine loading, but not the hereditary anomaly. Steroids, salicylates, and other agents used for rheumatoid arthritis have had no favorable effect on ochronotic arthritis. The missing enzyme, homogentisic acid oxidase, was first isolated by La Du and his co-workers (1958), but its therapeutic use appears to be impractical at present.

3. *Familial Dysautonomia (Riley-Day Syndrome)*

Of this condition, first described in 1949 by Riley *et al.*, more than 80 cases are now known. Only recently, however, was evidence obtained by Smith and his co-workers (1962) of an inborn error of metabolism. The disease occurs almost exclusively in persons of Jewish extraction, the defect being transmitted by an autosomal recessive gene.

Characteristic features are: reduced or absent tear formation during crying, excessive perspiration on excitement, transient blotching of the skin, cold hands and feet, poor motor coordination, difficulty in swallowing, relative indifference to pain, and a striking emotional lability. Many of these children are subject to breath-holding spells in infancy and to vomiting attacks as they become older. There may be periods of hypertension with excitement. These symptoms usually attract attention during infancy, the most troublesome being the violent emotional responses to minor stimuli and the injuries caused by the defective pain sense. Mental impairment is not a part of the picture, but education may be seriously interfered with. Most patients have survived but a few have died in unexplained shock. No abnormalities of the nervous system have been found at autopsy.

According to Smith *et al.* a partial block exists in the conversion of dopamine to norepinephrine (Fig. 3). They studied the urines of 11 children with this disorder and found that the end product of the epinephrine-norepinephrine pathway, vanillyl mandelic acid (VMA), was present in only one half of the usual quantity, whereas the end product of an alternate pathway, homovanillic acid (HVA), was present in double the normal quantity. The ratio HVA/VMA was nearly four times as high in dysautonomic patients as in control subjects. Two first cousins of dysautonomic children showed intermediate values. Four subjects were tested with infusions of norepinephrine and responded with elevations of blood pressure ten times as great as in control subjects. It was postulated that the block had had the effect of rendering these subjects hypersensitive to norepinephrine.

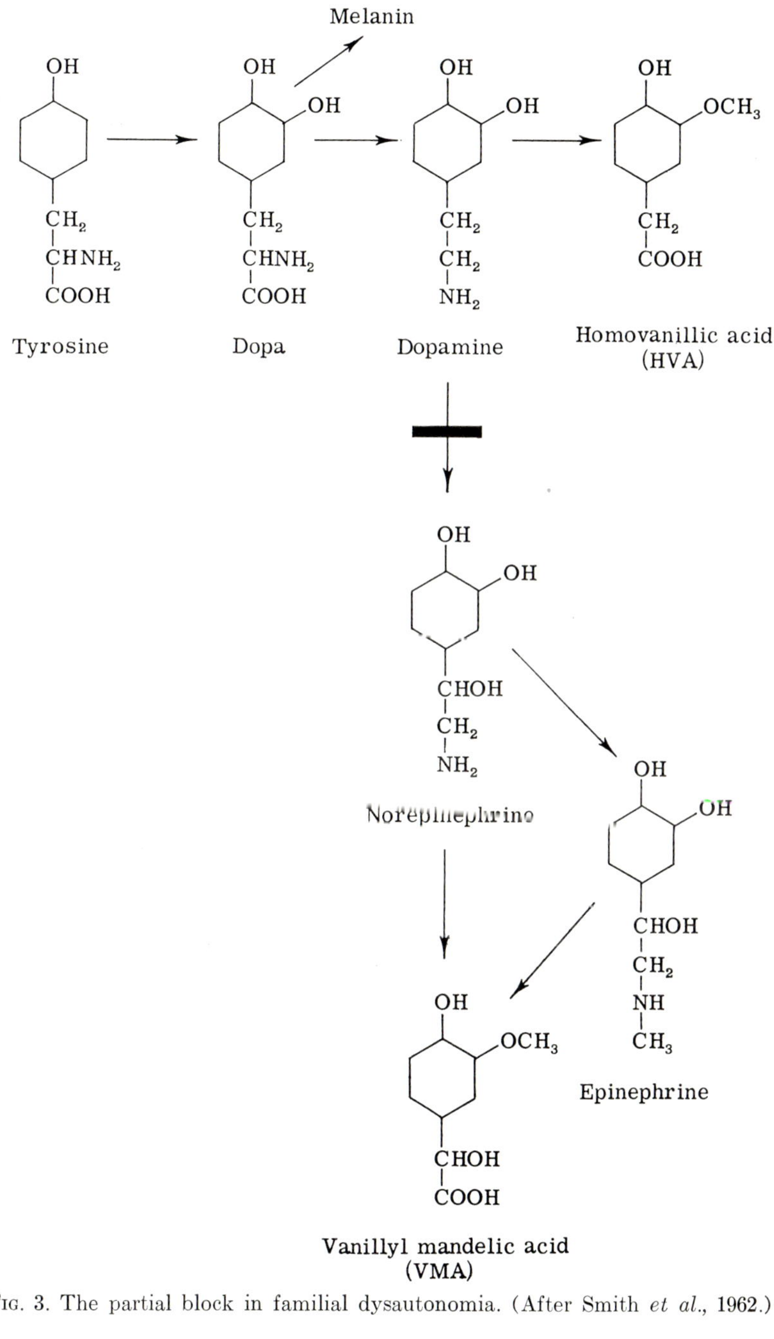

FIG. 3. The partial block in familial dysautonomia. (After Smith *et al.*, 1962.)

4. *Familial Goitrous Cretinism*

This condition differs from sporadic athyreosis and from endemic cretinism caused by iodine deficiency in that an enzymic defect is present in one or another step in the formation or degradation of thyroid hormone, which is formed by the iodination of tyrosine. This group of disorders may therefore be regarded as inborn errors of tyrosine metabolism.

The clinical picture, that of cretinism, will not be described here. Clinically the group caused by congenital metabolic errors is characterized by the presence of goiter in addition to classical signs of thyroid deficiency; in one subgroup there is an association with eighth nerve deafness.

The production of thyroid hormone is a complicated enzymically controlled process involving: (*a*) the trapping of inorganic iodide by the thyroid gland; (*b*) the "organification" of the iodide with tyrosine to form monoiodotyrosine (MIT or T_1) and diiodotyrosine (DIT or T_2); (*c*) coupling of these compounds to form triiodothyronine (T_3) and thyroxine (T_4); (*d*) these four iodinated amino acids are combined with protein to form thyroglobulin. The degradation of thyroxine in turn involves proteolysis and specific deiodination which split off iodine from the four iodinated amino acids. The process may go awry at several different points due to congenital enzymic defects (Fig. 4). Stanbury (1960) recognizes six distinct types of goitrous cretinism, one of which appears to be a heterogeneous group that may be further subdivided:

1. An exceedingly rare condition of which only a single case has been well studied in which the thyroid gland has difficulty in trapping iodide.

2. A group in which there is difficulty in forming the iodinated tyrosines. The administration of thiocyanate or perchlorate to such patients causes a prompt excretion of the trapped inorganic iodide.

3. A subdivision of this group is characterized by its association with eighth nerve deafness, by the prominence of the goiter at birth, and by a milder degree of thyroid deficiency. The intellectual development may be unimpaired in this group and the excretion of iodide following thiocyanate or perchlorate is less striking.

4. A group characterized by a coupling defect, as a result of which the gland contains only T_1 and T_2; T_3 and T_4 are absent. This is identified by giving labeled iodine and chromatographing the iodinated tyrosines obtained from a biopsy of the gland.

5. A group in which deiodination of the iodotyrosines is defective and iodotyrosines are excreted in the urine; this defect is readily identified by a load test in which T_2 is given by vein. This defect has been found to be widespread in a nomadic group of people known as tinkers, a group with a high rate of consanguinity located in western Scotland. Cases with this defect have also been observed in Holland.

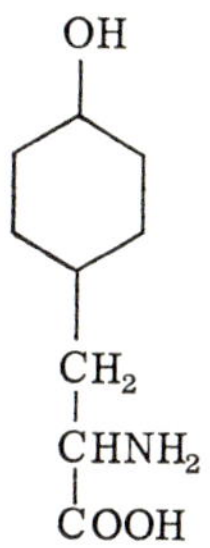

Tyrosine

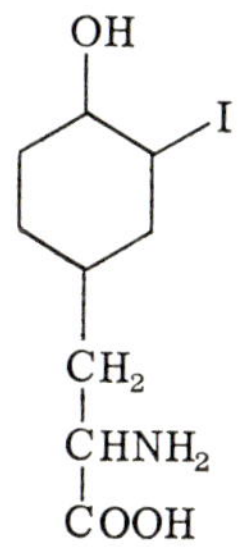

Monoiodotyrosine (T_1)

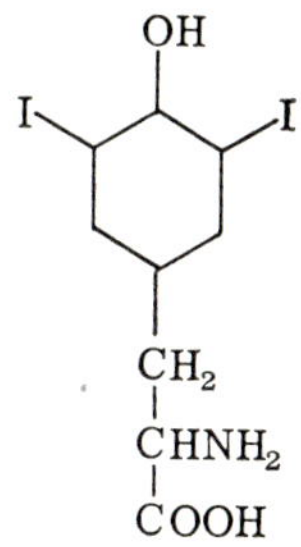

Diiodotyrosine (T_2)

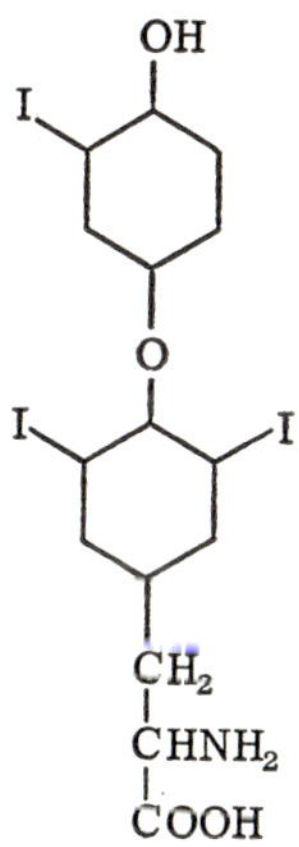

Triiodothyronine (T_3)

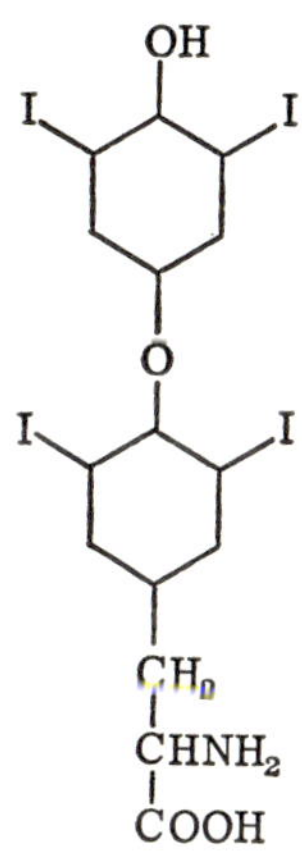

Thyroxine (T_4)

FIG. 4. Tyrosine and its iodinated derivatives.

6. This appears to be a miscellaneous group characterized by the presence in plasma of iodinated polypeptides.

In only three of these six groups are there sufficient data to establish the pattern of inheritance with reasonable certainty. Inheritance appears to be autosomal dominant in the group associated with nerve deafness and a mild difficulty of forming iodinated tyrosine derivatives from iodide. The coupling defect and the deiodinase defect appear to be inherited as autosomal recessives.

All forms of goitrous cretinism respond to thyroid therapy. A tendency to malignancy has, however, been noted in these disorders.

5. Phenylketonuria

Phenylketonuria was the first disease in which Garrod's concept between gene, enzyme, and clinical manifestation was clearly demonstrated. Since its original description by Fölling in 1934 the biochemical abnormality has been defined, the mode of inheritance demonstrated, and an effective treatment developed. The most important problem, the immediate cause of the brain damage, remains; its solution may throw light on processes of mental development in general.

Phenylketonuria is transmitted by a single autosomal recessive gene. It is seen particularly in northern European populations and is extremely rare among the Japanese, Negro, and Jewish populations. In America the incidence of the disease is about 1 in 25,000 births, the incidence of heterozygotes in the population being about 1 in 80. Approximately 1 % of institutionalized mental defectives suffer from this disorder (Jervis, 1954).

The most striking clinical manifestation is the mental deficiency which usually becomes apparent in infancy. There is a wide range of intelligence among untreated patients, the majority have an intelligence quotient below 20 and only about 2 % above 60; one or two instances of normal mentality have been described. Convulsions are seen in about one fourth of the cases and petit mal seizures in another 25 %. Most patients, even without seizures, exhibit abnormalities in the electroencephalogram. Other neurological abnormalities encountered include muscular hypertonicity, exaggerated reflexes, tremors, and hyperkinesis. A dermatitis resembling eczema occurs in 15 to 20 % of the cases. There is a disorder of pigment metabolism commonly present with blond hair, blue eyes, and fair skin in families normally more heavily pigmented; this defect may, however, be very slight. Many of these patients have a characteristic musty or mousy odor caused by phenylacetic acid in the urine.

Fölling, the discoverer of the disease, identified phenylpyruvic acid in the urine as the substance giving a green color with ferric chloride and he subsequently demonstrated that phenylalanine metabolism was at fault, there being high levels of this amino acid in the blood and urine. Jervis (1947, 1953) pinpointed the defect and demonstrated that the liver of these patients failed to hydroxylate phenylalanine to tyrosine (Fig. 2). The enzyme phenylalanine hydroxylase, since studied by Mitoma *et al.* (1957), is now known to consist of two fractions (Kaufman and Levenberg, 1959), a labile one found only in the liver and a stable one which is found in many parts of the body; only the former is lacking in phenylketonuria. Cofactors are reduced TPN (triphosphopyridine nucleotide), oxygen, and what appears to be a folic acid derivative.

As the result of the block in a major pathway of phenylalanine metabolism, phenylalanine accumulates in the blood and spinal fluid and is excreted in excess in the urine, transamination is stimulated, and large amounts of deaminated derivatives are found in the urine including phenylpyruvic acid, phenyllactic acid, phenylacetic acid (Jervis, 1950), *o*-hydroxyphenylacetic acid (Armstrong *et al.*, 1955), phenylacetyl glutamine (Woolf, 1951), and phenylethylamine. Abnormalities of tryptophan metabolism are regularly encountered in phenylketonuria; these are secondary and, like the phenylalanine metabolites, return to normal with appropriate therapy. There is diversion of tryptophan metabolism from the nicotinic acid pathway and the serotonin pathway (Pare *et al.*, 1958), more of the tryptophan being transaminated with the formation of indolepyruvic, indolelactic, and indoleacetic acids (Armstrong and Robinson, 1954); indoxyl compounds are also formed and there is indicanuria. This abnormality is thus similar to that found in Hartnup disease (Section II,H). The cause of the abnormal tryptophan metabolism is not established; it is believed by some that the increased transamination of phenylalanine has a stimulating effect on other transaminations; another possibility is that abnormal metabolites of phenylalanine exert an inhibitory influence on steps in the serotonin and nicotinic acid pathways, thus making more tryptophan available for transamination.

The deficit of the tyrosine fraction normally formed from phenylalanine tends to reduce the level of plasma tyrosine, making less tyrosine available for the formation of melanin and of the catecholamines—epinephrine and norepinephrine. A contributory effect interfering with melanin formation is an inhibition of the enzyme tyrosinase by the increased concentration of phenylalanine (Dancis and Balis, 1955), tyrosinase being effective at two steps in the pathway of melanin synthesis. It has been shown that hair pigmentation can be increased either by reducing the phenylalanine intake or by giving tyrosine (Snyderman *et al.*, 1955). The defect of epinephrine formation has been demonstrated both by a reduced level in the blood and by an increased sensitivity to the administration of epinephrine (Cawte, 1957). An early study suggested that thyroxine formation was defective in phenylketonuria, but it has not been possible to confirm this.

The nature and cause of the brain damage in this disease are not well understood. There appears to be a delay in myelination, but typical histological changes are not apparent. It is, however, clear that either phenylalanine or one of its metabolites, when present in excess, seriously interferes with brain function, eventually producing irreversible change. The continued administration of phenylalanine to monkeys and to rats has produced such damage (Waisman and Harlow, 1960). *In vivo* studies designed to demonstrate the toxicity of administration of particular phenylalanine metabolites have not given impressive evidence; *in vitro* studies by Tashian (1961) have, however, shown that several metabolites had an inhibiting effect on glutamic acid decarboxylase. An inhibitory effect of phenylalanine upon the uptake of other amino acids by brain slices has also been demonstrated (Neame, 1961). Finally, the damage has been attributed to a low level of serotonin in the brain, the result of the secondary disturbance of tryptophan metabolism (McKean *et al.*, 1962). Further work is needed to explain the brain damage. It is, however, quite clear that it is a result of the metabolic anomaly and that it can be prevented by early correction of the anomaly.

The diagnosis of phenylketonuria can be made with a fair degree of certainity by testing fresh, slightly acidified urine with a few drops of 5 % ferric chloride solution; a bluish-green color develops as a result of chelation of the ferric ion with the enol group of phenylpyruvic acid. False positives may be obtained when the urine contains bile, certain metabolites of chlorpromazine, or in histidinemia due to the presence of imidazolepyruvic acid. The diagnosis must be confirmed by a plasma phenylalanine determination. Because of the desirability of instituting treatment early, widespread efforts at early detection of the disease are being made. The ferric chloride test is usually, but not invariably, positive at 3 weeks of age. Test papers (Phenistix) impregnated with ferric chloride can be put in the diaper and have served to identify several cases in mass surveys. An earlier diagnosis can be made by measuring the plasma phenylalanine level which may be elevated by the third or fourth day of life. This can be done accurately by means of column chromatography, a procedure that would be attempted only in a known phenylketonuric family. Efforts to devise microchemical assays for plasma phenylalanine that can be applied in mass screening are now being studied and it is to be hoped will in time be practicable (Guthrie, 1962; Scheel and Berry, 1962).

The detection of heterozygote carriers of the phenylketonuric gene has been studied by several groups of investigators and it appears that the heterozygote can be detected with a fair degree of certainty by several procedures—by an elevated level of plasma phenylalanine, by a phenylalanine load test in which the rise of plasma phenylalanine is observed (Hsia and Payne, 1957), or by a phenylalanine load after which the plasma tyrosine level is followed (Jervis, 1960), the heterozygote showing a lesser rise of tyrosine than the normal control. Anderson and his co-workers (1962) have employed a double load of phenylalanine followed by plasma tyrosine measurement and found almost no overlapping between controls and heterozygotes.

Treatment with a diet low in phenylalanine will regularly control the biochemical abnormalities, secondary as well as primary (Armstrong *et al.*, 1955, 1957; Bickel *et al.*, 1954; Woolf *et al.*, 1958). Dermatitis clears, hair pigment returns, there is improvement noted in the electroencephalogram and, depending on the age of the patient, in the nervous symptoms. Seizures often come to an end. The effect on the intelligence is closely related to the age at which the corrective diet is instituted. If this is done in early infancy normal mentality can be confidently expected, the results being less satisfactory the longer the delay (Horner and Streamer, 1959; Knox, 1960). Significant improvement in intelligence may be observed if treatment is started before the age of 2 years and occasionally in children older than this. In a controlled study of adults no effects on intellectual function could be demonstrated, although there was improvement in other symptoms.

How long therapy should be continued is not clear at the present time. It has been suggested that this may safely be dispensed with at 3 to 4 years when myelination is virtually complete. This has been done in several cases without obvious relapse (Horner and Streamer, 1962), but longer follow-up information is needed.

Several proprietary preparations[1] are available for feeding phenylketonuric patients. These are casein hydrolyzates treated to remove most of the phenylalanine. Patients with phenylketonuria are not all alike and must be regulated individually to prevent unduly high phenylalanine levels. Since phenylalanine is essential for growth a certain amount must be included in the diet; in our experience 20 to 30 mg/kg/day is adequate for the growth of these infants as contrasted with 45 to 90 mg required by the normal child. A certain amount of milk or solid food may be given to supply the needed phenylalanine, a quantity which must be adjusted from time to time by following the blood phenylalanine. The ferric chloride test is not a satisfactory guide in regulating these patients, for it becomes positive only when the plasma phenylalanine has risen to 15 or 20 mg/100 ml a dangerously high level. It is our practice not to permit the plasma phenylalanine to rise above 10 mg/100 ml and it is often possible to maintain it at lower levels.

6. *Tyrosinosis and Tyrosyluria*

Tyrosinosis is a metabolic defect, presumably congenital, characterized by the persistent excretion in the urine of the keto derivative of tyrosine, *p*-hydroxyphenylpyruvic acid. A single case of this anomaly was described by Medes (1932) and an extensive survey by Blatherwick (1934) failed to reveal others. Medes' patient was a 49-year-old man with myasthenia gravis, but since other cases of myasthenia gravis have not exhibited this metabolic anomaly the association appears to be fortuitous and the condi-

[1] Ketonil (Merck) and Lofenelac (Mead Johnson) are used chiefly in the United States. Cymogran (Allen and Hanbury) and Minafen (Trufood) are available in England.

tion itself asymptomatic. The patient with tyrosinosis excreted continuously large amounts of the keto acid (1.6 gm/day). This could not be abolished by a tyrosine-free diet but could be increased by giving a high protein diet or a tyrosine supplement. Under the latter circumstances free tyrosine appeared in the urine as well as several other metabolites—*p*-hydroxyphenyllactic acid, *p*-hydroxyphenylacetic acid, and dihydroxyphenylalanine (dopa). Medes suggested a metabolic block caused by deficiency of the enzyme *p*-hydroxyphenylpyruvic oxidase which converts *p*-hydroxyphenylpyruvic acid to homogentisic acid (Fig. 2).

Cases designated as tyrosinosis, but differing in some respects from Medes' patient, have been reported by Felix *et al.* (1951) in Germany, by Sakai and Kitagawa (1957) in Japan, and by Berry and Sutherland (1960) in the United States. In the two former reports, liver disease was present; the German patients had also been splenectomized. The infant with "tyrosinosis" studied by Berry and Sutherland exhibited no other pathological findings; the tyrosinosis was, however, intermittent.

The term "tyrosyluria" is applied to conditions associated with an increased urinary output of tyrosine or of certain tyrosine derivatives which give the Millon reaction, notably, *p*-hydroxyphenylpyruvic acid, *p*-hydroxyphenyllactic acid, and *p*-hydroxyphenylacetic acid. Tyrosyluria occurs in a number of conditions associated with reduced activity of the hepatic enzyme *p*-hydroxyphenylpyruvic oxidase. The formation of this enzyme may be defective, as may be the case in liver disease, or a cofactor responsible for its activation may be wanting; one such cofactor is vitamin C, another, apparently a quinone, is thought to overcome the inhibitory effect upon the enzyme of its substrate (Zannoni and LaDu, 1959, 1960).

Tyrosyluria is seen in both experimental and clinical scurvy and responds readily to treatment with vitamin C. It has been described in a group of patients with both malabsorption and pernicious anemia, in whom it responded to large doses of vitamin C (Boscott and Cooke, 1954). The effect of vitamin C on the tyrosyluria of severe liver disease appears to have been studied on only three subjects. In two adults with liver disease and splenectomy and in a child with a malignant hepatoma it failed to respond to large doses of vitamin C.

Levine and co-workers (1941) observed tyrosyluria in premature infants fed on generous protein intakes and were able to induce it in full-term infants by loading with protein or tyrosine; in both situations it was readily controlled by vitamin C. The enzyme *p*-hydroxyphenylpyruvic oxidase, though normally present in fetal liver, appears to be in an inactive form. Kretschmer (1959) has shown that the addition of ascorbic acid or 2,6-dichlorophenolindophenol to fetal liver will restore its activity fully.

A more persistent form of neonatal tyrosyluria was described in 14 infants by Bloxam (1960); this failed to respond to adequate vitamin C supplementation but disappeared spontaneously around 2 months of age. In the patient reported by Berry and Sutherland the tyrosyluria which had been observed during the first 2 months of life had disappeared at the age of 3 months but was found to have recurred when the child was studied at 8 months and at 1 year of age. In none of these patients were there symptoms attributable to the tyrosyluria. Menkes and Jervis (1961), however, reported an infant with marked neonatal tyrosyluria which persisted more than 4 months and was associated with seizures, spasticity, and failure to thrive, symptoms which responded to a low phenylalanine and tyrosine diet and the administration of vitamin C. At 7 months no evidence of the metabolic anomaly remained.

G. Hyperhistidinemia

A new error of amino acid metabolism was described in 1961 by Ghadhimi and his co-workers in two siblings. The urine of these subjects gave positive ferric chloride and Phenistix reactions, suggesting phenylketonuria, but the plasma phenylalanine level was within normal limits and further investigation revealed large amounts of histidine in the plasma and urine.

The clinical features of this disorder are not as yet well defined. A familial incidence has been observed. The intelligence does not seem to be adversely affected. However, 4 of the 5 reported cases—all of them in young children—have shown retardation in speech development.

The three groups of investigators who have studied this condition have accumulated information that pinpoints the metabolic error as a deficiency of the enzyme histidine-α-deaminase (histidase), which converts histidine to urocanic acid (imidazoleacrylic acid). This is the first step in what is quantitatively the major pathway of histidine degradation (Fig. 5). The block in the urocanic acid pathway causes the accumulation of histidine in the plasma and its increased output in the urine. In addition more histidine is degraded by transamination to its keto derivative, imidazolepyruvic acid; it is this metabolite which is responsible for the positive ferric chloride and Phenistix tests. Imidazolelactic acid and imidazoleacetic acid are also found in the urine in excess. On the other hand, urocanic acid and its metabolite, imidazolone propionic acid, are not detectable in the urine of these subjects. The administration of histidine or of a high protein diet brings out the anomaly in a striking way; there is prolonged elevation of the plasma histidine level and a sharp rise in the excretion of histidine and imidazolepyruvic acid but no increase in urocanic acid and its derivatives such as occurs in the normal subject. The enzymic defect has been demonstrated in the skin by LaDu and his co-workers (1962), who reported that finger tip specimens from two patients when incubated with histidine failed

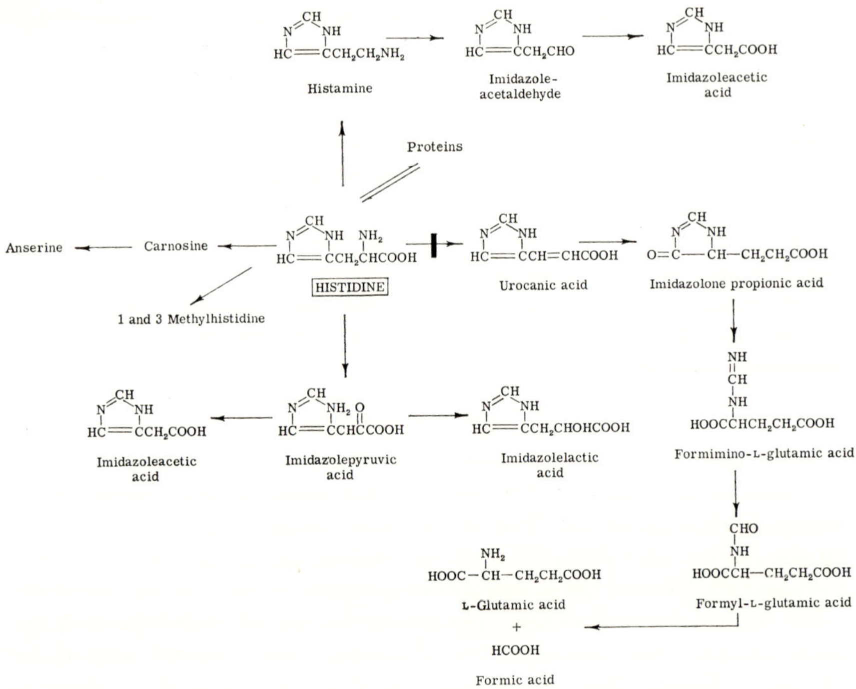

FIG. 5. The catabolism of histidine showing the block in hyperhistidinemia.

to produce urocanic acid in marked contrast to normal controls. Ghadhimi studied the parents of his two histidinemic subjects and found that they excreted in the urine a higher proportion of a histidine load than control subjects.

The significance of urocanic acid and the further metabolites along this pathway is incompletely understood. The keratinized epithelium of the skin is said to be rich in this substance and Zenisek and his co-workers (1955), who have studied its absorption spectrum, regard it as a physiological sun screen. A further metabolite along this pathway is formimino glutamic acid (FIGLU) which is concerned in the transfer of single carbon units in the synthesis of purines and pyridines. Auerbach *et al.* (1962) have reported a secondary disturbance of tryptophan metabolism in the subject studied by them—a reduction in plasma serotonin comparable to that observed in phenylketonuria. Further studies of this metabolic anomaly will be awaited with interest.

H. Hartnup Disease (Hereditary Pellagra)

This condition, described in 1956 by Baron *et al.* (1956) in England, has since been recognized in Holland and Germany (Weyers and Bickel,

1958). Some 15 cases are now known. The original report, entitled "Hereditary pellagra-like skin rash with temporary cerebellar ataxia, constant renal aminoaciduria and other bizarre biochemical features," describes the chief characteristics of the disease. It is a familial condition but the mode of inheritance is not established.

The skin rash, seen in most cases, is indistinguishable from that of nutritional pellagra, indeed several cases had been so diagnosed; the eruption is red and scaly and may be dry or weeping; it is largely confined to exposed surfaces and is seen most frequently in summer. Central nervous symptoms are inconstant and are reversible; their appearance is often associated with an exacerbation of the rash and they may be precipitated by intercurrent infections. A cerebellar ataxia has been observed in several cases. Other features have been chronic headaches, delirium, fainting attacks, and emotional instability. Although some patients have been subnormal mentally others are of average or even superior intelligence.

A conspicuous biochemical anomaly is a renal aminoaciduria with a highly specific pattern. Certain amino acids are excreted in quantities 6 to 10 times the normal, whereas others are found in normal or virtually normal amounts. Excreted in excess are the three branched chain amino acids: leucine, isoleucine, and valine; the aromatic amino acids: tryptophan, phenylalanine, and tyrosine; the hydroxy amino acids: serine and threonine; also histidine and alanine. Unaffected are the sulfur amino acids including taurine; the diamino acids: lysine, cystine, arginine, and ornithine; also proline and glycine. The dicarboxy amino acids—aspartic and glutamic—are not affected, but their amines asparagine and glutamine are excreted in great excess. The aminoaciduria is entirely attributable to increased renal clearance, the plasma amino acid concentrations being normal. The clearance rate of histidine is increased to such an extent that it approaches the glomerular filtration rate, tubular reabsorption being virtually nil.

A disturbance of tryptophan metabolism, not completely elucidated, appears to be responsible for the pellagra-like symptoms (Jepson and Spiro, 1960). The excretion of indole derivatives in the urine tends to be increased whereas the metabolites along the nicotinic acid pathway (Fig. 6) tend to be decreased. As yet it has not been possible to establish the existence of an enzymic block in the nicotinic acid pathway. There is evidence for defective absorption of tryptophan from the upper gut with bacterial cleavage and oxidation of the unabsorbed tryptophan and subsequent absorption of indole and indoxyl derivatives. An oral load of tryptophan will serve to demonstrate this derangement, leading to very little increase in nicotinic acid metabolites but to a great excess of indole and indoxyl derivatives. The defect can be even more readily demonstrated by the administration of D-tryptophan, for in contrast to the L-isomer the absorption of D-tryptophan is almost completely blocked, most of it being recoverable unchanged in the feces (Milne *et al.*, 1960). Impaired gastrointestinal absorption of phenylalanine, tyrosine, valine, and leucine as well as of tryptophan was demonstrated by Asatoor *et al.* (1963). It is possible that the renal and gastrointestinal disturbances in absorption of tryptophan are manifestations of a more widespread disturbance in cellular transport of this amino acid.

Therapy with nicotinic acid appears to have been successful in cases of Hartnup disease, though it is difficult to be certain of this because of the spontaneous fluctuation of the symptoms. The therapy should be maintained because of the limited formation of nicotinic acid from dietary tryptophan.

I. Congenital Tryptophanuria with Dwarfism

Another anomaly of trytophan metabolism was described by Tada *et al.* (1963) in a physically and mentally retarded child. This child also had photosensitivity and cerebellar ataxia similar to that which occurs in Hartnup disease. However, the aminoaciduria was limited to tryptophan, and there was no increase in indican or indoleacetic acid excretion. Loading with tryptophan resulted in high plasma levels for a prolonged period of time; there was an associated large increase in tryptophan excretion with little increase in the kynurenine excretion. A block in the conversion of tryptophan to kynurenine was postulated.

J. Other Abnormalities of Tryptophan Metabolism

In at least three of the inborn errors of amino acid metabolism—phenylketonuria, maple syrup urine disease, and histidinemia—secondary disturbances of tryptophan metabolism, somewhat similar in nature, have been described. That these are indeed secondary is shown by their disappearance when the primary disturbance is corrected by diet. The disturbance in tryptophan metabolism, best studied in the case of phenylketonuria, consists in an increased urinary output of the nonhydroxylated indole derivatives indoleacetic acid and indolelactic acid. This is attributed to interference with the hydroxylation of tryptophan and the formation of 5-hydroxytryptophan and 5-hydroxytryptamine (serotonin). The chief 5-hydroxy derivative found in urine—5-hydroxyindoleacetic acid (5 HIAA)—is diminished in phenylketonuria as is the serotonin level of blood plasma. It has been postulated that a deficiency of serotonin in the brain is responsible for the cerebral dysfunction in these various states, a view by no means established.

Bessman and Tada (1959) showed further that indicanuria was markedly increased in phenylketonuria, diminishing when the plasma phenylalanine level was controlled by diet. This, too, may be attributed to interference with some other pathway of tryptophan metabolism.

Several reports deal with abnormal tryptophan metabolism in other mental disturbances. A reduced ability to metabolize tryptophan along the 5-hydroxy pathway was found by Sutton and Read (1958) in a single case of infantile autism. In Mongolism three reports describe defective conversion of hydroxykynurenine to xanthurenic acid. However, the most recent study (McCoy, 1962) failed to confirm this.

Normally only about 1 % of the dietary tryptophan undergoes 5-hydroxylation, but in carcinoid this pathway is actively used by the tumor cells and as much as 60 % of the ingested tryptophan may be so metabolized, the plasma serotonin level being markedly elevated (Sjoerdsma *et al.*, 1956). This aberration limits the tryptophan available for other needs and when the intake is borderline, pellagra or protein deficiency may be precipitated (McNeely and Jones, 1946). Removal of the tumor restores the serotonin to normal levels, its persistence at high levels indicating that this has not been accomplished.

In states of malabsorption the metabolism of tryptophan is altered similarly to that found in Hartnup disease (Sleisinger, 1961). There is increased production of indole and indoxyl derivatives by the intestinal bacteria; these are absorbed and excreted in the urine. These changes can be abolished by the use of appropriate oral antibiotics.

Pyridoxine deficiency results in impaired metabolism of tryptophan along the nicotinic acid pathway (Snyderman *et al.*, 1953) beyond kynurenine (Fig. 6), resulting in an abnormal excretion of kynurenine, hydroxykynurenine, and xanthurenic acid after a load test of tryptophan. This has been observed in patients with pyridoxine-responsive anemias (Bottomley, 1962) and in individuals under treatment with isoniazid, which is a pyridoxine antagonist (Price *et al.*, 1957). A similar metabolic abnormality has been seen in a number of disease entities including congenital aregenerative anemia, leukemia, neoplastic diseases, diabetes, eczema, and in a family described by Anderson (1962) which appears to be symptom-free.

K. Metabolic Errors Involving the Urea Cycle

Three such metabolic errors are now known, each one giving rise to mental deficiency in early life. These are hyperammonemia, citrullinemia, and argininosuccinemia. The location of the metabolic block in these conditions is shown in the accompanying diagram (Fig. 7).

1. Hyperammonemia

This condition was described by Russell and his co-workers (1962) in two children who were first cousins. Clinically there was failure to thrive in

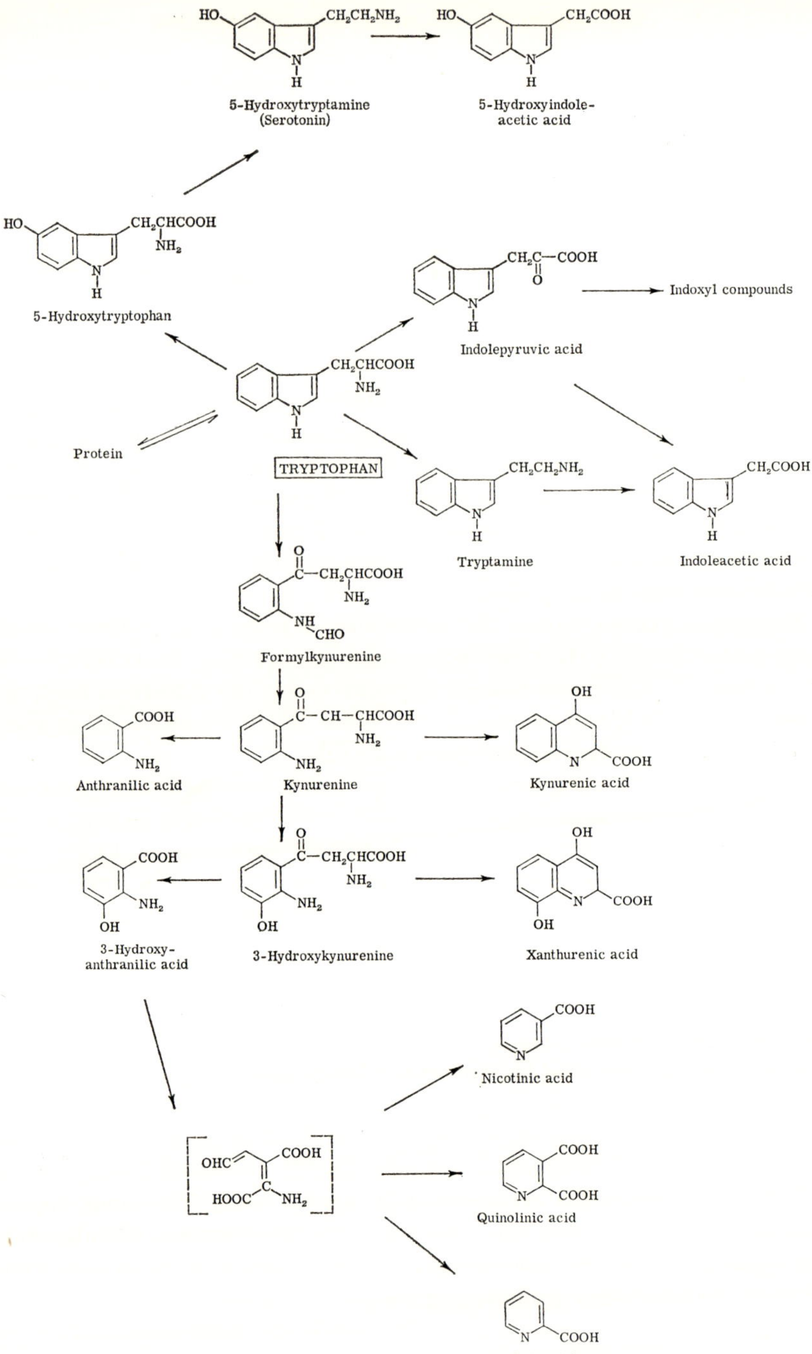

FIG. 6. The metabolic pathways of tryptophan.

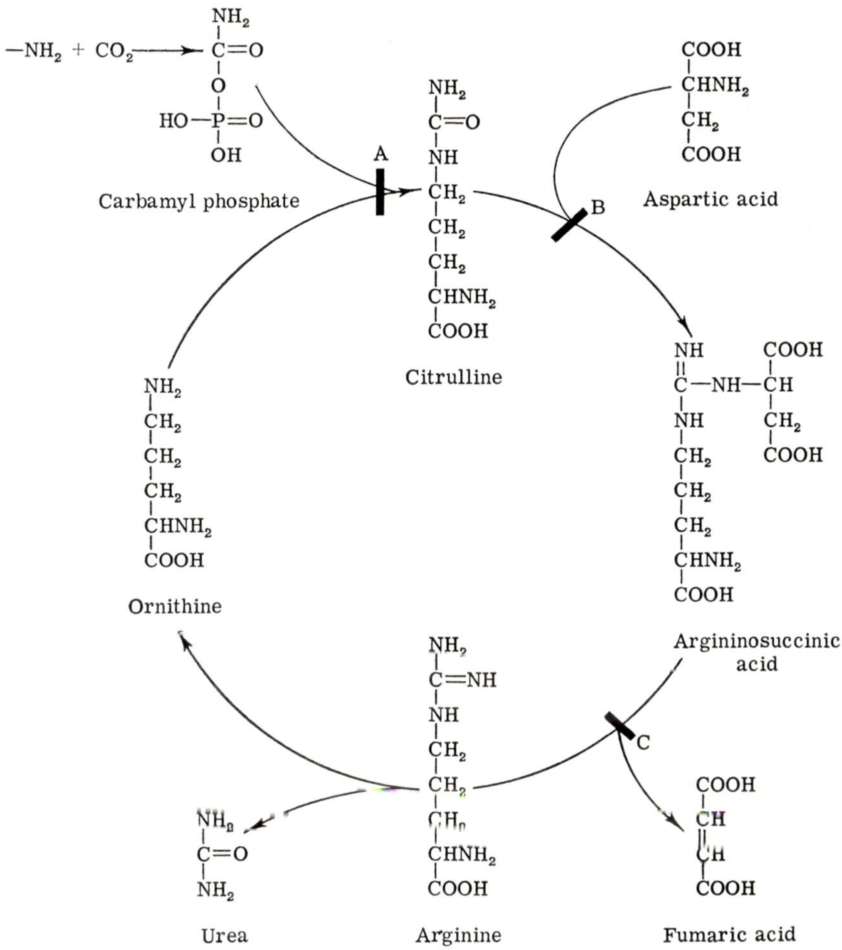

FIG. 7. The Krebs-Henseleit urea cycle, showing the block in hyperammonemia (A), citrullinemia (B), and argininosuccinemia (C).

infancy with evidence of progressive cerebral degeneration. The most striking chemical abnormality was the greatly increased blood ammonia level, higher than that encountered in hepatic coma. The ammonia level in the cerebrospinal fluid was also markedly elevated. A moderate amino-aciduria was present and an elevated level of glutamine in the spinal fluid. The abnormal chemical findings responded to protein restriction. A biopsy of the liver in one patient revealed a markedly reduced activity of the enzyme ornithine transcarbamylase, which participates in the formation of citrulline from ornithine and carbamyl phosphate.

2. Citrullinemia

McMurray and his co-workers (1962) have reported a single instance of a metabolic anomaly characterized by a markedly increased output of L-citrulline in the urine, a 50-fold increase in the concentration of this amino acid in the blood, and a 60-fold increase in the cerebrospinal fluid. The patient was a mentally defective child of 18 months, the parents being first cousins. The urea excretion was normal from which it appeared that the urea cycle, in the liver at least, was not significantly impaired. The possible impairment of this cycle in the brain remains to be investigated.

An increased excretion of citrulline in the urine has also been reported in human as well as in canine cystinuria (Milne *et al.*, 1962). The explanation for this association is not clear.

3. Argininosuccinemia

Nine cases of this rare familial condition are now known, 5 in the United Kingdom (Allan *et al.* 1958; Levin *et al.* 1961; Carson and Neill, 1962), 2 in Israel (Wallis *et al.* 1963) and 2 in the United States (Armstrong *et al.* 1964). Clinically there is mental defect which may not be detected in early infancy, but in one affected patient convulsions appeared at the age of 5 days, and there were episodes of unconsciousness, abdominal distention, enlargement of the liver, diarrhea, and edema. A dry skin has been observed. The nails are brittle as is the hair, which exhibits a specific type of trichorrhexis nodosa. Nodular swellings on the hair exhibit a frayed appearance, the frayed edges when treated with acridine orange showing a red fluorescence. The striking chemical anomaly is the presence in the urine, blood plasma, and spinal fluid of an unusual ninhydrin-reacting substance which Westall (1960) has identified as argininosuccinic acid (ASA), an intermediate in the Krebs-Henseleit urea cycle. Its concentration is highest in the spinal fluid.

The nature of the metabolic block is still unclear. The urea cycle in the liver appears to function normally but it has been postulated that it may not do so in the brain where the enzyme argininosuccinase, which splits argininosuccinic acid into arginine and fumaric acid, may be deficient. Westall and Tomlinson (1963) have demonstrated a deficiency of this enzyme in the red blood cells. The hair anomaly may likewise result from the anomalous metabolism of arginine since this amino acid is well represented in keratin structures.

Therapeutic measures have met with little success. A low protein diet and a synthetic diet low in arginine were found to decrease the excretion of the abnormal metabolite but were not well tolerated. The administration of ornithine and citrulline increased the output of ASA.

L. Oculo-Cerebo-Renal Dystrophy (Lowe Syndrome)

This syndrome, first described by Lowe and his co-workers (1952), consists of mental deficiency, ocular changes, muscular hypotonia, and a generalized renal aminoaciduria. Inheritance is as a sex-linked recessive, it being seen only in males and transmitted only by female carriers (Richards *et al.*, 1960). The most common ocular changes are cataract and buphthalmus. In addition to aminoaciduria there is renal glycosuria, phosphaturia, and acidosis attributable to defective ammonia formation. Bone changes are inconstant, rickets being present in some cases and osteoporosis in others. The mental defect is severe. The presence of ocular anad cerebral symptoms in addition to the renal has suggested a more widespread metabolic anomaly, perhaps of amino acids, but at the present time the evidence for this is only fragmentary. Two groups of investigators (Schoen and Young, 1959; Andrews *et al.*, 1962) have noted an increased excretion of glutamine in the urine. Another report describes abnormal amino acid excretion in the urine following an ornithine load (Schwartz *et al.*, 1962). Further study of these cases will be awaited with interest.

M. Anomalies of Sulfur Amino Acid Metabolism

1. Cystinosis (Cystine Storage Disease)

This familial disorder is characterized by the deposition of cystine crystals throughout the body. They are found chiefly, but not exclusively in phagocytic cells of the reticuloendothelial system. During life these crystals can be demonstrated in the cornea by slit lamp examination, in smears of the bone marrow, and, in some instances at least, by smears of the buffy coat of the blood where they are found within the leucocytes (Korn, 1960).

The typical clinical picture is that of the De Toni-Fanconi syndrome, characterized by resistant rickets, dwarfism and acidosis (Bickel *et al.*, 1952). This is associated with generalized aminoaciduria of renal origin, renal glycosuria, and renal hyperphosphaturia (glucoaminophosphate diabetes). The rickets is caused by the excessive loss of phosphate from the kidney. Abnormal loss of citrate in the urine has also been reported. The condition declares itself in infancy. Relatively few of these children grow up, most of them falling prey to intercurrent infections. The marked disturbance of tubular reabsorption is usually attributed to the deposition of cystine in these structures, which is apparent at autopsy. There is also an anatomical abnormality, a markedly narrowed proximal tubule known as the "swan neck" deformity, the significance of which is not altogether clear (Darmady and Stranack, 1957).

It has been thought that all cases of the Fanconi syndrome were caused by cystinosis and it is certainly true that the great majority seen in early childhood fall into this category. However, it is now clear from autopsy studies that cystinosis is not always responsible for this clinical picture. A few instances have been described in which cystinosis failed to give rise to the typical disturbance of renal function.

Cystinosis should not be confused with cystinuria, a condition to which it bears no relation, despite the fact that in one family tree (Eberlein, 1953) both abnormalities were found. In cystinuria the reabsorption defect is confined to the four diamino acids—cystine, lysine, arginine, and ornithine—in contrast to the widespread reabsorption defect in cystinosis. There is no deposition of cystine in the tissues in cystinuria. In cystinuria the plasma level of cystine is reduced, whereas in cystinosis it is somewhat elevated (Brigham *et al.*, 1960).

The cause of the widespread deposition of cystine has been obscure. Worthen and Good (1961) attribute it to deficiency of the enzyme cystine reductase which converts cystine to cysteine; they have found reduced levels of this enzyme in the blood in two patients studied. This important finding has recently been confirmed by Mahoney and Trump (1962).

Some benefit can be obtained in controlling the rickets by large doses of vitamin D, but the general condition shows little improvement. A promising therapeutic approach is reported by Clayton and Patrick (1961)—the administration of dimercaprol (BAL) or penicillamine. This was based on the view that the excess of disulfide groups exchanged with the thiol groups of sulfhydryl enzymes, tending to inactivate them. They were able to demonstrate in cystinosis elevated levels of succinic dehydrogenase in the liver and of pyruvate in the blood plasma. With the above-mentioned therapy temporary improvement in aminoaciduria and in the plasma pyruvate were obtained and, in one case, a marked resumption of growth.

2. Cystathioninuria

Harris *et al.* (1959) reported the excretion of large amounts of cystathionine by an elderly imbecile and of smaller amounts by her brother and one of his children. Post-mortem examination of the patient revealed abnormal amounts of cystathionine in the liver and kidneys. A metabolic block in the cleavage of cystathionine has been postulated (Fig. 8).

A second family with this metabolic anomaly has been studied by Frimpter *et al.* (1963). Large amounts of cystathionine were found in the blood and urine of a man with normal intelligence but with mental peculiarities, and smaller amounts in the urine of his sister and her four children. This patient is deaf with organic ear defects. The relationship of the metabolic defect to the mental peculiarities and the ear defects remains to be elucidated. Cystathionine has been found in the urine of pyridoxine deficient rats (Hope, 1957) and in the urine of an infant with vitamin B_6

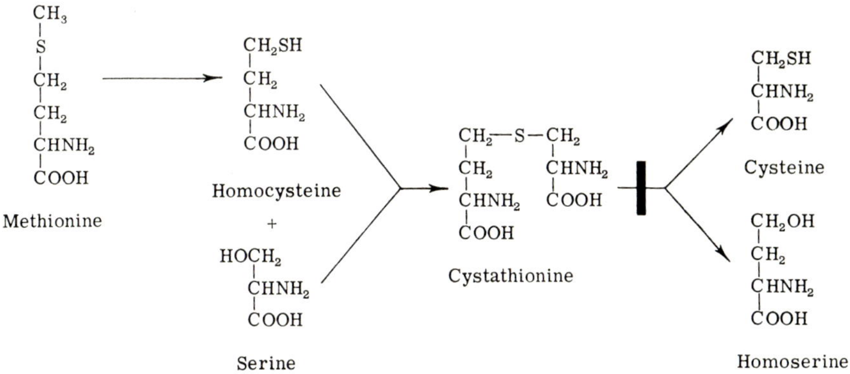

FIG. 8. The metabolic block in cystathioninuria.

deficiency (Scriver and Hutchison, 1963); appropriate therapy abolished the excretion in the infant. The amount of cystathionine excreted by Frimpter's patient was reduced when large amounts of this vitamin were administered.

3. *Homocystinuria*

A screening test of mentally defective institutionalized children in northern Ireland brought to light two siblings with a new anomaly of amino acid metabolism—homocystinuria. A number of other instances of this disorder have since been discovered in England (Carson *et al.*, 1963). A distinctive clinical picture was present in these subjects which, in addition to mental deficiency, included fine, dry, fair hair, bilateral dislocation of the lens of eye and hippus; the fingers and toes were unusually short and there was pes cavus. The urine showed an excess of a substance at first thought to be cystine because of a positive reaction with cyanide. Further investigation indicated that it was not cystine but homocystine. Some instances have shown marked elevation of the plasma methionine (Gerritsen and Waisman, 1964).

N. Anomalies of Branched Chain Amino Acid Metabolism

1. *Maple Syrup Urine Disease* (Branched chain keto aciduria)

This condition was first described by Menkes and his co-workers in 1954; they reported four siblings who died during the first 6 weeks of life with a progressive neurological disorder; their urine exhibited a characteristic odor resembling that of maple syrup. The disease was later identified as a disorder of the metabolism of the branched chain amino acids (Westall *et al.*, 1957), the exact location of the metabolic block being subsequently located by observations made in three different laboratories. More than 25 cases of this disorder are now known. It appears to be transmitted as an autosomal recessive. The disease has been described in widely separated parts of the United States, in England, Germany, Japan, and Greece (Jepson *et al.*, 1958; Mackenzie and Woolf, 1959; Müller and Schreier, 1962; Tada *et al.*, 1963). With one exception untreated patients have all succumbed in early infancy; one patient is known to be alive at the age of 8 years, grossly retarded and with severe neurological symptoms. One other patient exhibited retrogression of the symptoms and biochemical abnormalities in late infancy; it is a question whether this case represents a variant of this disease (Morris *et al.*, 1961).

The symptoms usually start soon after birth; they may be evident before the end of the first week; there is anorexia, failure to thrive, hypertonicity, and at times convulsions and a semicomatose state. With the development of symptoms the characteristic odor of the urine becomes apparent; it has been described as "malty," "like caramelized sugar," or like maple syrup. The anomaly reported by Jepson *et al.* (1958) is presumably the same entity, although their patient also suffered from phenylketonuria. At times other secretions exhibit this odor; it has been described in perspiration and in the cerumen of the ear.

The basic metabolic defect involves the branched chain amino acids—(Fig. 9) (Dancis *et al.*, 1959; Menkes 1959a,b, Mackenzie and Woolf, 1959).

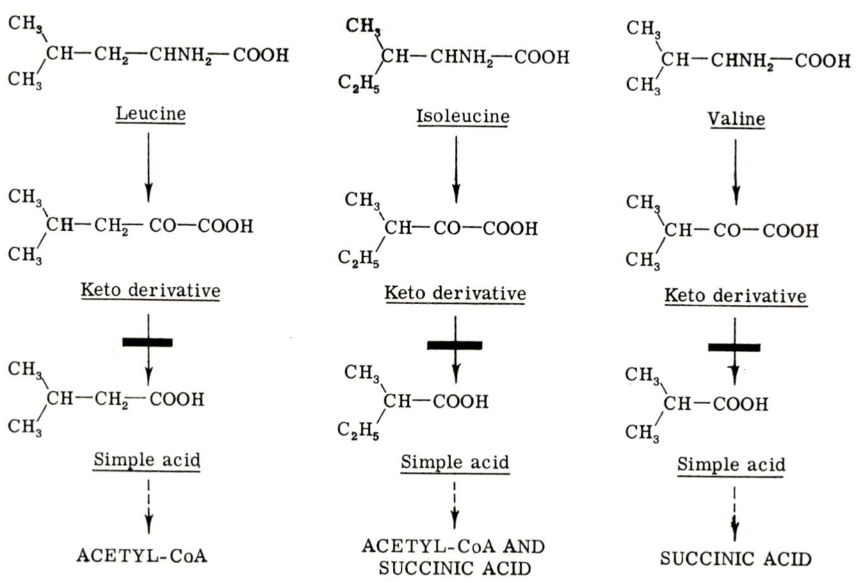

FIG. 9. The catabolism of the branched chain amino acids.

These are transaminated in a normal manner with the formation of the corresponding keto acids; the keto acids, however, are not further degraded by oxidative decarboxylation and they, as well as the amino acids themselves, accumulate in the blood and are excreted in excess in the urine. An excess of branched chain acids has also been demonstrated in the spinal fluid. The defect of the decarboxylating enzyme has been demonstrated by Dancis *et al.* (1960a) in the white blood cells and in skin fibroblasts in culture (Dancis *et al.*, 1963). Hydroxy derivatives of the branched chain amino acids have been identified in the urine (Jepson *et al.*, 1958). A secondary alteration of tryptophan metabolism similar to that observed in phenylketonuria, Hartnup disease, and several other conditions has also been observed in maple syrup urine disease; this consists in increased formation of tryptophan derivatives formed by transamination, notably, indoleacetic acid and indolelactic acid. When a corrective diet low in branched chain amino acids is instituted the abnormality of trytophan metabolism disappears.

The finding of alloisoleucine in the plasma and urine of these children is of some interest (Norton *et al.*, 1962). In the original analysis of plasma by column chromatography an abnormally large peak was observed in the region of methionine and an abnormality of methionine metabolism was postulated. However, improved techniques employing a more sensitive resin revealed two peaks in this region, a normal methionine peak and a large unknown peak which was subsequently shown to be alloisoleucine. The alloisoleucine is formed from isoleucine, apparently by enolization of the keto derivative, α-keto-β-methylvaleric acid, and subsequent reamination which results in the formation of both isoleucine and alloisoleucine (Fig. 10). The administration of isoleucine causes a prompt rise in the plasma isoleucine followed by a delayed rise in alloisoleucine; the latter persists for a long time indicating a poor renal clearance for this material. The alloisoleucine apparently exists in the plasma in the L-form; adminis-

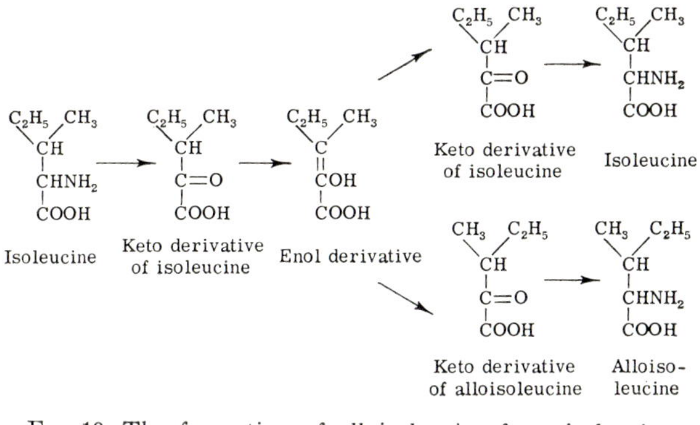

FIG. 10. The formation of alloisoleucine from isoleucine.

tration of L-alloisoleucine results in a prolonged high plasma level, while a comparable dose of D-alloisoleucine is rapidly excreted in the urine (Norton *et al.*, 1963).

The metabolite responsible for the peculiar odor has not been definitely identified; it appears to be related particularly to isoleucine, since it is brought out by a load of isoleucine, but not by a load of the other two branched chain acids. The exact cause of the neurological damage is also obscure. A direct toxic effect of the branched chain acids or of their derivatives is one possibility. Toxic neurological symptoms develop readily after a lapse of dietary control or after loading with branched chain acids, notably leucine. Tashian (1961) has shown in rat brain homogenates that the activity of the enzyme glutamic acid decarboxylase is markedly inhibited by the keto derivatives of leucine and valine. A second possibility is that the elevated levels of the branched chain amino acids interfere competitively with the uptake of other amino acids by the brain cell. It has also been suggested that the secondary abnormality of tryptophan metabolism is the cause of the brain damage; increased transamination and decreased 5-hydroxylation may be responsible for reduced serotinin levels in the brain.

It is difficult to relate the pathological changes to the biochemical abnormalities. There is a defect of myelin formation throughout the brain. This is accompanied by areas of spongy change, astrocytosis, and a decrease of oligodendrocytes (Silverman *et al.*, 1961). Pathological changes in the neurons have not been demonstrated.

Dietary restriction of the three branched chain amino acids will correct the biochemical abnormalities and has resulted in striking clinical benefit. When this is done sufficiently early in the disease it appears that the damage to the nervous system may be prevented, the benefit being less striking the longer therapy is delayed. The writers have had under treatment one such patient who developed symptoms at 6 days of age and whose diet was corrected at the age of 14 days. This patient, now 4 years of age, is in good physical health and shows no neurological symptoms; she is, however, somewhat retarded mentally, and has had a poor growth performance. It is possible that the retardation is related to nutritional factors because of prolonged restriction of methionine undertaken to reduce what had been regarded as a high methionine level. During the past year her growth and intellectual development have been accelerated. A second child, treated by Dent and Westall (1961) from the age of 5 days, and not deprived of methionine, is said to be developing normally at the age of 2 years. The importance of early diagnosis in these patients cannot be overemphasized. It may well be that if one waits for the urine to show the characteristic abnormality—a positive reaction for keto acids with dinitrophenyl hydrazine—some damage to the nervous system may already have occurred. An earlier diagnosis can be made by following the plasma levels of the branched chain acids by column chromatography or by identifying the enzyme deficiency in the white blood cells or skin fibroblasts. Such measurements are likely to be carried only when there are siblings.

The problem of dietary control is not a simple one since natural foods are all well supplied with branched-chain amino acids. It is therefore necessary to supply the major portion of the diet as a synthetic mixture, the nitrogen of which is furnished as a mixture of pure amino acids in which the branched chain acids are kept low, their intakes being adjusted at intervals to maintain normal plasma levels. This necessitates frequent analyses of plasma during the early months of life. We have found that it is possible to give these patients a small amount of milk which will furnish the bare minimum quantities of the branched acids which are needed for growth; the bulk of the diet must, however, be synthetic. In the initial period, however, when the diagnosis is first made and the branched chain acids are markedly elevated, it is desirable to reduce their concentration promptly by using a completely synthetic diet for a week or two.

A difficulty in the management of these patients has been that of intercurrent infections. At such times food intake may be curtailed and the patient be forced to live on his own tissues which contain an abundance of branched chain acids; as a result their levels rise promptly in the blood and there may be a recrudescence of symptoms and of the maple syrup odor. It is therefore of the greatest importance to protect these children from infection and to maintain their food intake during such infections.

With early detection and facilities for dietary regulation it appears possible to protect the nervous system in these patients. The recognition of heterozygote carriers of the trait will go far to make early diagnosis possible. Heterozygotes have been shown to respond to a leucine load with exceptionally high and sustained plasma leucine levels (Lonsdale *et al.*, 1963).

2. Hypervalinemia

A high plasma level of valine as well as increased urinary excretion was found by Tada (1963) in a retarded Japanese infant, associated with nystagmus and hyperkinesis. Symptomatic improvement resulted from a low valine intake. Increased valine was found in the parents' urine.

3. Leucine-Induced Hypoglycemia

In 1956, Cochrane *et al.* described the first cases of leucine-induced hypoglycemia. This appeared to be a distinct clinical entity which could be separated from the heterogeneous group of idiopathic hypoglycemias. These individuals responded to a load test of casein or of L-leucine by a further depression of their blood glucose values to less than half the control levels. A number of other amino acids have been studied for this effect with equivocal results; positive tests have been obtained less frequently in such patients after loading with isoleucine and valine. The keto acid derivative of leucine, α-ketoisocaproic acid, is the only other compound which has consistently produced hypoglycemia in these patients (Cochrane, 1960).

The hypoglycemic effect seems to be mediated through insulin. An increase in circulating insulin has been demonstrated concomitant with the fall in blood sugar after leucine loading in the majority of patients in whom this determination was carried out (DiGeorge and Auerbach, 1960; Grumbach and Kaplan, 1960). In normal dogs, the administration of L-leucine potentiates the action of exogenous insulin by exaggerating and prolonging the fall in blood sugar (Reaven and Lucas, 1962).

It has become increasingly clear that the leucine-sensitive hypoglycemias are not a homogeneous group. This condition is present in adults with islet cell tumors of the pancreas and disappears after surgical removal of such tumors (Flanagan *et al.*, 1961). Histological examination of the pancreas of these children has revealed different findings ranging from no discernible pathology to hyperplasia of the islet cells and abnormalities of the beta cells of the pancreas. Conversely, there are on record cases with hyperplasia of the islet cells who are not leucine-sensitive. Although the presence of leucine sensitivity is not useful in ascertaining the etiology of a given case of hypoglycemia, it may help in the management. The symptoms of these children are alleviated by a low protein diet which is the only practical way of providing a low leucine intake.

4. Other Leucine-Sensitive States

The tolerance for leucine is specifically reduced in several other pathological states. Gopalan and Srikantia (1960), in studying the diets which led to pellagra in Hyderabad, observed its striking frequency on a diet of millet as contrasted with rice, despite the fact that millet provides more tryptophan and nicotinic acid. This untoward effect was traced to the higher leucine content of millet. It was found that the oral administration of 5 gm L-leucine would increase the urinary excretion of *N*-methyl nicotinamide, both in normal subjects and in pellagrins. A somewhat larger dose of leucine, fed to two pellagrins produced marked deterioration of their symptoms. The untoward effect of leucine remains somewhat controversial.

In maple syrup urine disease, a condition discussed elsewhere, we have called attention to the particular untoward effects of this amino acid when given in excess.

A third condition in which leucine appears to be particularly upsetting to the metabolic balance is idiopathic hyperglycinemia (Nyhan *et al.*, 1961).

O. β-Aminoisobutyric Aciduria

β-Aminoisobutyric acid (BAIB) is found in normal urine in small amounts, but in certain individuals it is found in relatively large amounts (Crumpler *et al.*, 1951). The incidence of the high excretors varies from 5 to 15 % in Caucasians and Negroes, but in the Mongoloid race it is more common, from 40 to 45 % being so affected (Berry, 1960). The high excretion trait appears to be transmitted as an autosomal recessive.

In certain diseases the excretion of BAIB may be markedly increased, although this is not invariably so. High excretion occurs under conditions of tissue destruction. It has been observed in neoplastic disease, particularly in leukemia, in tuberculosis, liver disease, march hemoglobinuria, and after X-irradiation. It can serve to some extent as an index of tissue destruction. In thalassemia major, the performance of splenectomy has been observed to cause a sharp reduction in the excretion rate (Hillicoat, 1962).

Although it is possible in excised tissues to produce BAIB from valine, present evidence, chiefly from loading studies, points to thymine as the important source (Fink *et al.*, 1951). It is an intermediate in a pathway of thymine catabolism, being ultimately derived from DNA (see Fig. 11).

The cause of high excretion in the hereditary group is not altogether clear. There is no evidence of increased tissue catabolism. It is not a renal anomaly, for the renal clearance of BAIB is almost complete, there being virtually no tubular reabsorption. It is possible that there is a block in some metabolic step beyond BAIB or that some alternate pathway of

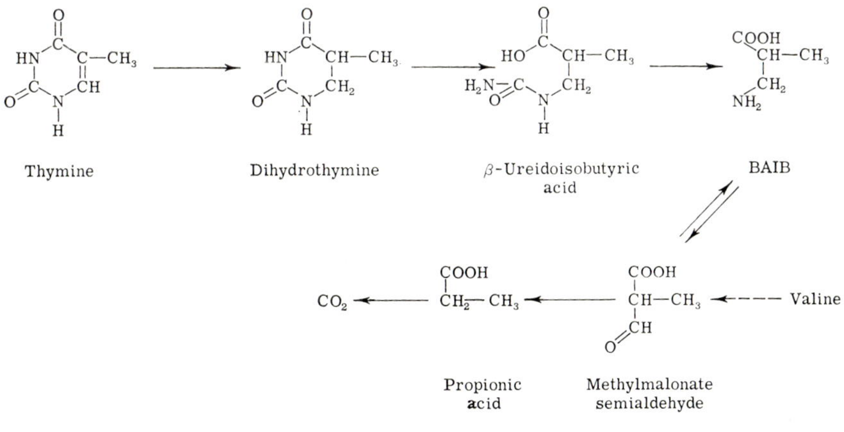

FIG. 11. The formation of β-aminoisobutyric acid.

thymine catabolism is less efficiently used in these subjects (Yates and Armstrong, 1962).

III. Acquired Disturbances of Amino Acid Metabolism

A. Premature and Newborn Infants

Premature and newborn infants tend to have a generalized aminoaciduria which is more marked in the premature. As much as 10 % of the total urinary nitrogen may be α-amino nitrogen; in the adult this ratio does not exceed 2 % (Simon, 1911). Some of this aminoaciduria of early life is due to immature renal function; a lower rate of tubular reabsorption of amino acids has been demonstrated (Sereni *et al.*, 1955). However a certain portion of the aminoaciduria is metabolic in origin since there are elevated levels of amino acids in the plasma (Norton *et al.*, 1963). Hydroxyproline and homocitrulline have been recovered from both the plasma and urine of this age group; these amino acids have not been observed in the plasma of normal adults. The plasma tyrosine levels tend to be higher during the newborn period. This aminoaciduria tends to disappear in several months, but persists longer in the premature than in the normal newborn.

B. Tissue Destruction

Aminoacidemia and aminoaciduria may occur in a number of conditions associated with excessive tissue destruction. These include severe burns (Eades *et al.*, 1955), the immediate postoperative state (Eades *et al.*, 1955), diabetes when complicated by ketosis (Gray and Illing, 1952), severe infections, leukemia (Kelley and Waisman, 1957), and progressive muscular dystrophy (Ames and Risley, 1948).

C. Liver Disease

Liver disease is frequently accompanied by an aminoaciduria which is metabolic in origin. In the milder cases this is a reflection of impaired deamination of amino acids by the liver; in acute massive liver necrosis, it is secondary to breakdown of liver tissue. Three patterns of aminoaciduria in liver disease have been described (Dent and Walshe, 1951). There may be a small rise in many amino acids; this is seen frequently in infectious hepatitis. There may be an increased excretion of certain amino acids—of cystine alone or in combination with β-aminoisobutyric acid, methylhistidine, or phosphoethanolamine. This pattern is seen in chronic liver disease such as cirrhosis or infiltration with fat or tumor (Wu *et al.*, 1955). In acute liver failure, there is a massive generalized aminoaciduria.

The aminoaciduria of hepatolenticular degeneration (Wilson's disease) may be both renal and metabolic. It is renal early in the course of the disease, presumably due to the deposition of copper in the renal tubule, and it may in addition become metabolic when liver involvement ensues. A renal aminoaciduria has been seen in unaffected members of such families (Uzman and Hood, 1952).

D. Kwashiorkor

A disturbance of amino acid metabolism characterized by an abnormal pattern of free amino acids in the fasting plasma is regularly associated with severe protein deficiency (Holt *et al.*, 1960; Holt *et al.*, 1963; Arroyave *et al.*, 1962; Vis, 1963). In incipient protein deficiency the total free amino acid concentration is little affected, but the concentration of individual amino acids deviates from the normal, certain ones being increased and others decreased. Concentrations of several of the unessential amino acids—glycine, alanine, proline, histidine, serine, taurine and aspartic acid—tend to be elevated, whereas the levels of the essential amino acids as well as tyrosine, arginine, citrulline and α-amino-n-butyric acid tend to be depressed. In more severe depletion states all the plasma amino acids are depressed; however, those which in the milder states were elevated, are less depressed than the others (Fig. 12A, B). Particularly conspicuous are the depressed levels of the branched chain amino acids. It is not possible at

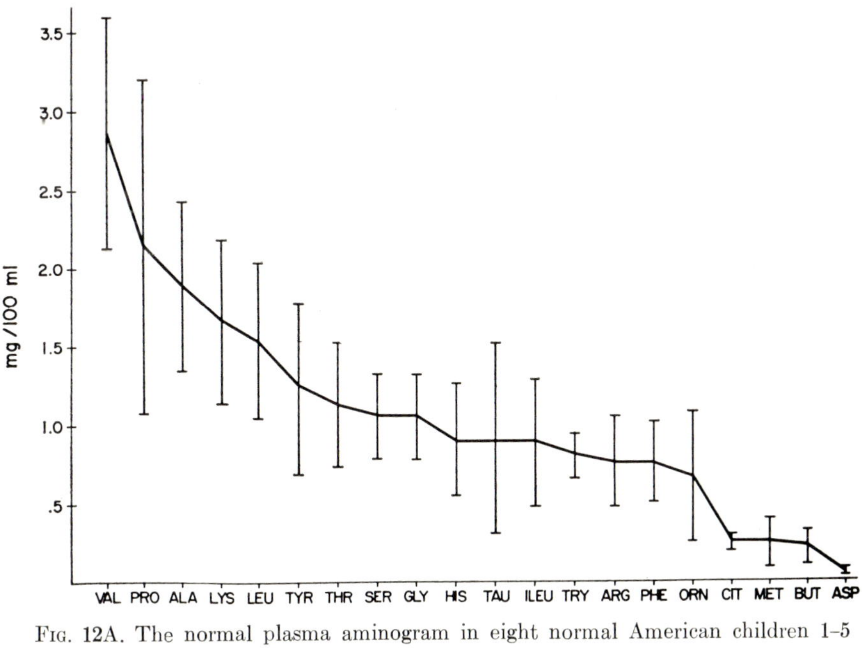

FIG. 12A. The normal plasma aminogram in eight normal American children 1–5 years old (average ±1 standard deviation).

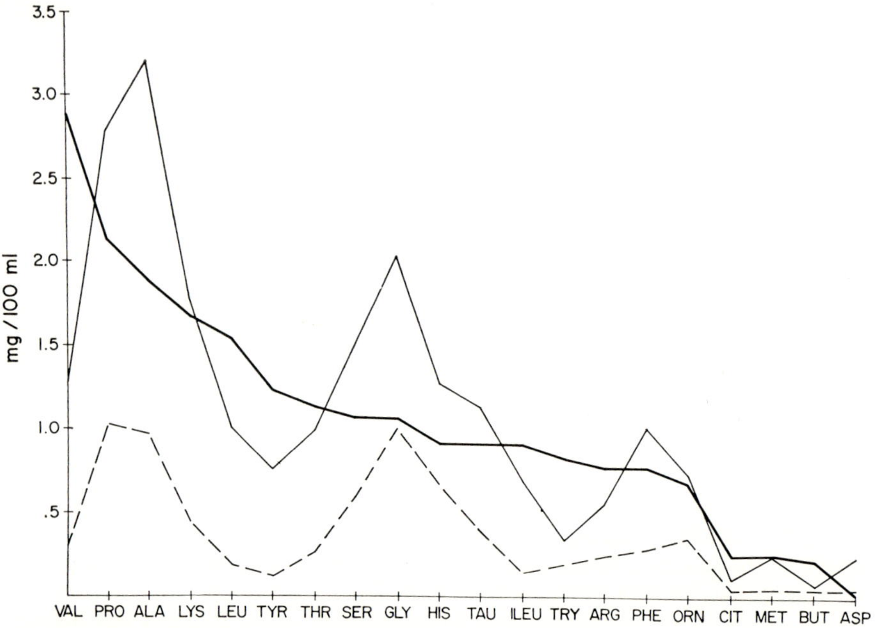

FIG. 12B. The plasma aminogram in incipient and advanced protein deficiency. The thin continuous line represents mild deficiency, the dotted line severe deficiency. The heavier line is the mean normal aminogram shown in Fig. 12A.

the present time to explain more than a few features of the kwashiorkor pattern, but it is clear that there is a widespread disturbance of amino acid metabolism. When protein food is restored in adequate quantities the abnormal plasma pattern corrects itself, sometimes rapidly; there may, however, be a delay in the restoration of synthetic processes resulting in elevated levels of amino acids and aminoaciduria.

REFERENCES

References are presented below in sections corresponding to the sections of the text. In order to provide a more ample bibliography for each disease, references additional to those cited in the text have been included and titles of papers have been retained.

I. CLASSIFICATION OF ANOMALIES OF AMINO ACID METABOLISM

Andrews, B. F., Bruton, O. C., and Knoblock, E. C. (1961). Aminoaciduria in salicylate intoxication. *Am. J. Med. Sci.* **242,** 411.

Bessman, S. P., and Baldwin, R. (1962). Imidazole aminoaciduria in cerebromacular degeneration. *Science* **135,** 789.

Bickel, H., Baar, A. S., Astley, R., Douglas, A. A., Finch, E., Harris, H., Harvey, C. C., Hickmans, E. M., Philpott, M. G., Smallwood, W. C., Smellie, J. M., and

Teall, C. G. (1952). Cystine storage disease with aminoaciduria and dwarfism (Lignac-Fanconi disease). *Acta Paediat.* **42,** Suppl., 90.

Borden, A. L., Brodie, E. C., Wallraff, E. B., Holbrook, W. P., Hill, D. F., Stephens, C. A. L., Jr., Johnson, R. B., and Kemmerer, A. R. (1952). Amino acid studies and clinical findings in normal adults and rheumatoid arthritis patients treated with ACTH. *J. Clin. Invest.* **31,** 375.

Choremis, C., Kiossoglou, K., Maounis, F., and Basti, B. (1959). Amino-acid tolerance curves and aminoaciduria in Cooley's and sickle-cell anemias. *J. Clin. Pathol.* **12,** 245.

Chowers, I., Czaczkes, J. W., Ehrenfeld, E. N., and Landau, S. (1962). Familial aminoaciduria in osteogenesis imperfecta. *J. Am. Med. Assoc.* **181,** 771.

Clarkson, T. W., and Kench, J. E. (1956). Urinary excretion of amino acids by men absorbing heavy metals. *Biochem. J.* **62,** 361.

Cusworth, D. C., Dent, C. E., and Flynn, F. V. (1955). The aminoaciduria in galactosemia. *Arch. Disease Childhood* **30,** 150.

Darling, S., Mortensen, B., and Søndergaard, G. (1960). Lactosuria and aminoaciduria in infancy, a new inborn error of metabolism. *Acta Paediat.* **49,** 281.

Davidson, L. A. G., Flear, C. T. G., and Donald, K. W. (1960). Transient aminoaciduria in severe potassium depletion. *Lancet* **I,** 911.

Dent, C. E. (1947). Aminoaciduria in the Fanconi syndrome: study making use of techniques based on paper partition chromatography. *Biochem. J.* **41,** 240.

Dent, C. E., Heathcote, J. G., and Joron, G. E. (1954). The pathogenesis of cystinuria. *J. Clin. Invest.* **33,** 1210.

de Vries, A., Kochwa, S., Lazebnik, J., Frank, M., and Djaldelti, M. (1957). Glycinuria, a hereditary disorder associated with nephrolithiasis. *Am. J. Med.* **23,** 408.

Dignam, W., and Harper, H. A. (1954). The causes of histidinuria in normal pregnancy. *Am. J. Obstet. Gynecol.* **68,** 110.

Doolan, P. D., Harper, H. A., Hutchin, M. E., and Alpern, E. L. (1956). The renal tubular response to amino acid loading. *J. Clin. Invest.* **35,** 888.

Dustin, J. P., Moore, S., and Bigwood, E. J. (1955). Chromatographic studies on the excretion of amino acids in early infancy. *Metab. Clin. Exptl.* **4,** 75.

Emslie-Smith, D., Johnstone, J. H., Thomson, M. B., and Laine, K. G. (1956). Aminoaciduria in acute tubular necrosis. *Clin. Sci.* **15,** 171.

Fanconi, G. (1936). Der Frühinfantile nephrotisch-glykosurische Zwergwuchs mit hypophosphatamischer Rachitis. *Jahrb. Kinderheilk.* **147,** 299.

Frimpter, G. W., Bass, A., Furth, E., Harwith, M., and Thompson, D. D. (1961). Cystinuria: amino acid clearances and the occurrence of a cystine peptide. *Proc. 53rd Meeting Am. Soc. Clin. Invest.* **40,** 1040.

Froesch, E. R., Prader, A., Wolf, M. P., and Labhart, A. (1959). Die hereditäre fructose Intoleranz. *Helv. Paediat. Acta* **14,** 99.

Gross, J. B., Comfort, J., and Ulrich, J. A. (1957). Abnormalities of serum and urinary amino acids in hereditary and non-hereditary pancreatitis. *Trans. Assoc. Am. Physicians* **70,** 127.

Harrison, H. E. and Harrison, H. C. (1954). Experimental production of renal glucosuria, phosphaturia and aminoaciduria by injection of maleic acid. *Science* **120,** 601.

Harrison, H. E., Chisholm, J. J., Jr. and Harrison, H. C. (1958). Congenital renal tubular acidosis. *A.M.A. J. Diseases Children* **96,** 558.

Hooft, C., Laey, P. D., Herpol, J., De Loore, F., and Verbick, J. (1962). Familial hypolipidaemia and retarded development without steatorrhea. *Helv. Paediat. Acta* **17,** 1.

Hosein, E. A., Smart, M., and Hawkins, K. (1960). Aminoaciduria in rats after treatment with the ethyl ester of gamma-butyrobetaine. *Can. J. Biochem. Physiol.* **38,** 837.

Jonxis, J. H. P., and Huisman, T. H. J. (1954). Aminoaciduria and ascorbic acid deficiency. *Pediatrics* **14,** 238.

Jonxis, J. H. P., Smith, P. A., and Huisman, T. H. J. (1952). Rickets and aminoaciduria. *Lancet* **II,** 1015.

Joseph, R., Ribierre, M., Job, J. C., and Girault, M. (1958). Maladie familial associant des convulsions a début très précise, une hyperalbuminorachnie et une hyperaminoacidurie. *Arch. Franç. Pediat.* **15,** 374.

Käser, H., Cottier, P., and Antener, I. (1961). Die Glucoglycinurie ein neues familiares Syndrom. *Helv. Paediat. Acta* **16,** 586.

Katz, E. J., and Hasterlik, R. J. (1955). Aminoaciduria following total body irradiation in the human. *J. Natl. Cancer Inst.* **15,** 1085.

Maggioni, G. (1960). "L'aminoaciduria normale e patologica del bambino." Fondazione Domenico Ganasseni, Milano.

Martin, E., Milhaud, G., and Doret, J. P. (1954). Analyse chromatographique de l'hyperamino-acidurie consecutive a l'administration de cortisone. *Colloque sur les acides amines, Lausanne, 1953.* p. 249.

Millichap, J. G., and Ulrich, J. A. (1962). Amino acid excretion with epilepsy. *Proc. Staff Meetings Mayo Clin.* **37,** 307.

Milne, M. D., Asatoor, A. M., Edwards, K. D. G., and Laughridge, M. B. (1961). Intestinal defect in cystinuria. *Gut* **2,** 323.

Parkes, W. E., and Neill, D. W. (1953). Acute nitrobenzene poisoning with transient aminoaciduria. *Brit. Med. J.* **I,** 653.

Ronzoni, E., Roberts, E., Frankel, S., and Ramasarma, G. B. (1953). Influence of administration of ACTH on urinary amino acids. *Proc. Soc. Exptl. Biol. Med.* **82,** 497.

Rothstein, A., and Berk, H. (1949). Aminoaciduria in uranium poisoning. *J. Pharmacol. Exptl. Therap.* **96,** 179.

Schendel, H. E., Antonis, A., and Hansen, J. D. L. (1959). Increased aminoaciduria in infants with kwashiorkor fed natural and synthetic diets. *Pediatrics* **23,** 662.

Scriver, C. R. (1962). Hereditary aminoaciduria. *Prog. Med. Genetics* **2,** 83.

Sereni, P., McNamara, H., Shibuya, M., Kretchmer, N., and Barnett, H. L. (1955). Concentrations in plasma and rate of urinary excretion of amino acids in premature infants. *Pediatrics* **15,** 575.

Spencer, A. G., and Franglen, G. T. (1952). Gross aminoaciduria following lysol burn. *Lancet.* **I,** 190.

Steinberg, A. H., Szeinberg, A., and Cohen, B. E. (1957). Aminoaciduria and hypermetabolism in progeria. *Arch. Disease Childhood* **32,** 401.

Uzman, L. L. and Hood, B. (1952). The familial nature of the aminoaciduria of Wilson's disease (hepatolenticular degeneration). *Am. J. Med. Sci.* **223,** 392.

Visakorpi, J. K. (1960). Aminoaciduria in various types of rickets. *Ann. Paediat. Fenniae* **6,** Suppl. 16.

Wallace, I. R., and Jones, J. H. (1960). Familial glomerulonephritis and aminoaciduria. *Lancet* **I,** 94.

Weaver, J. A., and Neill, D. W. (1954). Aminoaciduria in pernicious anemia and subacute combined degeneration of the cord. *Lancet* **I,** 1212.

Williams, R. D. B., and Tang, I. L. (1960). Mental defect, quadriplegia and ichthyosis. *Am. J. Diseases Children* **100,** 924.

Wilson, V. K., Thompson, M. L., and Dent, C. E. (1953). Aminoaciduria in lead poisoning. *Lancet* **II,** 66.

Woolf, L. I., and Giles, H. McC. (1956). Urinary excretion of amino acids and sugar in the nephrotic syndrome. *Acta Paediat.* **45,** 489.

Worthen, H. G., and Good, R. A. (1958). The de Toni-Fanconi syndrome with cystinosis. *A.M.A. J. Diseases Children* **95,** 653.

SECTION II. THE CONGENITAL ANOMALIES

A. Hyperglycinemia

Childs, B., Nyhan, W. L., Borden, M., Bard, L., and Cooke, R. E. (1961). Idiopathic hyperglycinemia and hyperglycinuria: a new disorder of amino acid metabolism. I. *Pediatrics* **27,** 522.

Cochrane, W., Scriver, C. R., and Krause, V. (1963). Hyperglycinemia-hyperglycinuria syndrome in a newborn infant. *Abstr. 33rd Meeting, Soc. Pediat. Res.* p. 1021.

Nyhan, W. L., Borden, M., and Childs, B. (1961). Idiopathic hyperglycinemia: a new disorder of amino acid metabolism. II. The concentrations of other amino acids in the plasma and their modification by the administration of leucine. *Pediatrics* **27,** 539.

Nyhan, W. L., and Childs, B. (1962). Hyperglycinemia: the turnover rate of glycine, pool size and the formation of serine. *Abstr. 32nd Meeting Soc. Pediat. Res.* p. 3.

Nyhan, W. L., Chisholm, J. J., Jr., and Edwards, R. V., Jr. (1963). Idiopathic hyperglycinuria. III. Report of a second case. *J. Pediat.* **62,** 540.

Tada, K., Yoshida, T., Morikawa, T., Minakawa, A., Wada, Y., Ando, T., and Shimura, K. (1963). Idiopathic hyperglycinemia. *Tohoku J. Exptl. Med.* **80,** 218.

B. Primary Hyperoxaluria and Oxalosis

Archer, H. E., Dormer, A. E., Scawen, E. F., and Watts, R. W. E. (1957). Primary hyperoxaluria. *Lancet* **II,** 320.

Archer, H. E., Dormer, A. E., Scawen, E. F., and Watts, R. W. E. (1958). The aetiology of primary hyperoxaluria. *Lancet* **I,** 175.

Burke, E. C., Bagenstoss, A. H., Owen, C. A., Jr., Power, M. H., and Lohr, O. W. (1955). Oxalosis. *Pediatrics* **15,** 383.

Crawhall, J. C., Scawen, E. F., and Watts, R. W. E. (1959). Conversion of glycine to oxalate in primary hyperoxaluria. *Lancet* **II,** 806.

Dunn, H. G. (1955). Oxalosis. *Am. J. Diseases Children* **90,** 58.

Elder, T. D., and Wyngaarden, J. B. (1959). The synthesis and turnover of oxalic acid in normal and hyperoxaluria subjects. *J. Clin. Invest.* **38,** 1001.

Frederich, E. W., Rabkin, M. T., and Smith, L. H. (1962). Primary hyperoxaluria: a defect in glyoxylate metabolism. *J. Clin. Invest.* **41,** 1358.

Godwin, J. T., Fawler, M. F., Dempsey, E. F., and Henneman, P. H. (1958). Primary hyperoxaluria and oxalosis. Report of a case and review of the literature. *New Engl. J. Med.* **259,** 1099.

Grossman, W. (1938). Current urinary stone wave in Central Europe. *Brit. J. Urol.* **10,** 46.

Joly, J. S. (1940). "Stone and Calculous Disease of Urinary Organs." Mosby, St. Louis, Missouri.

C. Hyperprolinemia

Schaefer, I., Scriver, C. R., and Efron, M. L. (1962). Familial hyperprolinemia, cerebral dysfunction and renal anomalies occurring in a family with hereditary nephropathy and deafness. *New Engl. J. Med.* **267,** 51.

D. Hydroxyprolinemia

Efron, M. L., Bixby, E. M., Pallatao, L. G., and Pryles, C. V. (1962). Hydroxyprolinemia associated with mental deficiency. *New Engl. J. Med.* **267,** 1193.

Fowler, D. I., Norton, P. M., Cheung, M. W., and Pratt, E. L. (1957). Observations on the urinary amino acid excretion in man: the influence of age and diet. *Arch. Biochem. Biophys.* **68,** 552.

Oye, I. (1962). Free hydroxyproline in human blood serum. *Scand. J. Clin. Lab. Invest.* **14,** 259.

Prockop, D. J., and Sjoerdsma, A. (1961). Clinical significance of urinary hydroxyproline in man. *J. Clin. Invest.* **40,** 843.

E. Marfan's Syndrome

Prockup, D. J., and Sjoerdsma, A. (1961). Significancy of urinary hydroxyproline in man. *J. Clin. Invest.* **40,** 196.

Sjoerdsma, A., Davidson, J. D., Udenfriend, S., and Mitoma, C. (1958). Increased excretion of hydroxyproline in Marfan's syndrome. *Lancet* **II,** 904.

F. Anomalies of Aromatic Amino Acid Metabolism

1. Albinism

Barnicot, N. A., and Birbeck, M. S. C. (1958). The electron microscopy of human melanocytes and melanin granules. *In* "The Biology of Hair Growth" (W. Montagna and R. A. Ellis, eds.), p. 239. Academic Press, New York.

Cooke, J. V. (1952). Familial white skin spotting ("partial albinism") with white forelock. *J. Pediat.* **41,** 1.

Cromartie, R. I. T., and Harley-Mason, J. (1957). Melanin and its precursors. *Biochem. J.* **66,** 713.

Fitzpatrick. T. B. (1960). Albinism. *In* "The Metabolic Basis of Inherited Disease" (J. B. Stanbury, J. B. Wyngaarden, and D. S. Fredrickson, eds.), p. 428. McGraw-Hill, New York.

Garrod, A. E. (1908). Inborn errors of metabolism. I. *Lancet* **II,** 73.

Knox, W. E. (1958). Sir Archibald Garrod's "Inborn errors of metabolism." III. Albinism. *Am. J. Human Genet.* **10,** 249.

2. Alcaptonuria

Black, R. L., Lawney, J. F., and Duffy, P. M. (1954). *A.M.A. Arch. Internal Med.* **93,** 75.

Garrod, A. E. (1923). "Inborn Errors of Metabolism," 2nd ed. Frowde, London.

Knox, W. E. (1958). Sir Archibald Garrod's "Inborn errors of metabolism." II. Alkaptonuria. *Am. J. Human Genet.* **10,** 95.

LaDu, B. N. (1960). Alcaptonuria. *In* "The Metabolic Basis of Inherited Disease" (J. B. Stanbury, J. B. Wyngaarden, and D. S. Fredrickson, eds.), p. 394 McGraw-Hill, New York.

LaDu, B. N., Zannoni, V. G., Laster, L., and Seegmiller, J. E. (1958). The nature of the defect in tyrosine metabolism in alcaptonuria. *J. Biol. Chem.* **230,** 251.

3. Familial Dysautonomia

Hutchinson, J. H., and Hamilton, W. (1962). Familial dysautonomia in 2 siblings. *Lancet* **I,** 1216.

Luft, R., and von Euler, U. S. (1958). Two cases of postural hypotension showing a deficiency in release of norepinephrine and epinephrine. *J. Clin. Invest.* **32,** 1065.

Riley, C. M. (1957). Familial dysautonomia. *Advan. Pediat.* **9,** 157.

Riley, C. M., Day, R. L., Greeley, D. M., and Langford, W. S. (1949). Central autonomic dysfunction with defective lacrimation. Report of 5 cases. *Pediatrics* **3,** 468.

Smith, A. A., Taylor, T. J., and Wortis, S. B. (1962). Abnormal catecholamine metabolism in familial dysautonomia. *New Engl. J. Med.* **268,** 705.

4. Familial Goitrous Cretinism

Stanbury, J. B. (1960). Familial goiter. *In* "The Metabolic Basis of Inherited Disease" (J. B. Stanbury, J. B. Wyngaarden, and D. S. Fredrickson, eds.), p. 273 McGraw-Hill, New York.

Stanbury, J. B., and McGirr, E. M. (1957). Sporadic or non-endemic familial cretinism with goiter. *Am. J. Med.* **22,** 712.

5. Phenylketonuria

Anderson, J. A., Gravem, B. S., Ertel, R., and Fisch, P. (1962). Identification of heterozygotes with phenylketonuria on basis of blood tyrosine responses. *J. Pediat.* **61,** 603.

Armstrong, M. D., and Robinson, K. S. (1954). On the excretion of indole derivatives in phenylketonuria. *Arch. Biochem. Biophys.* **52,** 287.

Armstrong, M. D., and Shaw, K. N. F. (1955). Studies on phenylketonuria. III. The metabolism of *o*-tyrosine. *J. Biol. Chem.* **213,** 805.

Armstrong, M. D., and Tyler, F. H. (1955). Studies on phenylketonuria. I. Restricted phenylalanine intake in phenylketonuria. *J. Clin. Invest.* **34,** 565.

Armstrong, M. D., Shaw, K. N. F., and Robinson, K. S. (1955). Studies on phenylketonuria. II. The excretion of *o*-hydroxyphenylacetic acid in phenylketonuria. *J. Biol. Chem.* **213,** 797.

Armstrong, M. D., Low, N. L. and Bosma, J. F. (1957). Further observation on the effect of phenylalanine restricted diet on patients with phenylketonuria. *Am. J. Clin. Nutr.,* **5,** 543.

Berry, H. K. (1962). Use of micro method for serum phenylalanine in management of phenylketonuric patients. *Clin. Chem.* **8,** 172.

Bickel, H., Gerrard, J. W., and Hickmans, E. M. (1954). Influence of phenylalanine intake on the chemistry and behavior of a phenylketonuric child. *Acta Paediat.* **43,** 64.

Cawte, J. E. (1957). Attempt to determine reaction to adrenaline in phenylketonuria. *J. Mental Deficiency Res.* **1,** 111.

Dancis, J., and Balis, M. E. (1955). A possible mechanism for disturbance in tyrosine metabolism in phenylpyruvic oligophrenia. *Pediatrics* **15,** 63.

Fölling, A. (1934). Ueber Ausscheidung von Phenylbrenztraubensäure in dem Harn als Stoffewechselanomalie in Verbindung mit Imbecillität. *Z. physiol. Chem.* **227,** 169.

Fölling, A., and Closs, K. (1938). Ueber das Vorkommen von L-Phenylalanin in Harn und Blut bei *Imbecillitas phenylpyruvica. Z. physiol. Chem.* **254,** 115.

Fölling, A., Closs, K., and Gamnes, T. (1938). Vorläufige Schlussvolgerungen aus Belastungversuchen mit Phenylalanine an Menschen und Tieren. *Z. Physiol. Chem.* **256,** 1.

Garrod, A. E. (1909). "Inborn Errors of Metabolism," 1st ed. Oxford Univ. Press, London and New York.

Guthrie, R. (1962). quoted by Scheel and Berry (1962).
Horner, F. A., and Streamer, C. W. (1959). Phenylketonuria treated from earliest infancy. *A.M.A. J. Diseases Children* **97,** 345.
Horner, F. A. and Streamer, C. W. (1962). Termination of dietary treatment of phenylketonuria. *New Engl. J. Med.* **266,** 79.
Hsia, D. Y.-Y., and Paine, R. S. (1957). Phenylketonuria: detection of heterozygote carrier. *J. Mental Deficiency Res.* **1,** 53.
Jervis, G. A. (1947). Studies of phenylpyruvic oligophrenia: the position of the metabolic error. *J. Biol. Chem.* **169,** 651.
Jervis, G. A. (1950). Excretion of phenylalanine and derivatives in phenylpyruvic oligophrenia. *Proc. Soc. Exptl. Biol. Med.* **75,** 83.
Jervis, G. A. (1953). Phenylpyruvic oligophrenia deficiency of phenylalanine-oxidizing system. *Proc. Soc. Exptl. Biol. Med.* **82,** 514.
Jervis, G. A. (1954). Phenylpyruvic oligophrenia (review with extensive bibliography). *Res. Publ. Assoc. Res. Nervous Mental Disease* **33,** 259.
Jervis, G. A. (1960). Detection of heterozygotes for phenylketonuria. *Clin. Chim. Acta* **5,** 47.
Jervis, G. A., Block, R. J., Bolling, D., and Kanse, L. (1940). Phenylalanine content of blood and spinal fluid in phenylpyruvic oligophrenia. *J. Biol. Chem.* **134,** 105.
Kaufman, S., and Levenberg, B. (1959). Further studies on the phenylalanine hydroxylation cofactor. *J. Biol. Chem.* **234,** 2683.
Knox, W. E. (1960). Phenylketonuria (a review). *In* "The Metabolic Basis of Inherited Disease" (J. B. Stanbury, J. B. Wyngaarden, and D. S. Fredrickson, eds.), p. 321. McGraw-Hill, New York.
McKean, C. M., Schanberg, S. M., and Giarman, N. J. (1962). A mechanism of the indole defect in experimental phenylketonuria. *Science* **137,** 604.
Mitoma, C., Auld, R. M., and Udenfriend, S. (1957). On the nature of the enzyme defect in phenylpyruvic oligophrenia. *Proc. Soc. Exptl. Biol. Med.* **94,** 634.
Neame, K. D. (1961). Phenylalanine as inhibitor of transport of amino acids in brain. *Nature* **192,** 1731.
Pare, C. M. B., Sandler, M., and Stacy, P. S. (1958). Decreased 5-hydroxy tryptophan decarboxylase activity in phenylketonuria. *Lancet* **II,** 1099.
Scheel, C., and Berry, H. K. (1962). Comparison of serum phenylalanine levels with growth in Guthrie's inhibition assay in newborn infants. *J. Pediat.* **61,** 610.
Sherwin, C. P., and Kennard, K. J. (1919). Toxicity of phenylacetic acid. *J. Biol. Chem.* **40,** 259.
Snyderman, S. E., Norton, P. M., and Holt, L. E., Jr. (1955). Effect of tyrosine administration in phenylketonuria. *Federation Proc.* **14,** 450.
Tashian, R. E. (1961). Inhibition of brain glutamic acid decarboxylase by phenylalanine, valine and leucine derivatives. *Metab. Clin. Exptl.* **10,** 393.
Waisman, H. A. (1960). Advances in the understanding of the metabolic abnormalities in phenylketonuria. *In* "Mental Retardation" (T. W. Bowman and H. V. Martner, eds.), p. 319. Grune & Stratton, New York.
Waisman, H. A., and Harlow, H. F. (1960). Phenylketonuria in infant monkeys. *Am. J. Diseases Children* **100,** 710.
Woolf, L. I. (1951). Excretion of conjugated phenylacetic acid in phenylketonuria. *Biochem. J.* **49,** IX.
Woolf, L. I., Griffiths, R., Moncrieff, A., Coates, S., and Dellestone, F. (1958). The dietary treatment of phenylketonuria. *Arch. Disease Childhood* **33,** 31.

6. Tyrosinosis and Tyrosyluria

Berry, H. K., and Sutherland, B. S. (1960). "Tyrosinosis" in an infant. *Am. J. Diseases Children* **100,** 571.

Blatherwick, N. R. (1934). Tyrosinosis: a search for additional cases. *J. Am. Med. Assoc.* **103,** 1933.

Bloxam, H. R. (1960). An inborn defect in the metabolism of tyrosine in infants on a normal diet. *Biochem. J.* **77,** 320.

Boscott, R. J., and Cooke, W. T. (1954). Ascorbic acid requirements and urinary excretion of *p*-hydroxyphenylacetic acid in steatorrhea and macrocytic anemia. *Quart. J. Med.* **23,** 307.

Felix, K., Leonardi, G., and Glasenapp, I. (1951). Ueber Tyrosinosis. *Z. Physiol. Chem.* **287,** 141.

Kretschmer, N. (1959). Enzymatic patterns during development. *Pediatrics* **23,** 606.

Levine, S. Z., Gordon, H. H., and Marples, E. (1941). A defect in the metabolism of tyrosine and phenylalanine in premature infants. II. Spontaneous occurrence and eradication by vitamin C. *J. Clin. Invest.* **20,** 209.

Medes, G. (1932). A new error of tyrosine metabolism: tyrosinosis. The intermediary metabolism of tyrosine and phenylalanine. *Biochem. J.* **26,** 917.

Menkes, J. H., and Jervis, G. A. (1961). Developmental retardation associated with an abnormality of tyrosine metabolism. *Pediatrics* **28,** 399.

Sakai, K., and Kitagawa, T. (1957). An atypical case of tyrosinosis. *Tokyo Jikeikai School Med. J.* **4,** 1.

Zannoni, V. G., and LaDu, B. N. (1960). Studies on the defect of tyrosine metabolism in scorbutic guinea pigs. *J. Biol. Chem.* **235,** 1651.

Zetterstrom, R. (1963). Tyrosinosis. *Ann. N. Y. Acad. Sci.*, **III,** 220.

G. HYPERHISTIDINEMIA

Auerbach, V. H., DiGeorge, A. M., Baldridge, R. C., Tourtelotte, C. D., and Brigham, M. P. (1962). Histidinemia. *J. Pediat.* **60,** 487.

Ghadhimi, H., Partington, M. W., and Hunter, A. (1961). A familial disturbance of histidine metabolism. *New Engl. J. Med.* **265,** 221.

Ghadhimi, H., Partington, M. W., and Hunter, A. (1962). Inborn error of histidine metabolism. *Pediatrics* **29,** 714.

LaDu, B. N., Howell, R. R., Jacoby, G. A., and Zannoni, V. G. (1962). Enzymatic defect in histidinemia. *Biochem. Biophys. Res. Commun.* **7,** 398.

Zenisek, A., Kral, J. A., and Hais, I. M. (1955). Sunscreening effect of urocanic acid. *Biochim. Biophys. Acta* **18,** 589.

H. HARTNUP DISEASE

Asatoor, A. M., Craske, J., London, D. R., and Milne, M. D. (1963). Indole production in Hartnup Disease. *Lancet* **I,** 126.

Baron, D. N., Dent, C. E., Harris, H., Hart, E. W., and Jepson, J. B. (1956). Hereditary pellagra-like skin rash with temporary cerebellar ataxia, constant renal amino-aciduria and other bizarre biochemical features. *Lancet* **2,** 421.

Colliss, J. E., Levi, A. J., and Milne, M. D. (1963). Stature and nutrition in cystinuria and Hartnup disease. *Brit. Med. J.* **I,** 590.

Jepson, J. B., and Spiro, M. J. (1960). Hartnup disease. *In* "The Metabolic Basis of

Inherited Disease" (J. B. Stanbury, J. B. Wyngaarden, and D S. Fredrickson, eds.), p. 1338. McGraw-Hill, New York.

Milne, M. D., Crawford, M. A., Girao, C. B., and Longbridge, L. W. (1960). The metabolic disorder in Hartnup disease. *Quart. J. Med.* **29,** 407.

Weyers, H., and Bickel, H. (1958). Photodermatose mit Aminoacidurie, Indolacetonie und cerebralen Manifestationen (Hartnup Syndrome). *Klin. Wochschr.* **36,** 893.

I. Congenital Tryptophanuria with Dwarfism

Tada, K., Ito, H., Wada, Y., and Arakawa, T. (1963). Congenital tryptophanuria with dwarfism ("H" disease-like clinical features without indicanuria and generalized aminoaciduria). A probable new error of tryptophan metabolism. *Tohoku J. Exptl. Med.* **80,** 118.

J. Other Abnormalities of Tryptophan Metabolism

Anderson, J. A. (1962). Familial defect in the metabolism of tryptophan *Abstr. 10th Intern. Congr. Pediat., Lisbon, 1962* p. 273.

Andrews, B. F., Bruton, O. C., and Knoblock, E. C. (1962). A study of aminoaciduria in infants and children. *Abstr. 10th Intern. Congr. Pediat., Lisbon, 1962*, p. 296.

Bessman, S. P., and Tada, K. (1959). A new abnormality of tryptophan metabolism in phenylketonuria. *A.M.A. J. Diseases Children* **98,** 563.

Bottomley, S. S. (1962). Pyridoxine responsive anemia. *J. Am. Med. Assoc.* **180,** 85.

Haverback, B. J., Dyce, B., and Thomas, H. (1960). Indole metabolism in the malabsorption syndrome. *New Engl. J. Med.* **262,** 754.

McCoy, E. E. (1962). Studies of 3-OH kynurenine metabolism in the Mongol. *Abst. 32nd Meeting Soc. Pediat. Res.* p. 120.

McKean, C. M., Schanberg, S. M., and Giarman, N. J. (1962). Mechanism of the indol defect in phenylketonuria. *Science* **137,** 604.

McNeely, R. G. D., and Jones, N. W. (1946). Secondary pellagra caused by multiple argentaffin carcinoma of the ileum and jejunum. *Gastroenterology* **6,** 443.

O'Brien, D., and Groshek, A. (1962). The abnormality of tryptophan metabolism in children with Mongolism. *Arch. Disease Childhood* **37,** 17.

Price, J. M., Brown, R. R., and Larson, F. C. (1957). Quantitative studies on human urinary metabolites of tryptophan as affected by isoniazid and deoxypyridoxine. *J. Clin. Invest.* **36,** 1600.

Sjoerdsma, A., Weissbach. H., and Udenfriend, S. (1956). A clinical, physiologic and biochemical study of patients with malignant carcinoid (argentaffinoma). *Am. J. Med.* **20,** 526.

Sleisinger, M. A. (1961). Clinical and metabolic studies in nontropical sprue. *New Engl. J. Med.* **265,** 49.

Snyderman, S. E., Holt, L. E., Jr., Carretero, R., and Jacobs, K. (1953). Pyridoxine deficiency in the human infant. *J. Clin. Nutr.* **1,** 200.

Sutton, H. E., and Read, J. H. (1958). Abnormal amino acid metabolism in a case suggesting autism. *A.M.A. J. Diseases Children* **96,** 23.

K. Metabolic Errors Involving the Urea Cycle

1. Hyperammonemia

Russell, A., Levin, B., Oberholzer, V. G., and Sinclair, L. (1962). Hyperammonaemia. A new instance of an inborn enzymatic defect in the synthesis of urea. *Lancet* **II,** 699.

2. *Citrullinemia*

McMurray, W. C., and Mohyuddin, F. (1962). Citrullinuria. *Lancet* **II,** 352.

McMurray, W. C., Mohyuddin, F., Rossiter, R. J., Rathburn, J. C., Valentine, G. H., Koelger, S. J., and Zarfas, D. E. (1962). Citrullinuria: A new aminoaciduria associated with mental retardation. *Lancet* **I,** 138.

Milne, M. D., London, D. R., and Asatoor, A. M. (1962). Citrullinuria in cases of cystinuria. *Lancet* **II,** 49.

3. *Argininosuccinemia*

Allan, J. D., Cusworth, D. C., Dent, C. E., and Wilson, V. K. (1958). A disease, probably hereditary, characterized by severe mental deficiency and a constant gross abnormality of amino acid metabolism. *Lancet* **I,** 182.

Armstrong, M. D., Yates, K. M. and Stemmermann, M. G. (1964). *Pediatrics,* **33,** 280.

Levin, B., Mackay, H. M. M., and Oberholzer, V. G. (1961). Argininosuccinuria—an inborn error of amino acid metabolism. *Arch. Disease Childhood* **36,** 622.

Wallis, K., Beer, S. and Fischl, J. (1963). A family affected by arginino succinic aciduria. *Helvet. Paed. Acta.* **18,** 338.

Westall, R. G. (1960). Argininosuccinic aciduria. Identification and reactions of the abnormal metabolite in a newly described form of mental disease, with some preliminary metabolic studies. *Biochem. J.* **77,** 135.

Westall, R. G., and Tomlinson, S. (1963). *Proc. 5th Intern. Congr. Biochem. Moscow, 1961* **103,** 364.

L. Oculo-cerebro-renal Dystrophy

Andrews, B. F., Bruton, O. C., and Knoblock, E. C. (1962). A study of aminoaciduria in infants and children. *Abstr. 10th Intern. Congr. Pediat., Lisbon, 1962,* p. 296.

Lowe, C. U., Terry, M., and McLaughlan, E. A. (1952). Organic aciduria, decreased renal ammonia production, hydrophthalmus and mental retardation; clinical entity. *Am. J. Diseases Children* **83,** 164.

Richards, W., Donnell, G. R., Wilson, W. A., and Stowens, D. S. (1960). Oculo-cerebro-renal syndrome of Lowe. *Am. J. Diseases Children* **100,** 707.

Schoen, E. J., and Young, G. (1959). Lowe's syndrome. *Am. J. Med.* **27,** 781.

Schwartz, R., Hall, P. W., III, and Gabuzda, G. J. (1962). The metabolism of ornithine and other amino acids in the cerebro-oculo-renal syndrome. *Am. J. Diseases Children* **104,** 484.

M. Anomalies of Sulfur Amino Acid Metabolism

1. *Cystinosis*

Bickel, H., Smallwood, W. C., Smellie, J. M., Board, H. S., Hickmans, E. M., Harris H., Astley, R., Teall, C. G., Douglas, A. A., Philpott, M. G., Harvey, C. C., and Finch, E. (1952). Cystine storage disease with aminoaciduria and dwarfism (Lignac-Fanconi disease). *Acta Paediat.* 42: Suppl. 90.

Brigham, M. P., Stein, W. H., and Moore, S. (1960). The concentrations of cysteine and cystine in human blood plasma. *J. Clin. Invest.* **39,** 1633.

Clayton, B., and Patrick, A. D. (1961). Use of dimercaprol or penicillamine in treatment of cystinosis. *Lancet* **II,** 909.

Darmady, E. M., and Stranack, F. (1957). Microdissection of the nephron in disease. *Brit. Med. Bull.* **13,** 21.

Eberlein, W. (1953). Aminoaciduria in childhood: cystinuria and cystinosis. *Am. J. Med. Sci.* **225,** 677.

Korn, D. (1960). Demonstration of cystine crystals in peripheral white blood cells of a patient with cystinosis. *New Engl. J. Med.* **262,** 545.

Mahoney, C. P., and Trump, B. F. (1962). Studies in cystinosis. *Abstr. 32nd Meeting Soc. Pediat. Res.* p. 44.

Parker, W. S., Prader, A., and Fanconi, G. (1955). Further observations on cystine storage disease. *Pediatrics* **16,** 228.

Worthen, H. G., and Good, R. A. (1958). The DeToni-Fanconi syndrome with cystinosis. *A.M.A. J. Diseases Children* **95,** 653.

Worthen, H. G., and Good, R. A. (1961). The pathogenesis of cystinosis. *Am. J. Diseases Children* **102,** 494.

2. Cystathionuria

Frimpter, G. W., Haymovitz, A., and Horwith, A. (1963). Cystathionuria. *New Engl. J. Med.* **268,** 333.

Harris, H., Penrose, L. S., and Thomas, D. H. H. (1959). Cystathionuria. *Ann. Human Genet.* **23,** 442.

Hope, D. B. (1957). L-Cystathione in the urine of pyridoxine-deficient rats. *Biochem. J.* **66,** 486.

Scriver, C. R., and Hutchison, J. H. (1963). The vitamin B_6 deficiency syndrome in human infancy: biochemical and clinical observations. *Pediatrics* **31,** 2401.

3. Homocystinuria

Carson, N. A. J., Cusworth, D. C., Dent, C. E., Field, C. M. B., Neill, D. W., and Westall, R. G. (1963). Homocystinuria: A new inborn error of metabolism associated with mental deficiency. *Arch. Disease Childhood* **38,** 425.

Gerritsen, J. and Waisman, H. A. (1964). Homocystinuria—an error in the metabolism of methionine. *Pediatrics.* In press.

N. Anomalies of Branched Chain Amino Acid Metabolism

1. Maple Syrup Urine Disease

Dancis, J. (1959). Phenylketonuria and maple syrup urine disease. *Bull. N.Y. Acad. Med.* **35,** 427.

Dancis, J., and Levitz, M. (1960). Maple syrup urine disease. *In* "The Metabolic Basis of Inherited Disease" (J. B. Stanbury, J. B. Wyngaarden, and D. S. Fredrickson, eds.) p. 473. McGraw-Hill, New York.

Dancis, J., Levitz, M., Miller, S., and Westall, R. G. (1959). Maple syrup urine disease. *Brit. Med. J.* **I,** 91.

Dancis, J., Hutzler, J., and Levitz, M. (1960a). Metabolism of the white blood cells in maple syrup urine disease. *Biochim. Biophys. Acta* **43,** 342.

Dancis, J., Levitz, M., and Westall, R. G. (1960b). Maple syrup urine disease; branched chain keto-aciduria. *Pediatrics* **25,** 72.

Dancis, J., Hutzler, J., and Levitz, M. (1961). Tissue distribution of branched chain keto acid decarboxylase. *Biochim. Biophys. Acta* **52,** 60.

Dancis, J., Janson, V., Hutzler, J., and Levitz, M. (1963). The metabolism of leucine in tissue culture of skin fibroblasts. *Biochim. Biophys. Acta* **77,** 523.

Dent, C. E., and Westall, R. G. (1961). Studies in maple syrup urine disease. *Arch. Disease Childhood* **38**, 259.

Holt, L. E., Jr., Snyderman, S. E., Dancis, J., and Norton, P. M. (1960). The treatment of a case of maple syrup urine disease. *Federation Proc.* **19**, 10.

Jepson, J. B., Smith, A. J., and Strang, L. B. (1958). An inborn error of metabolism with urinary excretion of hydroxy acids, keto acids and amino acids. *Lancet* **II**, 1334.

Lane, M. R. (1961). Maple syrup urine disease. *J. Paediat.* **58**, 80.

Lonsdale, D., Mercer, R. D., and Faulkner, W. R. (1963). Maple syrup urine disease—report of two cases. *Am. J. Diseases Children* **106**, 258.

Mackenzie, D. Y., and Woolf, L. I. (1959). Maple syrup urine disease—an inborn error of the metabolism of valine, leucine and isoleucine associated with gross mental deficiency. *Brit. Med. J.*, **I**, 90.

Menkes, J. H. (1959a). Maple syrup urine disease; isolation and identification of organic acids in the urine. *Pediatrics* **23**, 348.

Menkes, J. H. (1959b). Maple syrup disease; investigations into the metabolic defect. *Neurology* **9**, 826.

Menkes, J. H. (1960a). Maple syrup disease. *Maandschr. Kindergeneesk.* **28**, 54.

Menkes, J. H. (1960b). The pattern of urinary alpha-keto acids in various neurological disorders. *A.M.A. J. Diseases Children* **99**, 500.

Menkes, J. H., Hurst, P. L., and Craig, J. M. (1954). A new syndrome; progressive familial cerebral dysfunction associated with an unusual urinary substance. *Pediatrics* **14**, 462.

Morris, M. D., Lewis, B. D., Doolan, P. D., and Harper, H. A. (1961). Clinical and biochemical observations on an apparent nonfatal variant of branched chain keto aciduria (maple syrup urine disease). *Pediatrics* **28**, 918.

Müller, W., and Schreier, K. (1962). Die Ahorn-Sirup-Krankheit. *Deut. Med. Wochschr.* **87**, 2479.

Norton, P. M., Roitman, E., Snyderman, S. E., and Holt, L. E., Jr. (1962). A new finding in maple syrup urine disease. *Lancet* **I**, 26.

Norton, P. M., Roitman, E., Snyderman, S. E., and Holt, L. E., Jr. (1963). Further studies on alloisoleucine in maple syrup urine disease. *Federation Proc.* **22**, 549.

Patrick, A. D. (1961). Maple syrup urine disease. *Arch. Disease Childhood* **38**, 269.

Silverman, J., Dancis, J., and Feigin, I. H. (1961). Neuropathological observations in maple syrup urine disease: branched chain ketoaciduria. *Arch. Neurol.* **5**, 351.

Smith, A. J., and Strang, L. B. (1958). An inborn error of metabolism with the urinary excretion of α-hydroxy butyric acid and phenylpyruvic acid. *Arch. Disease Childhood* **33**, 109.

Snyderman, S. E. (1962). Maple syrup urine disease. *In* "Pediatrics," 13th ed. (L. E. Holt, Jr., R. M. McIntosh, and H. L. Barnett, eds.), Appleton, New York.

Tada, K., Wada, Y., and Okamura, T. (1963). Maple sugar urine disease. *Tohoku J. Exptl. Med.* **79**, 142.

Tashian, R. E. (1961). Inhibition of brain glutamic decarboxylase by phenylalanine, leucine and valine derivatives; a suggestion concerning the neurological defect in phenylketonuria and branched chain keto aciduria. *Metab., Clin. Exptl.* **10**, 393.

Waisman, H. A., Gerritsen, T., Boggs, D. E., Palidora, V. J., and Harlow, H. F. (1962). Mental retardation in monkeys. II. Branched chain aminoaciduria and ketoaciduria. *Am. J. Diseases Children* **104**, 488.

Westall, R. G., Dancis, J., and Miller, S. (1957). Maple sugar urine disease—a new molecular disease. *A.M.A. J. Diseases Children* **94**, 571.

2. Hypervalinemia

Tada, K. (1963). Idiopathic Hypervalinemia. *Tohoku J. Exper. Med.* **81,** 46.

3. Leucine-Induced Hypoglycemia

Cochrane, W. A., Payne, W. W., Simpkiss, M. J., and Woolf, L. I. (1956). Familial hypoglycemia precipitated by amino acids. *J. Clin. Invest.* **35,** 411.

Cochrane, W. A. (1960). Studies in the relationship of amino acids to infantile hypoglycemia. *A.M.A. J. Diseases Children* **99,** 476.

DiGeorge, A. M., and Auerbach, V. H. (1960). Leucine-induced hypoglycemia—a review and speculations. *Am. J. Med. Sci.,* **240,** 792.

Flanagan, G. C., Schwartz, T. B., and Ryan, W. G. (1961). Studies on patients with islet-cell tumor, including the phenomenon of leucine-induced accentuation of hypoglycemia. *J. Clin. Endocrinol. Metab.* **21,** 401.

Grumbach, M. M., and Kaplan, S. L. (1960). Amino acid and alpha-keto acid induced hyperinsulism in the leucine-sensitive type of infantile and childhood hypoglycemia. *J. Pediat.* **57,** 346.

Reaven, G., and Lucas, C. (1962). The use of insulin in the production of L-leucine induced hypoglycemia in normal dogs. *J. Clin. Invest.* **41,** 654.

4. Other Leucine-Sensitive States

Gopalan, C., and Srikantia, S. G. (1960). Leucine and pellagra. *Lancet* **I,** 954.

Nyhan, W. L., Borden, M., and Childs, B. (1961). Idiopathic hyperglycemia; a new disorder of amino acid metabolism. II. The concentrations of other amino acids in the plasma and their modification by the administration of leucine. *Pediatrics* **27,** 539.

Truswell, A. S., Goldsmith, G. A., and Pearson, W. N. (1963). Leucine and pellagra. *Lancet* **I,** 778.

O. β-Aminoisobutyric Aciduria

Berry, H. K. (1960). Excretion of β-amino isobutyric acid (BAIB). *Metab., Clin. Exptl.* **9,** 373.

Crumpler, H. R., Dent, C. E., Harris, H., and Westall, R. G. (1951). β-amino isobutyric acid (α-methyl-β alanine): a new amino acid obtained from human urine. *Nature* **167,** 307.

Fink, K., Henderson, R. B., and Fink, R. M. (1951). Beta-amino isobutyric acid—a possible factor in pyrimidine metabolism. *Proc. Soc. Exptl. Biol. Med.* **78,** 135.

Hillicoat, B. C. (1962). Altered urinary excretion of beta amino-isobutyric acid after splenectomy. *Lancet* **I,** 74.

Sandler, M., and Pare, C. M. B. (1954). Starvation aminoaciduria. *Lancet* **I,** 495.

Yates, K., and Armstrong, M. D. (1962). Studies on β-amino isobutyric acid excretion by familial high excretors. *Federation Proc.* **21,** 268.

SECTION III. ACQUIRED DISTURBANCES OF AMINO ACID METABOLISM

A. Premature and Newborn Infants

Fowler, D. I., Norton, P. M., Cheung, M. W., and Pratt, E. L. (1957). Observations on the urinary amino acid excretion in man: influence of age and diet. *Arch. Biochem. Biophys.* **68,** 452.

Gerritsen, T., Lipton, S. H., Strong, S. M., and Waisman, H. A. (1961). On the isolation and identification of homocitrulline from urine. *Biochem. Biophys. Res. Commun.* **4,** 379.

Norton, P. M., Roitman, E., Snyderman, S. E., and Holt, L. E., Jr. (1963). Unpublished data.

Sereni, F., McNamara, H., Shibuya, M., Kretchmer, N., and Barnett, H. L. (1955). Concentrations in plasma and rate of urinary excretion of amino acids in premature infants. *Pediatrics* **15,** 575.

Simon, S. (1911). Zur Stickstoffverteilung im Urin des Neugeborenen. *Z. Kinderheilk.* **2,** 1.

B. Tissue Destruction

Ames, S. R., and Risley, H. A. (1948). Aminoaciduria in progressive muscular dystrophy. *Proc. Soc. Exptl. Biol. Med.* **68,** 131.

Carver, M. J., Dutch, S. J., and Wittson, C. L. (1961). Aminoaciduria, in representative neuromuscular disorders. *Metab., Clin. Exptl.* **10,** 582.

Eades, C. H., Jr. (1955). Aminoaciduria in surgical conditions. ONR Report NONR-105-004.

Eades, C. H., Jr., Pollack, R. L., and Hardy, J. D. (1955). Thermal burns in man. IX. Urinary amino acid patterns. *J. Clin. Invest.* **34,** 1756.

Gray, C. H., and Illing, E. K. B. (1952). Plasma and urinary amino acids in diabetes. *J. Endocrinol.* **8,** 44.

Kelley, J. J., and Waisman, H. A. (1957). Quantitative plasma amino acid values in leukemic blood. *Blood* **12,** 635.

Nardi, G. L. (1954). Essential and nonessential amino acids in urine of severely burned patients. *J. Clin. Invest.* **33,** 847.

C. Liver Disease

Dent, C. E., and Walshe, J. M. (1951). Amino acid metabolism in liver disease. *In* "Liver Disease," a Ciba Foundation Symposium, p. 22. McGraw-Hill (Blakiston), Philadelphia, Pennsylvania.

Uzman, L. L., and Hood, B. (1952). Familial nature of the aminoaciduria of Wilson's disease. *Am. J. Med. Sci.* **223,** 392.

Wu, C., Bollman, J. L., and Butt, H. B. (1955). Changes in free amino acids in plasma during hepatic coma. *J. Clin. Invest.* **34,** 845.

D. Kwashiorkor

Arroyave, G., Wilson, D., de Funes, C., and Behar, M. (1962). The free amino acids in blood plasma of children with kwashiorkor and marasmus. *Am. J. Clin. Nutr.*, **11,** 517.

Arroyave, G. (1962). The estimation of relative nutrient intake and nutritional status by biochemical methods: Proteins. *Am. J. Clin. Nutr.* **11,** 447.

Holt, L. E., Jr., György, P., Pratt, E. L., Snyderman, S. E., and Wallace, W. M. (1960). "Protein and Amino Acid Requirements in Early Life," p. 14. N. Y. Univ. Press, New York,

Holt, L. E., Jr., Snyderman, S. E., Norton, P. M., Roitman, E., and Finch, J. (1963). The plasma aminogram in kwashiorkor. *Lancet.* **2,** 1343.

Maggioni, G. (1960). "L'Aminoaciduria normale e patologica del bambino."Fondazione Domenico Ganassini, Milan.

Norton, P. (1960). Plasma amino acids in Kwashiorkor. *Abstr. 5th Intern. Congr. on Nutr. Washington, D.C., 1960,* p. 7.

Schendel, H. E., Antonis, A., and Hansen, J. D. L. (1959). Increased amino aciduria in infants with kwashiorkor fed natural and synthetic diets. *Pediatrics* **23**, 662.
Tuttle, S. G., Swendseid, M., Friedrich, B., and Griffith, W. H. (1962). Effect of low protein diet and starvation on men over 60. *Federation Proc.* **21**, 395.
Vis, H. L. (1963). "Aspects et méchanismes des hyperaminoaciduries de l'enfance." Arscia Eds., Brussels.

CHAPTER 19

Physical Injury and Its Effects on Protein Metabolism

D. P. CUTHBERTSON

Rowett Research Institute, Bucksburn, Aberdeen, Scotland

I. Introduction

Physical injury results in some damage to, or loss of, body protein, both locally at the site of the injury and also generally, as part of the reaction to injury. First, injury may cause loss of protein by physical means: by separating tissue from the rest of the body or by destroying it. Second, for defense and in order to initiate repair of the damaged tissues, the injury is followed locally by an inflammatory reaction in which the participants are fundamentally structures made of protein. Third, there is a general response to injury, the most prominent feature of which is an increased output of nitrogen and sulfur and consequent loss of body protein, which can persist for several days after the injury has been inflicted. This general metabolic response may or may not be preceded by a state of shock at the time of the injury or soon after.

The time sequence of these events is presented in Table I, which demonstrates that the general and the local responses to injury overlap and cannot be considered as independent aspects of the effects of injury on the body. Consequently, a comprehensive picture of the metabolic changes following physical injury demands discussion of the local inflammatory change, of the condition called shock, and finally of the general metabolic response. Only then can the complexity of the problem be appreciated.

II. The Nature of the Inflammatory Reaction

The inflammatory reaction is characterized by local changes—swelling, redness, heat, pain, and loss of function. John Hunter (1794) believed that inflammation could be caused by any injury. The deleterious agents might be pressure, friction, heat, cold, corrosion, or "air in wounds." He correctly regarded inflammation as an adaptive defensive mechanism which tends to restore the injured part to its normal function. A hundred years later Burdon-Sanderson (see Florey, 1962) described inflammation as "the succession of changes which occurs in a living tissue when it is injured, provided that the injury is not of such a degree as at once to destroy its structure and vitality." Menkin (1940) pointed out that this reaction is elicited in higher animals by the presence of microorganisms but also by nonviable irritants and that the disturbances in the local physiology of inflamed tissues are closely associated with the immunological significance of the inflammatory reactions. Florey (1962) has properly emphasized that inflammation is a process and not a state: the affected area undergoes continuous change.

It is mainly to Cohnheim (1882) that we are indebted for a description of the march of events following injury; his classical picture is the basis for most chapters in textbooks of pathology. By macroscopic and microscopic examination, the character of the successive changes have now been worked out in great detail and more recently with the aid of the electron microscope. This has essentially been a study of localized changes in body protein, for the bulk of the cellular and intercellular substances involved are protein in nature.

A. The Local Response to Injury

Florey (1962) points out that, while the nervous system can modify the manifestations of inflammation, there is good experimental evidence that all the really essential phenomena can take place in the absence of nervous connections, as was maintained by Cohnheim. Florey (1961) states that it now seems likely that the swelling or edema, which is an important feature in many natural inflammations, is due to substances of the nature of histamine, 5-hydroxytryptamine (serotonin), and bradykinin, liberated from the tissues and causing increased permeability. Indeed the early reactions of tissues to injury are dominated by vascular changes, primarily exudation from the small blood vessels. This permeability change, which is remarkably similar for a variety of injuries, chemical and thermal burns, bacterial toxins, and bacterial infections, is biphasic: but only the transient initial stage is, according to Miles (1962), due to substances like histamine, bradykinin, and certain proteases, and occurs during the immediate acute phase of local anaphylaxis. It is characterized by fluid exudation only, leucocytes rarely emigrating from the blood. By direct test, this early exudative reaction proves to be associated with an increase of tissue resistance to infection of at least about threefold. The second, major exudative phase, which rises to a maximum in from 3 to 12 hours according to the nature and intensity of the injury, and declines a few hours later to low values, brings both soluble and cellular antimicrobial factors from the blood into the tissues, and the local tissue response may include the liberation or activation of further antimicrobial factors.

These local vascular changes are responsible for the classical signs of inflammation— swelling, redness, heat, pain, and loss of function. With the passage of protein into the tissue fluids the osmotic balance is disturbed and the fluid that escapes does not return to the blood vessel. The effect is exaggerated by the fact that the blood proteins in the inflammatory tissue fluid may break down by proteolysis and so increase the number of molecules that can exert osmotic pressure, thus further reducing the effective intravascular osmotic pressure. There arises swelling due to the exuded protein-containing fluid. Often inflammatory exudate contains enough fibrinogen to clot and form masses of fibrin, which can be readily seen when the exudate comes out on to a surface.

The sense of heat is due to blood from the interior passing more rapidly and in greater amount through the dilated vessels of the affected part. The phenomenon of increased heat of an inflamed part is only appreciated in lesions of the surface. As we shall see later, there is evidence that the metabolic processes of inflamed tissues are increased. Sevitt (1962) has found that the critical temperature at which the capillaries leak protein is 41°–44°C and this correlates with the critical temperature for the formation of bradykinin. The critical temperature for capillary stasis is 47°–48°C. Inflammation occurring in tissues which become particularly tense under local structures and pressure produces pain.

TABLE I

Succession of Changes Following Physical Injury

Time	Local changes	Blood changes	Changes in metabolism	Changes in hormonal activity
A few minutes after injury	Fleeting arteriolar contraction. Coagulation of escaping blood. Dilation of small blood vessels—redness and heat	Changes concerned with the defensive reactions of hemostasis, inflammation, and repair. Shortly after injury there is an increase in polymorphonuclear leucocytes and platelets, but a fall in eosinophils and lymphocytes. After severe injury, increased coagulability of blood, with decreased "suspension stability" of red cells (erythrocyte sedimentation rate) which tend to clump.	1. Primary reaction to the emergency attributable to stimulation of the sympathetic system. Heart rate, temperature and blood sugar raised. There may be primary shock akin to syncope (fainting)	Increased secretion and excretion of adrenaline and noradrenaline, and of cortical steroids
Beginning within a few hours after severe injury and generally complete in 24 hr	Increased capillary permeability and exudation; swelling. Polymorphs and monocytes migrate into the exudate. Exudate coagulates. There may be thrombosis of vessels in vicinity of damage		2. May progress into "secondary" or "traumatic" shock, a period of circulatory failure due to loss of circulating volume of blood through hemorrhage or fluid loss into the tissues or from the surface. In the early stages there is hyperglycaemia which later declines. This is the "ebb" period with progressive fall in body temperature and oxygen consumption. There is diminished urinary secretion lasting for some 24 hr and azotemia (increased blood nonprotein nitrogen content)	Concentrations of cholesterol and ascorbic acid in adrenal cortex decline. These changes are brief in moderate injuries but continue longer after more severe injuries. For a few days the secretion of aldosterone is inhibited by an excessive antidiuretic hormone output from

Maximal in first few days following injury, then declining	Increased local metabolism. Increased lymph flow, helping to remove inflammatory exudate	Changes occur in concentration of plasma proteins: albumin falls and fibrinogen and globulin increase. Proteolytic activity may appear in plasma, causing rapid digestion of fibrin and possibly of other proteins	3. If the body withstands this period of shock, it enters the third phase—the period of "flow," the early phase of which may be termed "traumatic inflammation." Repair begins with fever, increased oxygen consumption, pulse and respiration; leucocytosis and formation of inflammatory exudate and increased urinary excretion of N, S, P, and K by the well-nourished organism. This third stage may be reached without all but the vestige of passage through (1) and (2), particularly after operation	posterior pituitary lobe, but this declines and water metabolism gradually returns to normal

Increased blood flow associated with inflammation has a parallel in increased flow of lymph. Just as the finer blood vessels dilate and become more permeable to large molecules during inflammation, so the lymphatics undergo the same changes, which facilitate the removal of the excess fluid poured into the tissues from the blood vessels.

The details of the inflammatory reaction vary with the tissue. Pathogenic bacteria do not induce changes until they have passed certain barriers which are a first line of defense against their invasion. Thus the skin of animals is, in general, impervious to microorganisms, so that solution of its continuity is necessary for infection to occur. But the much more delicate mucous membranes of the respiratory, alimentary, and genitourinary tracts are more easily penetrated. Inflammation of mucous membranes is characterized by the phenomena common to all tissues. In addition, two features are to be noted that do not occur in other tissues, namely, the secretion of mucus and the desquamation or shedding of the epithelial cells of the mucous membrane, particularly of the small intestine.

The local effects of injury may also vary according to the noxious agent. Mechanical agents may damage cell membranes and disturb the spatial arrangement of enzymes and their substrates. Blockage of the blood supply to a portion of the body tissues causes an infarct in which the cells are deprived of oxygen and other substances that are necessary for the balance between synthesis and breakdown. Heat denatures cell enzymes, and cold may destroy their intracellular organization. Chemical substances may cause widespread damage to cell structures, or may merely inhibit the function of one enzyme; or, like radiation, they may injure chromosomes.

B. The Contribution of Blood Constituents to the Inflammatory Process

So far we have mainly outlined the local defensive reactions included in the term "inflammation." As MacFarlane (1962) has pointed out, it is not so well recognized that such damage, through physical, chemical, thermal, or toxic agents, may also produce striking changes in another compartment of the body, namely, the plasma and cells of the blood, and here the effects are primarily on proteins.

In addition to its part in hemostasis, blood coagulation provides a protective covering to a wounded surface, preventing infection and traumatic damage to the raw surface. The scab of dried blood does not become detached until the surface beneath has been covered by new epithelium.

Normal lymph clots like the plasma of the blood, and being derived from the tissue space fluid, it is probable that this fluid contains all that is required for coagulation. The fact that fibrin formation does not normally occur in the tissues suggests that there is no surface activation in the normal tissue spaces, such as occurs when the plasma comes in contact with damaged tissue. In severe inflammation, however, the concentration and amount of protein in tissue-space fluid increase, and may coagulate *in situ* forming a barrier of fibrin surrounding the inflamed area.

The reactions of the tissue components of the blood to injury consist of a rise in the number of polymorphonuclear leucocytes and platelets, and a fall in eosinophils and lymphocytes, a shortening of the coagulation time, and an increase in hemostatic efficiency indicated by a shortening of the bleeding time and an increase of capillary resistance to rupture by pressure.

III. The General Response to Physical Injury: Traumatic Shock

The author of this article has emphasized elsewhere (Cuthbertson, 1942) that the sequence of the general reactions to injury have to be considered in continuity as parts of a single process. As has long been known, no hard-and-fast rule can be drawn between what may be considered initial defense reactions and those which are concerned with subsequent repair of the damaged tissues, but for purposes of description it is easier to consider the events of healing and those involving the fully developed metabolic response to injury as occurring later than the first stages of inflammation and traumatic shock.

Three phases of general response to trauma can be recognized (Table I). Immediately after trauma or hemorrhage there is a sustained hyperglycemia. Such an effect is also observed when adrenaline is injected into a normal animal. This early effect of injury is presumed to be due to the release of sympathetic amines since the concentration of adrenaline in the adrenal medulla decreases greatly and the excretion of adrenaline and noradrenaline in the urine rises following surgical operation (see Section V for evidence).

Apart from the initial vasovagal disturbances (fainting) that sometimes follow receipt of trivial injuries by the conscious subject—one of the effects of sympathetic amines—the next systemic effect of serious injury is generally a diminution of metabolic activity with circulatory deficiency due to loss of circulating blood volume as the central feature—the period of traumatic shock. This is the period which the present author referred to as the "ebb" period and is associated with diminished metabolism. In severe but nonfatal injuries it is generally followed by a third period of increased metabolism or "flow" period in which some degree of "traumatic fever" can be detected (Cuthbertson, 1942).

Failure of oxygen transport by the blood is a major factor in the early stages of the individual's response to injury and for a particular injury its role will vary with the stage of the response, the tissue being studied, and with the environmental and other conditions pertaining at the time. A fall in oxygen consumption and in temperature are both consequences of the same reduction in cellular metabolism accompanying the phase of traumatic shock.

The later flow period, on the other hand, would seem to correspond to the period of local inflammatory defense and demolition in the surviving organism and may generally be accompanied by hyperglycemia, frequently, but not always, by a rise in body temperature, increased pulse and respiration rates, and the other signs of traumatic fever. It is later followed in severe injuries by a temporary anemia beyond that which would appear to be due to the actual blood loss. The injury eventually can cause depletion of body protein which is strangely more extensive than would apparently be warranted by the local damage.

The events which constitute this metabolic response to physical injury are gradually being elucidated as different facets of the reaction are being revealed through a variety of experimental techniques. How far the response is the resultant of a single chain reaction and how far it arises from a collateral series of such reactions is still a matter of exploration and conjecture. The author has attempted to outline the succession of changes in tabular form (Table I).

Views of the causation of shock will now be considered.

A. Early Theories of the Causation of Traumatic Shock

World War I saw the emergence of the concept of primary and secondary wound shock: the primary akin to syncope (fainting), and the secondary, a widespread depression in functional activity of many organs and tissues remote from the site of serious injury, characterized by sustained reduction in blood pressure and circulatory failure.

In 1910 Dale and Laidlaw drew attention to the reduction in blood pressure which results from the injection of histamine into cats. Together with Richards (Dale *et al.*, 1919) they investigated the phenomenon more fully during the later stages of World War I, and it was their view that these effects of histamine bear a closer resemblance to those occurring during wound shock than was at first expected. It was found that the low pressure resulting from histamine injection was not due to relaxation of the arteries but to a dilatation of the capillaries, with, in addition, an escape of plasma into the tissue spaces—a condition which Malcolm (1905) had surmised from clinical observation of patients in shock.

During World War I the French observer Quénu (1918) and his colleagues came to the view that everything that favored absorption at the site of injury favored the development of secondary shock. Thus arose the toxemic theory. The bacteriological studies of Wright and Colebrook (1918) on gas gangrene added further definition to the idea of toxemic factors of bacterial origin, whereas Bayliss and Cannon (1919) developed the concept of a toxic substance produced and liberated from the tissues. One possible source of shock-producing toxins could be bacteria present as normal inhabitants of the intestinal tract. Damage to the mucosal lining of the intestinal tract has been noted to follow certain types of injury and it is thus possible that the intestinal wall becomes more permeable to these bacterial toxins. The evidence regarding this theory will now be reviewed.

B. The Role of Endotoxins in the Production of Shock

Fine *et al.* (1960) believe that a toxin—probably a bacterial endotoxin—is liberated into the blood in hemorrhagic shock. In their view liberation of this toxin is the apparent cause of failure to respond to treatment by transfusion. A toxin with similar properties is held by Fine *et al.* (1960) to be present in the blood of an animal in tourniquet shock, and also in nonbacteremic shock complicating an induced peritonitis. The source of this toxin is believed by this group to be such Gram-negative bacteria as may be present in the animals' tissues and the bacteria normally present in the gastrointestinal tract. These endotoxins are believed to be constantly available and become free to inflict injury on the peripheral circulation and to paralyze blood flow resulting in death when the reticuloendothelial system's endotoxin-neutralizing potential is destroyed. Thomas (1956) has explained this action by the capacity of the endotoxins to potentiate the local action of adrenaline or noradrenaline.

By virtue of their capacity to induce adrenergic blockade, dibenamine and reserpine given in advance provide protection against shock. Thus, instead of intermittent hazard from whole bacteria sporadically invading the intestinal mucosa from the alimentary tract, Fine and his colleagues propose a continuous and continuing hazard from microbial endotoxin which, if not dealt with, leads to a fatal accumulation.

More recent studies have, however, failed to provide evidence for the presence of circulating bacterial endotoxin in experimental hemorrhagic shock (McClusky *et al.*, 1960). Nagler and Zweifach (1961) likewise cast doubt on the endotoxic theory and Janoff *et al.* (1961) find evidence which suggests that bacterial endotoxins are not the lethal factor in blood harvested from animals in shock induced by occlusion of the superior mesenteric artery.

Nevertheless, intestinal lesions are commonly associated with experimental and clinical shock induced by various methods in laboratory animals and in man. The lesions in man are said to bear little resemblance to the usual inflammatory lesions found in the small or large intestines. In man, severe surgical shock is occasionally associated with an extensive necrosis of the intestinal mucosa which is a sequel to an acute ischemia, this in turn being due to operative or postoperative shock and not attributable to local bacterial action. The evidence which appears to point to ischemia (Penner and Bernheim, 1939; Bruce, 1954), and the occurrence of gastroduodenal ulcers after various forms of shock (Klemperer *et al.*, 1940; Lillehei *et al.*, 1948) are both well established. This may be a more frequent phenomenon than is presently assumed, as subsequent healing may be rapid in those which survive the serious injury. Werle *et al.* (1942) record hemorrhage into the intestinal lumen of dogs reinfused with blood after a period of prolonged hemorrhagic hypotension, and Freeman *et al.* (1938) describe superficial necrosis of the duodenal mucosa as a postmortem finding in dogs dying of hemorrhagic shock.

The importance of adequate oxygenation in the maintenance of an intact gastrointestinal mucosa is emphasized by Palmer and Buchanan (1953) and Palmer and Sherman (1958). Extensive damage to the superficial gastrointestinal mucosa can occur in conditions other than shock, when there exists either sustained vascular engorgement or ischemia. The gastrointestinal lesions produced by prolonged parasympathetic stimulation in the dog resemble those occurring in shock (Wener *et al.*, 1948, 1949) and the repeated injection of pilocarpine causes gastric ulceration in the fasting rat (Basu Mallik, 1955). Abnormal parasympathetic action is probably the cause of gastrointestinal ulceration following brain injury (Cushing, 1932). Similar mucosal lesions occur in the alimentary tract of dogs and rabbits injected with pitressin (Berg, 1942; Crane, 1954); these workers attributed the damage to the action of gastric juice on cells rendered anoxic by arterial spasm. Ischemia of 15 minutes' duration is sufficient to cause necrosis of the alimentary epithelium (Crane, 1954).

We found that at death the cells of the intestinal epithelium become detached soon after the cessation of respiration, although the intestines still exhibited peristaltic movement (Badawy *et al.*, 1957; Fell, 1961). The rapid onset of this change suggested to us that the phenomenon might also occur in the living animal as a result of anoxia. Previous attempts to do this in the rabbit by inducing transitory vascular changes in the intestines were unsuccessful (Fell, 1961). To test the hypothesis that the cell shedding is due to local anoxic liberation of histamine or 5-hydroxytryptamine, the histamine liberator, compound 48/80, was injected into the normal unanesthetized rat. It caused severe respiratory embarrassment, cyanosis, and vascular congestion of the gastrointestinal tract. Associated with this change, and secondary to it, extensive erosions occurred in the intestinal epithelium, with hemorrhage into the intestinal lumen. Usually the rats recovered quickly, and regeneration of the epithelium was completed at between 2 and 3 hours after the injection, but some rats relapsed into a condition resembling traumatic shock. It was suggested that local anoxia due to vascular stasis is the cause of damage to the epithelium, which undergoes a rapid autolytic change. The damage to intestinal epithelium produced in these experiments is very similar to that which occurs as an early post-mortem change. With regard to the mechanism of action, intraperitoneal injections of histamine liberator caused neither severe shock nor intestinal cell shedding, although considerable displacement of histamine had apparently taken place, as evidenced by the clinical state and the mast-cell disruption. When given intravenously, compound 48/80 caused both severe anaphylactoid shock and intestinal damage. Although this agent liberates both histamine and 5-hydroxytryptamine in the rat, its failure to act intraperitoneally suggests that the damage to intestinal mucosa is due neither to the release of histamine and 5-hydroxytryptamine nor to the subsequent actions of these substances, but may be an independent consequence of the anaphylactoid shock which results from intravenous injection. This view is supported by the time of onset of the intestinal lesions, since the initiation of histamine-liberation shock precedes the earliest intestinal damage by more than 10 minutes. Our experiments do not eliminate local anoxic histamine release as a cause of the intestinal lesions, for the histamine content of the intestine is relatively unaffected by the action of histamine liberators (Feldberg and Talesnik, 1953).

As we have seen, one school of thought considers that the absorption of endotoxins from the intestine may be a cause of irreversibility in shock (Fine *et al.*, 1952, 1959, 1960; Ravin *et al.*, 1960; Schweinburg and Fine, 1960). Consequently, the rapid loss of integrity of the intestinal epithelium in abnormal vascular states may have some bearing on the pathogenesis of irreversible shock. Thus, we have observed typical epithelial shedding in the duodenum of shocked rabbits 2 hours after heavy bleeding from the ear vein (Fell *et al.*, 1961). However, intestinal damage does not necessarily lead to irreversible shock, for many of our surviving rats after treatment with compound 48/80 showed intestinal lesions while in the stage of recovery. From the work of Fine and his colleagues it would appear that the difference between our rats that recovered and those that passed into a state resembling traumatic shock might lie in the degree of damage sustained by the reticuloendothelial system during the initial reaction to compound 48/80, with consequent loss of capacity to deal with endotoxins coming from the intestine.

C. Other Aspects of Shock

1. Effects of Injury on Energy Transformations

Damage to the cell can impair the function of intracellular structures involved in energy transformations. Thus "cloudy swelling," which is a common cellular reaction to various types of cell damage, involves: (*a*) an alteration in the mitochondria followed by partial inhibition of phosphorylation coupled with oxidation; as a consequence, anaerobic glycolysis increases; (*b*) there is alleged to be increased protein synthesis; (*c*) at least in some cases there is also increased imbibition by the cell. According to Fonnesu (1960), mitochondrial alteration is the primary change in cloudy swelling. In these cells the available ~ P is decreased but is still compatible with life. There must be some additional factor which determines whether, in the liver, for example, the process goes on to fatty degeneration.

In the animal as a whole injury at first leads to a reduction of energy expenditure as muscle work, a fall in the temperature of the body surface and internal organs without evidence of increased heat loss, and a decreased respiratory exchange. If the injury is sufficiently severe these changes progress until the animal dies; otherwise they persist for a time and are followed shortly by increased oxidation and heat production and then recovery generally ensues.

One of the difficulties is to determine whether the biochemical changes seen after injury represent a primary metabolic change or are secondary to a decrease in circulating blood volume and blood pressure with consequent compensatory changes in skin and muscle. The work of Whiteley *et al.* (1953) and Stoner and Threlfall (1960) shows that, while the biochemical changes in the skin and kidney are mainly due to changes in blood supply, this is not primarily the case in liver, brain, and uninjured muscle. In shock due to limb ischemia, these latter tissues are apparently adequately oxygenated and in consequence the metabolic changes cannot be attributed to hypoxia. The secretion of the adrenal medulla during, and for a time after, limb ischemia mobilizes glycogen stores and glucose precursor sources, possibly including protein (Popják, 1944). The shortening of the survival time by adrenodemedullation would be accountable by the destruction of this protective mechanism. Survival time is also shortened by raising the environment temperature, largely because of its deleterious effect on the hemodynamic system (Stoner, 1958). But the occurrence of adrenal medullary and cortical hormones in increased amounts in the blood after injury (see Section V) does not mean that these hormones necessarily act in this way.

If the organism is unable to withstand the injury, oxygen transport fails during the latter part of the survival period and the resulting tissue hypoxia leads to the terminal changes in energy (heat) production and carbohydrate metabolism. As shock progresses the animal mobilizes energy sources: liver glycogen is depleted and blood sugar may eventually fall to hypoglycemic levels. Fat mobilization is apparently manifested by ketonemia until hepatic metabolism is impaired, then ketone levels decline. At this stage rising amino acid nitrogen may be an indication of failure of hepatic metabolism (Bloom, 1961). This metabolic sequel is discussed in more detail later.

2. *Impairment of the Nervous System*

Impairment of the nervous system plays an important part in the development of shock. Hypoxia, hormonal factors, toxic substances, and nociceptive afferent impulses all probably participate in bringing about the functional and biochemical changes in the central nervous system during shock (Kovách and Fonyó, 1960). But the oxygen consumption of the brain and its oxygen requirement do not decrease during shock; in fact they rise, in spite of the fact that the circulation to the head is reduced. Brain hexokinase activity increases up to the final stage of shock and the energy-rich P content of the brain is increased, even when severe reflex deficiencies, adynamia, and coma set in (Kovách and Fonyó, 1960).

3. *Circulatory Changes*

It has been shown that in states of shock, experimentally produced by losses of large volumes of fluid from the circulation either as a result of hemorrhage or as a result of swelling at the site of physical injury, death from the circulatory collapse can, under certain conditions, be prevented by administration of whole blood, plasma or appropriate fluids, electrolytes, and plasma proteins. It has been suggested that the release of K^+ into the blood after hemorrhage, muscle trauma, arterial occlusion, or asphyxia may be an important factor in shock (Millican, 1960).

The common finding of low plasma osmolarity for several days after trauma is explained by water retention in excess of electrolytes. Since renal function may be otherwise normal, present evidence suggests persistent antidiuretic hormone secretion (ADH) as the cause of the oliguria or restricted urine output (Wynn, 1960). Even in uncomplicated cases excessive ADH secretion may last for several days and minor complications may further prolong the antidiuretic phase. It would seem that the supraopticohypophyseal system is made more sensitive in some way by trauma so that a variety of stimuli provoke ADH secretion and the primary function of osmoregulation by the system is seriously impeded or lost. A secondary consequence of water retention is the inhibitory effect of the increased body fluid volume upon aldosterone production (see Section V). This leads to renal salt loss and aggravates the hypotonicity of the body fluids (Wynn, 1960).

IV. The General Disturbance of Protein Metabolism following Physical Injury

A. *The General Picture*

Apart from the actual loss of blood, exudate, or tissue substance due to the injury, the increased heat production and the increased urinary excretion of nitrogen, sulfur, phosphorus, K^+, and creatine constitute the most obvious energy and material loss to the body following physical injury (Cuthbertson, 1942, 1960a). My earliest studies showed that, in cases of bone or joint injury, and most probably in other forms of serious injury, the curves of the urinary excretion of nitrogen and sulfur run parallel and that of phosphorus follows closely, as shown in Figs. 1 and 2. Partition of the N- and S-containing catabolites demonstrated that the increase in nitrogen was mainly due to an increase in the urea excreted,

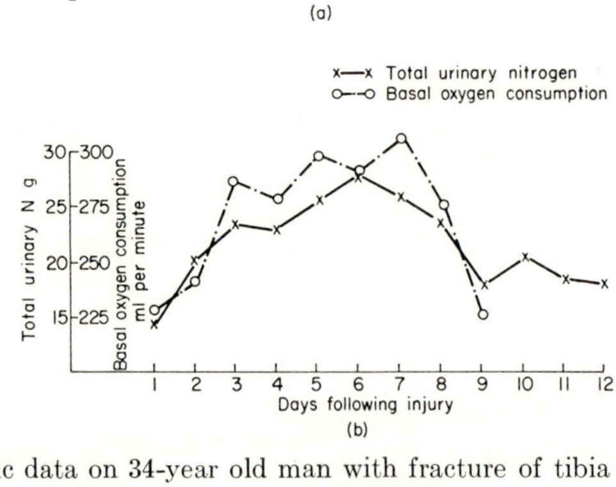

FIG. 1. Metabolic data on 34-year old man with fracture of tibia due to direct violence. No anesthetic required. Daily nitrogen intake 15.36 gm. [Redrawn from Fig. 1 of Cuthbertson (1932), by permission of the editors.]

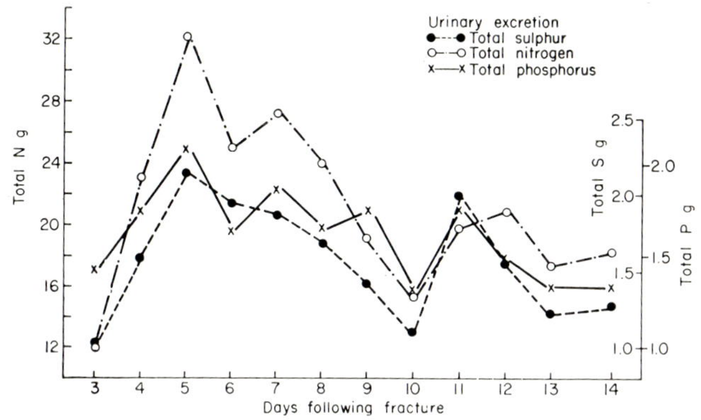

FIG. 2. Daily urinary excretion of total N, S, and P of youth following fracture of both bones of one leg. Daily nitrogen intake 11.7 gm. [Drawn from metabolic data in Table VII of Cuthbertson (1930).]

and that the increase in urinary sulfur was due to a similar proportional increment in the excretion of inorganic sulfate.

This reaction to injury is a much more active process than the catabolism due to disuse (Cuthbertson, 1929), a process which Schønheyder *et al.* (1954) ascribe to diminished synthesis of tissue protein along with unchanged rate of breakdown. Following severe accidental injury the excessive urinary loss in man was observed to reach a maximum about the third to eighth day and the maximum daily loss of nitrogen might exceed 23 gm (Cuthbertson, 1930, 1932). The values were observed to fall thereafter, but there was evidence in man that even 6 weeks after a fracture there might be a definite negative N balance (Cuthbertson, 1932, 1942), probably mainly due to disuse at this stage. While there were certain grounds for considering that the extent of the injury and the nitrogen loss were closely related, there was, on occasions, a lack of correlation between them. Muscle was suspected to be the source of this nitrogen and sulfur loss. Definite traces of creatine and occasionally traces of heat-coagulable protein were excreted, notably during the time of the maximum excretion of the nitrogenous catabolites. While the excretion of ammonia fluctuated, that of amino acids and uric acid generally rose (Cuthbertson, 1931). The fecal excretion of nitrogen was not appreciably affected by injury to skin, bone, or muscle.

Many of the nitrogenous and other constituents of blood exhibit changes as the result of injury. The observations of Gillette *et al.* (1958b) on the effect on goats of massive wounding indicate that both in untreated and penicillin-treated animals increases in blood nonprotein nitrogen, urea, creatine, creatinine, and plasma Mg levels occur. Unwounded animals that had received an intravenous lethal dose of the endotoxin of *Escherichia coli* showed similar, but lesser, effects. The plasma Na^+, K^+, and the amino-N remained constant in all groups.

It is now generally recognized that a negative nitrogen balance results from many other forms of injury, including surgical operations, but the details may differ because of treatment and the degree of involvement of the different tissues. Particularly is this so after burns (Browne, 1943; Moore *et al.*, 1950). Thus, Troell and Wretlind (1961) have pointed out that patients with extensive burns usually show weight loss along with negative nitrogen balance. The general deterioration produced by pain, infection, pyrexia, anemia, and wound seepage results in anorexia that further complicates the nutritional deficiency. Other special patterns have been observed in burned patients. Some 48 to 72 hours after a severe burn, Walker (1946) and Taylor *et al.* (1943) found an increase in the plasma undetermined nonprotein nitrogen, that is, the difference between the total nonprotein nitrogen and the sum of urea, uric acid, creatine, and α-amino nitrogen which seldom increased very much. Nardi (1954) found an aminoaciduria accompanying the increased urinary nitrogen loss following burns and other forms of trauma. It consisted of an increase in both the nonessential amino acids normally found in the urine, and threonine, valine, leucine, isoleucine, lysine, and methionine, essential amino acids not usually detected. Patients who were not burned severely enough to endanger their lives and patients undergoing minor surgical procedures showed only minimal increments in amino acid excretion with a preponderance of nonessential amino acids. Those dying of their burns or undergoing major surgery had a massive excretion of both essential and nonessential amino acids. Amino-aciduria has been observed as a prominent feature of other forms of injury (Chytil, 1956). Chytil suggests that the urinary loss of amino acids in rats arising from an injection of turpentine is the main factor causing the negative nitrogen balance in this form of injury, possibly from liver damage.

During the early stages of the generally depressed vitality following receipt of a serious wound there may be relative or absolute anuria (Cuthbertson, 1932). With a return to more normal conditions, after an injury such as a fracture of the femur, there is still a diminished volume of urine lasting for another day or two; thereafter the output slowly rises, often irregularly, to a volume practically double the amount passed during the early days. The maximum volume is not generally attained until about 2 days after the maximum excretion of nitrogen has appeared. There is thus a conservation of water during this period. The influence of injury on hormones regulating water output will be considered later.

In addition to these changes in urinary excretion and blood composition, early observations by the present author showed that there is generally, though not necessarily, a rise in body temperature and pulse rate resulting from severe injury (Cuthbertson, 1932). The general trend of this curve was parallel to that of the urinary excretion of nitrogen but sometimes preceded it in time. The basal consumption of O_2 in these experiments was found to rise in parallel with the increased excretion of nitrogen (Fig. 1). Florey (1954) has pointed out that whenever tissue destruction is severe, a rise in body temperature is usual and that this may occur in traumatic tissue damage or after an internal hemorrhage.

B. The Nature of the Disturbance in Protein Metabolism

In any consideration of the literature on this subject it is very necessary to distinguish studies of the early changes in metabolism from studies made during the later stages.

1. Studies of Amino Acid Metabolism after Trauma

Changes in blood amino acid levels have been described shortly after injury. Engel *et al.* (1943) showed that hemorrhagic shock in the rat is characterized by rising blood levels of amino nitrogen and considered that these were due partly to an increased breakdown of protein in the peripheral tissues (Russell *et al.*, 1944a) and partly to a decreased ability of the liver to dispose of amino acids, because of the accompanying anoxia and decreased metabolic activity of this tissue (Engel *et al.*, 1944; Russell *et al.*, 1944b). It was thus implied, but not proved, by Engel *et al.* (1943) that a depression in the rate of deamination of amino acids by the liver was an important feature of the metabolic picture accompanying shock. The failure of the liver to remove amino acids is confirmed by the observation of decreased urea formation in dogs during hemorrhagic shock (Van Slyke *et al.*, 1944). Indeed, the liver of the severely shocked animal may make a net contribution of amino acids to the plasma rather than effect their removal (Russell and Long, 1946). It is thus not surprising that infused solutions of amino acids are poorly tolerated by the injured dog (Burt, 1954). The liver defect is presumably one of deamination, since the low blood urea concentrations are not accompanied by any accumulation of ammonia in the liver of the shocked animal (Wilhelmi *et al.*, 1945).

Several studies on the immediate metabolic response of man to injury present a different aspect. In an examination of 177 battle casualties, Green *et al.* (1949) noted that the plasma amino nitrogen was not raised and indeed tended to bear an inverse relationship to the elevated blood sugar level. Hyperglycemia was a prominent feature of such injuries some 10 hours after wounding and its degree varied with the severity of the injury. Schreier and Karch (1954–1955) similarly observed in man that the serum amino acids were usually lower 3 to 4 hours after operation than before, although in 4 out of 5 patients the isoleucine and leucine values rose. There was an increased urinary output of all of eleven amino acids examined. In this connection, Eades *et al.* (1953) have reported the excretion of peptides in the urine of injured subjects. A further apparently discordant note is the observation in the battle casualties in the Italian campaign of a rise in blood urea in the first 48 hours after injury in those who had initially lost more than 40 % of their blood, irrespective of the extent of the injury (Grant and Reeve, 1951). The picture of a failure at the level of deamination in the injured experimental animal may thus not be applicable to injury in man.

2. Changes in Plasma Protein Metabolism

As already noted, injury causes a considerable disturbance in plasma protein concentrations apart from the immediate effects due to blood loss. In the studies already described, in which goats received massive wounds, Gillette *et al.* (1958a) observed that the phase of shock was accompanied by a relative increase of α_2-globulin concentration in the plasma. Animals receiving antibiotic treatment at the time of injury survived longer and from 24 hours onward showed marked changes in β-globulin concentration. The relative concentrations of the plasma proteins were not disturbed by shock induced by the intravenous administration of lethal doses of bacterial endotoxin.

The changes described in man generally relate to the postshock period. Cuthbertson and Tompsett (1935) noted that, in fracture and in operation cases, the general trend was a slight fall in plasma albumin combined with a marked rise in the globulin fraction as well as a frequently observed increment in fibrinogen level. In cases of burning, Bull (1958) has found essentially similar changes. A few days after a severe burn, the total plasma protein level is commonly low, but the change in the relative proportions of individual protein fractions is much more pronounced. The predominant early change is the continuing fall in the albumin fraction accompanied by a progressive reduction in total circulating albumin. An early change in the globulins is that the α_2-fraction commonly rises during the first day or two and the rise is accompanied by a very great increase in the proportion and the total quantity of γ-globulin. On the other hand, Yuile *et al.* (1953) found in their investigation of the turnover of plasma albumin and fibrinogen and of the protein of red blood cells in dogs with experimentally induced sterile inflammatory abscesses, a great increase in the rate of synthesis of fibrinogen and a less striking increase in the turnover rate of plasma albumin. They also observed a marked decrease in the rate of biosynthesis of the protein of the red cells, presumably mainly hemoglobin, and suggested that this latter finding may be connected with the anemia that often develops during inflammatory conditions.

In collaboration with Dr. A. S. McFarlane and his colleagues of the National Institute for Medical Research we found that the albumin, globulin and fibrinogen turnover rates immediately following fracture or cortisone implantation in rats were not sensibly affected as measured by the use of injected C^{14}-amino acids derived from a hydrolyzate of labeled algal protein or by use of I^{131}-proteins (McFarlane *et al.*, 1960). On the other hand, with the former labeled agent specific activities of free amino acid carbon were elevated by approximately 50 % after fracture or cortisone. The concentration of amino acids was correspondingly reduced in the plasma samples so that the specific activity change might be attributed mainly to a change in the amino acid pool size. It was then noted that random fluctuations, which are normally evident in free amino acid specific activities, were greatly increased after fracture or cortisone, as if the mechanism which normally maintains the level of amino acids was disturbed. While Fleck and Munro (1963) have confirmed the absence of an effect of injury in the rat on albumin catabolism, Davis *et al.* (1959) have found in man that I^{131}-albumin turnover was increased after burning injury. Sterling and his colleagues (1955) studied 10 patients following operation and concluded that there was a reduction in the half-life of albumin. However, they listed a discrepancy between the observed plasma radioactivity and the proportion of radioactivity retained, and they suggest that the rate of complete degradation of albumin to free amino acids may not in fact be increased after injury.

C. The Origin of the Nitrogen Lost by Man and the Rat

In the author's experience (Cuthbertson, 1932) the maximum daily loss of nitrogen of patients may exceed 23 gm per day following a fracture of the leg and in 10 days more nitrogen than is the equivalent of that present in the liver may be lost. Flear and Clarke (1955) report similar findings in untransfused burn cases. Grant and Reeve (1951) give a few reports on casualties with very large injuries and very great blood loss. In 7 days in one of the cases studied there was an accumulated negative balance of 173 gm nitrogen (1080 gm protein) and in another case 199 gm nitrogen (1250 gm protein) in 8 days. From observations of the fractured limbs of man and the rat and from the scale of the loss, I considered the loss to be more general than local and to represent a general catabolic effect (Cuthbertson, 1932, 1942). It has been suggested that blood lost into the area of trauma may account for it (Dudley, 1959), but this does not appear fully to fit the facts although it may well be a factor (Campbell and Cuthbertson, 1963).

Born (1962) has pointed out that, although there is a reduction in the total quantity of lymphoid tissue and in the number of lymphocytes following injury, this cannot account for the marked loss of nitrogen following an injury such as a fracture. The evidence points to muscle but of this we have no real proof.

In the rat the author and his colleagues noted that fracture of the femur by open operation under ether anesthesia caused a definite urinary loss of nitrogen, sulfur, phosphorus, K^+, and creatine, and there was but little change in Na^+ and creatinine—if anything a depression of the former (Cuthbertson *et al.*, 1939). This was more extensive than the disturbance caused by simple incision of skin and muscle without fracture. The wastage of body substance observed from the increased excretion of these catabolites could not be accounted for by the loss of muscle substance from the site of injury or from the injured limb, and the muscles adjacent to this fracture lost relatively more K^+ than nitrogen. Thus there appeared to be, in addition, a generalized increase in catabolism. It was found that additional carbohydrate exercised a definite sparing effect on the general loss of tissue substances, but did not appear to prevent appreciably the local wastage.

The experiments of Levenson *et al.* (1960) are of great importance. They administered N^{15}-glycine to rats daily for 4 days just prior to burning and killed animals, 1, 3½, and 6½ days later. They also fed the labeled glycine in one experiment at the height of the negative N balance. In the latter experiment the rate of incorporation of N^{15} in both burned animals and unburned controls was least in the skin and carcass protein but greatest in the liver, gastrointestinal tract, and the plasma proteins. In each case the N^{15} incorporated in the tissue proteins of the burned animals was equal to, or greater than, that in the corresponding tissue proteins of the control rats. If it is assumed that incorporation of N^{15} signifies synthesis, then anabolism was occurring as actively in the burned rats and so the increased nitrogen excretion cannot be explained as antianabolism following injury, at least during the postinjury stage examined. In the experiment in which the N^{15}-glycine was given for 4 days prior to trauma, the decline of N^{15} concentration in the tissues was faster in the injured group than in the control group. The protein content of the very active organs, e.g., the liver, changed but little despite their markedly increased turnover rates. The protein content of the carcass, on the other hand, fell despite its much slower inherent turnover rate and indeed accounted mathematically for most of the extra urinary nitrogen loss. This suggests that the integrity of certain organs is maintained at the expense of metabolically less active tissues, e.g., skeletal muscle. Absence of change in the protein content of liver has also been noted after burning (Calloway *et al.*, 1955) or fracture (Fleck and Munro, 1963) inflicted on rats.

Some light is also thrown on the origin of the N lost after injury by nutritional studies. When rats were given a protein-free diet for some time prior to fracture, they did not show the characteristic sharp increase in the urinary excretion of nitrogen. Only a slight and transient rise in nitrogen output occurred (Munro and Cuthbertson, 1943). A larger loss of nitrogen was observed by my colleagues Munro and Chalmers (1945) with animals on a 10 % protein diet; on a 25 % diet the loss was greater still. This greater nitrogen loss at the 25 % level seemed to depend on the rat having gained weight during the prefracture period. These observers also noted that if the femurs of rats on the 10 % diet were broken and the animals transferred to the 25 % diet, the nitrogen loss was similar to that obtained when the 10 % diet was used throughout, and was much less than the nitrogen loss observed when the 25 % diet had been eaten throughout. It was concluded that the amount of nitrogen lost after injury is modified by the level of protein intake during the prefracture period. The effect of feeding a high protein diet during the prefracture period is lost if the animal is transferred to a nonprotein diet after fracture, and it was suggested by these investigators that their findings were due to changes in the amount of labile (deposit) protein in the body at different levels of protein intake. The contribution of labile protein stores to the nitrogen lost after injury is discussed in more detail in Chapter 10 of this book, where it is suggested that operative injury may release labile protein stores whereas accidental injury, by its severer impact on the nervous system and prolongation of effect, as in fracture and burning, may result in a more extensive change in protein metabolism which is not confined to this limited compartment of body protein. In the rat, Fleck and Munro (1963) found that the protein content of the liver and of the plasma was not reduced after injury, even in spite of extensive loss of body nitrogen. They concluded that the labile body protein dissipated after injury must be located at other sites, possibly in muscle.

This effect of previous level of protein intake on the metabolic response of the rat to injury has been confirmed by Calloway *et al.* (1955); in man, it has been demonstrated that a similar lack of responsiveness to injury occurs in chronically undernourished subjects (Browne, 1944). Apart from suggesting the bodily origin of the nitrogen lost after trauma as in part labile body protein, these findings are of some importance in elucidating the mechanism by which injury influences metabolism; evidence to be considered later shows that adrenocortical hormones may play a part in the metabolic response to injury, and it is of interest that the increased nitrogen output caused by adrenocortical hormone administration is uninfluenced by the level of protein in the diet (Silber and Porter, 1953; Goodlad and Munro, 1959), whereas the data presented above show that the reaction to injury is virtually abolished on a protein-free diet. It is thus an oversimplification to regard the metabolic reaction to injury as an expression of adrenocortical overactivity.

D. The Increased Heat Production Consequent to Injury

The basal consumption of oxygen in my experiments made on fracture cases in 1932 paralleled the increased excretion of nitrogen (Cuthbertson, 1932). These early experiments also demonstrated that there is generally, although not necessarily, a rise in body temperature resulting from moderate to severe injury (Cuthbertson, 1932, 1936). The general trend in temperature rise in fracture cases was parallel to that of the urinary excretion of nitrogen but preceded it in time (Fig. 1). With few exceptions did the rise of temperature exceed 2°C. In certain cases with pronounced alterations in the urinary excretion of catabolites, little or no change in either pulse rate or temperature was detected. These observations are in agreement with those of Malcolm (1893), who noted that there may be evidence of increased nitrogen metabolism after a major operation yet the temperature remains at a low level.

Recently, using a form of the Benzinger-Kitzinger gradient-layer calorimeter constructed by my colleague Pullar, we have examined the rise in heat output of the rat when fed at an adequate or at a low level of protein intake and then subjected to fracture (Cairnie *et al.*, 1957). Since a low protein intake is known to suppress the metabolic response to injury in the rat, the relationship of extra heat output to the change in protein metabolism can be evaluated. The average total metabolism of the rats on a high protein diet remained steady at 145 kcal/kg$^{0.75}$/24 hr (66.1 kcal/rat/24 hr) prior to injury and rose to a maximum of 152 kcal/kg$^{0.75}$/24 hr (69.0 kcal/rat/24 hr) on the third day after injury, which, according to our previous investigations on rats, is the time of the maximum excretion of nitrogen. This rise was primarily due to a rise in the minimum metabolic rate, which was increased by 7% on the third day after injury.

The increase in urea output after trauma shows that more protein is being deaminated. The heat obtained from the nonnitrogenous residue can account fully for the observed increment in energy metabolism. If the nonnitrogenous residues had been stored, less heat would have been dissipated. If, on the other hand, a general acceleration of body metabolism had occurred, as on exposure to cold, fat would have been the preferred source of the extra energy. Consequently, this study established that protein metabolism is uniquely increased in response to trauma. In confirmation of this, rats fed on a protein-free diet before fracture did not show the increment in heat output noted in animals fed at a normal level of protein intake before injury. Since the protein-depleted rat fails to show an increased N output after injury, this provides further evidence that the increased heat output after injury is due to augmented combustion of protein only, and that carbohydrate and fat do not participate. Indeed, these experiments on protein-depleted animals demonstrate that the heat increment after injury normally arises only where there is oxidation of protein, and that in the absence of this increased oxidation injury *per se* does not lead to increased heat output.

Recent reexamination (Cairnie *et al.*, 1957) of my collected early data on man revealed that, during the first week after injury, the energy liberated by the oxidation of protein, when calculated as equivalent to the nitrogen excreted, can account for the rise in basal oxygen consumption observed and provides no evidence of a corresponding increment in oxidation of carbohydrate and fat. More recently, Cope *et al.* (1953) have made multiple but brief measurements of "basal" oxygen consumption by burned patients from the period beginning soon after trauma and extending through convalescence. The majority of severely burned patients increased their metabolic rates to plus 30–80% early after injury. The oxygen consumption rate gradually declined but continued to be elevated for as long as 2 months in many patients. Although Cope and his colleagues considered that fever probably played a role, the temperature rises were irregular and did not correlate with the metabolic rate. Following major elective operations there was usually a rise of 10 to 15% in basal oxygen consumption lasting from 3 to 5 days with a return to preoperative levels by the end of 1 week. Kinney and Moore (reported by Zimmermann, 1962) have shown that elective surgery of the abdomen is followed by a predictable increase in energy expenditure. In a 45% burn studied by Rabelo *et al.* (see Zimmermann, 1962), the heat production averaged 160% of the calculated normal and increased during the first 10 postburn days concomitant with the elevation of body temperature. With an overwhelming injury, such as an 85% burn, a progressive metabolic deterioration may occur. Comparison with pyrogen administration indicated that the metabolic response to burn injury is much greater than that caused by fever alone and, according to Zimmermann, represents a greater increase than is associated with any other form of injury.

It may be that this elevation of metabolism following severe injury—this "traumatic fever"—is a reaction of the body to enhance the healing process by accelerating the reactive processes which involve both inflammation and repair, and also to provide solutes for the elimination of the previously retained water. Caldwell *et al.* (1959) conclude that the rise of energy expenditure after burning is associated with evaporation of water from the wound and is independent of the thyroid, at least in the rat.

E. The Metabolic Reaction to Injury in Other Species

1. Studies on Poikilotherms

The discussion has so far been restricted to injury in homotherms, but in a most interesting series of experiments on the aquatic poikilothermic crustacean, the shore crab, Needham (1955) found that amputation of appendages in fasting specimens caused an immediate increase in nitrogen output which then declined exponentially during the next 12 days or so. Superimposed on this curve were cyclic fluctuations in output for a period of about 6 days. The excess output in these fasting creatures over the first 10 days was proportional to the normal output before operation. Particularly in the second week after injury, the feeding of carbohydrate reduced output and the giving of protein increased the short-term oscillation in output. Excess nitrogen output was related to the amount of damaged tissue remaining but not to the amount of tissue removed; nevertheless, it did not come entirely from the demolished tissues. Removal of the sinus gland region of the eye stalks caused a tenfold increase in nitrogen output over that due to limb amputation. After excision of portions of the body wall the pattern of output differed from that following amputation of the appendage and resembled that following skin and bone trauma in man. These fluctuations, or cycles of nitrogen output, are of interest since I also recognized two cycles in the metabolic response to injury of man (e.g., Figs. 1 and 2) and of the rat (Cuthbertson *et al.*, 1939).

Needham (1958) has also found that, after amputation of the tail, earthworms show increased excretion of nitrogen. Both ammonia plus amino nitrogen and urea nitrogen are affected and the increased output is proportional to the degree of injury and to the amount of tissue amputated and is higher if healing is defective. The response is proportional to the normal output of a species under the appropriate dietary regimen. Feeding reduces both normal output and the excess nitrogen output due to injury.

2. The Effect of Surgery on the Nitrogen Metabolism of the Ruminant

R. M. Campbell and Cuthbertson (see Cuthbertson, 1960a) have briefly reported that the ruminant, despite its markedly different digestive system, behaves like the rat and man in exhibiting a rise in urinary nitrogen output following operation under "Nembutal" anesthesia.

V. Physical Injury and Endocrine Function

A. Changes in Secretory Activity of Adrenal Medulla and Cortex

The immediate effect of injury or operation on the secretory activities of the adrenal medulla and the adrenal cortex has been extensively explored.

The sympathetic nervous system and adrenal medulla are not vital to survival in normal conditions, but can greatly facilitate adaptation of the organism to many exceptional situations. The sympathectomized animal cannot maintain its normal blood pressure, blood sugar level, or body temperature under stress conditions; survival capacity is impaired and the range of adaptation is reduced (W. B. Cannon, 1939). Secretion of additional noradrenaline and adrenaline plays an important part in the immediate response to traumatic injury, when the organism cannot maintain normal homeostasis. In middle-aged patients undergoing surgery, Pekkarinen (1960) found that, on the first and second postoperative days, the circulating levels of adrenaline and noradrenaline and of total 17-hydroxycorticosteroids (17-OHCS) were of the order which would promote adaptation to the stress of operation. However, the amounts of noradrenaline released into the circulation were not as large as those used in therapeutic shock treatment, nor was the amount of adrenaline excessively large, and in most cases the amounts excreted in the urine fell within normal limits by 3 days after operation, though an elevated level could occasionally persist for some 7 days. It has been suggested that, if the sympathetic nervous system and adrenal medulla became exhausted as the result of 3 or 4 days of strong stimulation, this might be a decisive factor in the patient's recovery (Uotila and Pekkarinen, 1951).

The pituitary-adrenocortical system is now considered to be even more important for homeostasis than the sympathetic nervous system and adrenal medulla. After major operations, the mean total 17-OHCS secretion rises, in some cases tenfold. Direct evidence of this occurrence during operation has been observed by cannulation of the adrenal vein (Hardy *et al.*, 1957). In severely burned cases, the increased adrenocortical activity can persist for as long as 16 or 17 days (Hume *et al.*, 1956), but following operations without complications the rise in blood 17-OHCS level is often over in 24 hours (Cooper and Nelson, 1962). Moore (1957) found the rise in plasma free 17-OHCS to be more extensive than the response to maximal doses of ACTH (adrenocorticotropic hormone) given over a similar period of time. However, Cooper and Nelson (1962) noted an increment in plasma ACTH coincident with the elevation in plasma corticosteroids after operation, which implies that part at least of the adrenocortical excitation may arise from stimulation by ACTH. Operations performed on dogs or human subjects under conditions of hypothermia (artificial cooling of the body) gave rise to less increase in the plasma level of 17-OHCS than did operations at normal body temperature (Hume *et al.*, 1956).

The secretion of adrenaline, noradrenaline, and corticosteroids involves the controlling influence of the central nervous system over the adrenal medulla, the sympathetic nervous system, and the adrenal cortex (von Euler, 1955; Fortier, 1956; Long, 1956; Hume, 1953; Hardy, 1955; Woodbury, 1958). The secretion of ACTH is partly regulated by neurotropic stimuli from the hypothalamus to the anterior lobe of the pituitary and partly by the systemic humoral changes in metabolism due to operation (Sayers, 1950; Fortier, 1956; Harris, 1956). Adrenaline can cause liberation of ACTH through this hypothalamic control. Stimulation of certain areas of the hypothalamus causes the selective secretion of adrenaline from the adrenal medulla or noradrenaline from the sympathetic nervous system (Long, 1956; von Euler, 1956, 1958). The normal sequence of endocrine response to injury can thus be set out as follows (Born, 1962):

Adrenaline → hypothalamus → anterior pituitary gland →
ACTH → adrenocortical secretion

The work of Llaurado and Woodruff (1957) indicates that the postoperative rise in serum potassium is most probably due to a circulating tropic hormone, elaborated by the diencephalon in response to changes in intracranial blood pressure. This appears to be the stimulus to postoperative aldosteronism, which is now regarded by these workers and by Moore (1957) as part of the normal pattern of response of the adrenal cortex to injury.

These endocrine activities are under feedback control. There is a reciprocal arrangement between the pituitary and the adrenal cortex, whereby, when an adequate or excessive amount of hydrocortisone is circulating in the human body, the production of ACTH is inhibited. However, it will be noted that Cooper and Nelson (1962) observed an increase in plasma ACTH level coincident with that in corticosteroid level after surgery, indicating that trauma may alter feedback control over ACTH secretion. The secretion of aldosterone is controlled mainly by the fluid and electrolyte state of the body and only to a slight extent by ACTH.

To what extent can these changes in the secretory activity of the adrenal gland account for the metabolic changes after injury? It is well known that body water, electrolytes, mesenchymal tissue, lymphoid tissue, the vascular system, and blood pressure are partly regulated by adrenocortical hormones, and these hormones have a profound effect on the metabolism of carbohydrate, fat, and protein (see Chapters 9 and 10). Administration of cortisone to normal animals over short periods results in increased excretion of nitrogen due to loss of carcass protein, concurrent with which there is accumulation of protein in the liver, gastrointestinal tract, and urogenital organs (Silber and Porter, 1953). To what extent can the increased excretion of N and other metabolic sequelae of injury be attributed to increased activity of the adrenal cortex? Several pieces of evidence demonstrate that the catabolic response to injury is not merely a specific example of adrenocortical overactivity.

1. The rise in plasma corticosteroids after uncomplicated surgery is often limited to 24 hours (Cooper and Nelson, 1962), whereas the increased N output attains its maximum several days later.

2. As discussed in Section IV,C, the response of protein metabolism to injury is almost completely obliterated by prior protein depletion, whereas the N loss provoked by cortisone remains unchanged in protein-depleted animals (Silber and Porter, 1953; Goodlad and Munro, 1959).

3. Implantation of cortisone subcutaneously into rats leads to an increased output of N in the urine. The effect of implanting 25-mg tablets of the hormone is similar in magnitude and shape of response curve to the metabolic response to a fracture of the rat's femur (Campbell *et al.*, 1954). Response curves of greater magnitude were observed after fracture combined with implantation of 25 mg of cortisone, as also from the implantation of 50-mg tablets of the steroid. However, most of the response to fracture or to the implantation of 25 mg of cortisone in this series of experiments appeared to result from the incision through the skin and the subsequent manipulation. The additional response to a third 25-mg tablet was very slight.

4. In the same experiments (Campbell *et al.*, 1954), it was shown that urinary N output does not rise after trauma to salt-maintained adrenalectomized rats, which would favor the view that the metabolic response to injury depends on intact adrenal glands. However, an increase in N output does occur in adrenalectomized rats maintained on low levels of cortisone not themselves sufficient to provoke such a response (Campbell *et al.*, 1954). This observation confirms earlier findings (Noble and Toby, 1946–1948; Ingle *et al.*, 1947, 1951). The response of the cortisone-maintained animal to trauma has been explained by Ingle (1951) and Selye (1954) in terms of the theory (Selye and Collip, 1936) that adrenocortical hormones play a "permissive" role in allowing some metabolic reactions to take place.

5. Repeated and increasing traumatization of rats in the Noble-Collip drum leads to adaptation to this type of injury, so that the adapted animal can survive several times the amount of injury which is fatal to the un-

adapted animal (Noble, 1942–1943). The adapted rat differs in a number of its metabolic responses from control animals (see Hrůza and Chytil, 1959, for review). The adapted animal shows changes indicative of increased catabolic processes; the raised urea levels in the blood may well represent increased glucocorticoid secretion. Hrůza (1962) argues that glucocorticoids are necessary for the establishment of this adaptation but not for its maintenance, since adrenalectomy of the adapted animal does not reduce the blood urea level.

It is thus apparent from this discussion that the mode of action of trauma on protein metabolism remains to be answered; it is clear that the adrenal cortex is involved, and the fact that it secretes more hormones for a short period after injury must have some metabolic consequences, but it is patently not the sole mediator of the metabolic response to injury. In this connection, it may be noted that the protein catabolic response to fracture is less in female rats than in male rats (Campbell and Cuthbertson, 1963). Female animals survive hemorrhage better than males (Goldstein and Harwood, 1956) and it has been said that women endure operation better than men. It is therefore of some interest that the excretion of adrenaline, noradrenaline, and of adrenocortical hormones is twice as high in men as in women.

B. The Influence of Other Hormones on the Metabolic Response to Injury

Suppression of the flow phase of excessive protein breakdown does not of necessity promote regeneration, as we observed in experiments in which N retention was induced with administration of anterior pituitary growth hormone (Cuthbertson *et al.*, 1941). Thyroid extract, which has a catabolic effect on protein metabolism, stimulated wound healing (Barclay *et al.*, 1944). Paradoxical responses of various tissues to these hormones are discussed at length in Chapter 10.

Pearce *et al.* (1960) observed that rats on a normal diet responded to administration of testosterone with an acceleration of healing of skin incisions which was not sustained, whereas an early but temporary impairment in rate of healing occurred when cortisone was injected. Wound healing was seriously impeded by feeding a low-protein diet and this was further inhibited by both hormones. During recovery from protein depletion, wound-healing remained impaired but was responsive now to testosterone. These investigators have suggested that testosterone increases the early formation of mucopolysaccharides and that these were present in minimal amounts in protein-depleted rats and in those treated with cortisone. An alternative explanation of the action of hormones on protein metabolism in different tissues and at different levels of protein intake is offered in Chapter 10 where it is demonstrated that testosterone administered at low levels of protein intake can impair deposition in some tissues while at the same time promoting protein deposition in muscle.

The action of synthetic anabolic agents has also been explored. Gilder *et al.* (1961) report that 17-ethyl-19-nortestosterone reduced losses of lean body tissue and fat in patients during their recovery from gastrointestinal operations. However, 19-nortestosterone had little effect on the N balance of male patients undergoing lung resections for tuberculosis (Webb *et al.*, 1960).

VI. Nutrition of the Injured

A. *General Aspects*

In approaching the nutritional problem presented by the catabolic response to injury, it is first necessary to assess the magnitude of the loss of body protein. The amounts of tissue (including blood) protein lost after various forms of trauma are enumerated in Table II, which has been compounded from the published data of Grant and Reeve (1951) and of

TABLE II

ASSESSMENT OF PROTEIN LOSS (GM) DURING FIRST 10 DAYS FOLLOWING SEVERE INJURY OR OPERATION TO A 70-KG MAN; ALSO FOR COMPARISON THE PROTEIN LOSS IN TYPHOID FEVER UNTREATED BY SPECIFIC THERAPY

Source of loss	Simple fracture of femur or both bones of leg	Muscle wound equivalent to 4 hands in volume	35% burn involving in considerable part full thickness of skin	Gastrectomy	Typhoid fever untreated by specific therapy
Loss of tissue	—	500–750	500	5–60	—
Hemorrhage or exudate (gm)	Unknown (as internal may be as much as 200 gm)	150–400	150–400	20–180	—
Catabolic phase (gm)	580 (may rise to 860)	650	600	525–650	675 (approx.)
Approximate estimate of disuse atrophy (gm)	100	100	150	100	50
Total protein lost (gm)	680 (may rise to 960 or even 1360)	1650–1900	1400–1650	650–990	725

the present author, and from discussion with Dr. J. P. Bull and Professor A. Wilkinson. For comparison, the loss of body protein caused by typhoid fever (Müller, 1884) has been included in the table. The observations of Clarke and his colleagues (Clarke, 1952; Clarke *et al.*, 1955) at the Birmingham Accident Hospital indicate that the magnitude of the blood loss associated with even closed major fractures and with multiple injuries is frequently underestimated, and can be masked by reutilization of the blood protein. Whereas Levenson *et al.* (1949) had found that transfusion of blood to normal individuals is followed by excretion of the nitrogen released by breakdown of the transfused erythrocytes, Flear and Clarke (1955) noted in cases of trauma that the nitrogen released by breakdown of transfused red cells is not excreted but is made available to the rest of the body. Consequently, the nitrogen balance of injured subjects receiving a transfusion, though still negative, is more favorable than that of untransfused patients with injuries. The pattern of nitrogen excretion further suggested that the catabolic response to injury was postponed by transfusion.

From Table II it is apparent that the catabolic response of the well-nourished subject after injury represents a considerable loss of body nitrogen. While the patient's metabolism is still in the catabolic phase, and the extent of the nitrogen loss is largely a measure of his nutritive state at the time of injury, there would appear to be no special need for concern at the loss of N over the first few days, provided that steps are immediately taken to replace rapidly and adequately the blood or plasma lost as a result of the accident or operation, and that as soon as possible the patient be encouraged to eat as much of a good diet as possible (Cuthbertson, 1957). In the U.S.S.R. (Biely, 1962), vitamin B_{12} is being used with effect in the treatment of severe trauma (see later).

B. The Special Problem of the Burned Patient

According to Artz and Reiss (1957) balance studies *per se* in burned cases yield no information about the current state of the dynamic processes of protein metabolism, and Bull (1958) has also taken up this view, which is perfectly logical in the absence of other information showing the extent to which anabolism and catabolism are respectively being influenced, although the trend is often quite obvious. Artz and Reiss point out that increased urinary excretion of nonprotein nitrogenous catabolites is characteristic of the severely burned, though this may not be evident in the first few postburn days because of poor renal function. The high excretion of urinary nitrogen is nearly always evident in 3 to 5 days and tends toward a maximum between the first and second postburn weeks. Thereafter, according to Artz and Reiss, it decreases progressively and is often low after a month, sometimes very low. According to these authors, there is a rough correlation between the severity of the burn and the amount of nitrogen excreted. Preburn body weight is positively correlated with nitrogen excretion and, like other injuries, the increased nitrogen excretion does not usually occur in previously malnourished patients.

Artz and Reiss state that the nitrogen loss in exudate tends to be high in the early postburn period but declines rapidly as the wound heals. In third-degree burns with extensive granulating wounds, the exudate nitrogen is well correlated with the amount of suppuration. The loss of 5 to 7 gm of nitrogen into the exudate per day is not uncommon in such burns. Early wound closure is thus of great importance. Reduction in food intake consequent on the loss of appetite contributes to the negativity of the nitrogen balance, but this is less decisive in the early period when there is increased catabolism.

A chronic anemia is invariably associated with deep extensive burns and is due partly to blood loss from the wounds and donor sites, and partly to impaired hematopoesis. James *et al.* (1951a,b) have recorded decreased hemoglobin synthesis as measured by the incorporation of N^{15}-labeled glycine into the heme of hemoglobin. An important practical consequence of this anemia is that unless the hemoglobin level is above 70% of normal by the second week, skin grafts are unlikely to take.

A pronounced hypoalbuminemia and hyperglobulinemia are generally present and continue, but in diminishing degree, until healing is complete. Their extent is usually directly related to the severity of injury. Vigorous intravenous therapy with plasma or albumin tends to diminish the hypoalbuminemia but cannot compensate for it completely, presumably mainly because of the loss of relatively albumin-rich fluid into the wound. Large intakes of protein generally fail to bring the level back to normal with greater speed. The increment in serum globulin is due mainly to an increase in the γ-glubulin fraction (Prendergast *et al.*, 1952).

After a month, the well-treated patient returns to nitrogen equilibrium or even positive balance. Artz and Reiss (1957) state that patients may store nitrogen consistently for months after complete replacement of previous losses. They consider that this unusual phenomenon is undoubtedly due to endocrine changes, but the nature of these changes is not understood.

In patients having burns of small extent, and in cases where most of the wound involves only the partial thickness of the skin, serious nutritional problems are unlikely to arise. If nausea and vomiting occur at all, these complications in such patients cease within a few days and the patient can then ingest increasing quantities of food. Artz and Reiss (1957) recommend that such patients should be encouraged to take a diet of 100 to 125 gm protein and about 3500 kcal, which is really the proportion of protein that one would expect in a well-balanced diet of this scale. The more severe the injury, the more critical the nutritional therapy becomes.

The protagonists of the concept that the catabolic response is nature's expression of the best way to provide energy and building stones for wound healing, and might therefore be regarded as a reflex which should not be thwarted by therapy, point out that excellent healing of wounds can occur at a time when the nitrogen balance is negative, whereas poor healing has been reported in debilitated patients who fail to respond to burns by increased nitrogen excretion. On the other hand, there are those who oppose this view and maintain that the paramount need is to keep the body composition as normal as possible by avoiding all possible loss through prompt provision of adequate nutrients from the beginning. Artz and Reiss (1957), while conceding that it is undisputed that excellent healing occurs in the presence of strongly negative nitrogen balance, point out that persistent negative nitrogen balance results in serious debilitation, increased susceptibility to bacterial infection, poor granulation tissue, poor graft take, and delay of healing at the donor sites. They consider that, if a laissez-faire attitude is assumed, disastrous results may follow.

Artz and Reiss (1957), like Wilkinson (1956), Moore and Ball (1952), and the author, consider that it is unwise to institute immediate forced-feeding programs to offset all nitrogen losses. The first-named authors consider that, though it is theoretically possible to achieve nitrogen equilibrium in the early postburn period, there are no indications that this is desirable. They draw attention to the fact that, during the first 5 to 10 postburn days, seriously burned patients often have nausea, vomiting, and gastrointestinal atony as manifested by abdominal distension and decreased peristalsis. Acute gastric dilatation is not uncommon and forced feeding may be followed by severe gastric dilatation.

During the first two days after injury it may be necessary to rely on intravenous feeding if nausea and vomiting occur. Thereafter the patient should be encouraged, but not forced, to ingest increasing amounts of food. Good intakes of a high-protein, high-calorie diet should be attainable within 7 to 10 days after injury. Where there is anorexia, as in severe burns, it may be kinder to tube-feed than to allow the patient to deteriorate.

Levenson and Watkin (1959), reviewing the protein requirements of burned children, state that they can be maintained in excellent nutritional condition when they are given dietary intakes 1–1½ times their "normal requirements" (as outlined by the National Research Council, see Food and Nutrition Board, 1958). They further state that severely burned women have been kept in good nutritional states by supplying them with about 2–4 gm protein/kg body weight and about 45 kcal/kg. Young adult males on a dietary intake similar to females show greater metabolic disturbances for comparable injury (see Section V,A); it seems that a daily intake of 3–4 gm protein/kg and 50–60 kcal/kg are desirable for this group. Bull (1958) has recommended even up to 8 gm protein/kg for children.

Bull (1958) reports that in burned patients who have had adequate transfusion, no difficulty arises in obtaining apparent positive nitrogen balance—the exudate loss not being considered as part of the catabolic loss. Soroff *et al.* (1961) have more recently studied the nitrogen requirements for equilibrium (including exudate loss) in 11 burned male patients during the later part of the acute catabolic phase (from 7 to 17 days after burning), the late catabolic phase (30–40 days after), the anabolic phase (60–70 days) and the late convalescent phase (90–100 days). The nitrogen intake varied from 1.0 to 28.8 gm/m^2 (normal resting adult, 5.3 gm), whereas the total energy intake remained constant at 3500 kcal/m^2 (normal resting adult, 1300 kcal/m^2). Isolated days of positive nitrogen balance occurred during the acute catabolic phase in the patients on high nitrogen intake but not in others. The cumulative nitrogen balance became positive in 9 of the 11 patients during the late catabolic phase and became progressively more positive during the subsequent phases. Regression analysis of the N balance on N intake showed a progressive decrease in the nitrogen requirements for equilibrium and obligatory nitrogen loss as the postburn course progressed to convalescence. The amounts of N/m^2 body surface/day estimated to be necessary for equilibrium during the four phases previously mentioned are 21–26, 13–16, 3–9 and 3–7 gm, respectively (surface area of normal adult, ca. 1.85 m^2). The N requirements of convalescents were similar to the requirements of controls. While this experiment suggests that diet can suppress the catabolic phase it must be noted that the experimental collection did not begin until the seventh day after burning, when the height of the catabolic phase would be passing off.

Inputs of 0.5 to 1 gm N/kg/day and 40 to 60 kcal/kg/day have been given to the severely injured but such a level of nitrogen is probably unnecessarily high. In a recent study of burned cases, Troell and Wretlind (1961) consider that an intake of 0.12–0.18 gm N/kg and 40–60 kcal/kg body weight is needed to maintain a positive nitrogen balance after the acute catabolic phase has passed. Nitrogen equilibrium was achieved with the lower N intake at the higher energy level. In case of anorexia they recommend complemental feeding of an emulsion of casein hydrolyzate, glucose, and peanut oil containing 4000 kcal/liter, to be given by nasogastric tube, but there is no proof that intact milk and other proteins could not be digested.

Blocker *et al.* (1955) have reported in a short series of cases that the continuous forced-feeding of high-caloric (45–85 kcal/kg), high-nitrogen (2–4 gm/kg) formulas through polyethene tubing by intranasal drip is successful in combating the negative balance after the first week of injury if interspersed with several rest pauses. During the subacute stages these patients can apparently utilize large amounts of high-nitrogen, high-calorie dietary supplements provided that alimentation is not impaired. Their radioactive studies using S^{35}-L-methionine indicate a greatly accelerated rate of protein catabolism which prevents equilibrium being reached until the healing phase is approached.

C. Practical Aspects of the Nutritional Problem

Every device that can stimulate appetite in the repair phase should be used. This is generally easy if there is no concomitant infection, but anything approaching forced feeding of protein-rich, energy-rich concentrates may cause nausea during the first few days. On occasions tube or intravenous feeding may be necessary if the patient is deteriorating through lack of adequate food. For those about to undergo operation it is strongly advised that within the time available every effort should be made to improve the nutritive state of the body if debilitated, and after the first few days following trauma advantage should be taken of the heightened synthetic process in the reconstructive stage so that nothing is lacking for optimal convalescence. Protein is used in this apparently profligate manner after injury only by previously well-nourished persons and in males to a greater degree than females. Moreover, this reaction to injury appears to be self-terminative. After it has continued for a certain length of time the loss of nitrogen gradually diminishes, though disuse atrophy may continue. Ultimately, provided that an adequate diet is given, a positive nitrogen balance can be established, even if the disease or injury continues.

Moore and Ball (1952) are in agreement with the author in warning against attempts to cover up the inevitable short-term nitrogen loss by giving unnecessary and poorly utilized proteins and protein hydrolyzates intravenously immediately after operation. Like the present writer, they consider that a deficiency in intake of energy-yielding food for the first few days after the trauma should not give rise to undue concern: they agree that every effort should be made thereafter to provide a diet of sufficient protein and energy-yielding nutrients as to obtain a positive nitrogen balance at the earliest possible moment.

Due notice must here be taken of the work at Cleveland of Abbott and his colleagues, who hold the view that much of the nitrogen deficit and loss of body weight which seems to characterize the first 5 to 10 days after surgical operation is in reality a result of the semistarvation regimens commonly employed, either because the patients are unable, or are not permitted, to have adequate food. Abbott *et al.* (1957) state that a combination of protein hydrolyzate solution containing sugar and fat emulsion (P. R. Cannon *et al.*, 1954), administered intravenously, is utilized preoperatively and postoperatively and will act against, or prevent the loss of, body weight and the nitrogen deficit. They advise that parenteral feeding should be used when the oral intake is grossly inadequate for more than a few days. But today little use is, in fact, made of protein hydrolyzates. Abbott's earlier observations had mainly been confined to soft tissue operations but his group (Spencer *et al.*, 1959) have more recently studied the nitrogen metabolism of a few patients undergoing orthopedic operations and found that they were in negative nitrogen balance when the intake of protein and energy was well below the levels recommended by Rhoads (1952), namely, protein present as 25 % of the energy and the energy at 30 to 40 kcal/kg/day. On the other hand, when two orthopedic patients were maintained on a high protein and calorie intake they were either in positive nitrogen balance or showed transient and minimal deficits after operation. These investigators found no correlation between the nitrogen deficits and the urinary excretion of 17-hydroxycorticosteroids. It should be emphasized that anesthesia protects the subjects from the most severe noxious stimuli of trauma. Accidental injury seems to induce a much greater catabolic response.

Levenson and Watkin (1959) have also commented on the conclusions of this Cleveland group as expressed by Holden *et al.* (1957). The view of the latter is that: "The major portion of the nitrogen deficit reported to be the result of operative trauma, in fact, is the result of poor nutritional intake." Levenson and his colleague have countered by stating: "But this is not true of a patient who has had a severe accident or of one seriously ill with an infectious disease: the nitrogen lost by these patients is greater than that lost by healthy subjects on comparable diets including those losses probably due to bed rest and fever (Wilkinson *et al.*, 1950). This is so when the diets are low in both protein and calories, low in protein but not in calories, 'normal' in protein and calories, and finally, high in protein and calories. But the net loss is less when the food intake is kept high than when the intake is low" (Werner *et al.*, 1949; Levenson *et al.*, 1959).

Levenson and Watkin consider that there is pretty general agreement that no "strenuous" attempt should be made to overcome completely the early increased urinary nitrogen excretion in the first few days following severe injury. It is their feeling—although they, like most, cannot substantiate this with incontrovertible proof—that the negative balance should be "reasonably minimized early by providing fairly rapidly, a full and 'adequate' diet." Unless there has been specific damage to the gastrointestinal tract they consider that there is seldom need for parenteral supplementation. When there is a serious deficiency in the capacity to ingest, digest, or absorb sufficient food over such a period of time as to jeopardize convalescence or even survival, intravenous alimentation should be adopted if tube feeding into the digestive tract be ruled out. This has been reviewed by Cuthbertson (1960b). Considerable amounts of protein hydrolyzate or amino acid mixture, electrolytes, vitamins, and carbohydrate can, as we have seen, be given, and to this armamentarium fat emulsions may now be added.

The optimal level of feeding still awaits definition and is a continuing challenge. But in stating this, we must not forget that Wilkinson (1956) has shown that even after operations as severe as gastrectomy and excision of the rectum, provided the initial state of the patient is eventually made good and that blood loss in excess of 400 ml is replaced by transfusion, the complete stoppage of water for 2 days and food for 4 days does not impair recovery. It should be remembered, however, that patients in a warm hospital environment do not tolerate the stoppage of water for as long as do those in cooler hospital environments.

VII. Conclusions

Physical injury results in damage to and loss of body protein. In the course of the first ten days following receipt of a severe injury, this may amount in total to some 1.5 kg protein or nearly 12 % of the total body content of protein (Table II). Provided there is not at once destruction of tissue vitality, a local and general defensive reaction sets in from the action of the agent whether it be physical, chemical, thermal, microbial, or toxic and such reactions are included under the term "inflammation." Many of the important aspects of the local defensive features of inflammation can be produced by substances of the nature of histamine, 5-hydroxytryptamine, and bradykinin, liberated from the tissues and causing increased permeability.

Injury also causes general responses in the body, both immediate and delayed (Table I). A description is given of the events involving first the emergency reaction, then the shock or "ebb" period and later the general disturbance of protein metabolism consequent on injury, the main characteristics of which in severe injuries in the well-nourished subject are an excessive loss of nitrogen, sulfur, phosphorus, potassium, and creatine in the urine, together with some fever. This constitutes the "flow" period following the "ebb." During the first 10 days or so there is thus an excessive catabolic loss, reaching a maximum about the third day with a secondary peak about the eighth day following an injury such as fracture of a long bone of the leg. The loss is, in part, proportional to the previous level of protein intake but is also related to the severity of the injury. This reaction is less in the female than in the male.

The loss of nitrogen and sulfur is greater than can be accounted for by local losses or by diminished food intake. The writer has long maintained that this reaction is associated with the fact that, since injury leads to loss of functional activity and this in turn to diminished food and water intake, there is set in motion a chain of events—a primitive reflex—to meet both the exigencies of repair and of maintenance, for the necessity to heal is urgent. For regeneration there is a need for energy, water, and the substrate for healing the breached tissues and these are apparently preferentially drawn from the body reserves. Nevertheless, we must at the same time admit that the mechanism underlying this extraordinary increase in protein catabolism after accidental injury is as yet imperfectly understood and remains a challenge. It can in part be overcome, or at least masked, by high intakes of a well-balanced diet. Muscle would appear to be a major source of the excessive excretion of nitrogen, sulfur, phosphorus, potassium, and creatine in the urine. Increased adrenocortical secretion occurs, but is not essential for this catabolic response.

The nutritional treatment of the injured is discussed in respect of dietary intake of protein and energy, but first blood or plasma losses must be made good. Most observers hold that the evidence then indicates that during the first few days following the injury in the previously adequately nourished person it is probably unwise to push the intake of a well-balanced diet beyond appetite as this may fail during this early postinjury phase in severe injuries. Further, kidney function may not be normal for a day or two. Thereafter the patient should be encouraged, but not forced, to take as much as he can of a well-balanced diet, relatively rich in protein. Rehabilitation may take months.

References

Abbott, W. E., Krieger, H., Holden, W. D., Bradshaw, J., and Levey, S. (1957). *Metab. Clin. Exptl.* **6,** 691.

Artz, C. P., and Reiss, E. (1957). "The Treatment of Burns." Saunders, Philadelphia, Pennsylvania.

Badawy, A. M., Campbell, R. M., Cuthbertson, D. P., and Fell, B. F. (1957). *Nature* **180,** 756.

Barclay, T. H. C., Cuthbertson, D. P., and Isaacs, A. (1944). *Quart. J. Exptl. Physiol.* **32,** 309.

Basu Mallik, K. C. (1955). *J. Pathol. Bacteriol.* **70,** 315.

Bayliss, W. M., and Cannon, W. B. (1919). Rept. Spec. Invest. Comm. on Surgical Shock and Allied Conditions. *Spec. Rept. Ser. No. 26,* VIII, p. 19. Med. Res. Comm. H.M.S.O., London.

Berg, M. (1942). *A.M.A. Arch. Pathol.* **33,** 636.
Biely, J. (1962). *Nature* **196,** 523.
Blocker, T. G., Jr., Levin, W. C., Nowinski, W. W., Lewis, S. R., and Blocker, V. (1955). *Ann. Surg.* **141,** 589.
Bloom, W. L. (1961). *Metab. Clin. Exptl.* **10,** 171.
Born, G. V. R. (1962). *In* "General Pathology" (H. Florey, ed.), 3rd ed., p. 284. Lloyd-Luke, London.
Browne, J. S. L. (1943). *In* "Metabolic Aspects of Convalescence," Trans. 4th Conf., p. 97. Josiah Macy Jr. Foundation, New York.
Browne, J. S. L. (1944). *In* "Metabolic Aspects of Convalescence," Trans. 6th Conf., p. 67. Josiah Macy Jr. Foundation, New York.
Bruce, J. (1954). *Gastroenterologia* **81,** 74.
Bull, J. P. (1958). *Proc. Nutr. Soc.* (*Engl. Scot.*) **17,** 114.
Burt, R. L. (1954). *J. Lab. Clin. Med.* **44,** 702.
Cairnie, A. B., Campbell, R. M., Pullar, J. D., and Cuthbertson, D. P. (1957). *Brit. J. Exptl. Pathol.* **38,** 504.
Caldwell, F. T., Osterholm, J. L., Sower, N. D., and Moyer, C. A. (1959). *Ann. Surg.* **150,** 976.
Calloway, D. H., Grossman, M. I., Bowman, J., and Calhoun, W. K. (1955). *Surgery* **37,** 935.
Campbell, R. M., and Cuthbertson, D. P. (1963). Unpublished data.
Campbell, R. M., Sharp, G., Boyne, A. W., and Cuthbertson, D. P. (1954). *Brit. J. Exptl. Pathol.* **35,** 566.
Cannon, P. R., Frazier, L. E., and Hughes, R. H. (1954). *J. Lab. Clin. Med.* **44,** 250.
Cannon, W. B. (1939). "The Wisdom of the Body," 2nd ed. Norton, New York.
Chytil, F. (1956). *Českoslov. Gastroenterol. Výživa* **10,** 244.
Clarke, R. (1952). *Ann. Roy. Coll. Surg. Engl.* **11,** 87.
Clarke, R., Topley, E., and Flear, C. T. G. (1955). *Lancet* **I,** 629.
Cohnheim, J. (1882). "Lectures on General Pathology." The New Sydenham Soc., London (transl. into English 1889).
Cooper, C. E., and Nelson, D. H. (1962). *J. Clin. Invest.* **41,** 1599.
Cope, O., Nardi, G. L., Quijano, M., Rovitt, R. L., Stanbury, J. B., and Wight, A. (1953). *Ann. Surg.* **137,** 165.
Crane, W. A. J. (1954). *J. Pathol. Bacteriol.* **67,** 379.
Cushing, H. (1932). *Surg., Gynecol. Obstet.* **55,** 1.
Cuthbertson, D. P. (1929). *Biochem. J.* **23,** 1328.
Cuthbertson, D. P. (1930). *Biochem. J.* **24,** 1244.
Cuthbertson, D. P. (1931). *Biochem. J.* **25,** 236.
Cuthbertson, D. P. (1932). *Quart. J. Med.* [N.S.] **25,** 233.
Cuthbertson, D. P. (1936). *Brit. J. Surg.* **23,** 505.
Cuthbertson, D. P. (1942). *Lancet* **I,** 433.
Cuthbertson, D. P. (1957). *Z. Tierernähr.* **12,** 259.
Cuthbertson, D. P. (1960a). *In* "The Biochemical Response to Injury" (H. B. Stoner and C. J. Threlfall, eds.), pp. 193–216. Blackwell, Oxford.
Cuthbertson, D. P. (1960b). *Surg. Gynecol. Obstet.* **110,** 105.
Cuthbertson, D. P., and Tompsett, S. L. (1935). *Brit. J. Exptl. Pathol.* **16,** 471.
Cuthbertson, D. P., McGirr, J. L., and Robertson, J. S. M. (1939). *Quart. J. Exptl. Physiol.* **29,** 13.
Cuthbertson, D. P., Webster, T. A., and Young, F. G. (1941). *J. Endocrinol.* **2,** 459.
Dale, H. H., and Laidlaw, P. P. (1910). *J. Physiol.* (*London*) **41,** 318.

Dale, H. H., Laidlaw, P. P., and Richards, A. N. (1919). Rept. Spec. Invest. Comm. on Surgical Shock and Allied Conditions. *Spec. Rept. Ser. No. 26*, VIII, 28. Med. Res. Comm. H.M.S.O., London.
Davis, J. W. L., Rickets, C. R., and Bull, J. P. (1959). *Lancet* **I,** 346.
Dudley, H. A. F. D. (1959). Personal communication.
Eades, C. H., Pollack, R. L., and Hardy, J. D. (1953). *J. Clin. Invest.* **34,** 1756.
Engel, F. L., Winton, M. G., and Long, C. N. H. (1943). *J. Exptl. Med.* **77,** 397.
Engel, F. L., Harrison, H. C., and Long, C. N. H. (1944). *J. Exptl. Med.* **79,** 9.
Feldberg, W., and Talesnik, J. (1953). *J. Physiol.* (*London*) **120,** 550.
Fell, B. F. (1961). *J. Pathol. Bacteriol.* **81,** 251.
Fell, B. F. (1961). Unpublished data. See Fell *et. al.* (1961).
Fell, B. F., Boyne, R., and Cuthbertson, D. P. (1961). *J. Pathol. Bacteriol.* **82,** 445.
Fine, J., Frank, H., Schweinburg, F., Jacob, S., and Gordon, T. (1952). *Ann. N.Y. Acad. Sci.* **55,** 429.
Fine, J., Rutenburg, S., and Schweinburg, F. B. (1959). *J. Exptl. Med.* **110,** 547.
Fine, J., Frank, E. D., Ravin, H. A., Rutenburg, S. H., and Schweinburg, F. B. (1960). *In* "The Biochemical Response to Injury" (H. B. Stoner and C. J. Threlfall, eds.), pp. 377–396. Blackwell, Oxford.
Flear, C. T. G., and Clarke, R. (1955). *Clin. Sci.* **14,** 575.
Fleck, A., and Munro, H. N. (1963). *Metab. Clin. Exptl.* **12,** 783.
Florey, H. (1954). Lectures on Pathology. Watson & Viney, London.
Florey, H. (1961). *Nature* **192,** 905.
Florey, H. (1962). *In* "General Pathology" (H. Florey, ed.), 3rd ed., p. 21. Lloyd-Luke, London.
Fonnesu, A. (1960). *In* "The Biochemical Response to Injury" (H. B. Stoner and C. J. Threlfall, eds.), pp. 85–104. Blackwell, Oxford.
Food and Nutrition Board (1958). "Recommended Dietary Allowances (Revised)," Report No. 500 Natl. Acad. Sci.—Natl. Research Council, Washington, D.C.
Fortier, C. (1956). *Abstr. 20th Intern. Physiol. Congr., Brussels* p. 490.
Freeman, N. E., Shaffer, S. A., Schecter, A. E., and Holling, H. E. (1938). *J. Clin. Invest.* **17,** 359.
Gilder, H., Moody, F. G., Cornell, G. N., and Beal, J. M. (1961). *Metab. Clin. Exptl.* **10,** 134.
Gillette, R. W., Mansberger, A. R., Jr., Oppenheimer, J. H., Caldwell, B. F., and Neff, E. G. (1958a). *Surgery* **43,** 747.
Gillette, R. W., Oppenheimer, J. H., Mansberger, A. R., Jr., Johnston, C. E., and Kookootsedes, G. J. (1958b). *Surgery* **43,** 740.
Goldstein, M. S., and Harwood, T. C. (1956). *Abstr. 20th Intern. Physiol. Congr., Brussels* p. 352.
Goodlad, G. A. J., and Munro, H. N. (1959). *Biochem. J.* **73,** 343.
Grant, R. T., and Reeve, E. B. (1951). *Med. Res. Council Spec. Rept. Ser.* **277,** 309.
Green, H. N., Stoner, H. B., Whiteley, H. J., and Eglin, D. (1949). *Clin. Sci.* **8,** 65.
Hardy, J. D. (1955). "Surgical Physiology of the Adrenal Cortex." Thomas, Springfield, Illinois.
Hardy, J. D., Turner, D. M., and Jackson, M. (1957). *Surgery* **42,** 194.
Harris, G. W. (1956). *Abstr. 20th Intern. Physiol. Congr., Brussels* p. 508.
Holden, W. D., Krieger, H., Levey, S., and Abbott, W. E. (1957). *Ann. Surg.* **146,** 563.
Hrůza, Z. (1962). *Physiol. Bohemoslov.* **11,** 206.
Hrůza, Z., and Chytil, F. (1959). *Physiol. Bohemoslov.* **8,** 307.
Hume, D. M. (1953). *Ann. Surg.* **138,** 548.

Hume, D. M., Nelson, D. H., and Miller, D. W. (1956). *Ann. Surg.* **143,** 316.

Hunter, J. (1794). "A Treatise on Blood, Inflammation and Gunshot Wounds." Nicol, London.

Ingle, D. J. (1951). *Recent Progr. Hormone Res.* **6,** 159.

Ingle, D. J., Ward, E. O., and Kuinzenga, M. H. (1947). *Am. J. Physiol.* **149,** 510.

Ingle, D. J., Meeks, R. C., and Thomas, K. E. (1951). *Endocrinology* **49,** 703.

James, G. W., Purnell, O. J., and Evans, E. I. (1951a). *J. Clin. Invest.* **30,** 181.

James, G. W., Purnell, O. J., and Evans, E. I. (1951b). *J. Clin. Invest.* **30,** 191.

Janoff, A., Nagler, A. L., Baez, S., and Zweifach, B. W. (1961). *J. Exptl. Med.* **114,** 205.

Klemperer, P., Penner, A., and Bernheim, A. I. (1940). *Am. J. Digest. Diseases* **7,** 410.

Kovách, A. G. B., and Fonyó, A. (1960). *In* "The Biochemical Response to Injury" (H. B. Stoner and C. J. Threlfall, eds.), pp. 129–160. Blackwell, Oxford.

Levenson, S. M., and Watkin, D. M. (1959). *Federation Proc.* **18,** 1155.

Levenson, S. M., Birkhill, F. R., Maloney, M. A., and Bell, J. A. (1949). *Ann. Surg.* **13,** 723.

Levenson, S. M., Malm, O. J., Haynes, B. W., and Lounds, E. A. (1959). In preparation.

Levenson, S. M., Braasch, J. W., Mueller, H., and Crowley, L. (1960). See Levenson and Watkin (1959).

Lillehei, C. W., Dixon, J. L., and Wangensteen, O. H. (1948). *Proc. Soc. Exptl. Biol. Med.* **68,** 125.

Llaurado, J. G., and Woodruff, M. F. A. (1957). *Proc. Univ. Otago Med. School* **35,** 19.

Long, C. N. H. (1956). *Ann. Rev. Physiol.* **18,** 409.

McClusky, R. T., Zweifach, B. W., Antopol, W., Benacerraf, B., and Nagler, A. L. (1960). *Am. J. Physiol.* **37,** 245.

McFarlane, A. S., Campbell, R. M., and Cuthbertson, D. P. (1960). Unpublished data. See Cuthbertson (1960a, p. 202).

Macfarlane, R. G. (1962). *In* "General Pathology" (H. Florey, ed.), 3rd ed., p. 216. Lloyd-Luke, London.

Malcolm, J. D. (1893). "The Physiology of Death from Traumatic Fever." Churchill, London.

Malcolm, J. D. (1905). *Lancet* **II,** 922.

Menkin, V. (1940). "Dynamics of Inflammation." Macmillan, New York.

Miles, A. A. (1962). *Abstr. Intern. Symposium on Injury, Inflammation and Immunity* Sect. 11, p. 18 (Miles Laboratories, Inc., U.S.A.).

Millican, R. C. (1960). *In* "The Biochemical Response to Injury" (H. B. Stoner and C. J. Threlfall, eds.) pp. 269–290. Blackwell, Oxford.

Moore, F. D. (1957). *Recent Progr. Hormone Res.* **13,** 511.

Moore, F. D., and Ball, M. R. (1952). "The Metabolic Response to Surgery." Thomas, Springfield, Illinois.

Moore, F. D., Langohr, J. L., Ingebretsen, M., and Cope, O. (1950). *Ann. Surg.* **132,** 1.

Müller, F. (1884). *Centr. klin. Med.* **5,** 569.

Munro, H. N., and Chalmers, M. I. (1945). *Brit. J. Exptl. Pathol.* **26,** 396.

Munro, H. N., and Cuthbertson, D. P. (1943). *Biochem. J.* **37,** xii.

Nagler, A. L., and Zweifach, B. W. (1961). *J. Exptl. Med.* **114,** 195.

Nardi, G. L. (1954). *J. Clin. Invest.* **33,** 847.

Needham, A. E. (1955). *J. Embryol. Exptl. Morphol.* **3,** 189.

Needham, A. E. (1958). *J. Exptl. Zool.* **138,** 369.

Noble, R. L. (1942–1943). *Am. J. Physiol.* **138,** 346.

Noble, R. L., and Toby, C. G. (1946–1948). *J. Endocrinol.* **5,** 303.
Palmer, E. D., and Buchanan, D. P. (1953). *Ann. Internal Med.* **38,** 1187.
Palmer, E. D., and Sherman, J. L., Jr. (1958). *A.M.A. Arch. Internal Med.* **101,** 1106.
Pearce, C. W., Foot, N. C., Jordan, G. L., Low, S. W., and Wantz, G. E. (1960). *Surg., Gynecol. Obstet.* **111,** 274.
Pekkarinen, A. (1960). *In* "The Biochemical Response to Injury" (H. B. Stoner and C. J. Threlfall, eds.), pp. 217–268. Blackwell, Oxford.
Penner, A., and Bernheim, A. I. (1939). *A.M.A. Arch. Pathol.* **27,** 966.
Popják, G. (1944). *J. Pathol. Bacteriol.* **56,** 485.
Prendergast, J. J., Fenichel, R. L., and Daly, B. M. (1952). *A.M.A. Arch. Surg.* **64,** 733.
Quénu, E. (1918). *Rev. Chir.* **56,** 204.
Ravin, H. A., Rowley, D., Jenkins, C., and Fine, J. (1960). *J. Exptl. Med.* **112,** 783.
Rhoads, J. E. (1952). *Federation Proc.* **11,** 659.
Russell, J. A., and Long, C. N. H. (1946). *Am. J. Physiol.* **147,** 175.
Russell, J. A., Long, C. N. H., and Engel, F. L. (1944a). *J. Exptl. Med.* **79,** 1.
Russell, J. A., Long, C. N. H., and Wilhelmi, A. E. (1944b). *J. Exptl. Med.* **79,** 23.
Sayers, G. (1950). *Physiol. Rev.* **30,** 241.
Schønheyder, F., Heilskov, N. S. C., and Olesen, K. (1954). *Scand. J. Clin. Lab. Invest.* **6,** 178.
Schreier, K., and Karch, H. L. (1954–1955). *Arch. Klin. Chir.* **280,** 516.
Schweinburg, F. B., and Fine, J. (1960). *J. Exptl. Med.* **112,** 793.
Selye, H. (1954). *J. Clin. Endocrinol.* **14,** 122.
Selye, H., and Collip, J. P. (1936). *Endocrinology* **20,** 672.
Sevitt, S. (1962). *Abstr. Intern. Symposium on Injury, Inflammation and Immunity* Sect. 1, p. 15 (Miles Laboratories, Inc., U.S.A.).
Silber, R. H., and Porter, C. C. (1953). *Endocrinology* **52,** 518.
Soroff, H. S., Pearson, E., and Artz, C. P. (1961). *Surg., Gynecol. Obstet.* **112,** 159.
Spencer, G. E., Jr., Krieger, H., Abbott, W. E., and Levey, S. (1959). *J. Bone Joint Surg.* **41A,** 381.
Sterling, K., Lipsky, S. R., and Freedman, L. J. (1955). *Metab. Clin. Exptl.* **4,** 343.
Stoner, H. B. (1958). *Brit. J. Exptl. Pathol.* **39,** 635.
Stoner, H. B., and Threlfall, C. J. (1960). *In* "The Biochemical Response to Injury" (H. B. Stoner and C. J. Threlfall, eds.), pp. 105–128. Blackwell, Oxford.
Taylor, F. H. L., Levenson, S. M., Davidson, C. S., and Adams, M. A. (1943). *New Engl. J. Med.* **229,** 855.
Thomas, L. (1956). *J. Exptl. Med.* **104,** 865.
Troell, L., and Wretlind, A. (1961). *Acta Chir. Scand.* **122,** 15.
Uotila, U., and Pekkarinen, A. (1951). *Acta Endocrinol.* **6,** 23.
Van Slyke, D. D., Phillips, R. A., Hamilton, P. B., Archibald, R. M., Dole, V. P., and Emerson, K., Jr. (1944). *Trans. Assoc. Am. Physicians* **58,** 119.
von Euler, U. S. (1955–6). *Ann. Rept. Stress* **5,** 125.
von Euler, U. S. (1956). "Noradrenaline: Chemistry, Physiology, Pharmacology and Clinical Aspects." Thomas, Springfield, Illinois.
von Euler, U. S. (1958). *Acta Endocrinol.* Suppl. 38, 36.
Walker, J., Jr. (1946). *Am. J. Med. Sci.* **209,** 413.
Webb, W. R., Doyle, R. S., and Howard, H. S. (1960). *Metab. Clin. Exptl.* **9,** 1047.
Welbourn, R. B. (1957). *Irish J. Med. Sci.* p. 401.
Wener, J., Hoff, H. E., and Simon, M. A. (1948). *Gastroenterology* **11,** 904.
Wener, J., Hoff, H. E., and Simon, M. A. (1949). *Gastroenterology* **12,** 637.

Werle, J. M., Cosby, R. S., and Wiggers, C. J. (1942). *Am. J. Physiol.* **136,** 401.
Werner, S. C., Habif, D. V., Randall, H. T., and Lockwood, J. S. (1949). *Ann. Surg.* **130,** 688.
Whiteley, H. J., Stoner, H. B., and Threlfall, C. J. (1953). *Brit. J. Exptl. Pathol.* **34,** 365.
Wilhelmi, A. E., Russell, J. A., Engel, M. G., and Long, C. N. H. (1945). *Am. J. Physiol.* **144,** 674.
Wilkinson, A. W. (1956). *Lancet* **II,** 428.
Wilkinson, A. W., Billing, B. H., Nagy, G., and Stewart, C. P. (1950). *Lancet* **1,** 533.
Woodbury, D. M. (1958). *Pharmacol. Rev.* **10,** 275.
Wright, A. E., and Colebrook, L. (1918). *Lancet* **I,** 763.
Wynn, V. (1960). *In* "The Biochemical Response to Injury" (H. B. Stoner and C. J. Threlfall, eds.), pp. 291–311. Blackwell, Oxford.
Yuile, C. L., Lucas, F. V., Jones, C. K., Chapin, S. J., and Whipple, G. H. (1953). *J. Exptl. Med.* **98,** 173.
Zimmermann, B. (1962). *Surg., Gynecol. Obstet.* **114,** 228.

CHAPTER 20

Protein Metabolism and Tumor Growth

G. A. J. Goodlad

Department of Physiology and Biochemistry, St. Salvator's College, University of St. Andrews, Fife, Scotland

I. Introduction

Normal growth, which is the orderly multiplication of cells, has to be distinguished from the growth of tumors (neoplasms), in which the cells proliferate without control and serve no apparent useful purpose. The unrestrained cell multiplication of tumors can, however, vary over a wide spectrum from slow-growing, benign neoplasms to rapidly multiplying, malignant tumors. Benign (simple, innocent) tumors are characterized by their local nature, being without tendency to spread to other parts of the body, and in consequence generally do not lead to a fatal issue. On the other hand, the malignant tumor has a propensity for spreading throughout the host's tissues, both locally by invasion and at a distance by lymphatic and vascular spread, with the consequent development of secondary deposits of the tumor (metastases) and eventually the death of the host if effective treatment is not applied. The malignant tumor thus has a much more profound general action on the host. This chapter will be concerned essentially with protein metabolism in animals bearing malignant tumors.

Malignant tumors can arise from the cells of all tissues; carcinomas are of epithelial origin, the term "sarcoma" being reserved for tumors arising from connective tissue and other mesodermal structures. When the histological structure of the malignant cell is compared with that of its parent cell, there is usually seen to be some degree of loss of differentiation, that is, the specialized structures of the parent cell tend to disappear and in consequence tumor cell morphology shifts toward the same undifferentiated pattern irrespective of origin, the most malignant cells having an undifferentiated (anaplastic) structure. This picture is encountered again in the fine structure changes revealed by electron microscopy (Bernhard, 1958; Mercer, 1961). In a well-differentiated tumor, cell structure may resemble that of the tissue of origin. On the other hand, in more anaplastic tumors the intracellular pattern tends toward that of the embryonic type in which the cytoplasm has a high content of freely dispersed ribosomes and no well-defined system of endoplasmic reticulum (Birbeck and Mercer, 1961). This final picture occurs irrespective of whether the tumor is derived from a tissue possessing a secretory type of cell with a well-defined granular endoplasmic reticulum (α-cytomembranes), or from cells with a low cytoplasmic ribonucleic acid content and a smooth, agranular type of reticulum (β-cytomembranes). Granular endoplasmic reticulum is especially abundant in cells secreting proteins, such as the pancreatic acinar cell, and the implication to be drawn from loss of this structure from the tumor cell is that some aspect of the cell's capacity to form proteins has been affected. This is also implied by the disappearance of individual enzymes from cells which have undergone malignant transformation; enzymes are specific types of protein, and impairment of capacity to form them indicates a change in the pattern of protein synthesis. It can thus be seen that the main characteristic of the tumor cell is a loss of specialized function and a redirection of its metabolic machinery toward reduplication (Rusch, 1954). It would consequently be predicted that the reactions of protein metabolism in the tumor cell would be much altered as compared with their tissue of origin, first because of a change in the types of protein synthesized within the tumor cell, second because of loss of some of the enzymes concerned with the intermediary metabolism of amino acids, and finally because the quantitative requirement for amino acids is intensified by growth. The account, which follows later in this chapter and is summarized in Fig. 1, shows that these predictions are indeed fulfilled. It is, however, disappointing to note that work to date throws little light on the reasons for the loss of capacity of the tumor cell to make specialized cell proteins. This may partly arise because our understanding of the control of protein synthesis in the normal cell is still being worked out (see Chapter 6) and this knowledge is a prerequisite to exploration of the loss of control in the malignant cell.

The presence of a tumor can also affect protein metabolism in the tissues of the host. Malignant tumors affect tissues in their vicinity by infiltration, but they also have general effects on the body, the extent of which varies widely with different tumor types. As Begg (1958) has pointed out, "It would appear that a tumor both concentrates metabolites to the detriment of the host and releases substances into the circulation which produce structural and functional changes in the tissues of the host." The tumor thus acts on the tissues of the host in two ways: first, by drawing on nutrients in competition with the host and second, by depressing certain functions of the host's tissues. Figure 1 displays some of the more frequently encountered effects of experimental tumors on the protein metabolism of the host, such as the well-known wasting of tissues (cachexia), particularly reflected in loss of muscle protein, the increased breakdown of plasma proteins by the tumor which is associated with a reduction in plasma protein concentration and an increase in their turnover, and an anemia which stimulates the development of new centers of blood cell formation (extramedullary hematopoiesis) in the liver and spleen.

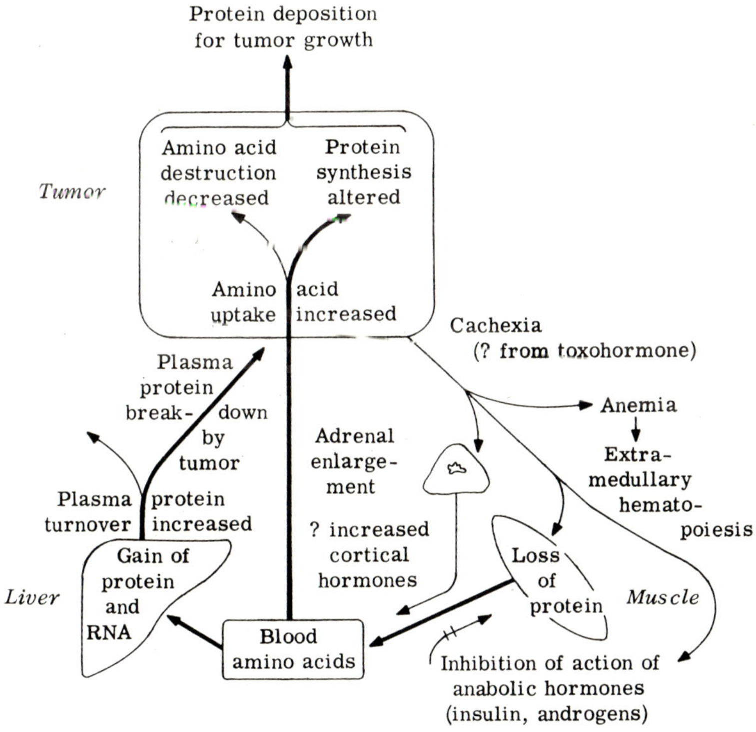

FIG. 1. A composite diagram of changes in protein metabolism which have been observed in animals bearing malignant tumors.

Factors regulating protein metabolism in the normal animal (Chapter 10) include the level of energy available to the tissues, and the action of hormones favoring either nitrogen retention (anabolic hormones) or nitrogen loss (catabolic hormones). Disturbances of the action of these regulating mechanisms have been observed in some tumor-bearing animals and subjects (Fig. 1), and would be expected to cause nitrogen losses from the tissues of the host. Thus, impairment in the metabolism and utilization of energy has been noted in patients with malignant tumors (Fenninger and Mider, 1954), but has not been explored as a cause of the disturbances in the protein metabolism of the host and will not be further discussed in this article. With regard to hormonal factors affecting protein metabolism, there is evidence in some tumor-bearing animals of impaired production or action of anabolic hormones (androgens and insulin). Some animals show enlargement of the adrenal cortex, suggesting that hypercortical activity may contribute to loss of body protein. Deposition of protein and ribonucleic acid in the livers of some tumor-bearing animals along with loss of carcass protein is compatible with overactivity of the adrenal cortex, but other evidence to be discussed later indicates that the changes in the protein metabolism of the host may not be due to any oversecretion of glucocorticoids.

Tumors on their part are not entirely autonomous, being influenced to a greater or lesser degree both nutritionally and hormonally by the host. From what has been said, it is thus apparent that there exists a complex interrelationship between the metabolism of the tumor and that of the host. Various aspects of this interrelationship will now be discussed, dealing first with protein metabolism in the tumor cell and then with the impact of the tumor on the protein metabolism of the host.

II. The Protein and Amino Acid Metabolism of Tumors

A. Nutrition of Tumors

The questions here to which we must seek answers are, (*a*) Do tumors need the same range of amino acids as those required by normal tissues? (*b*) Do tumors have a special capacity to obtain amino acids from the host, such as amino acid concentrating mechanisms and direct utilization of plasma proteins? (*c*) How is the capacity of the tumor to compete with the host for nutrients affected by nutritional and other factors in the host?

1. The Amino Acid Requirements of Growing Tumors

The amino acids essential for the growth of tumor cells under *in vitro* conditions show little variation from tumor to tumor, and include all those known to be essential for the growth of normal cells in tissue culture. In tissue culture, the basic requirement of malignant cells would appear to be: L-arginine, L-histidine, L-lysine, L-threonine, L-valine, L-leucine, L-isoleucine, L-phenylalanine, L-tyrosine, L-tryptophan, L-cysteine, L-methionine, and L-glutamine, while L-asparagine has been found to be essential for the growth of some tumors (Eagle, 1955; McCoy *et al.*, 1956, 1959a,b; Haley *et al.*, 1961). Glycine was found to be essential for the growth of the Novikoff hepatoma (McCoy *et al.*, 1959b) and both glycine and L-serine enhanced the growth rate of the Walker 256 carcinoma and the Jensen sarcoma (McCoy *et al.*, 1956, 1959a).

Tumor tissue must thus possess mechanisms for the biosynthesis of its other constituent amino acids. However, there is a certain amount of evidence to suggest that formation of glutamic and aspartic acids and alanine from intermediates of carbohydrate metabolism occurs to only a limited extent in tumors (Kit and Graham, 1956; Busch, 1955) and Henderson and LePage (1959b) believe that the tumor growing rapidly in the animal body may require an external supply of these amino acids. In this connection, an important part of the glutamate requirement of the tumor may be derived from glutamine (Roberts and Simonsen, 1960). Tumor cells are relatively impermeable to glutamic acid but readily permeable to its amide (Roberts *et al.*, 1956). Inside the tumor cell, glutamine is rapidly metabolized, the main products being glutamic acid and carbon dioxide (Roberts *et al.*, 1956; Coles and Johnstone, 1962). Glutamine has also been shown to be necessary for the *in vitro* incorporation of amino acids into protein in Ehrlich ascites cells (Rabinovitz *et al.*, 1956). Coles and Johnstone (1962) believe that its action in this respect is not just that it is required as a constituent of newly formed protein but that it may also serve as a source of energy for this process under certain conditions.

2. *Uptake of Amino Acids and Proteins by Tumor Cells*

As might be expected of a rapidly growing tissue, tumor cells under both *in vitro* and *in vivo* conditions have been shown to have a high capacity for the concentration of amino acids (Christensen and Riggs, 1952; Christensen and Henderson, 1952; Wiseman and Ghadially, 1955). Wiseman and Ghadially (1958) have actually advanced a hypothesis to explain the various aspects of tumor growth such as infiltration of normal cells and the anorexia and cachexia of the host based on this property of tumors. It would appear from inhibitor and from kinetic studies on tumor cells that uptake of different amino acids occurs by more than one system (Tenenhouse and Quastel, 1960; Johnstone and Scholefield, 1961). In this respect the tumor cell is no different from the nonmalignant cell, as discussed in detail in Chapter 4. A possible role of phospholipids in the uptake of amino acids by Ehrlich ascites cells has been suggested by Galy and his colleagues (1960). Tumor cells can also accumulate small peptides, though at a much slower rate than they can accumulate amino acids (Christensen and Rafn, 1952).

Although there is convincing evidence to suggest that plasma proteins are utilized by tumor tissue as a source of nitrogen to a greater extent than by normal tissues (Babson and Winnick, 1954; Busch and Greene, 1955; Henderson and LePage, 1959a; Busch *et al.*, 1961), there are conflicting views as to whether they are taken up by the tumor cell in an intact form. Direct utilization of plasma proteins without prior hydrolysis to amino acids was regarded for a time as a normal method by which tissues in general could obtain part of their protein supply, the evidence being largely due to the studies of G. H. Whipple and his colleagues; this is not now regarded as a likely mechanism for cell nutrition (see Chapter 8). However, evidence suggesting that tumor cells might be capable of taking up foreign proteins under conditions in which nonmalignant cells could not was presented by Cohen *et al.* (1961), who used immunohistochemical techniques to identify the protein in the cells. Thus, rat hepatoma cells but not liver parenchymal cells were shown to take up intact rabbit albumin and globulin. The appearance and distribution of the inclusions in the hepatoma cell suggested that the mechanism of uptake was different from that by which Kupffer cells of non-neoplastic liver take up foreign proteins. More recent studies by Ghose *et al.* (1962), using indirect or direct fluorescent tracing methods, have shown that direct uptake of foreign plasma proteins is not restricted to hepatomas but that cells of other types of primary and transplantable tumors also possess this property.

Although some radioactive studies have suggested that plasma protein may be incorporated into tumor cell protein in an unhydrolyzed or only partly degraded form (Babson and Winnick, 1954; Busch *et al.*, 1961), Campbell and Stone (1957b) concluded from an extensive study of all the factors involved in such experiments that there is no evidence for a direct transfer of serum albumin to tumor protein other than at the free amino acid level. The work of Henderson and LePage (1959a) also supports the view that plasma proteins undergo hydrolysis before utilization by the tumor. They made use of the fact that free glycine is a precursor of the purine nucleus of nucleic acids as well as of tissue proteins, and examined the ratio of tumor nucleic acid radioactivity to tumor protein radioactivity after administering either glycine-2-C^{14} or plasma protein labeled with glycine-2-C^{14} to mice bearing a lymphosarcoma. If the plasma proteins are hydrolyzed to free amino acids before utilization by the tumor cell, the radioactive glycine will have to pass through the free amino acid pool of the tumor tissue and will thus label the purines of the tumor nucleic acids as well as the tumor proteins. If, on the other hand, the plasma proteins are used directly for tumor protein formation, no significant labeling of purines should occur. In fact, the ratio of radioactivity in tumor nucleic acid to radioactivity in tumor protein was the same, irrespective of whether radioactive glycine was given as the free amino acid or in the form of labeled plasma protein, thus demonstrating that the plasma proteins had been hydrolyzed to their constituent amino acids before utilization. This experiment seems a compelling demonstration of the need for prior conversion to free amino acids before utilization in the cell. In order to explain the special ability of the tumor to utilize the plasma protein as an amino acid source, Henderson and LePage have suggested that either the tumor may have an increased permeability to plasma protein compared with other tissues, or that there may exist a high local concentration of proteolytic enzymes near the tumor which could make these proteins more easily available to the tumor, or that there may be a combination of these two circumstances. There is some experimental support for this idea of a localized area of proteolytic activity near the tumor. Benz and Lehmann (1959) found in the rat that the tissues surrounding a growing Walker 256 carcinoma had an increased catheptic activity, and tumor cells growing *in vitro* have been shown to secrete proteolytic enzymes into the medium (Holmberg, 1961). M. Green and Miller (1961), however, found that, whereas denatured hemoglobin and native insulin-I^{131} were hydrolyzed readily when present in the culture medium of Walker tumor cells, no hydrolysis of native rat plasma protein occurred.

Later work has suggested another mechanism of utilization of serum albumin by tumors. It has been proposed that amino acid derivatives, rather than free amino acids, are the end products of intracellular protein catabolism and that these may participate directly in protein anabolic processes without being degraded further to the free amino acid stage (Walter, 1960; Chin *et al.*, 1959). Penn (1961) has claimed that coenzyme A peptides are the primary products of albumin degradation by rat liver mitochondria and has further demonstrated the reincorporation of these products into protein. Since this same worker (Penn, 1962) has shown that mitochondria from Novikoff hepatoma also possess an active coenzyme A-dependent system which can hydrolyze albumin, he has suggested that this may be a factor involved in the utilization of plasma proteins by tumors.

From this discussion it would appear that tumors, like other tissues, utilize free amino acids as a main source of nitrogen for protein synthesis. To this end they have elaborated highly efficient mechanisms for the concentration of amino acids which enable them to compete successfully with the tissues of their hosts for available amino acids. However, tumors can also use plasma proteins to a greater extent than normal tissues. The available evidence is in favor of the view that plasma proteins must first undergo hydrolysis before utilization, which may occur both extracellularly, at a site in the immediate environment of the tumor, and intracellularly by a tumor mitochondrial enzyme. The relative importance of these two proteolytic systems cannot yet be assessed.

3. Factors in the Host Influencing Tumor Growth

Tumor-bearing animals fed a protein-containing diet were actually observed to excrete less nitrogen and consequently appeared to be in a more positive nitrogen balance than nontumorous controls (F. R. White, 1945; Bach, 1949). Analysis of the tumors of such animals, however, showed them to contain more nitrogen than was retained during the experimental period (Mider *et al.*, 1948). Thus part of the tumor protein must have been obtained at the expense of the tissues of the host. This ability of the tumor to draw directly on its host tissue proteins is also illustrated by the observation that tumor growth in animals maintained on a protein-free diet was still 75 % of the rate observed in protein-fed animals (F. R. White, 1945).

Observations such as these led Mider to describe the growing tumor as a "nitrogen trap" (Fenninger and Mider, 1954; Mider, 1955). In the present section an attempt will be made to discuss the effects of dietary, hormonal, and other factors on the competition between the tumor and its host for available nitrogen.

a. Dietary Protein and Tumor Growth. The growth pattern of most transplanted tumors may be divided into an initial latent period when little tumor growth is apparent, followed by a phase of more rapid growth. In some cases it would appear to be the former stage which is more sensitive to dietary changes. J. W. Green *et al.* (1950) found a significant increase in the latent period of the growth of the Walker 256 carcinoma in rats fed a protein-free diet and Allison and his colleagues (Allison, 1956) found the latent period of growth of a transplanted sarcoma to be dependent on the protein content of the diet. A dietary deficiency of lysine was shown to affect the growth of a mammary carcinoma in mice only during the early stages of tumor growth (F. R. White, 1945). However, Babson (1954) found that the latent period of development of the Flexner-Jobling carcinoma and the sarcoma R1 in rats was unaltered by diet although their growth rates did vary with protein intake.

Once the tumor has attained a certain size, it becomes less dependent on the diet and is able to utilize the tissues of its host to meet its requirements. Not all tumors develop this capacity to the same degree. Babson (1954) compared the growth of the Flexner-Jobling carcinoma and the sarcoma R1 in rats fed a protein-free diet. The latter tumor was much less affected by the low protein intake and, as might be anticipated, produced a greater weight loss in its host. However, the converse has also been observed. High levels of protein intake have been shown to have an adverse effect on the growth of some tumors (Allison *et al.*, 1954, 1961; Rosen *et al.*, 1961). Allison and his colleagues found that there was an optimal level of dietary protein for the growth of certain transplantable tumors, retardation by a high protein intake being attributable to the excess of methionine consumed (Allison *et al.*, 1956). Thus in rats bearing sarcomata, the giving of a high protein diet or a methionine supplement favored growth of the host's tissues at the expense of the tumor. On the other hand, Ghadially and Wiseman (1956) studied rats bearing the RD3 sarcoma and found that methionine supplementation retarded normal somatic growth but stimulated tumor growth. The effects of methionine supplements on tumor or somatic growth may be a consequence of increased adrenocortical activity, a known response to administration of excessive amounts of methionine (Munro and Mukerji, 1962).

b. Hormones and Tumor Growth. Certain tumors, generally found in organs in physiological or anatomical association with the endocrine system, have an absolute dependence on hormones for their growth (Noble, 1957). Examples of this type of tumor are estrogen-dependent mammary carcinoma and androgen-dependent carcinoma of the prostate. However, the growth of a much wider range of tumors has been shown to vary with modifications in the hormonal status of the host.

Since the action of androgens on protein metabolism is to cause nitrogen retention by the animal body (Chapters 9 and 10), androgen administration might be expected to reverse the flow of amino acids from carcass to tumor. However, Begg (1951b) observed no effect on the growth of the Walker 256 carcinoma when testosterone was given to rats.

On the other hand, adrenocortical hormones, which are known to promote mobilization of carcass protein in normal animals (Trémolières *et al.*, 1955), cause marked inhibition of the growth of Walker 256 carcinoma (Rosen *et al.*, 1961; Ingle *et al.*, 1950; Ingle, 1960) and of a mouse lymphosarcoma, but the Murphy-Sturm lymphosarcoma and the Novikoff hepatoma were not markedly affected (Rosen *et al.*, 1961). Suppression of the growth of Walker 256 carcinoma by cortisol is accompanied by a marked increase in the level of alanine-α-oxoglutarate transaminase activity in the tumor (Rosen *et al.*, 1961), which suggests that the hormone may act directly on the tumor. The situation is, however, confused because, although growth of the Walker 256 carcinoma is inhibited by excess corticosteroids, adrenalectomy also leads to inhibition of growth of this tumor (Ingle and Baker, 1951; Talalay *et al.*, 1952; Slawikowski, 1960).

Insulin and glucagon have also been shown to have an inhibitory effect on tumor growth (Salter *et al.*, 1958; Goranson *et al.*, 1959) and, when given in combination, the action of these two hormones is greater than the sum of their separate actions. Glucagon is known to cause an increase in nitrogen excretion (Salter *et al.*, 1957) and, since it increases the body weight loss of tumor-bearing animals, this hormone accordingly has an adverse action on protein metabolism in both the tumor and the host.

c. Tumor Growth in the Pregnant Host and in the Host with Regenerating Tissue. There is little effect on the growth of experimental tumors when the host animal is pregnant (Paschkis *et al.*, 1956). Partial hepatectomy has no inhibitory effect on growth of some tumors, and can actually enhance the growth of others (Paschkis *et al.*, 1955; Trotter, 1961; Leaders *et al.*, 1962). However, Straube and Hill (1956) observed a decrease in tumor growth in mice bearing an ascitic tumor during liver regeneration. The stimulant effect of partial hepatectomy on the growth of the Novikoff hepatoma in rats has been ascribed by Leaders *et al.* (1962) to the increased production of erythropoietic factor in the regenerating liver; they observed that a preparation of this factor was capable of stimulating the growth rate of this tumor.

From this survey of the factors in the host which influence tumor growth, it is apparent that the protein level in the diet can be a potent factor in tumor development during the early stages of induction, but is less important in the established tumor. In the latter case, there is an accompanying breakdown of the tissues of the host which will provide an alternative source of nitrogen for the tumor. Since different tumors vary in their response to a given hormonal stimulus, more data are required before any general conclusions can be made about the sensitivity of tumors as a class to the action of various hormones, such as can be made in the case of normal tissues (Chapters 9 and 10).

B. Protein Biosynthesis in Tumors

1. Protein Biosynthesis in the Cytoplasm

Studies on the uptake of labeled amino acids by the various subcellular fractions of tumor cells would appear to indicate that the basic reactions of protein biosynthesis in these cells are at least qualitatively similar to those of normal tissue (Chapter 6). Thus the microsomal protein is found to be the most highly labeled fraction of the tumor cell in both *in vitro* and *in vivo* experiments (Campbell *et al.*, 1957; Campbell, 1958) and Littlefield and Keller (1957), working with Ehrlich ascites tumor cells, found that the ribosomal component of the microsomes is the initial site of incorporation of amino acids into protein. The involvement of amino-acid-activating enzyme systems and transfer ribonucleic acid in the incorporation of amino acids into tumor microsomal protein may be inferred from the work of Greengard and Campbell (1959). Mitochondria isolated from tumor tissue, like those of normal tissues, can incorporate amino acids into protein by an energy-dependent process in the absence of microsomal and cell-sap fractions (Campbell and Greengard, 1959).

As discussed in Section I, the more rapidly growing anaplastic tumors tend to lose the structural elements of the parent tissue cell and to exhibit an embryonic type of cell with free ribosomes in place of organized endoplasmic reticulum. This implies that the cell has undergone a change in the pattern of protein synthesis, and Mercer (1961) has commented that the anaplastic cell is "devoting much of its synthetic machinery towards the formation of protein needed for cell division and the building of further cellular constituents."

On the other hand, some of the more differentiated tumors possess a very similar appearance to their tissues of origin. In these cases, where a well-defined system of endoplasmic reticulum persists, there is evidence that the neoplastic cell can still elaborate and secrete specific proteins. Thus plasma cell tumors have a well-differentiated system of granular endoplasmic reticulum (Askonas, 1961a) along with the ability to synthesize specific globulins which are secreted into the serum of their host (see Section IV,A). Since the physicochemical characteristics of these globulins have been found to be constant over 24 transplant generations (Potter and Fahey, 1960), it is apparent that in this particular tumor, at any rate, the control of protein synthesis is genetically stable despite a rapid rate of cell division (Askonas, 1961a).

2. Protein Biosynthesis in the Nucleus

It would appear that the synthesis of nuclear proteins is increased in tumor-bearing animals. Campbell and Stone (1957a) found under *in vitro* conditions that the incorporation of radioactive amino acids into nuclear proteins was much greater with slices from liver tumor than with slices from normal liver. With Krebs II mouse ascites tumor incubated *in vitro*, the nuclear protein fraction became more heavily labeled than that of any of the cytoplasmic fractions when incubation was continued for more than 30 minutes (Campbell *et al.*, 1957). This enhanced uptake was not observed in nuclei prepared from slices of regenerating liver.

The uptake of L-arginine-U-C^{14} into the histones of the Walker 256 carcinoma, the Jensen sarcoma, and the Ehrlich ascites tumor under *in vitro* conditions has also been shown to be higher than into the histones of nontumor tissue (Starbuck and Busch, 1960). An increased rate of uptake of amino acids by histones of tumors compared with that into histones of normal tissue has also been confirmed under *in vivo* conditions (Holbrook *et al.*, 1960; Butler and Lawrence, 1960; Busch *et al.*, 1959). In an attempt to discover whether the increase in labeling of the histone fraction was due to an increased synthesis of all histones or to certain specific ones, Davis and Busch (1959) separated the histones of nuclei from normal and tumor tissue of rats bearing the Walker 256 carcinoma, previously injected with L-lysine-U-C^{14}. They found that 30–37 % of the lysine incorporated into the acid-soluble proteins of tumor nuclei appeared in a cationic protein fraction not found in nontumor tissue. The presence of a similar highly labeled histone fraction was subsequently demonstrated in other types of malignant tumor (Davis and Busch, 1960). This particular histone fraction appears to be chromosomal in origin (Byvoet and Busch, 1961).

Thus protein biosynthesis in tumor cells does not apparently differ fundamentally from that in nonmalignant cells in that amino-acid-activating enzymes and a ribosomal system for amino acid assembly into proteins are present in both instances. No significant difference in the relative concentrations of the components of this system in neoplastic and normal cells has so far emerged, although there is a tendency for malignant cells to show a decrease in the amount of granular endoplasmic reticulum and an increase in free ribosomes in the cytoplasm. Protein synthesis in the nuclei of tumor cells appears to be quantitatively more important than in the nuclei of normal tissue cells and the presence in the former of a histone fraction with a rapid rate of synthesis which is absent from the nuclei of nontumor cells indicates a possible abnormality of protein biosynthesis in malignant cells. The findings which have been discussed do not provide any information on the means whereby the tumor cell redirects its metabolic pathways to the processes of reduplication at the expense of the synthesis of enzymes with specialized functions. Progress in this field must await further knowledge of the mechanism of the control of protein formation in normal mammalian cells.

C. Catabolism of Proteins and Amino Acids in Tumors

That the proteins of tumors are in a state of constant turnover has been shown by several workers. Babson and Winnick (1954) inoculated rats with a mince of Walker tumor which had been labeled previously with L-tyrosine-C^{14} and L-leucine-C^{14}. They found that there was a release of radioactivity from the tumor at such a rate as to suggest that the biological half-life of tumor protein was about 5 days. Moldave (1956) has also demonstrated the *in vivo* breakdown of Ehrlich ascites cell protein and Greenlees and LePage (1955) have shown that the protein of transplanted mammary carcinoma was available to some extent to the host.

In an attempt to exclude the possibility that cell death in tumors growing *in vivo* might be quite considerable and account for a large part of the observed protein breakdown, Miller and his colleagues grew Walker 256 carcinoma cells labeled with L-lysine-6-C^{14} (Jordan *et al.*, 1959) or L-leucine-1-C^{14} (M. Green and Miller, 1961) in tissue culture, where cell death is negligible, and followed the transfer of radioactivity from the cell to the medium. From their results they concluded that the Walker 256 carcinoma was capable of catabolizing its own protein at a constant rate of 12.4 % per day during the period studied.

The capacity of tumor cells to degrade proteins has been explored. Tumor cells grown *in vitro* have been found to secrete proteolytic enzymes into the culture medium to an extent which reflects the malignancy of the tumor (Holmberg, 1961). Green and Miller (1961) found that homogenates prepared from Walker 256 carcinoma cells grown in tissue culture were effective in hydrolyzing a large number of di- and tripeptides although they could not find any evidence for the occurrence of a trypsin-like enzyme, carboxypeptidase or cathepsin A. A variety of tumors growing *in vivo* were also shown by Wu and Bauer (1963) to possess significant peptidase activity. Moreover these workers found that the activities in various types of hepatomas were similar to or higher than that of normal liver and concluded that in the process of malignant transformation peptidase activities are retained or actually increased.

Enzymes involved in amino acid metabolism other than those hydrolyzing peptide bonds have also been studied in tumor cells. Before discussing this further it is well to point out that variations in enzyme content have been found to occur in different regions of a growing tumor. This is admirably illustrated by the work of Malmgren and Sylven (1959), who studied the peptidase and arginase levels of serial histological sections of three different solid mouse tumors, sarcoma 37, a transplanted mammary carcinoma, and a melanoma, and found that the cells in the peripheral region had at least three to four times as much arginase and peptidase activity per cell as the central regions of older but still viable cells, while necrotic regions of the tumor had no activity at all. Similarly, histidine decarboxylase activity in growing mouse ascites tumor was found to be related to the frequency of tumor cell mitosis and not to the total number of cells (Kahlson *et al.*, 1962).

Just as there is no special structural feature which can characterize a cancer cell as distinct from a normal cell, so there would not appear to be any particular pattern of enzymic activities common to all tumors by which they might be distinguished from nontumorous tissue. Greenstein (1954), however, from an extensive study of the occurrence of various enzymes, including several involved in amino acid metabolism in neoplastic and non-neoplastic tissue, concluded that tumors as a class differ less in enzymic content from one another than they do from their tissues of origin. This would require, as Weinhouse (1960) has pointed out, that those enzymes which are involved in special functional activities in a normal tissue should tend to become deleted in tumors derived from that tissue.

Liver normally has the highest concentration of most of the enzymes involved in the catabolism of the amino acids. It would therefore be expected that a highly undifferentiated hepatoma would have reduced amounts of these enzymes. This has indeed been found for the Novikoff hepatoma. Auerbach and Waisman (1958) found that in this tumor tryptophan pyrrolase, tyrosine transaminase, phenylalanine hydroxylase, threonine dehydrase, serine dehydrase, cysteine desulfhydrase, histidase, and *p*-hydroxyphenylpyruvate oxidase were either absent or present only in small amounts. The Morris hepatoma 5123, on the other hand, is a well-differentiated tumor and has been found to contain some of the above enzymes at significant levels of activity (Pitot *et al.*, 1961; Pitot and Morris, 1961). Thus complete deletion of amino-acid-catabolizing enzymes would not appear to be a prerequisite for malignancy. However one major difference in the behavior of some of these enzymes in this hepatoma from that in normal liver was observed. Induction of serine and threonine dehydrase by feeding high protein diets and tyrosine transaminase and tryptophan pyrrolase by administration of their respective substrates occurred in liver but not in hepatoma. Pitot and Morris (1961) therefore concluded that the cause of the malignancy in this case was related to the possibility that certain repressor mechanisms governed by substrate and products were partly or completely inoperative in the hepatoma cell.

Thus, tumor cells do possess proteolytic enzymes and it would also appear that there is a continual turnover of intracellular protein. However, defects in enzyme systems involved in the further catabolism of amino acids have been observed. This is of particular significance in hepatomas since their tissue of origin has a high content of these enzymes. Amino acids resulting from proteolysis in tumor tissues would not be expected to be degraded to the same extent as in normal tissues and would therefore be more readily available for further protein synthesis, a situation compatible with the demands of a rapidly growing tissue.

III. Protein Metabolism in Precancerous Tissue

Hepatocarcinogenesis, induced in rats by the prolonged administration of certain azo dyes, dimethylnitrosamine, 2-aminofluorene, and various other substances (Miller and Miller, 1953; Berman, 1958), offers an opportunity to study alterations in metabolism during the development of neoplasia. Hepatomas resulting from these treatments may originate in either parenchymal or bile duct cells with subsequent formation of different types of tumor. Weber (1961) has discussed the variety of cell types which may occur in liver during treatment with these carcinogens and the importance of cellular population on biochemical function.

Within a few hours of commencement of administration of various carcinogens, an inhibition of amino acid incorporation into protein has been observed (Emmelot and Benedetti, 1961; Hultin *et al.*, 1960; Arrhenius and Hultin, 1962). Brouwers and Emmelot (1960) concluded that this inhibition was exerted at the microsomal level, a finding of interest in view of the fact that carcinogenic azo dyes and aminofluorene bind preferentially with the protein of the membrane portion of the endoplasmic reticulum (= microsomes in homogenates), at least during early stages in the period of administration (Hultin, 1956, 1960; Fiala and Fiala, 1959; Westrop and Green, 1960). It is this fraction which contains the *N*-demethylases and hydroxylases which catabolize the carcinogens and it has been suggested that the reaction products of these enzymes may act preferentially with microsomal proteins (Emmelot and Benedetti, 1961; Hultin *et al.*, 1960). Structural alterations in the endoplasmic reticulum, involving a marked reduction in particle-associated portions of the endoplasmic reticulum, and hypertrophy of the agranular or smooth form of this membrane system, were observed during the first few hours of treatment with dimethylnitrosamine or 3′-methyl-4-dimethylaminoazobenzene (Emmelot and Benedetti, 1961; Porter and Bruni, 1959). However, this period of inhibition of microsomal protein synthesis is only transient and is followed by a period marked by an increased incorporation of amino acids into microsomal protein (Arrhenius and Hultin, 1962). This secondary stimulation of amino acid incorporation has been interpreted as being mainly due to an increased sensitivity of the liver to the action of glucocorticoids (Mukherjee *et al.*, 1963) since it is accompanied by a large deposition of hepatic glycogen (Mukherjee *et al.*, 1963) and does not occur in adrenalectomized animals (Arrhenius and Hultin, 1962). The significance of these early changes in the development of neoplasia in the livers of carcinogen-treated rats is not clear.

At a later stage in the feeding of these carcinogens there comes a period of intensive cell proliferation (Laird and Barton, 1960). Laird and Barton have suggested that this production of new cells may be compensating a loss of some functional capacity of the liver which is inactivated or destroyed by the carcinogen. Burke and Miller (1956, 1959a,b) studied amino acid metabolism in the isolated perfused liver preparation from normal rats and rats which had been maintained for 2–3 months on a diet containing 3′-methyl-4-dimethylaminoazobenzene. There appeared to be an over-all decrease in pathways of amino acid catabolism in the precancerous liver. This was demonstrated by the finding of decreased urea production from arginine or aspartic acid and a decrease in production of $C^{14}O_2$ when L-lysine-6-C^{14} was added to the perfusate. The incorporation of labeled L-lysine-6-C^{14} into both liver and plasma proteins was increased in the precancerous liver. Similar results have been obtained with livers from rats which had been fed 2-acetylaminofluorene (Burke and Miller, 1960). The impaired urea synthesis in these livers is believed to be due to a decrease in arginase activity. When either glutamine or ammonium carbonate was added to the perfusate, however, normal or enhanced levels of urea production were observed (Burke and Miller, 1959b). Burke and Miller have suggested that this result may be explained by the existence of a pathway of urea formation from glutamine-amide N independent of the Krebs-Henseleit cycle. In the perfused regenerating liver there is a similar picture of decreased amino acid catabolism, with a decreased rate of urea synthesis (Burke, 1962). Burke has shown, however, that the reduced ureogenesis in this case is not due to any defect in the enzymes of the Krebs-Henseleit cycle. The degradation of tryptophan by the precancerous liver would also appear to be impaired since the levels of hepatic tryptophan pyrrolase and kynureninase were decreased in rats fed 3′-methyl-4-dimethylaminoazobenzene (Petasnick *et al.*, 1961).

The alterations observed in protein metabolism in the precancerous state are thus similar to those in the fully developed malignant cell. There would appear to be a generalized deletion of enzymes concerned with amino acid catabolism and an increased incorporation of amino acids into protein associated with intensive cell proliferation. The significance of the structural changes observed in the endoplasmic reticulum which occur in the early stages is not clear but may reflect a reorganization of the control of protein synthesis in the premalignant cell.

IV. The Protein and Amino Acid Metabolism of the Tumor-Bearing Host

A. Alterations in Nitrogenous Constituents of Blood

The mobilization of tissue protein in the tumor-bearing animal to meet the requirements of the neoplasm might be expected to result in alterations in the nitrogenous constituents of blood. Babson (1954) did indeed note a progressive rise in the level of nonprotein N in the blood of tumor-bearing animals throughout the period of tumor growth. Relatively large increases in blood amino acid concentration have also been shown to accompany tumor growth, the magnitude of the increase depending on the size of the tumor (El Mehairy, 1950; Wu and Bauer, 1960). Plasma glutamic acid appears to be especially elevated in tumor-bearing animals (McHenry, 1955).

In view of the increased capacity of tumors to utilize plasma proteins as a source of amino acids, alterations in the plasma proteins of the tumor-bearing host might be expected. This has been shown to be the case. Mider *et al.* (1950), in a study of plasma proteins in a large series of cancer patients, found evidence of a fall in plasma albumin level even where the cancerous lesion was relatively localized. An early change in plasma albumin has also been observed by Hradec (1958), who noted an increased turnover of serum albumin as early as 3 days after implantation of experimental tumors in rats, before any tumor growth was apparent. This evidence indicates that the increased rate of turnover cannot be attributed to utilization of albumin by the tumor. In a study of albumin synthesis by liver slices of tumor-bearing rats, Hradec also found that the liver tends to compensate for this increased turnover by an increased rate of synthesis. Allison *et al.* (1961) showed that the fall in serum albumin observed in rats bearing the Walker 256 carcinoma could be prevented by feeding diets of high protein content.

As well as undergoing a quantitative change during tumor growth, qualitative changes in the plasma albumin fraction of tumor-bearing animals have also been reported. Berlinguet and Loiselle (1959) found that the albumin fraction of the serum of rats bearing the Walker 256 carcinoma had a different electrophoretic mobility from normal rat serum albumin and contained a much higher polysaccharide content. An antigen to anti-Walker antibody also appeared to be associated with this fraction.

The globulin fractions of the plasma proteins of tumor-bearing subjects also undergo alterations. An increase in the plasma globulin fraction of cancerous patients was observed by Mider *et al.* (1950), the alteration in the α-globulin component being more pronounced than that of either the β or γ-components. Increases in one or more specific components of the α-globulin fraction of the serum of experimental tumor-bearing animals have been frequently reported (Bernfeld and Homburger, 1955; Campbell *et al.*, 1959; Darcy, 1960a,b). These increases in serum α-globulin components are not characteristic of neoplastic growth alone, since similar increases occur in the serum of pregnant rats or rats with regenerating livers. Darcy has concluded that, in tumor growth, the increased α-globulin component does not originate at the site of the tumor, but he suggests that the tumor may selectively absorb it from the surrounding fluids. The significance of these α-globulin components is not know. The carbohydrate and lipid moieties of the various globulin fractions in the plasma of rats bearing the Walker 256 carcinoma have also been found to be increased (Allison *et al.*, 1961).

In mice bearing plasma-cell tumors or in humans with multiple myeloma, other abnormalities in the globulin fractions of plasma have been observed. Cryoglobulins have been found in the plasma of patients with multiple myeloma (Putnam and Miyake, 1956) and increased globulin levels, which may be due to an increase in β- or γ-globulin components, are frequently associated with these conditions (Potter and Fahey, 1960). These abnormalities in globulin production are not apparently due to an alteration in globulin synthesis in the tissues of the host since different types of plasma cell neoplasms in mice have been shown to produce different but characteristic globulins which appear in the serum (Potter and Fahey, 1960; Askonas, 1961b; Askonas and Fahey, 1961).

B. Alterations in Nitrogenous Constituents of Urine

Malignant growths cause wasting of the tissues of the host. Despite this, the total urinary N of the tumor-bearing subject or animal is not usually increased and may even be decreased due to the action of the tumor in retaining nitrogen (F. R. White, 1945; Bach and Maw, 1953). Bach and Maw (1953) studied the excretion of ammonia, urea, creatinine, and creatine in rats bearing the Jensen sarcoma during the growth of the tumor. They found that the levels of ammonia, urea, and creatinine in the urine of the tumor-bearing animals were not significantly different from control levels but that there was a very considerable increase in creatine excretion. Since this increase in creatinuria occurred at stages of tumor growth where there was no decrease in muscle creatine concentration, they concluded that an increased endogenous synthesis of creatine was associated with tumor growth. Bauer and Wu (1959), however, could not demonstrate an increase in creatine excretion in rats bearing the Walker 256 carcinoma, even when there was a decrease in total body creatine.

Increased excretion of allantoin (Begg, 1958; Wu and Bauer, 1962) and of uric acid (Wu and Bauer, 1962) has been reported in tumor-bearing animals, the extent of the increases being related to tumor size. Wu and Bauer (1962) suggest that the increased amounts of allantoin and uric acid in the urine arise from necrotic processes occurring in the tumor and also from the increased turnover of nucleic acids known to occur in the tissues of tumor-bearing animals (Payne *et al.*, 1952).

There appears to be little departure from the normal pattern of urinary amino acids in cancerous patients save for a tendency for lowered excretions of glycine and histidine (Eades and Pollack, 1954). Wu and Bauer (1960) found an increased excretion of a nonprotein-bound form of amino acids in rats bearing large Walker 256 carcinomas.

Proteinuria is not a common feature of tumor growth. However patients with multiple myeloma or mice bearing plasma-cell tumors, as well as showing abnormalities in their serum globulin components, may also excrete proteins of a type known as Bence-Jones proteins. These proteins are characterized by being precipitated between 45° and 55°C and redissolved as the temperature is further increased. They have a molecular weight of about 45,000 and vary in their physicochemical properties and structure from patient to patient (Putnam and Stelos, 1953). Since Bence-Jones proteins possess antigenic groupings common to both serum γ-globulins and the myeloma serum globulins, the relationship between these different groups of proteins is a matter of some interest. As a result of studies with isotopically labeled glycine carried out in patients with multiple myeloma, it has been shown that the urinary Bence-Jones proteins become labeled very rapidly and reach a higher specific activity than the serum myeloma protein, a finding which shows that the Bence-Jones protein is not a breakdown product of the serum globulins (Putnam and Hardy, 1955; Hardy and Putnam, 1955). The possibility that Bence-Jones protein might be a precursor of the serum myeloma protein has also been investigated by Askonas and Fahey (1961). As a result of their studies on the *in vitro* rates of incorporation of C^{14}-amino acids by a mouse plasma-cell tumor into Bence-Jones and myeloma proteins, these workers conclude that, while both proteins are formed by the tumor, there is no evidence for a precursor relationship.

C. Alterations in Protein Content of Host Tissues during Tumor Growth

As has already been discussed, the growing tumor can cause mobilization of the protein of the tissues of its host to provide for its own protein anabolism. Sherman *et al.* (1950) made a study of the nitrogen content of various tissues of rats bearing the Walker 256 carcinoma and found that muscle was the tissue which contributed most nitrogen to the tumor. Not all the muscles of the tumor-bearing animal were equally depleted; cardiac muscle showed a decrease in nitrogen content only in the terminal stages of tumor growth (cf. the resistance of cardiac muscle to protein depletion). In cancerous patients there is evidence of a fall in certain muscle protein components. Ritter (1958) separated muscle protein into three main fractions by electrophoretic techniques and found that cancerous patients had decreased amounts of two of these fractions. Lymphatic tissue (Sherman *et al.*, 1950), seminal vesicles (Babson, 1954), testes (Claus *et al.*, 1962), and thymus (Begg, 1951a) are other tissues which have been reported as undergoing atrophy during the growth of certain tumors.

Not all the tissues of the tumor-bearing animal suffer a depletion of protein. Renal nitrogen content did not appear to alter during the growth of the Walker 256 carcinoma in rats (Sherman *et al.*, 1950), while the spleen and liver have actually been shown to increase in nitrogen content in the presence of a growing tumor (Sherman *et al.*, 1950; Yeakel and Tobias, 1951; Stewart and Begg, 1953; Clark and Goodlad, 1960) and hypertrophy of the adrenals is frequently observed in tumor-bearing animals (Begg, 1958). The increase in the nitrogen content of spleen observed during the growth of the Walker 256 carcinoma was partly attributable to the occurrence of extramedullary hematopoiesis (Sherman *et al.*, 1950), a not uncommon finding in the tissues of tumor-bearing animals (Begg, 1958), possibly due to anemia. Woodruff and Symes (1962), however, have suggested the possibility of splenomegaly in tumor-bearing mice being associated with an immunological response to a tumor antigen.

The increase in liver nitrogen does not persist throughout the whole period of tumor growth and the liver undergoes a decrease in protein content during the terminal stages (Sherman *et al.*, 1950). There is some evidence which suggests that part, at any rate, of the increase in liver protein content may be due to the formation of new liver cells (Annau *et al.*, 1951; Hemingway, 1960; Clark and Goodlad, 1960). Hepatomegaly in rats bearing the Walker 256 carcinoma appears to be dependent to some extent on the thyroid since there was no liver enlargement in thyroidectomized or thiouracil-treated tumor-bearing rats (Girkin and Kampschmidt, 1961).

The relationship of these shifts in body protein to the effects of other agents on body protein distribution are discussed in more detail in Chapter 10.

D. Alterations in Intermediary Protein Metabolism of Host Tissues during Tumor Growth

At the present time there is insufficient information available on the intermediary metabolism of protein in the tissues of the tumor-bearing animal to explain the gross changes in protein content discussed above. However, certain alterations in specific proteins and enzymes concerned in protein and amino acid metabolism have been observed in the tissues of the host.

Wu and Bauer (1960) found a decrease in the total free amino acid concentration of muscles of rats bearing the Walker 256 carcinoma at a stage of tumor growth when the carcass was undergoing extensive loss of protein. This might be interpreted in terms of a rapid translocation of amino acids away from muscle or of a decrease in the ability of the muscle cell to accumulate amino acids. A similar decrease in the concentration of glutamine and certain amino acids was noted in the leg muscles of rats bearing an ascitic form of the Murphy lymphosarcoma (Roberts and Simonsen, 1960).

The amino acid levels of the liver, brain, and kidney of tumor-bearing animals have not been observed to alter significantly, though a decrease in free glutamine has been reported in the case of liver and kidney (Wu and Bauer, 1960; Roberts and Simonsen, 1960). Wu and Bauer believe that the fall in tissue glutamine level might be related to the requirements of the tumor for this amide for protein synthesis (see Section II). However, they also found that the synthesis of hepatic glutamine was impaired.

Moreover, there would appear to be a derangement in glutamic acid metabolism in the tissues of the tumor-bearing animal (McHenry, 1955). Both aspartate-α-oxoglutarate and tyrosine-α-oxoglutarate transaminases have been found to be elevated (J. M. White *et al.*, 1954; Goodlad and Clark, 1962; Auerbach and Waisman, 1958), whereas alanine-α-oxoglutarate transaminase has been found to be decreased in the livers of tumor-bearing animals (Beaton *et al.*, 1957; Goodlad and Clark, 1962). Dyer and his colleagues (1961) found an elevation in serum aspartate-α-oxoglutarate transaminase in rats bearing various transplanted tumors, presumably resulting from an increased destruction of the tissues of the host.

There is a reduction in the content of certain porphyrin-containing enzymes (catalase, tryptophan pyrrolase, and cytochrome c) in the livers of tumorous animals. Thus a decrease in the level of hepatic catalase has been one of the most consistent findings accompanying tumor growth of various kinds (Greenstein, 1954) and a fall in tryptophan pyrrolase activity in the livers of tumor-bearing animals has been reported, at least during certain stages of the growth of the tumor (Wood *et al.*, 1956; Kawachi *et al.*, 1961). There is also a progressive decrease in the cytochrome c content of various tissues of rats bearing the Walker 256 carcinoma (Kampschimdt *et al.*, 1959a), although an increase in hepatic cytochrome oxidase activity has been found to accompany the growth of this tumor (Greene and Haven, 1957). In the case of catalase, data obtained by chromatographic (Greenfield and Price, 1954) and immunological (Nishimura *et al.*, 1962) techniques suggested that there was a decreased rate of synthesis of this enzyme. There is no reason to believe that synthesis of tryptophan pyrrolase is impaired in the tumor-bearing animal, even although the level is low, for it has been observed in hepatoma-bearing rats that substrate induction of this hepatic enzyme still occurs on administration of tryptophan (Pitot and Morris, 1961). Since an increase in hepatic protoporphyrin and urinary coproporphyrin has been reported in tumor-bearing animals (Ono *et al.*, 1956), there would appear to be no defect in porphyrin biosynthesis but rather a failure in its utilization.

Several workers have reported increased levels of catheptic activity in various tissues of tumorous animals (Feinstein, 1950; Babson, 1956; Migliarese, 1956; Benz and Lehmann, 1959). Since the function of these enzymes is obscure, interpretation of this finding is difficult.

It is obvious, therefore, that on the evidence at present available on changes in the intermediary metabolism of proteins and amino acids in the tissues of the tumor-bearing host, no explanation can be offered regarding the means whereby the tumor brings about its effect on these tissues. Defects in the metabolism of glutamate and in the utilization of porphyrins would appear to be the most characteristic of the changes so far reported but the significance of these abnormalities is not clear.

V. Mechanism of Tumor Action on the Protein Metabolism of the Host

A. The Tumor as a Nitrogen Trap

Wiseman and Ghadially (1958) have postulated that the enhanced ability of the neoplastic cell to concentrate amino acids causes a continued deprivation of the cells of the tissues of the host of the precursors of their proteins, as a result of which these normal cells eventually suffer a lack of one or more essential amino acids, protein synthesis ceases, and tissue protein breakdown occurs. This theory, however, does not explain the protein anabolism observed in the liver during certain stages of tumor growth, although competition between tumor and host presumably contributes to the picture.

Since tumor growth affects certain endocrine organs (Section IV,C), the possibility exists that a resultant hormonal imbalance may account to some extent for the accompanying alterations observed in the protein metabolism of the host.

B. Effect of Tumor on Hormonal Balance of the Host

In the tumor-bearing animal it would appear that there is an impaired production or utilization of certain hormones with an anabolic action on protein metabolism. Mention has already been made of the atrophy of the testes and the seminal vesicles observed in male tumor-bearing rats (Claus *et al.*, 1962; Babson, 1954) which would suggest the impairment of androgen production. The tumor-bearing animal is not diabetic but it shows an impaired hepatic glycogen deposition to a glucose-load test (Goranson, 1955) which can be restored to normal by the administration of insulin (Begg, 1955), suggesting that there is either an impaired production of or a decreased sensitivity to this hormone.

On the other hand, other endocrine organs are affected by tumor growth in such a manner that a resulting increase in protein catabolism might be expected. Hypothyroidism, as is well known, leads to growth inhibition and there is some evidence to suggest that thyroid function may be impaired in tumorous animals (Claus *et al.*, 1962). There is also evidence to suggest that the adrenals of the tumor-bearing animal may be hyperfunctional, at least during some stages of tumor growth in experimental animals, with respect to both glucocorticoid and aldosterone production (Begg, 1958; Hilf *et al.*, 1960; Kavetsky *et al.*, 1962; Rechcigl *et al.*, 1961). Increased urinary excretions of 17-hydroxycorticosteroids and 17-oxosteroids have also been reported in cancer patients (Werk and Sholiton, 1960; Kavetsky *et al.*, 1962). Corticosteroid administration has a similar action to that of a growing tumor on protein distribution in the tissues of the animal body in that it causes a breakdown of carcass protein and a deposition of liver protein (Silber and Porter, 1953; Goodlad and Munro, 1959). However, Clark and Goodlad (1960) observed differences between the action of a growing tumor and the administration of cortisone on liver composition. Livers of rats bearing the Walker 256 carcinoma exhibited a much greater increase in ribonucleic acid content relative to protein than was observed in the hormone-treated rats. Liver deoxyribonucleic acid was increased in tumor-bearing rats but not in rats treated with cortisone. It was therefore concluded that the alterations in liver composition in rats bearing the Walker 256 carcinoma were effected by a different mechanism from that by which cortisone exerts its effect.

Certain other findings in the livers of tumor-bearing rats are not consistent with the idea that these animals are in a hypercortical condition. Hepatic alanine-α-oxoglutarate transaminase activity is increased following adminstration of corticotropin or corticosteroids and decreased following adrenalectomy in the normal animal (Rosen *et al.*, 1959; Harding *et al.*, 1961; Miyabo, 1959) while as has already been mentioned (Section IV,D) there is a fall in the level of this enzyme in the tumor-bearing animal. Increased liver mitosis has been observed in the livers of tumorous animals (Annau *et al.*, 1951; Hemingway, 1960) though corticosteroids are known to inhibit this process (Hemingway, 1961). Finally Ingle (1960) found that steroid glycosuria could not be established by injection of cortisone into rats bearing a well-established Walker 256 carcinoma and suggested that the tissues of tumor-bearing animals showed a decreased sensitivity to the action of corticosteroids.

These findings may be explained by the fact that the livers of rats bearing the Walker 256 carcinoma have an increased ability to inactivate corticosteroids by virtue of a rise in Δ^4-steroid hydrogenase activity (Goodlad and Clark, 1961). Evidence in favor of the level of hepatic steroid inactivation regulating the activity of certain corticoid-dependent enzymes in the liver has been presented by Lin *et al.* (1959). Urquhart *et al.* (1959) have shown a close correlation between the level of this enzyme system in the liver and adrenal size in several physiological and experimental conditions and have suggested that it is the level of this enzyme which determines adrenal size rather than the reverse. An increase in the rate of inactivation of glucocorticoids by the liver would cause a lowering of their level in the blood. Since the rate of secretion of corticotropin is under a negative feedback control, there would be an increased output of this pituitary hormone which would promote adrenal enlargement. Hence the increased hepatic Δ^4-steroid hydrogenase activity observed may also explain the occurrence of adrenal hypertrophy in tumor-bearing animals. In this connection the finding that testosterone implantation prevented adrenal hypertrophy in tumor-bearing rats (Begg, 1951b) is of interest in view of the fact that this hormone causes a depression in hepatic Δ^4-steroid hydrogenase activity (Yates *et al.*, 1958). That the breakdown in carcass protein which occurs in the tumor-bearing animal may also be unrelated to any increase in adrenal secretion might be inferred from the finding that the presence of a growing tumor actually reduces the carcass weight loss following cortisone administration (Ingle, 1960).

Thus the evidence presented suggests that in the tumor-bearing animal there may be a decreased production of certain hormones with a protein-anabolic action (insulin, androgens). The enlarged adrenals which are a characteristic of a tumor-bearing animal, might suggest an increased secretion of glucocorticoids, the main protein-catabolic group of hormones. However, certain findings indicate that the alterations in protein metabolism in the host are not due to the animal being in an hypercortical state.

C. Secretion of Humoral Agents by the Tumor

Whatever may be the mechanism by which the tumor brings about the various alterations in the tissues of its host, studies on the effect of administration of extracts of tumor tissue of varying degrees of purity have established that the tumor contains factors capable of producing similar systemic changes to that produced by the growing tumor.

In a search for extracts of tumors which, on injection, would produce the characteristic depression of liver catalase, Nakahara and Fukuoka (1948) prepared an active fraction from human tumors which they termed "toxohormone." The presence of a similar factor or factors in various tumors was confirmed by Greenfield and Meister (1951). These latter workers claimed that the active principle was contained in material which was nitrogenous in nature, dialyzable, resistant to acid, and thermostable. There is some evidence that the factor may also occur in a nondialyzable form (Nakahara and Fukuoka, 1954; Yunoki and Griffin, 1960). Yunoki and Griffin (1960, 1961) have further purified toxohormone preparations by the use of column chromatographic techniques. They isolated several proteolipids of rather low molecular weight (4200–6400) all with arginine as N-terminal amino acid and all possessing toxohormone activity. It appeared that both the polypeptide and lipid components were necessary for biological activity.

Partially purified toxohormone preparations are also effective in evoking other systemic changes associated with tumor growth. Kampschmidt and his co-workers (Kampschmidt *et al.*, 1959b, 1960) found that partially purified extracts prepared from the Walker 256 carcinoma or Flexner-

Jobling carcinoma when injected into normal rats resulted in a decreased concentration of plasma iron, inhibition of liver and kidney catalase activity, and increases in the weights of liver, spleen, and adrenals. Similar extracts prepared from Jensen sarcoma or Novikoff hepatoma significantly altered the spleen weight and plasma iron concentration, the only changes which were observed in the host during early periods of growth of these two tumors. The mitochondrial fraction of the tumor cell appeared to be the source of the factor(s) causing the various systemic changes (Kampschmidt and McCoy, 1960).

Normal tissues contain substances which have "toxohormone-like" properties, though at much lower concentrations than in tumor tissue (Day *et al.*, 1954; Kampschmidt *et al.*, 1960). Other effects in normal rats produced by administration of extracts and breis of tumors which are of similar nature to that evoked by the growing tumor are stimulation of liver mitosis (Malmgren, 1956), increased creatine excretion (Tedeschi and Mangiantini, 1957), increased plasma lactic dehydrogenase (Riley *et al.*, 1960), and decreased hepatic tryptophan pyrrolase (Kawachi *et al.*, 1961).

To date no evidence has been obtained for the actual secretion of any factors by the tumor into the blood stream of its host. However, McLaughlin (1960) has claimed that the plasma of patients with cancer possesses a factor (or factors) absent from the plasma of noncancerous individuals which is capable of stimulating liver regeneration in rats.

Recent work by Kampschmidt and his co-workers (Kampschmidt and Schultz, 1963; Kampschmidt and Upchurch, 1963) has raised doubts as to the significance of toxohormone in tumor host relationships. These workers have found that toxohormone activity is due to bacterial contamination of the tumor and that toxohormone cannot be extracted from bacteria-free tumors. However, in the absence of bacterial infection tumor growth still caused changes in the host. Kampschmidt and Upchurch (1963) have pointed out that, "Even if all the effects on the host now attributed to the tumor were found to be due to bacterial or viral contamination, their study would still be important, since they occur extensively not only in transplantable animal tumors but also in the human cancer patient."

References

Allison, J. B. (1956). *Am. J. Clin. Nutr.* **4,** 662.

Allison, J. B., Wannemacher, R. W., Jr., Hilf, R., Migliarese, J. F., and Crossley, M. L. (1954). *J. Nutr.* **54,** 297.

Allison, J. B., Wannemacher, R. W., Jr., Hilf, R., and Hetzel, C. A. (1956). *J. Nutr.* **59,** 27.

Allison, J. B., Wannemacher, R. W., Jr., Parmer, L. P., and Gomez, R. (1961). *J. Nutr.* **74,** 176.

Annau, E., Manginelli, A., and Roth, A. (1951). *Cancer Res.* **11,** 304.

Arrhenius, E., and Hultin, T. (1962). *Cancer Res.* **22,** 823.
Askonas, B. A. (1961a). *In* "Protein Biosynthesis" (R. J. C. Harris, ed.), pp. 363–378. Academic Press, New York.
Askonas, B. A. (1961b). *Biochem. J.* **79,** 33.
Askonas, B. A., and Fahey, J. L. (1961). *Biochem. J.* **80,** 261.
Auerbach, V. H., and Waisman, H. A. (1958). *Cancer Res.* **18,** 543.
Babson, A. L. (1954). *Cancer Res.* **14,** 89.
Babson, A. L. (1956). *Science* **123,** 1082.
Babson, A. L., and Winnick, T. (1954). *Cancer Res.* **14,** 606.
Bach, S. J. (1949). *Biochem. J.* **44,** xxv.
Bach, S. J., and Maw, G. A. (1953). *Biochim. Biophys. Acta* **11,** 69.
Bauer, J. M., and Wu, C. (1959). *Proc. Am. Assoc. Cancer Res.* **3,** 5.
Beaton, G. H., Curry, D. M., and Veen, M. J. (1957). *Arch. Biochem. Biophys.* **70,** 288.
Begg, R. W. (1951a). *Cancer Research* **11,** 341.
Begg, R. W. (1951b). *Cancer Research* **11,** 406.
Begg, R. W. (1955). *Can. Cancer Conf.* **1,** 237.
Begg, R. W. (1958). *Advan. Cancer Res.* **5,** 1.
Benz, G., and Lehmann, F. E. (1959). *Helv. Physiol. Pharmacol. Acta* **17,** 380.
Berlinguet, L., and Loiselle, J. M. (1959). *Can. J. Biochem. Physiol.* **37,** 1401.
Berman, C. (1958). *Advan. Cancer Res.* **5,** 55.
Bernfeld, P., and Homburger, F. (1955). *Cancer Res.* **15,** 359.
Bernhard, W. (1958). *Cancer Res.* **18,** 491.
Birbeck, M. S. C., and Mercer, E. H. (1961). *Nature* **189,** 558.
Brouwers, J. A. J., and Emmelot, P. (1960). *Exptl. Cell Res.* **19,** 467.
Burke, W. T. (1962). *Cancer Res.* **22,** 10.
Burke, W. T., and Miller, L. L. (1956). *Cancer Res.* **16,** 330.
Burke, W. T., and Miller, L. L. (1959a). *Cancer Res.* **19,** 149.
Burke, W. T., and Miller, L. L. (1959b). *Cancer Res.* **19,** 622.
Burke, W. T., and Miller, L. L. (1960). *Cancer Res.* **20,** 658.
Busch, H. (1955). *Cancer Res.* **15,** 365.
Busch, H., and Greene, H. S. N. (1955). *Yale J. Biol. Med.* **27,** 339.
Busch, H., Davis, J. R., Honig, G. R., Anderson, D. C., Nair, P. V., and Nyhan, W. L. (1959). *Cancer Res.* **19,** 1030.
Busch, H., Fujiwara, E., and Firszt, D. C. (1961). *Cancer Res.* **21,** 371.
Butler, J. A. V., and Lawrence, D. J. R. (1960). *Brit. J. Cancer* **14,** 758.
Byvoet, P., and Busch, H. (1961). *Nature* **192,** 870.
Campbell, P. N. (1958). *Advan. Cancer Res.* **5,** 97.
Campbell, P. N., and Greengard, O. (1959). *Biochem. J.* **71,** 148.
Campbell, P. N., and Stone, N. E. (1957a). *Biochem. J.* **66,** 19.
Campbell, P. N., and Stone, N. E. (1957b). *Biochem. J.* **66,** 669.
Campbell, P. N., Greengard, O., and Jones, H. E. H. (1957). *Exptl. Cell Res.* **12,** 689.
Campbell, P. N., Kernot, B. A., and Roitt, I. M. (1959). *Biochem. J.* **71,** 155.
Chin, L. E., Tannenberg, W. J., and Moldave, K. (1959). *Proc. Soc. Exptl. Biol. Med.* **101,** 140.
Christensen, H. N., and Henderson, M. E. (1952). *Cancer Res.* **12,** 229.
Christensen, H. N., and Rafn, M. L. (1952). *Cancer Res.* **12,** 495.
Christensen, H. N., and Riggs, T. R. (1952). *J. Biol. Chem.* **194,** 57.
Clark, C. M., and Goodlad, G. A. J. (1960). *Brit. J. Cancer* **14,** 327.
Claus, J. L., Trunnell, J. B., and Llaurado, J. G. (1962). *Acta Endocrinol.* **40,** 584.

Cohen, S., Beiser, S. M., and Hsu, K. C. (1961). *Cancer Res.* **21,** 1510.
Coles, N. W., and Johnstone, R. M. (1962). *Biochem. J.* **83,** 284.
Darcy, D. A. (1960a). *Brit. J. Cancer* **14,** 525.
Darcy, D. A. (1960b). *Brit. J. Cancer* **14,** 534.
Davis, J. R., and Busch, H. (1959). *Cancer Res.* **19,** 1157.
Davis, J. R., and Busch, H. (1960). *Cancer Res.* **20,** 1208.
Day, E. D., Gabrielson, F. C., and Lipkind, J. B. (1954). *J. Natl. Cancer Inst.* **15,** 239.
Dyer, H. M., Gullino, P. M., Ensfield, B. J., and Morris, H. P. (1961). *Cancer Res.* **21,** 1522.
Eades, C. H., and Pollack, R. L. (1954). *J. Natl. Cancer Inst.* **15,** 421.
Eagle, H. (1955). *J. Exptl. Med.* **102,** 37.
El Mehairy, M. M. (1950). *Brit. J. Cancer* **4,** 95.
Emmelot, P., and Benedetti, E. L. (1961). *In* "Protein Biosynthesis" (R. J. C. Harris, ed.), pp. 99-124. Academic Press, New York.
Feinstein, R. N. (1950). *Cancer Res.* **10,** 93.
Fenninger, L. D., and Mider, G. B. (1954). *Advan. Cancer Res.* **2,** 229.
Fiala, S., and Fiala, A. E. (1959). *Brit. J. Cancer* **13,** 236.
Galy, W. L., Wolin, H. L., and Zajac, I. (1960). *Cancer Res.* **20,** 1508.
Ghadially, F. N., and Wiseman, G. (1956). *Brit. J. Cancer* **10,** 570.
Ghose, T., Nairn, R. C., and Fothergill, J. E. (1962). *Nature* **196,** 1108.
Girkin, G., and Kampschmidt, R. F. (1961). *Am. J. Physiol.* **200,** 61.
Goodlad, G. A. J., and Clark, C. M. (1961). *Brit. J. Cancer* **15,** 833.
Goodlad, G. A. J., and Clark, C. M. (1962). *Nature* **195,** 186.
Goodlad, G. A. J., and Munro, H. M. (1959). *Biochem. J.* **73,** 343.
Goranson, E. S. (1955). *Can. Cancer Conf.* **1,** 330.
Goranson, E. S., Cinits, E., and Hercz, A. (1959). *Cancer Res.* **19,** 512.
Green, J. W., Benditt, E. P., and Humphreys, E. M. (1950). *Cancer Res.* **15,** 620.
Green, M., and Miller, L. L. (1961). *Cancer Res.* **21,** 1034.
Greene, A. A., and Haven, F. L. (1957). *Cancer Res.* **17,** 613.
Greenfield, R. E., and Meister, A. (1951). *J. Natl. Cancer Inst.* **5,** 997.
Greenfield, R. E., and Price, V. E. (1954). *J. Biol. Chem.* **209,** 355.
Greengard, O., and Campbell, P. N. (1959). *Biochem. J.* **72,** 305.
Greenlees, J., and Lepage, G. A. (1955). *Cancer Res.* **15,** 256.
Greenstein, J. P. (1954). "Biochemistry of Cancer," 2nd ed. Academic Press, New York.
Haley, E. L., Fischer, G. A., and Welch, A. D. (1961). *Cancer Res.* **21,** 532.
Harding, H. R., Rosen, F., and Nichol, C. A. (1961). *Am. J. Physiol.* **201,** 271.
Hardy, S., and Putnam, F. W. (1955). *J. Biol. Chem.* **212,** 371.
Hemingway, J. T. (1960). *Nature* **185,** 106.
Hemingway, J. T. (1961). *Nature* **191,** 706.
Henderson, J. F., and LePage, G. A. (1959a). *Cancer Res.* **19,** 749.
Henderson, J. F., and LePage, G. A. (1959b). *Cancer Res.* **19,** 887.
Hilf, R., Burnett, F. F., and Borman, A. (1960). *Cancer Res.* **20,** 1389.
Holbrook, D. J., Jr., Irvin, J. L., Irvin, E. M., and Rotherham, J. (1960). *Cancer Res.* **20,** 1329.
Holmberg, B. (1961). *Cancer Res.* **21,** 1386.
Hradec, J. (1958). *Brit. J. Cancer* **12,** 290.
Hultin, T. (1956). *Exptl. Cell Res.* **10,** 71.
Hultin, T. (1960). *Acta Unio Intern. contra Cancrum* **16,** 1115.
Hultin, T., Arrhenius, E., Low, H., and Magee, P. N. (1960). *Biochem. J.* **76,** 109.

Ingle, D. J. (1960). *Endocrinology* **66**, 289.
Ingle, D. J., and Baker, B. L. (1951). *Endocrinology* **48**, 313.
Ingle, D. J., Prestrud, M. C., and Rice, K. L. (1950). *Endocrinology* **46**, 510.
Johnstone, R. M., and Scholefield, P. G. (1961). *J. Biol. Chem.* **236**, 1419.
Jordan, H. C., Miller, L. L., and Peters, P. A. (1959). *Cancer Res.* **19**, 195.
Kahlson, G., Rosengren, E., and Steinhardt, C. (1962). *Nature* **194**, 380.
Kampschmidt, R. F., and McCoy, T. A. (1960). *Proc. Soc. Exptl. Biol. Med.* **103**, 869.
Kampschmidt, R. F., and Schultz, G. A. (1963). *Cancer Res.* **23**, 751.
Kampschmidt, R. F., and Upchurch, H. F. (1963). *Cancer Res.* **23**, 756.
Kampschmidt, R. F., Adams, M. E., and Goodwin, W. L. (1959a). *Arch. Biochem. Biophys.* **82**, 42.
Kampschmidt, R. F., Adams, M. E., and McCoy, T. A. (1959b). *Cancer Res.* **19**, 236.
Kampschmidt, R. F., Mayne, M. A., Goodwin, W. L., and Clabaugh, W. A. (1960). *Cancer Res.* **20**, 368.
Kavetsky, R. E., Samundgean, E. M., and Butenko, Z. A. (1962). *Acta Unio Intern. contra Cancrum* **18**, 115.
Kawachi, T., Fujii, S., Suzuki, T., Uesaki, N., and Yamamura, Y. (1961). *Gann* **52**, 219.
Kit, S., and Graham, O. L. (1956). *Cancer Res.* **16**, 117.
Laird, A. K., and Barton, A. D. (1960). *Nature* **188**, 417.
Leaders, F. E., Dixon, R. L., Osborne, J. W., and Long, J. P. (1962). *Proc. Soc. Exptl. Biol. Med.* **110**, 436.
Lin, E. C. C., Rivlin, R. S., and Knox, W. E. (1959). *Am. J. Physiol.* **196**, 303.
Littlefield, J. W., and Keller, E. B. (1957). *J. Biol. Chem.* **224**, 13.
McCoy, T. A., Maxwell, M., and Neumann, R. E. (1956). *Cancer Res.* **16**, 979.
McCoy, T. A., Maxwell, M., and Kruse, P. F., Jr. (1959a). *Cancer Res.* **19**, 591.
McCoy, T. A., Maxwell, M., and Kruse, P. F., Jr. (1959b). *Proc. Soc. Exptl. Biol. Med.* **100**, 115.
McHenry, E. W. (1955). *Can. Cancer Conf.* **1**, 149.
McLaughlin, E. D. (1960). *Surgery* **47**, 420.
Malmgren, R. A. (1956). *Cancer Res.* **16**, 232.
Malmgren, H., and Sylven, B. (1959). *Cancer Res.* **19**, 525.
Mercer, E. H. (1961). *Proc. Roy. Soc. Med.* **54**, 1057.
Mider, G. B. (1955). *Can. Cancer Conf.* **1**, 120.
Mider, G. B., Tesluk, H., and Morton, J. J. (1948). *Acta Unio Intern. contra Cancrum* **6**, 409.
Mider, G. B., Alling, E. L., and Morton, J. J. (1950). *Cancer* **3**, 56.
Migliarese, J. F. (1956). *Proc. Am. Assoc. Cancer Res.* **2**, 133.
Miller, J. A., and Miller, E. C. (1953). *Advan. Cancer Res.* **1**, 339.
Miyabo, S. (1959). *Endocrinol. Japon.* **6**, 113.
Moldave, K. (1956). *J. Biol. Chem.* **221**, 543.
Mukherjee, T., Gustafsson, R. G., Afzelius, B. A., and Arrhenius, E. (1963). *Cancer Res.* **23**, 944.
Munro, H. N., and Mukerji, D. (1962). *Biochem. J.* **82**, 520.
Nakahara, W., and Fukuoka, F. (1948). *Japan. Med. J.* **1**, 271.
Nakahara, W., and Fukuoka, F. (1954). *Gann* **45**, 67.
Nishimura, E. T., Kobara, T. Y., Kaltenbach, J. P., and Wartman, W. B. (1962). *Arch. Biochem. Biophys.* **97**, 589.
Noble, R. L. (1957). *Pharmacol. Rev.* **9**, 367.
Ono, T., Wmeda, M., and Sugimura, J. (1956). *Gann* **47**, 171.

Paschkis, K. E., Cantarow, A., Stasney, J., and Hobbs, J. H. (1955). *Cancer Res.* **15**, 579.
Paschkis, K. E., Cantarow, A., and Stasney, J. (1956). *Proc. Am. Assoc. Cancer Res.* **2**, 138.
Payne, A. H., Kelly, L. S., Beach, G., and Jones, H. B. (1952). *Cancer Res.* **12**, 426.
Penn, N. W. (1961). *Federation Proc.* **20**, 388.
Penn, N. W. (1962). *Cancer Research* **22**, 388.
Petasnick, J. P., Brown, R. R., and Price, J. M. (1961). *Cancer Res.* **21**, 1400.
Pitot, H. C., and Morris, H. P. (1961). *Cancer Res.* **21**, 1009.
Pitot, H. C., Potter, V. R., and Morris, H. P. (1961). *Cancer Res.* **21**, 1001.
Porter, K. R., and Bruni, C. (1959). *Cancer Res.* **19**, 997.
Potter, M., and Fahey, J. L. (1960). *J. Natl. Cancer Inst.* **24**, 1153.
Putnam, F. W., and Stelos, P. (1953). *J. Biol. Chem.* **203**, 347.
Putnam, F. W., and Hardy, S. (1955). *J. Biol. Chem.* **212**, 361.
Putnam, F. W., and Miyake, A. (1956). *Arch. Biochem. Biophys.* **65**, 39.
Rabinovitz, M., Olson, M. E., and Greenberg, D. M. (1956). *J. Biol. Chem.* **222**, 879.
Rechcigl, M., Jr., Grantham, F., and Greenfield, R. E. (1961). *Cancer Res.* **21**, 238.
Riley, V., Lilly, F., Huerto, E., and Bardell, D. (1960). *Science* **132**, 545.
Ritter, F. (1958). *Münch. Med. Wochschr.* **100**, 1113.
Roberts, E., and Simonsen, D. G. (1960). *In* "Amino Acids, Proteins and Cancer Biochemistry" (J. T. Edsall, ed.), pp. 121-146. Academic Press, New York.
Roberts, E., Tanaka, K. K., Tanaka, T., and Simonsen, D. G. (1956). *Cancer Res.* **16**, 970.
Rosen, F., Roberts, N. R., Budnick, L. E., and Nichol, C. A. (1959). *Endocrinology* **65**, 256.
Rosen, F., Budnick, L. E., Solomon, D. K., and Nichol, C. A. (1961). *Cancer Res.* **21**, 620.
Rusch, H. P. (1954). *Cancer Res.* **14**, 407
Salter, J. M., Davidson, I. W. F., and Best, C. H. (1957). *Diabetes* **6**, 248.
Salter, J. M., de Meyer, R., and Best, C. H. (1958) *Brit. Med. J.* **II**, 5.
Sherman, C. D., Jr., Morton, J. J., and Mider, G. B. (1950). *Cancer Res.* **10**, 374
Silber, R. H., and Porter, C. C. (1953). *Endocrinology* **52**, 518.
Slawikowski, G. J. M. (1960). *Cancer Res.* **20**, 316.
Starbuck, W. C., and Busch, H. (1960). *Cancer Res.* **20**, 891.
Stewart, A. G., and Begg, R. W. (1953). *Cancer Res.* **13**, 560.
Straube, R. L., and Hill, M. S. (1956). *Proc. Am. Assoc. Cancer Res.* **2**, 150.
Talalay, P., Takano, G. M. V., and Huggins, C. (1952). *Cancer Res.* **12**, 838.
Tedeschi, G. G., and Mangiantini, M. T. (1957). *Ricerca Sci.* **27**, 2174.
Tenenhouse, A., and Quastel, J. H. (1960). *Can. J. Biochem. Physiol.* **38**, 1311.
Trémolières, J., Derache, R., and Lowy, R. (1955). *Ann. Nutr. Aliment.* **9**, 179.
Trotter, N. L. (1961). *Cancer Res.* **21**, 778.
Urquhart, J., Yates, F. E., and Herbst, A. L. (1959). *Endocrinology* **64**, 816.
Walter, H. (1960). *Nature* **188**, 643.
Weber, G. (1961). *Advan. Cancer Res.* **6**, 403.
Weinhouse, S. (1960). *In* "Amino Acids, Proteins, and Cancer Biochemistry" (J. T. Edsall, ed.), pp. 109–119. Academic Press, New York.
Werk, E. E., and Sholiton, L. J. (1960). *Cancer* **13**, 469.
Westrop, J. W., and Green, H. N. (1960). *Nature* **186**, 350.
White, F. R. (1945). *J. Natl. Cancer Inst.* **5**, 265.
White, J. M., Ozawa, G., Ross, G. A. L., and McHenry, E. W. (1954). *Cancer Res.* **14**, 508.

Wiseman, G., and Ghadially, F. N. (1955). *Brit. J. Cancer* **9**, 480.
Wiseman, G., and Ghadially, F. N. (1958). *Brit. Med. J.* **II**, 18.
Wood, S., Jr., Rivlin, R. S., and Knox, W. E. (1956). *Cancer Res.* **16**, 1053.
Woodruff, M. F., and Symes, M. O. (1962). *Brit. J. Cancer* **16**, 120.
Wu, C., and Bauer, J. M. (1960). *Cancer Res.* **20**, 848.
Wu, C., and Bauer, J. M. (1962). *Cancer Res.* **22**, 1239.
Wu, C., and Bauer, J. M. (1963). *Cancer Res.* **23**, 956.
Yates, F. E., Herbst, A. L., and Urquhart, J. (1958). *Endocrinology* **63**, 887.
Yeakel, E. H., and Tobias, G. L. (1951). *Cancer Res.* **11**, 830.
Yunoki, K., and Griffin, A. C. (1960). *Cancer Res.* **20**, 533.
Yunoki, K., and Griffin, A. C. (1961). *Cancer Res.* **21**, 537.

CHAPTER 21

Experimental Protein-Calorie Deficiency

B. S. Platt, C. R. C. Heard, and R. J. C. Stewart

Human Nutrition Research Unit, National Institute for Medical Research, Mill Hill, London, England

I. Introduction: Objectives, and Nomenclature and Definition of Syndromes in Man and Animals

A. Objectives

The main purposes of the investigations described in this chapter were: (*1*) to study the effects on animals of diets and regimens which in human subjects produce protein-calorie deficiency; (*2*) to devise methods of defining the nutritional values of such diets and regimens; (*3*) to attempt to equate the clinical and pathological conditions in man and experimental animals; (*4*) to determine the contribution of various non-nutritional factors, e.g., the effects of zymotic disease, toxic substances, and the age of the subject.

Most of this chapter is devoted to category (*1*). Our method of expressing the protein-calorie values of the diets and regimens is presented in Section I,B,2. We have had limited success with the objective set out in (*3*), because it is clear from the literature that clinicians do not agree about what they mean by the terms they employ. The contribution to protein-calorie deficiency of infections and infestations is discussed in Section XIV.

B. Nomenclature of Protein-Calorie Deficiency

1. Clinical Nomenclature

There is probably no group of diseases which includes clinical entities known by so many different names. Confusion about nomenclature has existed for decades. For example, forty years ago Utheim (1922), in discussing this subject, recognized the need for a classification which indicated etiology and pathogenesis. She concluded that "the attempt of forcing the chronic nutritional disorders into a system based on causative factors must be abandoned and the more natural grouping based on clinical manifestations substituted." Waterlow *et al.* (1960) also recognized that the names attached to a disease may be of two kinds, i.e., they may describe either a clinical picture *or* a process. About the clinical nomenclature, in discussing the term "kwashiorkor," these authors said: "It would be convenient if all clinicians could agree upon their usage of the word, but if they do not, it is fruitless to argue about it." The same could also be said about marasmus and, therefore, of the term "marasmic-kwashiorkor."

At the sixth meeting of the Joint F.A.O./W.H.O. Expert Committee on Nutrition (see F.A.O., 1962), in its revision of the international classification of diseases, it was proposed that a main group of nutritional diseases should be classified as hypoalimentation and that one subgroup should consist of protein-calorie deficiencies, two forms of which are (*a*) kwashiorkor (including marasmic kwashiorkor), and (*b*) marasmus (e.g., athrepsia, cachexia, extreme wasting).

Clinically, the cardinal feature of the syndrome at one end of the range of protein-caloric deficiency (kwashiorkor) is edema and that at the other end (marasmus) is wasting; a brief comment on these is given to illustrate the difficulties of defining the two syndromes. In their discussion on the prevalence of "protein-calorie malnutrition," Nicholls *et al.* (1961) state that the only syndrome "which is clear-cut enough is kwashiorkor, in which, however, the real diagnostic criterion is one single feature, edema." Kohman (1920) produced edema in rats fed on a diet low in protein; these animals were grossly underweight for their age. Véghelyi (1950) described a "nutritional edema without infection" occurring in the final stage of a condition in infants due to underfeeding. After loss of edema the infants averaged 0.58 (0.31–0.83) of the weight of "standard" Hungarian infants; but no "edema ever appeared in subjects whose calorie intake was inadequate." Other starving infants "could easily be made edematous by feeding them carbohydrates with or without fat." Would these cases, whether rats or humans, be called "kwashiorkor"?

"Marasmus" means "wasting" and in the older literature (see Jackson, 1925) attention is paid to the actual losses in body weight (final weight compared with maximum observed during life). But in incomplete (underfeeding) or partial inanition Jackson concludes that "the weight of infants may be stationary or merely retarded to a variable degree, depending upon the length and severity of the malnutrition." How many of the reported cases of marasmus have, in fact, lost weight and how many failed to gain weight? In this chapter the term "marasmus" will not be used in the strict sense of the word, meaning wasting, but will refer to subjects (animals or children) whose growth has been severely retarded, as well as those who have lost weight. The clinical picture of protein-calorie deficiency is discussed in detail in Chapter 22.

2. *Nomenclature of Protein-Calorie Value of Food*

It was recognized in the F.A.O (1957b) report on protein requirements that the nutritive value of the protein in food depends on the quantity of protein and its quality; the energy content of the food and therefore on the relative quantities of carbohydrate, lipids, and alcohol; the caloric intake in relation to the consumer's needs; the nature and proportions in a diet of many essential accessory food factors; and the dietary regimen, e.g., the timing of meals and the distribution of protein between various meals during the day. The following procedure takes account of all these factors in the measurement of the protein value of the diet consumed; this measurement is expressed as Net Dietary-protein Calories per cent (NDpCal%) (Platt *et al.*, 1961).

The first of these factors, the quantity, is the amount of crude protein, and is (conventionally) determined by measuring the total nitrogen (N) content and multiplying this by the factor 6.25; thus

$$\text{Protein concentration} = \frac{\text{intake N (gm)} \times 6.25}{\text{total food intake (gm)}} \times 100 \qquad (1)$$

Biological Value (BV) has long been used to measure the second factor, the quality of dietary protein. It expresses the fraction of N *absorbed* from the intestine which is retained in the body, i.e.,

$$\text{BV} = \frac{\text{retained N}}{\text{absorbed N}} \qquad (2)$$

Usually the N of the diet is not fully absorbed, and sometimes these digestive losses are quite large. Consequently, to obtain the value of the dietary protein to the consumer, the Net Protein Utilization (NPU) of the protein is determined.

$$\text{NPU} = \frac{\text{retained N}}{\text{intake N}} = \text{BV} \times \text{digestibility } (D) \qquad (3)$$

$$\left(D = \frac{\text{absorbed N}}{\text{intake N}}\right)$$

It has been known for many years (Platt and Miller, 1958) that the BV of protein falls as the proportion of protein in the diet is increased above maintenance levels; similarly, the NPU is reduced under these circumstances. To determine BV, the protein is conventionally fed at 8 or 10% concentration in the assay diet. It is thought, probably unjustifiably, that this so-called "standardized" condition provides a basis for obtaining comparable data on different proteins. However, the measurement of NPU at levels of dietary protein which are adequate for maintenance provides reasonably constant values for protein quality; measurements of NPU made under such conditions are termed Standardized Net Protein Utilization (NPU_{st}). At higher levels of protein intake, NPU and BV both give progressively lower values for protein quality and, in order to distinguish NPU values obtained under circumstances not necessarily obeying the conditions for the standardized measurement, the term Operative Net Protein Utilization (NPU_{op}) has been introduced (Platt and Miller, 1959). This allows evaluation of dietary protein quality under practical conditions.

From the product of quantity and quality of protein, we obtain the Net Dietary-protein Value (NDpV), which expresses the total utilizable protein content of a food or mixture of foods, in the form in which they are customarily eaten (Platt and Miller, 1959).

$$\begin{aligned} \text{NDpV} &= \text{protein concentration} \times \text{NPU} \\ &= \left[\frac{\text{intake N (gm)} \times 6.25}{\text{total food intake (gm)}} \times 100\right] \times \left[\frac{\text{retained N (gm)}}{\text{intake N (gm)}}\right] \qquad (4) \\ &= \frac{\text{retained N (gm)} \times 6.25}{\text{total food intake (gm)}} \times 100 \end{aligned}$$

Thus a diet which contains 10% crude protein with an NPU_{op} of 0.70 will have an NDpV of 7 gm per 100 gm diet.

In view of the interdependence of the two dietary factors—protein and metabolizable energy—it is necessary to express NDpV in terms of the calories which could be derived from the protein if it were all utilized for energy purposes, calculated as a percentage of the total metabolizable energy in the diet. This value is called Net Dietary-protein Calories per cent (NDpCal%) (Platt *et al.*, 1961).

$$\begin{aligned} \text{NDpCal\%} &= \frac{\text{retained protein (expressed in kcal)}}{\text{total metabolizable energy in food consumed}} \times 100 \\ &= \frac{\text{retained N (gm)} \times 6.25 \times 4 \text{ kcal}}{\text{total food intake (gm)} \times \text{(kcal/gm food)}} \times 100 \qquad (5) \\ &= \frac{\text{NDpV} \times 4}{\text{kcal/gm food}} \end{aligned}$$

For example, if the diet has an energy value per gram of 5 kcal and the

NDpV is 7 gm per 100 gm diet, then NDpCal% is 5.6, i.e., the protein of this diet is equivalent in nutritive value to that of a diet containing 5.6 % (expressed as kcal) of the reference protein.

This method of expressing the nutritive value of protein in diets can be compared directly with the statement of protein requirements in terms of energy.

$$\text{Protein requirements} = \frac{\text{reference protein requirement in kcal}}{\text{total calorie requirement}} \times 100 \tag{6}$$

[reference protein is assumed to be one containing 16% N and to be completely utilized (NPU_{op} = 100)]

Taking the amounts of reference protein and the total calorie requirements from the FAO reports (1957a,b), the protein needs in terms of NDpCal% can readily be calculated. Typical values for various age and sex groups range from 4.6 % for an adult man requiring 2960 calories per day and 34 gm per day of reference protein to 8.0 % for an infant and 9.5 % for a lactating mother (Platt *et al.*, 1961).

For protein anabolism a supply of energy is always needed; therefore, when the intake of food is restricted, the amount of dietary protein that can be utilized by an animal may be limited by the energy available rather than by the quantity and quality of the protein in the food eaten (D. S. Miller and Payne, 1961). The following expression (7) shows the relationship between the maximum amount of protein that can be retained by the body and energy metabolism.

$$\begin{aligned}&\text{Maximum protein retained (expressed in kcal)}\\ &= \frac{\begin{pmatrix}\text{metabolizable energy}\\ \text{in food consumed}\end{pmatrix} - \begin{pmatrix}\text{energy required for}\\ \text{basal metabolism}\end{pmatrix}}{E}\end{aligned} \tag{7}$$

where E = energy in kcal required for the synthesis of 1 kcal of protein = about 6 kcal.

From (5) and (7) the protein value of food is therefore

$$\text{NDpCal\%} = \frac{100}{E}\left(1 - \frac{\text{energy for basal metabolism}}{\text{metabolizable energy in food consumed}}\right) \tag{8}$$

This equation (8) applies only when the synthesis of protein is limited by the energy available; under other conditions equation (5) applies.

The significance of this interdependence of the protein value on the energy content of food is, in practice, twofold:

(a) A regimen supplying even a large amount of protein will not meet the needs of an individual unless the energy content of the food is adequate.

(b) The protein value to the consumer of a well-balanced diet which is fed in restricted amounts is reduced, not only in proportion to the caloric restriction, but also by a lowered efficiency of utilization of the protein in the diet.

3. The Concept of Protein-Calorie Deficiency as a Disease Syndrome

The term "protein-calorie deficiency" does link the two interdependent dietary factors, protein and calories; on this protein-calorie axis will turn future research in nutrition and more especially many developments in applied nutrition.

Platt (1956) listed undernutrition as a cause of protein malnutrition and in 1958 he regretted the growing tendency to equate the term kwashiorkor with protein malnutrition (1958c); the term, he said, "loses its acclaimed merit of 'neutrality' and there is a risk that much protein malnutrition (i.e. of the marasmic type) may be overlooked." Very few nutritionists include undernutrition (or overnutrition) in the term malnutrition. Rhodes (1959) reported that kwashiorkor was found in Jamaican children receiving diets deficient in calories, but Waterlow (1948) stressed that in kwashiorkor the diets were adequate or even of high caloric value; on the other hand, Altmann (1948) in South Africa, Ramalingaswami *et al.* (1948) in India, Trowell (1949) in Middle Africa, and Oomen (1953) in Indonesia suspected a deficient intake of calories in kwashiorkor. Waterlow *et al.* (1957) expressed the opinion that "in kwashiorkor the limiting factor is protein, whereas in marasmus it is calories." Another view was recently stated (Scrimshaw and Behar, 1961), as follows: "The clinical spectrum of severe protein deficiency ranges from deficiency accompained by an adequate or even excessive caloric intake to that accompained by a lack of calories so great that this lack becomes the limiting factor and few signs and symptoms of protein deficiency are apparent."

Some of these statements may be explained by the fact that some observers relate caloric intakes to the actual size of the individual, while others, correctly, observe the recommendation of the F.A.O. (1957a) that "the average minimum requirements [should be] compatible with a rate of growth characteristic of a representative well-fed group of children." The risk of simplifying the difference between these two conditions as being, on the one hand, due to protein deficiency and, on the other, to calorie deficiency is clear; the remedy would logically seem to be to give protein to the child suffering from kwashiorkor and foods to supply calories to the marasmic child. Apart from the fact that it is incorrect that the marasmic subject is suffering only from a shortage of calories, the giving of extra calories as carbohydrate, if the food were low in protein, would almost certainly convert the case of marasmus into one of kwashiorkor. Therefore, in assessing the caloric needs of a growing animal from its weight, consideration should be given to the effects of previous malnutrition, and in making adjustments for body size in cases of "emaciation resulting from prolonged undernutrition, the actual body-weight should be ignored and a 'target' or desired body-weight should be used instead" (F.A.O., 1957a).

The interplay of influences important in the understanding of the etiology and pathogenesis of various forms of protein-calorie deficiency disease has been outlined in the sixth report of the Joint F.A.O./W.H.O. Expert Committee on Nutrition (F.A.O., 1962). In simple, uncomplicated cases an infant is usually fed entirely at the breast for from 4 to 6 months; it does well so long as the mother has enough milk and then, partly because of its increased needs and partly because of the state of nutrition of the mother—often because she has to work hard as well as to nurse her baby—the child fails to get enough food and does not thrive; the mother recognizes this, begins to give the infant starchy paps and gruels and, as her anxiety over the baby's failure to progress increases, tends to "push" these foods, i.e., to force feed the infant. The success with which she produces fulminating protein-calorie deficiency will largely depend on the food preparation she uses and her assiduity in forcing it upon her offspring. It was with this knowledge of the natural history of the development of the various types of protein-calorie deficiency in the human subject that the experimental work described in this chapter was planned.

C. Forms of Protein-Calorie Deficiency Produced Experimentally

1. Mild or Marginal Protein-Calorie Deficiency

Observations have been made on eight litters, each of four dogs. Immediately after weaning, one pair of each litter was given a good diet (better than NDpCal = 8 %) and the other pair was fed on a diet of lower protein value (NDpCal = 6.3 or 6.8 %). The results obtained are shown in Fig. 1 (Platt *et al.*, 1962b).

All the measurements were made within a period of 14 days, after the dogs had been on diet for periods varying from 6 to 30 months. Almost all values (except the serum iron) were low for the dogs on the diet of low protein value. Those animals on the low protein-value diet were not only lighter in weight, they were also smaller, as will be seen from the accompanying illustration (Plate I). They all had diarrhea which persisted for weeks or even months, but they were not obviously sick animals. The disparity in size and weight between pairs of littermates tended to lessen with age since, in the deficient animals, although the *rate* of growth was reduced the period of growth was extended. Most of these animals were females and about half of these were on a diet of the same protein value as the mothers of the congenitally malnourished pups described later.

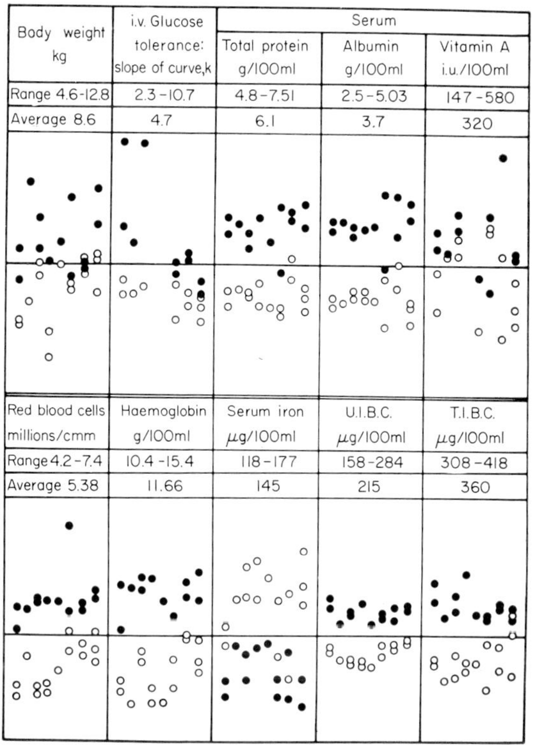

FIG. 1. Body weight, glucose tolerance, and values for some blood constituents of dogs maintained on diets of different protein value. Values for animals given diet of higher protein value: ●; values for animals given diet of lower protein value: ○. The horizontal line through the symbols shows the mean value, for all animals irrespective of diet, for each group of observations, the scale varies with the range; symbols in the same vertical are for animals of the same litter. UIBC: unsaturated iron-binding capacity; TIBC: total iron-binding capacity.

2. *Severe Protein-Calorie Deficiency*

The most marked differences in response to protein-calorie deficient diets [5 CLP (6.6), LP (3.5), and LP + CH (2.5) (see Section II,A)] were obtained in piglets which were only 10–12 days old when they were entirely maintained on the experimental diet. Usually they were taken from the sow at 5–7 days of age, put on to a gruel of "sow-replacer" weaner meal until 2–3 days before the start of the experiment, when they were given a mixture of "weaner meal" with an increasing proportion of the experimental diet. Less marked differences were found with older animals aged 25–30 days at the beginning of the experiment. When animals were started too young or at too low a weight, those on the LP (3.5) and LP + CH

PLATE I. Differences in size of two female littermate dogs, 5 months old, fed since weaning on isocaloric amounts of diets of different protein values; the smaller animal had NDpCal of 6.3%; the larger, 12.0%.

(2.5) diets did equally badly, and many died. Initial weights of 2–3 kg were desirable; when lighter animals (members of large litters) or heavier animals (members of small litters or older animals) were used, the characteristic differences due to the diets or regimens were progressively less marked (see Fig. 2). In addition, the success attending the production of the most marked and rapid changes following the addition of carbohydrate to the LP (3.5) diet depended in part on the assiduity with which the animal was offered the supplement, but an unexplained idiosyncrasy of the animal was sometimes a determining factor. Some trials with carbohydrate fed by stomach tube have been successful—with pigs as well as with dogs and rats.

3. Nutritional Dwarfism

We have not seen this "ill-defined clinical form of protein-malnutrition" (Nicholls *et al.*, 1961) in the progeny of well-fed animals. It is, however, found in the offspring of protein-calorie-deficient female rats (if the mother does not eat them) crossed with normal males, and in congenitally malnourished pups born of protein-calorie-deficient bitches crossed with stock sires, the offspring being reared on the same diet as was given to the

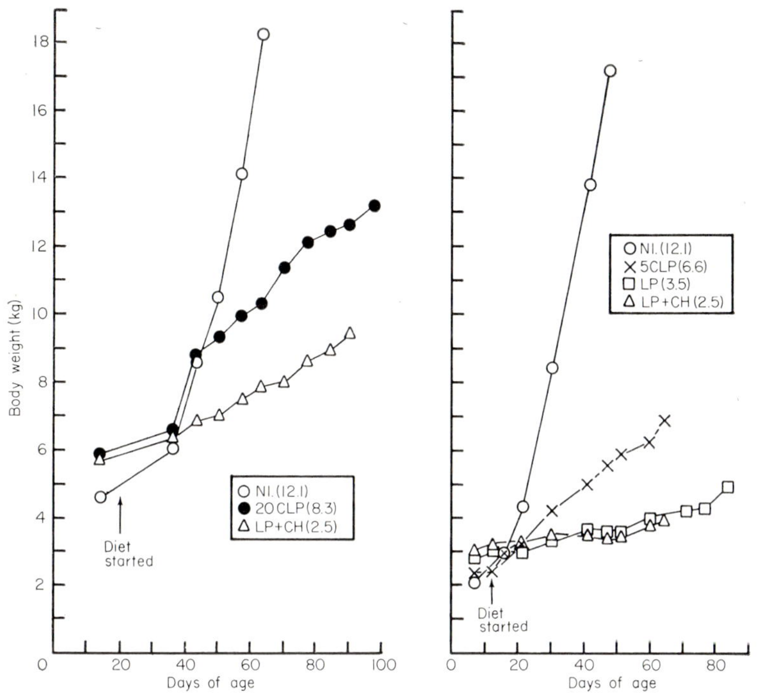

FIG. 2. Weight curves for two litters of pigs fed on diets of different protein values. The curves on the left are for pigs from a small litter (4 animals) started on a diet at 20 days of age (weight 5–6 kg); those on the right are for animals from a large litter (12 animals) started on diet at 11 days of age (weight 2–3 kg).

mothers. There is a high (50 %) mortality of such rat pups if they are weaned from the mother at 3 weeks; this can be eliminated if the young are allowed to run with the mother for 5 weeks before weaning. Similarly, there is a high infant mortality with the malnourished dog puppies and sometimes only 2 or 3 could be reared.

4. Equating of Syndromes in Man and Animals

Nicholls *et al.* (1961) say that any attempt to assess the prevalence of "protein-calorie malnutrition" is not at present practicable "owing to the indefiniteness of the various clinical pictures." We are, however, satisfied that we have produced clinical and pathological pictures in animals resembling most of those described in human populations. We can relate the changes to dietary factors and regimens with some precision and we can identify a number of environmental and other factors which modify the various syndromes.

Thus, protein deficiency may occur if a diet of good protein value is fed in insufficient amounts. Additional carbohydrate given to an animal on such a regimen would have a "protein sparing action" (Allison, 1958) by reducing the amount of protein metabolized for energy. On the other hand, the addition of carbohydrate to diets of low protein value will have a protein wasting effect, probably due in part to the lowered apparent digestibility, but also to the necessity of providing extra protein (enzymes) to metabolize the additional carbohydrate. This is well illustrated by work with piglets. Those given a low protein diet [LP (3.5)—see Section II] develop different degrees of incomplete pre- or marasmic-kwashiorkor. Increasing the caloric intake by supplementation with pure carbohydrate [LP + CH (2.5)] leads to the "sugar baby" type of protein-calorie deficiency. On the other hand, increasing the protein content but not the total calories of the diet [5 CLP (6.6)] produces the counterpart of nutritional marasmus in human infants. Marasmus is usually produced in rats given diets of low protein value *ad libitum*; an acute syndrome, sometimes with edema, can be induced by tube feeding more of these diets. Mild protein-calorie deficiency in dogs can be produced by feeding diets of appropriate protein-calorie values; bitches exhibiting evidence of such mild deficiency (crossed with normal males) produce offspring which suffer from "congenital malnutrition"; if reared on the deficient diet, the pups become examples of "nutritional dwarfism."

II. Diets Used in Experimental Studies

A. Pigs and Dogs

A brief description of the types of diets, with their protein values and abbreviations used in the text, is shown in Table I.

In the study of the effects of protein-calorie deficiency on the offspring of malnourished bitches, we usually supplemented the mothers' diet during pregnancy and lactation with carbohydrate (maize starch). The animals tended to be capricious about the amounts of the supplement they ate.

B. Rats

Rats quickly reduce their intake of food when given a diet of inadequate protein value. Having in mind the role of "forced feeding" in the development of the more florid forms of protein-calorie deficiency in human subjects, diets of low protein value have been tube-fed (Woodnott, 1961) in amounts up to about twice those eaten when offered *ad libitum*. Most of the observations on rats reported below are on "force-fed" animals; the diets used are not included in the table as they do not need detailed description. All the periods of forced feeding are of short duration—at the most 6

TABLE I

DESCRIPTIONS OF PRINCIPAL DIETS AND REGIMENS USED TO PRODUCE PROTEIN-CALORIE DEFICIENCY IN PIGS AND DOGS

Abbreviation used	Animal	Protein value (NDpCal%)[a]	Brief description of diet and/or regimen
LP (3.5)	Pig	3.5[b]	All vegetable "Gambia type" (Knowles, 1957; Friend *et al.*, 1961); two meals (each of 125 gm) per day
LP + CH (2.5)	Pig	2.5	LP (3.5) diet as above + 100 gm carbohydrate given between meals of LP (3.5) diet
5 CLP (6.6)	Pig	6.6[b]	LP (3.5) diet with 5 gm casein replacing 5 gm starch in every 100 gm diet; two meals (each of 125 gm) per day
20 CLP (8.3)	Pig	8.3[b]	LP (3.5) diet with 20 gm casein replacing 20 gm starch in every 100 gm diet; two meals (each of 125 gm) per day
Nl (12.1)	Pig	12.1	Commercial "weaner meal"; fed *ad lib.*
UK (8.6)	Dog	8.6	United Kingdom-type diet; fed *ad lib.*, or "pair fed" to litter mates given UK (6.8) diet
UK (6.8)	Dog	6.8	UK (8.6) diet modified by (*a*) omitting all the milk solids; (*b*) halving the meat; (*c*) restoring the calcium and the vitamins of the B complex; (*d*) adding cod-liver oil; fed *ad lib.*
B (12.0)	Dog	12.0	A complete diet; main ingredients breadcrumbs, meat, and casein; fed *ad lib.*, or "pair fed" to littermates given B (6.3) diet
B (6.3)	Dog	6.3	B (12.0) diet in which 16 gm casein in every 100 gm diet is replaced by dextrinized maize starch

TABLE I—*Continued*

Abbreviation used	Animal	Protein value (NDpCal%)[a]	Brief description of diet and/or regimen
Stock	Dog	9.4	Stock diet 41B (Bruce, 1958) + commercial dog food + skimmed milk powder

[a] Determined by assays on rats (Platt and Miller, 1959).

[b] The difference in NDpCal% of the LP (3.5) and 5 CLP (6.6) diet is much larger than between 5 CLP (6.6) and 20 CLP (8.3). This is because in the 5 CLP diet there are two factors: (1) amino acid supplementation and (2) an increase in concentration of protein. The effect of further increase in casein is relatively small because there is no further amino acid supplementation; also the increase in NPU is less at high than at lower concentrations of protein. Account must be taken of similar considerations in examining the differences between the protein values of the other diets listed.

days. It should be noted that the amounts of the diet given to the tube-fed rats, although twice as much as they would eat voluntarily, are still less than the average minimum requirements compatible with a rate of growth characteristic of a representative well-fed group of rats (see F.A.O., 1957a and Section I,B).

III. Body Size and Composition

A. Changes in Body Weight and Size

Dogs fed on the diets of low protein value grow more slowly than littermates on diets of high protein value (Section I,C, Plate I). Figure 3 shows, for pigs in a typical experiment, the dietary intake of calories and protein, the gain in weight, and the increase in bone length during the first 6 weeks on diet. It will be seen that all the experimental animals, including those fed the LP + CH (2.5) diet, consume less calories than a normally growing pig *of the same age*. The effect of adding protein without extra calories [5 CLP (6.6)] to the LP (3.5) diet is to promote bone growth relative to body weight; on adding extra carbohydrate [LP + CH (2.5)] without extra protein, body weight increases more than bone growth, which may even be reduced.

B. Weights of Organs and Tissues

The weights of the organs of the protein-calorie deficient pigs, as a percentage of body weight (Table II), conform more nearly to those found in the much younger normal pig of the same weight ("weight control") than to those of a normally fed animal of the same age ("age control").

Brody (1945) considers the parabola $y = ax^b$ "the most satisfactory empirical equation for relating part to part or to whole of structure and

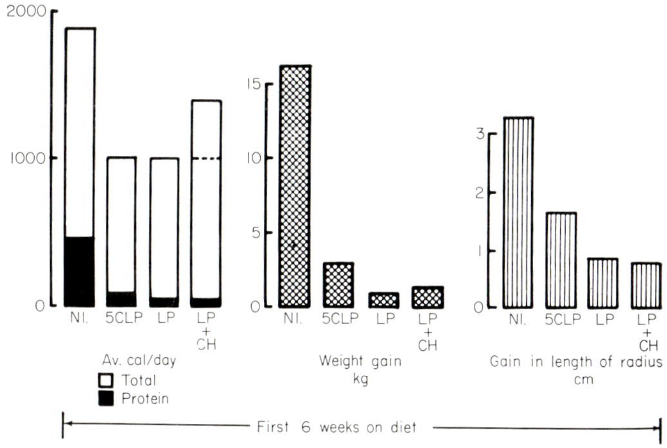

FIG. 3. Average daily intakes of total and protein calories, total gain in weight and in length of radius of pigs fed on Nl (12.1), 5 CLP (6.6), LP (3.5), and LP + CH (2.5) diets from 3 to 9 weeks of age. (From Platt, 1962.)

TABLE II
ORGAN WEIGHTS EXPRESSED IN TERMS OF BODY WEIGHT FOR PIGS FED ON VARIOUS DIETS

	"Age control"	Deficient animals		"Weight control"
Diet	Nl(12.1)	5CLP(6.6)	LP(3.5)	Nl(12.1)
Pig No.	90[a]	88[a]	92[a]	142
Age at death (days)	65	66	63	15
Weight (kg)	30.0	6.85	4.05	4.05
Spleen (gm/100 gm)	0.17	0.23	0.16	0.20
Liver (gm/100 gm)	2.47	2.76	4.05	3.56
Kidneys (gm/100 gm)	0.52	0.48	0.63	0.65
Brain (gm/100 gm)	0.24	0.92	1.48	1.09
Adrenals (mg/100 gm)	9.70	19.20	33.20	16.50

[a] Littermates.

function in animals of different size." Table III shows for various organs when a significant deviation in organ weight relative to body weight occurred. Because of the relatively small numbers of animals employed, numerical values have not been given for *a* and *b*.

C. Body Water and Extracellular Water

Trémolières (1958) has summarized observations on the effects of protein-calorie deficiency and rehabilitation on body compartments in humans. Abnormalities of fluid and electrolyte metabolism in malnutrition have

TABLE III

Deviations from the Normal Relationships of Organ Weight to Body Weight in Protein-Calorie Deficient Pigs[a]

	Diet			Recovery[b]
	LP(3.5)	LP + CH(2.5)	5CLP(6.6) or 20CLP(8.3)	
No. of animals	16	16	7	9
Spleen	− − −	− −	0	0
Kidney	− − −	− − −	− −	0
Thyroid	0	0	0	+++
Adrenal	++	0	0	0
Brain	+++	++	+	++
Total water	−	−	−	
Thiocyanate space	0	+	− −	

Key: +, ++, +++ = increase; −, − −, − − − = decrease ($P < 0.05$, 0.01, and 0.001, respectively); 0 = no significant change. Liver, heart, pancreas, and pituitary: no change on any diet. Thymus not included because in many LP (3.5) and LP + CH(2.5) animals it was so extremely atrophic.

[a] Normal relationship derived from 31 animals using $y = ax^b$ (Brody, 1945).

[b] Recovery: animal previously fed LP(3.5) or LP + CH(2.5) and then 5CLP(6.6) or 20CLP(8.3).

been reviewed by Eales (1961) for adults and by J. D. L. Hansen (1961) for infants. Widdowson *et al.* (1960), reporting on the impact of severe undernutrition on the chemical composition of the soft tissues of the pig, state that "the outstanding changes that take place in the chemical structure of the body of an undernourished man or animal are a loss of fat and protein and a gain in the proportion of water."

Four pigs of the same litter were fed on Nl (12.1), 5 CLP (6.6), LP (3.5), and LP + CH (2.5) diets. At 60 days of age their body water as a percentage of body weight was 62, 62, 67, and 70 % respectively. Body water as high as 80 % has been found in a pig fed the LP + CH (2.5) diet (Payne and Done, 1959). The effect of the diets on extracellular water (thiocyanate space) is similar to that on total body water (Godwin, 1960a).

In some animals, total body water, extracellular water, and glucose tolerance were determined simultaneously by injecting a solution of tritium-labeled water, thiocyanate, and glucose. The glucose tolerance curve was used to calculate "glucose space," which in most cases was the same as the "thiocyanate space," but in two animals, both on the LP + CH (2.5) diet, the thiocyanate space was high (53 and 47 % of body weight), while the glucose space was normal (33 and 27 %, respectively). This finding suggests that in these animals the thiocyanate is not confined to the extracellular compartment.

Measurements of total body water and extracellular water may be compared with body weight, as was done for organ weights (see Table III). Using this method of comparison (which is in effect using "weight controls"), the extracellular compartment of the severely deficient pigs is disproportionately large, while the total body water is below normal. Pigs on the 5 CLP (6.6) diet have thiocyanate space and total body water normally balanced in relation to each other, but low in relation to body weight.

D. *Composition of Tissues*

1. *Water, Fat, and Glycogen*

Table IV gives the composition of the liver on two different bases. First, the results are expressed as a percentage of fresh weight. Water content on this basis may actually be lower in the livers of deficient animals than in that of a normal animal, but fat and glycogen are increased. The water-, fat-, and glycogen-free residue (principally protein) is, however, very strikingly reduced in the severely deficient animals. If, as is shown in Table III, the total body water is reduced in deficient animals as compared with normal animals of the same weight, this reduction is probably slight compared with loss of weight of lean dry body tissue (excluding bone). Furthermore, the fact that the liver, compared with that of a "weight control," may be of normal weight could be due to retention of water and accumula-

TABLE IV

COMPOSITION OF LIVERS OF PIGS (LITTERMATES) FED ON DIETS OF DIFFERENT PROTEIN VALUE

Diet	NI(12.1)	5CLP(6.6)	LP(3.5)	LP + CH(2.5)
No. of animals	1	1	3	2
Age (days)	65	66	63–109	102
Weight (kg)	30.0	6.9	4.1–5.2	4.3–5.3
		A[a]		
Water	75	70	74	70
Fat	2	1	3	10
Glycogen	0.5	6.0	6.5	9.5
Residue	22.5	23.0	16.5	10.5
		B[b]		
Water	3.3	3.1	4.5	6.7
Fat	0.09	0.04	0.21	0.92
Glycogen	0.03	0.28	0.40	0.90

[a] Expressed as gm/100 gm fresh weight.

[b] Expressed as gm/gm (glycogen- and fat-free) dry weight.

tion of fat and glycogen, balancing out reduction in protoplasm. The second method of expressing tissue composition, which shows a better correlation with the protein value of the diet, is to give water, fat, and glycogen as a ratio of glycogen- and fat-free dry weight.

Figure 4 shows the proportion of fat, water, and glycogen in the skin, muscle, and liver in normal and deficient pigs. An increased proportion of water is found in all tissues in animals fed the LP (3.5) diet. Extra carbohydrate [LP + CH (2.5)] increases, and extra protein [5 CLP (6.6)] decreases the proportion of water found. Fat is lost from the skin of pigs fed the LP

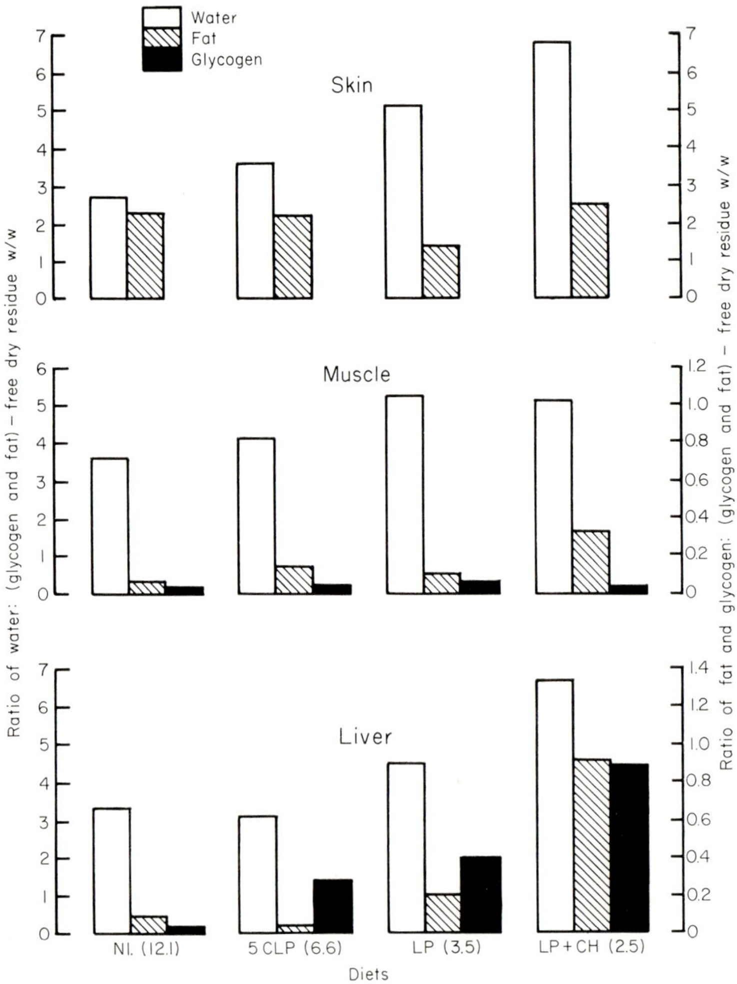

FIG. 4. Water and fat in skin; water, fat, and glycogen in muscle and liver of pigs fed on diets of different protein value.

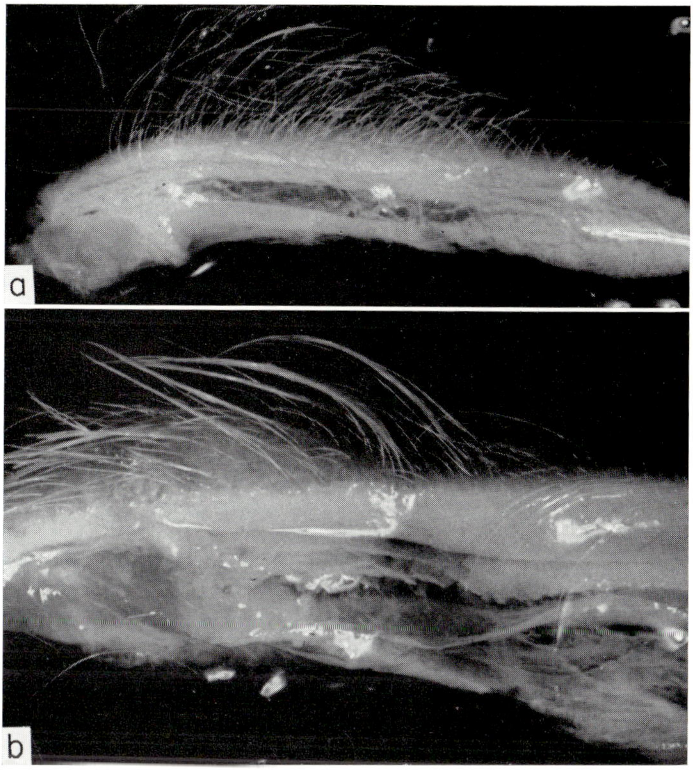

PLATE II. Photographs of skin, taken from corresponding sites from pigs given (*a*) diet LP (3.5) and (*b*) diet LP + CH (2.5) (for analyses see Fig. 1). Note that (*b*) is much more edematous than (*a*). (From Platt, 1958a.)

(3.5) diet; extra calories or protein prevents this (see Plate II). Accumulation of fat may occur to a slight extent in the liver of animals fed the LP (3.5) diet, but is much more pronounced in those fed the extra carbohydrate. The accumulation of fat which occurs in the liver of animals fed the extra carbohydrate is intensified if the level of caloric intake can be increased by feeding more LP (3.5) or carbohydrate, or both. In the animals given LP + CH (2.5) fat also accumulates in the muscle. Glycogen accumulates in the livers of all the protein-calorie deficient pigs; the concentration of glycogen is inversely related to the NDpCal% of the diet. Muscle glycogen is unaffected by protein-calorie deficiency. All analyses are of organs from animals fasted for 12–18 hours.

2. Sodium and Potassium in Muscle

As contrasted with normally developing pig muscle (Dickerson and Widdowson, 1960), that of protein-calorie deficient pigs retains more

TABLE V

SODIUM AND POTASSIUM IN THE MUSCLE OF PIGS (LITTERMATES)

Diet	Nl(12.1)	5CLP(6.6)	LP(3.5)	LP + CH(2.5)[b]
No. of animals	1	1	2	1
Age (days)	65	66	109	102
Weight (kg)	30.0	6.9	4.7	5.3
Na^{+a}	16	27	27	70
K^{+a}	60	53	52	49

[a] Milliequivalents/100 gm (glycogen- and fat-free) dry weight.
[b] This pig had a thiocyanate space of 47% and a glucose space of 28%.

sodium and less potassium (Table V). There are no corresponding changes in serum sodium and potassium. Similar changes have been found in our congenitally malnourished pups, and in malnourished infants (Smith and Waterlow, 1960).

IV. Blood

A. Anemia of Protein-Calorie Deficiency

The anemia of protein malnutrition in man is described (Woodruff, 1961) as being mild in degree and normocytic or slightly macrocytic. The majority of anemias in the tropics are said to be of the "iron-deficiency type," but under natural conditions only a minority can be attributed directly to an inadequate dietary intake of iron. It has been assumed that excessive loss of iron, or factors conditioning its absorption or utilization, must be responsible for the iron deficiency. However, treatment of some of these anemias with iron alone is not sufficient to raise the hemoglobin level above 11 to 12 gm/100 ml unless an adequate supplement of protein is also provided (Foy and Kondi, 1957). This suggests that deficiency of protein may often be responsible for anemias of the iron deficiency type.

A markedly reduced hemoglobin concentration (Hb) has always been found in pigs fed the LP (3.5) and LP + CH (2.5) diets (Knowles, 1957; Heard *et al.*, 1958). This is prevented by adding extra protein to the diet [5 CLP (6.6) and 20 CLP (8.3)] without change in iron intake. The anemia is also alleviated when pigs previously fed the LP (3.5) or LP + CH (2.5) diets are transferred to the diets of higher protein value. The restoration of hemoglobin levels is not as rapid as other signs of recovery (Table VI).

The figures shown in Table VI are for pigs in which Hb and packed cell volume (PCV) were both measured. This established the fact that under all the dietary conditions used the mean corpuscular hemoglobin concentration per cent remained unchanged at 29.3 %; therefore, in the majority of our animals only Hb was measured. Comparison of Hb with

TABLE VI

HEMOGLOBIN CONCENTRATION (HB), PACKED CELL VOLUME (PCV), AND MEAN CORPUSCULAR HEMOGLOBIN CONCENTRATION (MCHC) IN PIGS ON DIETS OF DIFFERENT PROTEIN VALUES

Diet[a]	No. of animals	Age (days)	Weight (kg)	Hb (gm/100 ml)	PCV (%)	MCHC (%)
Nl(12.1)	11	52	20.4	12.7	43	29.6
5CLP(6.6) or 20-CLP(8.3)	4	117	11.6	12.2	41	29.5
LP(3.5)	15	120	6.4	7.9	27	29.2
LP + CH(2.5)	9	90	4.8	8.0	28	28.6
Recovery[b]	9	172	11.9	10.2	—	—

[a] The daily intake of iron in pigs fed the LP(3.5), LP + CH(2.5), 5CLP(6.6), and 20CLP(8.3) diets was about 75 mg (Platt and Frankul, 1962). The pigs fed the commercial diet [(Nl(12.1)] *ad lib.* are included for comparison.

[b] 5CLP(6.6) or 20CLP(8.3) for 35 days following LP(3.5) or LP + CH(2.5) diet.

the red blood cell count (RBC) in protein-calorie deficient pigs showed that the anemia was normocytic and normochromic with a tendency toward the microcytic, hypochromic type.

The reticulocytes tend to be abnormally broad and thin in human protein malnutrition (see Woodruff, 1961, for discussion). No strictly comparable change has been observed in our animals, but when the red cells were suspended in isotonic formol-citrate for counting, the red cells of pigs fed on LP (3.5) and LP + CH (2.5) diets appeared crenated and misshapen; they also showed an increased resistance to hemolysis. Yoshimura (1961) has reported similar findings in humans.

The serum iron, unsaturated iron-binding capacity (UIBC), and total iron-binding capacity (TIBC) have been determined (Table VII) for three pigs fed on diets of widely differing protein value (Al-Rabii, 1962).

TABLE VII

HEMOGLOBIN (HB), RED CELLS (RBC), SERUM IRON, UNSATURATED (UIBC) AND TOTAL (TIBC) IRON BINDING CAPACITIES IN 3 PIGS (100-DAY-OLD LITTERMATES) FED DIETS OF DIFFERENT PROTEIN VALUES

Diet	Nl(12.1)	20CLP(8.3)[a]	LP + CH(2.5)[a]
Weight (kg)	43.0	12.4	9.5
Hb (gm/100 ml)	14.2	12.5	8.7
RBC ($\times 10^6$/mm^3)	6.6	6.4	4.1
Serum iron (μg/100 ml)	162	153	180
UIBC (μg/100 ml)	236	230	163
TIBC (μg/100 ml)	398	383	343

[a] Iron intake identical on these diets.

A fall in Hb and RBC is accompained by a fall in total and unsaturated iron-binding capacity, but a rise in serum iron. This same relationship has recently been reported in dogs exposed to a much milder degree of protein-calorie deficiency than the pigs (see Section I,C, Fig. 1). In two severely anemic protein-deficient pigs a comparison was made between treatment with iron and treatment with protein (Al-Rabii, 1962). An increase in Hb and RBC occurred only in the animal treated with protein. Serum iron was initially low in both animals, possibly because of a long history of poor appetite. The only effect of iron treatment was to increase the concentration of serum iron. Clearly the limiting factor in hemoglobin synthesis was protein.

B. Serum Proteins

Brock (1961) states that "in the present state of our knowledge, the earliest and most sensitive biochemical index of mild or impending protein deficiency is a drop in serum albumin into the marginal range." For his conditions and for children between 2 and 10 years of age, this marginal range is 2.75 to 3.50 gm albumin per 100 ml serum. The dogs on the poorer diets [UK (6.8) and B (6.3)] had serum albumin levels (determined by the methods used by Brock) within this marginal range.

In pigs fed on the LP (3.5) and LP + CH (2.5) diets, there is a very marked loss of serum albumin and the effect of extra protein in the diet [5 CLP (6.6) and 20 CLP (8.3)], without extra calories, in preventing and correcting this loss has been reported (Heard *et al.*, 1958; Friend *et al.*, 1961). These results are summarized in Table VIII.

TABLE VIII
MEAN SERUM PROTEIN LEVELS[a] IN PIGS ON DIETS OF DIFFERENT PROTEIN VALUE

Diet	No. of pigs	Age (days)	Weight (kg)	Protein fraction (gm/100 ml serum)				
				Total	Albumin	Globulin		
						α	β	γ
Nl(12.1)	15	91	37.2	6.21	2.58	1.39	1.00	1.33
5CLP(6.6)	4	92	8.8	4.79	1.75	1.37	0.70	0.97
LP(3.5)	8	128	7.3	3.98	0.72	1.36	0.65	1.24
LP + CH(2.5)	6	94	5.3	3.64	0.75	1.32	0.53	1.04
Recovery[b]	5	196	10.7	6.67	2.50	1.50	1.06	1.62

[a] As determined by electrophoresis (Friend *et al.*, 1961).

[b] 5CLP(6.6) or 20CLP(8.3) for 28 days or more following LP(3.5) or LP + CH (2.5) diet.

Payne and Done (1959) have shown that the reduction of serum albumin in the protein-calorie deficient pigs reflects an even greater reduction in extravascular albumin, so that the ratio of extravascular to intravascular albumin falls from a normal value of 2.2 to 0.8–0.7. Although Waterlow *et al.* (1960) found no change in the distribution of albumin in humans, Hoffenberg *et al.* (1962) have recently considered that an early effect of experimental protein deficiency in man is a preferential loss of extravascular albumin.

C. Transport

Many metabolites are known to be transported in the blood in the form of protein complexes. The effect of protein-calorie deficiency on the availability of these metabolites, through alteration in serum protein levels, has received scant attention. Olson and Vester (1960) have reviewed the role of nutrition in fat transport and close correlation between the concentration of serum albumin and protein-bound calcium was demonstrated by Gutman (1954). Serum vitamin A levels of pigs given adequate carotene supplies and having liver stores of the vitamin are dependent on the level of serum albumin and therefore on the protein value of the diet (Friend *et al.*, 1960, 1961). A correlation between serum vitamin A and albumin levels has also been demonstrated for humans (Woodruff and Stewart, 1962).

The occurrence of skin lesions similar to those attributable to zinc deficiency in one litter of protein-deficient pigs and of the anemia invariably associated with protein deficiency in our animals has led us to begin inquiries into the role of protein in the transport and utilization of zinc and iron. Platt and Frankul (1962) have shown that zinc is reduced in the serum and livers of protein-deficient pigs to about half the normal levels. Since about one third of the zinc in the serum is firmly bound to globulin and the remainder loosely bound to albumin (Vikhbladh, 1951), a reduction in these fractions might result in lowered serum zinc levels.

As is evident from Table VIII, β-globulin, as well as albumin, is reduced in protein-calorie deficient pigs. Three or four β-globulin fractions can be obtained by starch-gel electrophoresis of pig serum; the fraction corresponding in electrophoretic mobility with a sample of pure transferrin is reduced to a greater extent than is albumin in the serum from protein-deficient pigs. Feeding a diet of increased protein content [20 CLP (8.3)] results in a marked increase in transferrin and a rise in serum iron (Al-Rabii, 1962); at the same time the amount of iron in the liver is reduced (Platt and Frankul, 1962). The correlation between increase in transferrin and serum iron and decrease of liver iron suggests that the mobilization and transport of iron is also protein-dependent.

D. Serum Enzymes

Apart from the gross analysis of serum proteins, it is possible to study specific proteins in the form of enzymes. Such studies have, in the main, been related to the search for simple indexes of malnutrition and response to treatment (for reviews see Knox *et al.*, 1956; Waterlow 1959; Waterlow *et al.*, 1960).

A reduction in serum amylase and alkaline phosphatase was reported for pigs fed on the LP (3.5) and LP + CH (2.5) diets (Heard *et al.*, 1958); very low alkaline phosphatase values were thought to be a sign of a poor prognosis. Subsequent experience has suggested that the activity of the enzyme is, at best, a nonspecific measure of growth rate and may be equally low, whether growth is limited by calories [20 CLP (8.3)] or protein [LP + CH (2.5)]. Four dogs, fed diets of high or low protein value, were killed at 200 days. Serum alkaline phosphatase in the dogs fed on the UK (6.8) and B (6.3) diets was twice that found in the dogs receiving the diets of high protein value. In this case, the lower values were found in the better nourished animals because they had nearly attained full size.

Serum transaminases (glutamic-oxaloacetic and glutamic-pyruvic) may be double the normal values in pigs fed the LP (3.5) and LP + CH (2.5) diets, but not in those fed the 5 CLP (6.6) and 20 CLP (8.3) diets. Sometimes one enzyme is elevated more than the other, but no consistent picture has emerged, nor do we yet know the significance of these changes. Similar changes have been reported in malnourished infants; Waterlow (1959) has said that "this suggests that there may be an element of tissue destruction, in the liver or elsewhere."

The examples quoted (serum amylase always lowered; transaminases raised; alkaline phosphatase sometimes reduced and sometimes elevated) show that the pattern of enzyme change is related not only to the diet fed but also to the age and stage of development of the animal. It is not possible to state dogmatically that a given degree of protein-calorie deficiency will result in a known degree of serum enzyme change, nor that the serum enzyme changes are reliable indexes of the degree of protein deficiency.

V. Urine

A. Volume

Severely protein-calorie deficient pigs [LP (3.5) and LP + CH (2.5)] produce much less urine of higher specific gravity than do animals receiving extra protein [5 CLP (6.6)]. The deficient pigs also fail to respond to an oral water load (5 % body weight); little or no urine is produced in 3 hours, whereas normal animals excrete the full load. As shown in Section III,D,2, retention of water is accompanied by retention of sodium. Pituitary, thyroid, or adrenal insufficiency could cause poor response to a water load (Jadresic *et al.*, 1962). In our animals the first two are more apparent than the third (Sections VIII and IX). Response to a water load is restored within a week of feeding deficient pigs a diet of high protein value (Knowles, 1957).

B. Creatinine and Creatine

These were measured in pigs chiefly as checks on urine collection. Creatinine output varies with body weight. Creatine is excreted in appreciable amounts by the very young pigs, but diminishes with age on all diets.

C. Nitrogen: Total-, Urea-, and Ammonia-N

Platt (1954a, 1958a) and Platt and Heard (1958) have suggested that urinary urea nitrogen as a percentage of total urinary nitrogen reflects nutritional status in respect to protein, assuming that the subject has not, in the period immediately before the test, eaten meals of high protein content. The usefulness of this test has been confirmed [see Arroyave (1961); Dean and Whitehead (1963)]. Table IX shows the total, urea, and ammonia nitrogen excretion of pigs fed the Nl (12.1), LP (3.5), and LP + CH (2.5) diets. Total nitrogen excretion was less in the pig fed LP + CH (2.5) than in the animal fed the LP (3.5) diet alone and this was reflected by a lower value for urea-N:total N %. Nitrogen intake in the two animals was the same but fecal nitrogen was increased in the LP + CH (2.5) animal (Section VI,B,1). The increased excretion of ammonia ob

TABLE IX

URINARY NITROGEN IN PIGS FED VARIOUS DIETS FROM 19 DAYS OF AGE

Age (days)	Total nitrogen[a] (mg/24 hr)			Urea nitrogen (% total N)			Ammonia nitrogen (% total nitrogen)		
	Diet								
	Nl (12.1)	LP (3.5)	LP + CH (2.5)	Nl (12.1)	LP (3.5)	LP + CH (2.5)	Nl (12.1)	LP (3.5)	LP + CH (2.5)
39–43	9580	828	589	78	63	47	8	8	18
51–56	11290	984	623	79	64	30	7	12	35

[a] Over the period of study, the mean values for urine nitrogen:food nitrogen were: Nl (12.1), 43%; LP (3.5), 51%; LP + CH (2.5), 38%. The mean values for apparent digestibility of nitrogen over the same period were: Nl (12.1), 92%; LP (3.5), 80%, and LP + CH (2.5), 52%.

served in our protein-calorie deficient pigs is significant (see Section V,E,2) and has also been observed in malnourished humans.

D. Sodium and Potassium

Protein-calorie deficient pigs have more sodium and slightly less potassium in their muscles than normal animals (Section III,D,2). Analyses for sodium and potassium in pigs' urine have not given reliable results, probably because of contamination with food and feces. Urine from rats tube-fed a 0 % and a 15 % protein (casein) diet has, however, been analyzed and there is a lower Na^+/K^+ ratio in that of the deficient animals, due primarily to retention of sodium (Table X).

E. Other Metabolites

1. Proteins and Amino Acids

Proteinuria does not occur in protein-calorie deficient pigs and dogs. In rats, it is not greater than that found in normal animals. No abnormality of amino acid excretion is detected by two-way paper chromatography of deionized urine of protein-calorie deficient pigs and rats.

2. Keto Acids

Increased amounts of α-ketoglutaric acid have been found in the urine, of protein-calorie deficient rats (Doell and Stock, 1962). This would seem to be consequent on the increased ammonia excretion (Section V,C). Protein-calorie deficiency probably increases the activity of the renal enzymes which release ammonia from glutamine and glutamic acid. These enzymes are known to vary greatly in activity with changes in electrolyte and hormonal balance (Milne, 1957). Increased excretion of the ammonium ion would be at the expense of sodium, the retention of which has been observed (Section V,D).

No quantitative study of α-keto acids has been made in pig's and dog's urine but an elevation of pyruvic acid, and more particularly of lactic acid, has been noted in the blood of deficient pigs and rats. This has also

TABLE X

URINARY SODIUM AND POTASSIUM EXCRETION IN RATS TUBE-FED EQUAL QUANTITIES OF DIET

Diet	Nonprotein	Casein (15%)
Na^+ (meq./24 hr)	0.77	1.13
K^+ (meq./24 hr)	0.50	0.55
Na^+/K^+	1.54	2.05

been observed in humans and is thought to be related specifically to methionine deficiency (Edozien, 1959).

3. Glucose and Ketone Bodies

Glycosuria and urinary ketosis are not features of the impaired carbohydrate metabolism in protein-calorie deficient pigs and dogs (see Section IX). In rats (Platt *et al.*, 1962a) tube-fed isocaloric amounts of diet containing varying levels of casein, each feed was followed by a period of hyperglycemia the height and duration of which was inversely related to the protein value of the diet. The degree and duration of glycosuria was in accord with the blood sugar levels. Some animals tube-fed nonprotein diets died in a hyperglycemic state. Survivors appeared to adapt themselves to the carbohydrate loading and hyperglycemia and glycosuria diminished. Ketosis was not found in these rats.

4. Steroids

The excretion of 17-ketosteroids, 17-ketogenic steroids, and 17-hydroxycorticosteroids has been measured in pigs and rats (Durbin and Heard, 1962). The results are discussed in Section IX,E.

VI. Gastrointestinal Tract

A. Morphology

1. Alimentary Canal

Epithelial cells in scrapings from the buccal mucosa of the protein calorie deficient pigs and dogs stain poorly and, when compared with cells from normal animals, are distorted and fragmented (Squires, 1962). The number of abnormal cells is high in the animals given diets with low protein values and improving the diet leads to a prompt improvement in the condition of the cells. Squires (1961) has described changes in the buccal mucous membrane of children identical with his findings in experimental animals and there seems little doubt that these might be a reliable index of early malnutrition.

During one period vomiting was common among the deficient pigs [LP (3.5) and LP + CH (2.5)] and at death ulcers were found at the junction of the esophageal and gastric mucosa in 7 of 11 pigs (Plate III). Ulcers were occasionally found in the stomachs of pigs on a diet with a higher protein value [5 CLP (6.6)] and, although finally the occurrence of ulcers was related to the presence in the diet of uncooked haricot beans, a deficiency of protein greatly aggravated the effect. Steeping (steeping-water discarded) and cooking the beans eliminated this hazard. All the animals were without food for 12–18 hours before death, but there was

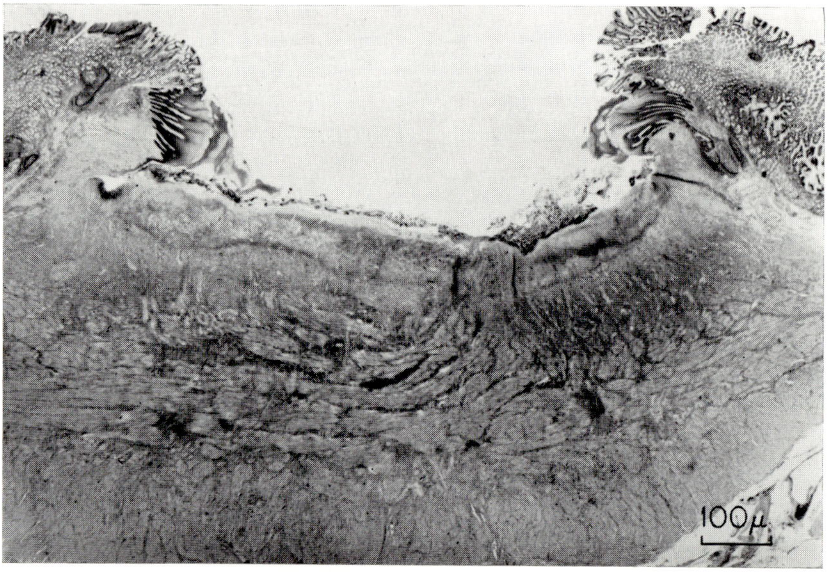

PLATE III. Gastric ulcer found in a pig on a low protein diet. Such ulcers are found in a region lined with squamous epithelium which surrounds the esophageal opening into the stomach. The sharp demarcation between squamous epithelium and secretory mucous membrane can be seen at the edge of the ulcer. (From Platt, 1958a.)

often a considerable bulk of incompletely digested food in the stomach and intestines of the deficient animals.

The intestines of protein-calorie deficient animals appear to be of small diameter and the mucous and muscle coats are thin. Lymphoid tissue is not obviously reduced, as might be expected from the severe loss of lymphoid tissue from the thymus and spleen. The tips of the intestinal villi are in some instances bare of epithelial cells. The changes are not thought to be due to autolytic or mechanical damage as the proportion of such villi is increased in the animals receiving the diets with the lower protein values and may be related to a reduced rate of cell production coupled with a normal rate of shedding. The epithelial cells are less columnar and the circular muscle coat thinner than in controls. Mucus-secreting cells may be reduced, but certainly not to the extent found in rats tube-fed a non-protein diet (Halder, 1962).

Atrophy of the stomach and intestine has been described in cases of human kwashiorkor and in severe cases the intestinal wall may become thin (tissue-paper intestine) and the mucous membrane smooth and atrophic (Passmore, 1947–1948).

2. *Pancreas, Exocrine Portion*

At death the pancreata of the protein-calorie deficient pigs often appear to be pale and wasted. Occasionally they are edematous and the lobules, separated by fluid, float like leaves in a semigelatinous mass.

The earliest histological change is a reduction in the concentration of the zymogen granules. Widespread degranulation occurs, but some cells containing zymogen granules are always present. The lobules are more widely separated and individual cells are shrunken and angular, with some vacuolation of the cytoplasm and hyperchromatic nuclei. In the most severe cases the acinar arrangement is disorganized; clumps of hyperchromatic nuclei practically devoid of cytoplasm and small necrotic foci are found.

Similar but less severe changes are found in malnourished pups whether born of normally fed or malnourished mothers. All the congenitally malnourished pups received a diet with a protein value of NDpCal % = 6.8, which might allow some "recovery" (see Section XI).

Changes similar to those described for pigs have been found in protein-deficient rats (Friedman and Friedman, 1946; Halder, 1962) and rabbits (Volk and Lazarus, 1960), and also in rats deprived of single essential amino acids (Véghelyi *et al.*, 1950; Sidransky and Farber, 1958, Sidransky and Clark, 1961).

The pancreata of the pigs and dogs do not show the cysts and fibrosis which are often found in malnourished human beings (Véghelyi *et al.*, 1950; Camain and Pierchon, 1954; J. N. P. Davies, 1948; Gillman and Gillman, 1951), but such changes have been observed in rats fed methionine- or tryptophan-deficient diets (Véghelyi *et al.*, 1950). As the fibrosis and cysts are not constant findings, even in man, it is probable that they indicate the presence of some additional factor or factors (see Section VII).

Véghelyi and Kemény (1962) have related the reduced enzyme production to the morphological changes in the pancreata of rats maintained on methionine-deficient diets. There is also evidence that deficiencies of single amino acids lead to a reduced production of specific pancreatic enzymes (Magee and Hong, 1959).

Changes in the exocrine pancreas similar to those brought about by protein deficiency can be produced by removal of the hypophysis (Bryans *et al.*, 1952) or thyroid gland (Townsend and Baker, 1959), but it is probable that dietary deficiencies have a direct effect on the pancreas (Véghelyi and Kemény, 1962) in addition to any deficiencies of hypophyseal or thyroidal hormones.

B. *Digestion*

1. *Apparent Digestibility of Nitrogen*

The protein of a diet, expressed in terms of NDpCal % of the diet, is available to the consumer only if digestive and absorptive functions are not impaired. Platt (1954a) showed that the apparent digestibility of nitrogen in malnourished Gambian children was low (mean value 58 %). A similar figure (56 %) was reported by Pellett (1957) for a pig which was on the LP (3.5) diet for 6 months. When the protein content of the pig's diet was raised, the apparent digestibility of nitrogen rose rapidly to 80 % within a week. After 2 weeks on the higher protein diet the pig was again offered the LP (3.5) diet; apparent digestibility of nitrogen fell only slowly. When young pigs are given the low protein diets, apparent digestibility also falls only slowly during the first weeks. The ability of the animals to digest and absorb nitrogen is rapidly regenerated on refeeding a better diet. The apparent digestibility of calories (mainly carbohydrate in pigs) is maintained at a high level for much longer than that of nitrogen (D. S. Miller and Payne, 1962), thereby exaggerating the carbohydrate/protein imbalance.

When littermate pigs were fed Nl (12.1), LP (3.5) and LP + CH (2.5) diets, the apparent digestibility of nitrogen at the end of 4 weeks was 92, 80, and 52 %, respectively. Heard *et al.* (1958) reported that the urinary excretion of nitrogen expressed as a percentage of nitrogen intake was 55 % for pigs fed the LP (3.5) diet and 41 % for animals fed the LP + CH (2.5) diet. The decreased urinary excretion of nitrogen in the animals fed the extra carbohydrate reflects not increased retention but decreased absorption of nitrogen. In fact, the animals fed the LP + CH (2.5) diet retain less nitrogen than their littermates fed the LP (3.5) diet alone.

Loss of protein into the alimentary canal has not been investigated in our animals; it has been shown to be a significant factor in the development of hypoalbuminemia associated with a variety of disorders (Dawson *et al.*, 1961; Waldmann, 1961). Cohen *et al.* (1962) have recently demonstrated that children suffering from kwashiorkor lose considerably more I^{131}-labeled polyvinylpyrrolidone into the gut than do normal children. It is generally believed that the albumin lost into the intestinal tract is hydrolyzed and reabsorbed. However, the markedly impaired apparent digestibility of protein in pigs fed the LP + CH (2.5) diet could be due to the extra carbohydrate aggravating the intestinal loss of proteins and impairing reabsorption.

2. *Enzymes*

Platt (1961a) has reviewed the relationship between malnutrition and the digestive process in the infant and young child and has indicated a number of aspects of this problem which lend themselves to experimental study in animals.

The offspring of malnourished mothers are probably immature and, in addition, show evidences of protein-calorie deficiency. This is likely to affect the development of digestive function both before and after birth. Studies of digestive enzymes in protein-calorie deficiency in growing animals must, therefore, be set against a knowledge of the normal changes in enzyme activity. For example, newborn pigs are known to be deficient in digestive enzymes (Bellis, 1957); they are able to hydrolyze lactose but not maltose or sucrose until they are about 10 days of age (Dollar *et al.*, 1957). Dextrin and starch are not well utilized until the animals are about 3 weeks old (Braude *et al.*, 1958).

Our pigs are weaned on to experimental diets when 2 to 3 weeks of age, and it is possible that the differences between experiments on pigs fed the same diets may be due to retarded enzyme synthesis in some cases. In an experiment started when the pigs were 19 days old, the animals' response to orally administered glucose and starch was tested when they were 6 weeks old. The rise in blood sugar following the test dose of starch was much less in animals fed the LP (3.5) and LP + CH (2.5) diets than in the normal animal and resembled the response of a 7-day-old normal pig (Braude *et al.*, 1958). However, the response to starch in these animals exactly paralleled the response to glucose, so that absorption rather than digestion must have been altered. As with apparent digestibility of nitrogen, the animals fed LP + CH (2.5) were more seriously affected than those fed the LP (3.5) diet alone.

Loss of pancreatic enzymes is also much more severe in children suffering from kwashiorkor than in those suffering from marasmus (Waterlow *et al.*, 1960). We have no direct measurements of these enzymes in pigs or dogs. Serum amylase has been reported to be diminished in LP (3.5) pigs to a greater extent than in those fed LP + CH (2.5) (Heard *et al.*, 1958).

The enzyme prolidase, found in intestinal mucosa but not in the pancreas, has been studied in protein-calorie deficient rats and pigs (Heard, 1957). Compared with normal animals of the same age, the *total* amount of this dipeptidase in the intestinal extract of deficient animals is drastically reduced, but it remains proportional to (body weight)$^{0.75}$, indicating that the synthesis of prolidase and probably therefore other intestinal enzymes is maintained in protein-malnourished animals at the level normal for their actual body weight.

Children suffering from kwashiorkor have, as a rule, a remarkable facility for retaining dietary nitrogen despite the loss of pancreatic enzymes (J. D. L. Hansen *et al.*, 1960). This may be due either to rapid restoration of these enzymes following improved diet, as suggested by the rapid rise in apparent digestibility, or to the undiminished activity of intestinal enzymes.

3. *Diarrhea*

The relationship between diarrhea and protein-calorie deficiency is confused in human subjects by the prevalence of infection. Diarrhea occurred in our pigs only when they were first removed from the sow and was not associated with protein-calorie deficiency. Vomiting was a feature in the terminal stages in some of our earliest work, but this was correlated with the presence of gastric ulcers (Section VI,A,1). In dogs fed on the UK (6.8) and B (6.3) diets, diarrhea is common and persists for many weeks, sometimes for nearly a year. Allison (1960) has also reported that protein-depleted dogs not exposed to infection have diarrhea. Balance studies show that diarrhea is not necessarily accompanied by impaired nitrogen absorption (Al-Rabii, 1962). As long ago as 1922 Utheim could write "That diarrhea may result from under-feeding has long been recognized," and she advocated that attention should be centered on the food requirement of the organism, the character of the stools being considered of secondary importance. Recent experience with children confirms that refeeding of the malnourished infant can and must be instituted despite diarrhea (Waterlow *et al.*, 1960), which "plays a minor role in regard to modifications of protein absorption and excretion" (Cravioto, 1958). The conclusion is that diarrhea is a symptom rather than a cause of protein-calorie deficiency and occurs in the presence or absence of infection.

VII. Liver

The livers of the protein-calorie deficient pigs show features similar to those frequently described for the livers of malnourished children and animals. They are often of a yellow-brown or even bright yellow color, have well-marked lobules, and are extremely friable.

Histological changes begin in the periportal area and spread with increasing severity and duration of the protein deficiency to the centre of the lobules. The most obvious change is the "vacuolated" and angular appearance of the cells; the cytoplasm is broken up by the accumulation of glycogen and fat and loses much of its basophilia. The cells appear to be enlarged as though distended by the inclusions and the sinusoids are relatively narrow. Some nuclei are enlarged and hypochromatic, others are small and hyperchromatic; the majority appear to be of normal size, having only one large prominent nucleolus. The number of double nucleated cells is increased, but no counts have been made. Karyolysis occurs in the most severely affected livers, in which small necrotic areas can also be found. Fibrosis is not a consistent finding.

Well-marked interlobular connective tissue develops in the livers of normally-reared pigs by about the 45th–60th day of age (White, 1938–1939); but in "minimal disease pigs" obtained by Caesarian section and reared apart from other pigs, the development of collagenous septa between the lobules may be delayed for 6 months or longer (Done, 1962). Godwin (1956) was, however, able to demonstrate increased amounts of fine connective tissue spreading among the cells of the portal areas of the protein-calorie deficient pigs. Other workers (Balfour, 1954; Ramalingaswami *et al.*, 1954) have not observed fibrosis of the liver in protein-depleted rats and we have not produced such a change in the depleted dog. Halder (1962) tube-fed a nonprotein diet to rats for 6 days and then allowed them access to a diet with a protein value of about NDpCal % = 3.5 for 3 weeks. There is in the liver of these animals proliferation of fine connective tissue fibers from the portal triads.

Excessive fat is an abnormal finding in the liver, whereas glycogen is stored, sometimes in large quantities, in the livers of normal, especially very young, animals. In the protein-calorie deficient pigs the relative unavailability of the glycogen, when needed to relieve insulin hypoglycemia (Section IX,C and D), can be regarded as abnormal. The animals examined by us were deprived of food for 12–18 hours before death; fat and glycogen were consistently higher in the livers of animals given the low protein diets than in littermates fed isocaloric quantities of a diet with a higher protein value (Fig. 4; Table IV), and the liver of the animal given the LP + CH (2.5) diet contained significantly more glycogen (Plate IV) and fat than that of animals on the LP (3.5) diet (Heard *et al.*, 1958).

Increasing the caloric intake of malnourished rats caused an increased deposition of fat (Lindan, 1954; Sidransky and Clark, 1961) and conversely, restricting the caloric intake, even when this involved further protein deficiency, reduced the amounts of fat and glycogen found in the liver (Ramalingaswami *et al.*, 1958). Sidransky and Clark (1961) also reported that the accumulation of fat in the livers of tube-fed rats was greater in animals given diets devoid of threonine than in those given diets devoid of valine.

In children suffering from protein-calorie deficiency the presence of fat globules is common. Glycogen accumulation in the liver is reported less often, but excessive amounts of glycogen occur in children (see Mukherjee and Nath, 1957), pigs (Heard *et al.*, 1958), rats (Wang *et al.*, 1949; Balfour, 1954; Halder, 1962), and dogs maintained on protein-calorie deficient regimens. It is possible that glycogen deposits have been overlooked or, more likely, mistaken for fat in paraffin sections. Fatty deposits are identified in paraffin sections of liver as clean-edged "spaces" within the cytoplasm;

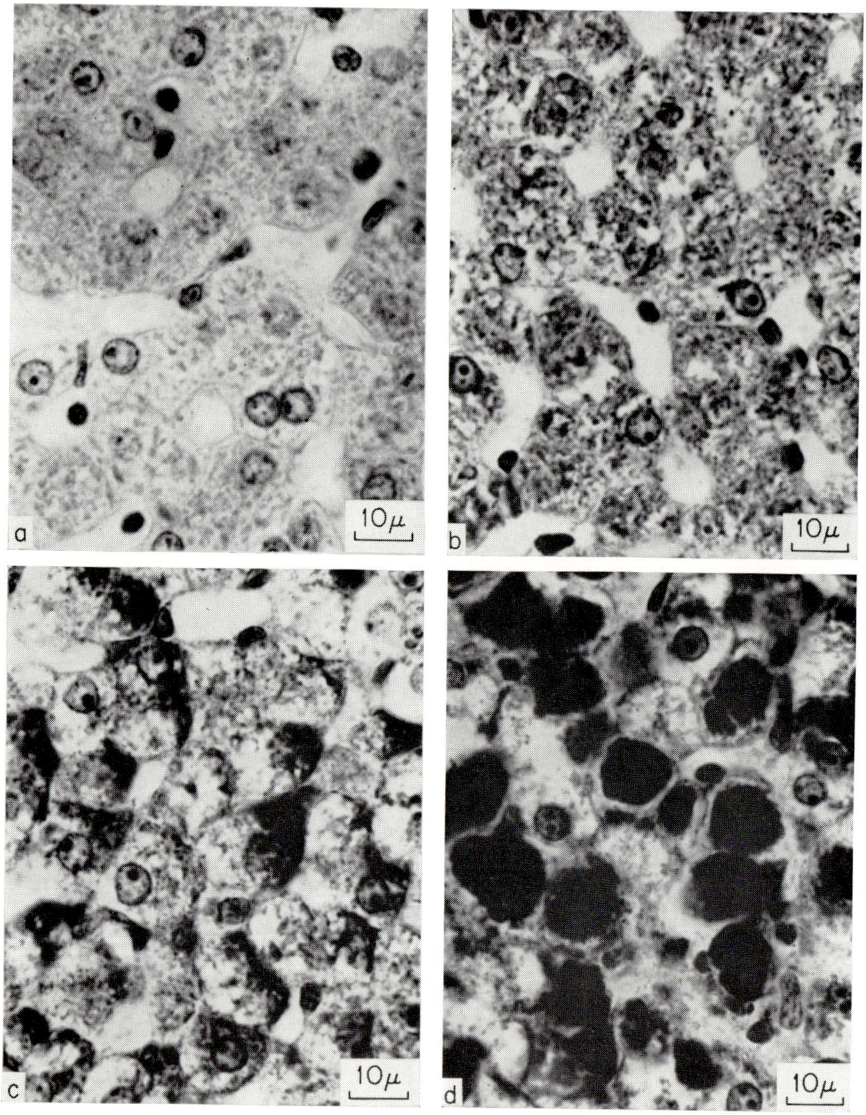

PLATE IV. Photomicrographs showing glycogen in the livers of pigs fed on diets of different protein values: (*a*) Nl (12.1), (*b*) 5 CLP (6.6), (*c*), LP (3.5), and (*d*) LP + CH (2.5). Note glycogen content increases as the protein value of the diet falls (see also Fig. 4).

these often fuse and the large single globule so formed forces the nucleus to the periphery of the cell (signet-ring cell), the nucleus becoming flattened and eccentric. Cells which originally contained glycogen can be recognized by the irregular spaces with ragged outlines and, in most cases, the nucleus remains central.

Alkaline phosphatase activity has been found to be increased in the liver of protein-depleted rats (Ramalingaswami *et al.*, 1954). In the malnourished pigs the alkaline phosphatase activity is increased when fat and glycogen accumulations are not excessive, but as the number of severely affected cells increases the activity is reduced.

The changes reported in the various experimental animals are similar to those found in human subjects suffering from severe protein-calorie deficiency, except that extensive fibrosis, which is so common in man, does not occur. Various workers (Brock, 1954) have suggested that, in man, viruses, alcohol, hepatotoxic alkaloids, excessive intakes of iron, genetic influences, and tropical parasites, notably malaria (Walters and Waterlow, 1954), play a part in the fibrotic changes in the liver.

VIII. Endocrine Glands

The changes described in the endocrine glands vary with the degree of deficiency produced and with the stage of maturation of the glands at the beginning of the experimental regimen. Composite reports are given to indicate that the endocrine glands are affected by protein-calorie deficiency and not to define the limits within which they may vary.

A. Hypophysis

Abnormalities in the pars anterior of protein-calorie deficient pigs (Godwin and Platt, 1960) include vacuolation of the cells and loss of cytoplasm. The reduced cell size results in a crowding together of nuclei, which are often hyperchromatic, and the identification of different cell types becomes difficult owing to the reduction of the cytoplasmic granules. In the most severely affected animals areas of necrosis and small cysts lined with ciliated epithelium can be seen.

Similar changes are found in the hypophyses of malnourished dogs and congenitally malnourished puppies, the reduction in cell size and hyperchromasia of the nuclei being particularly prominent. Large cysts and areas of necrosis were found in the hypophyses of two malnourished dogs, which produced short-lived litters. These changes resemble, though they are much less marked than, those found in Sheehan's syndrome (1937), i.e., in our experiments they may be related to a supply of "poor" blood and not, as in Sheehan's syndrome, a poor blood supply. However, since cysts lined with ciliated epithelium have been reported in apparently normal dogs and are considered to be remnants of Rathke's pouch, the changes cannot be unreservedly accepted as consequent on dietary deficiencies.

The consumption of food of inadequate quantity and/or quality has long been recognized as leading to atrophic changes in the hypophysis (Jackson, 1925) and it has been suggested that many changes in growth and maturation of protein-calorie deficient subjects can be related to reduced activity of the hypophysis (Mulinos and Pomerantz, 1940; Schwartz *et al.*, 1951; Srebnik and Nelson, 1962).

Srebnik and Nelson (1962) maintained rats for about 5 weeks, from the age of 40 or 114 days, on protein-free diets and reported on the histological appearance of the hypophyses and their content of hormones—follicle-stimulating (FSH), interstitial cell-stimulating (ICSH), growth (GH), and thyrotropic (TSH). The total amounts of all were found to be decreased in the younger of the deficient groups although the reduction in TSH was small. The thyrotropic hormone was not affected in the older group; the other hormones tested were reduced but less severely so than in the younger animals. The morphological changes found by these authors in rats resemble those already described for pigs and dogs. In the rats tube-fed a nonprotein diet for 6 days (Halder, 1962) the nuclei are large and hypochromatic, the cytoplasm is vacuolated, and chromophilic granules are reduced. The large pale-staining nuclei found in the short-term experiments are probably a feature of hyperactivity, the small hyperchromatic nuclei in the glands of more chronic cases indicating a nonproductive stage (Ham, 1957).

Mulinos and Pomerantz (1940) introduced the term pseudohypophysectomy "not only because of the resemblance between the effects of inanition and hypophysectomy but also because many of the effects of inanition are due to malnutrition of the hypophysis resulting in a diminished secretion of hormones." Reduced production of hypophyseal hormones has been reported (Leathem, 1958; Srebnik and Nelson, 1962), but the target glands are themselves malnourished and some direct effects undoubtedly occur.

The histological appearance of the hypophysis of human subjects suffering from malnutrition is often complicated by the effects of terminal infections (Tejada and Russfield, 1957). Cysts were found in hypophyses from cases of chronic malnutrition (Gillman and Gillman, 1951) and atrophy, with increased vacuolization, pycnosis (hyperchromasia of nuclei), cellular atrophy, and calcification, was reported by Zubirán and Gómez-Mont (1953), who also found that in 89 of 101 cases the number of eosinophilic cells was increased. However, the concentration of acidophilic cells varies greatly in different portions of the gland and unless extensive counting techniques (Rassmussen, 1929) are employed, actual percentages of cells are probably of less significance than the concentration of granules and the appearances of the nuclei of a representative sample of cells.

B. Pancreas, Endocrine Portion

According to the evidence reviewed by Haist (1944), during starvation, whether due to a complete absence of food or a reduced intake of a balanced diet, there is a reduction in the insulin content of the pancreas. Also when the dietary protein is provided as gelatin the insulin content of the pancreata is lower than when isocaloric quantities of casein were provided, and a good correlation is found between the concentration of granules in the beta cells and the insulin content of the pancreas.

In the protein-calorie deficient pigs there are marked changes in the endocrine portion of the gland, the islet cells are crowded together, counts showing that the number of cells per unit volume is increased because the size of the individual cells is reduced (Heard *et al.*, 1958). The aldehyde-fuchsin-positive granules (Gomori, 1950; Gabe, 1953) of the beta cells are reduced in number and intensity in the LP (3.5) pigs, and in the fulminating condition [LP + CH (2.5) pigs] there are often as many as 11 % of the cells completely degranulated. Halder (1962) tube-fed a protein-free diet to rats for 6 days and found a reduction in size of the individual islet cells and loss of aldehyde-fuchsin-positive granules in most of the cells. In contrast to this degranulation within the islets, there are in the protein-deficient pigs single cells, scattered among the atrophied acinar cells, which contain aldehyde-fuchsin-positive material. Whether these are cells developing insulin-producing properties is not known, neither has the condition or presence of insulin-producing cells within the duodenum (said to be present in normal swine) been investigated.

Selective phloxin staining of the alpha cells (Gomori, 1941) indicated that the ratio beta:alpha (β/α) cells is low in the deficient pigs (Stewart and Heard, 1959, 1960), but no attempt has been made to differentiate between α_1 or α_2 types (Hellerström and Hellman, 1960). The results of a few counts of the islet cells of young pigs suggest that, as in the rat (Mosca, 1956; Hellman, 1959), the ratio of β/α cells is lower in young animals than in adults and it seems likely that the low β/α cell ratio found in the protein-calorie deficient pigs is further evidence of immaturity, being similar to the continuation of the immature distribution in rats hypophysectomized at about 32 days of age (Mosca, 1957). Few other reports describing the effect of diet on alpha and beta cells are available. Rats fed a diet of starch, sugar, fat, and minerals (Verne and Hébert, 1953) suffered a severe loss of pancreatic alpha cells, but in these experiments the proportion of alpha cells in the control animals was 18 %—much lower than those found by other workers, for example, Mosca (1956) 25 %; Hellman (1959) 28 %; our own observations 26 %. In none of the protein-deficient pigs have degenerative changes (other than degranulation and reduced over-all size) been observed in the cells of the islets of Langerhans.

The malnourished dogs and congenitally malnourished pups show crowding of nuclei and reduced size of islet cells, sometimes to a severe degree, changes in the β/α cell ratio being less than in the pigs. Extensive degranulation of the islet cells is seen in congenitally malnourished pups which die before weaning (<6 weeks).

Although the changes in the exocrine pancreatic tissue in the protein-calorie deficient pigs (Section VI,A,2) are similar to those described in children dying of protein-calorie deficiency (J. N. P. Davies, 1948; Camain and Pierchon, 1954; Bras *et al.*, 1957), the cystic and fibrotic changes have not been found in pigs. As in the experimental animals, the morphological changes in the cells of the islets are, in comparison with the exocrine disturbance, slight. Bras *et al.* (1957) did not comment on the islet tissue; Camain and Pierchon (1954) described hypertrophy of the islets and a relative loss of alpha cells, and J. N. P. Davies (1948) reported a possible hypertrophy and hyperplasia of the islet tissue. The considerable reduction in the amount of exocrine tissue, by distorting the appearance of the pancreas, may give a false impression of the amount of islet tissue (Bryans *et al.*, 1952); hence the reports of hypertrophy of the islets. Similarly, the crowding of the nuclei may have led to reports of hyperplasia.

C. Adrenals

The adrenal glands of young pigs maintained on protein-calorie deficient diets are small for the age of the pigs, but large when considered as a percentage of the body weight. They are often fragile, but atrophy and extensive hemorrhage have not been observed.

The glands differ widely in histological appearance; most show some evidence of exhaustion, but in a few cases they are apparently normal. The demarcation between the cortical zones in the adrenals of the normal pig is relatively indistinct and the cells of the zona glomerulosa are less columnar than, for instance, those of a dog. The cells of the zona glomerulosa are small in the malnourished animals and are less columnar than in those of well-fed controls. Mitosis is seen but is not increased. The zona glomerulosa is decreased in thickness, but that of the fasciculata relatively increased. The lipid spaces within the spongiocytes are reduced in number, the cells consequently reduced in size; the cytoplasm and nuclei are hyperchromatic. The zona reticulosa does not appear to be affected, neither does the medulla of the gland. The distribution of adrenaline and noradrenaline, as well as the strength of their reactions observed histochemically, is within the range for normal pigs of the same age. Similar but less

severe changes are seen in the adrenal glands of the malnourished dogs; not all the adrenals of congenitally malnourished pups show abnormalities. Although McCarrison (1921) concluded that "the functional perfection of the adrenal glands is dependent on the quality of the food," there are few detailed accounts of the effects on the adrenal gland of protein-calorie deficiency, possibly because these are less consistent than for many other tissues. When young rats were deprived of protein the adrenals, although of low weight, were large in proportion to body weight (Leathem, 1958). However, the adrenal glands of adult rats maintained on a protein-free diet for 5–11 days suffered a greater proportional weight loss than did the body as a whole (Munro *et al.*, 1962). Great variation, from atrophy to hypertrophy, has also been reported in the adrenals of children suffering from inanition or protein deficiency (Jackson, 1925; Gillman and Gillman, 1951); there are reports of glands which were enlarged and hemorrhagic (Trowell, 1937); enlarged, fragile, and fatty (Bablet and Canet, 1952); markedly atrophic, with loss of lipid (Trowell *et al.*, 1954); necrotic, with patchy loss of lipid (Chatterji and Gupta, 1960); or even unaffected (Campbell, 1956). Fibrous tissue proliferation occurs in the adrenal glands of some malnourished children (Gillman and Gillman, 1951) but it is not a constant finding in human subjects (Zubirán and Gómez-Mont, 1953) and has not been seen in our protein-calorie deficient animals.

If the changes found in the adrenal glands were secondary to hypophyseal changes, more consistent effects, similar to those found after hypophysectomy, might be expected. Hypophysectomy in rats leads to a reduction in the zona fasciculata with a wide zona glomerulosa rich in lipid, whereas during starvation the cortex is hypertrophied and the lipid reduced (Sarason, 1943), observations which agree with those found for protein-calorie deficient pigs. Thus, it may be concluded that any explanation of the changes found must take account of the direct effects of protein-calorie deficiency on the adrenals themselves as well as of the indirect effects attributable to the hypophyseal secretions.

D. Thyroid

The state of activity of the thyroid gland is revealed by the shape of the follicular cells (Goormaghtigh and Thomas, 1934), the concentration of colloid-like droplets within the cells (De Robertis, 1942), the number of colloid "vacuoles" (Aron, 1930), and the staining reactions of the colloid. De Robertis (1949) has shown that in freeze-dried material vacuoles and basophilia of the colloid are not seen; it is, however, generally agreed that, following the usual fixatives, they correspond to some property of the activated gland, whereas lack of Aron vacuoles and an acidophilic colloid indicates a resting gland (Leblond, 1949).

In the thyroid glands of the protein-calorie deficient pigs (Plate V)

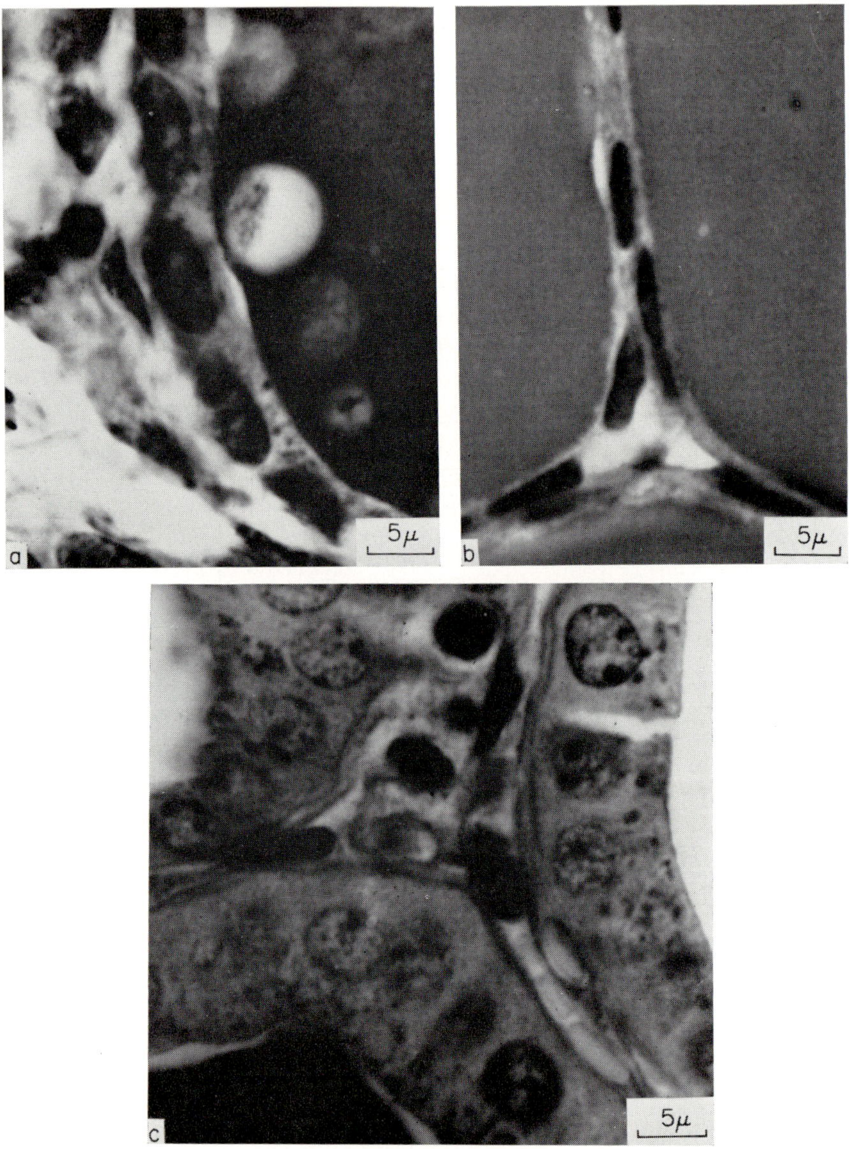

PLATE V. Photomicrographs of thyroid glands of pigs on (*a*) 5 CLP (6.6), (*b*) LP (3.5), and (*c*) "recovery" Nl (12.1) diets. Sections stained with PAS (periodic acid-Schiff technique). Note in (*a*) Aron "vacuoles" and low cuboidal cells with PAS-positive material within their cytoplasm; (*b*) no Aron vacuoles, elongated cells, and few if any PAS-positive granules; (*c*) marked resorption of colloid, tall columnar-type cells with many PAS-positive granules within their cytoplasm. (From Platt, 1962.)

the follicular cells are predominantly flat and elongated (epitheloid), there are very few droplets within the cytoplasm, the colloid is mainly acidophilic, and there are few Aron vacuoles (Stewart and Heard, 1960; Platt, 1962; Platt and Stewart, 1962a). There are differences between animals on the same diets and even within the gland of a single animal, but these are small when compared with those between littermates given diets of different protein values. The average length:height ratio (L:H) was calculated for 4 littermates of the same age and was found to vary from 1.6 (Nl 12.1) and 2.2 [5 CLP (6.6)] to 4.1 and 3.8 for the animals maintained on LP (3.5) and LP + CH (2.5) diets, respectively.

When, after a period of malnutrition, animals are given a diet [20 CLP (8.3) or Nl (12.1)] of higher protein value for 4 weeks, the glands show evidence of exceptional activity. The follicular cells become cubical (L:H 0.96–1.2), many colloid-like droplets are found within the cytoplasm, and there is evidence of collapse of the vesicles, presumably due to the rapid discharge of colloid. The appearance of the glands of the severely protein-calorie deficient pigs is in many ways similar to that found in simple colloid goiter, whereas glands of the recovery pigs are similar to glands which have been stimulated by thyrotropin (see Ham, 1957).

Thyroid glands from congenitally malnourished puppies are immature (see Cooper, 1925), the colloid content is low, cells are often found within the colloid vesicles which are irregular in shape and size, and there are large areas of intervesicular cells. Glands from dogs born of normal mothers and then maintained on low protein diets are similarly, but less severely, affected than are those of the malnourished pigs.

Plotnikova (1959) reported a reduced thyroid activity (I^{131} uptake) in rats maintained on a 3.5 % protein diet and the thyroid secretion rate of rats was progressively reduced by lowering the caloric intake or extending the period of caloric restriction (Grossie and Turner, 1962). Aschkenasy (1961a) found that, whereas thyroxine had approximately equal effects on the metabolism of normal and protein-depleted rats, thyrotropin was less effective in the latter group and suggested that protein deficiency had a direct effect on the thyroid gland. Skanse (1961), however, found that reduced I^{131} uptake and lowered response to TSH could follow ablation of the hypophysis.

No morphological changes are found in the thyroids of rats tube-fed a nonprotein diet for 6 days (Halder, 1962), but in rats maintained for long periods (up to 1 year) on calorically restricted regimens, changes similar to those described above for pigs have been reported (Nikitin, 1962).

There are few reports of the appearance of the thyroid glands of protein-malnourished children, but Trowell *et al.* (1954) refer to atrophy of the thyroid gland.

E. Gonads

As the experiments on pigs and congenitally malnourished puppies were terminated well before the animals reached sexual maturity, the testes and ovaries were not examined. Changes, however, do occur in these glands. Halder (1962) found in the testes of all male rats force-fed a nonprotein diet a great loss of interstitial cells, with a continuing growth of tubules. McCance (1962) made a similar observation in undernourished pigs and also found large cystic follicles in the ovaries of the females. During rehabilitation there was a sudden increase in the size of the external genitalia. McCance discusses the effects of undernutrition on the testes and ovaries and concludes that "at the right chronological age for their maturation, however, some development of the sexual glands evidently takes place even if the undernutrition is very severe."

IX. Carbohydrate Metabolism

A. Glucose Tolerance

Workers in the field of protein malnutrition in humans have been largely preoccupied with protein rather than carbohydrate metabolism, although impaired glucose tolerance has been reported (Holmes and Trowell, 1948; Slone *et al.*, 1961). Among the earliest observations made on pigs suffering from protein-calorie deficiency were: abnormally high levels of liver glycogen (Heard *et al.*, 1958), abnormal glucose tolerance, and changes in the islets of Langerhans suggestive of insulin deficiency (Stewart and Heard, 1959). In this early work, oral glucose tests appeared to indicate an abnormally high tolerance for glucose. When glucose was administered intravenously, the deficient pigs had a markedly impaired tolerance compared with normal animals of the same age (Heard and Stewart, 1960). The discrepancy between the oral and intravenous tests could be explained either by poor absorption of the glucose given by mouth (Section VI,B,1) or by the possibility that the oral test measures primarily hepatic glucose uptake and the intravenous test measures peripheral glucose uptake (Scow and Cornfield, 1954b). To avoid complications due to impaired absorption the intravenous test has been used in all our later work.

By glucose tolerance is meant the rate of removal of intravenously administered glucose from the blood. After initial equilibration of glucose in the extracellular space, its removal proceeds like a first-order reaction in that $dC/dt = kC$. Hence if the blood glucose concentration in mg/100 ml (C), at intervals after the injection of glucose, is plotted on a logarithmic scale, against time in minutes (t) the points fall on a straight line, the slope of which is a measure of glucose tolerance i.e., k, which is calculated from the equation:

$$k = \frac{2.3(\log C_1 - \log C_2)}{t_2 - t_1} \times 100$$

where C_1 and C_2 = glucose concentration at times t_1 and t_2, respectively (Ikkos and Luft, 1957).

In normal pigs k increases over the period 11–90 days of age from 1.0 to about 5.0. In pigs fed the LP(3.5) and LP+CH(2.5) diets k rises more rapidly, reaches a peak of 8 to 11 at 55 days of age, and then falls equally rapidly until at 100 days of age it is well below the normal values (Fig. 5), indicating a reduced capacity to tolerate glucose. Impaired glucose tolerance in deficient pigs at about 100 days was reported by Heard *et al.* (1961), which, as younger pigs had not at that time been tested, was interpreted as evidence of delayed maturity.

Puppies fed on a diet of NDpCal = 6.3 % also show an initially enhanced glucose tolerance compared with littermates on a diet of NDpCal = 12.0 %. By 6 months of age, glucose tolerance in dogs on protein-deficient diets [UK(6.8) and B(6.3)] falls below that of their littermate controls UK(8.6) and B(12.0), respectively (see Fig. 1). It would appear, therefore, that glucose tolerance is affected not only by the severe protein-calorie defi-

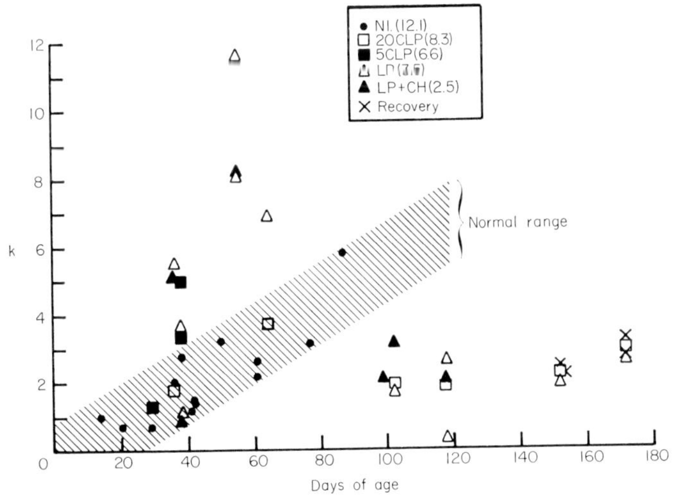

FIG. 5. Values for glucose tolerance (k, see Section IX,A) in pigs at various ages on diets of different protein values. Recovery (X) = 20 CLP (8.3) fed to animals 121 days of age which had been maintained from weaning on LP (3.5) or LP + CH (2.5).

ciency, as suffered by pigs, but also by the much milder forms of protein-calorie deficiency, as produced in dogs.

Impaired glucose tolerance suggests a deficiency of insulin which may be absolute or relative, for the slope of the intravenous glucose tolerance curve is a measure of the balance between insulin, on the one hand, and insulin antagonists or inhibitors, on the other. The latter include: growth hormone (Henneman and Henneman, 1960; Ikkos *et al.*, 1962); glucagon (Van Italie *et al.*, 1959); glucocorticoids (Papper *et al.*, 1956; Frawley *et al.*, 1959); and pituitary-adrenal-dependent inhibitors (Whitney and Young, 1957). Furthermore, impaired glucose tolerance is invariably associated with hypothyroidism (Kusama, 1957; Macho, 1960) and this association is independent of body size or food intake (Scow and Cornfield, 1954a).

B. Plasma Insulin-Like Activity

No measurements of insulin in the blood of protein-calorie deficient pigs are available. "Total" insulin-like activity, measured (Turner, 1962) by $C^{14}O_2$ output from rat epididymal fat pad, in young adult dogs (aged about 18 months) on the UK(6.8) diet gave values of 2600 microunits/ml serum: littermates fed the UK (8.6) diet gave 1150 microunits/ml. Despite the higher output of insulin by the UK(6.8) dogs, their glucose tolerance is impaired, which suggests excess insulin antagonism. In younger dogs (aged about 6 months) fed on the B(6.3) diet the total insulin-like activity is lower than that in adult dogs fed UK (6.8), but their glucose tolerance is somewhat greater than that of older dogs, which is compatible with incomplete development of insulin antagonism.

C. Insulin Tolerance

The pituitary occupies a key position in any consideration of insulin antagonism. The effect of changed activity of the anterior pituitary gland on carbohydrate metabolism is well established. Hypophysectomy or panhypopituitarism is associated with a marked sensitivity to insulin, as indicated by a rapid fall in blood glucose and a prolonged hypoglycemia following intravenous insulin (De Bodo and Altszuler, 1958). Conversely, acromegaly or treatment with growth hormone is associated with insulin resistance, as indicated by a delayed and less extensive fall in blood sugar, and a normal or increased rate of restoration to fasting blood sugar levels (Forsham and Mortimore, 1957).

Hypoglycemia has been reported as being a prominent cause of death in children suffering from kwashiorkor (Kahn and Wayburne, 1960), as well as during the early stages of treatment (Dupin, 1958). Stewart and Heard (1959) showed that pigs fed on the LP(3.5) and LP + CH(2.5)

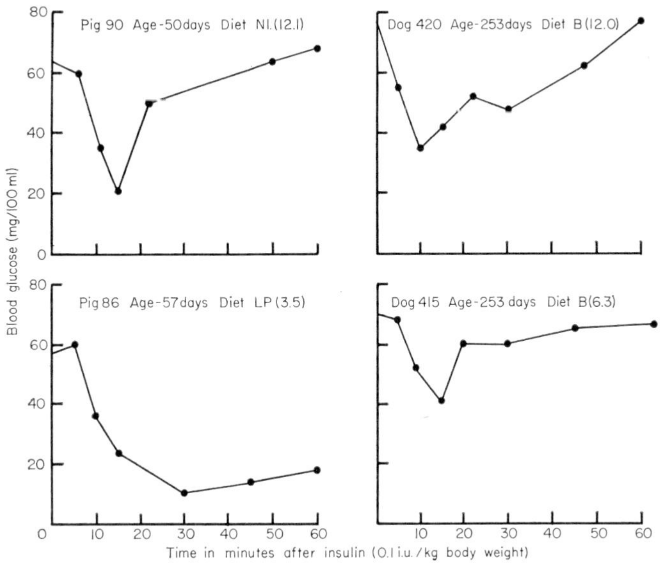

FIG. 6. Typical data for insulin tolerance in protein-calorie deficient pigs and dogs and in well-fed animals; insulin given intravenously.

diets had insulin sensitivity which closely resembled the condition found in hypophysectomized animals (see Fig. 6), and the blood sugar, following intravenous glucose, fell to hypoglycemic levels, again characteristic of pituitary insufficiency (Heard and Stewart, 1960).

Dogs suffering from a milder form of protein-calorie deficiency UK(6.8) and B(6.3), on the other hand, are insulin-resistant (Fig. 6) compared with dogs fed the higher protein diets, UK(8.6) and B(12.0) (Turner, 1962). In normal dogs growth hormone has a diabetogenic effect only in adult, nonpregnant animals (Young, 1941; Young and Korner, 1960). It seems possible that the retarded growth of the dogs on diets of NDpCal = 6.3 and 6.8 % results in a relative surplus of growth hormone production such that it is able to exert its anti-insulin effect. In the pigs, not only is growth stopped because of the protein-calorie deficiency, but the nutrition of the pituitary itself, and consequently hormone production, is drastically impaired.

D. Liver Glycogen Mobilization

The delayed recovery from hypoglycemia of protein-deficient pigs is not due to lack of glycogen in the liver, as abnormally large amounts may be present (see Fig. 4, Section III, and Plate IV). Durbin *et al.* (1960)

demonstrated that in pigs fed LP(3.5) and LP + CH(2.5) diets *from an early age* glucose-6-phosphatase could not be measured in the liver. This resembles the defect found in one form of glycogen-storage disease in man (Stetten and Stetten, 1960). Prolonged insulin hypoglycemia could be entirely due to this enzymic defect. Hepatic glucose-6-phosphatase activity is reduced in pigs started on LP(3.5) and LP + CH(2.5) diets *at a later age* but probably not sufficiently to be the limiting factor in glycogen mobilization (see Mukherjee and Nath, 1957). Furthermore, animals fed on the 5 CLP(6.6) diets have prolonged hypoglycemia but normal levels of hepatic glucose-6-phosphatase.

The initial response to profound hypoglycemia in normal animals is stimulation of the adrenal medulla, and is usually associated with evidence of adrenaline secretion (Himsworth, 1960). Histologically there is evidence of the presence of adrenaline in the medulla of protein-calorie deficient pigs (Section VIII,C). We were able to demonstrate a rapid and pronounced hyperglycemic response to adrenaline in these pigs in the fasting state but not when adrenaline was administered 30 minutes after insulin, when insulin hypoglycemia was most profound.

Glycogen and glucose-6-phosphatase have also been estimated in the livers of congenitally malnourished pups and in normally born pups tube-fed a 2% protein diet for 2 weeks. Glycogen was elevated and glucose-6-phosphatase was diminished in comparison with values for normal pups of similar age (90–100 days). Animals older than 6 months of age showed no difference in hepatic glycogen and glucose-6-phosphatase levels between dogs fed diets of high and low protein value.

E. Adrenal Function

Changes in adrenocortical activity could have a marked effect on carbohydrate metabolism. Impaired glucose tolerance without ketosis, glycogen accumulation in the liver, increased serum pyruvate and lactate levels, sodium retention with mild potassium depletion, and involution of the thymus are among the features of hyperadrenocorticalism (Frawley *et al.*, 1959). All are found in protein-calorie deficient pigs.

In pigs exposed to protein-calorie deficiency, the secretory activity of the adrenal cortex is less impaired than that of other endocrine organs. The deficient animals maintained a steady urinary corticosteroid excretion which was less than the output of the normal pig at the same age, but when expressed per kilogram of body weight was at least as great as that of the normal pig, and in the most severely malnourished animals was considerably greater than normal (Durbin and Heard, 1962). We have obtained similar results with rats fed on nonprotein diets. The metabolism of cortisone appears to be unaffected in the protein-deficient pigs. Thirty per cent of an oral dose of cortisone acetate (given according to weight) was excreted as increased 17-hydroxycorticosteroids (17-OHCS) in 24 hours irrespective of the diet.

Extremely reduced excretion of 17-OHCS has been reported for children suffering from kwashiokor (Castellanos and Arroyave, 1961). Lurie and Jackson (1962) found slightly reduced excretion of 17-OHCS in such children, but concluded "that there is rarely evidence of significantly reduced adrenal function in infantile malnutrition." Our experience with pigs suggests that low figures for steroid excretion need to be considered in relation to body weight.

From the work on man and animals, it may be concluded that the functional activity of the adrenal cortex is, relative to the other endocrine glands, maintained or increased in protein-calorie deficiency. From the work of Aschkenasy (1961b) it is clear that in many respects adrenalectomy affects rats fed on protein-deficient diets more severely than normal animals. At a time when dietary protein intake is inadequate and tissue protein has to be mobilized, the adrenal cortex has a vital homeostatic role to fill.

Extreme sensitivity to insulin and relative hyperactivity in the adrenal cortex appear incompatible; however, the anti-insulin activity of the adrenal in response to hypoglycemia is dependent upon an intact pituitary (De Bodo and Altszuler, 1958). The fact that the basal excretion of corticosteroids is maintained in our animals does not necessarily mean that this excretion could be further increased in response to stress.

F. Pregnancy

The impaired glucose tolerance induced in early life in bitches fed the UK(6.8) diet persists into adult life, although on the basis of human requirements the diet would be considered adequate for adults. They therefore enter pregnancy with a much lower glucose tolerance than bitches fed throughout life on a good diet. The animals fed on the UK(6.8) diet show no significant change in glucose tolerance throughout pregnancy, although once more the diet becomes inadequate with respect to protein-calorie value. Normal well-fed dogs show a slight increase in glucose tolerance in early pregnancy followed by a steady fall to a value, at the end of pregnancy, equal to that of the dogs fed the UK(6.8) diet (Turner, 1962). We believe that this is physiological and not due to any inadequacy of diet during pregnancy.

As shown above (Section IX,B and C), the impaired glucose tolerance of the animals on the UK(6.8) diet is associated with increased insulin antagonism. This, together with the dietary inadequacy during pregnancy, appears to take its toll of the fetus, for the offspring of the UK(6.8) mothers show evidence at birth of pancreatic damage (Section VIII,B) and (at 10 weeks of age) a glucose tolerance greatly in excess of normal, suggesting hyperinsulinism. This is also a feature of the offspring of diabetic mothers (Baird and Farquhar, 1962). Furthermore, the pups born of UK(6.8) mothers are underweight and show central nervous system lesions (see Section XI). Infants described by Hutchison *et al.* (1962) as suffering from congenital temporary diabetes mellitus are also underweight and show evidence of damage to the central nervous system.

G. Conclusions

Incomplete as this study of the complex interrelationships between the endocrine glands and nutrition may be, the point of major importance that has been established is that protein-calorie deficiency, ranging from severe in the young pigs to mild in the dogs, can induce changes of corresponding severity in the endocrine system. Both the dogs and the pigs maintain homeostatic control of fasting blood sugar levels, but an intravenous glucose load reveals that both pigs and dogs have an altered and precarious endocrine balance. That this is not achieved in the same way in the two groups of animals is revealed by the very different responses to insulin.

We suggest that diets of very low net dietary-protein value lead in the young pigs to an early diminution of insulin secretion, and to a compensatory decrease in the anti-insulin activity of the pituitary. Figure 5 suggests that these are not coincident in time of onset. Thyroid function is also reduced, but adrenocortical function is not seriously impaired. Not only do these suggested changes explain the glucose and insulin tolerance results, but they account for the increased liver glycogen and other features listed in Section III. Gillman and Gilbert (1958) showed that in baboons "a fatty liver did not emerge unless insulin was relatively or absolutely deficient and an adequate amount of adrenocortical hormone was available." Protein-deficient pigs fed the extra carbohydrate [LP + CH (2.5)] might be expected to have the imbalance between the beta cells and the adrenal cortex intensified, and indeed it is in these animals that most fat is found in the liver (Fig. 4, Section III).

Diets of higher but suboptimal protein value, fed to dogs, lead to stimulation of the beta cells either directly or by a relatively increased diabetogenic activity in the pituitary (see Section IX,C). Eventually an unstable balance is achieved between high insulin production and high levels of insulin antagonists. The relative deficiency of insulin leads to impaired glucose tolerance and the high level of antagonism to resistance to intravenous insulin. This imbalance may have no obviously serious consequences to the dogs themselves, but may have a profound effect on the developing fetus, particularly when the diet during pregnancy is itself inadequate.

X. Bones and Teeth

A. Bones

The observation that the bones of young malnourished subjects continue to increase in length, although at a reduced rate, even when there is great loss of body weight, has been confirmed for many species. Nevertheless, there are significant changes in the growth and structure of bones which develop during protein-calorie deficiency.

The bones of young pigs maintained on diets of low protein value have a reduced rate of growth, an effect which is intensified by feeding extra carbohydrate (Heard *et al.*, 1958). The remodeling of the bones conforms to the normal pattern of growth, but some modification of the intensity of reshaping occurs. This is best illustrated by the vertebrae, which are, for their age, small in the deficient animals; but the area of the transverse section of the vertebral canal is large in proportion to the total area of a transverse section of the bone. The canal accounts for 21 % of the total transverse area of the deficient bone, but only 16–17 % of the total area of the bones of "age" or of "size controls" (Platt and Stewart, 1962b). Radiographs reveal rarefaction, especially of the metaphysial regions of the long bones, and in these areas transverse trabeculation (Plate VI) is prominent and often more distinct than the usual longitudinal trabeculation (Stewart and Platt, 1958).

Histologically the epiphysial cartilages of the protein-calorie deficient animals are thin, the maturation of the cells retarded, and cells which have not attained full size are eroded by marrow elements. The columns of cells at the zone of penetration by marrow are consequently narrow and the columns of calcified cartilaginous matrix relatively wide. Chondrogenesis is reduced, osteoblastosis less so, and the deposition of bone around the broad cartilaginous columns leads to numerous and extensive transverse linkages. During the early stages of the deficiency osteoclasis is relatively enhanced; much of the primary and secondary spongiosa is destroyed (the rarefaction seen in radiographs, Plate VI), and the remnants are usually portions of the calcified cartilage linked together by bone (transverse trabeculae, Plate VII). In the more advanced stages, chondrogenesis, osteoblastosis, and osteoclasis cease; there is little primary or secondary spongiosa and the immature epiphysial cartilage is sealed off by a thin, often incomplete, border of bone.

The total mineral content of the bones of the protein-calorie deficient

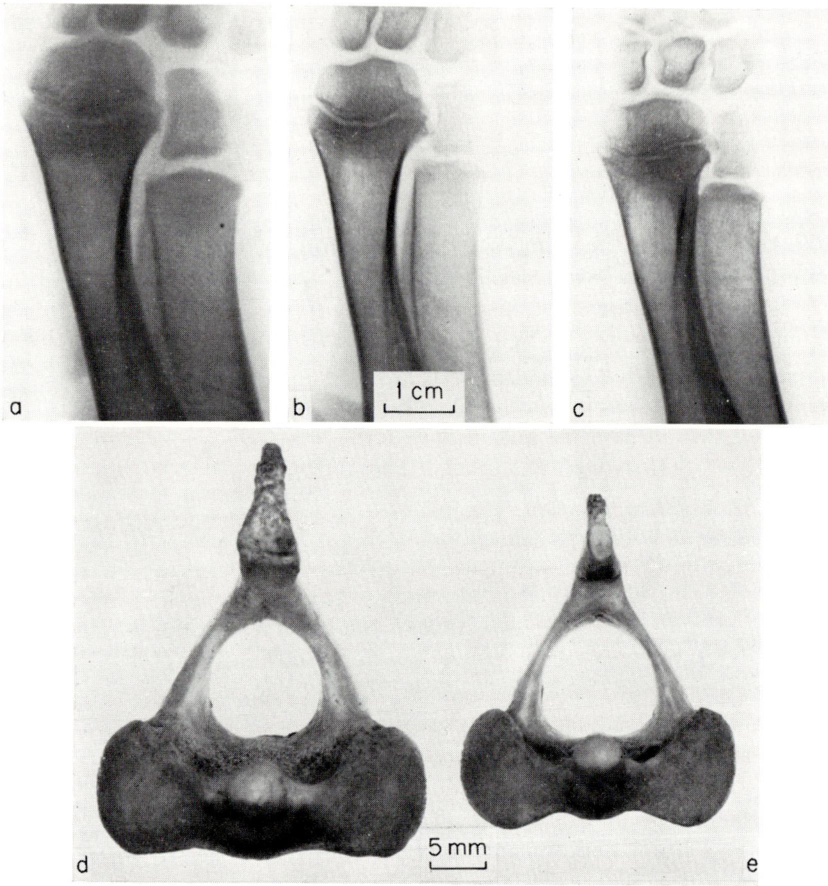

PLATE VI. Radiographs of distal ends of the radii of pigs with different degrees of protein-calorie deficiency after 112 days on (*a*) 5 CLP (6.6) diet, (*b*) LP (3.5) diet, and (*c*) LP (3.5) diet + Ca. Note the smaller size, lower density, and more numerous transverse trabeculae of (*b*) and (*c*) as compared with (*a*). The addition to the diet of a mixture of B vitamins and calcium (650 mg/day) as $CaCO_3$ did not bring about an increase in bone size or density [cf. (*b*) and (*c*)]. (From Platt and Stewart, 1962b.)

Photographs of the second cervical vertebrae of littermate dogs given diets of different protein value: (*d*) B(12) and (*e*) B (6.3). Note that (*e*) is smaller but the areas of the cross section of the vertebral canals of (*e*) and (*d*) are the same.

animals is always low when compared with that of bones from littermates receiving isocaloric quantities of a diet of higher protein value (Stewart and Platt, 1961). The ratio of ash:matrix is, in the short-term experiments, within the normal range for age, but in long-term experiments it would be expected that the proportion of incompletely mineralized bone would be low and the ratio of ash:matrix high for age.

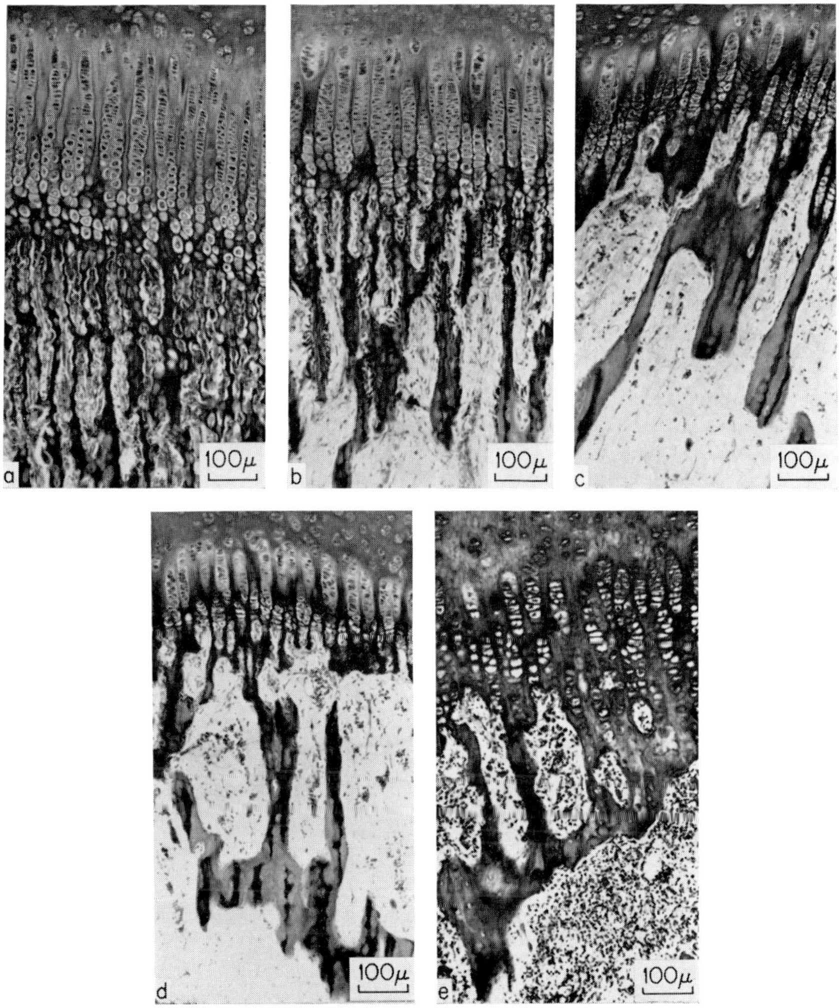

PLATE VII. Photomicrographs of the distal epiphysial cartilages and metaphyses of the radii of pigs (*a–d*) and of a malnourished child (*e*). The pigs' diets were: (*a*) Nl (12.1), (*b*) 5 CLP (6.6), (*c*) LP (3.5), and (*d*) LP + CH (2.5). Note that in the bones of the animals given diets with low protein values, the columns of proliferating and maturing cartilage cells are short and the individual cells are small; there are few normally placed primary but many transverse trabeculae and the amount of calcified cartilage within the trabeculae is excessive. The bone (*e*) from the malnourished child shows changes similar to those seen in the pig (*d*). [(a–d) from Platt and Stewart, 1962b.]

Additional calcium (650 mg daily as $CaCO_3$) does not prevent the retardation in growth, rarefaction, or development of transverse trabeculation. As the primary cause of the rarefaction is a reduced deposition of bone and cartilage, the condition should be regarded as osteoporosis (Platt and Stewart, 1962b).

Less pronounced differences of similar type are seen between 2 groups of dogs maintained on diets of NDpCal = 6.3 and 12 %, respectively; the effect of these diets on the vertebrae and vertebral canal is shown in Plate VI. The congenitally malnourished puppies exhibit retardation of growth, transverse trabeculation and modified remodeling. Some of these have been given diets with a higher protein value producing increases in the rates of growth (weight and bone length), as well as in density and structure of bone.

Similar histological appearances have been described in rats subjected to protein deprivation (Follis, 1950, 1958; Doyle and Porter, 1951; Frandsen *et al.,* 1954; Irving and Budtz-Olsen, 1955) or to caloric restriction (Silberberg and Silberberg, 1940; Handler *et al.,* 1947). Pratt and McCance (1960) found changes similar to those described above in the bones of pigs and cockerels which had been maintained on inadequate amounts of a balanced ration, but in these longer term experiments the ratio of calcium:collagen was higher than would be expected for age (Dickerson and McCance, 1961). Loss of calcified tissue without alteration in the ash:matrix ratio has been reported in adult (Fontaine *et al.,* 1950) and young (Estremera and Armstrong, 1948) rats subjected to protein deprivation.

Arrested growth lines, transverse trabeculae, and striations seen in metaphysial regions of the bones of children have, since 1924 (Harris, 1924–1925), been related to infections, starvation, or other episodes which could cause a reduced rate of growth. Platt and Stewart (1962b) have suggested on the basis of experiments with pigs and dogs that all transverse trabeculae are, directly or indirectly, evidence of protein malnutrition. They argue that during infection appetite often fails at a time when the needs for protein are high; that the protein value of a diet eaten in insufficient amounts is, as far as the consumer is concerned, reduced; that the widespread skeletal changes occurring after severe burns can be related to the severe loss of protein and that endocrine disturbances can lead to reduced protein anabolism. Transverse trabeculae, arrested growth lines, and lattice formation (although vitamin C may play a part in the last) have been described in malnourished children (Eliot *et al.,* 1927; Jones and Dean, 1956; Follis and Park, 1952; Dreizen *et al.,* 1956) and all of these conditions can probably be related to protein-calorie deficiency.

The bones of the protein-calorie deficient pigs exhibit all the histological (Higginson, 1954) and chemical (Walker and Arvidsson, 1954) changes described in the ribs of children dying of kwashiorkor, except that in the specimens of rib transverse linkages were less strong than in the radii of the deficient pigs. Through the cooperation of Dr. H. S. Platt (Department of Pathology, University College Hospital, Ibadan), we have been able to examine the distal epiphyses of the radii of a group of children dying from malnutrition and/or debilitating disease. Transverse trabeculation is common, and histologically the bones are remarkably similar to those from the protein-calorie deficient pigs [Plate VII, (*d*) and (*e*)], and it would appear that the pathogenesis is similar in the two species.

Experimental and clinical observations support the view that examination of the structure and growth of bone, especially the growing ends of the long bones, can provide valuable evidence in the assessment of the nutritional state of young subjects.

B. Teeth

Abnormalities have not been observed in the deciduous teeth of the protein-calorie deficient pigs. Eruption dates appear to be normal, the teeth are only slightly reduced in size, and no lines comparable to the transverse trabeculae of bone have been seen. Because of the relatively greater retardation in growth of the maxilla and mandible the teeth, especially the unerupted molars, are crowded. This discrepancy between tooth and bone size is probably related to the stage of development when the deficiency becomes established. On the other hand congenitally malnourished pups in which the deficient regimen was established when the tooth germs were laid down show retardation of equal degree in the growth of teeth and bones. The jaws and teeth are small, but development and the stage of eruption appear for the first 8–12 weeks of life to be normal for age. If this finding is confirmed on a larger series of pups, it would indicate a severe nutritional dwarfism.

The incisor teeth of rats maintained for 8 weeks on diets of low protein value have a reduced rate of growth and attrition (El-Maraghi *et al.*, 1964). Irving (1957) concludes that protein deficiency has relatively less effect on teeth than on bone, and that teeth appear to have a special facility for obtaining nitrogen even when it is being withdrawn from other tissues (Fritz and Burnett, 1953). Changes, which Bernick and Bavetta (1957) considered would prevent normal calcification, have, however, been found in the matrices of the teeth of tryptophan-deficient rats. McCance *et al.* (1961) maintained pigs for over a year on very inadequate amounts of a well-balanced diet and found that the teeth were crowded, tooth development, eruption, and absorption of deciduous dentine were all delayed, and there was extensive attrition of both deciduous and permanent teeth.

Discrepancies in the relative size of teeth and jaws in kwashiorkor have been reported by Trowell *et al.* (1954), who say: "The incongruous sight of a tiny child with its mouth crowded with teeth is not unusual." Infrequently other changes, such as caries and enamel hypoplasia, have been reported in protein-malnourished children. G. N. Davies (1956) drew attention to the high incidence of dental caries and defective matrix formation in an area in which the diets of pregnant and lactating women were known to be deficient in protein and calcium as well as in vitamins A, B, and C. The need for more intensive research on the effects on the offspring of maternal protein-calorie deficiency is obvious.

XI. Central Nervous System

Pigs born of well-nourished mothers and maintained on the LP(3.5) or LP + CH(2.5) diets are retarded in development, walk on tiptoe with a "hobble-skirt" gait and occasionally exhibit some incoordination of the hind legs (Platt *et al.*, 1964). Dogs born of well-nourished mothers and maintained on the B(6.3) or UK(6.8) diets are also retarded in growth, are more nervous than well-fed littermates, but show no definite evidence of neurological change. Rats tube-fed diets devoid of protein developed a spasmodic trembling of the head and forepaws after 3 or 4 days (Doell *et al.*, 1964a), but the signs of neurological involvement are most clearly seen in the dogs born of malnourished mothers (see Plate VIII).

Eight dogs which received the UK(6.8) diet from 6 weeks of age produced eleven litters, a total of 57 pups. Normal males were used for mating. Neonatal deaths among the pups were high (about 50 %) and the development of the survivors was retarded. The degree of abnormality varies widely, even within litters; the more obvious signs, such as athetoid movements of the head and neck (see Plate VIII) and ataxic gait, usually develop after weaning, quickly reach a peak and then regress. One animal died during an epileptiform fit and two others were found dead or injured in circumstances also suggestive of fits. In some of the dogs the acute manifestations do not regress and at 3 months of age they still show incoordinated movements which are exaggerated by the presence of a strange dog. Animals which recover from the acute stage remain retarded, often run in narrow circles both in and out of their cages, are extremely nervous, and resent being handled.

Electroencephalograms (EEG) were obtained, using needle electrodes (Pampiglione *et al.*, 1960), and showed considerable diminution of normal rhythmic activity. The puppies with obvious signs of central nervous

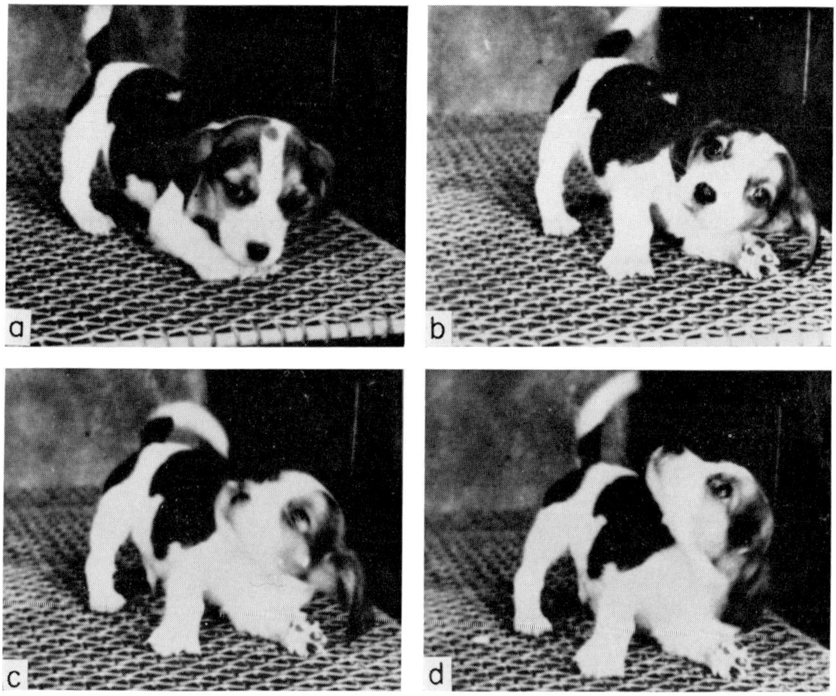

PLATE VIII. Photographs of a pup aged 9 weeks, reproduced from a cinematograph film showing athetoid movements of head and neck. Both the pup and its mother were maintained on diet UK (6.8). (From Platt, 1962.)

system involvement also show in the EEG an excess of irregular, slow activity of large amplitude and multifocal spikes and sharp waves (Fig. 7), particularly during sleep (Meyer *et al.*, 1961a). The severity of these EEG changes varies considerably between individual dogs, but similar abnormalities have not been seen in the EEG's of pups born of well-nourished mothers. Irregular, slow activity, and occasional discharges of large amplitude have also been observed in the EEG's of rats maintained on, or tube-fed, a nonprotein diet (Doell *et al.*, 1964a).

The growth of the central nervous system is less severely affected by protein-calorie deficiency than that of many other organs, so that the ratio of brain to body weight (Section III) is high in the deficient animals. At death, the space within the vertebral canal and around the spinal cord often appears to be reduced, but in no case have pressure effects, such as those seen in vitamin A deficiency (Mellanby, 1941), been observed.

Changes in the appearance of the nerve cells and neuroglial cells of the protein-calorie deficient animals have been found in the spinal

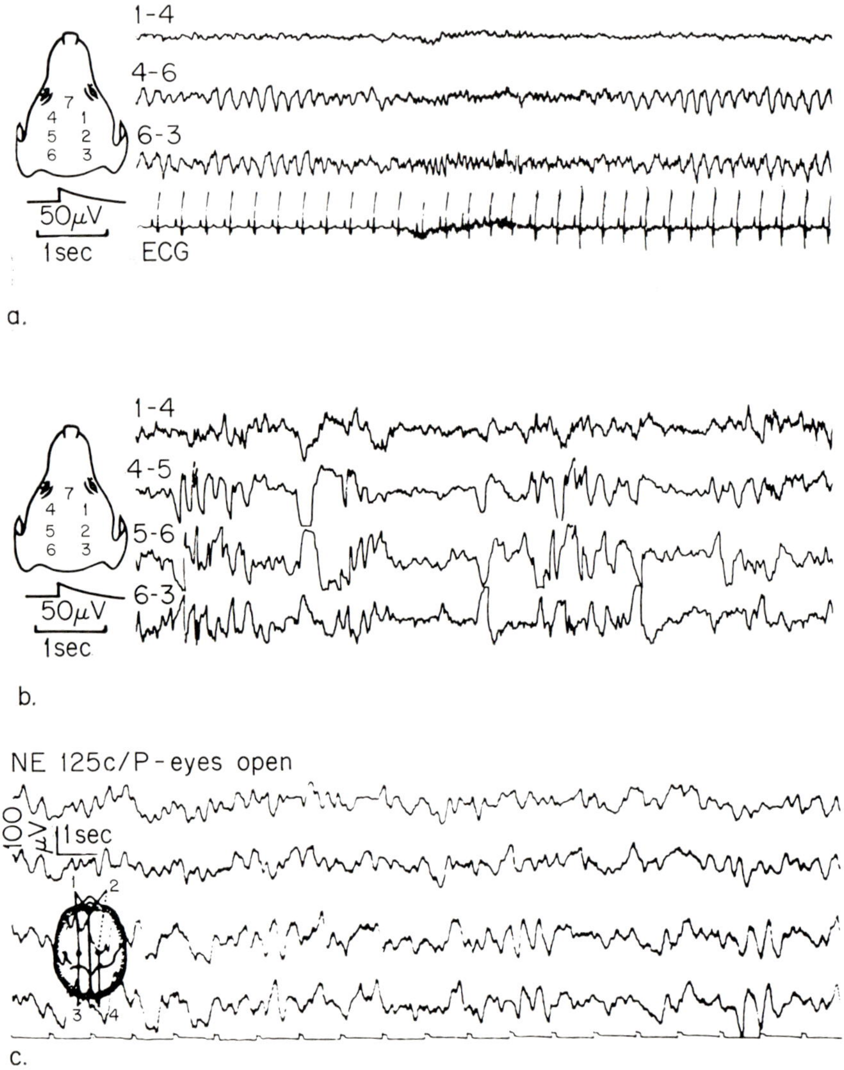

FIG. 7. Electroencephalographic (EEG) changes in protein-calorie deficiency: (*a*) EEG recordings (3 traces) from normal 3-month-old dog; (*b*) recordings from 3-month-old dog born of malnourished mother and weaned on to mother's diet [UK (6.8), see Section XI]; (*c*) EEG of child suffering from kwashiorkor; paper speed about half that in (*a*) and (*b*) [reproduced from Nelson (1959) with author's permission]. Note similarity of abnormalities in records (*b*) and (*c*); these are independent of any differences due to the location of the leads in dog and child.

cord, especially the lumbar and cervical enlargements, the medulla, and pons. The findings are less severe and more irregular in the higher levels and the following descriptions relate mainly to the spinal cord (Plate IX). Alterations include chromatolysis (Platt, 1961b), "foaming" of the cytoplasm (probably the vacuolation described in starvation) and an increase in the oligodendroglial nuclei, which are often clustered around damaged cells (Platt and Stewart, 1960). Activation of astrocytes, with an increase in the number and caliber of the astrocytic fibers (Meyer *et al.*, 1961b) and, in the more severe cases, Nissl's severe cell change, neuronal loss, and heavy fibrous gliosis also occur (Meyer *et al.*, 1961a). The neurofibrillar arrangement within the nerve cells becomes disorganized and the fibers lose much of their affinity for silver nitrate. Degeneration of the myelinated fibers is not observed, nor is any excess of fat, but in two of the congenitally malnourished puppies myelination of the peripheral nerve was thought to be delayed. Extensive demyelination has been reported in rats maintained on a protein-deficient regimen (Cuparencu *et al.*, 1959).

Most of the changes are probably reversible, but the more severe instances of gliosis and cell damage would be accepted as pathological in human material. Refeeding the depleted animals on a diet of high protein value leads to a prompt improvement in their condition, but at death, 1 to 3 months later, the central nervous system has not fully returned to normal; whether complete recovery would occur if the treatment were prolonged is not known.

The severity of the changes can be increased by reducing the age at which the deficiency is established, reducing the protein value of the diet, or increasing the duration of the experiments. Thus congenitally malnourished pups show more severe changes than those born of well-nourished mothers; pigs maintained on the LP(3.5) or LP + CH(2.5) diets are more severely affected than those given the 5 CLP(6.6) diet and in rats tube-fed a low protein diet for only 6 days the changes are usually limited to chromatolysis and "foaming" of the cells (Doell *et al.*, 1964b).

Cammermeyer (1961) considers the dog unsuitable for many tests of nervous system function because in a group of supposedly healthy adult dogs (history unknown) he found "numerous, widespread and variable neuropathological changes"; many of these were thought to be residual effects of meningoencephalitis. Cammermeyer's findings emphasize the need for adequate controls in any observations involving the central nervous systems of dogs. In our preliminary experiments the central nervous system of the dogs kept free of infection and maintained on diets with high protein values [B(12)] show little proliferation of astrocytes.

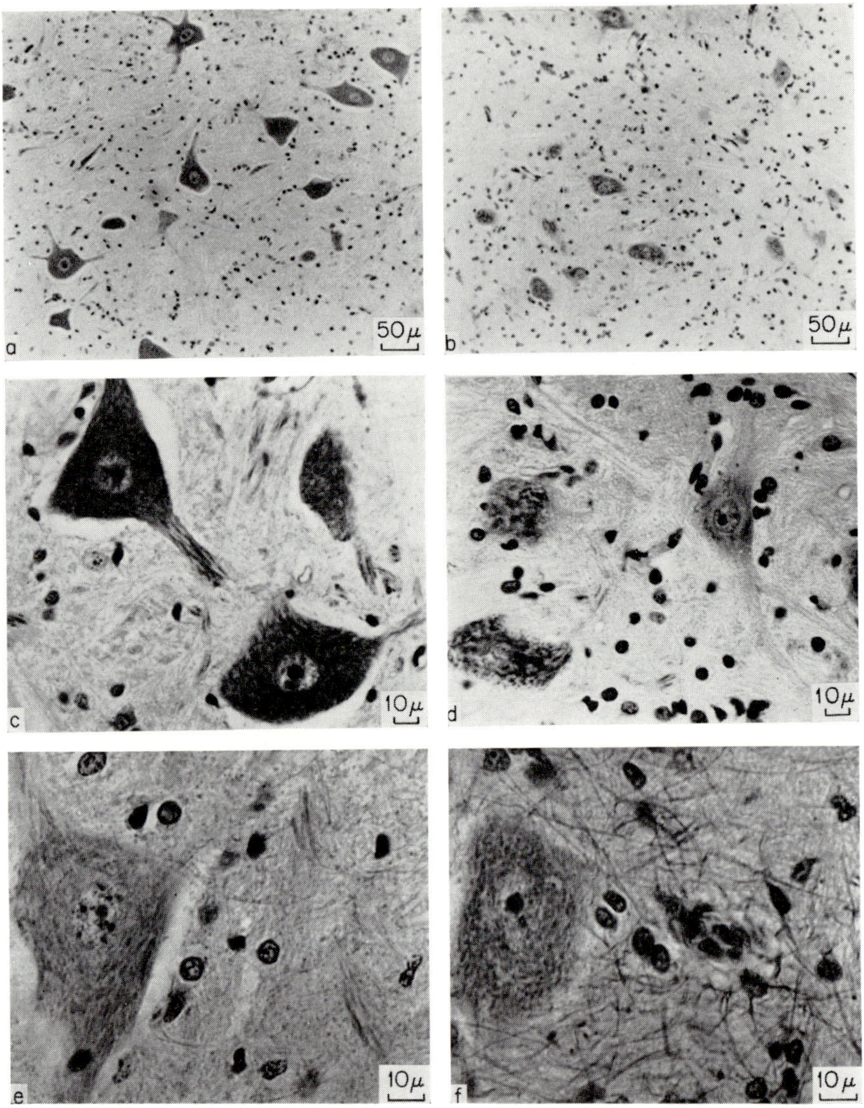

PLATE IX. Photomicrographs of the anterior horn of the spinal cord of pigs and dogs, stained (*a–d*) with Einarson's gallocyanin; (*e*) and (*f*) with Mallory's phosphotungstic acid hematoxylin. (*a*) and (*c*) from pig given diet Nl (12.1); (*b*) and (*d*) from pig given diet LP + CH (2.5). Note that (*b*) and (*d*) show increase in neuroglial cells and loss of Nissl substance. (*e*) from 3-month-old dog, born of normal mother, and maintained on stock diet; (*f*) from 3-month-old pup, born of malnourished mother, and maintained on diet UK (6.8). Note increase in number and caliber of astrocytic fibers in (*f*) as compared with (*e*). [(*e*) and (*f*) from Platt, 1962.]

Changes such as distention of the lateral ventricles, with tearing of the septum pellucidum and cellular meningitis, are not observed in either the well-fed or malnourished dogs.

Although the mental changes, misery, apathy, irritability, and drowsiness are such constant findings in children suffering from protein-calorie deficiency (Trowell, 1937; Clark, 1951; Brock and Autret, 1952; Kahn, 1954; Udani, 1960), there appear to be no reports of histological examinations of the central nervous system (Trowell *et al.*, 1954; Meyer, 1958). The effects of inanition have been reviewed by Jackson (1925), who reports that the nerve cells of animals, and especially the motor cells of the spinal cords, undergo atrophic degeneration, chromatolysis, cytoplasmic vacuolation with disorganization of the neurofibrils and, in severe cases, actual loss of some cells. Secondary diffuse changes in the myelinated fibers are also said to occur. Andrew (1941), when examining the spinal cords of mice starved from the 65th day of age, placed more emphasis on the increase in neuroglial cells. He also described (1940), in the brains of human subjects suffering from inanition, phagocytosis, vacuolation, and loss of cytoplasm, so that nerve cell nuclei appeared to be as "naked as neurologlial nuclei."

Nelson (1959) and Nelson and Dean (1959) reported evidence of retarded development of the nervous system and the presence of localized or focal disturbances in the EEG's of children suffering from kwashiorkor (Fig. 7), findings similar to those described for congenitally malnourished puppies. In many of the patients the abnormalities had not disappeared when they were discharged from hospital as cured and Nelson recalls that some years ago 58 % of a group of apparently normal West African soldiers had abnormal EEG's (Gallais *et al.*, 1951a,b; Mundy-Castle *et al.*, 1953), which it was thought might be due to malnutrition in early life. Abnormalities in the electroencephalograms of patients suffering from protein malnutrition have also been reported from Algiers (Sarrouy *et al.*, 1953), and Ceylon (Engel, 1956). Engel was able to follow up patients discharged from hospital as cured of protein malnutrition and found that in 2 of the 25 tested the EEG remained abnormal.

Impressed by the mental changes in malnourished infants, Udani (1960) conducted neurological examinations and found "motor system involvement, weakness, hypotonia, diminished or absent deep tendon reflexes"; the neurological signs receded during treatment (4–6 weeks) but tremors and rigidity often occurred during this period. The appearance of coarse tremors during recovery from kwashiorkor have also been investigated by Kahn and his associates (Kahn and Falcke, 1956; Kahn, 1954, 1957), who likened the condition to one of pseudo-Parkinsonism.

The acute exacerbations of nervous symptoms which occur when congenitally malnourished pups are weaned from malnourished mothers (Section XIII) on to a diet having a protein value of NDpCal = 6.8 %, may be a recovery syndrome similar to that described by Kahn (1954) and Udani (1960). This is not unexpected, for the diet has a protein value which is considered adequate for a young growing child (4–9 years of age).

There is, then, clinical and experimental evidence that the function of the brain is affected in protein-calorie deficiency and that lesions of the central nervous system may persist when physical disabilities have apparently disappeared.

XII. Other Tissues

A. *Eyes*

The growth (fresh weight) of the eye is relatively unaffected by protein deficiency (Limson and Jackson, 1932), cystine or lysine deficiency (Lafon, 1939), or inanition (Jackson, 1925; Addis *et al.*, 1936), the weight being related to age rather than body weight (McMeekan, 1940). McLaren (1958) fed rats from weaning to 6 months of age on diets with different protein values and found that even on a cassava diet the weight of the eyeball was only about 25 % less than that of well-fed littermates. The proportions of water to dry matter did not vary.

1. *Lens*

Cataract was noted in rats (Curtis *et al.*, 1932) and pigs (Cartwright *et al.*, 1945) which had been maintained on diets devoid of trytophan. Schaeffer and Murray (1950) demonstrated by means of delayed supplementation experiments that the cataractous changes were due not to a specific deficiency of tryptophan but to a lack of protein synthesis. McLaren (1959) noted cataracts only in pigs which had been maintained on diets low in protein for longer than 1 year. The fibers of the lens were swollen and separated by vacuoles; later the fibers degenerated and the interfibrillar clefts filled with an amorphous deposit; bladder or hydropic cells were also reported.

Cataracts were not found in the eyes of protein-depleted rats (McLaren, 1960) but the lenses of such animals contained reduced amounts of soluble (GSH) and protein-bound (PSH) sulfhydryl groups (Bagchi, 1957a). Supplementing the deficient diet with methionine led to normal GSH and PSH concentrations in the lenses, whereas the addition of methionine sulfoximine (a methionine antagonist) to the diet brought about a condition similar to that of protein deficiency (Bagchi, 1957b).

In the center of the lens of the eyes of some congenitally malnourished pups there was an opaque spiral (probably cellular infiltration), confirmed by Dr P. A. Gardiner. There was an error in refraction which disappeared as the animals grew older, possibly another indication of recovery on a diet of NDpCal = 6.8 % (see Section XI). Gardiner and MacDonald (1958) had earlier reported changes in refraction in rabbits maintained on protein-deficient diets.

2. *Cornea*

Vascularization of the cornea was found in rats which had been maintained on diets deficient in amino acids (Sydenstricker *et al.*, 1946), and McLaren (1959) made a similar observation in rats given diets containing 3 % or less of vegetable protein. On examination with a slit lamp the areas surrounding the vessels were hazy and histologically the basal cells were swollen (probably edematous), with a clear cytoplasm and pycnotic nuclei. The corneal epithelium was reduced in thickness, with some keratinization of the superficial layers. Opacity of the cornea has not been seen in the pigs, but occurs in some of the congenitally malnourished puppies. Changes in the eyes of the offspring of protein-deficient rats include "marked disorganization of the lens, poorly developed iris and ciliary body, immaturity of the corneal collagen and changes in the zone of separation of the eyelids" (McLaren, 1960).

Protein deficiency has aroused less interest with regard to abnormalities in eyes than has hypovitaminosis; Oomen (1958) has suggested that protein deficiency may be involved and there is experimental evidence to support this, at least in relation to riboflavin (Day *et al.*, 1934) and vitamin A (Friend *et al.*, 1961).

B. Skin, Hair, and Nails

The investigations in this unit on the effects of protein-calorie deficiency on the hair, skin, and nails of rats and pigs have been reported and discussed by Godwin (1962). The pigs maintained on the protein-calorie deficient diets had an unkempt appearance; the hair was sparse, fine, and long; the skin was dry and scaly, wrinkled, and cracked. In some animals papules and encrustations were found on the skin, especially of the abdomen and hocks. The changes were most severe in the LP + CH (2.5) group and did not occur in the normal animals. Platt and Frankul (1962) demonstrated a reduced transport of zinc in the animals deprived of protein and the hyperkeratosis was possibly due to zinc deficiency brought about in this instance by lack of dietary protein.

1. *Skin*

Apart from the areas of hyperkeratosis, it was found (Godwin 1960b) that when pigs were given rations of different protein values the epithelium was thinner in those given the diets with the lower protein value. In pigs on the LP + CH(2.5) diet the epithelium was often reduced to a single cell thickness, the rete pegs disappeared and the keratin layer was flaky (Plate X). A surface layer of periodic acid-Schiff (PAS)-positive material was present, similar to that previously reported in the protein-depleted rat.

The dermis of the deficient animals was disorganized, the collagen being foamy and broken up with many, probably edematous, spaces; fibroblasts were shrunken and there was some increase in histiocytes. Skins of protein-deficient rats (Godwin, 1956) also showed thinning of the epithelium and disorganization of the dermis, with poorly formed and fragmented collagen. Evan's blue injected intradermally leaves the site of the injection and spreads more rapidly in the malnourished than in the well-fed controls, indicating a loose and more permeable structure. There was also in the deficient rats some atrophy of the sebaceous glands.

Dermatoses are often, but not invariably, found in human subjects suffering from protein-calorie deficiency and such variations have added to the complications of nomenclature. Skin lesions developing in children deprived of essential fatty acids (A. E. Hansen *et al.*, 1958) are similar in many respects to those found in kwashiorkor; recently, Naismith (1962) has demonstrated an interaction between fatty acids and protein utilization, suggesting that in the dermatoses of kwashiorkor more than protein deficiency may be involved.

Histological changes in the skins of protein-calorie malnourished

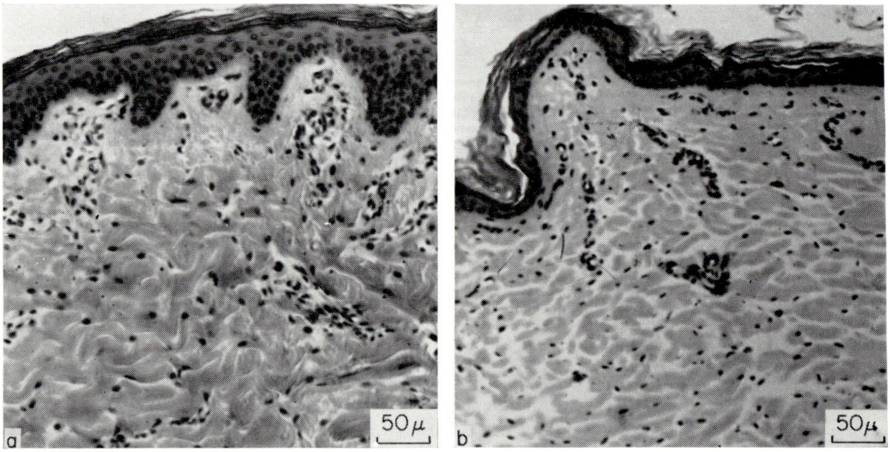

PLATE X. Photomicrographs of the skins of pigs given (*a*) diet 5 CLP (6.6) and (*b*) diet LP + CH (2.5). Note reduced thickness of epidermis and separation of collagen fibers in dermis in (*b*).

patients include thinning of the epidermis, loss of rete pegs, and patches of hyperkeratosis or parakeratosis; the horny layer may be reduced or absent. Edema of the dermis with separation and fragmentation of the collagen fibers is common, sebaceous glands are seen less frequently than in normal skin, and there is often an infiltration of histiocytes (Berte, 1954; Rangam *et al.*, 1962).

2. *Hair*

An intense sulfhydryl reaction was noted in the hair follicles of deficient rats and was related to an extension of the usual SH-containing zones (Nagchaudhuri, 1953). Evidence was presented to suggest that the fault was a reduced conversion of SH to S—S groupings in the deficient animals. The pigment obtained from the black hair of normal Africans and the red hair of kwashiorkor subjects gave different spots on a chromatogram (Nagchaudhuri and Platt, 1954); that from red hair was shown to behave like material obtained from oxidized black hair. Platt and Nagchaudhuri (1954) and Nagchaudhuri (1953) conclude that amino acid- or protein-deficient diets brought about not only an oxidative change in the pigment of hair, but a reduction in the cystine content of black hair. The changes in the cystine content of the hair from malnourished subjects were confirmed by Bigwood and Robazza (1955).

3. *Nails*

Although the nails are often long in the malnourished pigs this should be regarded as reduced attrition rather than increased growth, for Godwin (1959) showed that the rate of nail growth fell in rats given diets with a low protein value.

The rate of nail growth varies with age in the human subject (Hamilton *et al.*, 1955); the fastest rates are found in the 10–14-year olds, the slowest in those over 20 years, with the younger children (1–10 years) forming an intermediate group. In a preliminary investigation, Janes and Mittwoch (1962) compared the rates of growth of the nails of adult laboratory workers and a group of underprivileged children. The children aged 7–12 years were short and underweight for their age. Although the differences were not statistically significant, the lower rate of nail growth was found in the children and not, as would have been expected in a healthy population, in the adult group. The result suggests that with larger groups and adequate controls nail growth may be a good indicator of nutritional status.

C. *Spleen*

The spleens from the protein-calorie deficient animals are small and in many cases are little more than shrunken shells. Malpighian corpuscles are reduced in size, as is the red pulp, but the iron content (Prussian blue reaction) appears to be increased, especially in the spleens of animals given the LP + CH(2.5) diet. The reduction of the lymphoid tissue and red pulp gives to the splenic trabeculae an undue prominence. Atrophy of the spleen has been described as occurring during starvation (Jackson, 1925), as well as in human beings (Bablet and Normet, 1937; Trowell *et al.*, 1954) and animals (Sidransky and Clark, 1961; Halder, 1962) suffering from protein-calorie or amino acid deficiencies.

D. Kidney

At death the kidneys do not appear to be abnormal in color, shape or texture but are reduced in size, whether considered on an age or weight basis (Section III). Some loss of cell cytoplasm occurs in the kidneys of all the protein-calorie deficient animals [5 CLP(6.6) and LP(3.5) and LP + CH(2.5)], but is only severe in the LP(3.5) and LP + CH(2.5) groups.

The glomeruli are small, although no specific cell changes have been observed. Loss of cytoplasm with consequent reduction of cell size occurs in the proximal convoluted tubules, some cells show cloudy swelling, and the brush borders are less strongly PAS-positive than those of well-fed controls. Loss of cell cytoplasm is also seen in the cells of the loops of Henle and collecting tubules; nuclei in these regions appear to be devoid of cytoplasm. The changes vary considerably from animal to animal and even within a single kidney, and in no case are they as widespread as, nor do they approach the severity of, those seen in the liver and pancreas.

The changes described are amongst those reported by Jackson (1925) in the kidneys of animals subjected to starvation and in human patients suffering from protein-calorie deficiency (Bablet and Normet, 1937; Trowell *et al.*, 1954).

E. Thymus

The reduction in size of the thymus gland of the protein-calorie deficient animals is most impressive and in severe cases only vestiges of the gland are to be found.

Thymus glands that have been examined exhibit a great reduction in lymphoid tissue, especially in the cortex. The stroma appears to be increased, but this is probably only relative due to lymphoid loss. Hassal's corpuscles, although occasionally enlarged are usually normal.

Atrophy of the thymus may be so severe in children dying from "infantile edema" that the gland cannot be recognized (Vint, 1936–1937). Similar severe atrophy has been observed in starvation (Jackson, 1925) and amino acid deficiency (Sidransky and Clark, 1961) in animals.

Burnet (1962) has recently ascribed an important role to the thymus in antibody production. J. F. A. P. Miller (1961, 1962) reported that deaths due to common laboratory infections were higher in a group of mice thymectomized 1 to 16 hours after birth than in nonthymectomized mice and that there was a depression in the immune response to both viral and bacterial antigens. The effects of protein-calorie deficiency on the thymus, which may be regarded as being equivalent to partial and sometimes complete thymectomy, are not only interesting but probably important in relation to some aspects of the interrelationship between malnutrition and zymotic disease.

F. Muscle

The muscles of the protein-calorie deficient animals are usually pale and wasted. Histologically there is a reduced quantity of sarcoplasm, hence an apparent increase in nuclei; occasionally this increase is real. Transverse striations are usually present but often blurred. Muscles of the pigs fed on the LP + CH(2.5) diet show edema, which causes separation of the fibers from one another, and the myofibrils of some fibers, usually those which are larger than average, are prominent as though separated by fluid or other material.

Recently Van Bogaert *et al.* (1961) have described edema of the intermysial spaces, a foamy change in the sarcoplasm, eosinophilia, and dissolution of muscle fibers in children dying of kwashiorkor. Although the changes seen in our animals are of the same type, they are never as severe as those described by Waterlow (1948, Plate XI).

XIII. Reproduction

Young growing subjects need food of high protein value; for this reason they are likely to be specially susceptible to protein-calorie deficiency. However, accepting internationally agreed data for protein and calories, women (and presumably females of all species) need toward the end of pregnancy and during lactation food having as high, if not a higher, nutritive value (measured as NDpCal%) than the young. If the female has, during reproduction and lactation, a protein-calorie deficient diet, evidences of malnutrition might be expected to occur in her offspring.

At some stages of development the rat fetus shows considerable susceptibility to malnutrition. Depriving the mother of food for 24 hours (Bruce, 1962) or 48 hours (McClure, 1962) may cause resorption or abortion of the embryo. When pregnant rats are given a diet devoid of protein for 3 days the offspring show abnormalities the distribution and severity of which vary with the stage of development of the embryo at the time of protein restriction (Halder and Platt, 1962). These observations are pertinent to the prevalence of deterioration in or aberrations of appetite during pregnancy in humans (Taggart, 1961).

Rats maintained on a low-protein "Gambia-type" diet produce fewer and smaller litters (of lower birth weights) and with a higher neonatal mortality than do normal rats. These effects of the deficient diet can be prevented by additions of fishmeal (Platt and Miller, 1957). Similar results have been obtained by us using diets in which casein is the only dietary variant. Several examples of the effects on pups, the offspring of bitches suffering from mild protein-calorie deficiency, have been given: nutritional dwarfism (Section I), changes in the endocrine glands (Section VIII) and carbohydrate metabolism (Section IX), as well as neurological symptoms (Section XI). These pups, like the rats, are small at birth and the proportion of neonatal deaths is high.

Relatively less attention has been given to the mothers than to the offspring in the study of protein-calorie deficiency. As shown in Fig. 1 and Section IX, even mild protein-calorie deficiency results in many biochemical abnormalities and it is through such changes that the effect on the embryo must be mediated. The mothers are not outwardly abnormal, but during lactation the mammary glands are often difficult to distinguish from those of a nonlactating bitch and there is little doubt that shortage of milk contributes to the changes observed in the offspring. Similarly, the young of rats subjected to protein malnutrition are often found at death to have no curds in the stomach or upper part of the small gut, whereas in the stomachs of normal suckling rats the remains of milk taken 20 hours previously are often found (Platt, 1954b).

Hormonal deficiency or imbalance, such as occurs in protein-calorie deficiency, is known to have severe effects on the developing mammary gland, and therefore changes in the development of the gland might be expected in the protein-calorie deficient animals (Platt, 1958b). Plate XI shows the appearance of mammary glands of virgin rats maintained on diets of high and low protein values, respectively. The gland from the well-fed animal contains many rudimentary alveoli which could develop during pregnancy; in contrast, the gland from the deficient animal contains few rudimentary alveoli and it would not be surprising if the latter were unable to produce a normal supply of milk.

Platt (1961a) has reviewed his own and other work on the significance of curd and whey proteins in infant digestion; the casein remains for many hours in the stomach, while the whey proteins pass rapidly into the duodenum. It has been shown (Lutz and Platt, 1958; Lutz, 1958) that, in the milk of women suffering from protein-calorie deficiency, the curd:whey ratio is variable and usually much increased. The increase is related to the severity of the deficiency which was assessed qualitatively from dietary

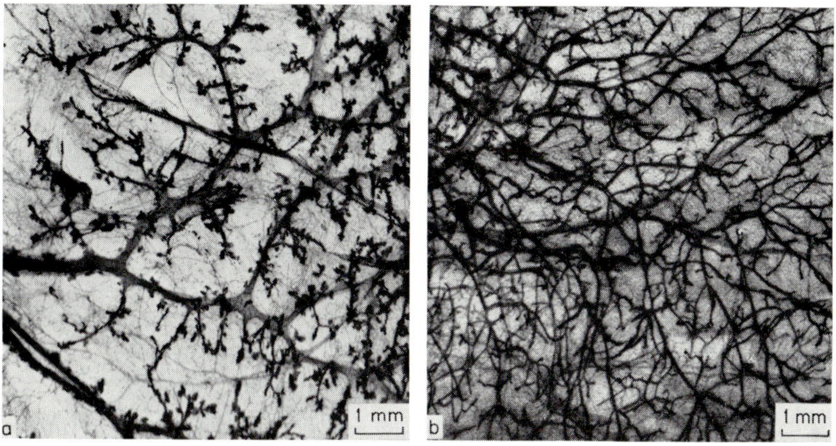

PLATE XI. Photographs of preparations of mammary glands from virgin rats fed (*a*) on stock diet and (*b*) on diet of low protein value (cassava + 3% casein). (From Platt, 1958b.)

and clinical information and quantitatively from measurement of the serum albumin, urinary urea nitrogen:total urinary nitrogen per cent (from a sample of first morning urine) and weight/height ratios.

Breast feeding has been referred to as "a device by which provision is made for a transition between complete dependence on the mother before birth and complete independence after weaning from the breast" (Platt, 1955). The experiments on animals suggest that early malnutrition during the development of the mammary gland may seriously reduce the efficiency of such a transition. It was on the basis of the experimental work that those responsible were urged to give the fullest consideration to the nutrition of the mother, especially in the underdeveloped countries (Platt, 1957a).

Even in advanced civilizations a relationship probably exists between maternal malnutrition and abnormalities in the offspring. Murphy and Bowes (1939) reported on the food habits of 545 mothers who had each produced a malformed child; about 35 % of the group took no milk during pregnancy and the diets of 40 % were considered to be inadequate. Although there was no proof that dietary inadequacies or anemia (possibly secondary to a poor diet (see Section IV,A)) were predisposing causes, the authors concluded: "It is of interest, however, that both characteristics should have been observed with unusual frequency."

XIV. Nutrition and Zymotic Factors in Disease

The term "zymotic factors" is used here to include all living organisms that produce disease, i.e., it includes all infections and infestations. The interplay between malnutrition and zymotic factors has been discussed in previous publications (Platt and Wadsworth, 1956; Platt, 1957b, 1958a).

Attention has been drawn above to the occurrence of diarrhea in protein-calorie deficiency and to the increased loss of protein which may develop in an already depleted subject (Section VI,B). The answer to the question as to which appears first or which is the predisposing factor, the malnutrition or gastroenteritis, is not yet clear, and many authors content themselves with postulating an interdependence or synergism of the two factors.

Also, in the light of the new role attributed to the thymus (Section XII,E), another feature of the interrelationship between protein-calorie deficiency and infection may be revealed. Wissler *et al.* (1946) showed that in laboratory animals a severe protein deficiency caused a diminished antibody response. However, Pretorius and de Villiers (1962) found no statistically significant differences in antibody production between normal children and children suffering from kwashiorkor during the first 10 days of treatment, but from their data it can be seen that low antibody titers were more commonly found in the malnourished group. The question of resistance to infection and protein intake is considered in detail in chapter 23.

Results obtained on the protein value of two Nigerian diets (Dema, 1959; Platt *et al.*, 1961) illustrate the interrelationship in the rat between malnutrition and infection with *Plasmodium berghei* (Table XI). The first (Kanuri) diet, when fed unrestrictedly to an uninfected rat, has a protein value of 8.0 NDpCal%. The same diet also fed in unrestricted amounts to an infected animal has a value of 4.3 NDpCal%. The animal which is infected eats less food and when an uninfected animal is fed (i.e., pair-fed) the same amounts of food as are eaten by the malarious animal the value to the consumer, obtained by assay, is reduced from 8.0 to 6.0 NDpCal%. This is the effect of caloric restriction on the protein value to the consumer of a diet, which is mentioned in Section I,B,2. The difference between the value for the uninfected animal on restricted

TABLE XI

EFFECT OF INFECTION WITH *Plasmodium berghei* ON THE PROTEIN VALUES TO THE CONSUMER (RAT) OF TWO NIGERIAN DIETS

Treatment	NDpCal%	
	Kanuri	Otukwang
Uninfected, unrestricted	8.0	4.2
Malarious, unrestricted	4.3	2.2
Uninfected, pair-fed	6.0	3.4

amounts of food and the value for the malarious animal on low, but unrestricted, intakes cannot yet be fully explained; part of the explanation may lie in the fact that the degree of restriction has been underestimated. It may, however, be noted that the proper control for the malarious animal fed unrestricted amounts of food would be an animal with the same body temperature as the infected animal. Similar results were obtained with the second (Otukwang) diet.

"There are many disease states to which protein-calorie and associated nutritional deficiencies may contribute; the study of these for the understanding of the pathogenesis as well as for the purposes of improving therapy, prophylactic and curative, would be most profitable" (Platt, 1962).

XV. Summary

A method has been developed of expressing the nutritive values of diets in terms of the two interdependent dietary factors, protein and calories. This method has been used to characterize the diets fed to animals in the experimental production of various types of protein-calorie deficiency, ranging from chronic marasmus to an acute florid form resembling classical kwashiorkor ("sugar baby" type), with many intermediate and mixed syndromes between these extremes.

Insufficiency of dietary protein has been shown to be a factor in the production of all forms of protein-calorie deficiency, but it is the addition of carbohydrate to the diet of a protein-deficient animal that is the determining factor in the production of the acute syndrome.

Changes have been found in all the tissues of the body, the histology of many of them has been described, and they are often accompanied by biochemical changes. The outstanding feature of them all is their mutual interdependence. In the experimental animal they can be attributed to the diet or the dietary regimen since, contrary to most experience with protein-calorie deficiency in human subjects, the complicating effects of zymotic disease factors have not been present.

Many of the observations should assist in the interpretation of protein-calorie deficiency disease syndromes in the human subject and point the way for future lines of clinical research, for example, on the role of the thymus in the infective process and on the conditions usually thought to be due to infection, e.g., diarrhea. Much has been written about the synergistic relationship between infection and malnutrition. The work reported in the foregoing account shows, however, how a zymotic factor may contribute to the development of protein-calorie deficiency. Furthermore, it is suggested that much of the pathology of an infectious process may be basically the same as that of protein-calorie deficiency.

Protein-calorie deficiency of the pregnant and lactating mother may affect the offspring at birth and in its subsequent development. Congenitally malnourished animals develop neurological disturbances; they also show other pathological evidences of protein-calorie deficiency. The importance of these observations in the elucidation of the etiology of various diseases, e.g., certain forms of diabetes, has still to be explored.

References

Addis, T., Poo, L. J., and Lew, W. (1936). *J. Biol. Chem.* **115,** 111.

Allison, J. B. (1958). *Ann. N.Y. Acad. Sci.* **69,** 1009.

Allison, J. B. (1960). Cited in Hansen *et al.* (1960).

Al-Rabii, H. A. (1962). Personal communication.

Altmann, A. (1948). *Clin. Proc.* **7,** 32.

Andrew, W. (1940). *Arch. Neurol. Psychiat.* **43,** 1188.

Andrew, W. (1941). *Am. J. Pathol.* **17,** 421.

Aron, M. (1930). *Compt. Rend. Soc. Biol.* **103,** 148.

Arroyave, G. (1961). *Federation Proc.* **20,** Suppl. 7, 39.

Aschkenasy, A. (1961a). *Compt. Rend. Soc. Biol.* **155,** 454.

Aschkenasy, A. (1961b). *Ann. Nutr. Aliment.* **15,** B165.

Bablet, J., and Canet, J. (1952). *Ann. Inst. Pasteur* **83,** 595.

Bablet, J., and Normet, L. (1937). *Bull. Acad. Méd. (Paris)* [3] **117,** 242.

Bagchi, K. (1957a). *Proc. Nutr. Soc. (Engl. Scot.)* **16,** xxiv.

Bagchi, K. (1957b). Ph.D. Thesis, London University.

Baird, J. D., and Farquhar, J. W. (1962). *Lancet* **I,** 71.

Balfour, B. (1954). *Rept. 2nd Inter-African Conf. on Nutr., Gambia, 1952* p. 120.

Bellis, D. B. (1957). *Proc. Nutr. Soc. (Engl. Scot.)* **16,** 98.

Bernick, S., and Bavetta, L. A. (1957). *J. Dental Res.* **36,** 142.

Berte, M. (1954). *Rept. 2nd Inter-African Conf. on Nutr., Gambia, 1952* p. 111.

Bigwood, E. J., and Robazza, F. (1955). *Voeding* **16,** 143.

Bras, G., Waterlow, J. C., and DePass, E. (1957). *W. Indian Med. J.* **6,** 33.

Braude, R., Dollar, A. M., Mitchell, K. G., and Porter, J. W. G. (1958). *Proc. Nutr. Soc. (Engl. Scot.)* **17,** xv.

Brock, J. F. (1954). *Ann. N.Y. Acad. Sci.* **57,** 696.

Brock, J. F., ed. (1961). "Recent Advances in Human Nutrition." Little, Brown, Boston, Massachusetts.

Brock, J. F., and Autret, M. (1952). *In* "Kwashiorkor in Africa," Nutr. Studies No. 8. Food and Agriculture Organization of the United Nations.

Brody, S. (1945). "Bioenergetics and Growth." Reinhold, New York.

Bruce, H. M. (1958). *Studies Fertility* **9,** 90.

Bruce, H. M. (1962). Personal communication.

Bryans, F. E., Kinash, B., Ashworth, M. A., and Haist, R. E. (1952). *Diabetes* **1,** 358.

Burnet, F. M. (1962). *Brit. Med. J.* **II,** 807.

Camain, R., and Pierchon, M. (1954). *Rept. 2nd Inter-African Conf. on Nutr., Gambia, 1952* p. 146.

Cammermeyer, J. (1961). *J. Neuropathol.* **20,** 386.

Campbell, J. A. H. (1956). *Arch. Disease Childhood* **31,** 310.

Cartwright, G. E., Wintrobe, M. M., Buschke, W. H., Follis, R. H., Jr., Suksta, A., and Humphreys, S. (1945). *J. Clin. Invest.* **24,** 268.

Castellanos, H., and Arroyave, A. (1961). *Am. J. Clin. Nutr.* **9,** 186.

Chatterji, A., and Gupta, P. C. S. (1960). *Bull. Calcutta School Trop. Med.* **8,** 148.
Clark, M. (1951). *E. African Med. J.* **28,** 229.
Cohen, H., Metz, J., and Hart, D. (1962). *Lancet* **I,** 52.
Cooper, E. R. A. (1925). "The Histology of the More Important Human Endocrine Organs at Various Ages." Oxford Univ. Press, (Humphrey Milford), London and New York.
Cravioto, J. (1958). *Am. J. Clin. Nutr.* **6,** 495.
Cuparencu, B., Rosenberg, A., and Doghie, V. (1959). *Studii Cercetari Med.* **10,** 267.
Curtis, P. B., Hauge, S. M., and Kraybill, H. R. (1932). *J. Nutr.* **5,** 503.
Davies, G. N. (1956). Conference on Diet and Oral Health (Chairman, J. F. Volker), School of Dentistry, University of Alabama.
Davies, J. N. P. (1948). *Lancet* **I,** 317.
Dawson, A. M., Williams, R., and Williams, H. S. (1961). *Brit. Med. J.* **II,** 667.
Day, P. L., Langston, W. C., and Cosgrove, K. W. (1934). *J. Nutr.* **7,** Suppl. 1, 12.
Dean, R. F. A., and Whitehead, R. G. (1963). *Lancet* **I,** 188.
De Bodo, R. C., and Altszuler, N. (1958). *Physiol. Rev.* **38,** 389.
Dema, I. S. (1959). Ph.D. Thesis, London University.
De Robertis, E. (1942). *Anat. Record* **84,** 125.
De Robertis, E. (1949). *Ann. N.Y. Acad. Sci.* **50,** 317.
Dickerson, J. W. T., and McCance, R. A. (1961). *Brit. J. Nutr.* **15,** 567.
Dickerson, J. W. T., and Widdowson, E. M. (1960). *Biochem. J.* **74,** 247.
Doell, B. H., and Stock, M. (1962). Personal communication.
Doell, B. H., Friend, C. J., Pampiglione, G., and Platt, B. S. (1964a). In preparation.
Doell, B. H., Stewart, R. J. C., and Platt, B. S. (1964b). In preparation.
Dollar, A. M., Mitchell, K. G., and Porter, J. W. G. (1957). *Proc. Nutr. Soc.* (*Engl. Scot.*) **16,** xii.
Done, J. (1962). Personal communication.
Doyle, M. D., and Porter, T. E. (1951). *J. Nutr.* **45,** 29.
Dreizen, S., Currie, C., Gilbey, E. J., and Spies, T. D. (1956). *Am. J. Roentgenol.* **76,** 482.
Dupin, H. (1958). "Etude des carences protidiques observées chez l'enfant en pays tropical (kwashiorkor)." Librairie Arnette, Paris.
Durbin, P. A. J., and Heard, C. R. C. (1962). *Proc. Nutr. Soc.* (*Engl. Scot.*) **21,** xxviii.
Durbin, P. A. J., Heard, C. R. C., and Platt, B. S. (1960). *Proc. Nutr. Soc.* (*Engl. Scot.*) **19,** xxxv.
Eales, L. (1961). *In* "Recent Advances in Human Nutrition" (J. F. Brock, ed.), p. 198. Little, Brown, Boston, Massachusetts.
Edozien, J. C. (1959). *Nature* **184,** 1150.
Eliot, M. M., Souther, S. P., and Park, E. A. (1927). *Bull. Johns Hopkins Hosp.* **41,** 364.
El-Maraghi, N., Platt, B. S., and Stewart, R. J. C. (1964). In preparation.
Engel, R. (1956). *Electroencephalol. Clin. Neurophysiol.* **8,** 489.
Estremera, H. R., and Armstrong, W. D. (1948). *J. Nutr.* **35,** 611.
F. A. O. (1957a). "Calorie Requirements," Nutr. Studies No. 15. Food and Agriculture Organization of the United Nations.
F.A.O. (1957b). "Protein Requirements," Nutr. Studies No. 16. Food and Agriculture Organization of the United Nations.
F.A.O. (1962). Joint F.A.O./W.H.O. Expert Committee on Nutrition. *World Health Organ. Tech. Rept. Ser.* No. 245.
Follis, R. H., Jr. (1950). *In* "Metabolic Interrelations," Trans. 2nd Conf., pp. 221–257. Josiah Macy Jr. Foundation, New York.

Follis, R. H., Jr. (1958). "Deficiency Disease." Thomas, Springfield, Illinois.
Follis, R. H., Jr., and Park, E. A. (1952). *Am. J. Roentgenol.* **68,** 709.
Fontaine, R., Mandel, P., and Guise, A. (1950). *Compt. Rend. Soc. Biol.* **144,** 1397.
Forsham, P. H., and Mortimore, G. E. (1957). *In* "Biochemical Disorders in Human Disease" (R. H. S. Thompson and E. J. King, eds.), pp. 494–543. Academic Press, New York.
Foy, H., and Kondi, A. (1957). *J. Trop. Med. Hyg.* **60,** 105.
Frandsen, A. M., Nelson, M. M., Sulon, E., Becks, H., and Evans, H. M. (1954). *Anat. Record* **119,** 247.
Frawley, T. F., Kistler, H., and Shelley, T. (1959). *Ann. N.Y. Acad. Sci.* **82,** 868.
Friedman, S. M., and Friedman, C. L. (1946). *Can. Med. Assoc. J.* **55,** 15.
Friend, C. J., Heard, C. R. C., Platt, B. S., Stewart, R. J. C., and Turner, M. R. (1960). *Proc. Nutr. Soc.* (*Engl. Scot.*) **19,** xxxiv.
Friend, C. J., Heard, C. R. C., Platt, B. S., Stewart, R. J. C., and Turner, M. R. (1961). *Brit. J. Nutr.* **15,** 231.
Fritz, I., and Burnett, G. W. (1953). *J. Dental Res.* **32,** 647.
Gabe, M. (1953). *Bull. Microscop. Appl.* [2] **3,** 153.
Gallais, P., Miletto, G., Corriol, J., and Bert, J. (1951a). *Méd. Trop.* **11,** 128.
Gallais, P., Bert, J., Corriol, J., and Miletto, G. (1951b). *Electroencephalol. Clin. Neurophysiol.* **3,** 110.
Gardiner, P. A., and MacDonald, I. (1958). *Proc. Nutr. Soc.* (*Engl. Scot.*) **17,** xx.
Gillman, J., and Gilbert, C. (1958). *Brit. Med. J.* **I,** 57.
Gillman, J., and Gillman, T. (1951). "Perspectives in Human Malnutrition." Grune & Stratton, New York.
Godwin, K. O. (1956). Ph.D. Thesis, London University.
Godwin, K. O. (1959). *J. Nutr.* **69,** 121.
Godwin, K. O. (1960a). *Abstr. 5th Intern. Congr. on Nutr., Washington, D.C.* Abstr. No. 219, p. 48.
Godwin, K. O. (1960b). *Proc. Nutr. Soc.* (*Engl. Scot.*) **19,** xxxviii.
Godwin, K. O. (1962). *World Rev. Nutr. Dietet.* **3,** 103.
Godwin, K. O., and Platt, B. S. (1960). *Proc. Nutr. Soc.* (*Engl. Scot.*) **19,** x.
Gomori, G. (1941). *Am. J. Pathol.* **17,** 395.
Gomori, G. (1950). *Am. J. Clin. Pathol.* **20,** 665.
Goormaghtigh, N., and Thomas, F. (1934). *Am. J. Pathol.* **10,** 713.
Grossie, J., and Turner, C. W. (1962). *Proc. Soc. Exptl. Biol. Med.* **110,** 631.
Gutman, A. B. (1954). *In* "Metabolic Interrelations," Trans. 5th Conf., 1953 p. 262. Josiah Macy Jr. Foundation, New York.
Haist, R. E. (1944). *Physiol. Rev.* **24,** 409.
Halder, K. (1962). Ph.D. Thesis, London University.
Halder, K., and Platt, B. S. (1962). *Proc. Nutr. Soc.* (*Engl. Scot.*) **21,** xxxii.
Ham, A. W. (1957). "Histology." Pitman Med., London.
Hamilton, J. B., Terada, H., and Mestler, G. E. (1955). *J. Gerontol.* **10,** 401.
Handler, P., Baylin, G. J., and Follis, R. H., Jr. (1947). *J. Nutr.* **34,** 677.
Hansen, A. E., Haggard, M. E., Boelsche, A. N., Adam, D. J. D., and Wiese, H. F. (1958). *J. Nutr.* **66,** 565.
Hansen, J. D. L. (1961). *In* "Recent Advances in Human Nutrition" (J. F. Brock, ed.), pp. 218–225. Little, Brown, Boston, Massachusetts.
Hansen, J. D. L., Schendel, H. E., Wilkins, J. A., and Brock, J. F. (1960). *Pediatrics* **25,** 258.
Harris, H. A. (1924–1925). *J. Anat.* **59,** 94P.
Heard, C. R. C. (1957). *Proc. Nutr. Soc.* (*Engl. Scot.*) **16,** xi.

Heard, C. R. C., and Stewart, R. J. C. (1960). *Acta. Endocrinol.* Suppl. 51, 1277.
Heard, C. R. C., Platt, B. S., and Stewart, R. J. C. (1958). *Proc. Nutr. Soc.* (*Engl. Scot.*) **17**, xli.
Heard, C. R. C., Durbin, P. A. J., and Platt, B. S. (1961). *Proc. Nutr. Soc.* (*Engl. Scot.*) **20**, xx.
Hellerström, C., and Hellman, B. (1960). *Acta Endocrinol.* **35**, 518.
Hellman, B. (1959). *Acta Endocrinol.* **31**, 80.
Henneman, D. H., and Henneman, P. H. (1960). *J. Clin. Invest.* **39**, 1239.
Higginson, J. (1954). *Metab., Clin. Exptl.* **3**, 392.
Hill, K. (1959). *Trans. Roy. Soc. Trop. Med. Hyg.* **53**, 217.
Himsworth, R. L. (1960). *Nature* **185**, 694.
Hoffenberg, R., Saunders, S., Linder, G. C., Black, E., and Brock, J. F. (1962). *In* "Protein Metabolism, an International Symposium" (F. Gross, ed.), p. 314. Springer, Berlin.
Holmes, E. G., and Trowell, H. C. (1948). *Lancet* **I**, 395.
Hutchison, J. H., Keay, A. J., and Kerr, M. M. (1962). *Brit. Med. J.* **II**, 436.
Ikkos, D., and Luft, R. (1957). *Acta Endocrinol.* **25**, 312.
Ikkos, D., Luft, R., Gemzell, C. A., and Almquist, S. (1962). *Acta Endocrinol.* **39**, 547.
Irving, J. T. (1957). *Vitamins Hormones* **15**, 291.
Irving, J. T., and Budtz-Olsen, O. E. (1955). *Brit. J. Nutr.* **9**, 301.
Jackson, C. M. (1925). "The Effects of Inanition and Malnutrition upon Growth and Structure." Churchill, London.
Jadresic, A., Carrasco, M., and López, E. (1962). *Lancet* **I**, 401.
Janes, M., and Mittwoch, A. (1962). Personal communication.
Jones, P. R. M., and Dean, R. F. A. (1956). *J. Trop. Pediat.* **2**, 51.
Kahn, E. (1954). *Arch. Disease Childhood* **29**, 256.
Kahn, E. (1957). *Central African J. Med.* **3**, 398.
Kahn, E., and Falcke, H. C. (1956). *J. Pediat.* **49**, 37.
Kahn, E., and Wayburne, S. (1960). *Proc. Nutr. Soc. South Africa* **1**, 21.
Knowles, C. B. (1957). *Proc. Nutr. Soc.* (*Engl. Scot.*) **16**, ix.
Knox, W. E., Auerbach, V. H., and Lin, E. C. C. (1956). *Physiol. Rev.* **36**, 164.
Kohman, E. A. (1920). *Am. J. Physiol.* **51**, 378.
Kusama, J. (1957). *Shinshu Med. J.* **6**, 376.
Lafon, M. (1939). *Ann. Physiol. Physicochim. Biol.* **15**, 1.
Leathem, J. H. (1958). *Recent Progr. Hormone Res.* **14**, 141.
Leblond, C. P. (1949). Discussion in De Robertis (1949), p. 333.
Limson, M., and Jackson, C. M. (1932). *J. Nutr.* **5**, 163.
Lindan, O. (1954). *Rept. 2nd Inter-African Conf. on Nutrition, Gambia, 1952* p. 129.
Lurie, A. O., and Jackson, W. P. U. (1962). *Clin. Sci.* **22**, 259.
Lutz, P. (1958). Ph.D. Thesis, London University.
Lutz, P., and Platt, B. S. (1958). *Proc. Nutr. Soc.* (*Engl. Scot.*) **17**, iii.
McCance, R. A. (1962). *Lancet* **II**, 671.
McCance, R. A., Ford, E. H. R., and Brown, W. A. B. (1961). *Brit. J. Nutr.* **15**, 213.
McCarrison, R. (1921). "Studies in Deficiency Disease." Frowde, Hodder, & Stoughton, London.
McClure, T. J. (1962). *J. Reprod. Fertility* **4**, 241.
Macho, L. (1960). *Arch. Intern. Physiol.* **68**, 713.
McLaren, D. S. (1958). *Brit. J. Nutr.* **12**, 254.
McLaren, D. S. (1959). *Brit. J. Ophthalmol.* **43**, 78.
McLaren, D. S. (1960). *World Rev. Nutr. Dietet.* **2**, 29.
McMeekan, C. P. (1940). *J. Agr. Sci.* **30**, 276, 387, and 511.

Magee, D. F. and Hong, S. S. (1959). *Am. J. Physiol.* **197,** 27.
Mellanby, E. (1941). *J. Physiol.* (*London*) **99,** 467.
Meyer, A. (1958). *In* "Neuropathology" (J. G. Greenfield, *et al.*, eds.), p. 286. Arnold, London.
Meyer, A., Pampiglione, G., Platt, B. S., and Stewart, R. J. C. (1961a). *Excerpta Med.* (*Amst.*) No. 39, Abstr. 17.
Meyer, A., Stewart, R. J. C., and Platt, B. S. (1961b). *Proc. Nutr. Soc.* (*Engl. Scot.*) **20,** xviii.
Miller, D. S., and Payne, P. R. (1961). *J. Nutr.* **75,** 225.
Miller, D. S., and Payne, P. R. (1962). Personal communication.
Miller, J. F. A. P. (1961). *Lancet* **II,** 748.
Miller, J. F. A. P. (1962). *Nature* **195,** 1318.
Milne, M. D. (1957). *In* "Biochemical Disorders in Human Disease" (R. H. S. Thompson and E. J. King, eds.), pp. 177–241. Academic Press, New York.
Mosca, L. (1956). *Quart. J. Exptl. Physiol.* **41,** 433.
Mosca, L. (1957). *Quart. J. Exptl. Physiol.* **42,** 267.
Mukherjee, K. L., and Nath, R. L. (1957). *Bull. Calcutta School Trop. Med.* **5,** 170.
Mulinos, M. G., and Pomerantz, L. (1940). *J. Nutr.* **19,** 493.
Mundy-Castle, A. C., McKiever, B. L., and Prinsloo, T. A. (1953). *Electroencephalol. Clin. Neurophysiol.* **5,** 533.
Munro, H. N., Hutchison, W. C., Ramaiah, T. R., and Neilson, F. J. (1962). *Brit. J. Nutr.* **16,** 387.
Murphy, D. P., and Bowes, A. D. (1939). *Am. J. Obstet. Gynecol.* **37,** 460.
Nagchaudhuri, J. (1953). Ph.D. Thesis, London University.
Nagchaudhuri, J., and Platt, B. S. (1954). *Rep. 2nd Inter-African Conf. on Nutr., Gambia, 1952* p. 215.
Naismith, D. J. (1962). *J. Nutr.* **77,** 381.
Nelson, G. K. (1959). *Electroencephalol. Clin. Neurophysiol.* **11,** 73.
Nelson, G. K., and Dean, R. F. A. (1959). *Bull. World Health Organ.* **21,** 779.
Nicholls, L., Sinclair, H. M., and Jelliffe, D. B. (1961). "Tropical Nutrition and Dietetics." Ballière, London.
Nikitin, V. N. (1962). *In* "The Development of Homeostasis," Symposium, Prague, 1960 (E. F. Adolph, Chairman), p. 191. Academic Press, New York.
Olson, R. E., and Vester, J. W. (1960). *Physiol. Rev.* **40,** 677.
Oomen, H. A. P. C. (1953). *J. Indonesian Med. Assoc.* **3,** 229.
Oomen, H. A. P. C. (1958). *Proc. Nutr. Soc.* (*Engl. Scot.*) **17,** viii.
Pampiglione, G., Stewart, R. J. C., and Platt, B. S. (1960). *Proc. Nutr. Soc.* (*Engl. Scot.*) **19,** ix.
Papper, S., Saxon, L., Prout, T. E., and Alpert, H. C. (1956). *J. Lab. Clin. Med.* **48,** 13.
Passmore, R. (1947–1948). *Trans. Roy. Soc. Trop. Med. Hyg.* **41,** 189.
Payne, P. R., and Done, J. (1959). *Proc. Nutr. Soc.* (*Engl. Scot.*) **18,** vi.
Pellett, P. L. (1957). *Proc. Nutr. Soc.* (*Engl. Scot.*) **16,** x.
Platt, B. S. (1954a). *Rept. 2nd Inter-African Conf. on Nutr., Gambia, 1952* pp. 153–159.
Platt, B. S. (1954b). *Rept. 2nd Inter-African Conf. on Nutr., Gambia, 1952* pp. 291–292.
Platt, B. S. (1955). *Brit. Med. J.* **I,** 179.
Platt, B. S. (1956). "Lectures on the Scientific Basis of Medicine," Vol. 4, pp. 145–166. Oxford Univ. Press (Athlone). London and New York.
Platt, B. S. (1957a). *In* "Human Protein Requirements and Their Fulfilment in Practice," Conference, Princeton, 1955 (J. C. Waterlow and J. M. L. Stephen, eds.), p. 57. F.A.O., W.H.O., and Josiah Macy Jr. Foundation, New York.
Platt, B. S. (1957b). *Am. J. Trop. Med. Hyg.* **6,** 773.

Platt, B. S. (1958a). *Trans. Roy. Soc. Trop. Med. Hyg.* **52,** 189.
Platt, B. S. (1958b). *Ann. Soc. Belge. Méd. Trop.* **38,** 425.
Platt, B. S. (1958c). *Proc. Nutr. Soc. (Engl. Scot.)* **17,** xl.
Platt, B. S. (1961a). *Federation Proc.* **20,** Suppl. 7, 188.
Platt, B. S. (1961b). *In* "Chemical Pathology of the Nervous System," *3rd Intern. Neurochem. Symposium, Strasburg, 1958* (J. Folch-Pi, ed.), pp. 114–118. Pergamon Press, New York.
Platt, B. S. (1962). *Proc. Roy. Soc.* **B, 156,** 337.
Platt, B. S., and Frankul, W. (1962). *Proc. Nutr. Soc. (Engl. Scot.)* **21,** vii.
Platt, B. S., and Heard, C. R. C. (1958). *Proc. Nutr. Soc. (Engl. Scot.)* **17,** ii.
Platt, B. S., and Miller, P. J. (1957). *Proc. Nutr. Soc. (Engl. Scot.)* **16,** ix.
Platt, B. S., and Miller, D. S. (1958). *Proc. Nutr. Soc. (Engl. Scot.)* **17,** 106.
Platt, B. S., and Miller, D. S. (1959). *Proc. Nutr. Soc. (Engl. Scot.)* **18,** vii.
Platt, B. S., and Nagchaudhuri, J. (1954). *Proc. Nutr. Soc. (Engl. Scot.)* **13,** ix.
Platt, B. S., and Stewart, R. J. C. (1960). *Proc. Nutr. Soc. (Engl. Scot.)* **19,** viii.
Platt, B. S., and Stewart, R. J. C. (1962a). *Proc. Nutr. Soc. (Engl. Scot.)* **21,** v.
Platt, B. S., and Stewart, R. J. C. (1962b). *Brit. J. Nutr.* **16,** 483.
Platt, B. S., and Wadsworth, G. R. (1956). *Proc. Nutr. Soc. (Engl. Scot.)* **15,** 103.
Platt, B. S., Miller, D. S., and Payne, P. R. (1961). *In* "Recent Advances in Human Nutrition" (J. F. Brock, ed.), pp. 351–374. Little, Brown, Boston, Massachusetts.
Platt, B. S., Halder, K., and Doell, B. H. (1962a). *Proc. Nutr. Soc. (Engl. Scot.)* **21,** vi.
Platt, B. S., Heard, C. R. C., Stewart, R. J. C., and Al-Rabii, H. A. (1962b). *Proc. Nutr. Soc. (Engl. Scot.)* **21,** xxix.
Platt, B. S., Pampiglione, G., and Stewart, R. J. C. (1964). In preparation.
Plotnikova, Ju. I. (1959). *Probl. Endokrinol i Gormonoterap.* **5,** 34; see *Abstr. World Med.* **27,** 680 (1960).
Pratt, C. W. M., and McCance, R. A. (1960). *Brit. J. Nutr.* **14,** 75.
Pretorius, P. J. and de Villiers, L. S. (1962). *Am. J. Clin. Nutr.* **10,** 379.
Ramalingaswami, V., Menon, P. S., and Venkatachalam, P. S. (1948). *Indian Physician* **7,** 229.
Ramalingaswami, V., Sriramachari, S., and Patwardhan, V. N. (1954). *Indian J. Med. Sci.* **8,** 7.
Ramalingaswami, V., Sriramachari, S., and Patwardhan, V. N. (1958). *Indian J. Pathol. Bacteriol.* **1,** 104.
Rangam, C. M., Bhagwart, A. G., and Gupta, J. C. (1962). *Indian J. Med. Res.* **50,** 184.
Rassmussen, A. T. (1929). *Am. J. Pathol.* **5,** 263.
Rhodes, K. (1959). Discussion in Hill (1959), p. 231.
Sarason, E. L. (1943). *A.M.A. Arch. Pathol.* **35,** 373.
Sarrouy, C., Saint-Jean, M., and Clausse, J. (1953). *Algerie Méd.* **57,** 584.
Schaeffer, A. J., and Murray, J. D. (1950). *A.M.A. Arch. Ophthalmol.* **43,** 202.
Schwartz, C., Scott, E. B., and Ferguson, R. L. (1951). *Anat. Record* **110,** 313.
Scow, R. O., and Cornfield, J. (1954a). *Am. J. Physiol.* **179,** 39.
Scow, R. O., and Cornfield, J. (1954b). *Am. J. Physiol.* **179,** 435.
Scrimshaw, N. S., and Behar, M. (1961). *Science* **133,** 2039.
Sheehan, H. L. (1937). *J. Pathol. Bacteriol.* **45,** 189.
Sidransky, H., and Clark, S. (1961). *A.M.A. Arch. Pathol.* **72,** 468.
Sidransky, H., and Farber, E. (1958). *A.M.A. Arch. Pathol.* **66,** 135.
Silberberg, M., and Silberberg, R. (1940). *A.M.A. Arch. Pathol.* **30,** 675.
Skanse, B. (1961). *Acta Endocrinol.* **38,** 166.
Slone, D., Taitz, L. S., and Gilchrist, G. S. (1961). *Brit. Med. J.* **I,** 32.
Smith, R., and Waterlow, J. C. (1960). *Lancet* **I,** 147.

Squires, B. T. (1961). *Proc. Nutr. Soc.* (*Engl. Scot.*) **20,** xix.
Squires, B. T. (1962). *Proc. Nutr. Soc.* (*Engl. Scot.*) **21,** xxxiii.
Srebnik, H. H., and Nelson, M. M. (1962). *Endocrinology* **70,** 723.
Stetten, D., and Stetten, M. R. (1960). *Physiol. Rev.* **40,** 505.
Stewart, R. J. C., and Heard, C. R. C. (1959). *Proc. Nutr. Soc.* (*Engl. Scot.*) **18,** x.
Stewart, R. J. C., and Heard, C. R. C. (1960). *Acta Endocrinol.* Suppl. 51, 1313.
Stewart, R. J. C., and Platt, B. S. (1958). *Proc. Nutr. Soc.* (*Engl. Scot.*) **17,** v.
Stewart, R. J. C., and Platt, B. S. (1961). *Proc. Nutr. Soc.* (*Engl. Scot.*) **20,** xlvi.
Sydenstricker, V. P., Hall, W. K., Hock, C. W., and Pund, E. R. (1946). *Science* **103,** 194.
Taggart, N. (1961). *Proc. Nutr. Soc.* (*Engl. Scot.*) **20,** 35.
Tejada, C., and Russfield, A. B. (1957). *Arch. Disease Childhood* **32,** 343.
Townsend, S. F., and Baker, B. L. (1959). *Proc. Soc. Exptl. Biol. Med.* **100,** 33.
Trémolières, J. (1958). *Bull. Soc. Sci. Hyg. Aliment.* **46,** 247.
Trowell, H. C. (1937). *Arch. Disease Childhood* **12,** 193.
Trowell, H. C. (1949). *Trans. Roy. Soc. Trop. Med. Hyg.* **42,** 417.
Trowell, H. C., Davies, J. N. P., and Dean, R. F. A. (1954). "Kwashiorkor." Arnold, London.
Turner, M. R. (1962). M.Sc. Thesis, London University.
Udani, P. M. (1960). *Indian. J. Child Health* **9,** 103.
Utheim, K. (1922). *J. Metab. Res.* **1,** 803.
Van Bogaert, L., Radermecker, M. A., Janssen, P., and Gatera, F. (1961). *Acta Neuropathol.* **1,** 363.
Van Italie, T. B., Felber, J. P., Hoet, J., and Renold, A. E. (1959). *Diabetes* **8,** 94.
Véghelyi, P. V. (1950). *Ann. Paediat.* (*Basel*) **175,** 349.
Véghelyi, P. V., and Kemény, T. T. (1962). *In* "The Exocrine Pancreas," Ciba Foundation Symposium, pp. 329–349. Churchill, London.
Véghelyi, P. V., Kemény, T. T., Pozsonye, J., and Sos, J. (1950). *Am. J. Diseases Children* **79,** 658.
Verne, J., and Hébert, S. (1953). *Compt. Rend.* **236,** 2170.
Vikhbladh, I. (1951). *Scand. J. Clin. Lab. Invest.* **3,** Suppl. 2, 9.
Vint, F. W. (1936–1937). *E. African Med. J.* **13,** 332.
Volk, B. W., and Lazarus, S. S. (1960). *Am. J. Pathol.* **37,** 121.
Waldmann, T. A. (1961). *Lancet* **II,** 121.
Walker, A. R. P., and Arvidsson, U. B. (1954). *Metab., Clin. Exptl.* **3,** 385.
Walters, J. H., and Waterlow, J. C. (1954). *Med. Res. Council, Spec. Rept. Ser.* **285.**
Wang, C. F., Hegsted, D. M., Lapi, A., Zamcheck, N., and Black, M. B. (1949). *J. Lab. Clin. Med.* **34,** 953.
Waterlow, J. C. (1948). *Med. Res. Council, Spec. Rept. Ser.* **263.**
Waterlow, J. C. (1959). *Federation Proc.* **18,** 1143.
Waterlow, J. C., Bras, G., and De Pass, E. (1957). *J. Trop. Pediat.* **2,** 189.
Waterlow, J. C., Cravioto, J., and Stephen, J. M. L. (1960). *Advan. Protein Chem.* **15,** 131.
White, E. G. (1938–1939). *J. Anat.* **73,** 365.
Whitney, J. E., and Young, F. G. (1957). *Biochem. J.* **66,** 648.
Widdowson, E. M., Dickerson, J. W. T., and McCance, R. A. (1960). *Brit. J. Nutr.* **14,** 457.
Wissler, R. W., Woolridge, R. L., Steffee, C. H., and Cannon, P. R. (1946). *J. Immunol.* **52,** 267.
Woodnott, D. P. (1961). *J. Anim. Tech. Ass.* **xii,** 59.

Woodruff, A. W. (1961). *In* "Recent Advances in Human Nutrition" (J. F. Brock, ed.), pp. 415–433. Little, Brown, Boston, Massachusetts.
Woodruff, A. W., and Stewart, R. J. C. (1962). *Proc. Nutr. Soc.* (*Engl. Scot.*) **21**, iv.
Yoshimura, H. (1961). *Federation Proc.* **20**, Suppl. 7, 103.
Young, F. G. (1941). *Brit. Med. J.* **II**, 897.
Young, F. G., and Korner, A. (1960). *In* "Diabetes" (R. H. Williams, ed.), pp. 216–232. Harper (Hoeber), New York.
Zubirán, S., and Gómez-Mont, F. (1953). *Vitamins Hormones* **11**, 97.

CHAPTER 22

Clinical Aspects of Protein Malnutrition

FERNANDO VITERI, MOISÉS BÉHAR, GUILLERMO ARROYAVE

Institute of Nutrition of Central America and Panama (INCAP), Guatemala, Central America

AND

NEVIN S. SCRIMSHAW

The Department of Nutrition and Food Science, Massachusetts Institute of Technology, Cambridge, Massachusetts

I. Introduction

Clinical protein deficiency still occurs widely. Severe protein malnutrition in children was first described by Czerny and Keller in Germany in 1906 under the name *Mehlnährschaden* and by Correa, in Yucatan, Mexico, in 1908 under the name *Culebrilla*. Reports of a severe protein deficiency syndrome in children appeared successively in French West Africa, Japan, Kenya, Italy, Haiti, France, Cuba, Indo-China, and South Africa, but the name now most widely used, kwashiorkor, stems from the 1933 and 1935 descriptions by Dr. Cecily Williams from the African Gold Coast, now the Republic of Ghana. Subsequent reports, in the order of publication, came from French Equatorial Africa, Tanganyika, El Salvador, Belgian Congo, Costa Rica, Guatemala, Honduras, Venezuela, Morocco, Chile, Colombia, China, Uruguay, Nigeria, Brazil, Rhodesia, Jamaica, Hungary, Greece, Indonesia, India, Philippines, Fiji, Curaçao, Spain, and Ceylon (Scrimshaw and Béhar, 1959; Trowell *et al.*, 1954a).

Interest in the problem was increased by the first F.A.O./W.H.O. Joint Expert Committee on Nutrition which met in Geneva in 1949, (F.A.O./W.H.O., 1952) and led to a survey of ten countries and territories in Africa south of the Sahara by Brock and Autret (1952). Their report adopted the name "kwashiorkor" and concluded that the various other local names referred to a single disease widely prevalent throughout Africa. The subsequent joint W.H.O./F.A.O. surveys of Autret and Béhar (1954) in Central America and Waterlow and Vergara in Brazil (1957) established the essential identity of protein deficiency disease as it occurred in Latin America.

Hundreds of reports of severe malnutrition in children have appeared since from various countries and it is clear that the syndrome is common to nearly all technically underdeveloped countries in the world. Since the authors have had most of their experience with protein-calorie malnutrition in Central America, it is inevitable that their emphasis or interpretation to some of the findings described would differ slightly from those of workers in other parts of the world.

Protein malnutrition tends to be manifest primarily in young children and is a major factor in the high mortality rate of children in the 1- to 4-year age group which in technically underdeveloped areas is 20 to 60 times greater than in the United States and western Europe. Most of the excessive mortality of preschool children characterizing these areas is due largely to the interaction between protein malnutrition and infection, which is discussed in Chapter 23.

Even when the child does not die with the signs and symptoms of kwashiorkor, death from infection may occur only because of an underlying malnutrition which is largely due to protein deficiency. Frank cases of kwashiorkor are thus only a small part of the total problem of protein deficiency in children. The prevalence of kwashiorkor has been compared with the visible portion of an iceberg which has six sevenths of its bulk concealed underneath the surface of the water, but a more correct simile would be an undersea mountain with only a protruding tip.

Very young children are less likely to suffer from protein deficiency because of the probability that they will be receiving breast milk when the parents cannot afford other food sources or would not use them properly. In some countries a new aspect of the problem is the migration of mothers to large cities in which even low-income mothers are resorting to formula feeding with frequently tragic results. As a child reaches school age he is also less likely to develop protein deficiency, because his requirements for protein per kilogram of body weight have decreased, the taboos against his sharing fully in the family diet no longer apply, and he is able to take some initiative in meeting his own food needs. For these same reasons, primary protein deficiency is not common in adults, although it does occur in special situations, such as those prevailing in some prisons and other institutions, arising during war and civil disturbance, or resulting from indulgence in food fads and bizarre habits.

Apart from young children, kwashiorkor is more likely to occur in adolescents or lactating women, because their protein needs per kilogram of body weight are increased. It has also been seen in alcoholics who have avoided the classical vitamin deficiencies associated with alcoholism by the use of enriched bread or vitamin concentrates, but whose diet has not contained adequate protein. Less severe protein malnutrition is common in elderly individuals who, for economic reasons, debilitation, or poor dentition, fail to purchase, prepare, or consume a diet adequate in protein. Secondary protein deficiency is commonly seen in adults in association with chronic infections, metabolic, or neoplastic disease.

II. Protein-Calorie Malnutrition in Children

A. *Epidemiology*

The multiple factors responsible for protein malnutrition can be conveniently classified as those relating to the agent, the host, and the environment.

1. *Agent Factors*

The agent is, of course, protein acting not by its presence as does the agent of an infectious disease, but by its absence when it is not sufficiently available to the body. The most pertinent characteristic of the protein is its Biological Value, which depends on both the relative amounts of essential amino acids present and their biological availability following digestion. Protein malnutrition is frequently observed with protein intakes which would be adequate if the Biological Value of the protein were good.

The proportion of calories in the diet is another major factor influencing protein deficiency. If dietary carbohydrates and fat are inadequate to supply needed calories, protein will be used for this purpose and Net Protein Value correspondingly reduced. The physical nature of the food is another factor affecting protein utilization; for example, finely ground cereals are better utilized than coarse preparations.

The concentration of protein in the diet is also of practical significance. If the quantity is too low, especially when Biological Value is also poor, it becomes impossible to consume enough food to fulfill the requirements for protein. In tropical and subtropical areas young children after weaning are frequently fed starchy gruels and starchy fruits or roots, with which they can satisfy their appetite and caloric requirements without fulfilling their protein requirements. Methods of evaluating the protein quality of the diet are discussed at greater length in the Introduction to Part II of this treatise and also in Chapters 12 and 21.

2. Host Factors

Age influences the appearance of protein-calorie malnutrition in several ways. Since nutritional requirements are directly related to the rate of growth, the younger the child, the greater the protein requirement per unit of body weight. This is one of the reasons why overt malnutrition develops more rapidly and frequently in young children and is correspondingly rare in adults.

Since small children are not able to masticate foods such as meat unless they are specially prepared, the young child tends to be denied many of the foods available to the rest of the family and therefore the diet is customarily limited to gruels and other foods which are easy to swallow but poor in protein.

Individuals, of course, vary genetically in their ability to utilize protein and in their requirements for essential amino acids. Consequently, they differ in the adequacy with which the protein in a given diet will fill their needs. However, there is no evidence of consistent racial or sexual differences in the development of protein deficiency which are not attributable to cultural and other environmental influences.

Among the pathological conditions of the host influencing the development of protein malnutrition are such noninfectious processes as the malabsorption syndromes, metabolic abnormalities of endocrine origin, such as hyperthyroidism and hypopituitarism, psychological disturbances resulting in severe anorexia, and an increasingly recognized number of metabolic disorders resulting from enzymic defects of genetic origin. Relatively more important, however, because of their greater frequency are the infectious diseases. In a high proportion of cases of protein-calorie malnutrition, infectious diarrhea or another infectious process can be identified as an important precipitating or aggravating factor. The effects of infections and other stress on protein metabolism are discussed in Chapter 23.

3. Environmental Factors

The environment can act either through the agent of protein deficiency by affecting its availability, or on the host, by influencing either food consumption or nutritional requirements.

Among the environmental factors limiting the availability of protein are (*a*) physical and geographical factors influencing food production, such as climate and characteristics of the soil; (*b*) biological factors, such as the varieties of plants or animals which can be supported in a given area, and local plant or animal diseases which may interfere with their production; (*c*) socioeconomic factors, such as knowledge of good agricultural practices, the provisions for land distribution or tenancy, and availability of the funds or credit required for efficient agricultural practice; population density relative to food resources; efficiency of conservation, processing and storage of foods, and adequacy of food distribution.

Of the environmental factors affecting food consumption, the socioeconomic factors are of the greatest importance. The purchasing power of the population and their habits, practices, beliefs, and taboos regarding food, particularly the feeding of young children, are often the most important factors in the occurrence of clinical protein deficiency. Environmental factors affecting the protein requirement of the host, such as temperature and other climatic characteristics, seem to be of relatively little importance, although some nitrogen loss occurs in sweat.

B. Classification

Protein deficiency may result in a range of clinical syndromes as a consequence of at least the following variables: (*a*) degree of adequacy of caloric intake, (*b*) age of the subject, and (*c*) magnitude and duration of the deficient diet.

1. Degree of Adequacy of Caloric Intake

Diets low in protein, but with adequate or even excessive calories, may be responsible for protein deficiency. This is observed when children are fed diets predominant in cereal flours, starchy roots and fruits, or gruels prepared from different refined starches. Children fed in this manner quickly reduce their caloric intake as a result of anorexia, one of the early signs of protein deficiency. When special care is taken, however, children can be forced to consume such a diet in large enough amounts to meet full calorie requirements. Under such circumstances the resulting clinical picture is that described in Europe earlier in the century under the name *Mehlnährschaden* (Czerny and Keller, 1906) and recently in the West Indies as "sugar baby" (Jelliffe *et al.*, 1954). These children conserve their subcutaneous tissue and may even appear obese.

At the other extreme are seen children whose caloric and protein intake are nearly equally deficient. This is brought about by feeding insufficient amounts of foods such as milk which have an adequate protein-calorie ratio or by giving very dilute cereal gruels, sweetened water, and other rations inducing starvation. Under these circumstances when the calorie deficiency predominates, the children rapidly lose their lipid reserve, and their own tissues, especially their muscles, furnish both protein and additional calories. The resulting clinical picture is one of emaciation or marasmus.

The two syndromes, the sugar-baby type of kwashiorkor and marasmus, are the extremes of a continuous spectrum since every degree of relative deficiency of protein and calories can be found. The large majority of cases of kwashiorkor show some loss of subcutaneous fat and tissue wasting as well as the hypoproteinemia, edema, and other signs and symptoms of protein deficiency. They might more properly be referred to as cases of kwashiorkor of the marasmic type.

2. Age

The younger a child, the greater his rate of growth and associated nutrient requirements. Furthermore, age influences the type of diet an individual is given. Before weaning, protein and calorie depletion are likely to occur from insufficient breast milk, overdiluted cow's milk, or cereal water. During and immediately after the weaning period, the child is likely to be fed a diet more deficient in protein than in calories. Accordingly, marasmus is more often seen in children under 1 year of age and kwashiorkor is most common in the second and third years.

3. Magnitude and Duration of the Deficient Diet

The nature and severity of the deficiency will determine whether there is time for muscle wasting to occur and for marked retardation in growth and maturation to become apparent before the acute crisis. Long-standing dietary deficiency in protein and calories will result in progressive emaciation upon which kwashiorkor may be superimposed at any time. An excellent classification of the types of protein-calorie malnutrition was provided in the report of the third session of the Joint Expert Committee on Nutrition (F.A.O./W.H.O., 1952). However, a classification of protein-calorie malnutrition is now possible which indicates the many possible combinations of protein and calorie deprivation through which children may pass in reaching a terminal state.

Figure 1 attempts to portray these schematically for the age range in

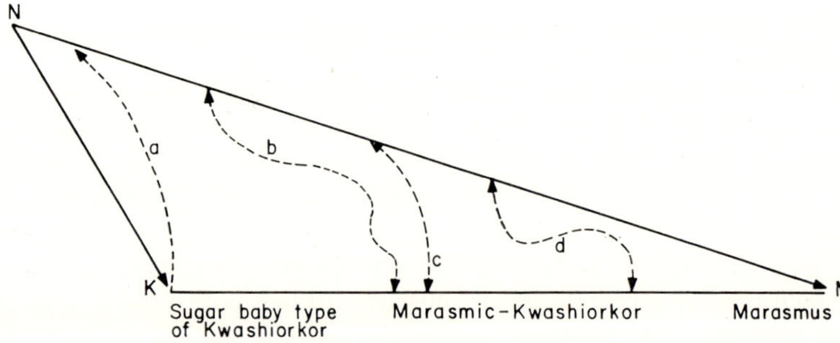

FIG. 1. Schematic representation of the development of the different types of protein malnutrition in children. (Reprinted from *Science* **133,** 2045, 1961, by permission.)

which protein deficiency is most common—the preschool child. The normal child (*N*) may develop kwashiorkor, following the line *N–K* or marasmus along *N–M*. This last line is longer because under field conditions most children take a longer time to develop this type of malnutrition.

The speed with which a child develops protein-calorie malnutrition depends both upon the magnitude of the deficiencies and upon his age. For a small infant, it may take only a few weeks to go all the way to *M*; for an older child, this process usually takes many months. Adverse socio-economic factors and infectious processes are complications which aggravate the deficiency and may cause a child to drop fairly quickly from line *N–M* toward the line *K–M* which represents the continuous spectrum between kwashiorkor and marasmus. The place in the spectrum which they eventually reach depends primarily on how far they have gone along the line *N–M*.

A continuous and smooth transition from simple undernutrition to the kwashiorkor type is indicated by line *c*. The abrupt development of a sugar-baby type of kwashiorkor is represented by line *a*. A child may also respond to an initial stress and then partially recover as indicated by line *b*; he then may continue toward marasmus (*M*) parallel to line *N–M* until a new stress takes him all the way to line *K–M*, at which time he will clinically manifest the marasmic type of kwashiorkor. In some cases, particularly in children already close to marasmus, the history indicates that they have developed edema, lost it, and acquired the appearance of simple marasmus before finally being brought to the hospital again with edema (line *d*).

The same lines *b*, *c*, and *d*, taken in the reverse direction, from the line *K–M* to the line *N–M*, indicate a possible path of recovery with the treatment. In recovery, children first return to a point on the line *N–M* close to the one from which they departed. Recovery to point *N*, i.e., normal height-weight for age, is a slower process and may no longer be possible. The further the child has gone toward marasmus, the more difficult it will be to return entirely to normal, especially for older children who have experienced prolonged deficiencies.

On the basis of this scheme for interpreting the different but closely interrelated types of protein-calorie malnutrition, the classification system graphically represented in Fig. 2 is suggested. The white apex represents the zone of normality and includes children who, although below average weight for their age, are still considered within the normal range.

The stippled area, parallel to the line *N–M*, includes the undernourished children in which the only apparent manifestation of malnutrition is retardation in growth and maturation. Weight can be used as a basis for classification into degrees. Following recommendations of Gómez *et al.* (1956b), we distinguish, for practical public health purposes, between

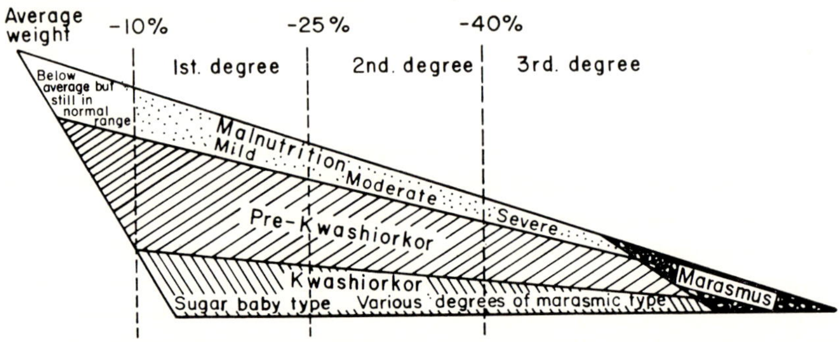

FIG. 2. Classification of the different types of protein malnutrition in children based on the concept represented in Fig. 1. The classification developed by Gomez *et al.* (1956b) is based on degree of weight lost between the dashed lines. (Reprinted from *Science* **133,** 2045, 1961, by permission.)

first (from 10 to 25 % below average weight for age), second (from 25 to 40 % below) and third (more than 40 % below average weight for age). When a child is severely underweight, but not markedly retarded in height for age, nutritional deficiency is recent. When height for age is as retarded as weight for age, this suggests a deficiency of long duration or the sequelae of a previous period of inadequate growth even if the child is presently receiving an adequate diet.

Cases classified as marasmus will fall into the black area. We diagnose as marasmus those children who have lost nearly all their subcutaneous tissue, have relatively little remaining muscle mass, and no edema. These are children who are extremely atrophic, the type referred to as only "skin and bones."

Within the kwashiorkor area of Fig. 2 parallel to the base of the triangle, occur all gradations from the sugar-baby type to the very marasmic type. The most consistent clinical criterion for classifying malnourished children as having kwashiorkor is the presence of frank edema. In the area between kwashiorkor and simple malnutrition should be classified those children who already show early psychic changes, slight edemas, and some of the other signs and symptoms of kwashiorkor to a mild degree. Even though such children are not yet severely enough affected to be classified as having kwashiorkor, it is evident they will very soon become frank cases of this syndrome if their situation does not change. They are sometimes referred to as representing cases of pre-kwashiorkor.

With this scheme every type of case within the wide range of syndromes due to varying degrees of protein and calorie malnutrition can be classified. It also takes into consideration the previous history of the case.

As an example, in Fig. 2 we have classified 2 cases seen by us which were in similar condition at the time of admission to the hospital but had very different histories; both reached a similar position in the triangle, but by different paths.

Although nutrients other than protein and calories have not been considered in the pathogenesis of these syndromes, deficiencies of vitamins and/or minerals associated with the basic syndrome of protein-calorie malnutrition are almost universal in field situations. These deficiencies, while a very important component of the patient's disease, are considered to be superimposed on the basic syndrome. They vary in their nature and severity from region to region and from case to case in the same region. Vitamin A and riboflavin deficiencies are particularly often associated with protein deficiency. On the other hand, when the protein-calorie malnutrition results in a severe impairment of growth and development and general metabolic activities, other specific deficiencies may not be as evident as the low intake of these nutrients would indicate. The variability from region to region of the specific deficiencies associated with the various types of protein-calorie malnutrition has contributed to the general confusion in classifying and characterizing these syndromes.

C. Chronic Protein-Calorie Malnutrition in Children

A large majority of the preschool children in areas where the acute protein malnutrition of kwashiorkor is prevalent go through a period of chronic protein deficiency. Although the importance of the milder forms has been recognized only in recent years, the much larger number of children affected give them great public health significance. In the schematic triangle (Fig. 2) such cases would be classified in the area parallel to the line N M. Frequently the only manifestation of nutritional deficiency in these children is a retardation of growth and development, which is usually not recognized, and such children are considered normal. The lack of adequate growth standards for populations other than European and North American makes the assessment difficult even for the trained field and health workers. It is known, however, that the size of the children at birth in most technically underdeveloped areas does not differ significantly from that of European and North American children and that their growth rates are similar during the first 4 to 6 months of life when their mother's milk covers their nutritional requirements (Sénécal, 1958; Ramos-Galván and Cravioto, 1958). Subsequently, proper supplementation to breast milk is not provided, rate of growth slows down, and by the time weaning is complete, growth is frequently at a standstill (Waterlow, 1948). During this period some of these children develop kwashiorkor, usually because of the stress of superimposed infection. The majority manage to survive this period of chronic malnutrition; after the age of 3 to 5 years they share in the family diet, suffer fewer infections, and have relatively lower nutritional requirements. When they enter school at the age of 6 or 7, they are small, but do not appear malnourished even though they may be two to four years retarded as judged by height, weight, and bone age.

Other characteristics of these cases which are more difficult to attribute with certainty to nutritional deficiencies are reduced physical activity, mental apathy, frequent episodes of ill-defined sickness often associated with anorexia and diarrhea, and a higher fatality rate from common infectious diseases. In direct contrast to normal children who are chacteristically very active, restless, curious, and mischievous, it is surprising to see these 2-, 3-, or 4-year-old children remain quiet for long periods. They seldom are interested in what goes on in their environment and are more apt to cry than to smile.

Recent studies carried out by Ramos-Galván *et al.* (1960) in Mexico have shown a strong correlation between retardation in growth and psychomotor development, as measured by conventional psychometric tests.

Other clinical signs frequently observed in these children include a thin, dry skin with areas of crazy pavement or "furfuraceous" desquamation. Even when there are no definite alterations, the skin lacks normal turgidity, softness, and elasticity. The hair is also likely to be thin and dry, mildly depigmented, and lacking in the luster and softness of normal hair. Often it is sparse and can be easily pulled out without pain.

After their second or third year, the signs and symptoms tend to disappear, but the retardation in growth remains. They look like well-nourished but younger children to whom Jelliffe (1959) applies the term "nutritional dwarfism." These children have suffered a period of lower growth rate, but have adapted themselves and are no longer suffering from protein deficiency at the time they are examined.

D. Pre-Kwashiorkor

This is a term used for children who, although they cannot be considered as kwashiorkor cases, will soon be typical cases of kwashiorkor if their situation is not corrected promptly. Clinically, these children have more marked signs and symptoms than described in the previous section. Apathy and irritability are more severe, appetite further reduced, and slight edema may be present. Biochemically, they present to a lesser degree some of the alterations characteristic of kwashiorkor. For example, the serum albumin which is still within normal levels in the chronically malnourished children is somewhat reduced in the pre-kwashiorkor cases. If the liver is biopsied, it is likely to show some degree of fatty change. These are incipient cases of kwashiorkor and many of them could be called mild kwashiorkor. In practice, it is impossible to establish sharp limits between chronic protein-calorie malnutrition and pre-kwashiorkor and even more difficult between pre-kwashiorkor and kwashiorkor.

E. Kwashiorkor

1. Clinical Characteristics

Pitting edema is the clinical characteristic in which the diagnosis of this type of protein malnutrition is based. It is soft, painless, and is partially localized by gravity. It is present in all cases, although its degree varies considerably from very mild in only the feet and ankles to severe and generalized. The eyelids may be swollen shut, movement of the extremities may be limited, and the peritoneal cavity frequently contains fluid. Severely edematous cases may lose as much as one third of their weight in a few days when the edema disappears during the initial phase of recovery. In emaciated children the edematous extremities are a sharp contrast with the shoulders and upper torso, which are extremely thin due to atrophy of muscles and loss of subcutaneous fat.

The degree of edema is not necessarily indicative of the clinical severity of the case nor of the degree of hypoalbuminemia, although total serum protein and albumin are always low when edema is present.

Retardation in growth and development has also been described as a constant feature of kwashiorkor. The degree of retardation is highly variable, and is determined by the history of the child. Previously well-nourished children who develop kwashiorkor within a few weeks as a result of sudden change from an adequate diet to one extremely poor in proteins, but not in calories, are not markedly retarded and after recovery, are close to the weight, height, and developmental stage expected for their age. Children who develop the syndrome after many months or years of deprivation are extremely retarded in growth and development, and are classified as having marasmic kwashiorkor.

Psychic changes are a clinical characteristic of kwashiorkor. Most cases present some degree of apathy which, in some instances, can be so severe that the child is lethargic and completely indifferent to his environment. They are also irritable and cry very easily at a slight stimulus with a cry which is weak and monotonous.

The skin lesions are characteristic although not always present. Whole areas of hyperpigmented and hyperkeratotic epidermis peel off in large scales, leaving a tender skin which can easily be infected. Lesions develop in two different ways. They may look like second-degree burns with erythema, and in some cases even phlyctenas, but these areas become progressively dry, hyperkeratotic, and hyperpigmented, and desquamation follows. Second, small areas of hyperkeratosis and hyperpigmentation with apparently normal surrounding skin grow progressively and become confluent, ending in areas of dermatosis very similar to those indicated above.

In infants and small children the skin lesions usually start and are more frequently localized in the perineal region. In older children they frequently start in the extremities. The face, neck, and lower trunk are affected next, and lesions are present in the upper trunk in only very few cases. The distribution of these lesions suggests that either chronic irritation of the skin or the presence of edema favors their development. The skin not affected by these lesions is usually thin and dry. Skin lesions of this type are not present in all cases, and their presence and severity have no correlation with the severity of the case. Hemorrhagic lesions of the skin, petechiae, and even ecchymosis are observed in extremely severe cases, and carry with them a bad prognosis.

The hair becomes dry, fine, brittle, sparse, and can be pulled out easily. Where the hair has been curly, it becomes straight. Black hair turns reddish, yellow, or even white, and in persons with blond hair, severe depigmentation is frequently observed. The color changes may be present in the whole hair, or stripes of depigmented hair may alternate with normal or less affected ones, producing the "flag sign" which reveals the nutritional history of the child. The changes in color are less constant than the alterations in texture and implantation of the hair and they do not correlate with the severity of the deficiency syndrome (Waterlow, 1948). We have observed extremely severe cases with black hair, and very mild cases with stripes of white hair. Changes similar to those described in the hair may be observed in the eyelashes and even in the nails, although much less frequently, probably because of their slower growth.

Anorexia may be severe enough to interfere with dietary therapy, forcing the use of small and frequent feedings, or even tube feeding during the first days. After a few days of adequate treatment, the appetite increases and may even become exaggerated.

Diarrhea is almost constant in the recent history, and is present in most cases at admission to the hospital. It is usually nonspecific in character, with undigested foods in the feces, and is promptly corrected by diet without any special treatment. In some cases a severe infectious diarrhea is present and should be promptly treated to prevent serious electrolyte imbalance and dehydration. Even when clinical diarrhea is not present, the stool volume is very large and stays so for many days or even weeks as in intestinal malabsorption.

Vomiting is frequent during the first days of hospitalization, particularly if the children are forced to eat.

Many other signs and symptoms of specific nutritional deficiencies may be present, such as those related to vitamin A, B-complex vitamins, or iron. They are observed with varied frequency from region to region, and should be considered as concomitant or superimposed deficiencies rather than part of the basic protein-calorie deficiency syndrome.

2. Hematological Alterations

Anemia is an almost constant accompanying sign in kwashiorkor, but many different types of anemia occur depending upon the secondary factors involved. The mild normocytic type observed has been interpreted as being mainly the result of hemodilution (Soto *et al.*, 1956; Metcoff *et al.*, 1957). A microcytic hypochromic type is usually the result of hookworm infestation, but in some areas may be related to a dietary deficiency of iron resulting from prolonged breast feeding, supplemented only with foods like sweet potatoes and plantains which have a very low iron content (Trowell and Simpkiss, 1957). A megaloblastosis, attributable to folic acid deficiency, has been observed in a certain proportion of cases (Walt, 1959; Lien-Keng and Ooang, 1959; MacIver and Back, 1960; Adams and Scragg, 1962). Recently, it has been suggested that some cases of megaloblastic anemia in kwashiorkor may be due to vitamin E deficiency (Majaj *et al.*, 1963). A normochromic and normocytic type has been interpreted as directly related to the deficiency of proteins.

These several different types of anemia have been observed in the same area (Scrimshaw *et al.*, 1956; Lien-Keng and Tumbelaka, 1960; Walt *et al.*, 1962) and anemias of a mixed type which are sometimes difficult to characterize are common. When, for example, they are due to a simultaneous deficiency of folic acid and iron, the peripheral blood cells are a mixture of small and large, although pale, red cells. A Price-Jones curve, or at least a careful study of a smear of peripheral blood will indicate this situation which is not revealed by the mean corpuscular volume.

Serum vitamin B_{12} levels of children with kwashiorkor have been found to be the same as in well-nourished children of the same age. Low serum copper and iron levels have been described in Central American children with kwashiorkor who were not infected with hookworm and whose diets appeared to contain sufficient iron (Lahey *et al.*, 1958). Furthermore, iron deposits have been observed in the reticuloendothelial cells of the bone marrow and in the liver of similar cases. The reduced levels of the serum carrier proteins (ceruloplasmin and transferrin) have been interpreted as being responsible for the low serum values of these minerals.

The course of the anemia of kwashiorkor is complicated by the very rapid recovery of these cases from their severe protein deficiency. The rate of hematopoiesis may be so great as to create extra demands for some of the hematopoietic factors and these may become limiting. A typical hypochromic microcytic anemia, responding to iron, may appear after 2 to 3 weeks of dietary treatment with milk just at a time when the child is recovering well from the signs and symptoms of protein deficiency and is gaining weight rapidly. Obviously, iron stores become exhausted, and the iron provided in the diet, although adequate for a normal child of the same age, is insufficient for the extra demands created by the rapid hematopoiesis of the recovery period.

A similar situation has been observed for folic acid deficiency anemia (Ghitis, 1962); the megaloblastosis sometimes appears after admission when the child is otherwise recovering. The sudden appearance of a crisis of hypoplastic bone marrow toward the end of the third week under good treatment and adequate recovery, which has been described mainly in Africa (Lien-Keng and Tumbelaka, 1960; Foy *et al.*, 1961; Walt *et al.*, 1962), may also be related to a similar phenomenon. Another factor which may complicate the interpretation of these anemias even further is the fact that infections are so frequently observed in association with kwashiorkor. The so-called "anemia of infection" has also been extensively studied (Cartwright and Wintrobe, 1952).

One type of anemia which is believed directly related to protein deficiency has been observed in children with severe kwashiorkor coming from the highland areas of Guatemala where hookworm is not present and where neither B_{12} nor folic acid deficiencies are prevalent. It is not usually a severe anemia; the red blood cell count is approximately 3 millon and the hemoglobin value about 10 gm. It is a normocytic or slightly macrocytic anemia with a mean corpuscular volume between 90 and 110 μ^3. The bone marrow is severely hypoplastic, especially for the erythroid series; the few elements of this series present are young erythroblasts, greater in size than the normal cells of this type but without the cytological characteristics of the megaloblasts.

When these children are treated with milk alone, the bone marrow rapidly becomes hyperplastic and a good reticulocyte response is observed in about a week. Unless iron is given, the anemia changes in a few weeks to a microcytic, hypochromic type because of the rapid increase in red blood cell count which is not accompanied by a parallel increase in hemoglobin. This type of anemia appears to be due to an arrest in erythrocyte formation at a very early stage in hematopoiesis because of the severe deficiency of proteins.

3. Pathology

A child with marasmic kwashiorkor who loses edema before dying will usually be considered a case of marasmus. Furthermore, in kwashiorkor there is usually some degree of coexisting marasmus. It is not surprising, therefore, that considerable confusion exists as to the pathological changes which are due to protein deficiency and those resulting from coexisting inanition.

a. General Findings and Gross Appearance. The lesions found at necropsy vary with the type, severity, and course of protein-calorie malnutrition and the quality and duration of the treatment given before death. In kwashiorkor the striated muscle generally looks pale, grayish-pink, and atrophied. As part of the general atrophy observed in these children (Béhar *et al.*, 1956), the visceral organs are usually reduced in both size and weight. Only the liver is enlarged and this is due to gross fatty change.

Severe wasting of tissues and organs is characteristic of marasmus, but the liver is not fatty. The osseous prominences are distinct due to the loss of adipose tissue and the muscles are markedly atrophic and pale. The visceral fat is also absent or greatly reduced. All of the organs, including the liver, are reduced in weight.

b. Liver. In the acute stages of kwashiorkor all of the hepatic tissue shows a marked fatty change of the large droplet type (Gillman and Gillman, 1951; Waterlow, 1948; Davies, 1951; Béhar *et al.*, 1956; Chanda, 1958). Small fat droplets generally appear first in the periportal area of the hepatic lobule and spread progressively to the central vein. The small droplets coalesce to form large fat globules that displace the nuclei and other cell structures toward the cell membrane. In severe cases, several fatty cells may fuse to form a large cyst.

With treatment the fat disappears in reverse order, first from the central zone of the hepatic lobule and last from the periportal areas. This change is accompanied by an increase in the number of reactive cells, chiefly lymphocytes and plasmacytes, in the sinusoids. There is, in addition, frank evidence of hepatic cell regeneration represented by an increase in nuclear size and chromatin content. Polyploids and mitotic figures are numerous in such areas. Deposits of hemosiderin pigment (which disappear with treatment) are seen in portal tracts as well and in the Kupfer and hepatic cells.

In the fatty liver of kwashiorkor, increased reticulin fibers radiating from the portal tract may develop a starlike pattern that was originally described as incipient cirrhosis. Many investigators considered kwashiorkor a precirrhotic disease (Davies, 1951; Chanda, 1958). Back and Bras (1958) found no evidence of fibrosis in children 3½ to 5 years after recovery from kwashiorkor. Most pathologists now agree that the fatty change of kwashiorkor does not lead to cirrhosis (Brock, 1954; Hill, 1959; Umaña and Tejada, 1961; Rhodes, 1957).

c. Pancreas. As a result of a low protein intake, the pancreas shows a relative hypofunction and atrophy (Veghelyi et al., 1950; Davies, 1952; Waterlow, 1948). The most common lesions are loss of zymogen granules and atrophy of the pancreatic acini. In some cases, there is also dilation of the pancreatic ducts and an increase in the number of interacinar ducts. Fibrosis has been described by Trowell *et al.* (1954a) as a constant finding, although in our material (Béhar *et al.*, 1956) and that of other investigators (Bras *et al.*, 1957) it has appeared rarely. It is possible that the collapse of the reticulum secondary to the atrophy of the acini has sometimes been interpreted as fibrosis. Where true fibrosis is observed, it is possibly due to fibrocystic disease of the pancreas with secondary malnutrition. With dietary treatment the pancreatic changes disappear rapidly and, in most cases, there may even be a hyperplasia with an apparent increase in the number of acini lining cells. This hyperplasia is probably similar to the one seen clinically in the parotid glands during the recovery state.

The islets of Langerhans in children with either kwashiorkor or marasmus frequently appear histologically normal (Béhar *et al.*, 1956). However, Normet (1937) has reported atrophy of the islet cells and a decreased number of beta cells.

d. Gastrointestinal Tract. Deficiencies of protein, as well as other essential nutrients, produce early functional and morphological changes (Vitale *et al.*, 1960) which markedly improve with 6 to 12 days of adequate dietary treatment and soon disappear. The changes observed in kwashiorkor and marasmus are similar.

(*i*) *Stomach.* Grossly, the organ is distended; its wall is thin and lined with an atrophic mucosa. The rugae are small and relatively scarce. In some cases the stomach is not dilated and the mucosa, although atrophic, shows frank submucosal edema (Béhar *et al.*, 1956).

Histologically, there is an atrophy of the mucosa at the level of the fundus. The mucous glands of the pyloric portion of the stomach are apparently well preserved and in some cases seem relatively increased (Tejada and Restrepo, 1962).

(*ii*) *Intestine.* The jejunum and ileum are usually dilated. The number and size of the intestinal rugae are decreased and the wall is very thin, even transparent (Béhar *et al.*, 1956; Tejada and Restrepo, 1962). In a few cases, there is edema of the submucosa and the wall is thicker with enlarged mucosal rugae. The presence of this latter change has not been explained, but it is usually seen in children with severe edema.

Histologically, there is atrophy of the intestinal villi, and a decreased number and size of the chief columnar cells. The picture is quite comparable to that observed in patients with primary malabsorption.

e. Kidneys. The changes are not specific and no gross or histological distinction can be made between cases of kwashiorkor and marasmus.

Cloudy swelling is usually found in severe cases (Trowell *et al.*, 1954a). Fatty degeneration of the cells of the proximal and distal convoluted tubules is often observed (Béhar *et al.*, 1956; Campbell, 1956; Davies, 1956), mainly in association with fatty metamorphosis of the liver. Glomerular changes include swelling of the epithelium of Bowman's capsule, partial or total hyalinization of the glomeruli, and sometimes a deposition of fine eosinophilic material in the capsular space (Bras *et al.*, 1957) that has been interpreted as protein in nature.

f. Heart. Gross and histological changes are minimal but the actual weight of the heart is decreased in malnourished children with kwashiorkor (Béhar *et al.*, 1956).

g. Lungs. In fatal cases of kwashiorkor and marasmus investigated by Tejada *et al.* (1956) bronchopneumonia was seen in 67% of the cases. Intra-alveolar hemorrhages and focal or diffuse atelectasis were observed independently or in association with the pneumonia.

h. Pituitary Glands. When the pituitary glands of children with kwashiorkor and marasmus are studied histopathologically and differential cell counts are performed, the relative percentages of basophiles are usually within normal limits. Acidophiles tend to be decreased and sparsely granulated amphophile cells which are thought to be the source of trophic hormones (Russfield, 1957) are often abundant. Hypertrophic amphophiles, characteristic of an active pituitary, are also present. The presence of numerically abundant amphophiles suggests the production of trophic hormones in response to stress. Since many of the children studied were treated vigorously for several days before they died, it is possible that these changes represent, in part, a response to the improved diet.

i. Thyroid. The prevalence of goiter in areas where malnutrition is also present makes it very difficult to determine whether changes are due to malnutrition or to goitrogenic factors (Tejada, 1958). Ambrosius (1956) observed flat epithelium and large follicles full of colloid in malnourished children over 5 years of age. However, the histological pattern was the same in well-nourished and malnourished children under 5 years of age.

j. Parathyroids. No histopathological changes have been observed in the parathyroid glands of children with either kwashiorkor or marasmus (Béhar *et al.*, 1956).

k. Thymus. The thymus gland is so reduced in size that it is very difficult to find at necropsy. Histologically, the lymphoid tissue is replaced by fat (Béhar *et al.*, 1956).

l. Adrenals. Contradictory results have been reported. In Central American material the adrenals are usually small (Béhar *et al.*, 1956). In approximately one third of the cases, the cortex has a linear pattern; the nuclei are pycnotic, and abundant lipoid is present as droplets in the cytoplasm of the cells. In the other two thirds of the cases, the nuclei are large and vesicular with varying degrees of lipoid exhaustion and obscured linear cortical pattern. The first is compatible with hypoactivity and the second with hyperactivity. These probably represent different stages of malnutrition in which hyperactivity is followed by hypoactivity and finally exhaustion.

4. Physiology and Biochemistry

Protein deficiency in its severe form is known to produce many physiological and biochemical changes. Since the individual may have associated deficiencies of other nutrients, as well as superimposed infections, there is considerable confusion in the literature as to which of these changes are due to protein deficiency per se. Furthermore, kwashiorkor may be superimposed upon any degree of marasmus, and many biochemical studies have failed to distinguish between these two syndromes. Reports of physiological and biochemical changes, attributed to marasmus, are frequently based on studies in children in whom marasmus is the dominant clinical characteristic, but in whom some signs of kwashiorkor are also present. There are many biochemical changes characteristic of kwashiorkor which do not occur to an appreciable degree in marasmus. This conclusion is based on our observations and those of other workers who have utilized a strict definition of marasmus, but is contrary to the impression given by much of the published literature.

a. Water, Electrolytes, and Circulatory Changes. A major clinical feature distinguishing kwashiorkor from marasmus is edema. Although hypoalbuminemia is a constant finding, many other factors are involved in edema production. These include hemodynamic changes, disturbed water and electrolyte balance, and hormonal abnormalities. Right heart failure does not seem to be involved in producing edema since venous congestion and enlarged, tender liver are not present. Instead of cardiomegaly the heart, as judged by X-ray upon admission, is small. The pulse in kwashiorkor may be slow, normal, or fast. Although bradycardia occurs in both malnourished children and adults (Howarth, 1951; Keys *et al.*, 1950), it is more common in marasmus than in kwashiorkor. Both types of malnourished infants develop prolonged tachycardia and cardiac dilatation during recovery, indicating a meager cardiac reserve. During initial recovery from protein malnutrition, the increase in intravascular volume may precipitate a mild congestive heart failure in both children and adults. The edematous lungs sometimes seen at necropsy (Tejada, 1962) may be due to overzealous rehydration or to some degree of congestive heart failure, although bronchopneumonia is also almost always present.

In electrocardiographic (ECG) studies (Gopalan *et al.*, 1955; Schyns and DeMaeyer, 1956; Smythe *et al.*, 1962; Janssen and LeRoux, 1950; Belli-Cortes *et al.*, 1961) only sporadic instances of arrhythmia due to extrasinusal pacemakers have been noted. Consistently decreased voltage and T-wave abnormalities have been interpreted as indicative of abnormal myocardial function or ischemia. An associated prolongation of the QT interval and notching of the QRS complex are difficult to interpret. Smythe *et al.* (1962) have suggested that some sudden deaths in kwashiorkor are due to cardiac arrest. ECG studies have also revealed characteristic changes resulting from electrolyte imbalances, particularly hypokalaemia, disappearing with adequate electrolyte therapy. Anemia may also play a role in the abnormal circulatory phenomena observed.

Castellanos *et al.* (1959), studying the vectocardiographic records of malnourished infants, have demonstrated alterations in the loops of depolarization and repolarization and have interpreted them as a reflection of a delayed progression of the usual left to right ventricular ratio and to defects in repolarization.

The child with severe kwashiorkor shows signs of decreased cardiac output and even of shock which are usually complicated by marked electrolyte imbalances and sometimes severe infections. Kerpel-Fronius *et al.* (1954) have found the decreased cardiac output to be greater than expected from the lowering of the oxygen requirement. The anatomical distribution of edema in the distal parts of the body, even in the presence of severe dehydration, suggests that tissue anoxia as a result of decreased peripheral circulation may be an important factor in the physiopathology of edema.

Overhydration in terms of total body water per unit of body surface or body weight is found in all cases of kwashiorkor (R. Smith, 1960; Schnieden *et al.*, 1958) as well as in undernourished adults (Keys *et al.*, 1950), and involves all fluid spaces (intravascular, extracellular, and intracellular) (Gómez *et al.*, 1950, 1957; Frenk *et al.*, 1957; Kerpel-Fronius, 1960; Gopalan *et al.*, 1953; Keys *et al.*, 1950; Waterlow *et al.*, 1960; Ramos-Galván and Cravioto, 1958).

Kerpel-Fronius and Frank (1949) point out that skin, skeleton, kidney, and central nervous system, which suffer very little weight loss in malnutrition, are tissues that already contain large extracellular fluid volumes. The loss of protoplasmic mass and fat would result in a relative increase of total extracellular fluid. It also has been suggested that decreased tissue elasticity results in lower hydrostatic pressure outside the capillaries to balance the intravascular pressure, thus permitting an increased seepage of fluid into the interstitial spaces (Keys *et al.*, 1950; McCance, 1951). The osmolarity of the three major water compartments is reduced (Gordillo *et al.*, 1957; Soto *et al.*, 1958; Pille, 1957).

A marked potassium depletion has been demonstrated in kwashiorkor by balance studies (Hansen, 1956), tissue analysis (Frenk, 1958; Metcoff *et al.*, 1957; Waterlow and Mendes, 1957), and exchangeable potassium measurements (R. Smith and Waterlow, 1960). Moreover, the potassium depletion is more severe than that of nitrogen. In other words, the nitrogen loss accounts for only part of the potassium deficit. The accompanying diarrhea, which is almost a constant finding in kwashiorkor, is at least partly responsible. The decline in intracellular potassium concentration is believed to cause a defect in energy regulation. One of the results of this alteration would be a less efficient "sodium pump" resulting in an increase in both intracellular sodium and water (Metcoff *et al.*, 1960).

The fact that these changes involve primary intracellular potassium loss explains why the serum potassium concentration is not an adequate index of its depletion in malnutrition (Thompson, 1955). Serum potassium concentration, however, is most commonly low, especially in children who have had diarrhea (Hansen and Brock, 1954; Kaul, 1960; Politzer and Wayburne, 1957; Goldis, 1959). Severe depletion of this element in kwashiorkor is accompanied by a low cerebrospinal fluid potassium, particularly in the presence of diarrhea (Goldis, 1959).

Intracellular potassium loss has a detrimental effect on the physiology of the renal tubular cells (Holliday *et al.*, 1960; Rehlman and Schwartz, 1956), and Gordillo *et al.* (1959) have demonstrated this type of renal dysfunction in severe malnutrition. The principal effects are hypoosmolarity in urine, decreased glomerular filtration rate, lowered renal plasma flow, and impaired tubular function, as well as a defect in urine acidification. The latter has also been described by R. Smith (1959). In severe, prolonged potassium depletion these changes may become irreversible. As a consequence, this tubular defect influences not only the water and electrolytic composition of the malnourished child, but also determines the survival of the patient in some instances. Supporting indirect evidence is found in the fact that edema in kwashiorkor may be reduced by the administration of potassium even in the absence of protein in the diet (Hansen and Brock, 1954). Alterations described in renal function from causes other than potassium depletion include a hypoosmolar urine with decreased osmolar clearance, decreased renal plasma flow and glomerular filtration rate, increased free water clearance (Gordillo *et al.*, 1957), and impaired renal response to a water load (Schnieden *et al.*, 1958).

There is a great deal of speculation regarding the effects of abnormal hormone metabolism. Gopalan (1950) found an antidiuretic substance in the urine of undernourished adults and suggested that liver damage may be responsible for a decreased breakdown of antidiuretic hormone (Gopalan, 1955). Increased plasma antidiuretic activity has been found by Heller (1953).

Abnormally elevated ferritin levels which are corrected by treatment were reported by Srikantia (1958) in plasma of kwashiorkor children and two adults with malnutrition edema. Ferritin is stored in the liver and the spleen and has a marked antidiuretic effect by stimulating the posterior pituitary; therefore, liver damage could thus be responsible for increased antidiuretic hormone activity in kwashiorkor. There is, however, suggestive evidence against this hypothesis: (*1*) Gordillo *et al.* (1957, 1959), as well as Kerpel-Fronius *et al.* (1954), demonstrated increased free water clearances in malnourished children. (*2*) Gordillo *et al.* (1957) demonstrated renal response to hypertonic saline; (*3*) Cravioto and Frenk (cited by Gordillo *et al.*, 1957) and R. Smith and Waterlow (cited by Waterlow *et al.*, 1960) demonstrated an adequate response to antidiuretic hormone administration. Both (*2*) and (*3*) indicate that any increase in antidiuretic hormone activity must at least be moderate and capable of being overcome or augmented by other factors.

R. Smith (cited by Waterlow *et al.*, 1960) has found low aldosterone excretion in children with kwashiorkor while Lurie and Jackson (1962) reported increased excretion in some edematous malnourished children, but little or none in a few. The difficulty of accurate measurement of aldosterone is still a great obstacle for this type of research. In kwashiorkor sodium retention has been demonstrated mainly through balance studies which show a prolonged period of sodium loss during recovery (Hansen and Brock, 1954; Hansen, 1956). The chloride space in poorly nourished hypoalbuminemic children is expanded (Cheek, 1954) as is the sodium space and total exchangeable sodium (Gomirato-Sandrucci and Mussa, 1960). As a consequence, water retention occurs mainly in the extracellular spaces. Intracellularly, the potassium loss and its replacement by sodium and hydrogen ions, plus the suggested increase in free amino acids, produce an excess in cation concentration (Eckel *et al.*, 1954; Frenk *et al.*, 1957) which favors intracellular overhydration.

In protein malnutrition the elevated free amino acid concentrations complicate the classic mechanisms described by Darrow *et al.* (1948). A lack of adequate correlation between intracellular potassium decrease and intracellular sodium content has been recognized (Frenk *et al.*, 1957). To what extent sodium retention is due to the renal changes already discussed or to hormonal or cardiovascular dysfunction is still a moot question. Waterlow *et al.* (1960) propose from clinical impressions that sodium and water may be the first altered in a sequence that leads to the final overhydration of malnutrition.

Magnesium metabolism has recently been studied in kwashiorkor and both Montgomery (1960a,b, 1961) and Metcoff *et al.* (1960) believe that magnesium deficiency in this syndrome has wide physiological and biochemical implications. Magnesium deficit follows the pattern of potassium depletion and has been incriminated in the tetanic manifestations sometimes observed in kwashiorkor (Montgomery, 1960a,b; Back *et al.*, 1962).

Both the organic and inorganic phosphorus content of muscle have been found to be low (Metcoff *et al.*, 1960). In children surviving severe malnutrition there was an increase in both total and organic phosphorus during treatment, while in children who died, only inorganic phophorus rose, probably indicating a failure in energy metabolism. Phosphorus depletion in kwashiorkor has been observed by many other investigators. Hypophosphatemia and hypocalcemia (M. Carvalho *et al.*, 1945), increased phosphorus excretion with an abnormal increase in the fecal/urinary ratio (Holemans *et al.*, 1959) as well as an increase in lipid phosphorus content in feces—all revert to normal with recovery. Gomez *et al.* (1956a) not only described hypocalcemia and hypophosphatemia, but also reported cases of acute rickets, tetany, and fractures during initial phases of recovery. Phosphorus retention markedly increases during initial recovery and decreases as the child advances in nutritional repletion. Absorption increases only slightly (Waterlow and Wills, 1960).

b. Basal Metabolism. Basal oxygen consumption, as an index of general energy metabolism, has been shown to be reduced in adult starvation (Keys *et al.*, 1950). Recently, the oxygen consumption of marasmic children was noted to be low (Kerpel-Fronius, *et al.* 1954), and the decrease was proportional to the weight deficit. Kerpel-Fronius (1960) theorized that because of this fact these children adapt better to the tissue anoxia produced by the decreased cardiac output.

Montgomery (1962a), in a detailed study of the basal metabolic rate of malnourished children, found that it was subnormal when expressed in terms of surface area. However, when expressed in terms of body weight or of total body solids(as calculated from determinations of total body water) it was normal. When the subjects were severely marasmic, it tended to be high. These seemingly contradictory results have been explained on the basis of the relative preponderance in the malnourished child of tissues such as the central nervous system, heart, and kidney which under basal conditions account for a large proportion of the oxygen consumption. Determinations of thyroid function by I^{131} studies (Montgomery, 1962b) gave essentially normal results. The greater the weight deficit, the higher was the basal metabolic rate during recovery and the higher were the caloric requirements for weight gain. Thyroid evaluation by means of protein-bound iodine (Valderrabano Jolsen, 1954) yielded low values in only 43 % of "malnourished children," a decrease which correlated significantly with that of total serum proteins and albumin. In all cases there was an increase with nutritional therapy. The fraction of protein-bound iodine which is nonextractable in butanol showed an increase in severe "malnutrition" (Lifschitz, 1962). With treatment the ratio of butanol extractable to total protein-bound iodine reverted to normal or above. This points to the lack of specificity of total protein-bound iodine as an index of thyroid function when marked alterations in plasma protein are present.

c. Carbohydrate and Energy Metabolism. Because the main difference between the diets leading to kwashiorkor and marasmus is their carbohydrate content, several authors, particularly MacDonald (1960c) and Gillman *et al.* (1961), have suggested that altered carbohydrate and lipid metabolism are the determining factors. Nevertheless, carbohydrate metabolism has been studied less intensively than lipid and protein metabolism.

Glycemia has been found low (Chaudhuri, 1948; Pille, 1957; Gómez *et al.*, 1958; Slone *et al.*, 1961) and hypoglycemia has been suspected as a cause of death. Glucose absorption appears normal (Holmes and Trowell, 1948), although the handling of intravenous glucose and galactose loads has been observed to be abnormal. Glucose loads produce prolonged hyperglycemia in a number of cases (Niemeyer and Meneghello, 1950; Chaudhuri, 1948; Slone *et al.*, 1961; Holmes and Trowell, 1948), but Gómez *et al.* (1958) report severe hypoglycemia in the last stages of the test.

Several explanations have been proposed for these findings. The only experimental data on glycogen formation (Holmes and Trowell, 1948) indicate impaired synthesis, although Waterlow and Weisz (1956) have found increased glycogen concentration in liver in kwashiorkor. Gillman *et al.* (1961) using C^{14}-labeled glucose found a decrease in carbon dioxide production in kwashiorkor which indicated that its utilization was reduced. The same occurred with C^{14}-labeled acetate and pyruvate, but the delayed carbon dioxide production from acetate-1-C^{14} was less marked. Also C^{14} incorporation from pyruvate and acetate into blood glucose was greater and labeled blood glucose was present for a longer period of time in kwashiorkor than in control cases. This may mean that: (*a*) blood glucose utilization, or disappearance, is delayed; (*b*) glucogenesis is increased in kwashiorkor; (*c*) glycogen in the normal child is produced from available amino acids and fat from internal deposits diluting the radioactive carbon; or (*d*) the tricarboxylic acid cycle activity is depressed.

Although the origin of hypoglycemia is obscure, adrenal hypofunction may play a role (Ramachandran *et al.*, 1956; Castellanos and Arroyave, 1961). Lack of adequate growth hormone production by the hypophysis would sensitize the child to insulin and glycogen depletion and could be another mechanism for the production of hypoglycemia, although liver glycogen stores have been reported as normal and responses to subcutaneous administration of glucagon and epinephrine are either normal or difficult to interpret (Salazar de Sousa, 1959, 1960). There appears to be no hypoglycemic response to leucine in these cases.

Pyruvic and α-ketoglutaric acid levels have been found elevated in the blood of malnourished children, but not in urine (Edozien, 1959; Chaudhuri *et al.*, 1959) regardless of the type of malnutrition (edema present or absent). Cutroneo-Guarniere and Serrano (1960), however, found decreased pyruvic acidemia in marasmus and elevated values in kwashiorkor (carbohydrate dystrophy). Muscle analysis (Metcoff *et al.*, 1960; Frenk, 1961) for several intermediary metabolites of glucose oxidation and the Krebs cycle, in relation to intracellular ionic concentration, have yielded further support to altered energy metabolism in malnutrition.

Based on the data of Metcoff *et al.* (1960), the condensation step of acetate with oxaloacetate appears to be impaired. Studies by Gillman *et al.* (1961) with C^{14} suggest that both normally and in malnutrition infants metabolize glucose through the malic acid cycle with active CO_2 fixation. This conclusion is consistent with the previous findings of Waterlow (1958) of an increased malic dehydrogenase activity expressed per milligram of DNA (deoxyribonucleic acid)-phosphorus and of Burch *et al.* (1957), who found constant values before and after treatment in children with kwashiorkor, when the results were expressed per milligram of liver nitrogen.

Waterlow (1950, 1958) determined the activity of some of the enzymes involved in the tricarboxylic acid cycle and "electron transfer chain system" in the liver of children with kwashiorkor. Expressed per milligram of DNA-phosphorus, succinoxidase tended to decrease but cytochrome oxidase was normal. Glutamic dehydrogenase was also normal. Burch *et al.* (1957) reported stable levels of DPNH (reduced diphosphopyridine nucleotide)-dehydrogenase per milligram of tissue protein.

Metcoff *et al.* (1960) found very low levels of muscle glutamic dehydrogenase activity in malnutrition. From estimations of isocitrate, oxaloacetate, and α-ketoglutarate levels, it was concluded that in fatal cases a block exists in the enzymic step from α-ketoglutarate to succinate. The rise of CO_2 production from acetate following treatment of kwashiorkor (Gillman *et al.*, 1961) also suggests that the tricarboxylic acid cycle is impaired even in nonfatal cases.

Waterlow *et* al. (1960), studying different steps in oxidative phosphorylation in kwashiorkor, have found that in liver the incorporation of phosphorus in phospholipids is decreased, while respiration and oxidative phosphorylation are preserved. Metcoff *et al.* (1960) reported impaired production of organic phosphorus and accumulation of inorganic phosphorus in fatal cases of malnutrition and interpreted this as due to a block of oxidative phosphorylation.

Galactose utilization after an intravenous dose was abnormal in 2 of 11 patients studied by Niemeyer and Meneghello (1950). There was no correlation with degree of liver steatosis.

A number of cases of galactose intolerance have been described among kwashiorkor patients (Badr El-Din and Aboul-Wafa, 1961) with galactose abnormally present in their blood and severe diarrhea at the beginning of a milk regimen. The defect disappeared after a week of treatment with a galactose-free diet. One child, who died 12 hours after a galactose tolerance test, showed the highest value of the group studied.

d. Lipid Metabolism. Changes in lipid metabolism in protein malnutrition have been sought since the early recognition of the fatty liver of kwashiorkor (Williams, 1932). Waterlow (1948) believed initially that the intensity of liver steatosis paralleled the intensity of protein depletion. Because diarrhea, wasting, and lipid abnormalities occurred, early researchers were prompted to investigate abnormalities in serum lipids and fat absorption (Trowell *et al.*, 1954a).

A number of investigators (Van der Sar, 1951; Jayasekera *et al.*, 1951; Gómez *et al.*, 1956c; Robinson *et al.*, 1957) have noted varying degrees of steatorrhea and decreased fat absorption in kwashiorkor, which became normal with adequate diet therapy. A large percentage of the fecal fat is composed of fatty acids and neutral fats (Jayasekera *et al.*, 1951). Holemans and Lambrechts (1955) also reported increased fecal lipids. In acute kwashiorkor treated with a lipid-free diet we still find an average fecal loss of about a gram of fat per day.

Several factors may be involved in the production of steatorrhea. The pancreatic enzymes, trypsin, lipase, and amylase, have been observed to be markedly reduced (Veghelyi, 1950; Thompson and Trowell, 1952; Trowell *et al.* 1954a; Gomez *et al.*, 1954) and the intestinal wall is severely atrophied (Davies, 1948; Tejada, 1955). The fact that fatty acids as well as phospholipids are elevated in fecal fat point more toward an intestinal defect, but the simultaneous increase in neutral fats suggests an enzymic insufficiency. Absorption tests with vitamin A as a lipid marker (Arroyave *et al.*, 1959) and low chylomicronemia indicate a defect in fat absorption in kwashiorkor. Vitamin A palmitate is not absorbed in kwashiorkor but is absorbed by marasmic children. Vitamin A alcohol in kwashiorkor is absorbed better than the palmitate but not as well as the water-soluble vitamin A acetate. Simultaneous administration of pancreatic enzymes and bile salts, independently or together, did not improve the absorption of vitamin A palmitate (unpublished INCAP data, 1962).

Abnormalities in lipid transport have also been found. Cravioto (1958a) and Cravioto *et al.* (1959a) demonstrated a marked decrease of α-lipoproteins as well as a decrease in β-lipoproteins in the majority of patients.

Chatterjee and Chaudhuri (1961) agree on the decrease in α-lipoprotein but report β-lipoproteins to be initially increased (Chaudhuri, 1961). A deficiency of plasma protein for the transport of lipids in severe malnutrition has been suggested also by Arroyave *et al.* (1961a). They found that children with kwashiorkor experimentally receiving skim milk therapy with no vitamin A or carotenes showed not only a rise in albumin and total protein but also a corresponding rise in serum vitamin A and a concomitant decrease in liver vitamin A. In the majority of kwashiorkor cases there is a decrease in total blood lipids (Schwartz and Dean, 1957; Cravioto *et al.*, 1959a; Gómez Caballero, 1959; Mey *et al.*, 1961; Schendel, 1960), phospholipids (Trowell *et al.*, 1954a; J. Carvalho, 1947) and total cholesterol (DeMaeyer and Vanderborght, 1954; Mey *et al.*, 1961; Schendel, 1960; Carvalho, 1947). The initial decrease in serum cholesterol appears due to a frank lowering of the cholesterol ester fraction (DeMaeyer and Vanderborght, 1954; Schwartz and Dean, 1957), suggesting impaired esterification of cholesterol in the liver.

All the lipid fractions studied rose promptly with adequate dietary therapy, even in the absence of dietary fat. This has been interpreted as a reflection of restoration of the protein transport system and consequent mobilization of lipid stores (Dean, 1960; Matthew and Dean, 1960). There was a concomitant decrease in liver lipids (Waterlow, 1948; Burch *et al.*, 1957) and a rise in the proportion of the cholesterol esters to total cholesterol in blood (DeMaeyer and Vanderborght, 1954; Schwartz and Dean, 1957). This response (Schwartz and Dean, 1957) is not influenced by the type of protein given but the type of fat influences both the degree of cholesterol elevation and the duration of the initial hypercholesterolemia seen during therapy. Butterfat, as compared with vegetable oil, produced a higher initial cholesterol rise and a higher level of cholesterol during treatment (Matthew and Dean, 1960), suggesting that not only transport but also cholesterol metabolism may be affected during therapy.

The findings that the total serum lipids are low in kwashiorkor even when adipose depots are preserved favors a defect in fat transport from depots and/or in lipid synthesis and catabolism. The similarity between liver fatty acids and fat depot fatty acids reported by MacDonald (1960a,b) suggests that the liver fatty acids in kwashiorkor originate from the depot fat. MacDonald (1960a,b) has shown that the production of phospholipids in the fatty livers of children with kwashiorkor is not elevated. This suggests that the defect in lipid transport is related to a relative impairment in phospholipid metabolism.

Studies by Schwartz and Dean (1957) and Schendel and Hansen (1958) on phospholipid synthesis also point to a decreased production of this lipid fraction in kwashiorkor. The possibility of a deficiency of methyl donors as a cause of fatty liver (through poor synthesis of phospholipids) has been discarded because the administration of choline and methionine does not affect the liver steatosis (Auffret and Tanguy, 1950). Cholesterol esterification in the liver also seems to be reduced in kwashiorkor. The hepatic cholesterol accumulation reported by several authors (Ramalingaswami *et al.*, 1952) also points to a defect in the transport of lipid from the liver.

The pituitary and adrenal changes previously described and the decreased 17-hydroxysteroid excretion in kwashiorkor but not in marasmus (Castellanos and Arroyave, 1961) may, by influencing protein and lipid metabolism in adipose tissue, account for some of the differences between these two syndromes.

Both serum lipase and esterase are reduced in kwashiorkor (Srinivasan and Patwardhan, 1952). Whether these nonspecific lipolytic enzymes play any role in the abnormal lipid transport and metabolism is unclear, but the decreases in plasma esterase seemed to correlate well with liver steatosis.

Alterations in lipid synthesis and catabolism are beginning to be explored by the use of C^{14}-labeled acetate, pyruvate, and glucose (Gillman *et al.*, 1961). An increase in the synthesis of cholesterol, phospholipid and glyceride-glycerol from C^{14}-acetate, and an increased synthesis of phospholipids and glyceride-glycerol from pyruvate and glucose may also account for some of the lipid changes in kwashiorkor.

A deficiency of essential fatty acids in kwashiorkor has been suggested by the fact that children who respond to initial therapy show a rise in trienoic acids and a lowering of di-, tetra-, penta-, and hexaenoic acids (Schendel and Hansen, 1959b; Schendel, 1960). This is another aspect of lipid metabolism in kwashiorkor which needs to be explored further, since fat-free diets seem adequate for treatment of the syndrome.

e. Protein and Amino Acid Metabolism. The absolute nitrogen content of liver as per cent of body weight is normal (Méndez and Tejada, 1962). The decrease in protein concentration in liver tissue, expressed as gm/100 gm of fresh or dry weight is entirely accounted for by the increase in fat since on a fat-free basis the ratio of nitrogen to water is normal. The nitrogen content of the liver in kwashiorkor is low per unit of DNA-phosphorus (Waterlow, 1956; Waterlow and Weisz, 1956). DNA-phosphorus was used as a reference to calculate nitrogen content assuming that it expresses the amount of nitrogen per cell.

The increases in liver nitrogen per DNA-phosphorus during recovery, however, may be somewhat understimated due to an increase in DNA with increased mitosis (Ely and Ross, 1951). Arroyave *et al.* (1961b) and Standard *et al.* (1959), using urinary creatinine excretion in 24 hours to evaluate the total muscle mass, found it very low in kwashiorkor. On the basis of DNA-phosphorus muscle nitrogen is also decreased (Waterlow and

Mendes, 1957; Waterlow, 1956). Waterlow *et al.* (1960) postulate that in protein-calorie malnutrition as much as 50 % of the total body protein can be lost.

In protein depletion growth ceases and "labile proteins" are lost in appreciable amounts all through the body with the exception of the central nervous system. Proper measurements of body composition have been hampered by the lack of suitable methods (Brozek, 1961; Forbes, 1962) since the measures of partitioning the body into active tissue mass, fat-free solids, total body fat, and total body water involve assumptions that may not apply to a malnourished individual. Interpretation is hampered because (*a*) the malnourished child is overhydrated and in most instances has lost an appreciable amount of fat; (*b*) the rate of fat loss and decrease in muscle mass in malnutrition occur at different rates and with patterns which complicate the final picture (McFie and Welbourn, 1962); and (*c*) liver and muscle analysis in children, autopsy organ weights in relation to normals, and various experimental findings (Diener, 1961; Waterlow, 1956, 1959a; Waterlow and Mendes, 1957; Kerpel-Fronius and Frank, 1949; Tejada, 1955) support the concept that different organs respond to nitrogen loss by losing protein or weight at different rates.

In spite of its disadvantages, the cumulative nitrogen retention during recovery from kwashiorkor would give another estimate of the protein deficit (Forbes, 1962; Fomon and Owen, 1962). The rates of individual tissue losses, however, seem to be different. The proteins with specific functions such as enzymes, hemoglobin, and some plasma protein fractions seem to change in characteristic ways which are partly independent of total nitrogen loss.

In kwashiorkor, pancreatic enzymes have been found low (Veghelyi, 1948; Gómez *et al.*, 1954; Venkatesan, 1957; Thompson and Trowell, 1952). The physiological reason for this phenomenon, as well as for the pancreatic atrophy constantly observed pathologically, is probably the normally very high protein turnover of this organ. Whether this enzymic deficiency is partly responsible for the occurrence of diarrhea in malnutrition remains uncertain. The absorption of nitrogenous compounds seems to be little affected, even at the very start of treatment. Robinson *et al.* (1957) and Pretorius and DeVilliers (1961) have found no difference between the absorption of milk and protein hydrolyzates, which suggests that the mild decrease in absorption is due to intestinal wall defects. On the other hand, Sarrouy *et al.* (1957a) claim to have detected a polypeptide compound in the serum of malnourished children after administering a gelatin load which disappears with nutritional recovery.

Serum amylase (Dean and Schwartz, 1953; Scrimshaw *et al.*, 1956; DeMaeyer and Vanderborght, 1954; Senecal *et al.*, 1957; Smit and

Pretorius, 1957) has been found consistently low in kwashiorkor. In marasmus, however, it is often normal. Serum amylase and serum lipase which is also low (Srinivasan and Patwardhan, 1952) are believed to reflect pancreatic atrophy.

Serum pseudocholinesterase activity has also been found low (Dean, 1960; Arroyave *et al.*, 1958; Edozien, 1961; Burch *et al.*, 1957; Schendel and Hansen, 1962). This decrease may reflect decreased liver synthesis of the enzyme since it was also found low in this organ (Mukherjee and Sarkar, 1958; Waterlow, 1950, 1952, 1958). Serum pseudocholinesterase as a possible measure of marginal protein deficiency has been investigated by Arroyave *et al.* (1958) and Schendel *et al.* (1960); it appears that this enzyme's activity drops concomitantly with the albumin in malnutrition and therefore has little special value for this purpose.

Serum alkaline phosphatase is also low (Scrimshaw and Béhar, 1961; Scrimshaw *et al.*, 1956; Dean and Schwartz, 1953; Schwartz, 1956; Edozien, 1961; Gómez *et al.*, 1956a) in severe protein malnutrition. This drop may reflect decreased osteogenesis. Schwartz (1956) describes three alkaline phosphatase fractions, probably of different origin, which change at different rates during the initial repletion. As a result total alkaline phosphatase activity drops initially before rising to normal levels. The elevated alkaline phosphatase in the liver (Mukherjee and Sarkar, 1958) suggests a decreased inactivation or deficient excretion of the enzyme by this organ.

Some blood enzymes which have been found elevated in destructive processes of parenchymatous organs are also increased in kwashiorkor. Among these, Baron (1960) reports twofold elevations of serum glutamic-oxaloacetic transaminase and isocitric dehydrogenase, and a mild increase in glutamic-pyruvic transaminase. Cutillo *et al.* (1961) described elevated L-leucine amino peptidase in "dystrophic infants" and in one case of kwashiorkor. However, Edozien (1961) found no systematic changes in serum levels of glutamic-oxaloacetic transaminase, glutamic-pyruvic transaminase, and isocitric dehydrogenase. In the 3 cases out of 50 in whom these three enzymes were markedly elevated, other indexes of severe liver disease were also present. Only one showed a decrease in the serum levels of glutamic-oxaloacetic transaminase, glutamic pyruvic transaminase, and isocitric dehydrogenase following the administration of 2 gm methionine. Diwany *et al.* (1961) and C. E. Smith (1962) found no elevation of serum glutamic-oxaloacetic transaminase and serum glutamic-pyruvic transaminase in children with either marasmus or kwashiorkor.

Waterlow (1950) found decreased activity of cholinesterase expressed either as carbon dioxide output per milligram dry weight, or per unit of DNA-phosphorus (Waterlow, 1952) but the results are variable when expressed per milligram of fat-free solids (Waterlow and Patrick, 1954).

Mukherjee and Sarkar (1958) also found a very low activity of cholinesterase expressed as per milligram nitrogen.

Liver transaminase has been found normal or elevated if expressed on DNA-phosphorus (Waterlow, 1952, 1958) or on a fat-free solid basis in measurements made on liver tissue (Waterlow and Patrick, 1954). Burch *et al.* (1957) found the same values per milligram of liver nitrogen before and after recovery from kwashiorkor.

Liver catalase is reported to be elevated per milligram of nitrogen (Mukherjee and Sarkar, 1958). The fact that transaminase and glycolic acid oxidase (Burch *et al.*, 1957) are constant in liver in relation to either nitrogen or DNA-phosphorus and only slightly reduced on a fat-free solid basis is interesting, since they have active roles in protein anabolism or catabolism. That the protein anabolic potentiality is not lost is confirmed by the large nitrogen retention observed on repletion by nitrogen balance techniques (Robinson *et al.*, 1957).

Liver xanthine oxidase and D-amino acid oxidase have been described low per unit of nitrogen, probably due to both riboflavin and protein depletion (Burch *et al.*, 1957). However, another flavoenzyme, DPN (diphosphopyridine nucleotide)-cytochrome C reductase, seems to be preserved, perhaps due to its vital role (Waterlow and Patrick, 1954; Waterlow, 1958; Burch *et al.*, 1957).

The total serum proteins are low (Cofiño and Arguedas Klee, 1938; Trowell *et al.*, 1954a; Scrimshaw *et al.*, 1956) in kwashiorkor and considerably less so in marasmus (Scrimshaw *et al.*, 1956). Literally hundreds of studies have been published on the electrophoretic separations of serum proteins in kwashiorkor. Hypoalbuminemia has been a constant finding (Trowell *et al.*, 1954a; Scrimshaw *et al.*, 1956; Gómez *et al.*, 1950; Waterlow *et al.*, 1960). The globulins are less altered and therefore an inversion of the albumin/globulin ratio is frequent.

The hypoalbuminemia is a basic change in protein malnutrition while changes in specific globulin fractions vary, depending on areas where investigations are conducted (Scrimshaw *et al.*, 1956; Gómez *et al.*, 1957; Burgio and Giacalone, 1955; Souza *et al.*, 1958; Gupta and Pohowalla, 1958; Hellyer and Went, 1960; Edozien, 1960; Bollo and Montero, 1957; Awwaad and Abdel-Wahab, 1960, 1961; Awwaad *et al.*, 1961; Anderson and Altmann, 1951; Goswami and Barua, 1959; Dean and Schwartz, 1953; El-Gholmy *et al.*, 1960; Kulkarni *et al.*, 1960; Pille, 1957; DeMaeyer and Vanderborght, 1954). When measured, the α_1 and α_2 globulins have been found increased and the β-globulins frequently decreased. γ-Globulin serum concentrations are variable, but when expressed per unit of serum protein, they are consistently high. The rise in α-globulins has been thought to be related to cellular destruction and the high levels of γ-globulin to infection.

Serum protein separation by means of starch gel (Potgieter *et al.*, 1960; Beaton and Arroyave, 1962) and immunoelectrophoresis (Peetoom, 1962) have disclosed several abnormalities in kwashiorkor, few of which seem specific. The β-lipoprotein and the transferrin bands are markedly reduced. Elevated glycoprotein levels were found by Peetoom (1962), but this is a common accompaniment of increased tissue breakdown. Transferrin and coeruloplasmin are low in kwashiorkor (Lahey *et al.*, 1958), and hemoglobin synthesis and hematopoiesis appear to be decreased. Cerebrospinal fluid proteins are decreased when serum proteins become very low (Gómez *et al.*, 1957).

These serum protein abnormalities plus, in some cases, severe hepatic steatosis and impaired liver circulation are believed to be responsible for the occasionally elevated bromsulfalein, flocculation, and turbidity liver function tests in kwashiorkor (Kinnear and Pretorius, 1956; Gómez *et al.*, 1950; Niemeyer and Meneghello, 1950; Waterlow, 1948).

Investigations with I^{131}-labeled human serum albumin carried out in malnourished children during initial recovery in Mexico and Jamaica (Gitlin *et al.*, 1958; Cravioto, 1958b; Garrow and Waterlow, 1959) indicated that the relation of intravascular to extravascular albumin remained constant during recovery and the total albumin pool was markedly decreased. Cohen and Hansen (1962), on the other hand, have reported that the extravascular albumin pool is more markedly reduced than the intravascular pool. In the 1958–1959 studies cited the rate of catabolism of albumin determined by the plasma slope method was found not to be altered (as compared with normal children), the half-life being usually between 9 and 11 days.

Later work by Picou and Waterlow (1962) has clearly shown, however, that when albumin turnover rates are calculated from the daily urinary excretion of I^{131} the rate of catabolism of albumin is much less in the malnourished child than after recovery. The results of Cohen and Hansen (1962) agree entirely in that the albumin breakdown rate is slow in South African kwashiorkor cases before treatment (0.84 gm per day) compared with values after recovery (2.49 gm per day). The latter investigators kept the children on suboptimal diets in order not to change the clinical and physiological status. The discrepancy between earlier and later results are probably due to the method of study. In the earlier investigations, the labeled albumin was diluted by newly formed albumin, resulting in an increased slope of the plasma decay curve which gives an erroneous picture of more rapid breakdown. It has also been shown that the reduced total body albumin is due primarily to decreased albumin synthesis. The rate of synthesis is markedly enhanced, however, during the early phase of adequate protein therapy (Cohen and Hansen, 1962).

In marasmus plasma protein catabolism is accelerated, according to

Imperato *et al.* (1959a,b): 14.7 % turnover per day before treatment compared with 10.9 % after recovery.

In "dystrophic children" increased migration of albumin-I^{131} into the extravascular space has been demonstrated (Imperato *et al.*, 1959a,b). Cohen and Hansen (1962) described increased gastrointestinal protein loss in kwashiorkor by means of labeled albumin and γ-globulin. The gastrointestinal loss will shorten the half-life of albumin, but the resulting increased turnover of albumin may reflect losses via the gut and not necessarily increased catabolism in the body. Cohen and Hansen (1962) have reported, however, that the total gastrointestinal loss of protein must be small.

γ-Globulin turnover rates are not altered in kwashiorkor unless infection is a complication, in which case both the synthesis and catabolism of γ-globulin may be markedly increased. The latter suggests the preferential use of amino acids for the synthesis of this protein when infection is present even in severe protein depletion and may be contributory to the kwashiorkor (Cohen and Hansen, 1962).

Nitrogen balance studies by many investigators (Robinson *et al.*, 1957; Scrimshaw *et al.*, 1955; Gómez *et al.*, 1950; Dean, 1961; Hansen *et al.*, 1960) have conclusively demonstrated that the child with severe protein malnutrition retains an abnormally large proportion of the nitrogen even with a high protein diet. This increased retention slowly returns to normal as the body protein is progressively repleted.

Arroyave *et al.* (1962) have measured serum amino acid levels under conditions which included: (*a*) the strict diagnosis of marasmus and kwashiorkor by criteria previously cited, (*b*) all children studied before dietary treatment, and (*c*) the blood drawn after the same postabsorptive period. In kwashiorkor serum amino acid levels were low; in marasmus they were higher than in kwashiorkor. Other authors have not followed such rigid criteria and total serum amino acids have been found high (Sarrouy *et al.*, 1957a; Awwaad and Eisa, 1959; Awwaad and Abdel-Wahab, 1960), variable or normal (Corda, 1957; Cravioto *et al.*, 1959b; Misra, 1960; Schendel and Hansen, 1962) and low (Westall *et al.*, 1958; Edozien *et al.*, 1960; Norton, 1960).

The partition of blood amino acids has indicated that proline is absent in malnourished children (Sarrouy *et al.*, 1957a), that both essential and nonessential amino acids are decreased (Westall *et al.*, 1958) and that arginine, leucine, and valine are most reduced (Cravioto *et al.*, 1959b). According to Edozien *et al.* (1960), as well as Arroyave *et al.* (1962), the essential amino acids are more depressed in severe cases of kwashiorkor. Awwaad *et al.* (1962), studying only methionine content, found it low.

Inversion of the tyrosine/phenylalanine ratio has been an almost constant finding, suggesting a deficiency of phenylalanine hydroxylase. Arroyave *et al.* (1962) found a decrease of the arginine/ornithine ratio in association with an elevated phenylalanine/tyrosine ratio and decreased blood urea, suggesting an impairment in the Krebs urea cycle. Essential amino acids are markedly depressed except for phenylalanine and lysine which, due to their peculiar catabolic pathways, may not reveal their depletion by low serum levels.

Holt and co-workers (1962) have determined the plasma-free amino acids of children with kwashiorkor from many regions of the world and have found a very consistent pattern of alterations regardless of origin. The magnitude of the changes varies, however, apparently in relation to the severity of the condition.

Edozien *et al.* (1960) found increased levels of ethanolamine and amino-isobutyric acid in kwashiorkor and suggested a metabolic block in transmethylation which is minimized by giving methionine-supplemented diets, as well as in nucleic acid metabolism. Both substances disappeared from the serum following adequate dietary treatment.

Urinary nitrogen in kwashiorkor has been studied carefully by workers in Africa. Whitehead and Matthew (1960) and Dean (1961) reported increased excretion of total urinary nitrogen in children given glucose and water by mouth for 24 hours after admission to the hospital. Fractionation of this urinary nitrogen revealed a marked decrease in the nitrogen contributed by urea. Increases were seen in urinary ammonia, bound amino acid nitrogen, α-amino acid nitrogen, creatine, and creatinine nitrogen, and "unaccounted-for nitrogen." Uric acid was not abnormal, but phenolic compounds varied markedly.

Edozien and Phillips (1961), studying 24-hour urine samples of children with kwashiorkor who were receiving low protein diets, found decreased values for total urinary and urea nitrogen. The proportion of total urinary nitrogen contributed by urea was low; creatinine was normal; and ammonia, creatine, uric acid, and amino acid nitrogen were high. In total amounts per 24 hours, total nitrogen and urea as well as creatinine and uric acid were all low. Ammonia was slightly elevated and creatine and amino acid nitrogen markedly elevated. Paper chromatographic analysis of amino acids disclosed that this elevation did not involve elevation of essential amino acids. A great majority of cases demonstrated elevated excretion of ethanolamine and β-amino isobutyric acid.

Cheung *et al.* (1956) did not find increased urinary amino acid levels in kwashiorkor, and total urinary nitrogen, urea, and creatinine were all low. Depressed urinary urea excretion is paralleled by low plasma urea values (Dean and Schwartz, 1953; Edozien *et al.*, 1960), but uric acid excretion was low even in the presence of elevated serum levels. The low urea levels in blood and urine suggest decreased deamination and diminished Krebs cycle activity. Serum creatinine and creatine levels have been reported "normal" by several authors (Arroyave *et al.*, 1961b; Dean, 1961; Edozien, 1960). The excretion of urea nitrogen and creatinine rose progressively with therapy to nearly normal levels after an initial rise above them. The increase in the excretion of urea was paralleled by elevated blood urea levels, but the initial creatinine excretion peak was accompanied by a decrease in blood creatinine to subnormal levels. Apparently the "normal" blood creatinine levels reported prior to treatment are due to abnormally low creatinine clearance (Arroyave *et al.*, 1961b).

The high excretion of "unaccounted-for nitrogen" has been investigated. Dean (1960) reported four or more components which seem to be xanthine and uric acid derivatives or precursors, but he is still unable to identify more than half of the "unaccounted-for nitrogen." Whitehead and Arnstein (1961) and Whitehead (1962) have found a catabolite of histidine, imidazole acrylic acid, or urocanic acid to be part of this urinary nitrogen fraction. Oral administration of 5 gm L-histidine produced a sharp and elevated excretion of histidine in acute kwashiorkor as compared to the same child after treatment. A decreased hydroxylation of phenylalanine to tyrosine in kwashiorkor has also been shown by Whitehead and Milburn (1962).

Sarrouy *et al.* (1957a) found that after administering a test dose of either gelatin or a protein hydrolyzate the plasma amino acid increase during the first week of therapy was very low and occasionally delayed. After 2 to 3 months of treatment, it was high for gelatin and normal for protein hydrolyzates, but in both cases the elevation persisted longer than normal. The qualitative amino acid pattern after the test meals did not change from the fasting pattern during the first week of treatment. This observation has been confirmed by Arroyave *et al.* (1962). After a month of treatment, the rise in plasma amino acids reflected the composition of the protein hydrolyzate (rich in glutamic acid, lysine, and proline in that order); proline increased markedly and glutamic acid and lysine rose very little. Awwaad and Eisa (1959) and Awwaad *et al.* (1962) have suggested that the increased amino acid blood levels after protein feeding in kwashiorkor are correlated with the severity of the disease as judged by the degree of fatty liver. Schendel and Hansen (1962) found that plasma amino acid levels increase in the "recovering" child in proportion to the severity of kwashiorkor.

The high dietary protein therapy is thought to be responsible for the aminoaciduria observed during recovery from kwashiorkor (Schendel and Hansen, 1959a, 1960, 1962). Further support is given by the fact that the rise in plasma amino acids was greater the higher the protein intake and that this elevation correlated with the degree of aminoaciduria.

Cheung *et al.* (1955, 1956) described altered amino acid ratios in urine of children with kwashiorkor. Isoleucine was higher than leucine, phenylalanine higher than tyrosine, and threonine excretion was low. Corda (1957) found sharp increases in urinary elimination of α-amino nitrogen after a load of glycine in both marasmic and edematous children. Vis *et al.* (1958) reported elevated glycine, alanine, histidine, serine, taurine, and β-amino isobutyric acid in malnourished children during the first 10 to 12 days of therapy.

Evidence of increased recycling of body amino acids in kwashiorkor has been provided by studies of Picou and Waterlow (1961) and Garrow (1957) in which S^{35}-labeled methionine was used. These workers reported a rapid increase in S^{35}-albumin and a slower disappearance rate than observed with I^{131}-albumin.

It is tentatively concluded that:

1. There is marked net protein loss in states of protein malnutrition.
2. The organs most rapidly affected are probably those which have a rapid turnover rate of their protein.
3. Gastrointestinal protein leakage occurs in kwashiorkor, mainly due to a marked intestinal mucosal atrophy.
4. The body tends to adapt to this situation by decreasing the catabolic rate of certain proteins, such as albumin. γ-Globulin catabolism is not necessarily decreased.
5. As a consequence of nitrogenous losses, several blood proteins and enzyme systems suffer because of decreased synthesis, along with resulting abnormalities of the general metabolic functions.
6. Blood amino acids, as judged by modern techniques, are low in kwashiorkor before treatment.
7. Although urinary urea and creatinine are diminished, other nitrogen fractions in kwashiorkor tend to be high. The fractions elevated are amino acid nitrogen, purine derivatives, and abnormal protein and amino acid metabolites: imidazole-acrylic acid, ethanolamine, and β-amino isobutyric acid. The aminoaciduria apparently does not involve increased elimination of essential amino acids.
8. Protein synthesis occurs in severe kwashiorkor, but in general at a slower rate than normal. An adaptive reduction in enzyme activity occurs which can be corrected by a few days of adequate therapy. In the meantime, there is slight impairment in amino acid incorporation into tissues as evidenced by elevated blood amino acid levels and aminoaciduria in the initial phases of therapy.

f. Vitamins. In kwashiorkor the serum levels of fat-soluble vitamins are found consistently decreased and the incidence of vitamin A deficiency signs is high. Poor vitamin intake and defective absorption and transport of vitamin A explain these findings. Vitamin E (as well as vitamin A and carotene) rises with adequate therapy (Béhar *et al.*, 1956; Trowell *et al.*, 1954b). Vitamin D reserves may become inadequate during the initial recovery, leading to mild signs of rickets (Gómez *et al.*, 1956a).

Signs of water-soluble vitamin deficiency in kwashiorkor vary with local dietary patterns. Red and white blood cells as well as serum levels of thiamine, riboflavin, and vitamin C are often reduced. In tissues, Burch *et al.* (1957) and Béhar *et al.* (1956) report low normal levels of riboflavin, niacin, and total pyridine nucleotides, which rise with adequate therapy.

The lack of a severe deficiency of these vitamins, in spite of low intake, may reflect a decreased demand. Russo and Balsamo (1959) and Gerbasi (1960) have also reported riboflavin deficiency in kwashiorkor. Pyridoxine deficiency has been suggested by increased xanthurenic acid and decreased pyridoxic acid excretion (Theron *et al.*, 1961; Abbassy *et al.*, 1959). Vitamin B_{12} in serum has been found normal or high (Macdougall and Ross, 1960) and folic acid serum levels variable (Satoskar *et al.*, 1962).

g. Central Nervous System Changes. The child with kwashiorkor suffers various mental changes which have been described in a previous section. Investigations using the Gesell and other psychomotor tests have revealed that the general performance of the malnourished child is poor, but particularly disturbed was language development, followed by social, adaptative, and motor development (Geber and Dean, 1956; Robles *et al.*, 1959; Cravioto and Robles, 1963). During the acute illness there was a poor correlation between these psychological abnormalities and the age or somatic development of the child. However, after initial recovery, and in field trials, Cravioto and Robles (1963) have found a high correlation between somatic growth, retardation of maturation, and delayed psychological development.

Severely malnourished children develop slowing of the EEG (electroencephalogram) frequencies (Lo Jacono, 1957) and localized abnormal wave patterns in temporal and posttemporal areas (Nelson, 1959; Nelson and Dean, 1959), whose severity correlates with the degree of edema (Sarrouy *et al.*, 1957b). These abnormalities coincide with the abnormal behavior and may require several months to disappear (Lo Jacono, 1957; Engel, 1956).

Hypothalamic and autonomic nervous system abnormalities are suggested by the severe anorexia, poor temperature regulation capacity, almost complete cessation of perspiration, and faulty gastrointestinal motility which are characteristic of kwashiorkor.

Spasticity, coarse tremor affecting the extremities and tongue, and Parkinsonian movements have been described during initial recovery in both India (Udani, 1960) and South Africa (Nelson and Dean, 1959).

F. *Marasmus*

The continuous spectrum of deficiency which includes both kwashiorkor and marasmus has already been described. The characteristics attributed to marasmus in the literature are often those of marasmic kwashiorkor. It seems clear that when clinical criteria are strictly applied there are many biochemical differences between the two syndromes.

Serum protein determinations as well as serum enzymes studies consistently demonstrate higher levels of total protein, albumin, amylase, and pseudocholinesterase in marasmus when compared to kwashiorkor. Vitamin A absorption is normal in marasmus in contrast to a very depressed absorption in kwashiorkor. This implies a better fat absorption in marasmus and the conservation of enzymic systems in the digestion and transport of lipids through the intestinal wall as well as sufficient lipid transport protein.

Serum amino acids under basal conditions are higher in marasmus than in kwashiorkor, although both are lower than normal. This suggests that amino acid availability and turnover rate of proteins are less affected in marasmus. The excretion of 17-hydroxycorticosteriods is also higher in marasmus than in kwashiorkor.

This biochemical picture of marasmus strongly suggests that the child afflicted with this condition catabolizes muscle proteins to preserve liver and other vital organ proteins. Also, due to the severe caloric restriction and the chronicity of the situation which leads to the development of marasmus, the child utilizes fat and protein to meet its caloric requirements. This, of necessity, requires that fat transport mechanisms be adequate and that the complex enzyme systems required for gluconeogenesis be functional.

The biochemical picture of kwashiorkor, on the contrary, suggests that the child with this syndrome must adapt itself to an acute situation where amino acid rather than caloric restriction is the primary limitation and therefore, neither fat transport mechanisms nor muscle catabolism are essential for fulfilling the caloric needs which take precedence over any other physiological requirements. Without either endogenous amino acids from muscle catabolism or exogenous ones from the diet, the liver suffers an acute and drastic protein depletion.

The role of the adrenal hormones and other endocrine secretions becomes very important in these two syndromes. The fact that glucocorticoid excretion is high in marasmus and low in kwashiorkor is of importance and explains many of the differences between these syndromes. It seems that when glucocorticoid secretion fails kwashiorkor develops rapidly. The exact mechanism of the failure of glucocorticoid production which occurs when the caloric intake greatly exceeds that of protein is still unknown.

III. Protein Malnutrition in Adults

To an even greater extent than in children, protein malnutrition in adults is associated with caloric deficiency. Therefore, the typical kwashiorkor syndrome is seldom seen in adults, although occasional reports of this syndrome appear in the literature especially in conjunction with chronic debilitating diseases, gastrointestinal absorption defects, food faddism, and alcoholism. Sometimes an adolescent with chronic infection or a lactating woman will show acute kwashiorkor-like syndromes.

Hunger or famine edema is seen in areas of the world where chronic severe protein-calorie malnutrition is endemic, or after a population group is exposed to severe dietary restriction for several months.

Stanier and Holmes (1954) have studied a group of severely malnourished adults in East Africa and have described a clinical picture of edema, weakness, breathlessness, and palpitations. Their diet was markedly lower in protein than in calories. Biochemically, the serum proteins of their subjects were low at the expense of albumin, the γ-globulin levels usually being high. Albumin levels were consistently below 3 gm ($\bar{x}$ 1.3 and 1.7 in two groups studied). Low serum pseudocholinesterase and lipoprotein values were also noted. When adults of this type were given a diet rich in protein or calories, they retained large amounts of nitrogen for very long periods of time (Holmes *et al.*, 1954).

Studies carried out by Goswami and Barua (1959) in India on patients with hunger edema also demonstrate serum albumin values of about 2 gm. γ-Globulins were relatively elevated (38.7 % of total protein). Venkatachalam *et al.* (1954) reported decreased oxygen consumption expressed on active tissue mass, as did Keys *et al.* (1950) in starvation. Electrolyte and water changes, except for occasional elevated serum potassium, are very similar in these patients to those described in malnourished children (Srikantia *et al.*, 1953; Gopalan, 1950) as are electrocardiographic abnormalities, renal physiological alterations, and antidiuretic activity in urine. Serum protein ranged from 2.9 to 6.8 gm/100 ml, and albumin from 1.1 to 4.5 gm/100 ml. Serum amylase was low when albumin decreased below 2.2 gm, but serum lipase was normal. Liver biopsy studies revealed fatty infiltration in some cases but in others the liver appeared normal (Mukherjee and Werner, 1954).

In a comparative study of malnourished adults with edema and children with kwashiorkor, it has been found that liver nitrogen, iron, and catalase, cholinesterase, and alkaline phosphatase activities suffer almost identical changes. Serum proteins and cholinesterase are also reduced to the same degree (Mukherjee and Sarkar, 1958).

Investigations performed at the Hospital de Enfermedades de la Nutrición in Mexico City (Zubirán *et al.*, 1955) on patients with different degrees and clinical pictures of malnutrition, either primary or secondary to other illnesses, indicate many endocrinological alterations. These may explain the appearance of sexual disturbances, skin pigmentation, low basal metabolic rate, gynecomastia, asthenia, hypotension, and other symptoms which suggest endocrine malfunction in states of impaired nutrition. They report general endocrine hypofunction, most clearly defined in the gonads. In men, no hyperestrogenism was observed. Hypoadrenalism and hypopituitarism was strongly suspected. These investigators theorize that the tissues respond subnormally to hormonal stimulation and that hormone inactivation is impaired. This would allow more circulating hormones to reach the pituitary and cause a suppression of trophic hormone production.

While tropical ulcer is considered by many to be an infectious disease, deficiencies of vitamin A (Martin, 1955; I. G. Thomson, 1956), vitamin B complex (Gelfand, 1948; Manson-Bahr, 1954), essential fatty acids (Martin, 1955; Thomson, 1956), and protein (Gelfand, 1948; Martin, 1955) have been implicated. It is probable that this syndrome represents a synergism between various nonspecific infectious agents and protein deficiency with possibly other nutrient deficiencies also involved.

Acknowledgments

The authors wish to express their appreciation to Drs. Carlos Tejada, Moon Chung, and Luis Mosovich for their assistance in the preparation of this chapter.

References

Abbassy, A. S., Zeittoun, M. M., and Abouiwfa, M. H. (1959). *J. Trop. Pediat.* **5,** 45.

Adams, E. B., and Scragg, J. N. (1962). *J. Pediat.* **60,** 580.

Ambrosius, K. (1956). *Bol. Med. Hosp. Infantil (Mex.)* **13,** 897.

Anderson, C. G., and Altmann, A. (1951). *Lancet* **I,** 203.

Arroyave, G., Feldman, R., and Scrimshaw, N. S. (1958). *Am. J. Clin. Nutr.* **6,** 164.

Arroyave, G., Viteri, F., Béhar, M., and Scrimshaw, N. S. (1959). *Am. J. Clin. Nutr.* **7,** 185.

Arroyave, G., Wilson, D., Mendez, J., Béhar, M., and Scrimshaw, N. S. (1961a). *Am. J. Clin. Nutr.* **9,** 180.

Arroyave, G., Wilson, D., Béhar, M., and Scrimshaw, N. S. (1961b). *Am. J. Clin. Nutr.* **9,** 176.

Arroyave, G., Wilson, D., Funes, C. de, and Béhar, M. (1962). *Am. J. Clin. Nutr.* **11,** 517.

Auffret, C., and Tanguy, F. (1950). *Bull Med. A.O.F.* **7,** 43.

Autret, M., and Béhar, M. (1954). "Síndrome Policarencial Infantil (Kwashiorkor) and Its Prevention in Central America," Nutr. Studies No. 13. Food and Agriculture Organization of the United Nations.
Awwaad, S., and Abdel-Wahab, E. M. (1960). *Arch. Pediat.* **77,** 85.
Awwaad, S., and Abdel-Wahab, E. M. (1961). *Clin. Chem.* **7,** 694.
Awwaad, S., and Eisa, E. A. (1959). *Arch. Pediat.* **76,** 395.
Awwaad, S., Eisa, E. A., and Attia, O. (1961) *J. Egypt. Med. Assoc.* **44,** 437.
Awwaad, S., Eisa, E. A., and El-Essawy, M. (1962). *J. Trop. Med. Hyg.* **65,** 179.
Back, E. H., and Bras, G. (1958). *West Indian Med. J.* **7,** 160.
Back, E. H., Montgomery, R. D., and Ward, E. E. (1962). *Arch. Disease Childhood* **37,** 106.
Badr El-Din, M. K., and Aboul-Wafa, M. H. (1961). *J. Trop. Med. Hyg.* **64,** 110.
Baron, D. N. (1960). *J. Clin. Pathol.* **13,** 252.
Beaton, G., and Arroyave, G. (1962). *Proc. Can. Federation Biol. Soc.* **5,** 9.
Béhar, M., Arroyave, G., Tejada, C., Viteri, F., and Scrimshaw, N. S. (1956). *Rev. Col. Med. Guatemala* **7,** 221.
Belli-Cortes, E., Argüello Gil, H., Baez Diaz, C., and Arcia, R. (1961). *Arch. Inst. Cardiol. Mex.* **31,** 328.
Bollo, F. A., and Montero, R. (1957). *Rev. Chilena Pediat.* **28,** 6.
Bras, G., Waterlow, J. C., and DePass, E. (1957). *West Indian Med. J.* **6,** 33.
Brock, J. F. (1954). *Ann. N.Y. Acad. Sci.* **57,** 696.
Brock, J. F., and Autret, M. (1952). "Kwashiorkor in Africa," Nutr. Studies No. 8. Food and Agriculture Organization of the United Nations.
Brozek, J. (1961). *Science* **134,** 920.
Burch, H. B., Arroyave, G., Schwartz, R., Padilla, A. M., Béhar, M., Viteri, F. and Scrimshaw, N. S. (1957). *J. Clin. Invest.* **36,** 1579.
Burgio, G. R., and Giacalone, O. (1955). *Pediatria* (*Naples*) **63,** 630.
Campbell, J. A. H. (1956). *Arch. Disease Childhood* **31,** 310.
Cartwright, G. E., and Wintrobe, M. M. (1952). *In* "Recent Advances in Internal Medicine" (W. Dock and E. Snapper, eds.), Vol. V, p. 165. Year Book Publishers, Chicago.
Carvalho, J. deM. (1947). *Hospital* (*Rio de Janeiro*) **32,** 25.
Carvalho, M., Pinto, A. G., Schmidt, M. M., Potsch, N., and Costa, N. (1945). *J. Pediat.* (*Rio de Janeiro*) **11,** 395.
Castellanos, H., and Arroyave, G. (1961). *Am. J. Clin. Nutr.* **9,** 186.
Castellanos, A., Azan, L., Junco, J. A., and Taquechel, N. (1959). *J. Pediat.* **54,** 330.
Chanda, N. K. (1958). *Brit. Med. J.* **I,** 1263.
Chatterjee, K., and Chaudhuri, J. N. (1961). *Indian J. Pediat.* **28,** 195.
Chaudhuri, K. C. (1948). *Acta Paediat.* **36,** 110.
Chaudhuri, K. C. (1961). *Indian J. Pediat.* **28,** 409.
Chaudhuri, K. C., Chaudhuri, J. N., and Ghosal, S. P. (1959). *Indian J. Pediat.* **26,** 399.
Cheek, D. B. (1954). *Pediatrics* **14,** 193.
Cheung, M. W., Fowler, D. I., Norton, P. M., Snyderman, S. E., and Holt, L. E., Jr. (1955). *J. Trop. Pediat.* **1,** 141.
Cheung, M. W., Fowler, D. I., Norton, P. M., Snyderman, S. E., and Holt, L. E., Jr. (1956). *Bol. Med. Hosp. Infantil* (*Mex.*) **13,** 217.
Cofiño, E., and Arguedas Klee, G. (1938). Paper presented at *5th Congr. Med. Centroam., San Salvador.*
Cohen, S., and Hansen, J. D. L. (1962). *Clin. Sci.* **23,** 351.
Corda, R. (1957). *Ann. Ital. Pediat.* **10,** 1.

Correa, J. P. (1908). *Rev. Med. Yucatan (Merida)* **3,** 86.
Cravioto, J. (1958a). *Bol. Med. Hosp. Infantil (Mex.)* **15,** 805.
Cravioto, J. (1958b). *Bol. Med. Hosp. Infantil (Mex.)* **15,** 823.
Cravioto, J., and Robles, B. (1963). *In* "Mild-Moderate Forms of Protein-Calorie Malnutrition" (G. Blix, ed.), p. 115. [Symposia of the Swedish Nutrition Foundation I, Bastad, August 29–31, 1962.] Almqvist and Wiksells, Uppsala.
Cravioto, J., Peña, C. L. de la, and Burgos, G. (1959a). *Metab., Clin. Exptl.* **8,** 722.
Cravioto, J., Gómez, F., Ramos-Galván, F., Frenk, S., López Montano, E., and García, N. (1959b). *Pediat. Intern. (Roma)* **9,** 135.
Cutillo, S., DiToro, R., Lupi, L., and Ansanelli, V. (1961). *Pediatria (Naples)* **69,** 925.
Cutroneo-Guarniere, I., and Serrano, P. (1960). *Lattante* **31,** 121.
Czerny, A., and Keller, A. (1906). "Des Kindes Ernährung, Ernährungsstörungen und Ernährungsterapie," 1st ed., p. 62. Deuticke, Vienna.
Darrow, D. C., Schwartz, R., Iannucci, J. F., and Colville, F. (1948). *J. Clin. Invest.* **27,** 198.
Davies, J. N. P. (1948). *Lancet* **I,** 317.
Davies, J. N. P. (1951). *In* "Liver Injury," Trans. 9th Conf. Josiah Macy Jr. Foundation, New York. (Cf. Trowell *et al.*, 1954a.)
Davies, J. N. P. (1952). *Ann. Rev. Med.* **3,** 114.
Davies, J. N. P. (1956). *Am. J. Clin. Nutr.* **4,** 539.
Dean, R. F. A. (1960). *Clin. Chim. Acta* **5,** 186.
Dean, R. F. A. (1961). *Federation Proc.* **20,** 202.
Dean, R. F. A., and Schwartz, R. (1953). *Brit. J. Nutr.* **7,** 131.
DeMaeyer, E. M., and Vanderborght, H. (1954). *Ann. Soc. Belge Med. Trop.* **34,** 417.
Diener, K. A. (1961). *Bol. Med. Hosp. Infantil (Mex.)* **18,** 47.
Diwany, M., Tallat, M., Gabr, M., Mokhtar, N., and Rahman, Y. A. (1961). *Arch. Pediat.* **78,** 4.
Eckel, R. E., Pope, C. E. and Norris, J. E. C. (1954). *Arch. Biochem. Biophys.* **52,** 293.
Edozien, J. C. (1959). *Nature* **184,** 1150.
Edozien, J. C. (1960). *J. Pediat.* **57,** 594.
Edozien, J. C. (1961). *Pediatrics* **27,** 325.
Edozien, J. C., and Phillips, E. J. (1961) *Nature* **191,** 47.
Edozien, J. C., Phillips E. J., and Collis, W. R. F. (1960). *Lancet* **I,** 615.
El-Gholmy, A., El-Nabawy, M., Shukry, A. S., Ismail, S., and El-Hawary, M. F. (1960). *J. Trop. Med. Hyg.* **63,** 137.
Ely, J. O., and Ross, M. H. (1951). *Science* **114,** 70.
Engel, R. (1956). *Electroencephalol. Clin. Neurophysiol.* **8,** 489.
F.A.O./W.H.O. (1952). Join Expert Committee on Nutrition. *World Health Organ. Tech. Rept. Ser.* **72.**
Fomon, S. J., and Owen, G. M. (1962). *Pediatrics* **29,** 495.
Forbes, G. B. (1962). *Pediatrics* **29,** 477.
Foy, H., Kondi, A., and Macdougall, L. (1961). *Brit. Med. J.* **I,** 937.
Frenk, S. (1958). *Bol. Med. Hosp. Infantil (Mex.)* **15,** 789.
Frenk, S. (1961). *Federation Proc.* **20,** Suppl. 7, Pt. III, 96.
Frenk, S., Metcoff, J. Gómez, F., Ramos-Galván, R., Cravioto, J., and Antonowicz, I. (1957). *Pediatrics* **20,** 105.
Garrow, J. S. (1957). *In* "Amino Acid Malnutrition" (W. H. Cole, ed.), p. 14. Rutgers Univ. Press, New Brunswick, New Jersey.
Garrow, J. S., and Waterlow, J. C. (1959). *Clin. Sci.* **18,** 35.

Geber, M., and Dean, R. F. A. (1956). *Courrier* (*Paris*) **6**, 3.
Gelfand, M. (1948). "The Sick African Child," 2nd ed. Stewart Printing, Capetown.
Gerbasi, M. (1960). *Arch. Pediat.* **77**, 157.
Ghitis, J. (1962). Personal communication.
Gillman, J., and Gillman, T. (1951). "Perspectives in Human Malnutrition." Grune & Stratton, New York.
Gillman, J., Gillman, T., Scragg, J., Savage, N., Gilbert, C., Trout, G., and Levy, P. (1961). *S. African J. Med. Sci.* **26**, 31.
Gitlin, D., Cravioto, J., Frenk, S., López Montano, E., Ramos-Galván, R., Gómez, F., and Janeway, C. A. (1958). *J. Clin. Invest.* **37**, 682.
Goldis, G. (1959). *Monatsschr. Kinderheilk.* **107**, 371.
Gómez, F., Ramos-Galván, R., and Cravioto, J. (1950). *Bol. Med. Hosp. Infantil* (*Mex.*) **7**, 497.
Gómez, F., Ramos-Galván, R., Cravioto, J., and Frenk, S. (1954). *Pediatrics* **13**, 544.
Gómez, F., Ramos-Galván, R., Cravioto, J., Frenk, S., and Bolok, R. (1956a). *Bol. Med. Hosp. Infantil* (*Mex.*) **13**, 865.
Gómez, F., Ramos-Galván, R., Frenk, S., Muñoz, J. C., Chavez, R., and Vazquez, J. (1956b). *J. Trop. Pediat.* **2**, 77.
Gómez, F., Ramos-Galván, R., Cravioto, J., Frenk, S., Vázquez, J., and Peña, C. de la (1956c). *Lancet* **II**, 121.
Gómez, F., Ramos-Galván, R., Cravioto, J., Frenk, S., Janeway, C. A., Gamble, J. L., and Metcoff, J. (1957). *Pediatrics* **20**, 101.
Gómez, F., Ramos-Galván, R., Cravioto, J., and Frenk, S. (1958). *In* "Diseases of Children in the Subtropics and Tropics" (H. C. Trowell and D. B. Jelliffe, eds.), p. 474. Arnold, London.
Gómez Caballero, G. (1959). *Rev. Clin. Espan.* **75**, 244.
Gomirato Sandrucci, M., and Mussa, G. C. (1960). *Minerva Pediat.* **12**, 711.
Gopalan, C. (1950). *Lancet* **I**, 304.
Gopalan, C. (1955). *In* "Protein Malnutrition," Proc. Conf., Jamaica, 1953 (J. C. Waterlow, ed.), p. 173. Food and Agriculture Organization of the United Nations.
Gopalan, C., Venkatachalam, P. S., and Srikantia, S. G. (1953). *Metab., Clin. Exptl.* **2**, 335.
Gopalan, C., Srikantia, S. G., and Venkatachalam, P. S. (1955). *Indian J. Med. Res.* **43**, 15.
Gordillo, G., Soto, R. A., Metcoff, J., López, E., and García Antillón, L. (1957). *Pediatrics* **20**, 303.
Gordillo, P. G., López, M. E., and Soto, R. A. (1959). *Bol. Med. Hosp. Infantil* (*Mex.*) **16**, 495.
Goswami, P., and Barua, R. K. (1959). *Indian J. Pediat.* **26**, 253.
Gupta, P. S., and Pohowalla, J. N. (1958). *Indian J. Pediat.* **25**, 351.
Hansen, J. D. L. (1956). *S. African J. Lab. Clin. Med.* **2**, 206.
Hansen, J. D. L., and Brock, J. F. (1954). *Lancet* **II**, 477.
Hansen, J. D., Schendel, H. E., Wilkins, J. A., and Brock, J. F. (1960). *Pediatrics* **25**, 258.
Heller, H. (1953). *Proc. 19th Intern. Congr. Physiol., Montreal* p. 449.
Hellyer, G. C., and Went, L. N. (1960). *J. Trop. Pediat.* **5**, 109.
Hill, H. R. (1959). *Trans. Roy. Soc. Trop. Med. Hyg.* **53**, 217.
Holemans, K., and Lambrechts, A. (1955). *J. Nutr.* **56**, 477.
Holemans, K., Lambrechts, A., and Martin, H. (1959). *J. Trop. Pediat.* **4**, 129.
Holliday, M. A., Segar, W. E., Bright, N. H., and Egan, T. (1960). *Pediatrics* **26**, 950.

Holmes, E. G., and Trowell, H. C. (1948). *Lancet* **I**, 395.

Holmes, E. G., Jones, E. R., and Stanier, M. W. (1954). *Brit. J. Nutr.* **8**, 173.

Holt, L. E., Jr., Snyderman, S. E., Norton, P. M., and Roitman, E. (1962). *Federation Proc.* **21**, 2.

Howarth, S. (1951). *Med. Res. Council, Spec. Rept. Ser.* **275**, 238.

Imperato, C., Giovannelli, G., Battistini, F. D., Ghirardini, G., and Mazzeo, G. (1959a). *Lattante* **30**, 433.

Imperato, C., Giovannelli, G., Battistini, F. D., and Ghirardini, G. (1959b). *Pediatria (Naples)* **67**, 549.

Janssen, E., and LeRoux, J. S. (1950). *S. African Med. J.* **24**, 762.

Jayasekera, H. T. W., DeMel, B. V., and Collumbine, M. (1951). *Ceylon J. Med. Sci.* **8**, 1.

Jelliffe, D. B. (1959). *J. Pediat.* **54**, 227.

Jelliffe, D. B., Bras, G., and Stuart, K. L. (1954). *West Indian Med. J.* **3**, 43.

Kaul, K. K. (1960). *Indian J. Pediat.* **27**, 1.

Kerpel-Fronius, E. (1960). *J. Pediat.* **56**, 826.

Kerpel-Fronius, E., and Frank, K. (1949). *Ann. Paediat. (Basel)* **173**, 321.

Kerpel-Fronius, O., Varga, F., Kun, K., and Venoczky, J. (1954). *Acta Med. Acad. Sci. Hung.* **5**, 27.

Keys, A., Brozek, J., Henschel, A., Mickelsen, O., and Taylor, H. L. (1950). "The Biology of Human Starvation," Vol. 1. Univ. of Minnesota Press, Minneapolis, Minnesota.

Kinnear, A. A., and Pretorius, P. J. (1956). *Brit. Med. J.* **I**, 1528.

Kulkarni, B. S., Satoskar, R. S., and Chitre, R. G. (1960). *Indian J. Med. Res.* **48**, 488.

Lahey, M. E., Béhar, M., Viteri, F., and Scrimshaw, N. S. (1958). *Pediatrics* **22**, 72.

Lien Keng, K., and Ooang, O. (1959). *A.M.A. J. Diseases Children* **97**, 209.

Lien-Keng, K., and Tumbelaka, W. A. F. (1960). *Ann. Paediat. (Basel)* **194**, 257.

Lifschitz, F. (1962). *Bol. Med. Hosp. Infantil (Mex.)* **19**, 319.

LoJacono, F. (1957). *Pediatria (Naples)* **65**, 707.

Lurie, A. O., and Jackson, W. P. U. (1962). *Am. J. Clin. Nutr.* **11**, 115.

McCance, R. A. (1951). *Med. Res. Council, Spec. Rept. Ser.* **275**, 21.

MacDonald, I. (1960a). *Abstr. 5th Intern. Congr. Nutr., Washington, D.C., 1960* p. 18.

MacDonald, I. (1960b). *Metab., Clin. Exptl.* **9**, 838.

MacDonald, I. (1960c). *Arch. Disease Childhood* **35**, 449.

Macdougall, L. G., and Ross, G. I. M. (1960). *J. Pediat.* **57**, 589.

McFie, J., and Welbourn, H. F. (1962). *J. Nutr.* **76**, 97.

MacIver, J. E., and Back, E. H. (1960). *Arch. Disease Childhood* **35**, 134.

Majaj, A. S., Dinning, J. S., Azzam, S. A., and Darby, W. J. (1963). *Am. J. Clin. Nutr.* **12**, 374.

Manson-Bahr, P. H. (1954). "Manson's Tropical Disease," 14th ed. Cassell, London.

Martin, K. (1955). *Abstr. Trop. Diseases Bull.* **52**, 90.

Matthew, C. E., and Dean, R. F. A. (1960). *J. Trop. Pediat.* **5**, 135.

Méndez, J., and Tejada, C. (1962). *Exptl. Mol. Pathol.* **1**, 344.

Metcoff, J., Frenk, S., Gordillo, G., Gómez, F., Ramos-Galván, R., Cravioto, J., Janeway, C. A., and Gamble, J. L. (1957). *Pediatrics* **20**, 317.

Metcoff, J., Frenk, S., Antonowicz, I., Gordillo, G., and López, E. (1960). *Pediatrics* **26**, 960.

Mey, H. S., Smit, Z. M., Labuschagne, C. J., and Pretorius, P. J. (1961). *S. African J. Lab. Clin. Med.* **7**, 141.

Misra, R. P. (1960). *J. Indian Med. Assoc.* **34**, 481.

Montgomery, R. D. (1960a). *Lancet* **II,** 74.
Montgomery, R. D. (1960b). *Lancet* **II,** 264.
Montgomery, R. D. (1961). *J. Pediat.* **59,** 119.
Montgomery, R. D. (1962a). *J. Clin. Invest.* **41,** 1653.
Montgomery, R. D. (1962b). *Arch. Disease Childhood* **37,** 383.
Mukherjee, K. L., and Sarkar, N. K. (1958). *Brit. J. Nutr.* **12,** 1.
Mukherjee, K. L., and Werner, G. (1954). *J. Lab. Clin. Med.* **43,** 727.
Nelson, G. K. (1959). *Electroencephalol. Clin. Neurophysiol.* **11,** 73.
Nelson, G. K., and Dean, R. F. A. (1959). *Bull. World Health Organ.* **21,** 779.
Niemeyer, H., and Meneghello, J. (1950). *Am. J. Diseases Children* **80,** 898.
Normet, L. (1937). *Bull. Soc. Sci. Hyg. Aliment.* **25,** 153.
Norton, P. M. (1960). *Abstr. 5th Intern. Congr. Nutr. Washington, D.C., 1960* p. 7.
Peetoom, F. (1962). *Trop. Geograph. Med.* **14,** 193.
Picou, D., and Waterlow, J. C. (1961). *West Indian Med. J.* **10,** 36.
Picou, D., and Waterlow, J. C. (1962). *Clin. Sci.* **22,** 459.
Pille, G. (1957). *In* "Le controle du traitement du kwashiorkor au laboratoire de biochimie clinique. Importance des facteurs électrolytiques," pp. 171. Gouvernment Général de l'Afrique Occidentale Française, Direction Général de la Santé Publique, Dakar, J. G. Malochet S. A., Paris.
Politzer, W. M., and Wayburne, S. (1957). *Brit. J. Nutr.* **11,** 105.
Potgieter, G. M., Smythe, P. M., and Kench, J. E. (1960). *S. African Med. J.* **34,** 841.
Pretorius, P. J., and DeVilliers, L. S. (1961). *S. African J. Lab. Clin. Med.* **7,** 108.
Ramachandran, M., Venkatachalam, P. S., and Gopalan, C. (1956). *Indian J. Med. Res.* **44,** 227.
Ramalingaswami, V., Sriramachari, S., and Tulpule, P. G. (1952). *Lancet* **II,** 661.
Ramos-Galván, R., and Cravioto, J. (1958). *Bol. Med. Hosp. Infantil (Mex.)* **15,** 763.
Ramos-Galván, R., Perez-Navarrete, J. L., and Cravioto, J. (1960). *Bol. Med. Hosp. Infantil (Mex.)* **17,** 455.
Rehlman, A. S., and Schwartz, W. B. (1956) *New Engl. J. Med.* **255,** 195.
Rhodes, K. (1957). *West Indian Med. J.* **6,** 1 and 73.
Robinson, U., Béhar, M., Viteri, F., Arroyave, G., and Scrimshaw, N. S. (1957). *J. Trop. Pediat.* **2,** 217.
Robles, B., Ramos-Galván, R., and Cravioto, J. (1959). *Bol. Med. Hosp. Infantil (Mex.)* **16,** 317.
Russfield, A. B. (1957). *In* "Analytical Pathology" (R. C. Mellors, ed.), p. 293. McGraw-Hill, New York.
Russo, G., and Balsamo, V. (1959). *Minerva Pediat.* **11,** 1186.
Salazar de Sousa, C. (1959). *Pediat. Intern.* **9,** 53.
Salazar de Sousa, C. (1960). *Rev. Port. Med.* **33,** 222.
Sarrouy, C., Garcia, I., Cabannes, R., and Clausse, J. (1957a). *Presse Méd.* **65,** 1221.
Sarrouy, C., Saint-Jean, M., and Clausse, J. (1957b). *Nourrisson* **45,** 105.
Satoskar, R. S., Kulkarni, B. S., Mehta, B. M., Sanzgiri, R. R., and Bamji, M. S. (1962). *Arch. Disease Childhood* **37,** 9.
Schendel, H. E. (1960). *Abstr. 5th Intern. Congr. Nutr., Washington, D.C., 1960* pp. 6–7.
Schendel, H. E., and Hansen, J. D. L. (1958). *Metab., Clin. Exptl.* **7,** 731.
Schendel, H. E., and Hansen, J. D. L. (1959a). *S. African Med. J.* **33,** 871.
Schendel, H. E., and Hansen, J. D. L. (1959b). *S. African Med. J.* **33,** 1005.
Schendel, H. E., and Hansen, J. D. L. (1960). *S. African Med. J.* **34,** 791.
Schendel, H. E., and Hansen, J. D. L. (1962). *J. Pediat.* **60,** 280.

Schendel, H. E., Hansen, J. D. L., and Brock, J. F. (1960). *S. African Med. J.* **34,** 791.
Schnieden, H., Hendrickse, R. G., and Haigh, C. P. (1958). *Trans. Roy. Soc. Trop. Med. Hyg.* **52,** 169.
Schwartz, R. (1956). *J. Clin. Pathol.* **9,** 333.
Schwartz, R., and Dean, R. F. A. (1957). *J. Trop. Pediat.* **3,** 23.
Schyns, C., and DeMaeyer, E. M. (1956). *Folia Sci. Africae Centr.* **2,** 21.
Scrimshaw, N. S., and Béhar, M. (1959). *Federation Proc.* **18,** 82.
Scrimshaw, N. S., and Béhar, M. (1961). *Science* **133,** 2039.
Scrimshaw, N. S., Béhar, M., Guzman, M., Viteri, G., and Arroyave, G. (1955). *Federation Proc.* **14,** 449.
Scrimshaw, N. S., Béhar, M., Arroyave, G., Viteri, F., and Tejada, C. (1956). *Federation Proc.* **15,** 977.
Sénécal, J. (1958). *Ann. N.Y. Acad. Sci.* **69,** 916.
Sénécal, J., Pille, G., Vuylsteke, J., Dupin, H., and Aubry, L. (1957). *Bull. Med. A.O.F.* **2,** 140.
Slone, D., Taitz, L. S., and Gilchrist, G. S. (1961). *Brit. Med. J.* **I,** 32.
Smit, A. M., and Pretorius, P. J. (1957). *S. African J. Lab. Clin. Med.* **3,** 142.
Smith, C. E. (1962). *J. Pediat.* **61,** 617.
Smith, R. (1959). *Lancet* **I,** 764.
Smith, R. (1960). *Clin. Sci.* **19,** 275.
Smith, R., and Waterlow, J. C. (1960). *Lancet* **I,** 147.
Smythe, P. M., Swanepoel, A., and Campbell, J. A. H. (1962). *Brit. Med. J.* **I,** 67.
Soto, R. A., García, A. L., Gordillo, P. G., and Tellez, G E. (1956). *Bol. Med. Hosp. Infantil (Mex.)* **13,** 625
Soto, R. A., Gordillo, P G., Uribem, F., Juliao, E. H., and Meunier, G. (1958). *Bol. Med. Hosp. Infantil (Mex.)* **15,** 3.
Souza, S. B., Gondim, A., and Villela, R. (1958). *Bol. Inst. Puericult.* **15,** 218.
Srikantia, S. G. (1958). *Lancet* **I,** 667.
Srikantia, S. G., Venkatachalam, P. S., and Gopalan, C. (1953). *Metab., Clin. Exptl.* **2,** 521.
Srinivasan, P. R., and Patwardhan, V. N. (1952). *Lancet* **II,** 864.
Standard, K. L., Wills, V. G., and Waterlow, J. C. (1959). *Am. J. Clin. Nutr.* **7,** 271.
Stanier, M. W., and Holmes, E. G. (1954). *Brit. J. Nutr.* **8,** 155.
Tejada, C. (1955). *Rev. Col. Med. Guatemala* **6,** 1.
Tejada, C. (1958). *Federation Proc.* **17,** 66.
Tejada, C. (1962). Personal communication.
Tejada, C., and Restrepo, C. (1962). *Proc. 8th Pan Am. Congr. Gastroenterol., New York, 1962*
Tejada, C., Béhar, M., and Confiño, E. (1956). *Rev. Col. Med. Guatemala* **7,** 134.
Theron, J. J., Pretorius, P. J., Wolf, H., and Joubert, C. P. (1961). *J. Pediat.* **59,** 439.
Thompson, M. D. (1955). *Lancet* **I,** 1181.
Thompson, M. D., and Trowell, H. C. (1952). *Lancet* **I,** 1031.
Thomson, I. G. (1956). *Trans. Roy. Soc. Trop. Med. Hyg.* **50,** 485.
Trowell, H. C., and Simpkiss, M. J. (1957). *Lancet* **II,** 265.
Trowell, H. C., Davies, J. N. P., and Dean, R. F. A. (1954a). "Kwashiorkor." Arnold, London.
Trowell, H. C., Moore, T., and Sharman, I. M. (1954b). *Ann. N.Y. Acad. Sci.* **57,** 734.
Udani, P. M. (1960). *Indian J. Child Health* **9,** 103.
Umaña, R. C., and Tejada, C. (1961). *Rev. Col. Med. Guatemala* **12,** 170.
Valderrabano Jolsen, D. C. (1954). Thesis, Univ. Nacl. Autónoma de México.

Van der Sar, A. (1951). *Doc. Neerl. et Indonesia Morbus Trop.* (*Amsterdam*) **3,** 25.
Veghelyi, P. V. (1948). *Acta Pediat.* **36,** 483.
Veghelyi, P. V. (1950). *Ann. Pediat.* (*Basel*) **175,** 349.
Veghelyi, P. V., Kemeny, T., Pozsonyi, T., and Sos, J. (1950). *Am. J. Diseases Children* **79,** 658.
Venkatachalam, P. S., Srikantia, S. G., and Gopalan, C. (1954). *Metab., Clin. Exptl.* **3,** 138.
Venkatesan, T. V. (1957). *Indian Practitioner* **10,** 187.
Vis, H., Dubois, R., Loeb, H., Vincent, M., and Bigwood, E. J. (1958). *Ann. Soc. Belge Med. Trop.* **38,** 991.
Vitale, J. J., Hegsted, C. M., and Zamcheck, N. (1960). *Am. J. Clin. Nutr.* **8,** 156.
Walt, F. (1959). *J. Trop. Pediat.* **5,** 3.
Walt, F., Taylor, J. E. D., Magill, F. B., and Nestadat, A. (1962). *Brit. Med. J.* **I,** 73.
Waterlow, J. C. (1948). "Fatty Liver Disease in Infants in the British West Indies," Med. Research Council, Spec. Rept. Ser. No. 263. His Majesty's Stationery Office, London.
Waterlow, J. C. (1950). *Lancet* **I,** 908.
Waterlow, J. C. (1952). *Rept. 2nd Inter-African* (*C.C.T.A.*) *Conf. Nutr., Gambia, 1952* p. 176.
Waterlow, J. C. (1956). *West Indian Med. J.* **5,** 167.
Waterlow, J. C. (1958). *West Indian Med. J.* **7,** 44.
Waterlow, J. C. (1959a). *Federation Proc.* **18,** 1143.
Waterlow, J. C. (1959b). *Nature* **184,** 1875.
Waterlow, J. C., and Mendes, C. B. (1957). *Nature* **180,** 1361.
Waterlow, J. C., and Patrick, S. J. (1954). *Ann. N.Y. Acad. Sci.* **57,** 750.
Waterlow, J. C., and Vergara, A. (1957). "Protein Malnutrition in Brazil," Nutr. Studies No. 14. Food and Agriculture Organization of the United Nations.
Waterlow, J. C., and Weisz, T. (1956). *J. Clin. Invest.* **35,** 346.
Waterlow, J. C., and Wills, V. G. (1960). *Brit. J. Nutr.* **14,** 183.
Waterlow, J. C., Cravioto, J., and Stephen, M. L. (1960). *Advan. Protein Chem.* **15,** 131.
Westall, R. G., Roitman, E., Peña, C. de la, Rasmussen, H., Cravioto, J., Gómez, F., and Holt, L. E., Jr. (1958). *Arch. Disease Childhood* **33,** 499.
Whitehead, R. G. (1962). *Lancet* **II,** 203.
Whitehead, R. G., and Arnstein, H. R. V. (1961). *Nature* **190,** 1105.
Whitehead, R. G., and Matthew, C. E. (1960). *East African Med. J.* **37,** 384.
Whitehead, R. G., and Milburn, T. R. (1962). *Nature* **196,** 580.
Williams, C. D. (1932). *In* "Gold Coast Annual Report of the Medical Department for 1931–32," Appendix E. Government Printing Office, Accra.
Williams, C. D. (1933). *Arch. Disease Childhood* **8,** 423.
Williams, C. D. (1935). *Lancet* **II,** 1151.
Zubirán, S., Gomez-Mont, F., and Laguna, J. (1955). *Ann. Internal Med.* **42,** 1259.

CHAPTER 23

Protein Deficiency and Infective Disease

NEVIN S. SCRIMSHAW

Department of Nutrition and Food Science, Massachusetts Institute of Technology, Cambridge, Massachusetts

I. Introduction

The relationship between protein deficiency and infective disease must be viewed in terms of reciprocal effects. Severe protein deficiency interferes with many of the mechanisms by which the body normally resists disease of infectious origin. On the other hand, infections almost always have some adverse effect on protein nutritional status, although in a well-nourished individual this effect is of no serious consequence if the infection is not unduly prolonged.

When the host is not well nourished, the effects of either an infectious agent or a period of more severe protein deficiency may still be resisted, but the two together are likely to lead to more serious illness. This type of relationship between infection and protein malnutrition can be identified as *synergism* and is commonly seen when a bacterial or rickettsial infection is superimposed on moderate to severe protein deficiency. It is also the characteristic interaction when intestinal protozoa or helminths are involved.

Certain of the systemic viruses and protozoa are so dependent for their multiplication on normal enzyme systems and nutrient levels within the host cell that the alterations associated with severe protein malnutrition are more inhibitory to the agent than they are damaging to the host. Such an interaction can be referred to as *antagonism* of nutrition and infection. Even when antagonism can be demonstrated experimentally, protein malnutrition has no value as a therapeutic measure and is exceedingly dangerous. In order to produce antagonism, it must be so severe that mechanisms of resistance to potentially fatal secondary infections are disrupted.

Since the cause of death in viral infections is commonly a secondary bacterial complication, severe protein malnutrition leads to a high mortality in viral as well as in other infectious diseases. This is illustrated by the mortality rates from measles among preschool children in technically underdeveloped areas which are often several hundred times higher than in the United States and Western Europe—a difference noted even before the advent of chemo- and antibiotic therapy. While dietary restriction to produce antagonism is not practical, the specific administration of an antimetabolite which interferes with reproduction of the agent, but has little or no effect on the host, is the basis of both sulfonamide and antibiotic therapy.

Another aspect of the synergism between nutrition and infective disease which is of great public health importance is the effect of infection in worsening already existing protein malnutrition. As pointed out in other chapters, trauma, infection, and other physical and psychological causes of stress result in an increase in urinary nitrogen excretion and decrease in nitrogen balance. In the case of infective diseases several other factors are likely to make the situation even more grave. These include a decrease in appetite, the deliberate therapeutic restriction of dietary protein, and the administration of medication which interferes with absorption of nutrients from the gastrointestinal tract.

The high mortality which is seen in young children during the weaning and postweaning period in technically underdeveloped countries is due directly to the synergism of malnutrition and infective disease. This mortality, which in the 1- to 4-year age group varies from 20 to 60 times higher than in the United States and Western Europe, stems from the fact that the foods initially substituted for mother's milk by the populations of most technically underdeveloped areas contain too little protein relative to their carbohydrate content.

The almost universal occurrence of some degree of protein malnutrition among children of lower socioeconomic groups in technically underdeveloped areas is reflected in retardation of both physical and psychomotor growth and development. Because of poor sanitation, they are also subjected to more enteric and other diseases, but the severity of disease produced by known pathogens is greater because of the lower resistance of the children. Furthermore, there is growing evidence that "weanling diarrhea" is often caused by enteric agents which would not ordinarily produce disease in well-nourished children (Gordon *et al.,* 1963). The observed high mortality rates in this age group are largely accounted for when the possibility of the infection per se terminating fatally—because of reduced resistance resulting from malnutrition—is added to the chance that the infection will precipitate kwashiorkor.

During the years 1955 and 1956, the mortality of children 1 to 4 years of age in Guatemala averaged 42.7 per 1000, compared with 1.1 for the United States and only 1.0 for Sweden (W.H.O., 1958; Pan Am. Sanit. Bur., 1958). An investigation of each death occurring over a 2-year period in four highland villages with approximately the same mortality rate as the country as a whole revealed that nearly two fifths died with the signs and symptoms of kwashiorkor, approximately one third of acute diarrheal disease, and nearly all of the remainder from complications of such common childhood infections as measles, whooping cough, and chickenpox. The histories indicated that in every case the development of kwashiorkor was preceded by one or more recent episodes of diarrheal disease or other infection (Béhar *et al.*, 1958). The fatal infections were caused by the same agents which are seldom serious for better nourished children of middle and upper-income groups in this and other countries.

Although the proportion of deaths from kwashiorkor compared to those from acute infectious diseases varies somewhat from one technically underdeveloped area to another, the example of toddler mortality in Guatemala has general application and illustrates the significance of the interaction between malnutrition and infection for preschool children in such areas. Official vital statistics fail to reveal this type of information because deaths in Guatemala are usually not medically certified and kwashiorkor is not a commonly recognized cause of death. In the Guatemalan study most of the deaths from this disease were tabulated in the official records as due to "worms" or "dysentery."

Although the potential synergism of malnutrition and infection is not an important factor in the higher mortality rates in the first months after birth, it does play a progressively greater role in the first year of life as the proportion of infants adequately fed at their mother's breast decreases and weaning practices introduce both infections and an excess of carbohydrate relative to protein. At the other end of the life span of man, inability to work, edentia, neoplastic and degenerative disease as well as other factors often result in protein depletion severe enough to interfere with resistance to infection. That special circumstances can arise in which persons of any age may be subjected to a degree of protein malnutrition which is sufficient to reduce resistance to infection is well illustrated by the mortality rates from infective disease among prisoners of the Japanese in World War II (D. A. Smith and Woodruff, 1951).

The remainder of this chapter will summarize observations suggesting or confirming the various types of interaction between protein nutrition and infective agents and examine some of the determinants.

II. Evidence of the Effect of Protein Malnutrition on Infection

There is abundant and varied published evidence which demonstrates conclusively that protein malnutrition, if sufficiently severe or prolonged, has a profound effect on resistance to infections. Table I condenses the results of studies in which protein was clearly the major limiting dietary factor, and Table II lists those in which there was a multiple deficiency although protein deficiency appeared to predominate. The most common finding is that protein malnutrition is synergistic with infection; examples of protein deficiency interfering more with the multiplication of the agent than with resistance of the host are rare.

In general, mild degrees of protein malnutrition have no detectable effect on the incidence or severity of infection. The exact degree or duration of protein deficiency which will cause an observable effect varies with both the host and the agent under study. Thus, Miles (1951a) found that a 4 % protein diet greatly increased the mortality of rats from *Salmonella typhimurium* infection, while Metcoff *et al.* (1948) and Metcoff (1949) in equally well-designed experiments observed no effect. Since the diets and duration of deficiency appear similar, the difference in results is presumably attributable to strain differences in host or infecting agent or both. Effects may be detected in young animals at levels of protein restriction which do not seem to affect the resistance of older animals. For example, Foster and Cort (1931, 1932a,b, 1935) were able to infect young dogs fed a protein-deficient diet with the cat strain of hookworm, *Ancylostoma caninum,* but were unsuccessful in infecting older animals by the same procedure. Only when the deficiency was considerably prolonged were they able to infect older dogs.

In studying swine influenza virus in the mouse, Sprunt (1948, 1949) found that acute or chronic deficiency in immature mice was antagonistic to the virus while in older animals more prolonged protein deficiency made the host more susceptible to infection by the virus. The explanation proposed was that older animals fed a low protein diet were not affected until they became depleted; because young animals have no appreciable reserves, even acute deficiency was sufficient to deplete them to the point where response could be observed. On the other hand, periods of fasting as short as 24 to 48 hours immediately before or after infection of mice with mouse tuberculosis, *Staphylococcus aureus,* or *Klebsiella pneumoniae* were synergistic (Dubos, 1955a,b; Dubos *et al.*, 1955; S. M. Smith and Dubos, 1956).

Some reported results are difficult to interpret. For example, Ratcliffe

TABLE I

THE EFFECT OF PROTEIN DEFICIENCY ON SPECIFIC INFECTION

Agent	Host	Response	Reference	Remarks
		BACTERIA		
Mycobacterium tuberculosis	Mouse	Synergistic	Hedgecock (1955, 1958)	Greatest protection at 20% protein—decreased with less or more protein. Protective effect abolished by lard or methyl linoleate
M. tuberculosis	Mouse	Synergistic	Dubos and Pierce (1948)	Higher mortality with corn meal and gelatin than with bread and milk diet
M. tuberculosis	Mouse	Synergistic	Dubos *et al.* (1955)	Survival time of infected animals decreased by fasting 36–48 hr after infection; corrected by complete diet but not when protein omitted
M. tuberculosis	Mouse	Synergistic	Schaedler and Dubos (1956)	Greater mortality, more tissue bacteria with 7.5% gelatin diet than with 7.5% gelatin plus 18% casein in the diet
M. tuberculosis	Mouse	Synergistic	Dubos and Schaedler (1958)	Highest mortality on 8% casein; less with 8% casein and 12% amino acids or 20% casein
Mycobacterium fortuitum	Mouse	Synergistic	Schaedler and Dubos (1959a,b)	Mortality higher with 5–8% than with 15–20% casein
M. tuberculosis	Rat	No effect	Metcoff *et al.* (1949), Metcoff (1949)	Relatively mild deficiency (18% vs. 8% casein only up to 58 days)
M. tuberculosis	Rat	Synergistic	Koerner *et al.* (1949)	With 15 or 25% protein, 100% mortality in 150 days. With 40% protein, little or no progression
M. tuberculosis	Rat	Antagonistic	Ratcliffe (1951) (abstract)	With 40% casein slightly larger tubercles than with 6–8%, but only after 16 weeks
M. tuberculosis	Rat	Antagonistic	Ratcliffe (1954) (abstract)	Only after 150 days tubercles progressed more rapidly with 25% than 5% protein diet, 17% intermediate
M. tuberculosis	Rat	No effect	Lange and Simmonds (1923)	High resistance on both high and low protein diets
M. tuberculosis	Rat	No effect	Sriramachari and Gopalan (1958)	2, 5, and 18% diet groups all relatively resistant
M. tuberculosis	Hamster	Antagonistic	Ratcliffe (1951) (abstract)	40% casein resulted in larger tubercles and more secondary spread after 16 weeks than 6–8% casein
M. tuberculosis	Hamster	Synergistic	Ratcliffe and Merrick (1957)	Progress of disease beyond 12 weeks more rapid with 6% than 30% protein diet, 17% intermediate. Eventual arrest or regression of infection with 30% protein
M. tuberculosis	Guinea pig	Synergistic	Narasinga Rao and Gopalan (1958)	More rapid return to positive nitrogen balance with 18% than 2 or 5% protein

TABLE I—*Continued*

Agent	Host	Response	Reference	Remarks
M. tuberculosis	Guinea pig	No effect	Sriramachari and Gopalan (1958)	Level of protein did not significantly influence mortality and survival time of infected animals
M. tuberculosis	Guinea pig	Synergistic	Ratcliffe (1954) (abstract)	Less rapid progression with 17 and 25% than with 5% protein diet. After 154 days more healing with 25% than with 5%; 17% protein diet intermediate
M. tuberculosis	Human (adult)	Synergistic	Co Tui *et al.* (1954)	High protein diet beneficial to 12 patients with advanced disease and poor prognosis
Corynebacterium murium	Rat	No effect	Miles (1951a)	With 4, 6, 8, 12, and 16% casein diets, reduced survival compared with 16% protein stock diet but no effect of protein levels
Klebsiella pneumoniae	Mouse	Synergistic	Schaedler and Dubos (1959a,b)	Mortality higher with 5–8% than 15–20% casein
K. pneumoniae	Mouse	Synergistic	Schaedler and Dubos (1956)	Greater mortality, more tissue bacteria with 7.5% gelatin diet than with gelatin plus 18% casein
K. pneumoniae	Mouse	Synergistic	Dubos and Schaedler (1959)	Higher and quicker mortality with 5% casein or corn diet than with 20% casein
Pneumococcus type I	Rat	No statistically significant effect	Chen and Li (1930)	76% mortality with 30–50% whole wheat, 30% millet, 15% soya diet; 59% mortality with 66% wheat, 33% milk (43 animals in each group).
Pneumococcus type I	Rat	Synergistic	Wissler (1947b)	Mortality increased and agglutinin and leucocyte response decreased with 1.78% compared with 18–22% casein
Pneumococcus type I	Rabbit	Synergistic	Wissler (1947a)	Increased mortality, decreased agglutinin response and phagocytic activity with severe protein depletion; approximately 2% protein and 125–150 calories compared with 15% protein and 340 kcal/day
Pneumococcus type II	Mouse	Synergistic	Sako (1942)	Survival time doubled by 55 and 60% vs. 4.5% protein; 20% protein intermediate (100 animals/group).
Streptococcus hemolyticus B	Rabbit	Synergistic	Mannino and Schiraldi (1956)	More severe infection with 12.75% vs. 34.25% protein
Rheumatic fever	Human (children)	Synergistic	Warner and Winterton (1935)	Institutional incidence reduced to ⅓ following change from deficient to abundant protein
Staphylococcus aureus	Mouse	Synergistic	Schaedler and Dubos (1959a,b)	Mortality higher with 5–8% than with 15–20% casein
S. aureus	Mouse	Synergistic	Dubos and Schaedler (1958)	Highest mortality 8% casein; less with 8% casein and 12% amino acids or 20% casein
S. aureus	Mouse	Synergistic	Schaedler and Dubos (1956)	Greater mortality, more tissue bacteria with 7.5% gelatin diet than with 7.5% gelatin plus 18% casein in diet

TABLE I—*Continued*

Agent	Host	Response	Reference	Remarks
Staphylococcus albus	Mouse	Synergistic	Dubos and Schaedler (1959)	Higher and quicker mortality with 5% casein or corn diet than with 20% casein
Salmonella	Mouse	Varied with *Salmonella* resistance factor (SRF)	Hill *et al.* (1962)	5, 15, and 30% protein. Low protein synergistic with SRF, antagonistic without SRF at minimal vitamin levels. No protein effect with 10-fold vitamin levels
Salmonella typhimurium (*Bacillus pestis caviae*)	Mouse	Synergistic	Pritchett (1927)	Mortality lower with 15–17% milk (vitamin A content also increased by milk addition).
S. typhimurium	Mouse	Synergistic	Watson (1937), Watson *et al.* (1938)	Good controls; effect obtained with added skim milk
S. typhimurium ("hog cholera bacillus")	Rat	Synergistic	Chen and Li (1930)	84% mortality with 30–50% wheat-millet diet and 42% with 66% wheat, 33% milk (30 animals/group).
S. typhimurium	Rat	Synergistic	Miles (1951a)	Much less mortality with 16 than 4% protein diet; no correlation with agglutinin response
S. typhimurium	Rat	No effect	Metcoff *et al.* (1948), Metcoff (1949)	18% vs. 2% casein diet
Salmonella enteritidis	Rat	Synergistic	Robertson and Doyle (1936)	Soya vs. casein diet, each approximately 14% protein
"Enteritis"	Rabbit	Synergistic	Casady *et al.* (1961)	Lower mortality when 0.1 mg methionine added to approximately 22% protein stock ration with or without added lysine; effect abolished when 0.2 mg used
Borrelia persica	Rat	Synergistic	Guggenheim *et al.* (1951)	Longer duration, higher fatality on protein-free diet compared with 18% casein diet
RICKETTSIA				
Rickettsia prowazeki	Rat	Synergistic	Fitzpatrick (1948)	Well controlled; 18% vs. 5% casein
VIRUS				
Infectious hepatitis	Mouse	Synergistic	Ruebner and Bramhall (1959), Ruebner and Miyai (1961)	Mortality doubled by low protein compared with either casein or 12 amino acid supplement
Infectious hepatitis	Rat	Synergistic	MacCallum and Miles (1946)	Transmissible only to protein-deficient animals
Louping ill	Mouse and rat	Synergistic	Miles (1951b)	Virus disappeared more rapidly and higher antibody titers with 16% than 4% casein
Mouse encepalomyelitis (Theiler)	Mouse	No effect	Kearney *et al.* (1948)	36% vs. 9% casein (more paralysis when specific tryptophan deficiency).
Swine influenza	Mouse	Synergistic	Sprunt (1948)	Chronic deficiency in older animals
Swine influenza	Mouse	Antagonistic	Sprunt (1949) (abstract)	Acute or chronic deficiency in immature animals

TABLE I—*Continued*

Agent	Host	Response	Reference	Remarks
PROTOZOA				
Leishmania donovani	Hamster	Synergistic	Ritterson and Stauber (1949)	Earlier emaciation and death with 10–20% than with 40% protein
Trypanosoma gambiense	Rat	Antagonistic	Yeager (1960)	19.6% protein—survival time 6.5 days with casein, 9.9 days with gluten, and 7.6 days with gluten plus lysine
Trypanosoma congolense	Mouse	Antagonistic	Keppie (1953)	10, 20, 30% casein suppressed infection—confounded by "normal" control diets low in protein and deficient in several other nutrients
Giardia muris, Hexamitus muris, Trichomonas muris	12 carnivorous wild animals of 10 species	Synergistic	Hegner (1924)	Animals receiving carnivorous diets very hard to infect with protozoa
Giardia lamblia	Cats	No effect	Hegner (1924)	Cats difficult to infect with *G. lamblia* because of carnivorous diets
Trichomonas hominus	Rat	Synergistic	Hegner and Eskridge (1935)	Rats fed high protein diet eliminated organisms to greater degree than those fed a standard diet
HELMINTHS				
Nematodes				
Nippostrongylus muris	Rat	Synergistic	Donaldson and Otto (1946)	More worms with 9% vegetable vs. 11% vegetable plus 11% animal protein
Haemonchus contortus	Sheep	Synergistic	Ross and Gordon (1933)	Heavier infestation 5.3% vs. 3.1% protein instead of pasture
Cooperia punctata, Cooperia oncophora, Ostertagia ostertagi, Haemonchus contortus, Trichostrongylus axei	Cattle	Synergistic	Goldberg (1959)	Artificially infected and control animals fed two levels of dietary protein; more worms recovered and poorer growth of animals receiving the low level of protein compared with other three groups
Oesophagostomum radiatum	Cattle	No effect	Goldberg (1959)	In the above experiment, no marked effect of diet on number of intestinal lesions or number of larvae in the intestinal wall when infected experimentally with *O. radiatum*
Ancylostoma	Human (adults)	Synergistic	Demarchi (1958)	Recovery from anemia greater in 27 patients receiving 102 gm than in 27 with 65 gm protein

TABLE I—*Continued*

Agent	Host	Response	Reference	Remarks
Mixed intestinal nematodes	Human (adults)	Synergistic	Orr *et al.* (1931), Orr and Gilkes (1931)	Comparison of predominately cereal-eating with meat-eating African tribes
Cestodes				
Hymenolepis nana	Mouse	Synergistic	Larsh (1950) (abstract)	Nearly twice as many cysticercoids in 9% protein group than in controls on 22% protein
Hymenolepis diminuta	Rat	No effect	Chandler (1943)	More worms in animals on protein-free vs. stock diet with 18% casein
Trematodes				
Schistosoma mansoni	Mouse	Synergistic	DeMeillon and Patterson (1958)	Earlier and higher mortality with 12.8% than 20.2% protein diet. Also fewer eggs and male worms smaller with abnormal reproductive organs

(1951) found no difference during the first 16 weeks in resistance to tuberculosis in either rats or hamsters fed a 6–8 % casein diet compared with those consuming 40 %, but thereafter the tubercles were larger in the animals receiving the higher intake of protein. From a subsequent report by Ratcliffe and Merrick (1957), however, it was concluded that the progress of tuberculosis in the hamster was more rapid on a 6 % protein diet than on a 30 % protein diet and was intermediate with 19 % protein. Evidently the 40 % level of protein in his earlier study was too high for optimum response. When Ratcliffe (1954) fed diets containing 25, 17, and 5 % protein to both rats and guinea pigs infected with *Mycobacterium tuberculosis*, the results differed with the host species. The previous finding in rats of more rapid tubercle development after 150 days on the higher protein diets was confirmed, but in guinea pigs fed 25 % protein the infection was arrested or regressed compared to those receiving 17 % protein. These animals, in turn, resisted the infection better than those receiving only 5 %.

There is considerable evidence that malnutrition is associated with an increased frequency and severity of tuberculosis in man (Scrimshaw *et al.*, 1959). While some of the increase can be accounted for by poor sanitary conditions and crowding together of people, the disease, itself, has a more rapid course and higher case fatality rate in a malnourished host. Protein deficiency is usually a major part of the general malnutrition in such situations (Marche and Gounelle, 1950), although in none of the published studies can protein alone be singled out sufficiently to justify incorporation into Table I or II.

While protein deficiency is generally synergistic with both natural

TABLE II

THE EFFECT ON INFECTION OF INADEQUATE DIETS PREDOMINANTLY DEFICIENT IN PROTEIN

Agent	Host	Response	Reference	Remarks
BACTERIA				
Mycobacterium tuberculosis	Human (adults)	Synergistic	Leyton (1946)	Prisoners of war receiving supplementary food with 30 gm protein had higher plasma protein and 1.2% tuberculosis compared with 15%–19%
"Diarrhea"	Human (adults)	Synergistic	McKenzie (1940)	From analysis of 320 cases with 142 deaths, believes deficiency increases susceptibility to variety of intestinal pathogens. Severe diarrhea frequent in malnourished laborers although pathogen not usually found. Diarrhea incidence decreased when diet generally improved
"Diarrhea"	Human (children)	Synergistic	Garcia-Erazo (1960)	Diarrhea more severe in malnourished children
VIRUS				
Infectious hepatitis	Human (adults)	Synergistic	Findlay (1948)	Much more severe disease and mortality in malnourished individuals—protein therapy beneficial
Herpes simplex	Human (children)	Synergistic	Hansen (1961)	Severe disease and death from virus almost limited to children with protein malnutrition
PROTOZOA				
Entamoeba histolytica	Human (adults)	Synergistic	Westphal (1937)	Increased in World War I prison camps with greater malnutrition; decreased when more food was available
E. histolytica	Human (adults)	Synergistic	Elsdon-Dew (1949)	In Durban, South Africa, acute fulminating disease common in maize-eating Bantus, rare in Indians with curry and rice diet and Europeans receiving a relatively balanced diet
HELMINTHS				
Nematodes				
Haemonchus contortus	Sheep (lambs)	Synergistic	Fraser and Robertson (1933)	Fewer worms and eggs with milk or fresh grass plus oats than grazed pasture alone when placed on infected pasture
Ostertagia spp., H. contortus, Trichostronglyus axei	Sheep (lambs)	Synergistic	Fraser *et al.* (1939)	Artificially infected; more worms retained by poorly fed than well fed group

TABLE II—*Continued*

Agent	Host	Response	Reference	Remarks
		HELMINTHS—*Continued*		
Nematodes—*Continued*				
H. contortus, Trichostrongylus spp., Ostertagia circumcincta, Cooperia curticei	Sheep (lambs)	Synergistic	E. L. Taylor (1934)	Heavier mixed natural infection with straw or hay only than hay plus concentrates
H. contortus, Oesophagostomum columbianum	Sheep (lambs)	Synergistic	Laurence *et al.* (1951)	Artificial infection; increasing maize content of hay diet improved growth and decreased mortality; eventually reduced worm load
H. contortus, O. columbianum	Sheep (lambs)	Synergistic	Lucker and Neumayer (1947)	More worms; greater egg output; diet also poor in calcium, iron, and some vitamins
Bunostomum trigonocephalum	Sheep (lambs)	Synergistic	Lucker and Neumayer (1947)	More worms and greater egg output with hay diet than hay plus oats and corn; diet also poor in calcium, iron, and some vitamins
Mixed nematodes	Sheep and lambs	Synergistic	White and Cushnie (1952)	Supplementation of pasture diet of ewes with oats and linseed cake resulted in lower intensity of mixed nematode infection of their lambs (but no effect in ewes).
Ancylostoma caninum (dog strain)	Dog	Synergistic	Foster and Cort (1931, 1932a, 1935)	Age acquired resistance lost on poor diet; worm and egg load decreased when good diet used
A. caninum (cat strain)	Dog	Synergistic	Foster and Cort (1932b)	Only in young animals fed 35% ground peas as only protein source vs. good diet of "hog lungs, bread, and powdered milk"; older animals not infected despite poor diet
A. caninum (cat strain)	Dog	Synergistic	Foster and Cort (1935)	Older animals also infected when poor diets prolonged. Both younger and older animals were heavily infected
A. caninum (dog strain)	Cat	Synergistic	Foster and Cort (1932b)	Same diets as above. Infected cats gave significantly greater infestation percentages
Strongyloidiasis	Human (adults)	Synergistic	Adamson and Path (1960)	Claim frequently associated with malnutrition in Durban, but authors present no data
Cestodes				
Hymenolepis nana	Mouse	Synergistic	Larsh (1950)	Many more cysticercoids with 9% vegetable than 11% vegetable plus 11% animal protein
Nippostrongylus muris	Rat	Antagonistic	Chandler (1932)	Fewer worms and fewer eggs per worm on protein-free diet

and experimental helminth infections, there are exceptions and variations. More protein in the diet usually results in fewer worms and decreased egg production (Fraser and Robertson, 1933; Fraser *et al.,* 1939; E. L. Taylor, 1934; Lucker and Neumayer, 1947; Donaldson and Otto, 1946; Ross and Gordon, 1933) or cysticercoid formation (Larsh, 1950), and if the infection in animals on the deficient diet is severe enough to cause deaths, mortality is reduced when dietary protein is increased (Laurence *et al.*, 1951; DeMeillon and Patterson, 1958). In protein-deficient animals, mortality is earlier and higher in infection with *Schistosoma mansoni* even though the development of the male schistosomes was abnormal (DeMeillon and Patterson, 1958). Hegner (1924) showed many years ago that all ten species of carnivores which he studied were difficult to infect with such protozoa as *Giardia, Hexamitus,* and *Trichomonas* until placed on protein-deficient diets of vegetable origin.

Variability in response is not necessarily due to inadequate experimental design, to poor technique, or to uncertain results. There are many well-controlled studies to indicate that a wide variety of infections are synergistic in the different experimental hosts selected. While some of the studies, such as those of Dubos, are not well adapted to statistical analysis because experimental conditions are changed from one experiment to another, there is an overwhelming consistency in the results observed. Futhermore, most of the experimental studies cited do have adequate controls, although for some of the earlier ones the exact protein value of the diets employed and the extent of complicating deficiencies of vitamins and minerals is uncertain.

It is not surprising that in some of the studies reported the experimental diet did not appear to influence the course of the infections because the protein deficiency was not sufficiently severe or prolonged. There are a number of other circumstances under which wide variations in nutritional status may fail to exert a detectable influence on an infection. This is to be expected when the hosts under study have such a high natural resistance to the infection that none is seriously affected or, conversely such a low natural resistance that all succumb. It will also occur when the relative virulence of the agent is so great that all hosts are highly susceptible or so low that none is seriously threatened. In the studies of Stewart and Gordon (1953) lambs previously kept worm-free and fed two levels of dietary protein had so little resistance when first exposed to *Trichostrongylus colubriformis* that both groups became maximally infected.

A high natural resistance of rats to both bovine and human strains of *Mycobacterium tuberculosis* is the most likely explanation for the repeated failures to observe any difference between well-nourished and malnourished rats infected experimentally with these organisms (Lange and Simmonds, 1923; Metcoff *et al.*, 1949; Metcoff, 1949; Sriramachari and Gopalan, 1958) or to find reduced resistance on an excessively high protein intake (Ratcliffe, 1951). Since there is a tendency to use inbred lines and single cultures in experimental studies, these are not always as pertinent as epidemiological observations for an understanding of the over-all problem. In natural situations there is more likelihood that considerable variation in both the virulence of infectious agents in the environment and the resistance of individual hosts within a population exists. Under these circumstances, protein malnutrition has an effect on a significant number of the resulting agent-host interactions.

It is well recognized that in some instances, particularly when viral infections are involved, the deficiency will affect the agent more than the host with a resulting antagonism rather than synergism. It is incorrect to conclude that either those investigators who report synergism or those who find antagonism must be wrong. On the contrary, the interaction of nutrition and infection represents a complex biological situation. With severe protein deficiency the various factors usually result in synergism, but occasionally antagonism is observed if the protein deficiency is sufficiently severe and prolonged and the agent highly dependent upon the host metabolism.

In view of the marked effects of protein deficiency in experimental animals and farm animals and the strong indirect evidence for the importance of protein deficiency in resistance to infection of populations which comes from the morbidity and mortality statistics previously mentioned, it is surprising how few direct observations have been reported for man. Tuberculosis is the disease most frequently studied in relation to nutritional status, but most of the reports involve such general improvement of the diet that it is impossible to assume that protein was the principal factor, even though this is most likely to have been the case.

Co Tui *et al.* (1954) selected 12 patients with advanced tuberculosis and a poor diet and improved their dietary protein intake from 0.5–1.0 gm to approximately 2.0 gm per kg of body weight. While they described the results as beneficial as compared with previous experience with similar patients and considered that the patients were serving as their own controls, this type of experimental design lacks true controls and other dietary factors besides protein were implicated. An observation of Leyton (1946) involved Russian and British prisoners of war. The latter received the same prison diet but were given in addition supplementary food which supplied 30 gm of protein. While the difference in the incidence of tuberculosis, as determined by X-ray survey, was very great between the two groups, it is obvious that the dietary supplement was not the only factor involved.

The same reservation must be expressed regarding the comparisons of

Orr and Gilkes (1931) of illnesses among cereal-eating and meat-eating African tribes. There is always the possibility that nutritional factors other than protein as well as influences other than diet were also involved. Demarchi (1958) compared the effects of diets supplying 65 gm of protein with those containing 102 gm of protein daily on ancylostoma infections in man. The higher protein diet resulted in recovery from clinical anemia without a decrease in worm burden.

Warner and Winterton (1935) noted that after the diet was markedly improved there was a decrease in the incidence of rheumatic fever in children who had previously been consuming a seriously inadequate institutional diet. It is apparent from other types of studies of diet and rheumatic fever (Hess, 1920, 1932; Jacobs *et al.*, 1954) that such results will be obtained only when the diet is severely inadequate before supplementary feeding.

The same conclusions apply generally to the effect of protein deficiency in man when he is infected with a variety of pathogens; unfortunately, serious protein malnutrition is still widespread in the populations of technically underdeveloped areas, especially among preschool children. Both epidemiological observations and laboratory experiments have confirmed the fact that protein deficiency of a severity sufficient to be synergistic with infections occurs in large numbers of persons in technically underdeveloped areas and occasionally, particularly in the elderly, in the developed countries.

III. Determinants of the Effect of Protein Nutrition on Infection

A. Innate Characteristics of the Host and Agent

It is axiomatic that the reaction of the host to an infectious agent depends, in part, upon genetic make-up. There is such a wide genetic variation in resistance to disease that the same agent may be so lethal in one individual and innocuous in another that wide variations in protein nutritional status will have no detectable effect on resistance to the infection in either case. It is in between these extremes, however, that the severity of disease resulting from an infection may depend upon nutriture.

Irrespective of diet the reaction of individuals of a given genetic make-up to an infection will also be influenced by their physiological state and by pathological conditions already present. Such factors as age, sex, pregnancy, and lactation must be taken into account and the presence of endocrine disorders such as diabetes and Addison's disease, of accidental or surgical trauma, of neoplasms, and of coexisting acute and chronic infections will have a profound influence on the outcome of any combination of infectious agent and malnutrition in a susceptible host. Moreover, the host may have developed sufficient immunity from contact with the infecting agent that he is not susceptible unless the deficiency is sufficiently severe as to destroy acquired immunity.

When dealing with populations, however, there is usually sufficient range in innate characteristics that the effects of protein malnutrition and other nutritional deficiencies can be demonstrated. Not only do the innate characteristics of the host vary but also those of the agent. A given strain of a known pathogen may be so virulent as to be fatal even when the protein intake is adequate or so avirulent as to fail to produce disease even in the presence of severe protein deficiency.

Where the combined effects of malnutrition and infection are synergistic, it can be assumed that one or more of the mechanisms of host resistance have been affected. When the effect of protein deficiency in the host appears to be antagonistic, the innate characteristics of the agent are more likely to be of major importance. The protein requirements and, more specifically, the individual amino acid requirements of genetically different strains may vary greatly and may determine whether a given deficiency within the host's tissues will or will not affect the agent. Where deficiency exerts part of its effect by altering the flora of the gastrointestinal tract or oral cavity, sensitivity of enteric infecting agents to the metabolic demands or products of so-called normal microbial populations may determine whether or not disease is produced.

The vector or vehicle of the infection may also influence the effect of a given agent even when other factors appear to be the same. The greater relative prevalence of diphtheria of the skin in India and nasopharyngeal diphtheria in the United States and Western Europe is probably determined more by exposure to monsoon conditions in the former than any other single factor (Gordon, 1963).

B. Consequences of Protein Deficiency

1. Synergism

a. Antibody Formation. Anamnestic mechanisms have long received the greatest attention in efforts to evaluate the synergistic effects of protein malnutrition on most infections. This is due, in part, to the fact that methods for measuring antibodies have been well developed for nearly 50 years while good quantitative methods of measurement are still lacking for most of the other recognized or suspected mechanisms of resistance to infection. On the other hand, better methods might long since have been developed were it not for the erroneous tendency to assume that most resistance was accounted for by antigen-antibody mechanisms.

In 1940, Madden and Whipple demonstrated that dogs depleted of protein by being fed a low protein diet and subjected to repeated plasmapheresis had a reduced capacity to elaborate specific antibodies and

Miller *et al.* (1940) showed that these animals were much more susceptible to infection. Both changes were reversed by protein feeding. Cannon *et al.* (1943) showed that protein-depleted rabbits had a reduced capacity to produce agglutinins to *Salmonella typhosa* and *Salmonella paratyphi* and Wissler (1947a) reported a decreased resistance of protein-deficient rabbits to pneumococcal infection which could be restored by passive immunization. This immunization produced a lower agglutinin titer in malnourished than in well-nourished animals. Although the differences were less marked, similar results were obtained in rats normally resistant to a strain of pneumococcus (Wissler, 1947b). When an amino acid mixture was substituted for part of the carbohydrate in the diet, the capacity to produce antibodies was partially recovered.

Berry *et al.* (1945) described the agglutinin titers against *Bacterium tularense* and mixed typhoid-paratyphoid vaccine in rats as being less than one fourth in males and less than one half in females on a basal diet of corn and white flour as protein sources compared with those on the same diet to which 18 % casein had been added. The loss of response of rats to injected sheep erythrocytes or Friedlander's bacillus as antigens was directly correlated with the duration of protein depletion (Benditt *et al.*, 1949). The antibody response of protein-depleted rabbits to beef serum has been found to be greatly reduced in protein-depleted rabbits as compared with those fed either a commercial diet or repleted following protein deficiency (Gemeroy and Koffler, 1949). Working with *Nippostrongylus muris,* Donaldson and Otto (1946) observed that animals immunized by repeated infections of antigen lost a significant part of their immunity when fed a 9 % vegetable protein diet for 36 days.

It has been clearly demonstrated that the severe protein deficiency of kwashiorkor interferes with antibody formation in children following administration of diphtheria toxoid (Olarte *et al.*, 1956) or typhoid vaccine (Budiansky and DaSilva 1957). Children admitted to the hospital with kwashiorkor and immediately fed a high protein diet do not have impaired ability to form antibodies (Fernandez and Scrimshaw, 1958; Pretorius and DeVilliers, 1962). Under such circumstances, abundant protein is available for metabolic purposes.

A number of other investigators have reported failure to find any interference with antibody formation in individuals whom they considered to be suffering from protein malnutrition (Bieler *et al.,* 1947; Havens *et al.,* 1954; Dancis *et al.,* 1953; Larson and Tomlinson, 1952), but in each case it is clear that they were not dealing with a really severe degree of malnutrition. In contrast Krebs (1946) reports the single case of a 15-year-old girl, so deficient in protein that her plasma protein level was 3.1 gm and albumin 2.0 gm per 100 ml, who showed no response to typhoid vaccine. Cannon (1944) described 15 patients dying of terminal infections who had serum protein levels of less than 5 gm per 100 ml and showed diminished antibody response to typhoid vaccine. Similarly, Wohl *et al.* (1949) reported that 88 patients with serum albumin values below 4 gm per 100 ml had a lessened ability to respond to typhoid vaccine compared with 14 similar patients given special supplements of protein or protein hydrolyzates.

While a direct relationship is found between lowered resistance to infection and decreased antibody formation in some of the studies cited, there are also a number of examples in which resistance is diminished even though the ability to form antibodies appears unaffected (Miles, 1951a), as well as examples of diminished capacity to form antibodies when clinical resistance was unimpaired.

b. Phagocytic Response. Because severe protein depletion leads eventually to marked atrophy of liver, spleen, and bone marrow (Asirvadham, 1948; Béhar *et al.*, 1956; Cannon *et al.*, 1944; Trowell *et al.*, 1954), where most phagocytic cells are presumed to originate, normal production of phagocytes might be expected to be affected. Asirvadham showed leucocyte response in turpentine abscesses to be markedly reduced on 2% compared with 22% casein diets. Following 14 weeks of depletion, 18 days of refeeding were required for the normal leucocyte response to reappear. Guggenheim and Buechler (1946, 1948) concluded from their experimental studies in rats that protein deficient diets "invariably" impair leucocyte regeneration in rats and that the effect can be reversed by an improvement in dietary protein. They emphasized that both quantity and quality of the protein intake must be taken into account in evaluating this effect, and while the improvement in leucocyte response generally paralleled growth, the effect in promoting blood regeneration was greater than in promoting growth (Guggenheim and Buechler, 1949).

It is, of course, possible that the phagocytic activity of the leucocytes could be affected quite independently of their numbers. Although the method for determining the opsonocytophagic index is still poor, it appears from the work of Berry *et al.* (1945) that a protein-deficient diet reduced the capacity of rat leucocytes to ingest and destroy *Micrococcus candidus*, an effect which paralleled reduction in agglutinin response to *Bacterium tularense*.

Children with the severe protein deficiency of kwashiorkor often show little or no leucocytic response even to serious infections (Trowell *et al.*, 1954; Scrimshaw and Béhar, 1961). This is presumed to be a significant factor in the greatly reduced resistance of children with kwashiorkor to infection.

c. Altered Tissue Integrity. Another factor, long assumed to be significant in resistance to infection, is the barrier to invasion of infectious agents offered by the skin and mucous membranes. It is well known that these structures are seriously affected by various nutritional deficiencies of which the skin changes in kwashiorkor are but one example (Trowell *et al.*, 1954; Scrimshaw and Béhar, 1961). Unfortunately, no techniques have been developed for quantitative estimation of the role of tissue integrity alone as a factor in the resistance to infection.

The principal possible changes include not only increased permeability of the intestinal and other mucosal surfaces, but also the reduction or absence of mucous secretions, the accumulation of cellular debris and mucus to give a more favorable culture medium, loss of ciliated epithelium in the respiratory tract, fluid in edematous tissues, and interference with normal tissue replacement and repair.

d. Collagen Formation. Many studies have been done on the influence of nutrient deficiencies on the healing of superficial wounds and especially on the formation of collagen (Dunphy, 1957). Protein deficiency, particularly of methionine, not only interferes with the conversion of the procollagens, but also with the tensile strength of collagen fibers (Dunphy *et al.*, 1956; Dunphy, 1957). P. Taylor and Tejada (1962) have shown a marked impairment in the capacity of protein-deficient rats to wall off abscesses induced by croton oil. This is paralleled by a decreased resistance of these animals to spontaneous infections as compared with similar animals receiving an adequate diet.

e. Interference with Nonspecific Protective Substances. An observed variation in "bacteriocidal power" of body tissues or fluids may be unrelated to either antibodies or known specific substances. The so-called lysozymes fall into this category. Most of the studies reporting nutritionally induced loss of nonspecific bacteriocidal power deal with vitamin deficiencies, but protein deficiency may, under some circumstances, have the same effect. For example, Dawson and Blagg (1950) tested the saliva from patients with severe malnutrition compared with healthy controls against a variety of infectious agents and showed that individuals in the malnourished group differed from normal controls by having little or no antibacterial action.

Several substances which have been identified recently appear to be involved in the destruction of infectious agents through nonspecific mechanisms. Interferon is a protein which appears to be a normal constituent of animal cells and which is in some way involved with resistance to infection. It appears to be elaborated in a host cell under the stimulus of the presence of a virus and seems to prevent infection of the cell by a second virus by interfering with the formation of enough ATP (adenosine triphosphate) to supply the needs of the virus in addition to those of the host cell (Isaacs, 1961). In this case the deficiency apparently has more serious consequences for the agent than for the host.

Properdin is a euglobulin occurring in normal individuals and associated with natural resistance to many diseases of bacterial, viral, and even protozoan etiology (Pillemer *et al.*, 1956). Decreased levels of properdin have been shown to result in lowered resistance to many infections. It would seem inevitable that with severe protein deficiency, at least, levels of this substance would decrease as has been described in the case of pantothenic acid deficiency (von Wiss *et al.*, 1957), various types of leukemia (Hunter and Hill, 1958), and pregnancy (Homer *et al.*, 1961). Studies with C^{14}-labeled leucine have shown that the turnover of properdin in normal rabbits occurs in as short a period as 2 to 3 days (Keller *et al.*, 1957), but the effect of protein deficiency on this system has not been described.

f. Nonspecific Resistance to Bacterial Toxins. The resistance of the host to bacterial toxins does not depend entirely upon antibody formation. There is considerable evidence that deficiencies of the B-complex vitamins, ascorbic acid, and vitamin A reduce resistance to such bacterial toxins as those produced by diphtheria and tetanus organisms completely independently of any effect on antibody formation (Scrimshaw *et al.*, 1959). Brief fasting, as well as feeding deficient diets, reduced resistance of mice to the endotoxin of *Klebsiella pneumoniae* (Dubos *et al.*, 1955), but the effect of protein deficiency per se does not seem to have been studied.

g. Alterations in Intestinal Flora. Changes in bacterial flora may have either a favorable or an unfavorable influence on pathogenic agents invading the gastrointestinal tract. In guinea pigs a synthetic diet used by Lynch (1957) drastically altered the bacterial flora and intestinal mucosa, presumably contributing to the fulminating amebic dysentery which followed. In the experiments in which Hegner (1924) found a decreased susceptibility to *Giardia muris, Trichomonas muris,* and *Hexamitus muris* in rats fed high protein diets, conditions in the gastrointestinal tract obviously became less favorable for these pathogens, but whether the effect occurred because of alteration in competing intestinal organisms is not known.

Dubos and Schaedler have reported recently (1962) that rats in which lactobacilli make up the largest percentage of the fecal flora are more resistant to the lethal effects of even very large doses of *Staphylococcus aureus* or *Klebsiella pneumoniae* endotoxins, but are somewhat more susceptible to experimental infection with *Mycobacterium tuberculosis.* The availability of techniques for raising germ-free animals should encourage direct studies of the effect of the presence or absence of competing intestinal flora on enteric infections.

h. Altered Endocrine Balance of the Host. Altered endocrine function is almost certainly involved in some of the mechanisms already discussed. One effect of protein and choline deficiency is lowered adrenal function. Kinsell (1955) states that "it has been well established that the resistance of the Addisonian patient or the adrenalectomized animal to infection, in particular, and to stress, in general, is markedly diminished." Except for evidence that cortical hormones may have a direct inhibiting effect on certain bacterial endotoxins (Boyer and Chedid, 1953; Geller *et al.*, 1954), the mechanisms of corticoid action in infection are not clear.

Infection is a common and severe complication in diabetes, although notably less with long-acting insulin preparations. High nitrogen loss was once common as a result of the frequent short periods of ketosis associated with regular insulin. Pollack (1955) believes that maintenance of positive nitrogen balance with protamine zinc insulin, even with glycosuria, is the basis for the better resistance to infection characteristic of diabetes under modern management.

2. Antagonism

The mechanism of antagonism is presumably interference with the presence of chemical substances within the host which are essential to the maintenance, growth, or reproduction of the agent. The many studies of the nutrient requirements of pathogenic agents are beyond the scope of this chapter, but examples can be cited. For the growth of psittacosis virus in tissue culture, tyrosine, threonine, methionine, isoleucine, phenylalanine, tryptophan, leucine, valine, and cystine are required (Bader and Morgan, 1958). Interference with the relative availability of any one of these might, under some circumstances, affect the agent more than the host.

A striking example of natural antagonism is the discovery that virulent strains of *Salmonella typhosa* require cystine and tryptophan while avirulent mutants also need a purine (Formol *et al.*, 1954). A strain, avirulent in mouse peritoneal fluid which normally lacks purine, became virulent when xanthine was injected intraperitoneally or when the organism reverted to purine independence.

Further understanding of the possible mechanisms involved in the examples of antagonism cited in this chapter requires correlation of the specific biochemistry and physiology of the agent with information on the effects of protein malnutrition on specific enzyme systems and other aspects of metabolism.

IV. Summary

The interaction between protein malnutrition and infective disease is dynamic and is usually characterized by synergism. The synergistic effects lie both in the many different ways in which protein deficiency can interfere with mechanisms of resistance to infection and the extent to which infective disease reduces nitrogen retention and often results in negative nitrogen balance.

Where both infection and protein malnutrition are highly prevalent as among young children in most technically underdeveloped areas, this synergism leads to a high age-specific mortality and constitutes a major public health problem. Antagonism sometimes occurs when the deficiency affects the agent more seriously than the host. However, it is of no practical importance since protein deficiency is always synergistic with the secondary bacterial infections which usually cause death in those diseases, mainly viral, in which antagonism has been demonstrated. Problems of protein malnutrition and infective disease are interdependent in laboratory experiments, clinical management of patients, and in public health programs. This fact needs more widespread recognition.

References

Adamson, P. B., and Path, H. D. (1960). *S. African Med. J.* **34,** 383.

Asirvadham, M. (1948). *J. Infectious Diseases* **83,** 87.

Bader, J. P., and Morgan, H. R. (1958). *J. Exptl. Med.* **108,** 617.

Béhar, M., Arroyave, G., Tejada, C., Viteri, F., and Scrimshaw, N. S. (1956). *Rev. Colegio Méd. Guatemala* **7,** 221.

Béhar, M., Ascoli, W., and Scrimshaw, N. S. (1958). *Bull. World Health Organ.* **19,** 1093.

Benditt, E. P., Wissler, R. W., Woolridge, R. L., Rowley, D. A., and Steffee, C. H. (1949). *Proc. Soc. Exptl. Biol. Med.* **70,** 240.

Berry, L. J., Davis, J., and Spies, T. D. (1945). *J. Lab. Clin. Med.* **30,** 684.

Bieler, M. M., Ecker, E. E., and Spies, T. D. (1947). *J. Lab. Clin. Med.* **32,** 130.

Boyer, F., and Chedid, L. (1953). *Ann. Inst. Pasteur* **84,** 453.

Budiansky, E., and DaSilva, N. (1957). *O. Hospital (Rio de Janeiro)* **52,** 251.

Cannon, P. R. (1944). *J. Michigan State Med. Soc.* **43,** 323.

Cannon, P. R., Chase, W. E., and Wissler, R. W. (1943). *J. Immunol.* **47,** 133.

Cannon, P. R., Wissler, R. W., Woolridge, R. L., and Benditt, E. P. (1944). *Am. Surg.* **120,** 514.

Casady, R. B., Damon, R. A., and Suitor, A. E. (1961). *J. Nutr.* **74,** 120.

Chandler, A. C. (1932). *Am. J. Hyg.* **16,** 750.

Chandler, A. C. (1943). *Am. J. Hyg.* **37,** 121.

Chen, T. T., and Li, C. P. (1930). *Chinese J. Physiol.* **4,** 59.

Co Tui, Kuo, N. H., and Schmidt, L. (1954). *Am. J. Clin. Nutr.* **2,** 252.

Dancis, J., Obsorn, J. J., and Julia, J. F. (1953). *Pediatrics* **12,** 395.

Dawson, C. E., and Blagg, W. (1950). *J. Dental Res.* **29,** 240.

Demarchi, M. (1958). *Am. J. Clin. Nutr.* **6,** 415.

DeMeillon, B., and Patterson, S. (1958). *S. African Med. J.* **32,** 1086.

Donaldson, A. W., and Otto, G. F. (1946). *Am. J. Hyg.* **44,** 384.

Dubos, R. J. (1955a). *Bull. N.Y. Acad. Med.* **31,** 5.

Dubos, R. J. (1955b). *J. Exptl. Med.* **101,** 59.

Dubos, R. J., and Pierce, C. (1948). *Am. Rev. Tuberc.* **57,** 287.

Dubos, R. J., and Schaedler, R. W. (1958). *J. Exptl. Med.* **108,** 69.

Dubos, R. J., and Schaedler, R. W. (1959). *J. Exptl. Med.* **110**, 935.
Dubos, R. J., and Schaedler, R. W. (1962). *Am. J. Med. Sci.* **244**, 265.
Dubos, R. J., Smith, J. M., and Schaedler, R. W. (1955). *Proc. Roy. Soc. Med.* **48**, 911.
Dunphy, J. E. (1957). *Yale J. Biol. Med.* **29**, 550.
Dunphy, J. E., Udupa, K. N., and Edwards, L. C. (1956). *Ann. Surg.* **144**, 304.
Elsdon-Dew, R. (1949). *Am. J. Trop. Med.* **29**, 337.
Fernandez, N., and Scrimshaw, N. S. (1958). Unpublished data.
Findlay, G. M. (1948). *Monthly Bull. Ministry Health (Gt. Brit.)* **7**, 2 and 32.
Fitzpatrick, F. (1948). *Am. J. Public Health* **38**, 676.
Formol, S. B., Baron, L. S., and Spillman, W. (1954). *J. Bacteriol.* **68**, 117.
Foster, A. O., and Cort, W. W. (1931). *Science* **73**, 681.
Foster, A. O., and Cort, W. W. (1932a). *Am. J. Hyg.* **16**, 241.
Foster, A. O., and Cort. W. W. (1932b) *Am. J. Hyg.* **16**, 582.
Foster, A. O., and Cort, W. W. (1935). *Am. J. Hyg.* **21**, 302.
Fraser, A. H. H., and Robertson, D. (1933). *Empire J. Exptl. Agr.* **1**, 17.
Fraser, A. H. H., Thomson, W., Robertson, D., and George, W. (1939). *Empire J. Exptl. Agr.* **6**, 316.
Garcia-Erazo, T. H. (1960). Doctoral thesis, Fac. Med., Universidad de El Salvador.
Geller, P. E., Merrill, E. R., and Jawetz, E. (1954). *Proc. Soc. Exptl. Biol. Med.* **86**, 716.
Gemeroy, D. G., and Koffler, A. H. (1949). *J. Nutr.* **39**, 299.
Goldberg, A. (1959). *Am. J. Vet. Res.* **20**, 806.
Gordon, J. E. (1963). Personal communication.
Gordon, J. E., Chitkara, I. D., and Wyon, J. B. (1963). *Am. J. Med. Sci.* **245**, 345.
Guggenheim, K., and Buechler, E. (1946). *J. Immunol.* **54**, 349.
Guggenheim, K., and Buechler, E. (1948). *J. Immunol.* **58**, 133.
Guggenheim, K., and Buechler, E. (1949). *Blood* **4**, 958.
Guggenheim, K., Buechler-Czaczkes, E., and Halevy, S. (1951). *J. Infectious Diseases* **88**, 105.
Hansen, J. D. L. (1961). *S. African Med. J.* **35**, 131.
Havens, W. P., Jr., Bock, D. G., and Siegel, I. (1954). *Am. J. Med. Sci.* **228**, 251.
Hedgecock, L. W. (1955). *J. Bacteriol.* **70**, 415.
Hedgecock, L. W. (1958). *Am. Rev. Tuberc.* **77**, 93.
Hegner, R. W. (1924). *Am. J. Hyg.* **4**, 393.
Hegner, R. W., and Eskridge, L. (1935). *J. Parasitol.* **21**, 313.
Hess, A. F. (1920). "Scurvy, Past and Present." Lippincott, Philadelphia, Pennsylvania.
Hess, A. F. (1932). *New Engl. J. Med.* **207**, 637.
Hill, C. H., Colburn, R. W., and Schneider, H. A. (1962). *Federation Proc.* **21**, 277.
Homer, R. S., McNall, E. G., Oura, M., and Galino, M. (1961). *Am. J. Obstet. Gynecol.* **81**, 29.
Hunter, D., and Hill, J. M. (1958). *Am. J. Clin. Pathol.* **29**, 128.
Isaacs, A. (1961). *Sci. American* **204**, No. 5, 51.
Jacobs, A. L., Leitner, A. A., Moore, T., and Sharman, I. M. (1954). *Am. J. Clin. Nutr.* **2**, 155.
Kearney, E. B., Pond, W. L., Plass, B. A., Maddy, K. H., Elvehjem, C. A., and Clark, P. F. (1948). *J. Bacteriol.* **55**, 89.
Keller, H., Isliker, H., and Aebi, H. (1957). *Helv. Physiol. Pharmacol. Acta* **15**, C21.
Keppie, A. A. N. (1953). *Brit. Med. J.* **II**, 853.
Kinsell, L. W. (1955). *Ann. N.Y. Acad. Sci.* **63**, 240.

Koerner, T. A., Getz, H. R., and Long, E. R. (1949). *Proc. Soc. Exptl. Biol. Med.* **71,** 154.
Krebs, E. G. (1946). *J. Lab. Clin. Med.* **31,** 85.
Lange, L. B., and Simmonds, N. (1923). *Am. Rev. Tuberc.* **7,** 49.
Larsh, J. E., Jr. (1950). *J. Parasitol.* **36,** 45.
Larson, D. L., and Tomlinson, L. J. (1952). *J. Lab. Clin. Med.* **39,** 129.
Laurence, G. B., Groenewald, J. W., Quin, J. W., Clark, R., Ortlepp, R. J., and Bosman, S. W. (1951). *Onderstepoort J. Vet. Res.* **25,** 121.
Leyton, G. B. (1946). *Lancet* **II,** 73.
Lucker, J. T., and Neumayer, E. M. (1947). *Am. J. Vet. Res.* **8,** 400.
Lynch, J. E. (1957.) *Am. J. Trop. Med. Hyg.* **6,** 813.
MacCallum, F. O., and Miles, J. A. R. (1946). *Lancet* **I,** 3.
McKenzie, A. (1940). *E. African Med. J.* **17,** 30.
Madden, S. C., and Whipple, G. H. (1940) *Physiol. Rev.* **20,** 194.
Mannino, N., and Schiraldi, O. (1956). *Boll. Ist. Sieroterap. Milan.* **35,** 478.
Marche, J., and Gounelle. H. (1950). *Milbank Mem. Fund Quart.* **28,** 114.
Metcoff, J. (1949). *Am. J. Public Health* **39,** 862.
Metcoff, J., Darling, D. B., Scanlon, M. H., and Stare, F. J. (1948). *J. Lab. Clin. Med.* **33,** 47.
Metcoff, J., Darling, D. B., Wilson, D., Lapi, A., and Stare, F. J., (1949). *J. Lab. Clin. Med.* **34,** 335.
Miles, J. A. R. (1951a). *Brit. J. Exptl. Pathol.* **32,** 285.
Miles, J. A. R. (1951b). *Brit. J. Exptl. Pathol.* **32,** 295.
Miller, L. L., Ross, S. F., and Whipple, G. H. (1940). *Am. J. Med. Sci.* **200,** 739.
Narasinga Rao, B. S., and Gopalan, C. (1958). *Indian J. Med. Res.* **46,** 93.
Olarte, J., Cravioto, J., and Campos, B. (1956). *Bol. Med. Hosp. Infantil (Mex.)* **13,** 467.
Orr, J. B., and Gilkes, J. L. (1931). *Med. Res. Council, Spec. Rept. Ser.* **155,** 15–82.
Orr, J. B., MacLeod, J. J. R., and Mackie, T. J. (1931). *Lancet* **I,** 1177.
Pan Am. Sanit. Bur. (1958). "Summary of Four-Year Reports on Health Conditions in the Americas." Sci. Publ. No. 40, Washington, D.C.
Pillemer, L., Blum, L., Lepow, I. H., Wurz, L., and Todd, E. W. (1956). *J. Exptl. Med.* **103,** 1.
Pollack, H. (1955). *Ann. N.Y. Acad. Sci.* **63,** 311.
Pretorius, P. J., and DeVilliers, L. S. (1962). *Am. J. Clin. Nutr.* **10,** 379.
Pritchett, I. W. (1927). *J. Exptl. Med.* **46,** 557.
Ratcliffe, H. L. (1951). *Federation Proc.* **10,** 368.
Ratcliffe, H. L. (1954). *Federation Proc.* **13,** 441.
Ratcliffe, H. L., and Merrick, J. V. (1957). *Am. J. Pathol.* **33,** 107.
Ritterson, A. L., and Stauber, L. A. (1949). *Proc. Soc. Exptl. Biol. Med.* **70,** 47.
Robertson, E. C., and Doyle, M. E. (1936). *Proc. Soc. Exptl. Biol. Med.* **35,** 374.
Ross, I. C., and Gordon, H. McL. (1933). *Australian Vet. J.* **9,** 100.
Ruebner, B. H., and Bramhall, J. L. (1959). *Nature* **183,** 609.
Ruebner, B. H., and Miyai, K. (1961). *J. Lab. Clin. Med.* **58,** 627.
Sako, W. S. (1942). *J. Pediat.* **20,** 475.
Schaedler, R. W., and Dubos, R. J. (1956). *J. Exptl. Med.* **104,** 67.
Schaedler, R. W., and Dubos, R. J. (1959a). *J. Exptl. Med.* **110,** 921.
Schaedler, R. W., and Dubos, R. J. (1959b). *J. Exptl. Med.* **110,** 935.
Scrimshaw, N. S., and Béhar, M. (1961). *Science* **133,** 2039.
Scrimshaw, N. S., Taylor, C. E., and Gordon, J. E. (1959). *Am. J. Med. Sci.* **237,** 367.

Smith, D. A., and Woodruff, M. F. A. (1951). *Med. Res. Council, Spec. Rept. Ser.* **274.**
Smith, S. M., and Dubos, R. J. (1956). *J. Exptl. Med.* **103,** 109.
Sprunt, D. H. (1948). *Proc. Soc. Exptl. Biol. Med.* **67,** 319.
Sprunt, D. H. (1949). *Federation Proc.* **8,** 369.
Sriramachari, S., and Gopalan. C. (1958). *Indian J. Med. Res.* **46,** 105.
Stewart, D. F., and Gordon, H. McL. (1953). *Australian J. Agr. Res.* **4,** 340.
Taylor, E. L. (1934). *J. Comp. Pathol. Therap.* **47,** 235.
Taylor, P., and Tejada, C. (1962). Unpublished data.
Trowell, H. C., Davies, J. N. P., and Dean, R. F. A. (1954). "Kwashiorkor." Edward Arnold, London.
von Wiss, O., Weber, F., and Isliker, H. (1957). *Schweiz. Med. Wochschr.* **87,** 1430.
Warner, E. C., and Winterton, F. G. (1935). *Quart. J. Med.* **4,** 227.
Watson, M. (1937). *J. Hyg.* **37,** 420.
Watson, M., Wilson, J., and Topley, W. W. C. (1938). *J. Hyg.* **38,** 424.
Westphal, A. (1937). *Arch. Schiffs- u. Tropen-Hyg.* **41,** 262.
White, E. G., and Cushnie, G. H. (1952). *Brit. J. Nutr.* **6,** 376.
W. H. O. (1958). "Epidemiological and Vital Statistics Report," Vol. 11, p. 98. World Health Organization, Geneva.
Wissler, R. W. (1947a). *J. Infectious Diseases* **80,** 250.
Wissler, R. W. (1947b). *J. Infectious Diseases,* **80,** 264.
Wohl, M. G., Reinhold, J. G., and Rose, S. B. (1949). *A.M.A. Arch. Internal Med.* **83,** 402.
Yaeger, R. G. (1960). *Abstr. 5th Intern. Congr. Nutr. Washington, D.C., 1960* p. 72.

Author Index

The numbers in italics show the pages on which the complete reference is listed.

A

B

C

D

G

I

J

L

N

Q

R

T

Subject Index

E

L

M

N

O